Franz Anton Mesmer arrived in Paris from Austria in February 1778 and established a medical practice which hailed him as one of the most remarkable men of his time. His skill was in his theories and in his use of "animal magnetism" to cure disease. Mesmer began a school of thought which spread across Europe and America, and enjoyed, at various stages, favour as a respected scientific and medical therapy, and disrepute as the tool of occultists, mystics and charlatans.

Dr Gauld has created a detailed and scholarly history of the phenomena, practice and theory of mesmerism, hypnotism and multiple personality disorder. The book traces the course of the mesmeric and hypnotic movements and gives special attention to the ideas and influence of certain leading figures. It considers the theories which developed to explain the phenomena and the uses and shortcomings of the technique in medical practice. Throughout the book, case histories and anecdotal accounts provide a fascinating insight into this controversial subject. The book also touches on the social and intellectual issues which influenced the progress and development of mesmerism and hypnotism. The work is extensively annotated and referenced. In his epilogue, Gauld discusses modern approaches to hypnotism and multiple personality disorder and the role of hypnotism in clinical practice and offers some ideas for understanding these intriguing phenomena.

A History of Hypnotism is a wide-ranging and detailed history of the mesmeric and hypnotic movements and is an essential work of reference and scholarship in this field. A fascinating account, this important work will be of interest to psychiatrists, psychologists and medical historians, as well as to the general reader who wishes to learn more about this absorbing topic.

A history of hypnotism

A history
of hypnotism

Alan Gauld

CAMBRIDGE
UNIVERSITY PRESS

Published by the Press Syndicate of the University of Cambridge
The Pitt Building, Trumpington Street, Cambridge CB2 1RP
40 West 20th Street, New York, NY 10011–4211, USA
10 Stamford Road, Oakleigh, Victoria 3166, Australia

First published 1992

Printed in Great Britain at the University Press, Cambridge

A catalogue record for this book is available from the British Library

Library of Congress cataloguing in publication data
Gauld, Alan.
A history of hypnotism/Alan Gauld.
p.   cm.
Includes bibliographical references (p.   ) and index.
ISBN 0-521-30675-2 (hc)
1. Hypnotism–History.   2. Mesmerism–History.   3. Animal
magnetism–History.   I. Title
BF1125.G38   1992
154.7′09–dc20   92-5106
CIP

ISBN 0 521 30675 2 hardback

To my mother
Dorothy L. Gauld

Contents

Illustrations

Preface

This work is designed as a straightforward historical survey or panorama of the literature and alleged phenomena of mesmerism and of hypnotism. It is intended primarily for those interested in the phenomena, theory and practice of hypnotism. It is not an essay in social history or in biography, though I have sketched in some aspects of social, medical and intellectual history, and some biographical details, which could be of relevance to our understanding of the progress and development of the mesmeric and hypnotic movements.

There are several reasons why such a survey is appropriate at the present time: a wide interest in the problems and uses of hypnotism, and a rapidly-growing interest in, and increase in numbers of, cases of multiple personality disorder; the prevalence of over-simplified accounts of the history of both topics; the possibility that such a survey may assist one in interpreting the phenomena, or at any rate discourage one from propounding ideas which, as so often happens in this field, are revived cyclically by persons seemingly unaware of the previous cycles.

I have endeavoured to present readers, as impartially as I can, with enough information to make up their own minds on leading issues. Naturally I have comments of my own to make, but I have tried to treat each period or phase of the subject in its own right, and to avoid retrospective criticisms based on present-day knowledge and viewpoints. Where I introduce facts or consider-ations drawn from current work I hope it will be clear that I have done so to lay the groundwork for concluding reflections.

The survey begins in 1778 with Mesmer's arrival in Paris. Pre-mesmeric "mesmerism" and "hypnotism" are touched upon only insofar as they were discussed by later mesmerists and hypnotists. Those who are interested in mesmerism and hypnotism before Mesmer will find much useful material in W. E. Edmonston's *The Induction of Hypnosis* (New York, 1986). I devote Part I of my survey to the mesmeric or animal magnetic movement from 1778 to a little after 1850. The movement did not, as is often supposed, end then, but its history becomes less interesting and harder to trace. Part II follows the hypnotic movement from its origins out of mesmerism in the mid-nineteenth century, through its heyday towards the end of that century, to its partial decline in the

years before the First World War. I largely omit the period between the wars, but
in a longish epilogue I touch on some current approaches to the problems of
hypnotism and multiple personality disorder. In no way does this epilogue
purport to be a comprehensive or even balanced survey of contemporary
literature. Its main function is as a vehicle for some concluding reflections on
certain present issues in their historical setting.

I have used the terms "animal magnetism" and "mesmerism" more or less
interchangeably, although the balance of usage between them gradually shifted
somewhat from the former to the latter. By "animal magnetism", or simply
"magnetism" in contexts where this will not mislead, and "mesmerism", I
mean the *tout ensemble* of ideas and practices which passed under those names.
And by "magnetist" or "mesmerist" I shall generally mean someone who
adhered to those doctrines and practices, with an emphasis, perhaps, on the
practical side. A person in the process of mesmerizing or magnetizing I shall refer
to as a "mesmerizer" or "magnetizer", whilst "mesmerization" and "magnetization" are what the subject or patient correspondingly undergoes. The
peculiar state said to be induced in certain patients as the result of magnetization
is perhaps best called the "magnetic" state. Similarly, I shall use the terms
"hypnotism" for the *mélange* of ideas and practices current under that heading
in the later nineteenth century, and "hypnotist" for one involved in or adhering
to these ideas and practices. By "hypnotizer" I shall mean one in the process of
hypnotizing, whilst "hypnotization" is that process considered mainly from the
point of view of the individual being hypnotized. "Hypnosis" or the "hypnotic
state" is the supposed special state which some subjects are alleged to enter as
a result of being subjected to hypnotization. It should be noted that these usages
do not wholly reflect the current practice of the periods under survey. That
practice was rather variable. Most variable of all, perhaps, were the terms used
to denote the magnetic state and the state of hypnosis. "Magnetic sleep",
"somnambulism", "artificial sleep", "artifical somnambulism", "hypnotic
sleep", "induced sleep" (*sommeil provoqué*), and "sleep-waking" (*Schlafwachen*)
are among the commoner variants. But I do not think that confusion is likely to
arise. Curiously enough, the term "trance" (*transe*, *Verzückung*), which has such
a popular vogue today, was somewhat rarely used, being most often adopted by
mid-nineteenth-century English and American writers.

The literatures of both mesmerism and hypnotism are very extensive, and I
make no pretense that my survey is anything like complete. Indeed, difficulty in
locating certain books and periodicals, my ignorance of various relevant
languages, and my having less than a life-time to devote to the task, ensure its
incompleteness. None the less, my lists of references are fairly long, and I hope
that they may be of some use to subsequent explorers in the same fields. They are
not offered as bibliographies, though I have included various items by writers
mentioned in the text which are not themselves referred to in the text. In the

interests of economy of space I have sometimes omitted the later parts of a title when (as was often the case in earlier days) they appear to convey no additional information of use.

Two valuable sources of further bibliographical information are A. Crabtree's *Animal Magnetism, Early Hypnotism, and Psychical Research, 1766–1925* (New York, 1988), and the library catalogue of the Society for Psychical Research, by Theodore Besterman, published in five parts in the *Proceedings* of that society between 1927 and 1934. On animal magnetism in France A. Dureau's *Notes bibliographiques pour servir à l'histoire du magnétisme animal* (Paris, 1869) has still not been superseded, and may be supplemented by A. L. Caillet's *Manuel bibliographique des sciences psychiques ou occultes* (3 vols, Paris, 1912). Also of use are the three bibliographical articles published in the *Archives du magnétisme animal* (Vols 7 and 8, 1823) by E. F. d'Hénin de Cuvillers. Of bibliographies of hypnotism by far the most valuable is M. Dessoir's *Bibliographie des modernen Hypnotismus* (Berlin, 1888; *Nachtrag*, 1890), which incorporates articles as well as books. On mesmeric and hypnotic periodicals, M. A. Gravitz's useful article "Two centuries of hypnosis specialty journals", in *The International Journal of Clinical and Experimental Hypnosis* for 1987, may be consulted.

For permission to reproduce illustrations I have to thank the Mary Evans Picture Library, the Wellcome Institute Library, London, Clark University Press, and the C. und O. Vogt-Institut für Hirnforschung, Düsseldorf.

For helping me with the location and consultation of scarce items, and in many other ways, I have to thank N. Clark-Lowes and Eleanor O'Keeffe of the library of the Society for Psychical Research, and A. H. Wesencraft of the Harry Price Library, University of London. For the provision of references, and for the gift or loan of relevant material in original or photocopy, or for help in obtaining such material, I am indebted to Sylvia Pybus of the Sheffield City Libraries, and to Dennis Roughton, Ian Stevenson, John Shotter, John F. Schumaker and the late Eric Dingwall. For linguistic help I am grateful to Doreen Newham, Mildred Morant and Zofia Weaver. For permission to use certain passages already published in the *British Journal of Experimental and Clinical Hypnosis* (now *Contemporary Hypnosis*) I have to thank the editor, Brian Fellows. For help in photographic matters I am indebted to Sam Grainger and Steve Tristram. For provision of photographs I have to thank Hilary Evans, and Professor K. Zilles of the C. und O. Vogt-Institut für Hirnforschung, Düsseldorf. For his kindness in reading the proofs I am indebted to Brian Inglis. To Alison Litherland I owe the removal of many errors and infelicities from the text. Above all I am grateful to my family who have patiently endured my absence from the human race for over five years.

Prologue: Franz Anton Mesmer

Franz Anton Mesmer (1734–1815) is one of that select band of persons whose names have passed into common parlance. Of course, few of those who use the words "mesmerism" and mesmerize" know anything about Mesmer himself, but information concerning him is easily come by. He figures in numerous encyclopaedias and histories of psychology, psychiatry and medicine, often as a charlatan, but sometimes as a key figure, a man born before his time, a pioneer of hypnotism, psychotherapy, mental healing, dynamic psychiatry and the unconscious mind, even of education and of paediatrics.[1]

This posthumous recognition would perhaps have surprised Mesmer. For he thought of himself as first and foremost the discoverer of a physical force ("animal magnetism") with important applications in the cure of bodily disease. His influence on the history of psychology and psychiatry was almost a by-product of the activities which he regarded as his chief claim to fame.

The number of Mesmer's biographers has been proportionate to his fame, and it has been rather more than proportionate to the available materials.[2] Much of our knowledge about him comes from his own writings, especially the *Mémoire* of 1779 and the *Précis historique* of 1781. These writings seem to be generally reliable so far as facts are concerned. But owing to Mesmer's imperfect command of French, they were to an unknown extent corrected or even drafted by friends and pupils.[3]

Mesmer's early career falls outside the scope of this work, and I shall outline it only briefly. He was born in the village of Iznang, Swabia, on 23rd May 1734. His father was a forester to the bishop of Constance. Anton was a clever child, and after studying at the Jesuit universities of Dillingen and Ingolstadt, ended up in 1759 at the University of Vienna, where he took up medicine. In 1766 he published a doctoral dissertation which, though heavily derivative, has since become famous.[4] Its subject was the influence of the moon and planets on the course of disease.

This little treatise has sometimes been talked of as though it represented a dalliance with astrology. It is, however, essentially an essay in popular Newtonian physics. Mesmer gives an account of the conjoint action of the sun and moon in raising tides; adds that there are also tides in the atmosphere; and proposes that there are likewise tidal effects on bodily humours. He supports this

Figure 1. F. A. Mesmer. (From the Library of the Society for Psychical Research.)

suggestion with various rather unconvincing examples of patients whose illnesses fluctuated with the lunar cycle. Sometimes he talks as though he regards these phenomena as gravitational effects; but he also refers (following Newton) to another kind of force "which is the cause of universal gravitation and, very probably, the basis of all corporeal properties; which, indeed, in the smallest particles of the fluids and solids of our organism stretches, relaxes and disturbs the cohesion, elasticity, irritability, magnetism and electricity, a force which can, in this respect, be called *animal gravity*".[5]

Mesmer's doctoral dissertation does not seem to have been thought eccentric. And it must be remembered that the Vienna medical school was one of the foremost in Europe. Its tradition of sound clinical teaching and close study of individual cases had been established by Gerard van Swieten (1700–1772), a pupil of the celebrated Hermann Boerhaave of Leyden. It was continued by Anton Stoerck (1731–1803), a noted pharmacologist and toxicologist, who succeeded van Swieten as President of the Faculty, and was Mesmer's teacher and close friend. So when Mesmer entered practice, there seemed no reason why a prosperous career should not have lain before him. He married a wealthy widow, set up a palatial home, moved in good society, and devoted his leisure to music and the arts.[6] For a forester's son, still under forty, his prospects seemed remarkably good.

Mesmer was, however, one of those obstinate individualists whose ideas, though slow to form, are held with unshakeable conviction. He began to try out his theories about bodily tides in his clinical practice. In 1773 and 1774 he treated Miss Oesterline, a young lady who for several years had suffered from a convulsive ailment accompanied by the most cruel toothaches and earaches, and by delirium, mania, vomiting and fainting fits.[7] He attempted by the application of magnets[8] to set up in her body "a kind of artificial tide", presumably on the assumption that her symptoms were in some way due to interference with "natural" tides. At first she experienced burning and piercing pains, but soon her condition improved remarkably. Mesmer made certain discoveries. The force concerned could not be "mineral magnetism", for that does not affect nervous tissue. Furthermore, it is only possible to magnetize certain metals, whereas Mesmer found that he could "magnetize" "paper, bread, wool, silk, leather, stones, glass, water, different metals, wood, men, dogs ... to the point at which these substances produced the same effect on the patient as does the magnet".[9] Before long he abandoned the use of the magnet, substituting the downward "passes" of his hands over the patient's body with which his name is still linked. He came to see that the "animal magnetism" resided in himself, and acted to revive the impaired circulation (the "tides") of the patient's own magnetic fluid, restoring his nervous system to "harmony" with the universe.

Mesmer told Stoerck of his discoveries, and requested a commission of the Faculty to investigate and publicize them. To his pained surprise, Stoerck became distant and evasive. Neither then nor thereafter did Mesmer's Viennese colleagues offer him the slightest encouragement. Of good advice they gave him plenty, and its tenor may easily be guessed. Mesmer's response was threefold. Early in 1775 he published his discoveries in a pamphlet[10] which he sent to most of Europe's scientific institutions. He visited neighbouring countries – Hungary, Switzerland, Bavaria – and enjoyed some modest successes. He was made a member of the Bavarian Academy of Sciences, probably because of his exposure of the exorcist Gassner.[11] And he tried (not without apparent success) to achieve such cures as would force conviction on his timid or reluctant colleagues.[12]

It was Mesmer's pursuit of this last aim that precipitated his migration to Paris. He had had some striking results with cases of ophthalmia, and in January 1777 began to "magnetize" a young pianist of eighteen, Miss Paradis, who had been blind from the age of four. Under Mesmer's care her sight was partially restored. Her parents were at first overwhelmingly grateful;[13] but later, for obscure reasons, they insisted that Mesmer cease treating her. Bitter disputes followed, and the patient's vision again deteriorated.[14]

To add to Mesmer's troubles he was no longer on good terms with his wife. He decided to quit Vienna, and arrived in Paris in February 1778, to find that his reputation as a wonder-worker had preceded him.

MESMER IN PARIS

A question which has aroused some interest is that of why Paris proved so receptive to his ideas and methods of treatment. One school of thought[15] points to the existence there of a large class of wealthy, bored and indolent persons, over-fed, under-exercised, and prone to constipation, indigestion, hypochondria and the vapours. These individuals threw themselves with enthusiasm upon any therapeutic novelty. And a therapeutic novelty which (like animal magnetism) had the advantage of being much less unpleasant than orthodox remedies, and of being administered by a foreign physician of stately carriage and strange powers, was assured of success.

Another school of thought[16] focusses on the vogue which occultisms of many varieties then enjoyed among these same idle persons. According to the Baronne d'Oberkirch, a very sharp observer of French society

> The end of this very sceptical century is conspicuous for this incredible characteristic – the love of the marvellous ... It is certain that Rosicrucians, adepts, prophets and all that goes with them, were never so numerous or so much listened to. Conversation turns almost entirely on these matters; they fill all heads; they strike all imaginations, even the most serious ... Our successors will hesitate to believe it; they will not understand how people who doubt everything, even God, can add a complete faith in omens.[17]

Mesmer himself repudiated all mysticisms and occultisms. None the less, features of his doctrines and practices were bound to appeal to the occultists: cures seemingly miraculous; a mysterious "magnetic fluid"; the strange ambience of his dimly-lit salon; and in the background, perhaps, his Masonic affiliations – certain Masonic lodges were much given over to ceremonial magic. And of course occultism has always mingled readily with "fringe" medicine – witness the earlier success of St. Germain the deathless and the later success of Cagliostro.

A final school of thought emphasizes the prestige which science and the applications of science (such as balloon ascents and lightning conductors) had come to enjoy in late-eighteenth-century France. Though professional scientists were few, amateur scientists and popular scientific publications flourished. The end-product was hardly less occult than the illuminism which we have just been discussing. It seems safe, says Darnton

> ... to draw one conclusion from the pulp literature of the 1780's: the reading public of that era was intoxicated by the power of science, and it was bewildered by the real and imaginary forces with which scientists peopled

the universe. Because the public could not distinguish the real from the imaginary, it seized on any invisible fluid, any scientific-sounding hypothesis, that promised to explain the wonders of nature.[18]

To these factors one should perhaps add a degree of intellectual freedom not attainable in most other European capitals. As Vinchon remarks, the Parisians "thought more freely than the Viennese. In their city thought and behaviour were a good deal less repressed by official surveillance. Here one need not keep out of sight when joining a group connected with new ideas".[19]

Whatever the reasons for the vogue which Mesmer enjoyed in Paris, there is no doubt of the fact. Before long his modest rooms in the place Vendôme were invaded by hordes of patients. Towards the end of the year he hired larger apartments at the hôtel Bullion, rue Coq-Héron.

Patients were so numerous that, even with the help of assistants, Mesmer could no longer treat each one individually. He therefore devised a piece of apparatus – foreshadowed earlier[20] – for storing animal magnetism. This was a wooden tub (*baquet*), containing iron filings and bottles of "magnetized" water, from which protruded a number of moveable iron rods. Patients sat in groups round the the tubs, grasping the rods, with which they touched afflicted parts of their bodies. To promote free circulation of the magnetic fluid, they linked fingers, and the whole group was joined by a cord. One *baquet* was kept for the poor. Individual treatment was reserved for special cases.

It was part of Mesmer's doctrine that when the treatment began to break down harmful obstructions to the circulation of the patient's magnetic fluid, there would result a "crisis" of which the nature and severity were determined by the site and extent of the obstructions. Some patients would laugh, some cry; some would yawn, some sleep or faint; and some would pass into unpleasant convulsions. Several writers have left vivid word-pictures of Mesmer's clinic[21] – the spacious rooms, dimly lit and hung with many mirrors (animal magnetism was reflected by mirrors); walls decorated with astrological symbols; luxurious carpets; a background of harmonious music, skilfully adapted to the general mood; the patients round the *baquet*, amused, awe-struck or passing into crisis; and moving among the company the stately form of Mesmer in his lilac suit, occasionally directing a dose of animal magnetism at a patient by means of a metal wand or of his singularly potent index finger.[22] No wonder critics wrote him off as a charlatan. Yet an ordinary charlatan would have been content with the pickings at the hôtel Bullion. One observer calculated that Mesmer's earnings were at the rate of nearly 8000 livres a month, while his yearly outlay could not have been more than 20000 livres.[23] Now Mesmer certainly liked money. But he had another and even stronger motive. He wanted to be recognized – not just by the masses, but by establishment science and medicine – as the discoverer of a new physical force or principle with practical applications of benefit to all humanity.

He pursued this aim in a serious and systematic manner almost from the moment he entered Paris. He approached in turn the highly prestigious Royal Academy of Sciences; the Royal Society of Medicine, a recently founded body, which had an official responsibility for promoting medical research and for vetting patent medicines;[24] and its arch-enemy, the Faculty of Medicine of the University of Paris, described by a recent writer[25] as the most formidable opponent of change and medical progress in France. Finally, through a somewhat circuitous route, he obtained the sympathy of the Queen, Marie-Antoinette, and through her the ear of Maurepas, the minister of state, thus in effect appealing to the government over the heads of the scientific and medical establishments.

The details of the negotiations and arguments (which Mesmer recounts at somewhat tedious length in his *Précis historique*) are not worth describing. In general terms we may say that the negotiations with the Academy of Sciences, the Royal Society of Medicine, and Maurepas broke down for much the same reasons. On the one hand lay the reluctance of the scientific and medical establishments to risk the ridicule and loss of face which premature endorsement of heterodox ideas could bring upon them. On the other lay the proud, obstinate, suspicious, and curiously naive, personality of Mesmer. Mesmer had difficulty in conceiving that any reasonable person faced with the evidence, especially with an actual demonstration, could fail to accept his doctrines immediately. Thus anyone who was sufficiently fair-minded to watch and listen was liable to be thought a convert and later called a backslider. Those who were not converted were timid, or obtuse, or influenced by his enemies. In the negotiations with the Faculty of Medicine Mesmer was not himself directly involved, but the conservatism of that body ensured failure even without the help of his peculiar brand of tetchy paranoia.

Mesmer's attempts to gain official recognition ended with the publication in April 1781 of his extraordinarily petulant and conceited open letter to the Queen, a lady who had done her best for him. Thereafter, his hopes of recognition became increasingly centred on his pupils.

PUPILS AND THE SOCIETY OF HARMONY

Mesmer's first pupil had come to him three years previously, in September 1778, following an accidental meeting. This was Charles d'Eslon (1739 – 1786),[26] physician-in-ordinary to the King's brother, the Comte d'Artois. D'Eslon was interested by Mesmer's methods and results, underwent treatment himself, and began to assist at Mesmer's clinic.[27] He tried to interest the moribund Faculty of Medicine in Mesmer's claims, and for his pains was struck off the roll of *docteurs régents*. None the less, he continued to support Mesmer even when the latter was behaving with characteristic obstinacy. His pamphlets are among the best of those which put the case for animal magnetism. He writes well, argues calmly, assembles his facts with care.[28]

Like most who associated with Mesmer for any length of time d'Eslon eventually fell out with him. Mesmer had conflicting attitudes to his supposed discovery. He wanted on the one hand to be acclaimed as the discoverer of a novel force with important medical applications; this involved, or should have involved, his putting his findings and techniques fully and fairly before the world. But on the other hand he wished also to treat animal magnetism almost as a kind of patent medicine, the secret of which he was free to communicate or not as suited him. D'Eslon, a frank and open character, had basically the former attitude, and was therefore bound to come into conflict with Mesmer sooner or later. The breach came in August 1782. The disgruntled Mesmer had recently retreated to Spa, in Belgium, with various favoured friends and clients, and showed every intention of quitting France permanently. At this point d'Eslon spoke before the Faculty of Medicine, to defend himself for the third and last time against the deprivation of his status as *docteur régent*. He made it clear that he had now set up on his own as a practitioner of animal magnetism. Mesmer was highly incensed.[29] He went so far as to deny that he had fully instructed d'Eslon, a tactic which he used again in disputes with other pupils, leading to persistent rumours that he had an undivulged "secret".

D'Eslon soon had his own corpus of patients and his own pupils (in the autumn of 1784 he claimed that he had given instruction to 160 doctors[30]). He began to develop his own methods and ideas. In 1783 he briefly patched up the quarrel with Mesmer, but the two were soon on bad terms again.

The final rupture came in the spring of 1784 when d'Eslon, through influential friends, and tact, and other favourable circumstances, procured the establishment of two Royal Commissions to investigate animal magnetism as practised in his own clinic. Mesmer boiled over, addressing furious letters to various prominent persons and to the *Journal de Paris*.[31] He accused d'Eslon of breaking his solemn word, and hinted yet again that he had not been fully initiated. Mesmer's anger was understandable, for if the reports were favourable, the credit which should have been his would go in large part to d'Eslon, and if not both would be pilloried together.

Mesmer's second and third disciples of importance stepped conveniently into the breach created by his quarrel with d'Eslon in August 1782. They were Nicolas Bergasse (1750–1832),[32] a lawyer from a wealthy Lyons family, and Guillaume Kornmann, an Alsatian banker. Both had reason to be grateful to Mesmer – he had helped alleviate Bergasse's chronic ill-health, and had cured Kornmann's infant son of serious eye trouble and other maladies[33] – and both had accompanied him to Spa in July 1782. When news of d'Eslon's speech to the

Faculty reached Spa, Bergasse and Kornmann hatched a scheme to protect both
Mesmer's discoveries and his finances. A society of a hundred members was to
be formed, each member paying, in effect to Mesmer, a subscription of 100 louis
(2400 livres), and receiving in return instruction in animal magnetism.[34]
Subscriptions were opened in March 1783, and by the end of the year there were
forty-eight members. These included eighteen members of the nobility, two
knights of Malta, four doctors, two surgeons, a lawyer, eight bankers or
merchants, two abbés and three clergymen.[35] The number continued to grow.
Each member had to sign a declaration that he would neither teach nor practice
animal magnetism without Mesmer's written consent.

   At the beginning of 1784, the new "Lodge of Harmony" (later "Society of
Harmony") took rooms at the hôtel Coigny, rue Coq-Héron, and established a
clinic with four *baquets*, one of which was reserved for the poor.[36] Members
received instruction from Mesmer or (Mesmer's French not being always easy to
understand) from Bergasse. The latter prepared a guide for the adepts, entitled *La
théorie du monde et des êtres organisés* (1784). Notes of the former's lectures were,
to his great indignation, published by Caullet de Veaumorel as *Aphorismes de M.
Mesmer* (1785). He disavowed them, but they are undoubtedly correct in
substance.[37] Clinical teaching was supervized by Mesmer. The formal sessions of
the Society seem to have followed a masonic model.[38] On completing the
courses, members received a signed diploma, with permission to set up treatment
in a named town.

PROGRESS AND DISPUTES 1784 TO 1789

The early months of 1784 were for Mesmer a time almost of triumph. In the
quiet elegance of the Society's assembly room,[39] a distinguished company
acknowledged him as master. In the clinic, it was his judgment to which his
pupils deferred. In the streets of Paris, his name was heard wherever idlers
foregathered. To be sure, he was the butt of numerous satires and squibs, even
of plays;[40] but he was not, he could not be, ignored.

   In the provinces, animal magnetism spread with great rapidity, its progress
marked by a plethora of pamphlets.[41] By the autumn, the Royal Society of
Medicine was able to concoct a pamphlet of its own from the correspondence it
had received on the subject. "One cannot be unaware", says M. A. Thouret,[42]
the editor, "that it has spread in all the provinces, and that there are few towns
in which public treatments have not been established."[43] Among major cities
mentioned by the Royal Society's correspondents as centres of magnetism are
Nantes, Dijon, Marseille, Calais, Rouen, Clermont and Lyon. Animal magnetism, says Thouret, flourishes especially in Guienne, Brittany and the district

of Lyon; least in Provence and Languedoc; least of all in cities, such as Montpellier, with old-established universities. Foreign correspondents of the Society reported outbreaks of mesmerism in countries other than France, for instance Holland, Germany, Italy, Malta and Haiti. Among those who established clinics in the provinces, pupils of Mesmer (many of them medically qualified) were prominent; but quite a few clinics were set up by enterprising persons whom stories from Paris had induced to experiment for themselves.[44]

The advance of animal magnetism was largely at the expense of orthodox practitioners and was viewed with corresponding disfavour by the medical profession. According to a country doctor quoted by Thouret,[45] "Melancholics and hysterics daily demand from us the marvellous wand of M. Mesmer; they simply do not want to hear of other anti-spasmodics. Purgatives also are losing their reputations with other patients."

Not only the medical profession but also the government itself was alarmed by the spread of animal magnetism. Anything which undermined "official" medicine in France undermined a whole complex of institutions – professional, educational, charitable, ecclesiastical and military – enjoying old-established royal or governmental licence, patronage, endowment, support or tacit approbation.[46] To a government already weakened and under pressure, even such indirect threats caused a degree of apprehension. The apprehension was not lessened by the fact that mesmeric societies had, like certain masonic lodges, a reputation for democracy; persons of different ranks met there on terms of equality.[47] Some prominent members, like Bergasse and Lafayette,[48] held reformist political views. Mesmer himself was not politically active, but some features of his doctrine could be given a political gloss, especially his frequent talk of his patients' need to achieve "harmony" both with other individuals and with the universe at large. Court de Gebelin, in a pamphlet eulogizing Mesmer, says that magnetism "must reestablish the primitive harmony which reigned between man and the universe ... a new physical world must necessarily be accompanied by a new moral world."[49]

Thus when in March 1784 Breteuil, minister at the Maison du Roi, set up Royal Commissions to investigate the claims of animal magnetism, it is likely that he was in part actuated by other motives than those which had led d'Eslon to push the matter. The negative reports of the two commissions no doubt suited the government very well. The immediate outcome was, however, not the collapse of the magnetic movement but an intensification of the pamphlet war between Mesmer's supporters and his enemies. But soon the Paris Society of Harmony was riven by internal disputes which lessened its political significance.

The root problem was Bergasse. Bergasse had become the principal introductory lecturer at the hôtel Coigny. But his restless ambition, his abrasive character, his vanity and conceit, soon brought clashes with Mesmer.[50] He had always tended to think of the Society and its offshoots as a means of promoting his own political schemes.[51] He and his immediate friends constituted a cell of political schemers within the Society.

And this brings us to a second problem. Mesmer required each member of the Society to sign an undertaking not to divulge any of his teachings without permission. Since he was notoriously jealous of his discoveries, this would have effectively shackled any member who had ambitions to walk forth and spread the word. Now there were indeed various members who wished not just to practice animal magnetism but to teach it; some, such as Bergasse, mainly for political and ideological reasons, but others, such as the Puységur brothers, for purely humanitarian ones. And many seem to have thought themselves entitled to do so. The matter is somewhat obscure;[52] but it seems probable that the original members of the Society regarded their engagement to Mesmer as lasting only until one hundred members had each paid him their hundred louis. During the course of 1784, this target was exceeded,[53] and several influential members apparently thought that they were now totally free to teach and practice and (even worse) to modify what they had learned from Mesmer. Mesmer objected strongly and emphasized the letter of the signed agreements.[54] Disputes dragged on through much of 1784,[55] and culminated in May 1785, when Bergasse and his group were expelled, and rules were set down to govern any future daughter society. These rules reaffirmed Mesmer's predominant position.

After this the Paris Society, though somewhat emasculated, continued to grow until the eve of the Revolution, when it had 430 members;[56] various Societies of Harmony were founded at home and abroad, though their relation to the Paris Society is often obscure;[57] and Mesmer, tired perhaps of bickering, travelled round Europe a good deal. For a while, the French Revolution caused him financial and other problems, but he eventually settled in the neighbourhood of Lake Constance, where relations of his still lived.[58] He started no more clinics; unsettled no more medical faculties; practised animal magnetism mostly to help friends, neighbours and local persons; wrote several essays further formulating his ideas;[59] enjoyed a brief Indian summer of *rapprochement* with the German medical establishment; and died peacefully and in comfortable circumstances at Meersburg on 5th March 1815.

The details of Mesmer's biography are not of great importance for present purposes. But three questions require some preliminary remarks. These questions are as follows:

1   What, in brief, was Mesmer's theory?
2   What were his therapeutic practices?
3   What evidence is there that these practices worked?

MESMER'S THEORY

The first question has commonly been answered in terms of the twenty-seven confusing and incoherent propositions which Mesmer places at the end of his *Mémoire* of 1779, and the second by a paraphrase of the brief *Catéchisme du magnétisme animal* put out in 1784 for the benefit of members of the Society of Harmony.[60] Both questions, however, are better approached through the somewhat more systematic (if pirated[61]) *Aphorismes de M. Mesmer*.

Mesmer regarded his discovery of animal magnetism as a major contribution not just to physiology but to physics. "I dare to flatter myself", he exclaims in the *Mémoire* of 1799, "that the discoveries I have made will push back the boundaries of our knowledge of physics as did the invention of microscopes and telescopes for the age preceding our own."[62] Animal magnetism, in Mesmer's view, is a material influence which bodies, animate or inanimate, exercise upon each other through the mediation of a universal and extremely fine fluid (the "magnetic" fluid), which underpins the phenomena of animal life in the same sort of way as air is the vehicle of sound or the ether of light. This fluid penetrates and perfuses all material bodies, but some permit its passage more readily than others. Glass, iron, gold, steel, silver are good conductors, and so in the animal organism are nerves. In certain substances – for instance water, broken glass, trees – the magnetic fluid can accumulate, and be tapped by a suitable conductor. Mesmer gives elaborate instructions for magnetizing bottles of water and trees.[63] With a tree one starts from the branch-ends, stroking towards the trunk and then down to the roots; finally one embraces the trunk itself.

All this is far from clear and comprehensible, but greater obscurities follow. For the all-pervasive magnetic fluid is not static. Not merely does it have a tendency to flow from regions of low to high concentration, but by appropriate measures it can be set moving in one direction rather than another, and it appears also to possess numerous natural or spontaneous currents, great and small, macrocosmic or microcosmic. A small current, such as is constituted by the circulation of magnetic fluid through an individual's nervous and vascular systems, will not be totally insulated from larger currents or circulations, including ones on a cosmic scale. One upshot seems to be that the human organism will develop poles – points of inflow and outflow of magnetic fluid – somewhat analogous to the poles of a magnet and usually situated at the extremities (head, hands, fingers, feet). Mesmer claims that corresponding points of the left and right sides are generally of opposite polarity, but different polarities can be established by suitable magnetic procedures.[64]

In his last works, Mesmer propounds a somewhat subtler theory. The procedures of animal magnetism occasion not a *movement of* the fluid but a kind of vivifying *motion through* the fluid. This motion through the fluid is referred to metaphorically as that of "invisible fire" or "life-fire".[65] Mesmer's earlier ideas were on the whole more influential than his later ones.

THERAPEUTIC PRACTICES

Along with Mesmer's theory of the magnetic fluid goes a theory of disease. Disease arises when the circulation of magnetic fluid is obstructed. This may happen if, for instance, some organ or part becomes engorged and presses upon a neighbouring blood vessel, lymph duct, or nerve trunk. Symptoms do not necessarily appear at the site of the obstruction; in fact a considerable percentage of all ailments have causes in the abdominal viscera.

Mesmer's therapeutic practices[66] are closely tied to his theory. To cure most diseases it is only necessary to reestablish the blocked circulation. Nature has a tendency to do this anyway, and the task of the physician is to assist her. The magnetic influence will not touch the contents of the stomach and intestines, which are in effect extraneous to the body. If these are disordered, the appropriate remedies are mild antacids, emetics or purgatives (Mesmer recommended magnesia and cream of tartar). Inflammations may be treated by bleeding, and the system toned up with medicinal baths.[67] For the rest, the remedy of choice is animal magnetism, supplemented by the drinking of magnetized water. Animal magnetism is best begun with operator and patient seated facing each other, feet touching. When the left side of the one makes contact with the right side of the other, and vice versa, opposite poles are linked, and a kind of circuit completed. In effect a new whole has been created. "In a man on his own", says the Mesmer of the *Aphorismes*, "when one part suffers, his whole life force is turned upon it to destroy the cause of the sufferings. It is the same when two men act upon each other. Their whole united force acts upon the diseased part, with a strength proportional to the increase of mass."[68] In fact a chain of people, united by hands or a cord, can be therapeutically even more effective.

Next may be tried some general procedures for moving or injecting magnetic fluid. With hands on corresponding points of each side of the patient's body, the operator may lightly run his fingers over or just above the abdomen , or, in cases of general sickness, make passes down the whole body, starting from the head, and going over both the front and the back. If a localized pain or discomfort results, this marks the site of the obstruction. The operator should concentrate on this region, perhaps delivering an extra strong "jolt" of magnetic fluid with a metal wand or an outstretched finger. For particular diseases certain special treatments are appropriate. In epilepsy, one touches the top of the head with one hand and the nape of the neck with the other, but one also seeks in the viscera the obstructions which have caused engorgement of the brain. Eye problems are treated by application of the fingers or of an iron rod. Tumours, sores and ulcers of all kinds should be bathed with magnetized water. Asthma and other chest problems may be treated by passing one hand slowly down the chest and the other down the spine. Menstrual disorders yield to the palm of the hand placed on the vulva.

In favourable cases, magnetic procedures will produce results almost immediately. The patient will undergo a "crisis" or series of "crises", after which he will rapidly recover.[69] The symptoms of the crisis may include tears, laughter, gastric disturbances, coughing, loss of consciousness, and convulsions resembling those of epilepsy.[70] These occur when the reinvigorated stream of animal magnetism begins to break down the obstructions which have held it in check. "It is", says Mesmer in the *Mémoire* of 1799, "easy to conceive ... that the impulse to movement [in the magnetic fluid] has always to make an effort against resistance, and must be proportionate to that resistance in order to overcome it. This effort is called *crisis* ... the different forms which this effort of nature may take depend upon the diversity in structure of the organic parts or the viscera which suffer its effects, their connections and affinities, the various degrees and kinds of their resistance, and the stage of their development."[71] It is important to note that being "magnetized" should have no effect on a healthy person in whom the magnetic fluid will already be circulating freely.

Most descriptions of the procedures of magnetization derive either from the pamphlet mentioned above, or from the reports of the two Royal Commissions.[72] It is rather harder to find accounts from the patient's point of view. Clearly Mesmer had a very powerful effect on certain people. Upon being touched by him, the Comte de Puységur, a naval officer, passed out for an hour.[73] Another patient, at whose chest Mesmer merely pointed, immediately vomited.[74] More typical, perhaps, were the less dramatic experiences of Heinrich Schreiber, a young Freiburg historian, who happened in 1815 to be in Mesmer's neighbourhood in Meersburg, and kept a journal in which he recorded the following:

> I had suffered for some while with considerable hoarseness, with oppression of the chest, and coughing. I was therefore induced ... to let myself be magnetized by Mesmer. The first strokes produced in me a mild feeling of flowing warmth ... At the same time Mesmer sat directly opposite me, knee to knee, his flashing and deeply penetrating gaze fixed incessantly on me, stroking downwards, fairly near me, with both hands, or only with the right hand or the thumb thereof. He also pressed both thumbs together directly on the pit of the stomach. Gradually sweat appeared on my chest and back. I felt a burning in the pit of the stomach, then an inclination to vomit. There followed strong coughing with plentiful expectoration, and an extraordinarily powerful, and quite involuntary, respiration ... Gradually the sensation ceased to be painful. I began to have a great inclination to sleep ... So the crisis passed, and Mesmer amply compensated me for the discomfort with a masterly performance on the glass-harmonica.[75]

Mesmer was by then 81 years old, but clearly he had neither lost his gifts nor changed the practices which he had first developed forty years before.

Perhaps the most interesting account of Mesmer in action is that of a certain M. Bertrand de la Grézie, a somewhat hypochondriacal gentleman who was persuaded to visit Mesmer in 1784:

> Before beginning [Mesmer] forecast the effects of magnetism which I ought to feel, effects which should take the form of intestinal colics, copious evacuations, and excessive sweating. I awaited these crises with eagerness, and I was so biassed towards them, that if I had had a tense and delicate nervous system, I would have believed I felt them. Meanwhile Mesmer did not cease to ask me if I did not wish to be sick, if I did not feel movements in the stomach; but I was obliged to tell him, after several sessions, that I felt no effect from his operations.[76]

It is clear that during Mesmer's treatments the patient might be subjected to a barrage of direct and indirect suggestions; this barrage probably passed unnoticed by most contemporaries, perhaps even by the operator himself.

CURES

It remains for me to say a little about the third question which I raised above, the question, namely, of whether in Mesmer's hands the techniques of animal magnetism actually had any of the curative effects he claimed for them. This is a question which it is impossible to answer with any degree of certainty. The problem is not a shortage of materials. It is rather that so much of the literature is rabidly partisan. The protagonists of animal magnetism announce its merits with an almost evangelical fervour, and their opponents castigate them as fools, optimists, knaves or charlatans.[77] Neither party has time for such trivia as the proper verification of case histories,[78] the obtaining of medically attested statements concerning the state of the patient before and after treatment, and the likelihood of recovery in the ordinary course of events; while the possibility that magnetic procedures might be assessed by comparing the progress of a group of magnetically treated patients with that of "control" groups, was only very rarely entertained. In fact the first suggestion of the kind that I have come across was made by Mesmer himself, perhaps at the instigation of d'Eslon.[79]

For the time being, the best thing I can do is perhaps to take a few examples of the cures or ameliorations which Mesmer allegedly wrought. For this purpose I shall cite accounts by two medically qualified observers, Charles d'Eslon, whom we have already met, and M. Giraud, a Turin doctor who attended Mesmer's clinic at the hôtel Coigny and presented him with a *compte-rendu* of 31 patients from those attending the *baquet* reserved for the poor. All these patients were ones whom conventional medicine had failed (the d'Eslon ones had been his own patients).

Here, first, are three cases from the d'Eslon series.

1. Mlle ***, aged 35, had suffered for several years from a painful tumour, diagnosed as cancerous, in the lower part of her left breast. Several other tumours had formed around, and had grown together forming so great a swelling that the skin could scarcely withstand it. Two painful, lead-coloured prominences were joined to the original mischief, and at the tip of the breast had formed, breaking through, a blackish circle, the site of especial, shooting pains. Eventually the right breast too began to exhibit scattered lumps. Her general health was destroyed. She could find neither sleep nor rest. When Mesmer took on her case, she improved greatly. The secondary tumours disappeared; the principal one diminished considerably; the pains lessened; the patient could sleep and walk and travel in a vehicle, and her despair was replaced by calmness of mind.[80]

I have not discovered the fate of this patient; analogous cases suggest that her tumour would eventually have returned.

2. "When I presented the said *** to M. Mesmer, I thought he would refuse to treat her...her left eye was profoundly sunk into its socket, and probably softened. The right eye...was covered with a thick grey film, so that this lady was absolutely blind." Mesmer said that he could do nothing for the left eye. But after four or five weeks of treatment, the film had gone from the right eye, the eye had been brought back into position, and the patient could see very well.[81]

3. About September 1780 Mesmer undertook the treatment of M. Busson, physician to the Comtesse d'Artois. M. Busson suffered from an appalling nasal polypus,[82] which had displaced his right eye onto his temple and enlarged his right nostril as far as his cheek-bone. Hearing that Mesmer had procured the fall of another nasal polypus,[83] M. Busson begged d'Eslon to arrange a meeting with him. Mesmer began to treat Busson, and the polypus shortly fell beneath his hand. The eye returned to its socket, the nose recovered its normal size, and Busson believed himself cured. Word of this spread rapidly, and greatly alarmed M. Roussel de Vauzesmes, a young physician who was preparing a speech against d'Eslon for an assembly of the Faculty of Medicine. Vauzesmes claimed six leading physicians had denied that Busson's tumour was malignant. Unfortunately for M. Busson, for Mesmer, for Vauzesmes, and for the six leading physicians, it soon became apparent that malignancy was present, and Mesmer could do no more than soothe the pain of Busson's last days.[84]

The cases described by Giraud[85] suffer from the disadvantage that the case histories had to be obtained from the sufferers themselves, and were not medically authenticated. But the series has certain features of interest. There is a high proportion of rheumatic cases, and of cases of menstrual disorders, all successfully treated. Some of the other cases are at least curious. Case 10 is that of a man of sixty who had suffered for ten years from a perfect hemiplegia[86] of the left side, presumably as the result of a stroke. He was admitted to magnetic treatment on 28th May 1784. Magnetism produced sharp pains in the paralyzed parts. Before the end of June he could walk and move his arm. In case 12, again one of perfect hemiplegia following apoplexy, the time interval was much shorter. The stroke occurred in January 1784 (the patient was a man of thirty-three). Conventional remedies had had no effect by May, when magnetic treatment was begun; by the end of June the patient was capable of walking and making some arm movements.

CHARACTER AND INFLUENCE OF MESMER

Mesmer has been honoured as a pioneer by a variety of different groups, many of them at or beyond the fringes of scientific respectability. Not just psychiatrists and hypnotists, but theosophists, spiritualists, Swedenborgians, magicians, believers in occult physics, dowsers, and practitioners of radionics, radiaesthesia and psychotronics, have praised Mesmer's insights.[87] Naturally enough he has in consequence been often dismissed as a charlatan. Yet if being a charlatan involves conscious deceit, he was no charlatan. All the indications are that from about 1773 until his death in 1815, in public and in private, in theory and in practice, he adhered with bulldog tenacity to a body of doctrine which changed but little, and then only in inessentials. He looked upon himself as a scientist whose observations had led him to important theoretical and practical discoveries. But in what way his observations underwrote his theory or licensed his factual assertions remains obscure. He tells us little, for instance, about the observations which convinced him that animal magnetism can be stored, reflected or conducted by certain substances. My own suspicion is that the rationalistic physician, who had early on rejected the priesthood as a career, and had exposed the self-deception of the exorcist Gassner, had an inner tendency towards mysticism; his theory derived as much from the wordless communings with nature, of which he tells us in a rare passage of early autobiography[88], as from any observations he had made himself. He *felt* the pulsations of the cosmic fluid, suffusing and uniting all things, especially all living things, and he felt it flow with peculiar strength through him. He *knew* the theory to be true, and facts, observations, cases were simply fitted to it.

It is curious that Mesmer continues to enjoy a reputation as a pioneer of psychology and psychiatry. There seems to be a widespread covert assumption that because he has had a considerable influence on the history of psychology and psychiatry, he must in some sense have been a psychologist. In fact, there is little that is directly psychological in his writings, and what there is is mostly late, and concerned with the alleged phenomena of somnambulism and clairvoyance.[89] Nor did he regard animal magnetism as being a form of treatment specially suited to cases of mental disturbance. Of course, some of his patients may well have been suffering from hysterical or psychosomatic disorders.   Mlle de Berlancourt, for instance, one of his most celebrated successes,[90] was the precursor of a long line of young women who certainly suffered from such disorders, and whose improvements, relapses, and ultimate cures were recounted in loving detail by their attendant mesmerists. But Mesmer himself regarded most of these ailments as having a physical basis (e.g. obstructions in the viscera).

The urge to make a psychiatrist out of Mesmer is in some writers so strong that they present him as someone whose psychological intuitions, though unformulated, were so powerful that they in fact guided the therapeutic procedures which he thought he was basing on quite other principles.[91] All the indications are quite to the contrary. Mesmer seems to have been remarkably devoid of psychological insight. He was blinkered by his own single-mindedness and the paranoid tendencies which went with it. He could be amazingly insensitive, as can be seen time and again from his autobiographical writings and from his letter to Marie-Antoinette. If he carried conviction to his patients, it was not because of his psychological skills, but because he was himself, whole-heartedly, unswervingly, and unalterably, convinced of the truth of his own doctrines.

To understand Mesmer, and the impact which, almost accidentally, he has had on the history of psychology and psychiatry, one needs to look not at his writings, or at his therapeutic practices, but at his personality. I have met, talked to, and watched, several well-known "fringe" healers, of quite different persuasions, who share, I should suppose, some of Mesmer's characteristics: an imposing presence;[92] immense, but not oppressive, vitality; a benevolent appearance, reflecting a real goodness of heart;[93] unshakeable belief in some cherished set of doctrines; great obstinacy and pertinacity in propounding them; a rather paranoid reaction to criticism or incredulity; and a way of overcoming doubts and of sweeping patients along on a tide of enthusiasm and hope. Mesmer's influence on the history of psychological science owed little to his theory or to his particular therapeutic practices; it was mainly due to the personal qualities which made him a successful healer, and to the interest which his apparent successes stirred up in a number of high-minded, energetic and influential persons.

NOTES

1   On the last see Stone (1974).

2   Among biographical studies of Mesmer the most useful are those by Benz (1977); Buranelli (1976); Figuier (1881); Kerner (1856); Schurer-Waldheim (1930); Tischner (1928); Tischner and Bittel (1941); Vinchon (1936); Walmsley (1967); and Wohleb (1939). The best brief account is that by Gillispie (1980, pp. 261–289).

3   Bachelier d'Agès, d'Eslon and Bergasse were at different times his principal helpers.

4   Mesmer (1766). There is a French translation in Robert Amadou's edition of Mesmer's

works (Mesmer, 1971) which I shall use whenever possible, and an English one in Mesmer (1980). On Mesmer's sources, and the materials which he "lifted" from Richard Mead, see Pattie (1956).

5 Mesmer (1971), p. 40; cf. p.47, n15.

6 He was a friend and patron of Mozart; the first performance of *Bastien et Bastienne* took place at Mesmer's home in Vienna in 1768.

7 Mesmer (1779); see Mesmer (1971), p. 63.

8 These magnets were manufactured for Mesmer by Maximilian Hell, S.J., (1720–1792). The result was an acrimonious dispute over priority in the curative use of magnets. See the items reprinted in *Sammlung* (1778).

9 Mesmer (1971), p. 51.

10 Mesmer (1775); cf. Mesmer (1971), pp. 49–52, and Mesmer (1980), pp. 25–29.

11 J.J. Gassner (1727–1779). A useful account of Mesmer's encounter with Gassner will be found in Benz (1977), pp. 25–31. Materials on Gassner will be found in Semler (1776). For a detailed account of Mesmer's therapeutic procedures during his trip to Hungary in 1775 see Kerner (1856), pp. 18–50, Watts (1883), pp. 114–128. At this period he was using passes, magnets, and an electrical machine. There are even foreshadowings of the *baquet*. He dropped the use of magnets the following year.

12 A number of Mesmer's early cures are described in *Sammlung* (1778); cf. also Tischner (1928), pp. 37–54.

13 See the statement of Miss Paradis' father, Mesmer (1971), pp. 81–84.

14 She visited England in 1785; cf. *The Gentleman's Magazine*, 1785, 55, Pt.1, pp. 175–6. She played before their majesties at Windsor. She is said to have been able to distinguish colours by touch. On the "Mesmer-Paradis myth" see Pattie (1980–81).

15 E.g. Vinchon (1936), p. 69.

16 Figuier (1881), pp. 3–7; cf. in general Viatte (1928).

17 Oberkirch (1970), p. 495.

18 Darnton (1968), p. 23.

19 Vinchon (1936), p. 31.

20 See note 11.

21 Many of them seem in fact to derive from accounts of d'Eslon's clinic. For an amusing account, taken from Grimm, of Mesmer's establishment at the hôtel Bullion, see Wohleb (1939), pp. 48–50.

22 As a sample of contemporary scuttlebut about Mesmer I cite from Doppet (1784a) the following lines about Mesmer's famous finger:

> Le magnétisme alors de l'amour fait la fête.
> C'est-là que mon héros charmant, magnétisant,
> Sous un verrou fermé dissipe un feu naissant;
> Il fait trouver l'endroit pour fixer la cruelle;
> Jamais son heureux doigt ne trouve une rebelle.

There is more in the same strain.

23 Brack (1784); Figuier (1881), p. 60.

24 See Gillispie (1980), pp. 194–203.

25 Vess (1975), p. 17.

26 The *Dictionnaire de biographie française* gives his dates as 1750–1786.

27 For d'Eslon's account of his early relations with Mesmer see d'Eslon (1781), pp. 19–29; For Mesmer's account see the latter's *Précis historique* in Mesmer (1971), pp. 325–326. D'Eslon's name is sometimes spelled Deslon or Delon.

28 D'Eslon (1781; 1782; 1784).

29 On Mesmer's reception of the news see Bergasse (1786), pp. 35–38. For his denial that d'Eslon had been fully instructed in animal magnetism see his letter to M. Philip, Dean of the Faculty, and his exchange with d'Eslon, Mesmer (1971), pp. 229–233. Bergasse, p. 38, says that he drafted the letter to M. Philip on Mesmer's behalf.

30 D'Eslon (1784), pp. 38–39.

31 Mesmer (1971), pp. 236–243.

32 On Bergasse see especially Darnton (1968).

33 See *Cure operée par M. Mesmer* (1785).

34 Audry (1924), p. 24, quotes a popular song of the period:

> Dans l'ordre de l'harmonie,
> Des élus seront admis,
> Ils consacreront leur vie
> Au secours de leurs amis.
> Mais que chacun ait, je prie,
> La quittance de cent louis.

35 Vinchon (1936), p. 78; Brack (1784), p. 17; *Tableau des cent premiers membres* (1784), cited in Dureau (1869), pp. 32–33.

36 Brack (1784), p. 21. The other three, he says, brought in 300 louis a month.

37 See Mesmer's letter to the *Journal de Paris* of 6th January 1785, reprinted in Mesmer (1971), p. 259. Caullet de Veaumorel's reply is in the *Journal de Paris* for 9th January 1785. He maintains that the *Aphorismes* agree with the notebooks of several persons. Mesmer (1814), pp. 163–197 (on mesmerizing trees) is lifted straight from the *Aphorismes*; Bergasse (1786), p. 70, says that the teachings of the *Aphorismes* are essentially Mesmer's.

38 See Baron de Corberon's account in Darnton (1968), pp. 180–182.

39 On these see Brack (1784), pp. 23ff.

40 Among satirical verse and prose pieces are Doppet (1784a), quoted above; Doppet (1785); *Histoire véritable* (1785); *Prophétie du douzième siécle* (1785); Meltier (1787). Among plays Guigoud-Pigale (1784); Barré and Radet (1785); and Duval d'Esprémesnil (1784), which is a reply to the last.

41 The best source of information about these is still Dureau (1869).

42 Thouret (1785); reprinted in Burdin and Dubois (1841), pp. 190–236. M. A. Thouret (1748–1810) was an energetic and leading member of the Royal Society of Medicine, and a vigorous opponent of animal magnetism. His previous work on the topic, which contains many historical parallels (Thouret, 1784), was much admired. He was the first Dean of the École de Santé founded at Paris in 1794. The impartiality of his selection of correspondence for publication is called in question by Valleton de Boissière (1785) and by Bonnefoy (1785).

43 Burdin and Dubois (1841), p. 190.

44 From the membership list of the Paris Society of Harmony, published in the *Journal du magnétisme*, 1852, 11, pp. 131–139, 198–203, 240–243, 266–269, 496–503, 570–577, 626–631, we learn that pupils of Mesmer, a very high percentage of them medically qualified, were commissioned to set up magnetic treatments at Malta, St. Domingue (Haiti), Versailles, Bourbonne-les-Bains, Lyon, Amiens, Bordeaux, the United States, Quimper, Mâcon, Langon, Chartres, Rouen, Castres, Saint-Etienne en Forez, Bourg-en-Bresse, Saint-Macaire, Nantes, Dijon, Erné, Dinan, Beaune, Aix-en-Provence, Toulon and Constantinople. Treatments at La Sauve Entre-deux-mers, Toulouse, Montaubon, Grenoble, Beaubourg, Valence en Brie, Calais, Clermont, and Marseille are mentioned by

other writers, e.g. A. M. J. de Chastenet de Puységur (1786), pp. 190–191; Thouret (1785); Brack (1784), pp. 28–29.

45   Burdin and Dubois (1841), pp. 193–194.

46   Vess (1975), pp. 3–4.

47   Cf. Servan (1784a), p. 7.

48   Lafayette (no. 91 on the membership list of the Society of Harmony – see note 44 above) tried to rouse interest in mesmerism in the United States during a visit there in 1784. See Benz (1977), pp. 34–37.

49   Court de Gebelin (1783), pp. 42–43; Viatte (1928), Vol. I, p. 223.

50   On Bergasse's clashes with Mesmer see Corberon's account in Darnton (1968), p. 182. Bergasse's vanity is scathingly described from personal observation by Camille Desmoulins (Viatte, 1928, Vol. I, p. 228n).

51   Brissot, quoted by Viatte (1928), Vol. I, pp. 228n-229n, says of Bergasse: "Bergasse did not hide from me that in raising an altar to magnetism, he had nothing in view but making it an altar to liberty ... The time has come, he said to me, when France needs a revolution. But to wish to bring it about openly is to wish to fail; to succeed it must be wrapped in mystery; we must unite men under the pretext of experiments in physical science, but in reality to overthrow despotism."

52   See Bertrand (1826), pp. 51–61; Deleuze (1813), I, p. 29.

53   Duval d'Esprémesnil (1786); Mesmer (1971), pp. 262n-264n; Bergasse (1784) pp. 30n-34n. Mesmer is said by June 1785 to have received in subscriptions from the Paris and other Societies of Harmony a total of 343764 livres.

54   His position may be judged from his letter to Duval d'Esprémesnil, reprinted in Mesmer (1971), pp. 262–275; *Recueil de pièces* (1786), pp. 9–28. This is a bitter attack on Bergasse, who replied with Bergasse (1786). Mesmer may at this time have been contemplating setting up another Society in England, a venture which would have been impeded if his teachings had become too widely disseminated. He seems in fact to have visited London in August 1785. See Banks (1958), p. 866. (I am indebted to Patricia Fara for pointing out this reference.)

55   See especially Bergasse (1786).

56   See the list referred to in note 44 above.

57   There seem to have been societies in at any rate the following places: Metz, Nancy, Guienne, Bayonne, Bordeaux, Saint-Etienne, Saint-Domingue, Lyon, Grenoble, Amiens, Chartres, Turin, Malta, Berne, Strasbourg, Marseille, and Stockholm. See A. M. J. de Chastenet de Puységur (1807); Bergasse (1784), pp. 135–136; Deleuze (1813), I, pp. 24–25, 2, pp. 128–132; Mesmer (1971), p. 264n; Nicolas (1784); Tournier (1911); and the account of the Strasbourg Society in chapter 2 below. Of these societies, only those of Bordeaux, Saint-Etienne and Saint-Domingue are recorded as contributing to the funds of the main (Paris) Society (Mesmer, 1971, p. 264n).

58   Wohleb (1939) gives much useful information about Mesmer's wanderings and later days.

59   Mesmer (1799; 1812; 1814).

60   Mesmer (1971), pp. 225–227.

61   Cf. note 37 above.

62   Mesmer (1971), p. 294. There is a large literature from Thouret (1784) onwards tracing the intellectual ancestry of Mesmer's ideas back to the seventeenth century, sometimes with the implication that he is a conscious or unconscious plagiarist. I have omitted discussion of this topic as not being relevant to the purposes of this book. For convenient brief treatments see Benz (1977) and Podmore (1909, ch. 2). There is no doubt that ideas recognizably similar to Mesmer's are to be found scattered in the eighteenth century medical literature which he

had almost certainly studied. For instance Richard Mead (1673–1754) and Friedrich Hoffmann (1660–1742) had both speculated about the identity of "nervous fluid" or "nervous liquor" with the universal ether postulated by Newton, and doctrines about obstructions were prominent in the system of van Swieten's teacher, Hermann Boerhaave (1668–1738).

63 Mesmer (1785a), aphorisms CCCIV and CCV.

64 Cf.A. M. J.de Chastenet de Puységur (1807), pp. 134–136. The complex system of poles described by Paulet (1784), pp. 3–17, seems to be a satire or an invention.

65 Mesmer (1814); cf. Schott (1982). The later ideas are foreshadowed in Mesmer (1799).

66 A somewhat different account of Mesmer's therapeutic practices to that contained in the Aphorismes will be found in Système raisonné (1786), pp. 49–55.

67 On the orthodox remedies favoured by Mesmer see Mesmer (1785a), aphorisms CCCXII and CCCXIII; Moulinié (1785), p. 302; and Mahon (1784), p. 22.

68 Mesmer (1785a), aphorism CCXXXVIII.

69 The notion that a crisis must precede recovery is of course an old one.

70 Cf. d'Eslon (1781), p. 79.

71 Mesmer (1971), pp. 300, 301.

72 Some editions of Mesmer's Aphorismes have a note or appendix describing d'Eslon's methods which, in a general way, are of course very similar to Mesmer's. A detailed eyewitness account of d'Eslon's procedures is given by Jussieu (1784), in Burdin and Dubois (1841), pp. 146–163.

73 A. H. A. de Chastenet de Puységur (1783), pp. 11–12.

74 Watts (1883). p. 121.

75 Wohleb (1939), p. 112.

76 Grézie (1784). pp. 4–5; cf. Servan (1784a), pp. 39–40.

77 The extremes are well represented by, on the sceptical side, Paulet (1784a; 1784b; 1785), and, on the side of the believers, Hervier (1784), who sets Mesmer above Descartes and Newton.

78 For instance, I should be disinclined to accept unchecked any of Paulet's versions of the case histories of Mesmer's patients (Paulet, 1784a, pp. 152–168).

79 To the Paris Faculty of Medicine. See Mesmer's Précis historique, Mesmer (1971), p. 139. The proposal was rejected. Pattie (Mesmer, 1971, p. 320, n5) says Hell suggested such experiments to Mesmer.

80 D'Eslon (1781), pp. 194–196.

81 D'Eslon (1781), pp. 50–52.

82 A tumour as it were on a stalk. Nasal polypi are not usually malignant.

83 Bouvier (1784), p. 4.

84 D'Eslon (1782), p. 22; Mesmer (1781) in Mesmer (1971), pp. 156–157, 165–166.

85 Giraud (1784).

86 A paralysis of the muscles of one side of the body.

87 See, among many others, Bush (1847) and Sinnett (1892).

88 Kerner (1856); Watts (1883), pp. 142–143.

89 He devotes substantial parts of the Mémoire of 1799 to attempts to explain somnambulism clairvoyance, precognition, etc., in terms of his universal fluid; and he adopts not dissimilar views in the Mesmerismus of 1814. The ideas are to some extent foreshadowed in the section on nervous maladies in the Aphorismes. I shall touch on such speculations in a later chapter.

90 See Fournier-Michel (1781).

91 E.g. Ellenberger (1970), p. 69; Buranelli (1976), pp. 72–73; Schott (1982), pp. 208–212.

92 A description of Mesmer, from an identity card of 1798, says he was "aged 64, 1.76 m. in

height, hair and eyebrows brown, chin double, face full, forehead high, nose and mouth medium." (Buranelli, 1976, p. 187).

93 Testimonies to Mesmer's benevolence towards the sick and to the worthiness of his character are given by, for instance, A. H. A. de Chastenet de Puységur (1783), pp. 17–18; d'Eslon (1781), p. 79; Court de Gebelin (1783) in *Recueil des pièces* (1785), pp. 156–157; and Moulinié (1785), pp. 312–313. Hervier (1784), pp. 37–38, says it was Mesmer's custom to give poor patients something to live on between visits to the *baquet*. Deleuze (1813), I, p. 19n, tells the story of Mesmer's generosity to M. Nicolas, of Grenoble (author of Nicolas, 1784), a physician who could ill-afford the membership subscription to the Society of Harmony. Mesmer surreptitiously slipped him the requisite amount, thus, as it were, paying his left hand with his right hand. Deleuze had the story from Nicolas. Mesmer probably was generous to those whom he did not regard as his equals. A totally different view of his character is, as one might expect, presented by Bergasse (1786).

PART I

The successors of Mesmer

CHAPTER 1

The Royal Commissions and the pamphlet war

Mesmer had always been "news",[1] and the success of his clinic, together with the publication in 1779 of his *Mémoire sur la découverte du magnétisme animal*, made him something more than "news". He became the centre of a rapidly intensifying pamphlet war in which medical men and amateur *savants* of various kinds were the principal protagonists. In this war jealousy, gratitude, curiosity and (more rarely) public spirit were among the motives; while among the weapons satire, misrepresentation and plain lying were at least as common as reason and the judicious observation of facts.

Of Mesmer's enemies, the most persistent was J. J.Paulet (1740–1826), editor of the *Gazette de santé*, in which he attacked Mesmer with considerable ability and not much scruple.[2] I would be reluctant to accuse him of simply inventing his facts; let us just say that no *canard* was too gross or too preposterous for inclusion. His later pamphlets (1784a; 1784b) contain many gems from the same vein. A staunch defender was Charles d'Eslon, whom we have already met (d'Eslon 1781; 1782).

Following the publication in July 1783 of Court de Gebelin's *Lettre de l'auteur du* Monde primitif, the pamphlet war escalated. Antoine Court de Gebelin (1728–1784), a well-known antiquary, had sustained two wounds in his left leg. The leg had become swollen and covered in erysipelatous boils. His right leg had withered from lack of exercise, and other afflictions followed. In March 1783, a friend induced him to seek Mesmer's help. Mesmer visited him and said that the cause of the swellings was obstructions hindering the free circulation of the humours. He passed and repassed his hand over the swellings, and the next day Court was able to attend Mesmer's clinic. In a few weeks he had completely recovered and wrote the pamphlet just mentioned. In it he describes his own case, various cases from the literature, and some striking cases that had come under his own observation; he rejects the idea that the cures or ameliorations merely happened to coincide with the treatments, and he is unrestrained in his praise of Mesmer and his theory, and of animal magnetism and its benefits.

The case of Court de Gebelin was uncritically and enthusiastically taken up by Mesmer's supporters. The least critical and the most enthusiastic was perhaps

Father C. Hervier, librarian of the Grand Augustinians, whose *Lettre sur la découverte du magnétisme animal* (1784) describes his own illness, and his cure by Mesmer, and praises Mesmer as not just the equal but the superior of Descartes and Newton.[3] To this effusion Girard[4] replies that Hervier's fellow monks regard his ailment as imaginary. An inhabitant of Bordeaux accuses Hervier of self-contradictions,[5] and an anonymous "member of several academies" says that his views are irreligious and ought not to be allowed.[6] M. de Bourzeis, doctor in ordinary to the King, accuses Mesmer of killing a patient,[7] whilst a "doctor of the Faculty of Paris" informs M. Court de Jebelin (*sic*) that he had never really been ill.[8]. Court silenced this critic by dying (May 1784) of malignant disease.[9] His death unleashed a tasteless and quite unjustified howl of triumph from Mesmer's enemies.[10]

Not all of the pamphlet and periodical war was conducted on such a low level. I shall discuss certain more noteworthy pamphlets in the ensuing chapters. In August 1784 appeared the most important pamphlets of all – the reports on animal magnetism of the two Royal Commissions chosen respectively from the Paris Faculty of Medicine and the Royal Academy of Sciences, and from the members of the Royal Society of Medicine.[11]

THE REPORTS OF THE ROYAL COMMISSIONS

The former report is considerably the more interesting. The commissioners from the Faculty of Medicine were Majault, Sallin, J. Darcet (1727–1801), and J. I. Guillotin (1738–1814). The scientific commissioners were Benjamin Franklin (1706–1790), J. B. Laroi (c.1724–1800), J. S. Bailly (1736–1793), G. de Bory (1720–1801) and A. L. de Lavoisier (1743–1794). (Two of these commissioners, Bailly and Lavoisier, later met their deaths on the device named after their colleague Guillotin.) The report, drawn up by Bailly, is elegantly written, and breathes a tone of calm reasonableness and careful thought which sets it quite apart from most previous contributions to the animal magnetism debate.

The commissioners of the Royal Society of Medicine were P. I. Poissonnier, Caille, P. J. C. Mauduyt de la Varenne, F. Andry (1741–1829) and A. L. de Jussieu (1748–1836), the last of whom issued a separate and partially dissenting report. The others seem to have felt it their special duty to take a strong line against dubious medical practices, and their report is very much a case for the prosecution.

According to the Commissioners of the Academy and the Faculty, the prime question is not whether animal magnetism works, but whether it exists. For something that does not exist can produce no effects, such as cures. Some magnetists have thought its existence demonstrated by the fact that certain subjects may see the fluid emanating from the operator's fingers, or feel sensations of warmth or cold as his hand moves above their faces. In the former case, what is seen is the vapours of perspiration magnified by the "solar microscope"; in the latter the subjects are merely detecting air-currents or heat. Those who think they can detect the characteristic odour of the fluid when the operator's finger, or metal wand, pass beneath the nose, are really detecting the smell of sweaty fingers or of warm and sweat-impregnated iron. Nor can the cures so often attributed to animal magnetism be taken as proof of its reality, for nature on her own cures a large proportion of patients.[12]

The commissioners undertook some experiments of their own upon the alleged "momentary effects" of animal magnetism upon the "animal economy". They began by submitting themselves to magnetization at d'Eslon's clinic – d'Eslon was at all times candid and cooperative towards the commissioners who were investigating his claims. Since it is very easy to find, or imagine, all sorts of small pains or feelings of warmth, they took care not to be too attentive to what passed within them. Not surprisingly, they felt very little, and nothing that they thought could be unequivocally set down to animal magnetism.

Recognizing that animal magnetism was supposed to affect only those in indifferent health, the commissioners next presented d'Eslon with seven sick persons drawn from the ranks of "the people". When magnetized, three of these suffered assorted pains and twitchings. But unfortunately a number of other patients, from "society", and "gifted with more intelligence and more ability to give an account of their sensations", experienced very little, and the commissioners dismissed the statements of the poorer persons as the products of expectation and compliance.[13]

Noting that sceptics and children too young to realize what is supposed to happen rarely respond to magnetization, the commissioners organized a lengthy and quite careful series of experiments to test the proposition that persons susceptible to animal magnetism would feel its supposed effects when they merely imagined themselves subjected to it, and would fail to feel them if unaware that they were being magnetized. I can only give one brief example of these experiments.

The subject in this case was a serving-man whose eyes were covered by a formidable blindfold. He was induced to believe that he was being magnetized although he was not.

> Then he felt an almost general warmth, and movements in his stomach. His head grew heavy. Little by little he became drowsy, and he seemed on the point of going to sleep. Which proves, as we said above, that this effect related to the situation, to boredom, and not to magnetism.
>
> Magnetized next with his eyes uncovered, and the iron rod presented to his forehead, he felt prickings; when his eyes were rebandaged, and it was presented to him, he did not feel it at all; and when, though it was not presented, he was asked if he did not feel something on his forehead, he replied that he felt something go and come in the expanse of his forehead.[14]

Experiments such as this constituted *prima facie* evidence that the sensations and other feelings so often reported by magnetized subjects had psychological causes – presumably "imagination". Could imagination likewise be the cause of the disagreeable convulsions which so often occurred in Mesmer's and d'Eslon's treatments? More experiments of a design similar to the above suggested that it could. Furthermore, an unfortunate lad of 12, made to embrace four trees which he believed d'Eslon might have magnetized, felt progressively stronger effects at each, and at the fourth fell into convulsions. At no time did he in fact come closer than 24 feet to the tree which d'Eslon had actually magnetized. Similarly, a woman, one of d'Eslon's most susceptible subjects, fell into convulsions when she drank non-magnetized water which she thought had been magnetized, and was unaffected by magnetized water which she thought had not been magnetized.

The commissioners suggest additional causes for the convulsions. Touch and pressure may irritate the nerve plexuses of the lower stomach, which in turn communicate with all other parts of the body; pressure in that region may also facilitate the evacuations which were a common and valuable result of magnetization. The commissioners of the Royal Society of Medicine mention certain accessory causes arising from the ambience of the treatment: the warmth; the heavy and impure air; the serious, indeed impressive, scene; the gloomy reflections it is likely to inspire; and above all the effect upon sensitive persons of seeing others fall into convulsions.[15] "Everywhere", say the academic commissioners, "example works on the mind, mechanical imitation sets the body in play. We have a recent example in the young girls of Saint-Roch,[16] who, separated, were cured of convulsions from which they suffered again when reunited."[17]

The commissioners of the Royal Society of Medicine followed the fortunes of a number of patients at d'Eslon's *baquets* but their account is too brief to be of much value. These patients, they say,[18] fell into three classes: those whose ailments were marked and had a known cause; those whose ailments were slight, vague and without determinate cause; and melancholics. In four months they saw no patient of the first category either cured or notably relieved. Cures of such patients had, it is true, been reported. "But ... out of a multitude of sick persons ... nature will cure some, and in a space of time often less than that required for animal magnetism."[19] Cures of the other classes of patients may be attributed to the hope inspired by the treatment, the exercise involved in visiting the *baquets*, the renunciation of harmful remedies, or the fact that the patients were never really ill.

Finally, both sets of commissioners issued warnings of the danger to health posed by the convulsive crises, and in a "secret" report, also drafted by Bailly, the academic commissioners dwelt upon the moral dangers occasioned by the practice of animal magnetism. It is always men who magnetize women. The process involves knee to knee contact and application of the hand, even in the neighbourhood of the most tender parts of the body. The nervous systems of women, being more highly strung than those of men, are liable to be stirred without their fully realizing it. "In lively and sensitive women", says the report, " ... the end of the sweetest of emotions, is often a sort of convulsion; this state is succeeded by ... a sort of sleep of the senses ... ".[20] A scoundrel might exploit such sensitivity, and even a well-intentioned operator might be unable to resist the temptations which might arise.

THE PAMPHLET WAR

Copies of the reports were printed in a quantity[21] and distributed with a speed which suggest that the Government was exceedingly anxious to bury animal magnetism for good. And the first signs were encouraging. The reports received a great deal of publicity. They were even in part translated into English in *The Gentleman's Magazine*.[22] But the supporters of animal magnetism refused to be intimidated. A considerable number of replies to the reports appeared shortly. It is impossible to deal with them all separately, nor do they deserve it. I shall, however, attempt a digest of some of the leading themes of the better ones, for instance: the dissenting report of A. L. de Jussieu, a distinguished botanist who was one of the commissioners of the Royal Society of Medicine;[23] Charles d'Eslon's *Observations sur les deux rapports* (1784); the *Supplément aux deux rapports* (1784) by a group of d'Eslon's supporters; *Réflexions impartiales* (1784) by A. E. de Dampierre (1743–1824), a counsellor of the *parlement* of Burgundy; *Doutes d'un provincial* (1784) and *Questions du jeune docteur Rhubarbini de Purgandis* (1784) by J. M. A. Servan (1737–1807), advocate general to the *parlement* of Grenoble; *Analyse raisonée des rapports* (1784) by J. B. Bonnefoy (1756–1790), a Lyons surgeon; and *Remontrances des malades* (1785) by J. F. Fournel (1745–1820), a Paris advocate.

A frequently recurring theme is that of sheer hostility to the medical profession, at whose hands, or lancets, several of the authors seem to have suffered grievously. One gets a vivid picture of the barbarities and crudities of late eighteenth century medicine. The bitterness felt towards the medical profession by certain of its victims has of course no direct bearing upon the validity or otherwise of the two reports. But there is reason behind the bitterness. For even at the time it did not take too much thought to perceive that a good deal of contemporary medical treatment was mere ignorance and superstition. Furthermore, says Servan, throughout the history of medicine it has been doctors themselves who have been the chief opponents of every beneficial medical discovery.[24] Who are such persons – and nine of the commissioners were medical men – to pronounce upon the merits of animal magnetism, and why should we listen to their pronouncements?

A second major theme in the replies consists in what may be called giving the lie to particular claims or assertions made by the commissioners. For instance, d'Eslon states that the commissioners of the Royal Society of Medicine saw at his clinic three patients, two of whom they had presented to him themselves, whose maladies improved during the period of observation. He quotes a certificate, signed by two of the commissioners, concerning one of them, a girl of nine suffering from scrofula. Yet the commissioners deny that they saw any improvement in any malady of known cause. D'Eslon also states that convulsions are nothing like so common among his patients as the commissioners make out, only 20 patients in more than 500 having suffered from them; as for the supposed dangers to which the convulsions give rise, he declares that only five of his patients have died during the last three years, and that all the world knows these were in a desperate state when they came to him. Furthermore, crises take many forms other than convulsions – some patients cough and spit, others sleep, and others are agitated and troubled – so how can the commissioners use "imitation" to explain the supposed prevalence of convulsions at his clinic? Nor can the convulsions be explained in terms of the irritation of abdominal nerve plexuses by strong pressure of the operator's hands; the touches employed are always soft and light, never powerful.[25]

Several of these claims are confirmed by Bonnefoy as an eyewitness of Mesmer's procedures. At Mesmer's clinic he has seen only eight convulsive crises among more than two hundred patients, and he has observed crises of which the aftermath has been unmistakably beneficial. The commissioners, he adds, make the treatment rooms sound like places of darkness and horror which one should tremble to approach. Mesmer's clinic, and that at Lyon, are not like this at all. Windows and curtains are always open, weather permitting, and no-one observes silence. Tranquillity, cheerfulness, laughter, and varied and amusing conversation, make the time pass quickly.[26] The truth, I suspect, is that convulsions at first occurred not infrequently around Mesmer's *baquets*, but that he later on somewhat discouraged them because they aroused so much unfavourable comment.[27]

Perhaps the dominant theme in the replies to the reports is dissatisfaction with the commissioners' imagination-explanation of magnetic phenomena. The objections can be taken under two heads: criticisms of the imagination-hypothesis as applied to the pains, feelings of warmth, etc. to which the process of magnetization so often gave rise,[28] and criticisms of the hypothesis as applied to the alleged cures.

The critics of the reports are least effective on the former topic. Some believed they had evidence for phenomena which were demonstrably independent of the imagination. Jussieu, for instance, succeeded in producing sensations and convulsions in patients whom he believed unaware that he was magnetizing them; but the precautions which he took to exclude sensory cues were hopelessly inadequate.[29] Other critics argued that the circumstances of the commissioners' experiments were such as to confuse and baffle the subjects. According to Bonnefoy, the ignorant persons who served as subjects for the assembled commissioners would have been in a state of turmoil likely to obliterate all magnetic sensations. On the other hand, when they think they are being magnetized, although they are not, " ... they will carefully study their sensations, and if they have an ailing part it will reveal itself by pain as a result of the attention paid to it."[30]

Turning now to the imagination-hypothesis as applied to the cures: the commissioners of the Royal Society of Medicine denied that they had seen any cures among patients seriously ill with known diseases. The academic commissioners, on the other hand, declined, for reasons mentioned above, to investigate the alleged cures. Strictly speaking, therefore, neither set of commissioners propounded the imagination-hypothesis of the cures. Yet almost to a man, critics responded as if they had done so. And these critics had considerable justification. For when the academic commissioners have finished demonstrating that the sensations, convulsions, etc. occasioned by magnetic procedures were really due to "imagination", they remark:

> It remains to consider whether the crises or the convulsions ... can be useful in curing or improving the sick. Without doubt the imagination of patients often has a considerable influence on the cure of their ailments. The effect is only known through general experience, and has not been established by positive experiments; but it does not seem that one could doubt it.[31]

The implication seems clear that, insofar as the magnetic procedures result in cures, those cures may be ascribed to the power of the imagination. Thus although the commissioners had not investigated the alleged cures, they create the impression that the imagination-hypothesis is a sufficient explanation of them. And this was obviously unsatisfactory, for two reasons: first, it was quite conceivable that the sensations and convulsions should (in some sense) have been due to the imagination, whilst the cures were due to a physical influence directly linked to the magnetic "passes". And, second, readers were led to suppose that there was nothing about the alleged cures that required further investigation, whereas if some of them could have been genuine (which the academic commissioners did not altogether deny), public policy and public interest alike dictated that the question should have been further investigated.

The excuses offered by the academic commissioners for not investigating the alleged cures are specious to the point of casuistry. They remark that since nature cures diseases anyway, animal magnetism might reap the credit. But, as Servan justly observes,[32] this difficulty could have been approached by comparing the progress of magnetized and unmagnetized groups of patients, and it is hard to suppose that such a solution did not occur to scientists as distinguished as the academic commissioners. The commissioners also advanced as a reason for not studying the results obtained at d'Eslon's clinic the possibility that the distinguished patients who frequented it could have been annoyed at their questions.[33] This unworthy prevarication was, as we shall see, exposed as nonsense by the distinguished patients themselves, who came forward in large numbers upon publication of the reports to state publicly how much they had benefitted from the magnetic treatment.[34]

Several critics of the reports tackle the imagination-hypothesis of the cures by citing cures which, they believe, can neither be denied nor set down to imagination.[35] However, the case histories are usually so brief and so little substantiated as to be of little value.

The most interesting work of this kind is a substantial pamphlet, the *Supplément aux deux rapports* (1784), prepared, apparently, by a group of d'Eslon's supporters. It contains 115 case histories,[36] mostly written by the patients, though some are by medical men. The patients' accounts are, however, in most cases obviously derived from their doctors. Many of the patients were of high social standing, a fact no doubt reflecting d'Eslon's position as physician to the Comte d'Artois. As Podmore remarks, for persons of education and refinement, "it can scarcely have been a congenial task to enter into intimate and frequently repulsive details of their maladies and cure. It is evident, in fact, that in many cases nothing short of a conviction of the unfairness of the report, and a strong sense of gratitude to Deslon and his colleagues, would have induced them to come forward."[37]

The compilers of the *Supplément* present their certificates in four categories, which emphasize the shortcomings of the imagination-hypothesis. The first is that of cures and effects wrought on children; the second that of cures and improvements in persons who experienced neither special sensations nor convulsions; the third that of patients who experienced the sensations of cold, heat, pain, etc., but did not go into convulsions; the fourth (and smallest) that of patients subject to convulsions and crises.

Among the cases were

> ... ailments of every kind, long-standing ailments on which the art of medicine was exhausted over a long period of years; nevertheless one sees that a large number [about half] of these patients have been cured, nearly all the others greatly improved, very few [six] who have cause to complain of the ineffectiveness of magnetism, and not one who has found it harmful or even dangerous.[38]

Many were cases of rheumatic conditions or gout. For instance, for over a year Madame de la Perrière (category two), a farmer-general, had had rheumatic pains all over the body; her fingers had become bent, and nodules had formed all over her joints. Treatment with animal magnetism during the summer of 1783 removed these.[39] Sometimes relief was almost instantaneous. M. Chauvet, a priest (category three), had suffered continual and sometimes immobilizing rheumatic pains in the arm for over five years. In September 1783 he visited d'Eslon. The latter placed his hand on the patient's shoulder. Chauvet at once broke into a profuse perspiration on his left side, and his pain disappeared. A year later he writes that he no longer knows what rheumatism is.[40] M. Perruchot (category two), who had suffered from gout for three years, was laid up by a severe attack. D'Eslon treated him at home. The effect was so remarkable that in the evening he was able to go out paying calls. Let us trust that he passed the time abstemiously.[41]

Although the symptoms described were sometimes of a rather vague character, there are not a few accounts of cures of ailments which were at once serious and unmistakable. Magdelon Prin (category two), a portress, had suffered since she was fifteen from tumours the size of eggs on her leg and thigh. Ten weeks' treatment by d'Eslon saw their departure.[42] Dame Gaddart (category four), housekeeper to Madame d'Avignon, had suffered for seven years from a scirrhus (a hard tumour) nearly as big as a head, and from associated dropsy and visceral problems. After treatment by animal magnetism for nearly a year her general health was better and the tumour had altogether disappeared.[43]

M. Patillon, a doctor of the Faculty of Besançon, describes from his own practice a case that has modern parallels in the literature of hypnotism. A young girl of eleven (category two) had suffered since birth from a skin disease "which might be called leprous". Conventional treatments had been wholly ineffective. After fifteen days of animal magnetism the skin had changed from leaden colour to white, and the scabs had begun to fall off, leaving a healthy skin behind. Treatment was still in progress at the time of writing.[44]

The compilers of the *Supplément* have this to say about the imagination-hypothesis of the cures:

> If magnetism brings on *slumber*, it is my imagination which makes me *sleep*. If I am purged, better than I would be by *manna* and *double catholicon*, it is my imagination that purges me. If I render an abscess, a gathering, if sharp pains are calmed, if, beneath the hand which magnetizes me, a colic disappears; if at last, after ten years of privation of appetite, sleep and health, I recover these boons, I am indebted only to my imagination! It is imagination which cures this scrofulous infant, still sucking its nurse's breast! And when we believe that it is magnetism that calls an apoplectic, in a coma, back to life, we deceive ourselves again; it is his imagination which, triumphing over his inertness, brings back sensation and life.[45]

The compilers here pick out two of the three kinds of evidence most commonly held to count against the imagination-hypothesis; namely the alleged influence of animal magnetism upon unconscious persons, and upon very young children (the third was its alleged influence upon animals[46]). They later present twelve cases of the cure or amelioration of ailments in children. One or two of the cases are perhaps of some interest. A boy of two had his arm burned to the elbow, the skin being entirely destroyed. He was cured without permanent scarring after nine days of magnetic treatment, no drugs or ointments being used. A boy of six months, suffering from spasmodic breathing, inversion of the eyes, and greyness of the skin (possible symptoms, Podmore suggests, of retropharyngeal abscess, a very serious condition) was much improved after only one and a quarter hours of treatment. His breathing got better and his eyes resumed their natural position. Further treatment brought a discharge of pus from the nose, followed by a complete cure.[47]

After reading the *Supplément*, one is inescapably confronted with two questions:

1. Would all or many of these patients have recovered or improved as they did if they had not been subjected to the procedures of animal magnetism? Here one must remember that many of these had already exhausted all conventional remedies.

2. Is "imagination", as that is ordinarily conceived, a sufficient explanation of the reported cures and ameliorations? Here it must be remembered that many of the patients were suffering from ailments of an unmistakably physical character; that such ailments are not ordinarily cured by simply imagining them better; and that cures and ameliorations were apparently wrought on persons who were not sufficiently imaginative to feel the customary sensations of heat, pain, etc., or to pass into convulsions.

If one has to answer "no" to these questions, or if (like the present writer) one is simply unable to answer "yes", then one has, I think, to say that the commissioners failed to get to grips with issues that clearly merited investigation.

LE COLOSSE AUX PIÈDS D'ARGILE

Replies to the critics were not numerous and need not long detain us. Servan's *Doutes d'un provincial* attracted one or two answers of no great note.[48] The commissioners themselves did not respond. However, the pamphlet by Thouret (1785), quoted in the previous chapter, is in effect a continuation of the Royal Society of Medicine's war against animal magnetism, and that by Devillers (1784) is intended as a general reply to all the critics of the reports. Thouret stresses the rôle of emotion and motivation in producing the phenomena attributed to animal magnetism. Doctors have long known, he says, that extreme confidence, a burning desire to be cured, and a great hope of being so by a cause thought of as almost supernatural, can in certain maladies work beneficial changes. Pains may be removed or assuaged, tumours diminish or disappear, paralyzed limbs regain some movement. These effects have often been the result of strong emotion. Mesmer's cures are chiefly due to the enthusiasm he has aroused in credulous persons.[49]

Devillers was a medical man and a member of the Academy of Ville-Franche, Rouen and Marseille.[50] He has for the commissioners that sort of respect, almost reverence, which it is only proper that a member of the lesser academy should show towards members of the greater. His *Le colosse aux pièds d'argille* (1784) is a bluff and fiddlesticks reply to leading critics of the reports, especially Servan, Dampierre and Bonnefoy. He has little difficulty in exposing the rashness of some of their arguments. Yet somehow he never gets to grips with what seem, in retrospect, their central contentions. His treatment of the *Supplément aux deux rapports* is very curious.[51] The magnetic fluid does not exist. What, then, has cured the 54 patients who felt nothing during the magnetic treatment, and who were therefore presumably not excessively imaginative? Why, what else but the other remedies they were undoubtedly taking. Devillers does not seem troubled by the fact that in most cases these remedies had remained ineffective until the patients began attending d'Eslon's clinic. Indeed, the commissioners of the Royal Society of Medicine had advanced exactly the opposite argument to explain certain cures – they proposed that the cures were due to the patients' ceasing to take harmful remedies.[52] If the left barrel misses, the right one is bound to hit.

Devillers does, however, accept certain of the animal magnetic cures. He draws an extended parallel (which others had already drawn) between the convulsions undergone by magnetized patients and the extraordinary antics of the Jansenist *convulsionnaires* around the tomb of the deacon François de Pâris in the churchyard of St. Médard, Paris, in 1731–2, and subsequently elsewhere. These antics often resulted in the cure or amelioration of the *convulsionnaire's* ailments. Some of the cures were very remarkable, and are well-documented.[53] They present many analogies with mesmeric cures. We can hardly suppose, Devillers thinks, that even after death the deacon Pâris continued to exude animal magnetism from his cold and lifeless tomb, still less that he did so from earth scraped from that tomb, or from shreds of his former garments. What underlies both sets of convulsions and of cures must be imagination. This powerful faculty can correct a disorder in the workings of the body as readily as it can create one.[54] Devillers' position would have been greatly strengthened had he been able to cite copious examples of undeniable imagination-cures as well-authenticated as the cases adduced by the Jansenists and the animal magnetists.

The reports of the two commissions did not halt the advance of animal magnetism; they produced at most a slight stumble in its onward progress. On balance I am inclined to say, with Fournel,[55] that the magnetists proceeded with more candour than their opponents. It would be pleasant, therefore, if I could truthfully add that the reports failed to achieve their object because of the rational counter-arguments of their critics. All the indications, however, are that the critics had at best a marginal influence. The continued spread of animal magnetism was due much more to rumour and excitement working on the hopes and fears of the sick and on the mindless curiosity of the general populace,

than to any balanced assessment of the cogency of opposed arguments. Typical
was perhaps the sequence of events at Castres as described in a letter, dated 24th
December 1784, from the local correspondent of the Royal Society of Medicine,
a certain M. Pujol.[56] Publication of the two reports, says M. Pujol, produced at
first "a general calm on the subject of magnetism". But then a local doctor, who
had gone to Paris to learn Mesmer's secrets (the records of the Society of
Harmony suggest that he was a M. Malzac, the 151st member[57]), returned with
assorted pamphlets and catalogues of cures, and was fortunate enough to
achieve two or three quick successes. The fire of enthusiasm overcame the
coolest heads, and everyone talked of the ferocity of Mesmer's enemies, and of
the conspiracy against him. Soon Malzac had more patients than he could cope
with, and himself trained two assistants.

By this time the magnetists had acquired fresh ammunition to hurl at their
opponents. Certain new and very peculiar categories of "magnetic" phenomena
had came into prominence. To the great annoyance of Devillers, it was alleged
that these phenomena definitively disposed of the imagination-hypothesis.[58]
Since they emerged principally in the provinces, it is to developments in the
provinces that we must now turn.

NOTES

1 To assess the extent of Mesmer's fame before he came to Paris see e.g. *Sammlung* (1778);
   Buchner (1922); Tischner (1928), pp. 37–54.
2 Selections were published as Paulet (1780).
3 Hervier (1784), p. 11.
4 Girard (1784), pp. 8–9.
5 *Lettre d'un Bordelais* (1784).
6 *Remarques sur la conduite du Sieur Mesmer* (1784).
7 Bourzeis (1783).
8 *Lettre d'un médecin de la Faculté de Paris* (1784).
9 Mesmer had foreseen the fatal outcome, but none the less resumed treatment to alleviate
   Court's sufferings. See *Lettre sur la mort de M. Court de Gebelin* (1785), to which is added a
   report of the inquest.
10 Figuier (1881), pp. 194–196.
11 *Rapport des commissaires chargés par le Roi* (1784); *Rapport des commissaires de la Société
   Royale de Médecin* (1784). Both exist in various editions with varying pagination. Both are
   reprinted, along with the separate report of Jussieu, the "secret" report, and the later
   reports of 1826 and 1837, in the very convenient volume by Burdin and Dubois (1841), to
   which the ensuing page references relate.
12 Burdin and Dubois (1841), p. 41.
13 Burdin and Dubois (1841), p. 52. Compliance is nowadays a rather fashionable explanation
   of hypnotic phenomena!
14 Burdin and Dubois (1841), p. 59.
15 Burdin and Dubois (1841), p. 117.
16 Burdin and Dubois (1841), p. 80n.
17 Burdin and Dubois (1841), p. 80.

18 Burdin and Dubois (1841), p. 137.

19 Burdin and Dubois (1841), p. 137.

20 Burdin and Dubois (1841), pp. 95–96.

21 Figuier (1881), p. 230, states that the royal presses produced no less than 80 000 copies of the reports, but he does not give his source.

22 *The Gentleman's Magazine*, 1784, 54, Pt.2, pp. 944–946. A full English translation was published the following year as *Report by Dr. Benjamin Franklin, and other Commissioners* (1785); it is reprinted in Tinterow, ed., (1970).

23 On Jussieu see Gillispie (1980), pp. 151–156.

24 Servan (1784a), pp. 95–96.

25 D'Eslon (1784), pp. 27–29, 33, 40–41.

26 Bonnefoy (1784), pp. 60, 78–79.

27 Devillers (1784), p. 107.

28 The commissioners ascribe certain sensations felt by some patients to real physical causes; e.g. streams of light seen coming from the operators are set down to the "solar microscope" acting on "transpiration" from the operator's fingers. Yet as Podmore (1909, p. 65) remarks, the most superficial examination of the actual facts would have sufficed to convince them that the effects did not correspond to the causes alleged.

29 Burdin and Dubois (1841), pp. 160–165.

30 Bonnefoy (1784), p. 49. Cf. Servan (1784a), p. 42. Servan draws (pp. 46–47) an interesting parallel with the tests made on Bléton the dowser.

31 Burdin and Dubois (1841), p. 85.

32 Servan (1784a), p. 26.

33 Burdin and Dubois (1841), p. 34.

34 See *Supplément aux deux rapports* (1784).

35 Dampierre (1784), pp. 13–15; C***, G. (1784), pp. 6–11; Bouvier (1784), pp. 4–17.

36 The summary at the end says that there are 111 cases; but Podmore (1909), who gives a most valuable analysis of these case histories, points out (p.9) that the correct figure is 115.

37 Podmore (1909), pp. 10–11.

38 *Supplément aux deux rapports* (1784), pp. 3–4.

39 *Supplément aux deux rapports* (1784), p. 22.

40 *Supplément aux deux rapports* (1784), pp. 51–52.

41 *Supplément aux deux rapports* (1784), p. 28.

42 *Supplément aux deux rapports* (1784), p. 43.

43 *Supplément aux deux rapports* (1784), p. 76.

44 *Supplément aux deux rapports* (1784), p. 32.

45 *Supplément aux deux rapports* (1784), pp. 7–8.

46 Examples of the alleged magnetic cure of animals can be found even at this early date, though details and convincingness are alike lacking. See e.g. J.M.P. de Chastenet de Puységur (1784), pp. 9–10.

47 Podmore (1909), pp. 12–13 tabulates these twelve cases.

48 *L'evangile du jour* (1785); Paulet (1785).

49 Burdin and Dubois (1841), pp. 202–204.

50 He is not to be confused with Charles Villers, author of *Le magnétiseur amoureux* (see chapter 3). The *Dictionnaire de biographie française* identifies him with Charles-Joseph Devillers (1724–1810), member of the Academy of Lyon, and owner of a cabinet of curiosities which he exhibited publicly. Audry (1924), p. 73n, says that the two are different people.

51 Devillers (1784), pp. 172–173.

52 Burdin and Dubois (1841), p. 138.

53 Carré de Montgeron (1745–7).

54 Devillers (1784), p. 55.

55 Fournel (1785b), p. 3.

56 Gillispie (1980), pp. 288–289, who cites this letter from a manuscript, spells the name
    Payol. But Dr. Payol of Castres is obviously identical with the Dr. Pujol of Castres whose
    letters to the Royal Society of Medicine are cited in Thouret (1785), *apud* Burdin and Dubois
    (1841), pp. 195n, 201n-203n. Presumably the name given in the official publication of the
    Royal Society of Medicine is correct.

57 A letter from Malzac is quoted by Deleuze (1813), Vol. 2, pp. 132–133.

58 See e.g. Dampierre (1784).

CHAPTER 2

Puységur

The name which stands second to Mesmer's in histories of animal magnetism is that of Amand-Marie-Jacques[1] de Chastenet, Marquis de Puységur (1751–1825). An aristocrat of old family, an owner of large estates, a colonel in the Royal Corps of Artillery, he seems at this distance of time the very embodiment of the *ancien régime*. Puységur, however, was far from the popular stereotype of the French aristocrat. Well, though not deeply, educated in literature and the sciences, he was a liberal in politics, humane and philanthropic towards the less fortunate of all categories, and a well-wisher towards humanity in general. In character he was straightforward, conscientious, generous, brave, unselfconscious, unassuming, totally honest, and just a little naive. If my account of this admirable man taxes credulity, I can only reply that I base myself both upon the remarks of contemporaries and upon his own writings, and that he possessed two younger brothers who seem to have been scarcely less admirable than himself.

It was almost certainly through the middle brother, Antoine-Hyacinth-Anne de Chastenet de Puységur (1752–1809), known as the Comte de Chastenet, that Puységur became seriously interested in animal magnetism. This brother, a naval officer, was attacked by a dry asthma which defied the best efforts of the best doctors in Paris. Chance, he tells us,[2] led him to Mesmer in March 1780. Touched by Mesmer, he shortly lost consciousness, and on awakening an hour later was much improved. After three months of attendance at the *baquets* he was completely cured.

In the same year he began to experiment for himself, not without success.[3] His most interesting patient was a lady of Brest, aged 27, suffering from vomiting, convulsions and weakness so pronounced that on 9th January 1783 she received the last sacraments. However, by the end of February the Comte had cured her, a fact certified by several doctors and surgeons. In the same year, he introduced animal magnetism with notable success on board *le Fréderic-Guillaume*, the ship of which he was commander.[4] In 1784, he founded a magnetic society in Saint-Domingue (Haiti).[5] Magnetic practices spread rapidly among the black slaves and no doubt merged with the occult activities for which the island is still famous.

The name of the Comte de Chastenet is number seven on the original

Figure 2. The Marquis de Puységur. (From the Mary Evans Picture Library.)

members' list of the Paris Society of Harmony;[6] that of the youngest brother, the
Comte Jacques-Maxime-Paul de Chastenet de Puységur (1755–1848), called
the Comte de Puységur, is number three. Puységur himself (I shall so designate
only the Marquis) was the forty-sixth member, and probably enrolled about the
end of 1783. He later implied, however, that he had first become involved with
animal magnetism in 1781.[7] About April 1784, he completed his initiatory
courses, and set off for his estates at Buzancy near Soissons with the intention of
practising what he had learned; though he later hinted that at the end of his
course with Mesmer he was still not much wiser about the theory and practice
of animal magnetism.[8]

On visiting his estate manager, he discovered that the latter's daughter was suffering from a severe toothache. He jokingly asked if she wanted to be cured, and when she agreed set about magnetizing her. In less than ten minutes she was completely relieved of pain, and the pain did not return.[9] He had a similar success with the wife of a keeper, and, on 4th May, he tackled the case of a young peasant of twenty-three, named Victor Race.

Victor had been suffering for four days with an inflammatory condition of the lungs. He had pain in his side, was spitting up blood, and was greatly enfeebled by fever. After quarter of an hour of magnetization he fell into what seemed to be a sleep. He then began to talk, loudly, about his domestic worries. Possibly he was delirious. Puységur rose to the occasion:

> When I thought his ideas might affect him in a disagreeable way, I stopped them and tried to inspire more cheerful ones; ... At length I saw him content, imagining that he was shooting at a target, dancing at a festival, and so on. I fostered these ideas in him, and ... forced him to move about a good deal on his chair, as though to dance to a tune, which I was able to make him repeat by singing in my mind. By this means I produced in him an abundant sweat. After an hour of crisis,[10] I calmed him down and left the room. He was given a drink, and ... I made him take some soup. The following day, being unable to remember my visit of the previous evening, he told me of his improved state of health.[11]

Victor rapidly improved and by 8th May was almost better. During his convalscence it became apparent that he possessed some remarkable gifts – gifts, however, which he could only exercise when rendered "somnambulic" by the magnetic influence. He appeared to enjoy a peculiar "rapport" with Puységur. Merely by silent "willing", Puységur could make him do this or that, speak or not speak, notice or not notice other people in the room.[12] But Victor did not become an automaton. Though ordinarily a simple and tongue-tied peasant, he would, in the somnambulic state, converse in a fluent and elevated manner, expressing such sentiments of friendship and gratitude that, says Puységur, "we were unable to restrain tears of admiration and tenderness to hear the voice of nature express itself so frankly."[13] Victor assumed management of his own case, diagnosing and prescribing for his illness, and predicting its course. More than this: on being brought into contact with other patients, he seemed able to do the same for them.

Word of Puységur's success with Victor spread rapidly, and emboldened other ailing peasants to seek his help. It was impossible for him to treat them all, and he therefore made use of an expedient recommended by Mesmer – he magnetized a tree. For this purpose he selected a large elm.[14] Cords were hung from it, and stone benches set round it. Patients encircled it as if at a *baquet*, forming chains and holding the cords. Cures and alleviations were numerous. On the morning of 17th May, Puységur wrote to his brother the Comte de Chastenet, that one hundred and thirty patients were gathered for treatment.[15]

We have various contemporary accounts of the scene. One is a letter dated 13th June 1784 by M. Clocquet, a collector of taxes at Soissons.[16] Clocquet paints a kind of rustic idyll: the huge and venerable elm-tree, where young folk traditionally met for dancing and their elders for gossip; the spring of purest water bubbling forth at its feet; the chains of patients through whom passed the vivifying magnetic influence; Puységur and his brother directing the whole affair, so earning the universal gratitude and affection of their peasantry; above all certain gifted somnambules[17] moving through the throng to diagnose and prescribe.

It is apparent that already Victor Race had found a fair number of imitators and successors. According to mesmeric doctrine, once cured he should no longer have passed into a "crisis" state. Fortunately, other persons (mostly female) proved equally good "clairvoyant" somnambules, and Clocquet gives us this account of them:

> M. de Puységur, whom I shall henceforth call the Master, chooses ... several subjects, whom, by touching with the hands and presentation of a wand [of metal] he makes fall into a full crisis. [There is] an appearance of sleep during which the physical faculties seem suspended, but to the advantage of the intellectual ones. Their eyes are closed, they can hear nothing. They awaken only at the voice of the Master ... These patients in crisis, who are called doctors [*médecins*], have a supernatural power by which, through treating a patient who is presented to them, or even running a hand over his clothes, they feel which is the diseased organ, the affected part. They announce it and indicate pretty well the proper remedies ... A peculiarity not less remarkable ... is that these doctors who, for four hours, have treated patients ... remember absolutely nothing at all, when it has pleased the Master to awaken them ... The Master has the power, not only ... to make himself heard by the doctors in crisis, but [to] present his finger from a distance towards one of these doctors, still in crisis, and make them follow wherever he has wished ... But how does the Master awaken these doctors? It is sufficient for him to touch their eyes; or else he says to them: Go and embrace the tree.[18]

Much of the future development of the magnetic movement is latent in this passage, and not a little of that of hypnotism.

Our next account is from a letter to the administrator of Soissons.[19] The unknown writer does not paint quite such a heart-warming picture as Clocquet. In fact he makes the proceedings sound like a somewhat hysterical revivalist meeting. He implies that many of those present were just idly curious; that persons who were really ill were not cured and were even made worse; that certain of the somnambules were more interested than they should have been in the Puységur brothers; and that the brothers themselves were at best extraordinarily naive. Probably there are elements of truth in both accounts.

Puységur shortly published a list of 62 cures worked at Buzancy in a six-week period.[20] The case histories are very brief. There is a great predominance of fevers, difficult to interpret in modern terms. A sprinkling of cases seem to have involved fairly serious ailments. At any rate, Puységur deserves the credit for being (so far as I know) the first magnetizer to publish his results in a systematic manner.

His brother Maxime shortly followed suit. Summoned, about July 1784, to join his regiment at Bayonne, he at first kept quiet about his interest in mesmerism. However, when a soldier at drill was struck down by apoplexy, Maxime came forward and succeeded in reviving him. Thereafter he was approached by so many ailing persons that he had to magnetize three trees and train several helpers.

In a pamphlet setting forth his results[21], he prints certificates describing the cures of sixty persons, and gives a list of sixteen persons not yet cured, who were continuing to follow the treatment. Doctors present were invited to ascertain the states of the patients before and after treatment, and the certificates were deposited with the public notary of Bayonne.

The most interesting cases were perhaps those of two partial paralytics. The first was a surgeon with long-standing right hemiplegia, who had to walk with a stick, dragging his foot. He was cured in a month. The second was a monk, aged seventy-five, who had suffered from right hemiplegia since June 1783. He could walk with a stick, dragging his foot, but could raise his arm no higher than his chest. Gout was added to the other afflictions, but all yielded to the curative influence of animal magnetism.

By the end of 1784, Puységur's brand of magnetic procedures had made considerable progress throughout the country. The reasons for its spread are obvious enough: it largely cut out convulsions and other unpleasant forms of crisis, while the mysterious gifts, real or supposed, of the clairvoyant somnambules, added an element almost of the supernatural, the very rumour of which aroused both hope and curiosity. More obscure are the channels through which it spread. Though Puységur and his brother Maxime had each published pamphlets describing their activities, it seems likely that much of the spread was through correspondence and visits between interested persons. It is noteworthy also that many of animal magnetism's most forceful early exponents were military officers.[22] No doubt this was partly because such persons had a sufficiency of leisure and a ready supply of captive but willing subjects; but we may also suspect that news could travel relatively quickly from one officers' mess to another. There may also have been some overlap between military and masonic channels of communication.

What did Mesmer think of the new developments? There is not much information on this question. Puységur brought Victor Race to see him in the autumn of 1784, but Mesmer showed little interest.[23] However, chapter XIV of his *Aphorismes* (1785) discusses the phenomena of clairvoyant somnambulism, and we learn from several eyewitnesses that such phenomena were exhibited at his clinic before the end of 1784.[24] In his later books,[25] he goes into the theory of them in some detail. I know of nothing, however, which convincingly suggests that he had anticipated Puységur's discovery of the so-called "somnambulic" state. It is true that Mesmer's patients had often exhibited a tendency to fall into a peculiar sleep-like state. It is true also that Puységur's naval brother, the Comte de Chastenet, having guessed from his own experiments at Brest in 1783 the importance of the will of the operator in influencing the response of the patient, found that Mesmer too was in possession of this "secret" (characteristically Mesmer made the Comte promise to tell no-one[26]). However, none of this amounts to a full anticipation of Puységur.[27] To Puységur, then, the credit of the discovery seems to belong.

But in what sense and to what extent was the "discovery" of this "artificial somnambulism" really a discovery? Many modern workers – whose claims we cannot consider at this point – hold that the phenomena supposedly characteristic of "hypnosis" or artificial somnambulism can all be explained without postulating any "special state" of consciousness. If these workers are in the right, Puységur did not discover anything, though perhaps he unwittingly invented something. But if it comes to invention, Victor Race has as good a claim as Puységur to be the inventor. Victor continued to have problems; even in 1811, by which time he was married and the father of six children, he still suffered from a "nervous indisposition".[28] Was "artificial somnambulism" the product of Victor's psychological problems and of his peasant cunning, a product which "caught on" like the symptoms of any other hysterical epidemic but on a wider scale? To whom should a well-deserved monument be raised at Buzancy where it all began? To Puységur? To Victor? To both? It is hard to find clear answers.

CASES AND THEORY

Puységur left Buzancy about the middle of June 1784 to join his regiment at Strasbourg.[29] He was back by October, treating more cases, several of which he describes in detail. These cases run pretty much to a pattern, as do many other cases from the magnetic literature of the next year or two. Indeed, it was clearly Puységur who set this pattern through the book (*Mémoires pour servir à l'histoire du magnétisme animal*) which he published at the end of 1784. Patients would pass when magnetized into the "somnambulic" state (it must be emphasized that the majority did not, and were treated by passes and the clairvoyant prescriptions of those who did); in this state they diagnosed and prescribed for their own ailments, and quite often for the ailments of others; they were particularly prone to "see" such causes of trouble as inflammations, obstructions, worms, abscesses (*depôts*) and accumulations of blood or pus; they predicted, often with considerable accuracy, evacuations, bouts of pain, the arrival of suppressed periods, and the bursting of abscesses; above all they predicted the dates and times of renewals of symptoms and of cures. A couple of Puységur's cases will serve as examples.

Henri-Joseph-Claude Joly[30], aged 19, had suffered for ten years from severe and progressive deafness consequent upon an acute illness. He came to Puységur on Wednesday 13th October 1784, and proved an excellent subject. On Saturday 16th October[31] he announced in the somnambulic state that he had an abscess in the head. If it discharged into his throat he would die, but if it discharged by the nose he would be cured of his deafness. Next day he added that it would discharge by the nose in two instalments, one on Monday morning, the other later. On the Monday morning he discharged from his nose whitish matter of about the volume of an egg. There were no independent witnesses of this occurrence.

On Tuesday 19th October Joly woke from the somnambulic state with a bad headache and an hour later was discovered stretched out on the ground in Puységur's park choking and rattling as if in his death agony. Puységur gave him magnetized water to drink, calmed his convulsions and put him into the magnetic state. The following day, Wednesday 20th October, still in crisis, he stated that his abscess would discharge by the nose on Thursday evening.

On Thursday, after further attacks of choking and rattling, during one of which he passed spontaneously into the somnambulic state, a thick white substance, mixed with a little blood, was discharged from his nose before ten or a dozen witnesses. He shortly came to, and found his hearing restored. His case was attested in a statement signed on 4th November by a number of responsible persons who had previously known him.

Philippe-Hubert Viélet, aged 36, formerly a game-keeper, and now a schoolmaster, came to Puységur on 8th October 1784.[32] According to written consultations by various medical men, he had been suffering for four years from assorted ailments which one Dr. Jumilther succinctly diagnosed as follows:

> The original cause of the illness was an arrested perspiration which degenerated into a true inflammation of the chest. The bad treatment he received turned the inflammation into a cavity or abscess in the lungs, which is proved by the spitting of pus mixed with blood. The abscess renews itself from time to time. ... To this there is furthermore joined an asthmatic affliction which constricts his breathing.[33]

After some days, Viélet began to achieve clairvoyant somnambulism. He said that some of his symptoms were due to abscesses in the pilorus and the spleen, and that he had in addition a very dangerous abscess in the chest. The abscess in the spleen would discharge through the bowels on the evening of 28th October, and that in the chest through a gathering which he would spit out. According to the patient the former prediction was punctually fulfilled. Two days later, in the magnetic state, he announced that the abscess in his chest would discharge the same evening between nine and ten o'clock. At 9.15 p.m. he voided through the mouth a quantity of stuff as black as ink. After drinking some magnetized water he was fully restored and next day he was able to return home.

Eight days later he was back. Enforced overwork had brought on pains in his side and in the pit of his stomach. Puységur took him in and made him pass his nights in a state of magnetic sleep. On 16th November, Viélet's room-mate reported that during the night he had heard him wake up and begin to write. It turned out that he had written his own detailed medical history. He predicted that most of the abscess in the pilorus would discharge between nine and ten in the evening of the following day through an evacuation of the bowels. If he were fortunate enough to vomit he would get rid of the remainder. He ordained that he should be given two ounces of cream of tartar in the morning.

Puységur sent this document to the public notary at Soissons and awaited events. After frightful colics and repeated spasms, Viélet had the predicted evacuation, but he did not vomit. The result, he said in trance, was that the membrane which had covered his abscess remained behind. This membrane, which he could see attached to his nerves, caused him frightful spasms over the next few days.

Eventually, in the somnambulic state, Viélet asked for a consultation with another somnambule, Catherine Montenecourt, which duly took place.[34] Catherine ordained for Viélet various kinds of poultices on the stomach, and the membrane came away piece by piece, though not without the most frightful spasms. Whether any of these pieces were actually perceived in Viélet's stools, or whether the evidence for their detachment was purely clairvoyant, is not made clear. But either way the upshot was beneficial. On the evening of 30th November, after spasms and sufferings worse than any that had gone before, Viélet was able to indicate, in a feeble voice, that he was better. And infinitely better he was and remained, even though his health was not perfectly restored.

It is exceedingly difficult to know what to say of these cases, and of others like them which were soon appearing in the magnetic literature in increasing numbers. For the copious testimony printed by Puységur leaves little doubt that both Joly and Viélet had suffered for many years from incapacitating ailments which had resisted the best efforts of numerous doctors, and little doubt also that after a relatively brief period of magnetic treatment those ailments had very substantially improved. On the other hand, Joly's and Viélet's clairvoyant self-diagnoses lack any substantial confirmation, and the suggested anatomy ranges from the vague to the preposterous. It is true that in both cases purulent matter was undeniably discharged, and was discharged, furthermore, at the very times predicted. But we do not know – the relevant studies have not been carried out – how far self-suggestion might go towards producing similar effects.

To what, then, may we attribute the cures? To sheer chance – the recoveries just happened to coincide with the treatment? This does not seem likely when we reflect upon the length of time for which the ailments had proved intractable. To deliverance from poisonous drugs and barbaric methods of treatment? In neither case do the remedies previously administered, though often unpleasant, seem likely to have produced or prolonged the symptoms described. To magnetic "passes"? Conceivably these might have both psychological and physiological effects, but since nobody uses passes any more we lack relevant data. To the therapeutic force of the patients' own imaginations? Why was this not stirred into life by the numerous orthodox treatments which both patients had tried out? In short, we can dream up quite a lot of explanations of the cures; but none appears wholly convincing.

Puységur, of course, had few doubts about the explanation. The cures were due to the vivifying effects of animal magnetism, constantly monitored and reinforced by diagnoses and prescriptions from the somnambules. Puységur seems largely to have accepted Mesmer's view that disease results from impediments to the circulation of a quasi-magnetic fluid through the body, and that treatment must consist in reestablishing this circulation.. Whether the fluid is universal, or special to living organisms, he hesitates to say.[35] Puységur differs from Mesmer over the question of how the circulation of the fluid is to be reinstated. According to Mesmer, "magnetic passes" will do the trick. Puységur does not deny this, but thinks there is another and preferable way, namely the establishment of a *rapport* between the magnetizer and the patient. This involves the magnetic fluid of the operator and that of the patient mingling to establish a conjoint circulation directed by the operator's will. The operator can then move the patient's body pretty well as he can move his own, simply by willing the result he desires.[36] The patient is now in the somnambulic state, and in extreme cases may almost become an automaton controlled by the magnetizer. Rapport can be instituted, and the patient put into the magnetic state, by laying a hand or hands on his head or stomach, and strongly willing the desired result. It can thereafter continue even in the absence of contact or be transferred to another operator. *Croyez* and *veuillez* are the key words, says Puységur, "believe" and "will".[37] Patients might emerge from the somnambulic state spontaneously, or be aroused by having their eyes lightly rubbed.

The will of the magnetizer could not, however, always completely constrain his subject's behaviour. He might direct a somnambule to diagnose or prescribe, but the details of the diagnosis or prescription would come from the somnambule. Some somnambules would (like Viélet) spontaneously carry out activities relevant to the purpose in hand. The heightened creativity and enhanced faculties exhibited by somnambules are not easy for Puységur to accommodate. Take for instance the supposed ability of certain somnambules clairvoyantly to diagnose their own or other peoples' ailments. Where these clairvoyants diagnosed, as did Joly, by feeling appropriate pains or sensations in their own bodies,[38] no doubt some explanatory story could be cooked up in terms of the reciprocal flow of animal magnetism between somnambule and patient. But when somnambules began to claim that they could "see" diseased organs or even intestinal worms inside patients, difficulties crowded thick. Puységur toyed with the theory that clairvoyants may "see" accumulations of magnetic fluid around the obstructions to its flow.[39] But his more educated somnambules told him that "seeing" was not the right word to describe the experience; "knowing" would be less misleading.[40] At root, clairvoyance is a faculty of the soul, a faculty which we may suppose to be released in the somnambulic state when other faculties are in abeyance or under external control.[41]

This sort of explanation later became very common, and it is consistent with Puységur's views on the nature of the "will". Puységur is certain of only one thing about the will – that it is not a physical faculty but a faculty of the free and immaterial soul. The magnetic fluid is so to speak the last, most nearly spiritual, stage of matter, and the will can act directly upon it. Will can thus indirectly act upon other forms of matter, for instance living organisms.[42]

Puységur was, however, no mystic. He was opposed to the mysticism and illuminism that soon became, and remained, a strong undercurrent in the magnetic movement. He was essentially a practitioner, a therapist, whose prime concern was to alleviate suffering. So long as the treatment worked he was not too deeply concerned over its rationale. The fluidic theory at any rate serves to fix the magnetizer's attention.[43] He offered it simply to make some sense of what he had observed, and it must be admitted that he had observed some very curious things.

Most curious of all was the way in which certain somnambules would respond to unspoken commands or mental willings. A particularly striking example was the somnambule Madeleine whom he discovered in the autumn of 1784 and brought to Paris the following winter to demonstrate before friends. However, word got about and his house "soon became a public place, to which people resorted in the frame of mind they would have brought to a show by a tumbler."[44] The general procedure at séances was as follows. After a demonstration of ordinary phenomena, Madeleine, with eyes still shut, would be put through contact en rapport with individual volunteers. Each volunteer would designate some object with hand or eyes, and will Madeleine to fetch or touch it. She was successful in a considerable proportion of trials – provided that the operator's will-power was strong enough! Her eyes could be heavily bandaged without affecting the results.

Almost all such blindfolds can easily be circumvented. Still, some of the episodes were quite odd. Thus Puységur brought Madeleine to the house of a certain M. Mitouard who was himself put en rapport with her. M. Mitouard successfully made her stand, walk, sit, and pick up various objects. Then he stood stock-still in front of her, concentrating very hard. Suddenly Madeleine put her hand into his pocket and extracted three small nails – the very action which he had been willing her to perform. Until this incident there had been a good deal of gossip that Puységur was a confederate or a dupe. Thereafter the gossip declined;[45] but the reflections on his good faith had depressed him, and he was at a low ebb when he returned to Buzancy in April 1785.

THE STRASBOURG SOCIETY

Next month, just as Puységur was about to rejoin his regiment in Strasbourg, he was considerably cheered by a letter from the Comte de Luxelbourg (or Lutzelbourg), written on behalf of the freemasons of that city. The Comte requested Puységur, a fellow mason, to instruct them in the science of animal magnetism.[46] Puységur agreed, and on arrival set about the task with the energy and enthusiasm of an apostle. He gave a course of lectures,[47] and he established a Society, the *Société harmonique des amis réunis*, which was to become by far the most influential and the most respected of all the provincial societies. The Society was inaugurated on 25th August 1785. Mesmer was honoured as its perpetual President, but otherwise the statutes are purely Puységurian.[48] The society was run by a self-electing college of eighteen founders, who had to be aged at least twenty-five, domiciled at Strasbourg, of respectable status, of irreproachable character, of a healthy constitution, and if possible non-smokers. There were also trained associates (*associés initiés*), corresponding members (*initiés corres-pondans*) and associates under training (*associés élèves*). The treatment rooms were open from March to November. Treatments (always free) were given by founders and trained associates, and were carefully supervised and recorded. The period from December to February was occupied with courses of instruction and the preparation of printed volumes of case histories.

Among animal magnetists, the *Société harmonique* enjoyed in its own time and for some decades thereafter a deservedly high reputation. This reputation was principally due to three factors. The first was the serious and unselfish purposes which all recruits to the Society undertook to pursue. The welfare of patients was paramount, and experimentation for mere curiosity was frowned upon. The second was that many leading figures in the Society were highly respectable persons of some local standing. They included a sprinkling of physicians and surgeons. A group of medical men examined patients who presented themselves at the public treatment. Hostility between the magnetizers and the medical profession, so marked in Paris, was singularly absent at Strasbourg.[49] The third was the three printed volumes of the Society's annals,[50] which are among the most important documents of the early history of animal magnetism.

In 1789 the society had 188 members, of whom 31 were founders and 32 trained associates. There were 17 lady members – 14 more than the Paris Society of Harmony could boast.[51]

By the early months of 1785, it was already apparent that Mesmer had become something of a back number. Perhaps that was why he began to travel abroad. The demonstrations arranged by Puységur in Paris had made a great impression. According to J. F. Fournel, more than 500 persons had attended them, and they had received notice in the national and even foreign press. Every magnetizer was now trying to reproduce the new and startling phenomena of artificial somnambulism.[52]   Although many somnambules were simply pretenders, others, thought Fournel, were beyond suspicion: respectable mothers of families; sober men of known probity; persons of weak intellect; and children.[53] He had himself experimented successfully with such somnambules, and believed that in Paris and the provinces more than 6000 persons had done likewise.[54] The interest and excitement which these phenomena aroused[55] largely obliterated the sobering impression made by the reports of those academic and medical killjoys, the royal commissioners of 1784.

In Paris, this excitement quickly died away. Parisians had a reputation for volatile sensation-seeking, and the very month in which Puységur founded the Strasbourg Society saw the eruption of the most exciting court scandal for years, the extraordinary affair of Cardinal Rohan and the Queen's diamonds. In the provinces, however, interest in animal magnetism, though it began to cool, by no means abated. In 1786, Puységur observed that only in Paris was there no interest in animal magnetism.[56] The dedicated members of the Strasbourg Society, and of other provincial societies, continued their work in the firm conviction that it was of the greatest importance to mankind.

NOTES

1 I follow Ellenberger (1970, p. 70), who has looked carefully into the genealogy of the Puységurs, in giving this version of Puységur's Christian name, instead of the more usual Armand-Marie-Jacques. Puységur prints his name as Amand-Marie-Jacques on p. 422 of the 3rd (1820) edition of his *Mémoires*.

2 A. H. A. de Chastenet de Puységur (1783), p. 11.

3 A. H. A. de Chastenet de Puységur (1783), pp. 15n-16n.

4 Tissart de Rouvre (1785), pp. 413–415.

5 Société magnétique du Cap-Français (1828).

6 *Journal du magnétisme*, 1852, 11, p. 32.

7 A. M. J. de Chastenet de Puységur (1811), p. 54.

8 A. M. J. de Chastenet de Puységur (1807), pp. 28–31.

9 A. M. J. de Chastenet de Puységur (1786), p. 22. I have used principally the London (1786) edition of this book, Puységur's *Mémoires*, which was first published as A. M. J. de Chastenet de Puységur (1784b).

10 The state of somnambulism qualified as a "crisis" in the terminology of the magnetists.

11 A. M. J. de Chastenet de Puységur (1786), p. 23.
12 A. M. J. de Chastenet de Puységur (1786), pp. 23–24.
13 A. M. J. de Chastenet de Puységur (1786), p. 29.
14 This tree survived until 1940. Ellenberger (1970), p. 74.
15 A. M. J. de Chastenet de Puységur (1786), p. 27.
16 In A. M. J. de Chastenet de Puységur (1785), pp. 321–331.
17 Puységur at first referred to these somnambules as "doctors" because of the role they played in diagnosing and prescribing for his patients. By the autumn of 1784, he was using the term "somnambule" because of the apparent similarities between the behaviour of the "doctors" and that of persons liable to walk in their sleep. I shall follow Puységur's usage, but I do not, of course, wish to imply that the magnetic, or hypnotic, state really is at root the same as ordinary somnambulism.
18 In A. M. J. de Chastenet de Puységur (1785), pp. 324–327.
19 Taken from Montègre (1812) and printed as an appendix to Villers (1978) by the editor, M. Charles Azouvi. There is a long reply to Montègre's book in Deleuze (1813), Vol. 2, pp. 139–151. Audry (1924), p. 71, quotes part of another contemporary account by L. C. de Saint-Martin.
20 A. M. J. de Chastenet de Puységur (1784a), reprinted as A. M. J. de Chastenet de Puységur (1785).
21 J. M. P. de Chastenet de Puységur (1784).
22 This is borne out by the membership list of the Paris Society of Harmony. See Prologue, note 44.
23 A. M. J. de Chastenet de Puységur (1811), Foreword (unpaginated).
24 See e.g. Dampierre (1784), pp. 6–7; Bonnefoy (1784), p. 24.
25 Mesmer (1799; 1814).
26 A. M. J. de Chastenet de Puységur (1807), pp. 142–143.
27 This issue is discussed at greater length by Figuier (1881), pp. 264–268; cf. Pattie in Mesmer (1971), pp. 321–322, note 14.
28 A. M. J. de Chastenet de Puységur (1811), p. 119n.
29 A. M. J. de Chastenet de Puységur (1786), p. 34.
30 A. M. J. de Chastenet de Puységur (1786), pp. 54–71.
31 Puységur confuses matters (1786, p. 57) by calling 15th October a Saturday. In fact 16th October 1784 was a Saturday.
32 A. M. J. de Chastenet de Puységur (1786), pp. 83–112, 310–321.
33 A. M. J. de Chastenet de Puységur (1809a), p. 135.
34 A. M. J. de Chastenet de Puységur (1809a), p. 125.
35 A. M. J. de Chastenet de Puységur (1807), p. 163.
36 A. M. J. de Chastenet de Puységur (1786), p. 210.
37 On Puységur's methods of magnetizing see A. M. J. de Chastenet de Puységur (1807), pp. 157–178; (1811), pp. 14–15.
38 A. M. J. de Chastenet de Puységur (1786), p. 73.
39 A. M. J. de Chastenet de Puységur (1807), pp. 205–206.
40 A. M. J. de Chastenet de Puységur (1786), p. 285.
41 The most extensive exposition of it in English is Colquhoun (1836).
42 A. M. J. de Chastenet de Puységur (1786), pp. 244–249.
43 See his instructions on learning to magnetize, A. M. J. de Chastenet de Puységur (1807), pp. 157–178.
44 A. M. J. de Chastenet de Puységur (1786), p. 207.
45 A. M. J. de Chastenet de Puységur (1809b), pp. 11–13.

46 A. M. J. de Chastenet de Puységur (1807), pp. 108–109.

47 A. M. J. de Chastenet de Puységur (1807), pp. 116–157.

48 The statutes are printed in *Système raisonné* (1786), pp. 116–132. On the Strasbourg Society in general see Deleuze (1813), Vol. 2, pp. 184–199; Eckartshausen (1788–91), Vol. 2, pp. 286–317; Ueber das Magnetisiren in Strasburg (1787); Lévy (1976).

49 Thus a distinguished physician, J. F. Ehrmann, was at one and the same time professor in the faculty of medicine, an active magnetizer, and "Physician and Inspector of Patients to the [Magnetic] Treatment". See Lévy (1976), p. 204.

50 *Exposé de différentes cures* (1786); *Suite des cures* (1787); *Annales de la Société harmonique* (1789).

51 *Annales de la Société harmonique* (1789), f.i.

52 Fournel (1785a), pp. 8–9.

53 Fournel (1785a), pp. 9–10.

54 Fournel (1785a), p. 13.

55 On this general ferment see Carra (1785), p. 4.

56 A. M. J. de Chastenet de Puységur (1786), p. 385.

CHAPTER 3

Phenomena and speculations
1784–1789

In this chapter, I shall consider certain later developments in the pre-revolutionary phase of the magnetic movement. That these developments mostly took place away from Paris may be ascribed to two factors. The first is that the hostility of the Paris medical authorities must greatly have discouraged any Parisian doctor through whose mind flitted the idea of becoming a magnetic practitioner, whereas quite a number of provincial medical men were able to set up their own magnetic clinics with impunity, not to say profit. The second is that the presence of Mesmer in Paris until some time in 1785 constituted a formidable obstacle to change and development. Jealous as always of his own claims to priority, Mesmer initially gave a chilly reception to Puységur's discoveries, and was unlikely to give a warmer one to anyone else's.

The developments in question may be considered under two general headings, phenomena and speculations to which these phenomena gave rise. The phenomena may in turn be grouped into two categories: the cures supposedly wrought by animal magnetism, and the peculiar phenomena, chiefly psycho-logical, apparently promoted by or characteristic of the state of "artificial somnambulism".

CURES AND CASE HISTORIES

On the cures I shall touch only briefly. The Annals of the Strasbourg Society of Harmony contain a large number of accounts of cures, many of them medically attested, so I shall begin by looking at these.

Treatment at Strasbourg would usually begin with conventional mesmeric procedures; but frequently these procedures were supplemented, in the Puségurian tradition, by treatments (medicines, baths, magnetized water, even bleeding) prescribed by the patient himself in a somnambulic trance, or (more often) by a clairvoyant somnambule with whom he had been brought into contact.

The first two volumes of the Society's annals[1] contain between them accounts of 186 cures. The categories of ailment most commonly involved seem to have been as follows:

> Rheumatism and related ailments (26 cases)
> Assorted stomach ailments (26 cases)
> Fevers of various kinds (24 cases)
> Headaches and migraines (16 cases)
> Suppression of periods (11 cases)
> Ear troubles (10 cases)
> Choking and oppression (7 cases)
> Epilepsy and convulsions (7 cases)
> Eye troubles (6 cases)
> Dropsy (ascites) (5 cases)

The figures are to be taken as approximations only – the mixture of popular and archaic technical terms in which the cases are described is not easy to penetrate. Anyone who reads the case histories hoping for reports of miraculous recoveries will be disappointed. In fact a sceptic could be forgiven for remarking that many of the ailments concerned wax and wane,or get better of themselves, or respond readily to conventional remedies, or are susceptible to psychological influences. If magnetic procedures played any part in these cures, that part probably consisted of boosting the patient's morale and expectations, and inducing him to adopt a healthier régime.

Still, it has again to be said that many of the patients had been ill for a long time, had exhausted conventional remedies, and were desperate, and that against this background some of the cures are quite striking. I take a few sample cases:

Ambroise Fleuri, a soldier aged 19, had suffered from *grand mal* epilepsy from his earliest infancy. Conventional remedies had brought no benefit. His convulsive seizures were very violent and ordinarily lasted three hours. Magnetic treatment was undertaken by M. de Laulanié. The patient became somnambulic, and began to prescribe for himself and predict the onset of his attacks. Gradually these diminished in length and frequency. The patient said in trance that his troubles were due to an abscess the size of a broad-bean beneath the right nipple. His cure was certified by the surgeon-major of his regiment.[2]

Jean-Louis Broc, a soldier, was afflicted with a severe ophthalmia, for which he had unavailingly spent many weeks in hospital. He had pustules and ulcers, which caused him the most painful smarting, around the eyelids of both eyes. He was magnetized by M. des Chabert, his commanding officer. Becoming somnambulic, he prescribed for himself magnetized water and a sneezing powder and rapidly improved.[3]

It appears likely that one of these young men suffered from a true epilepsy, and the other from a serious eye infection; that both these problems had resisted conventional treatments for a relatively long period of time; and that both were cured or immensely improved after a relatively short period of magnetic treatment. Analogous remarks could be made concerning many of the other cases.

A far more complex case was that of Marie-Catherine Emmich, a dress-maker, aged 37. At the age of 24, following a violent fright, this lady was attacked by a raving fever, with fearful convulsions passing into states of faintness amounting to catalepsy. The cataleptic states persisted and became so frequent as to incapacitate her for the ordinary business of life. Magnetic treatment was begun on 31st January 1786 by the Baronne de Reich (née Baronne de Boecklin), of Kiensheim.

The patient's condition at this time is attested by a certificate from a Dr. Weiler. He describes her as dreadfully thin and enfeebled, barely able to take nourishment, afflicted by a continual cough with occasional spitting of blood, and undergoing palpitations of the heart. So swollen and hard, and so excessively sensitive to touch, was her abdomen that he concluded scirrhi were present, and believed the illness incurable.[4]

Under treatment the patient became somnambulic. She foresaw the course of her own sufferings, and prescribed fearsome courses of purgation, emetics and blood-letting. Her cure, which was attested on 8th May 1786 by the signatures of numerous distinguished persons who had attended the séances, can at any rate not be set down to the banishment of heroic remedies.

The Baronne's detailed diary of this case[5] highlights a change that was coming over the magnetic literature, a change that was to reach its climax in Germany in subsequent decades. Whereas earlier writers, such as d'Eslon, Puységur and his brothers, or the Marquis Tissart de Rouvre, had at first published only brief or fairly brief case reports, it now became the fashion to publish immensely detailed journals, often of book length, setting forth the symptoms, treatment, progress, relapses and final triumphant cure of a single patient. The vogue perhaps developed out of some of the longer case histories which Puységur published in his *Mémoires* (1784). Of course, only a few cases were suitable for a really extended description, and by and large they conformed to a certain general type. The patients suffered from all sorts of aches, pains, spasms and fevers, together with some complex of physical ailments, not clearly corresponding to the symptoms of any generally recognized disease, but containing hints or elements of several. The ebb and flow of such symptoms, and the constant battle with them, gave the diarists plenty of material. Above all, it was essential that the patients should become somnambulic, so that their clairvoyant self-diagnoses, predictions and prescriptions could be minutely chronicled and compared with the eventual outcome. Most such case records are tedious to the last degree.

Particularly influential then, and for decades to come, were the series of journals published by Tardy de Montravel, a captain of artillery, who, when we first hear of him, appears to have been stationed at Valence.[6] At 255 pages his *Journal du traitement magnétique de la Demoiselle N.* (1786) established a new record. The Demoiselle N. was a very poor young woman of 21 whom Tardy had encountered at the public *baquet* where he seems to have been a pupil. She had been wasting away for no obvious cause since she was 15 or 16, and three or four years later an "imprudence" brought about a total suppression of her periods. She suffered from a "slow fever", and had become feeble and exceedingly thin, which, combined with constant coughing of blood and pus had led to a diagnosis of consumption and a prognosis that death would shortly follow.

Tardy began to treat her daily on 31st March 1785. He concentrated on restoring her periods, to the suppression of which he attributed all her superficial symptoms. She became somnambulic and started to prescribe for herself. Magnetized water was her most favoured remedy. On 4th April she predicted the restoration of her periods for 15th May, a prediction which was fulfilled with great benefit to herself. Unfortunately, it appeared that the magnetic procedures which benefitted her also benefitted certain intestinal worms from which she believed herself to be suffering. One in particular waxed and grew strong. In trance she could see it very well. It was longer than her arm, and its head was like a serpent's, with two protuberant eyes, and a very large mouth. It was eventually slain by a course of vermifuges, and fearful were its dying spasms. Most regrettably, no part of this novel specimen emerged for scientific scrutiny.

Soon after her cure the unfortunate Demoiselle N. contracted smallpox, and this occasioned a continuation of Tardy's journal.[7] Later in the year, he went to Strasbourg and threw himself into the work of the society there. He began to treat a certain Madame B., and his published journal of the treatment[8] runs to 279 pages. For 15 months this lady had suffered from frequent palpitations, suffocations, depressions, and fainting fits, and from almost continual stabbing pains in the left side. She had entirely lost her appetite, her periods had become deranged, and she was thin and listless. Her doctors, not knowing how to define her malady, had unsympathetically labelled it "vapours". Under treatment, however, she became somnambulic, and thereafter Tardy had to battle not just with suppressed periods, but with a polypus on the heart. He was not daunted, and again brought matters to a successful conclusion.

Tardy was without doubt a conscientious and dedicated person for whom the welfare of his patients was a paramount consideration. Yet it would scarcely be unjust to say that it was not the medical but rather the broadly psychological aspects of the cases which interested him most. Both the Demoiselle N. and Madame B. exhibited when in the somnambulic state apparent powers of clairvoyance and prevision, and various other psychological peculiarities, from the study of which, and from the didactic utterances of the somnambules themselves, Tardy tried to draw far-reaching conclusions about man, the universe, and their interrelationships. I shall later on touch briefly on the phenomena and the speculations in question. Similar tendencies are manifest in other writers. Some writers went even further, and published what were in effect volumes of somnambulic teachings, generally of a mystical and religious character.[9]

From a psychological point of view, cases such as those written up in laborious detail by the Baronne de Reich and Tardy de Montravel are indeed not uninteresting, partly because they foreshadow much that was to come in the mesmeric and hypnotic movements, and partly because of the undoubtedly psychosomatic nature of many of the ailments treated. The psychological interest of these materials is, however, largely offset by the fact that the enthusiasm for collecting them led to a relative neglect of cases that, from a medical point of view, were of much greater interest – I mean cases in which genuine if commonplace ailments with an undoubtedly organic basis were apparently cured or alleviated by the procedures of animal magnetism. I have already given examples of such cases and indicated my belief that it is not altogether satisfactory to dismiss them as due to chance or to renouncing harmful remedies. I shall conclude this section by giving one further example. The magnetizer was J. P. F. Deleuze (1753–1835), from 1795 an assistant and later librarian at the Museum of Natural History, who was to become one of the best-known and most respected exponents of animal magnetism in early-nineteenth-century France. It appears that the subject did not become somnambulic, and that Deleuze must simply have used the traditional magnetic "passes".

About 1785, Deleuze undertook the treatment of a lady who had suffered for two years from a right hemiplegia so complete that she could only move her right hand by grasping it with her left. She also had severe speech and reading problems, not due to any paralysis of the tongue. Soon she could move her fingers and after fifteen days she moved her arm. Little by little her speech improved. Deleuze left her in this state, entrusting her husband with the task of completing the cure. But owing to the events of the Revolution he did not learn the upshot.[10]

This lady had undoubtedly sustained a serious lesion to the left hemisphere of the brain in the area supplied with blood by the middle cerebral artery. Her speech difficulties would nowadays be classified as a "Broca's aphasia". A substantial improvement in the condition of such a patient after a lapse of two years is unusual, though not unknown.[11] It seems at any rate fairly unlikely that the improvement was not connected with the treatment, though the exact relationship between them remains obscure.

PHENOMENA OF THE MAGNETIC STATE

I come now to the second major subdivision of the "phenomena" with which this chapter is to deal, namely that of the phenomena apparently promoted by or characteristic of the state of "artificial somnambulism". Many of these phenomena are conveniently listed by M. de Mouillesaux, joint secretary of the Strasbourg Society, in his *Appel au public sur le magnétisme animal* (1787).[12] I shall base myself on his list, adding occasional comments and examples, but shall separate those categories of phenomena which apparently involve extrasensory perception (ESP) in some form or another[13] from those that do not. Taking the latter first we find that they fall under eleven headings.

1. "To hear only the magnetizer, to have need of his will to be put into rapport with other people, that is to say to be able to converse with them, and touch or be touched by them without being upset."

2. "To be, in certain circumstances, disturbed by the presence of sceptics or scoffers about magnetism."

3. "To be unable to recollect, in the waking state, everything that had happened or been said during the somnambulic crisis, unless the express will of the magnetizer gives them the command and the power."

Puységur observed this phenomenon with his very first somnambule, Victor Race, and it was for many years afterwards widely confirmed by others.

4. "To hear, respond to, obey, in crisis, the commands, signs and even thought of the magnetizer."

Again this was noted by Puységur with Victor Race.

5. "To execute, in the waking state, with an irresistible impulse, and without being able to offer any reason, the magnetizer's will, without any indication of this will except a notification given during the somnambulic crisis. It makes no difference whether the command is to take place immediately or after several days."

We are dealing here, of course, with the phenomenon later known as "post-hypnotic suggestion", and Mouillesaux has himself a good claim to be called its discoverer.[14] In an article published in the Annals of the Strasbourg society in 1789,[15] he describes as follows an experiment which he carried out with a gifted somnambule, a young lady of twenty-two:

> I had told Md.lle * in crisis, to come the next day at nine in the morning to visit someone to whose house she never went ... Measures were taken and well kept to make sure that in her waking state she had no hint of this agreement ... [several named persons] were with me next day ... to observe the outcome. Shortly before nine o'clock we saw Md.lle * pass and repass several times in front of the house in question where we were. She seemed to us very preoccupied ... However, seeing her at length enter a church, we believed the experiment had failed. We were talking about it as we

breakfasted, when, to our great surprise, Md.lle * entered the room. Her embarrassment cannot be described ... She was soon reassured and told us that ever since she had got up she had had the idea of taking this step, that she even believed she had dreamed it, that ... an irresistible impulse had made her brush aside all opposition, and that she then found herself much relieved ...[16]

6. "Be put into a state of crisis [somnambulism] by any person receiving the commission or the power of the accustomed magnetizer."

7. "Be put into this state of crisis without being ill; but without doubt on account of simple or slight indispositions, or as a result of tendencies towards ailments kept at bay, or through the effect of other appropriate or reciprocal dispositions."

It was at first generally believed, following Mesmer, that only persons who were ill, i.e. persons in whom the circulation of magnetic fluid required unblocking, would pass into a "crisis", whether convulsive or somnambulic.[17] It was therefore with some surprise that Tardy de Montravel discovered that the Demoiselle N., even when not ill, would pass into the somnambulic state if he wanted her to, and that Mouillesaux found in Md.lle * a perfectly healthy young woman who exhibited all the phenomena of magnetic somnambulism. Both these gentlemen had to juggle with their theoretical systems in order to accommodate such facts.

8. Be put into, and awakened from, the somnambulic state by the verbal commands of the operator, without any use of touching or magnetic passes.

This, of course, is the central phenomenon of late-nineteenth-century hypnotism. It seems to have been Tardy de Montravel who first regularly obtained it. Here is his account:

In the afternoon of [15th September 1786], I put Madame B. in crisis without touching her ... and only by saying to her: *Sleep.* Without doubt her period problems had made her nerves more feeble and more susceptible and had completely fixed my ascendancy over her; for, at this single word, she fell asleep on the spot, and since that day I have not needed to use any other form of magnetism to put her in crisis; and likewise to bring her out of crisis it has since always been sufficient to say to her: *Wake up.*[18]

Tardy looked on these phenomena as further examples of the power of his will over his patient, and the practice, though revived from time to time, did not catch on.

9. "To preserve in the waking state the faculty possessed in crisis of giving consultations with intelligence and success to sick persons, and finally be able at need, after having agreed the matter with one's magnetizer, to put oneself in crisis in his absence and without his help...."

We have here foreshadowings of the possibility of self-hypnosis. In the shorter term this development cleared the way for the appearance of a class of "medical" clairvoyants, who later merged into the class of fortune-tellers in general.

10. "To have, in the state of somnambulism, an extension, a superiority of the senses and the understanding not possessed in the waking state."

Apparent enhancement of understanding was observed in Puységur's first subject, Victor Race, who was able to converse on equal terms with persons much above his ordinary intellectual and social level. Other somnambules developed (as Viélet had done) the gift of writing spontaneously. Quite soon the curious and the credulous were being amused or overawed by the writings and utterances of various classes of specialist somnambules. Some rhymed and versified spontaneously, some praised or expounded the truths of animal magnetism, some held forth on philosophical, quasi-scientific, or religious topics.[19] Tardy de Montravel describes a Catholic somnambule who, espying a Protestant, discoursed for half an hour, as if inspired, upon the topic of confession. The Protestant, touched and melting into tears, left, beating his breast.[20] Other somnambules professed different verities, and a school of Catholic writers grew up which attributed somnambulic phenomena to the activities of demons.[21]

The more sober magnetists deprecated the enthusiasm of some of their colleagues for the teachings of the somnambules. These enthusiasts, they felt, would simply find their own preconceptions reflected back to them. As for the supposed enhancement of faculty of which some writers took such an exalted view,[22] the more mundane Deleuze merely remarks:

> It is in no wise proven that in the state of somnambulism one has knowledge which one did not have in the waking state: one has simply sensations infinitely more delicate, a distinct recollection of everything that one has known or by which one has been affected, and a great facility for making new combinations of ideas; that is quite enough to produce very remarkable results.[23]

However, Deleuze presents very little evidence for the postulated enhancement of memory in the somnambulic state.[24]

11. If a subject in a state of magnetic somnambulism is given a command that offends his or her moral sensibilities he or she will wake up.

A problem that both Puységur and Tardy de Montravel had in the wake of the "secret report" of 1784, was this. A somnambulic patient *en rapport* with her magnetizer will at once obey his commands, even unspoken commands. Suppose the magnetizer were a scoundrel; he might then have some girl at his mercy! Puységur and Tardy both asked their favourite somnambules how they would react in such circumstances, and the former even tried a very mild experiment.[25] Both received essentially the same answers: any immoral or indecorous proposal that went beyond the bounds of a joke would cause the subject to wake in indignation. Indeed, it seemed that the somnambulic state might even produce a heightening of moral faculties. Tardy tells us of a lady of the town who, put into the somnambulic state, found that her magnetizer was looking at her too freely. She covered her breast with the greatest care.[26]

It does not seem to have occurred to either of these high-minded gentlemen that some of the magnetized young ladies might have developed an interest in their magnetizers. Yet the Demoiselle N. liked to have Tardy's hand and no-one else's on her knees. Many years later the middle-aged Puységur was surprised when a lady began to fear that he was inspiring too strong an attachment in her seventeen-year-old daughter. Puységur found this supposition flattering but incredible.[27]

The eleven categories of phenomena I have so far listed cover between them a strikingly high proportion of those that a century later came to be thought characteristic of "hypnosis". Indeed to those listed, several others might be added. The most obvious of them is that of sensory hallucinations, which were inadvertently produced on a large scale by the numerous magnetizers who induced their somnambulic subjects to see, smell or taste the magnetic fluid as it streamed in glowing filaments from the operator's fingers or wand, or added its sanative influence to glasses of magnetized milk or water.[28] The magnetizers naturally regarded the streams of fluid as having an objective existence. It would have been easy enough to have tested this hypothesis; but I have not found in this period even one reasonably conducted experiment to ascertain (for instance) whether somnambulic subjects could reliably distinguish glasses of magnetized water from glasses of unmagnetized water.[29]

Another kind of phenomenon *almost* discovered in this period was that of the susceptibility of good subjects to waking suggestions. Thus Mouillesaux engaged one day with the (waking) Md.lle * in a somewhat heated argument on matters magnetic. Putting his hand on her shoulder, though with no intention of magnetizing her, he told her that to convince her he would have to make her dumb. At once she became fixed where she stood, mouth open but unable to speak. He repeated this experiment on a number of subsequent occasions.[30]

A recurrent source of controversy has been that of markings (up to and including blistering) allegedly produced on the skin by suggestion. This development also is foreshadowed in the mesmeric literature. Tardy de Montravel describes the case of a certain Mlle. B., an ignorant country girl, who in the somnambulic state would preach the truths of Catholicism for somewhat extended periods of time. On 4th October 1785 she had been holding forth in the somnambulic state to a lady who remained unconvinced, and said to her:

> To prove that what I say does not come from myself, put ... your thumb on
> my arm, and there you will see the sign of him who speaks through me.
> There was no mark on her arm. A moment later she told this lady to remove

her thumb, and one saw on Mlle B.'s skin a cross clearly marked out by two
red strokes.[31]

For the most dramatic of all hypnotic phenomena – the painless carrying out of
surgical operations – I have discovered no clear precursor in the period before
the French Revolution. Animal magnetism was, however, often successful in
alleviating pain; and on one occasion Mesmer seems to have delivered an infant
painlessly. By so doing he brought down on his head the wrath of certain writers
who cited scripture (Genesis iii, 16) to prove that women ought to suffer during
childbirth[32] – a curious foretaste of arguments heard when chemical anaesthetics were first used by obstetricians.[33]

PARANORMAL PHENOMENA

This brings us to those phenomena on Mouillesaux's list which apparently
constituted evidence for ESP. Whether or not one believes in such phenomena,
there is no escaping the fact that they were, and for upwards of fifty years
remained, a central feature of the animal magnetic scene. Nor did they by any
means wholly disappear with the advent of hypnotism.[34] It is therefore necessary
to give them some attention, though we can say at once that, while there is a
smallish residuum of rather odd findings, much of the material from this early
period consists of second-hand or poorly reported anecdotes, together with a few
accounts of experiments in which the precautions taken were clearly quite
inadequate.

12. Respond to the unexpressed will of the magnetizer, even when he gives no
external indication of what it is.

This was (as we have seen) quite widely reported, but it is in most cases
impossible to be sure that the magnetizer gave no cue to the subject.

13. Fall into the somnambulic state in response to the unexpressed will of the
operator, even when operator and subject are separated by a wall, indeed by the
width of a town.

This phenomenon too was widely reported and believed in; Mesmer succeeded
in magnetizing a subject through a wall as early as 1775.[35] Magnetization at a
distance was later reported frequently, but I have not found an eighteenth-
century example that seems worth even a brief summary.[36]

14. "See into all one's interior, see the state of one's viscera, of one's nerves,
one's humours; the effect of remedies; foresee the effects they will have in the
short or long term."

15. "See equally well into other sick persons ... even give consultations (when
these are undertaken by the order and will of the magnetizer) concerning absent
and unknown persons, see them, and give a correct account of them."

Most of the treatments prescribed by somnambules were quite harmless. Not all, however: one of Puységur's somnambules prescribed fifteen leeches on the fundament for a luckless lady, and another prescribed for herself dangerously large doses of a violent purgative. Puységur went ahead, and the somnambule was cured.[37]

The further claim that somnambules may diagnose and prescribe for distant and unknown persons is not substantiated by any noteworthy case reports. The Demoiselle N. once diagnosed and prescribed for an unknown lady by holding a piece of glass which the latter had worn against her solar plexus (glass was supposed to accumulate the magnetic fluid very readily). Mouillesaux tells us that he has successfully repeated such experiments.[38]

16. "Foresee that one will have crises more clairvoyant, and determine the times...Foresee events in one's own and other peoples' lives; and, especially as regards health problems, foresee onsets, crises, accidents, their causes, their characters, their duration, their dangers, their treatments."

Especially common here were successful predictions of the date of return of suppressed periods. Such predictions were attributed to a clairvoyant insight into the impending course of the problem. Nowadays, we would see in them only the effects of self-suggestion (though what this involves is itself obscure). Of course, the successful prediction of a disease or accident which could neither have been foreseen at the time nor consciously or unconsciously brought about by the subject would be much more remarkable. There were one or two apparent cases of this. A certain M. de R.****** says that in June 1786 a clairvoyant somnambule told him that his youngest child, then six months old, would have smallpox in the summer of the following year. On 8th November 1786 she stated that the eruption would last from 8th to 10th July 1787, a prediction which was precisely fulfilled.[39]

17. "See the fluid...Appreciate at once, by the state and colour of the fluid... the state of health, strength and constitution of...bodies, and, when the bodies are minerals, recognize the kind by the emanation of the fluid; recognize also the properties, the usefulness of plants, by taste or smell."

I know from this period of no interesting examples of ostensible diagnosis from the fluid's state or colour, or of the recognition thereby of minerals or plants.[40] Such beliefs are of course popular with many kinds of occult practitioners.

18. "See persons present [sc. through closed eyes], as well as those who arrive during the crisis; attend to them, even to those one does not know at all; speak about their concerns, and also about their health; indicate and predict events for them, although they have never been put into rapport, and no request has been made on their behalf."

Within a relatively short time after Puységur's discovery of clairvoyant somnambulism there appeared somnambules who would give not just medical consultations but consultations upon all aspects of life, and this to persons previously unknown to them. This class of somnambules before long merged with the class of fortune-tellers,[41] and many no doubt were or became professionals. Few accounts of them have come down to us.[42]

19. See, hear, taste, smell, with parts of the body other than the normal sense organs, and especially with the pit of the stomach.

I do not know who first claimed to have observed this phenomenon in a mesmeric context. It may have been Tardy de Montravel, whose favourite somnambule, the Demoiselle N., began during April 1785 to complain in the somnambulic state that loud noises pained her in the pit of the stomach. Tardy tried whispering immediately above her stomach, and he found that no matter how quietly he whispered she could always hear him. He also noticed that when he asked her to examine a certain drug and comment on it she held it, and moved it about, in front of the pit of her stomach. He began to gather similar stories from other places.[43]

Tardy was immensely interested when he came across[44] a book by J.H.. Petetin (1744–1808), a well-known physician of Lyon, which appeared to confirm and amplify these findings on a considerable scale.[45] Petetin, who was not involved with animal magnetism, had been studying a number of women suffering from what he regarded as spontaneous catalepsy. He had a theory of "animal electricity" which closely resembled the theory of animal magnetism. Petetin found that his cataleptics would not respond to remarks directed at their ears, but would respond if the remarks were addressed to the pit of the stomach. The best way to speak to the subject's stomach was to place one of one's hands on it and whisper to the finger tips of the other hand. Since the human body conducts "animal electricity" the subject can hear and reply. Indeed the speaker might be at the end of a long chain of persons holding hands, of whom only the first had a hand on the patient. But if the chain included a non-conducting "link", such as a stick of wax or a piece of glass, the patient would not "hear". Petetin's cataleptics could also "see" and "taste" objects laid on or near the pit of the stomach, or other parts of the body, even when this was done in such a way that they had no opportunity to observe them by normal means.

The alleged phenomenon of "stomach-seeing" had a considerable influence on later theoretical speculations about the *modus operandi* of the magnetic fluid.

THE MAGNETIZERS OF LYON

One further category of phenomena that might perhaps be regarded as paranormal requires a brief mention in view of its enormous and in some quarters influential expansion during the nineteenth century. To deal with it I must first say a little about the history of animal magnetism in the city of Lyon, a history which was varied, interesting and filled with controversy.

There seems already to have been at Lyon some background interest in animal magnetism when, early in 1784, Pierre Orelut, a surgeon, established the city's first magnetic clinic. Orelut (who was executed in 1794 for his part in the Lyon rebellion) had attended the courses at the Paris Lodge of Harmony. He was joined at his clinic by three other Lyon surgeons who had been trained by Mesmer, namely J. B. Bonnefoy, L. J. P. Grandchamp, and M. Faissole. They seem to have adopted strictly mesmeric doctrines and practices, and in the summer of 1784 Orelut published a short pamphlet on their results.[46] Meanwhile, a Lyon apothecary, J. B. Lanoix (1740–1845[sic]),another pupil of Mesmer's, had set up a clinic with a surgeon named Dutreich, who had not been trained by Mesmer but had been initiated in another way.[47] There were probably several clinics in addition to these.

Some time in the spring of 1784, a group of interested persons (not trained by Mesmer) banded together to form a Lyon magnetic society called *La Concorde*. Among its founders or leading members were an artillery officer, the chevalier de Barberin, Dutreich (the surgeon mentioned above), P. P. A. de Monspey de Vallière (1739–1807, commandeur de Montbrison), and J. J. Millanois (1749–1794, a lawyer, who like Orelut was executed for his part in the Lyon rebellion).

Barberin and Monspey seem to have been the driving force and were innovators in doctrine and practice. They emphasized the importance of the magnetizer's preparation, state of mind, etc. They found that when they magnetized they produced in themselves a sort of "doubling" which enabled them to feel in their own bodies the ailments of those they were treating and to operate accordingly.[48] (Later, they found they could diagnose just from the sensations they felt in their own hands as these passed around the patient's body.) Furthermore, the magnetic fluid, according to their view, was so far subjected to the thought and will of the operator that magnetization at a distance was perfectly possible.[49]

The power of the Barberinist method of diagnosis was put to the test during the summer of 1784 in three experiments which became famous.[50] They were conducted at the Lyon Veterinary College with the cooperation of the director, and various students. All three were of a similar pattern, and I will take that of 22nd July 1784 as an example.[51] In the presence of a large number of witnesses two Barberinist initiates, Millanois and Dutreich, were given an ailing horse to diagnose. They announced that the animal was diseased in the lateral and anterior part of the right lung, and in the left lung all over but particularly towards the sixth of the true ribs, and that there were obstructions in the liver and spleen, especially the latter. These diagnoses were proved correct at autopsy; but critics alleged that the ailments named by the operators would have been found in many decrepit old horses.[52]

The activities of the Lyon magnetizers aroused a good deal of hostility, especially among more orthodox practitioners, and a lively pamphlet war ensued.[53] The Barberinist school, however, found some powerful allies in a local masonic lodge, the *Respectable loge de la Bienfaisance*. The *Bienfaisance* was one of a network of lodges whose members were much given to the practice of ceremonial magic. An important figure in this lodge was J. B. Willermoz (1730–1824), a martinist, that is, a disciple of the mystic and occultist Martinez de Pasqually (1710–1774). Also a member was L. C. de Saint-Martin (1743–1803), another martinist, who later wrote as the "Unknown Philosopher". Several prominent members of the *Concorde* (e.g. Monspey, Millanois) were also members of the *Bienfaisance*, and the occultists among the latter soon realized there was much to interest them in the magnetic practices of the former. Willermoz, in particular, worked so effectively for a *rapprochement* that, as Alice Joly puts it, "the *Concorde* very quickly became a magnetic branch of the *Bienfaisance*."[54]

Barberin himself, an influential figure of whom we know very little, did not join the *Bienfaisance* until 1786, perhaps because he left Lyon for Paris in June 1784. There he met up with Saint-Martin, and they between them worked out the doctrines and practices of what came to be called the "Barberinist" school of animal magnetists.[55] These, at least in the accounts we have,[56] are fairly obscure, though the influence of martinism is here and there apparent. Intelligent beings of all categories were created as a series of emanations from God. Man is the latest of these emanations, and there are thus intelligent beings nearer the divine than he is. Some of these beings have parted from God and are morally neutral or positively evil. Man too has fallen and since his fall has been imprisoned in a material body. He has two selves, a real "I", or residual spark of the divine, and a "Not-I", presumably associated with the body and baser instinctive tendencies. There are also two kinds of magnetic fluid, one corresponding more or less to the fluid of the mesmerists, which accumulates especially round "obstructions" in the body, and another, all-pervasive one, which flows from God, does not accumulate, but may be influenced and directed by human will. It appears to be through this second kind of fluid that magnetic fluid of the first kind is impelled or aided to break down "obstructions", so promoting the cure of disease. Successful magnetizing is thus essentially a matter of an active and properly motivated will, and magnetic passes, wands, etc. have only the function of concentrating the operator's mind. Magnetization at a distance does not depend on any intervening medium. Each real "I" emanates from and hence participates in the Divine and is not truly separate from any other real "I". If an operator meets difficulties he should therefore endeavour by religious exercises to raise himself nearer the Source from which he and all others have emanated, thus participating in its omnipresence and in its control over the fluid. Magnetizers of the Barberinist school enjoyed a high reputation for piety, purity of life, and freedom from mercenary motives.

The real "I" is not totally tied to the body, but exists in its own "atmosphere"

(I suspect this means a kind of vaporous envelope), which can reach out indefinitely. Wherever this "atmosphere" has reached, there the self is. Such "expansion" away from the body occurs particularly in the somnambulic state. There has come down to us a journal of Barberin's from 1785 recording his experiments with various somnambules of religious tendencies.[57] These ladies were able to "reach out" towards the Divine (this apparently involved some variety of "out-of-the-body experience"), and *en route* encountered and conversed with various "intermediate" spirits and guardian angels. One lady met the spirit of her deceased father. In addition they were, in this state, able to answer various religious and cosmological questions. It will come as no surprise that these answers reflected Barberin's own views.

Meanwhile, in Lyon, the members of the *Concorde* were treading a similar and probably not independent path. Why, they asked themselves, should not the clairvoyance of the somnambules, so remarkable in medical and mundane matters, extend to the realm of spiritual beings and bring back answers to the religious and metaphysical questions which Willermoz and his associates had hitherto approached by the intricate paths of ceremonial magic? Several suitable somnambules were quickly located. The most successful was Mlle. Gilberte Rochette, who soon developed the faculty of putting herself into the magnetic sleep whenever required. Detailed records of the séances with her remain. She saw and conversed with angels and saints, and she met and (with appropriate gestures) retailed messages from deceased members of her own family, and the families of Willermoz, Monspey, and other members of the *Concorde*.[58] She ended by marrying one of Willermoz's nephews.[59]

The mystical magnetizers of Lyon seem to have had a considerable and spreading influence in the years up to the French Revolution and even beyond. Much of this influence probably passed along masonic channels, but it surfaces occasionally. Thus we learn from Bergasse, that Barberin converted d'Eslon to his doctrines shortly before the latter's death.[60] Bergasse himself had for a period dealings with spiritualist somnambules whose pronouncements he took seriously.[61] Certain Swedish and German masonic lodges of Swedenborgian tendencies found somnambules through whom deceased persons communicated Swedenborgian views.[62] Somnambules of other religious persuasions delivered their teachings.[63] Mystical magnetism seems to have survived and flourished underground even into the Napoleonic era – at least if we are to judge by the amount of space Deleuze devotes to it in his *Histoire critique* of 1813.[64] In the congenial climate of German Romanticism it was to emerge and proliferate into many strange forms.

SPECULATIONS

The speculations of Barberin and Saint-Martin were only the most extreme of the various bizarre theories provoked by Puységur's discovery of the state of "artificial somnambulism". The peculiar phenomena of the somnambulic state, especially the ostensibly paranormal ones, rendered most previous theories of animal magnetism obsolete. It was no good continuing to say, with the commissioners, that the phenomena were all due to "imagination", for imagination can hardly be supposed to create totally new faculties, such as those of clairvoyant diagnosis, prediction and prescription. The best ploy of an imagination-theorist with regard to such phenomena is to deny that there is any solid evidence for them. Believers in the magnetic fluid tended to welcome reports of the odder sorts of somnambulic phenomena as making things difficult for the imagination-hypothesis. It is not, however, immediately obvious that their own theory can cope with them any better. The first question to ask from the points of view of both parties is therefore clearly that of whether or not some excuse can be found for dismissing the alleged phenomena at the outset.

A work that was very influential in this connection, both at the time and for some decades to come, was J. F. Fournel's *Essai sur les probabilités du somnambulisme magnétique* (1785). Fournel, a Parisian lawyer, argues that it is impossible to dismiss all the marvels of somnambulism as fraud and pretence. Though there certainly are pretended somnambules, many somnambules are persons beyond suspicion, deception by all of whom would be a moral impossibility. Furthermore, the phenomena of magnetic somnambulism find many parallels in those of natural somnambulism. Fournel quotes the *Encyclopaedia* article on somnambulism:

> Persons who are afflicted [with somnambulism], plunged into a profound sleep, walk, speak, write, and carry out different actions, as if they were wide awake; and sometimes even with more intelligence and precision ... [65]

He takes from the same article accounts of various natural somnambules who supposedly exhibited peculiar and even paranormal powers resembling those of magnetic somnambules.[66] We may safely conclude that there is no a priori reason for rejecting the alleged phenomena of magnetic somnambulism.

A further implication of Fournel's arguments is that a theory of magnetic somnambulism cannot be divorced from a theory of sleep and of natural somnambulism. For there is no sharp division between ordinary sleep, in which various purposeful movements may be made, and natural somnambulism, and between the latter and magnetic somnambulism, which is simply a kind of somnambulism reinforced by magnetic procedures. This generalization was accepted by most subsequent theorists of animal magnetism.

Just as influential as Fournel's pamphlet was Tardy de Montravel's *Essai sur la théorie du somnambulisme magnétique* (1785).[67] He believes in a universal fluid, infinitely subtle, and underlying many kinds of natural phenomena, but especially concentrated in the animal organism, whose activities it sustains and through whose nervous system it circulates. Obstructions to its circulation are a cause of illness, and the procedures of animal magnetism reinforce nature's attempts to break down the obstructions. Now the fluid is exceedingly elastic, and when impelled upon an obstruction is thrown back from it upon some closely connected part of the body. From there it is thrown back again upon the obstruction, and so on, like a kind of water-hammer in the neural tubes. Not surprisingly, convulsions and other kinds of "crisis" may result. When the fluid is reflected back from the obstruction on to the brain, the crisis takes the form of somnambulism. To understand why this should be so we must turn aside to consider Tardy's doctrine of "the sixth sense".

According to Tardy we possess a "sixth sense" which is normally kept in abeyance by the other five. It lacks a special organ, and can be regarded as a sort of interior sense of touch, penetrating all internal parts, and revealing all kinds of damage and disharmony. It also picks up neural disturbances from the five external senses, and transmits these to the soul or intellectual principle.

Now the "seat" of the sixth sense is not in the brain, but in the solar plexus and complex of nerves linked thereto. If the brain becomes numbed or stupefied for any reason – such as incessant bombardment by magnetic fluid "reflected" from obstructions – the five exterior senses are damped down, and the interior sense comes into its own. The patient, now somnambulic, becomes able to perceive his own internal malfunctionings. Since the interior sense reports direct to the soul, the somnambule will be able to judge the extent and likely progress of his malady, and see the probable effects of the remedies which "instinct" suggests. If the somnambule is in rapport with another person, in other words if the magnetic fluid is circulating between them, "one can regard all their nerves as the cords of two instruments tuned to the same pitch or to mutual harmony". Then any discord in the nerves of the patient touched will reverberate in those of the somnambule who touches him, and the somnambule's emancipated sixth sense will perceive it. Thus we may explain the ability of somnambules to diagnose the ailments of other people.

As for clairvoyance of external objects: when, in the somnambulic state, the external senses are suspended, the interior sense (seated in the solar plexus) will begin to communicate with external objects through the medium of the universal fluid and will pass impressions of them to the soul. Furthermore, the soul, being partly disengaged from matter, will begin to enjoy the faculties proper to it; it will rise above time and distance, and display a judgment more far-reaching and more certain.

It is interesting to note that in his *Mémoire* of 1799, Mesmer gives an account of somnambulic phenomena not dissimilar to Tardy's. He offers, however, no acknowledgment either to Tardy or to Puységur, but talks as if he had been the first to discover the phenomena and to proffer the explanations.

This kind of theorizing – arm-chair speculation by persons, often competent in their own fields, but amateurs in physics and physiology with little scientific knowledge and no appreciation whatever of scientific method – was to dominate the animal magnetic movement for the next half-century. Between such theories and the mysticism of Barberin and the martinists there is little to choose, and perhaps no way of choosing. Both are probably based in part upon the pronouncements of clairvoyant somnambules whose "clairvoyance" extended no further than picking up hints of their magnetizer's own beliefs and hopes. These "fluidic" explanations of somnambulic phenomena were less extensively criticized at the time than they might have been. Of those who did criticize them, one of the most acute was Charles Villers (1765–1815), a young artillery officer (yet another artillery officer!) who was stationed at Strasbourg from 1783 to 1786, and who joined the *Société harmonique* there. Later he became well-known as an expositor of Kant and a writer on German history and literature. In 1787, he set forth his own ideas on animal magnetism in the form of a novel, *Le magnétiseur amoureux*.[68] This work is a good deal better as a treatise on animal magnetism than as a novel.

Among the objections that Villers raises to the fluidic theory is the following: The magnetic fluid is supposed to be all-pervasive, immensely subtle and penetrating. It can, allegedly, traverse great distances, pass through the thickest walls, etc. Yet at the same time, its movement is supposed to be prevented by such physically insignificant things as the "obstructions" to its circulation through the viscera that figure so much in the magnetists' theories of disease.[69] He might have added that its flow, or the passage of undulations through it, must also be stopped by the relatively insubstantial nerve plexuses of the stomach regions, the surrogate "organ" that makes clairvoyant perception possible. There is a complete contradiction here.

Villers' own theory of these phenomena is at first sight almost as extravagant as Barberin's.[70] Magnetism is a direct effect of the soul of one individual on that of another, brought about by an energetic concentration of the magnetizer's will. Magnetic procedures are not necessary, though they may serve to fix the attention. Magnetization is facilitated by a unity of purpose between two individuals, especially a unity in the desire that the patient be cured. If the magnetizer attains a great ascendancy over the patient, the latter may become somnambulic, due to his brain becoming charged with a superabundance of vital force. Then the action of the patient's soul on his body may be strongly augmented by the magnetizer's will, leading to curative effects and the enhancement of faculties. The supposed evidence for somnambulic clairvoyance Villers dismisses as due to the direct influence which the magnetizer's soul can exercise over that of the somnambule. What the magnetizer strongly pictures (e.g. the fluid emanating from his own fingers) the somnambule will see.[71] Indeed, Villers asserts that such powers have been misused by young men for the purposes of seduction.[72] Perhaps he had good reason for saying this: according to one story he had an affair while at Strasbourg with Lorenza Feliciani, the beautiful wife of Cagliostro.[73]

To most modern writers, Villers' theory, which postulates a form of what might be called active-agent telepathy, will appear exceedingly far-fetched. Yet, if we set aside the matter of telepathy, Villers seems to be offering some penetrating and thoroughly modern insights. He is saying (in effect) that certain peculiar phenomena – those of hypnotism – may involve covert communication between operator and patient along channels that it is very difficult to detect; that through these channels the imagination of the operator may decisively influence that of the patient without either party fully realizing what is going on; and that an essential part of this process may be the tacit development of complementary motives and a kind of unexpressed agreement to cooperate in pursuing them.

With Villers, I reach the end of my brief account of animal magnetism in France prior to the Revolution, a period of remarkable richness during which many of the phenomena later thought characteristic of hypnosis were discovered or adumbrated, and some of the counters still used in today's theoretical games had already been brought into play. Though popular interest grew steadily less during and after 1785, animal magnetism continued to have active and enthusiastic supporters right up to the Revolution. The Revolution swamped it,[74] as it swamped so much else, and when the waves of that great cataclysm subsided magnetism did not speedily reemerge. Probably it continued underground – this at least was Deleuze's opinion,[75] and he was well-placed to know. But some of its personnel had vanished for ever, and others had more pressing matters to occupy their minds. The impetus of the movement was largely lost. Fortunately, however, a little of that impetus had been transmitted outside France. The next phase in the story centres on Germany. It was a phase not of extensive further innovation and discovery, but of amplification, consolidation and finally codification. It had, however, some features of considerable interest.

NOTES

1 *Exposé de différentes cures* (1786); *Suite des cures* (1787).
2 *Suite des cures* (1787), pp. 209–220.
3 *Suite des cures* (1787), pp. 226–240.
4 *Exposé de différentes cures* (1786), pp. 133–135.
5 *Exposé de différentes cures* (1786), pp. 162–250.
6 I have not been able to find anything more about Tardy's life than is contained in his own writings of the period 1785–1787. Deleuze (1813), Vol. 2, p. 162n, says he had to leave France during the Revolution, and died "a few years ago".

7 Tardy de Montravel (1786b).

8 Dureau (1869), p. 72, gives the name as Brown.

9 *Extrait du journal* (1787); Luetzelbourg (1788); Tschiffli (1789).

10 Deleuze (1813), Vol. 1, pp. 223-225. A somewhat similar case, in which both speech and movement had been lacking for nine months, was successfully treated by M. Malzac of Castres, and reported by him to the Guienne society. *Recueil d'observations* (1785), pp. 157-161.

11 Cf. Kolb and Whishaw (1990), pp. 720-721.

12 Pp. 81-84. Mouillesaux was director of posts at Strasbourg. He died in Paris in 1811, aged 72. There is a eulogy of him in Deleuze (1813), 2, p. 119.

13 We might, in a preliminary way, define extrasensory perception as the acquisition of factual information without the use of the known sense organs. In "telepathy" the information relates to the contents of another person's mind; in "clairvoyance" to some physical event or state of affairs; in "precognition" to an event or state of affairs that lies in the future.

14 On a number of occasions, Puységur dictated to somnambulic subjects the hour at which they were to waken up or fall asleep again.

15 Mouillesaux (1789).

16 Mouillesaux (1789), pp. 70-72.

17 See above, p. 13.

18 Tardy de Montravel (1787c), pp. 201-202.

19 *Système raisonné* (1786), pp. 69-76.

20 Tardy de Montravel (1786b), pp. 195-196.

21 Cf. Dingwall (1967), pp. 32n-33n.

22 See e.g. *Système raisonné* (1786), pp. 79-84.

23 Deleuze (1813), Vol. 1, pp. 179-180.

24 See Deleuze (1813), Vol. 1, pp. 221-222.

25 Tardy de Montravel (1786a), pp. 21, 59-61; Puységur (1786), pp. 152-153.

26 Tardy de Montravel (1787a), p. 82.

27 Puységur (1811), pp. 383-384; Tardy de Montravel (1786a), p. 19; Deleuze (1813), Vol. 1, pp. 204-206; Tardy de Montravel (1787c), pp. xviii, 43-45.

28 Mesmer himself produced hallucinations of smell and taste in a member of the Academy of Sciences. See *Précis historique* in Mesmer (1971), p. 107. The "seeing" of the fluid is mentioned, and explained away, in the report of the academic commissioners, *apud* Burdin and Dubois (1841), pp. 36-37. For other accounts of the seeing of the fluid see Bonnefoy (1784), pp. 23-26, and (especially) Tardy de Montravel (1786a), pp. 5, 63-64, 67-72, 78, 182-183; (1787a), pp. xxvii, 54-57, 78-80.

29 The academic commissioners found that a lady who drank non-magnetized water believing it magnetized fell into convulsions, and vice versa. Burdin and Dubois (1841), pp. 65-66.

30 Mouillesaux (1789), pp. 62-63.

31 Tardy de Montravel (1787b), p. 28.

32 Grézie (1784), pp. 23-24; Girard (1784), pp. 17-19.

33 Haggard (1932), pp. 320-327.

34 The classic work is Dingwall (1967-8); see also Podmore (1909).

35 Kerner (1856), pp. 28-29.

36 See e.g. Dampierre (1784), p. 18; Tardy de Montravel (1787c), pp. 184-188.

37 Puységur (1811), pp. 60-62, 353.

38 Tardy de Montravel (1786b), pp. 164-172; Mouillesaux (1789), pp.72-75.

39 R., M. de (1787); cf. Tardy de Montravel (1786b), p.180.

40 But see Tardy de Montravel (1786c), pp. 60-62; Mouillesaux (1789), pp. 75-76.

41 Cf. Devillers (1784), p. 39; Deleuze (1813), Vol. 1, pp. 284–285.

42 A curious account of such a performance by one of Puységur's somnambules will be found in Oberkirch (1970), pp. 391–393.

43 Tardy de Montravel (1786a), pp. 65, 71; (1786b), pp. 203–204; (1787c), pp. 64–65.

44 Tardy de Montravel (1787c), pp. 258 ff.

45 Petetin (1787b).

46 Orelut (1784).

47 Some cures from these clinics are described by Dampierre (1784), pp. 14–15; Gilibert (1784), pp. 58–62. Criticisms of these cases will be found in L'antimagnétisme martiniste (1784), pp. 17–23.

48 Joly (1938), p. 219; Amadou and Joly (1962), pp. 20–21. Cf. Dampierre (1784), pp. 8–9.

49 Cf. Audry (1924), p. 73 and Dampierre (1784), pp. 11–13.

50 Audry (1924), pp. 85–87.

51 Monspey and Barberin (1784).

52 Devillers (1784), p. 87.

53 I have already mentioned the pamphlet by Bonnefoy (1784) in favour of Orelut and the pupils of Mesmer, and that by Dampierre (1784) supporting Barberin. Janin de Combe-Blanche (1784) and Gilibert (1784) are generally on the side of the magnetizers. O'Ryan (1784) is violently hostile to the pupils of Mesmer, whose clinics are, he says, hysterical in tone, mephitic in atmosphere, and run by persons of great eccentricity. The Lyons College of Physicians, he adds, has never approved or permitted any magnetic demonstrations – he is here getting at the surgeons. Also sceptical or hostile are Pressavin (1784), the author of the Abrégé de l'histoire (1784), and the author of L'antimagnétisme martiniste (1784), who may have been Charles Devillers (1724–1804), on whom see note 54 to chapter 1.

54 Joly (1938), p. 218.

55 Joly (1938), p. 223.

56 On Barberin's views see Luetzelbourg (1786), pp. 141–145; Puységur (1807), pp. 143–148; Système raisonné (1786), pp. 55–63; Barberin (1818). Amadou (ed.) (1969), prints, pp. 81–132, two treatises of 1784 by Saint-Martin on animal magnetism.

57 Barberin (1818).

58 Joly (1938), p. 228.

59 Amadou and Joly (1962), p. 77.

60 Bergasse (1786), p. 99n. Cf. a letter of d'Eslon's addressed to the editors of the Journal de Paris, and published in the Journal du magnétisme 1848, 6, pp. 237–242. It is dated 4th March 1785. See also Gauthier (1845), pp. 236, 241–289, and Deleuze (1825), p. 328.

61 Bergasse also had for a period dealings with spiritualist somnambules whose pronouncements he took seriously. See Viatte (1928), Vol. 1, p. 230.

62 Lettre sur la seule explication (1788); Tschiffli (1789); Halldin (1787; 1816).

63 See e.g. Extrait du journal (1787); Dieu, l'homme et la nature (1788).

64 Deleuze (1813), Vol. 1, pp. 235–293.

65 Fournel (1785a), p. 40. The author of this article was Dr. Menuret.

66 A case used by Fournel (1785a, pp. 42–44), and cited again and again by others, was that of a young ecclesiastic in the household of an Archbishop of Bordeaux. This young man would compose music and write sermons with his eyes shut, and would correct errors he had made even when the Archbishop interposed a sheet of cardboard between his eyes and the manuscript. Yet on other occasions he was totally confused as to the nature of his surroundings, plunging, for instance, into his bed under the impression that it was a river, making swimming movements, and "rescuing" a bundle of bedclothes which he took for a drowning child.

67 I have used the second edition of 1787. See Tardy de Montravel (1787c), pp. 201–222; cf. Tardy de Montravel (1787a).
68 Reprinted in 1978, with a valuable introduction by M. Charles Azouvi.
69 Villers (1978), pp. 15–16.
70 A summary of it is given by Deleuze (1813), Vol. 2, pp. 105–107.
71 Villers (1978), pp. 219–220.
72 Villers (1978), pp. 140, 220n.
73 Azouvi, introduction to Villers (1978), p. xiv.
74 According to Puységur (1807), p. 152, the Strasbourg Society survived until 1792.
75 Deleuze (1813), Vol. 1, p. 35.

CHAPTER 4

Mesmerism in the German-speaking countries

I call this chapter "Mesmerism in the German-speaking countries" rather than just "Mesmerism in Germany" because I wish also to touch upon mesmerism in German-speaking Switzerland and (obliquely) in the German-speaking parts of the Austro-Hungarian Empire. In any case, Germany in the 1780s was not a single country but a "divinely ordained confusion" of more than three hundred states, ranging in size from territories so tiny that their rulers could hunt a deer from one frontier to its opposite in the course of an afternoon, to the substantial and powerful electorates of Prussia, Hanover, Saxony and Bavaria. Among the best administered states were the fifty-one free cities, such as Lübeck, Bremen, Hamburg, Cologne and Frankfurt. Much less efficiently run were the ecclesiastical territories, supervised by bishops and archbishops. The remaining states were ruled by assorted princes, princelings and aristocrats, even the pettiest of whom was likely to have such extravagant ideas of his status that he felt obliged to have his own court, his own Versailles, his own comic-opera army, his own Montespans and du Barrys, and his own expensive gala-days. Some rulers were no doubt "benevolent despots"; probably most were more despotic than benevolent; and not a few capriciously alternated between benevolence and despotism.[1]

Against such a complex and chaotic background it is almost impossible to trace in any detail the paths through which animal magnetism spread into Germany, or to assign satisfactory reasons why it should have taken root in some areas and not in others. Freemasonry, and its mystical offshoots the Rosicrucians and the Illuminati, flourished widely,[2] and various princes (mainly Protestant ones) were adepts or sympathizers.[3] We have already noted the links in France between freemasonry, mysticism and mesmerism; and it seems likely that in Germany, as in France, the masonic network was a channel for the initial spread of mesmerism. But the overt progress of mesmerism was through more orthodox pathways.

One matter in which the German states were, on balance, not backward, was that of higher education. At the end of the eighteenth century the German states possessed twenty-six universities,[4] several of which – notably Halle (Prussia),

Göttingen (Hanover) and Königsberg (Prussia) – enjoyed a high reputation for academic freedom and for work in the pure and applied sciences. J. H. Baas claims that in medical education "the ascendancy and the lead fell to Germany",[5] apparently on the ground that most of the important medical systems of the period were developed there. Regional or professional learned societies sprang up and began to publish their own transactions and journals. By the 1780s, the influence of the French Enlightenment was very strong, and was promoted by a number of rulers following the lead of Frederick the Great of Prussia. Berlin became a centre of rationalist intellectual debate. The "Berlin frame of mind" received its most sustained and characteristic articulation from C. F. Nicolai (1733–1811), a prolific bookseller, publisher, editor and writer, and a resolute and cantankerous foe of all forms of irrationality and superstition.[6] The principal organ of the Berlin enlightenment was the *Berlinische Monats-schrift*, founded in 1783 and edited by J. E. Biester (1749–1816) and F. C. Gedike (1754–1803).

Little enough in all this might seem favourable to the serious study of animal magnetism. But growing interest in the sciences, along with increased freedom of debate, aroused, perhaps, here and there the idea that one should be consciously open-minded in the study of natural phenomena, and the first stirrings of the romantic movement may in some quarters have lent an added interest to the strange phenomena of animal magnetism. And, as we shall see, before long certain findings in the biological sciences began to lend a specious plausibility to the speculations of the mesmerists. Whatever the explanation, the fact remains that by the turn of the century, animal magnetism was being taken up and investigated by serious, even distinguished, German scientists and medical men.

ANIMAL MAGNETISM IN SWITZERLAND

The means by which animal magnetism was reintroduced[7] into the German-speaking lands was, however, one not likely to commend it to the enlightened. J. C. Lavater (1741–1801), the Zürich pastor, is now remembered principally as a physiognomist. But in his own day he was a figure of international note, a preacher of great fervour and personal charisma, a poet and writer on religious topics, an indefatigable traveller and evangelist, a correspondent of almost anyone who was anyone in contemporary mysticism and its offshoots. Hearing something of animal magnetism, Lavater went in the summer of 1785 to Geneva, where he received practical instruction in Puységurian methods. On his return, he immediately applied these methods in an attempt to cure the long-standing and variegated maladies of his wife.

This lady was suffering from prolapse of the womb, bad colics, an enlarged abdomen, frequent vertigos, frightful migraines, rheumatic attacks, and suppression of the periods. Magnetism caused violent convulsions, headaches, prostrations, and copious perspirations, but also restored the periods, and regular bowel movements. Before long Frau Lavater became somnambulic, and diagnosed and prescribed for herself and others with considerable success. Lavater was exultant, and not just on his wife's account. Animal magnetism, with its laying on of hands and its remarkable cures, seemed at the very least to lend added credibility to New Testament miracles, and it was not hard to conceive its practitioners as possessing apostolical gifts, as touched with and transmitting divine influence. Lavater wrote and spoke extensively in public and in private about his wife's case and about other cases seen by him. Partly as a result of his activities, animal magnetism became a subject of interest and debate in Switzerland, even in medical circles.[8]

A central figure in this debate was J.H. Rahn (1749–1812), professor of medicine at Zürich. As editor of the *Archiv gemeinnütziger physischer und medizinischer Kentnisse* (Zürich, 1787–1791), he published a number of pieces both for and against animal magnetism. This was, so far as I know, the first time animal magnetism had received anything like a fair hearing in a medical journal. The most interesting contribution is an exchange of letters dated 1786–7 between Rahn himself and Dr. Christoph Scherb of Bischofzell.[9] Scherb reported on six of his own patients. Most of them felt various peculiar sensations during the process of magnetization, and one became somnambulic. Only three showed any lasting improvement. However, Scherb concludes from these and other cases that the methods of animal magnetism can sometimes bring about improvements where other methods have failed, and that the whole matter deserves the closest investigation.

Rahn's reply consists of a lengthy and learned historical and psychological analysis of the phenomena of animal magnetism. His key notion is that of a relationship between one person and another which he designates "sympathy". By this he means a largely involuntary tendency to feel what a person you are watching feels, weep when he weeps, laugh when he laughs, act as he acts. The power of sympathy is greatest in those suffering from certain sorts of nervous disorders. Sympathy can be aroused by purely imaginary situations, as when listening to stories or even telling oneself stories. Its explanation is to be sought in the laws of association. If one sees another human being acting in a certain way in a certain situation, or even merely reads about it, there comes to one's mind the idea of the feelings, etc., which one has undergone oneself in similar circumstances. But since this idea is conjoined with an impression originating from another person, it does not produce so much a *memory* as a peculiar (and involuntary) community of feeling with the other individual. Insofar as the feelings engendered already have preestablished associations with the neuromuscular apparatus, "sympathetic" movements will be aroused.

One can, in a general way, see how this kind of explanation might be applied to the more notorious phenomena of the mesmeric *baquet* – the peculiar sensations, and (particularly) the convulsions. But Rahn does not pretend to provide an explanation of *all* the phenomena of animal magnetism. He proposes, for instance, that the "inner vision" and heightened imagination of magnetic somnambules is due to a dampening down during illness of the intellectual side of our being with a correspondingly enhanced awareness of the state of our animal economy. The same phenomenon can be observed in fever, hypochondria, hysteria, catalepsy, and as the result of taking various drugs. There is nothing special about it, and we do not need to invoke a mysterious magnetic influence to explain it.

Rahn's arguments are moderate in tone, and backed by substantial erudition. He reported them in various other publications,[10] and according to Milt they dug the grave of animal magnetism in Switzerland,[11] despite the fact that Mesmer himself lived there in retirement for extended periods of time.

MESMERISM IN GERMANY

Outside Switzerland, Lavater's influence on the spread of animal magnetism was longer lasting. Among the confidants to whom he wrote concerning his magnetic experiences was Charles Frederick, Margrave of Baden (ruled 1738–1811), a man as deeply interested in spiritual and mystical matters as he was humane and practical in mundane affairs. At Lavater's suggestion, the Margrave sent a certain Dr. Grob for training at Strasbourg, and then established him as professor of magnetism at Rastatt, where a harmonic society on the Strasbourg model was shortly formed.[12] Charles Frederick also invited to his court Jean de Crook, Russian ambassador extraordinary in Germany, and a prominent member of the Strasbourg society. De Krook arrived at Carlsruhe in July 1786, and a lively interest in animal magnetism sprang up both in court circles and more generally. So great was this interest that the Margrave sent the professor of physics at Karlsruhe, J.L. Böckmann (1741–1802), to learn Puységur's methods at Strasbourg. Böckmann had not been convinced by what he had heard of Mesmer, but he was convinced by what he saw at Strasbourg, and on his return came forward at the Margrave's request as leader of the numerous and eminent individuals in Karlsruhe who wished to help their fellow-men by animal magnetism. The group did not include many members of the medical profession, and no institute or society of harmony was set up. But in July 1787, Böckmann established the first periodical of importance devoted to animal magnetism, his *Archiv für Magnetismus und Somnambulismus* (eight parts, 1787–1788). This is a useful but somewhat disappointing work – useful because it contains many articles by persons practising animal magnetism outside Baden, but disappointing because it tells us almost nothing of the remarkable things that we know from other sources were going on in Baden itself. Many interested persons were coming there to observe the phenomena and receive instruction. With increasing experience Böckmann came more and more to favour the mesmeric over the Puységurian system. He felt that the Puységurians placed too much emphasis on the psychological element in the cures, whereas he had himself obtained striking effects by the use of steel conductors and other mesmeric paraphernalia.

Animal magnetism lingered on in Karlsruhe through the 1790s, but Baden never again became an especial centre for its study and practice.[13]

It was through a visit to the free city of Bremen in July 1786 that Lavater most powerfully influenced the development of animal magnetism. A girl of nineteen had been suffering for three months from a nervous illness which had resisted the best efforts of two competent doctors, A. Wienholt (1749–1804) and H. W. M. Olbers (1758–1840).[14] The patient would suddenly fall down unable to move, and remain so, sometimes for as much as half an hour. Convulsions and long and painful coughing fits were shortly added. Lavater was asked for his advice by friends of the patient, and suggested animal magnetism. Wienholt and Olbers accordingly proceeded to magnetize the girl by Puységur's methods. By September she was completely better.

Encouraged, and not a little surprised, by this success, Wienholt and Olbers successfully treated two further patients by the same means. Both were young women suffering from complexes of refractory ailments, amongst which convulsions and fits of powerlessness figured prominently.

Early in 1787, these cases were reported in print by Dr. G. Bicker, a professional colleague whom Wienholt and Olbers had invited to witness the phenomena. The latter gentlemen shortly gave their own accounts,[15] and in October Wienholt described the three cases in a small book.[16] All the subjects became somnambulic. Two of them appeared often able to see with closed eyes, though Wienholt was not sure to what extent light was getting through the lids. Hearing and speech were heightened in trance, and one patient would issue practical and comprehensive instructions as to the handling of her case.

Upon awakening, the subjects would recollect nothing of what had happened. But in the magnetic state they could recall both the events of the waking state and the events of previous magnetic episodes. Indeed, their powers of memory were strikingly enhanced. Long-forgotten events from their earliest years would return to them unheralded. Two of the patients became quicker to learn and apter to retain new verbal material.

Sometimes patients in the magnetic sleep were able to predict the course of their illnesses, and the effect of different treatments, and they would also prescribe medicines and diets for themselves. One patient even prescribed (apparently by means of detailed description) medicines of which she had never heard.

Publication of the Bremen results aroused fierce antagonism. In Bremen itself, animal magnetism was attacked from the pulpit,[17] whilst in Berlin it was equally frowned upon by the rationalists of the *Berlinische Monatsschrift*. Indeed, Funck proposes[18] that Biester, Gedike and Nicolai hoped that if they could comprehensively shatter the absurd pretensions of animal magnetism, this would promote their larger aim of blowing Christianity to the winds; for Lavater was the most ardent and the most charismatic contemporary apostle of both.

The unfortunate Wienholt, who saw nothing supernatural in animal magnetism, found himself the target of acrimony which he had done nothing to deserve. Many of the more controversial articles were reprinted by J. H. Cramer in his *Magnetistisches Magazin für Niederteutschland* (eight parts, 1787–1788), a work which Wienholt regarded as prejudiced in the extreme.[19] For the next fifteen years he pursued his investigations and speculations in relative privacy, publishing nothing, though he did not hide his activities from interested professional colleagues. Among these were J. Heineken, a well-known Bremen physician, and the noted botanist, G. R. Treviranus (1776–1837), both of whom published on the subject.[20] Heineken's little book, which came out in 1800, foreshadows some of the notions which Wienholt himself developed in his monumental *Heilkraft des thierischen Magnetismus* (1802–6).

A name often paired with Wienholt's was that of Eberhard Gmelin (1751–1809), a physician of Heilbronn, in Württemberg. It is not clear how Gmelin became interested in animal magnetism, but it seems on geographical grounds likely that he would have heard of the activities of the Strasbourg Society, and also of the growing interest in Baden. He certainly visited Böckmann in Karlsruhe and observed striking phenomena there.[21] Gmelin first became known as an animal magnetist through the publication in 1787 of his small book *Ueber thierischen Magnetismus*. It was quickly followed by two very substantial works, *Neue Untersuchungen über den thierischen Magnetismus* (1789) and *Materialen für die Anthropologie* (two volumes, 1791–3). Gmelin held animal magnetism to be of value chiefly in cases of nervous illness. The "magnetic stuff" transferred from operator to patient is to be identified with the "nervous ether" and it has close affinities with electricity. It streams from points (fingers, metal rods), has conductors and insulators, possesses polarity, etc.

There were of course other persons who spread animal magnetism in Germany, and other centres from which it spread. For instance, we learn from the *Deutsche Zeitung* of 4th May 1787 that one Pichler had come from the Strasbourg Harmonic Society to Mannheim, Mainz and Frankfurt am Main, spreading the word as he went.[22] In 1788, Böckmann heard from correspondents that several doctors in Duisburg and Düsseldorf were experimenting with animal magnetism, as was the younger Unzer in Altona. Later still he heard of progress in Mainz, Frankfurt, Heidelberg and Mannheim.[23] But the major influences in early German animal magnetism remained Böckmann in Karlsruhe, Wienholt and his friends in Bremen, and Gmelin in Heilbronn. Many subsequent workers acknowledged an initial indebtedness to one or other of these.

It was Gmelin who was indirectly responsible for animal magnetism penetrating at last the citadel of German enlightenment, Berlin. Frederick the Great of Prussia had died in 1786. His successor Frederick William II was a man of a very different temperament, devoted to mysticism and the irrational. Biester began to have problems with the censor. His attitude to animal magnetism shifted somewhat, and in November 1789 he reported in the *Berlinische Monatsschrift*[24] a series of experiments supervised by Professor C.G. Selle (1748–1800), medical director of the charity hospital. Selle had become interested in animal magnetism through reading one of Gmelin's books. His experiments led him to a guardedly favourable conclusion. He concluded that the effects of animal magnetism are due to a quasi-electrical material in the nerves, set in motion by stroking. Biester concurred. Selle continued to experiment. Meanwhile, an interest in animal magnetism had grown up at court, and in his last illness (1797) Frederick William II received magnetic treatment.[25]

SCIENCE AND ANIMAL MAGNETISM IN GERMANY

By the early 1790s animal magnetism in Germany had become respectable enough to receive serious attention from scientists and medical men. Thus C.W. Hufeland (1762–1836), a well-known, and moderate, medical writer and editor of medical journals, who had attacked Mesmer and mesmerism at the time of the French Royal Commissions of 1784, had changed his views considerably by 1794:

> By good fortune [animal magnetism] came at last, where it should properly have been, in the hands of philosophical doctors, amongst whom I name with genuine respect Dr. Gmelin of Heilbronn ... It is ... surely worthy of remark, and redounds to the honour of the German nation, that, as soon as magnetism began to become trickery, it could no longer maintain itself on German soil, and that as soon as it returned there, it very quickly took on a more solid and philosophical aspect.[26]

Science in the last decade of the eighteenth century, and for some while thereafter, contained tendencies which seemed in a general way to favour some of the assumptions of the animal magnetists. Most important was the suggestion, recently reinforced by the experiments of Luigi Galvani (1737–1798), that the action of nerves and muscles was not quasi-hydraulic but electrical in nature. Galvani's findings sparked off controversies into which we need not enter in detail.

Among those who joined the debate was F. A.von Humboldt (1769–1859) of whose work Mary Brazier says, "The design of Humboldt's experiments and the clarity of his reasoning are a pleasure to study in the welter of acrimonious controversy that greeted Galvani's findings."[27] Humboldt concluded that Galvani had uncovered two genuine phenomena: bimetallic electricity and intrinsic animal electricity. Certain of Humboldt's experiments were cited again and again by animal magnetists. He was thought to have demonstrated that nervous activation could cross a gap of up to 5/4 line in a frog's crural nerve, and still cause constriction of the muscle. This "induction" would also take place if the two ends of the severed nerve were carried past each other so that the halves of the nerve ran parallel with a small separation. It is to be noted that the cut ends of the nerve were held in mid-air.[28] Animal magnetists took such experiments not just as disproving quasi-hydraulic views of the nature of nervous action, but as demonstrating that an activated nerve had round it a circle of influence or "nervous atmosphere" extending well beyond its anatomical boundaries.[29] By implication, the nervous system as a whole might be surrounded by an atmosphere passing altogether outside the body. Many magnetists supposed that this "atmosphere" would be disturbed by the approach of metal, and the literature of early German magnetism is full of accounts of the supposed sensitivity of somnambulic subjects to the presence of this, that or the other kind of metal.

It was not just with frogs that experiments on the electrical properties of the nervous system were conducted. Others used higher animals. Some even experimented on human beings. Galvani's nephew, G. Aldini (1762–1834), applied strong electrical currents to corpses and to the freshly severed heads of criminals, producing contortions and grimaces which certain spectators could not bear to look upon. "The fresher the head," says Brazier, "the more remarkable the grimace."[30]

Speculation and experiment on the role of electricity in neural function flowed into and sometimes merged with, sometimes remained distinct from, an older stream of speculation concerning a supposed "life-force", a notion which had in turn emerged from seventeenth and eighteenth century doctrines about the soul, for instance those of G. E. Stahl (1659–1734). Soul-doctrines were wearing a little thin by the end of the century, but what may be called the problem of the *differentia*, of that which differentiates an organized and ongoing living creature from a spent corpse, remained.

This problem was addressed in an influential article of 1796[31] by a person who, though not himself a devotee of animal magnetism, was to become one of the great heroes of the magnetists, namely J. C. Reil (1759–1813), professor of medicine at Halle and later Berlin, remembered today for the "Island of Reil", but best known in his own time for his enlightened approach to psychiatry,[32] and as first editor of the *Archiv für die Physiologie*, founded in 1796. According to Reil, the phenomena of the living body will in the end be explained in terms of physics and chemistry. However, he thought it probable that with the coarse ordinary matter of the body is blended a finer stuff, and that upon the consequent refining of coarse matter the possibility of "animal phenomena" depends.[33] As for the nature of this fine stuff and its relation to coarse matter, we can, Reil thinks, as yet say nothing definite. Later, under the influence of Schelling and nature-philosophy, he came to conceive life-force in accordance with "the galvanic schema".[34] And some of his ideas about life-force and "the ganglion system" (i.e. the autonomic nervous system) were, as we shall see, to have especial appeal to animal magnetists.[35]

Though findings and speculations of these kinds did not directly support the contentions of the animal magnetists, they did give the magnetists, in most peoples' eyes, if not a cloak of scientific respectability, at any rate materials from which it seemed likely that such a cloak could be manufactured. In 1797 came a kind of breakthrough with the publication in Reil's *Archiv* of Dr. J. N. Pezold's account of his experiments in animal magnetism.[36] Pezold was a Dresden medical man influenced by Gmelin. It cannot be said that these experiments were very exciting, but after the turn of the century further articles on animal magnetism began to appear in Reil's *Archiv* and in the very influential *Journal der practischen Arzneykunde und Wundarzneykunst*[37] edited by C. W. Hufeland. Although an attempt by A. W. Nordhoff to run a periodical devoted to animal magnetism[38] was a failure, there clearly was in Germany at this time a considerable interest, both medical and lay, in animal magnetism.[39] An Englishman who travelled there in 1803 and 1804 remarked that he heard "many very enlightened men of the universities talk of animal magnetism, nearly with the same certainty as of mineral magnetism...".[40]

WIENHOLT'S *HEILKRAFT DES THIERISCHEN MAGNETISMUS*

Interest was no doubt heightened by the publication in 1802–1806 of what must, I think, be rated in many ways the most impressive book in the whole literature of animal magnetism, Arnold Wienholt's *Heilkraft des thierischen Magnetismus*. Following the hostile reaction to his first book, Wienholt had shunned publicity for fifteen years. When at last he felt his results were firm enough, he began to publish them, and the three chubby though disorganized volumes in which he did so are at once a landmark in the history of animal magnetism and a quarry from which many subsequent writers extracted their materials.

The fifty-nine case histories[41] scattered across these volumes vary a good deal in length, but are on balance rather impressive. Wienholt chronicles successes and failures alike with the most laudable impartiality. When he chronicles a success one is correspondingly more tempted to accept the implication that magnetic treatment must have contributed to the favourable outcome, especially if, as in not a few cases, conventional remedies had been tried for an appreciable time without success.

According to Wienholt, animal magnetism may most usefully be employed in nervous illnesses in which there is no gross organic lesion. However, he does not mean cases of hysteria and hypochondria – he expressly states that with hysterics little benefit may be expected. By nervous diseases he seems to mean such complaints as epilepsy (he recounts some successes – one, Case I, with an epilepsy of 13 years standing – and some failures); other kinds of fits and convulsions; impairments of hearing, cause not clear (Cases IV, V and VI, much improved); amaurosis (blindness without known cause – Case VII, considerably improved); certain sorts of paralyses and muscular problems; and cases in which syncope and attacks of "powerlessness" were leading symptoms. With two cases of apparent mental retardation (XIV and XV) he was totally unsuccessful. Wienholt did not by any means confine his endeavours to nervous ailments. He tackled, unsuccessfully, but with some transient alleviation of suffering, two cases of malignant disease ((XVIII and XLVIII). In a dangerous case of pneumonia (Case XVI) he obtained a rapid improvement and ultimate recovery, which he attributed to the effects of animal magnetism in promoting sleep. Suffering himself from an excruciatingly painful leg wound (Case XLVI), he was magnetized by downward movements of the flat hand about an inch above the wound; pain was greatly alleviated and sleep promoted.

In only a small proportion of these cases did the patient become somnambulic; most of the treatments simply involved mesmeric passes (often, however, resulting in drowsiness or sleep).

Wienholt was not just a clinician, but a well-read and reflective person, intensely interested in the theoretical and philosophical issues. He holds a conception of the life-force somewhat resembling Reil's, but avoids any firm commitment as to the relationship between the life-force and electricity, magnetism, etc. In a long and erudite disquisition on sleep,[42] he explains the healing efficacy of sleep by supposing that sleep promotes a redistribution of life-force away from the brain and sense-organs to any other system where it might be needed. Can animal magnetism produce a similar redistribution of life-force? Undoubtedly. The magnetizer's life-force draws that of the patient away from brain and sensory systems to other parts. Ultimately the brain becomes depleted, producing drowsiness and sleep. That the magnetizer acts in such a way is further shown by the effect of contrary passes, which reawaken the patient by reversing the conduction of life-force.[43]

The magnetizer, however, does not act purely through redistributing the patient's life-force. He transmits life-force to the patient. This is suggested by the sensations which magnetization produces. And without postulating such a transmission it would be difficult to explain how a short and often incomplete sleep could rally the patient's forces for an extended period.

As for the phenomena of somnambulism, these occur when life-force is withdrawn from parts of the brain or from certain sense organs only. In such conditions "the cessation of many animal functions will bring others to an often striking level of activity, and the inactivity of the former will mostly appear the more profound, the stronger is the activation of the latter."[44]

With regard to the paranormal powers allegedly exhibited by certain somnambules, Wienholt offers two kinds of explanation. The first is contained in his "Seven lectures on the sphere of influence of living bodies" printed in the third volume of *Heilkraft*.[45] Here he takes up the idea, favoured by Reil, and supposedly established by Humboldt, that an active nerve has a "circle of influence" extending beyond the actual substance of the nerve. Following Gmelin, he further supposes that the whole body may be surrounded by a "life-atmosphere" produced by the conjoint life-force of all the organs of the body. He finds evidence for this life-atmosphere in phenomena that seem to suggest the externalization of a kind of tactile perception beyond the skin, and an externalization of force beyond the body; for instance the supposed "eyeless vision" of bats[46] and polyps, the power of the torpedo to stun other creatures, the power of snakes to fascinate their prey. In humans, comparable evidence is provided by such phenomena as sensitivity to the presence of metals, peculiar feelings, twitchings, etc., produced by the approach of a human hand, instantaneous sympathies and antipathies, the revivifying effect upon the elderly of contact with the young. Wienholt supposes further that the life-atmosphere, with its attendant form of perception (whose nature and laws are never made very explicit), extends inside the body-cavity, making possible some kind of awareness of events and conditions within it.

What may be called full-blooded clairvoyance – the alleged ability of certain somnambules, natural or magnetic, to read, discern objects, even perceive distant scenes, without the aid of the ordinary senses – Wienholt discusses in the "Seven lectures on somnambulism" to be found in volume three of *Heilkraft*.[47] Considerable parts of these lectures are devoted to presenting evidence for this somnambulic clairvoyance, and also for the supposed dermal and other vision of certain blind persons. I shall touch on such phenomena in the next chapter. Though Wienholt is well aware of the strength of the case for materialism, and cites evidence for "the dependence of the thinking and willing principle ... upon the condition of the brain",[48] he thinks that the facts of clairvoyance tend to disprove materialism and to open up other possibilities:

> If an individual can acquire visual perceptions without the assistance of the external organ, it is equally possible that, in another state, he may also be capable of thinking without the assistance of the brain, and that his

> imagination and his memory may not be indissolubly attached to this soft
> and so easily destructible mass ... [49]

Wienholt speculates further that the life-force may act as intermediary between the soul and the body. Now a very small redistribution of life-force, as in somnambulism, may, without any structural changes in the brain, bring about marked changes in the intellectual faculties, lead to apparently separate streams of memory (a somnambulic and a waking one), etc. If mere redistribution of life-force, and a change in its concentration, can produce such profound alterations in the mind, we may reasonably assume that the mental impairments caused by brain damage are the result of interference with the life-force, and "the dependence of the condition of the mind upon the body, might be merely apparent, and only indicate the change which has taken place in the intermediate organs."[50]

Wienholt's theoretical speculations were exceedingly influential on later German animal magnetism.

ANIMAL MAGNETISM IN BERLIN

For the next few years – let us say the period from 1806 to 1820 – the most important scene of events in German mesmerism was Berlin.[51] I must again emphasize that we are talking about a relatively respectable sort of animal magnetism, one whose leading supporters were accepted members of the medical and scientific communities. During the later part of this period a rather different approach – which, at the risk of some oversimplification, we may call the "mystical" – became prominent, and ultimately dominant, with a consequent increase in hostility from orthodox science.

The best starting point is perhaps the fact that in 1800 C. W. Hufeland moved to Berlin, succeeding Selle as Director of the medico-surgical college, doctor in charge of the Charité, and full member of the Academy of Sciences. Hufeland, as we have seen, began to open his *Journal der practischen Arzneykunde und Wundarzneykunst* to articles on animal magnetism – the first was by Dr. J. F. L. Lentin of Hanover, who had just paid a visit to the Bremen magnetizers.[52] Most of the contributions were case reports. Hufeland himself reported a case in 1809,[53] and prefaced it with a statement of his current views on the subject. Originally, he says, he did not deny the facts, but ascribed them to the power of imagination. But he had to take the investigations of men like Reil, Gmelin, Wienholt, Heineken and Schelling[54] seriously. These men had established the physical rather than psychical basis of the phenomena, and tied them into the range of higher natural powers of electricity and galvanism. He himself accepted that there exists an unknown rarefied medium, analogous to mineral magnetism, which unites living beings with each other, and which may be aroused by the procedures of animal magnetism. Its channels in the organism are the nerves, and it appears to be closely related to the principle of nervous activity. Hence powerful effects are obtained when magnetic manipulations follow the course of the nerves.

Hufeland was an influential writer and it is probable that no single article did more to enhance the respectability of animal magnetism. However, Hufeland had not himself magnetized the patient. The operator was a young Berlin surgeon, C. A. F. Kluge (1782–1844). Kluge had for some while been at work on a book, which appeared in 1811 under the title *Versuch einer Darstellung des animalischen Magnetismus als Heilmittel*.[55] It was intended as a guide for practising doctors, but delves quite deeply into history, phenomena, theory and practice, and contains numerous, and largely accurate, references to the literature. With the possible exception of Deleuze's *Histoire critique du magnétisme animal* (1813), it is the most useful of all books on animal magnetism[56] – so useful that I propose to devote the next chapter to a brief discussion of it. Though in most respects just a synthesis and compendium of data and ideas already current, it served to fix these ideas and to give them a form and definiteness which kept them part of the stock-in-trade of magnetic literature to the mid-century and beyond.

It is a curious fact that Mesmer, though German-speaking, is mentioned only somewhat rarely in the early German mesmeric literature, and when mentioned tends to be dismissed as a charlatan. It seems to have been widely assumed that he was dead. However, though he had kept out of the public eye for over twenty years, Mesmer was still alive and tolerably robust. In April 1809, Dr. J. A. Zugenbühler (1774–1855) of Glarus contributed to Hufeland's *Journal* a brief account of Mesmer's wanderings since the French Revolution, and of a visit to him at his present retreat in Frauenfeld, Switzerland.[57] The old leopard had in no way changed his spots. He still held firmly to his beliefs, and had beside him a completed manuscript which, however, no publisher would undertake at their own cost.[58]

After this a number of medical men and scientists interested in animal magnetism entered into correspondence with Mesmer, or visited him. They included L. Oken (1779–1851), professor of medicine in Jena, and a leading exponent of nature-philosophy, Reil, Kluge, and K. C. Wolfart, of whom more shortly. In December 1811, Reil, who had moved to Berlin in 1810, suggested that Mesmer should visit that city and demonstrate in a hospital there. Mesmer replied that his age did not permit him to undertake so long a journey. A suitably qualified person must come to him, learn his methods, and return to instruct others. Furthermore, he must be a person in some way officially commissioned by the government. Mesmer had not lost the obstinate wish that governments should bend the knee to him.

He got his wish, more or less. In March 1812, Reil showed the correspondence to Schuckmann, head of the department of culture and public education in the Prussian ministry of the interior.[59] Schuckmann was undisguisedly hostile. He would, he said, remain sceptical until Mesmer or one of his disciples had swallowed an ounce of arsenic before his eyes and been cured by magnetism. He himself would be happy to bear the cost of this experiment, which would permanently silence either the disbelievers or the apostle of superstition.[60] However, the State Chancellor, Hardenberg, was more sympathetic. Schuckmann therefore set up a commission, under Hufeland's chairmanship, to test the claims of animal magnetism. And, as head of the department of police, he issued an order restricting the practice of animal magnetism to medically qualified persons.[61]

In September 1812, a member of this commission, K. C. Wolfart (1778–1832) arrived in Frauenfeld and presented Mesmer with a certificate signed by Hufeland. (There was later much dispute as to whether or not Wolfart had any official government status in the matter, and because of this, and the outbreak of war, the committee did not issue its – favourable – report until 1816.) Wolfart's promising career in his native Hanau (Prussia) had came to an abrupt end through an affair with the wife of a business man. After various wanderings, he settled in Berlin about 1809. In 1811 he started *ΑΣΚΛΗΠΙΕΙΟΝ* a general medical and surgical journal which ran until 1814. In this he showed himself favourable to animal magnetism, and by the beginning of 1812 had established his own magnetic clinic, and was in correspondence with Mesmer. He kept a large house and gave hospitality to a large circle of literary friends. He seemed to have a gift for getting to know, indeed to treat, the "right" people. This may not have been deliberate policy, but in the end it helped him to achieve what quite soon became his leading ambition – a chair in the medical school which could be a forum for the teaching of animal magnetism.[62]

From the accounts of those who met Mesmer in old age,[63] it is quite apparent that he had lost none of his power to fascinate. Indeed, the main features of his character became easier to discern in tranquil retirement than they had been amidst the glitter and the strife of Paris in the 1780s. Visitors found him an active and very sociable old gentleman, in comfortable, though not affluent, circumstances, no longer in practice, but happy to magnetize friends or deserving cases free of charge, and possessing a great love of animals, and a strange power over them. He liked to talk, and over a simple but interesting meal would expound his system with the sort of obstinate conviction that by itself half-convinces a sympathetic listener. He would dwell, too, with much emphasis, upon the bigotry and short-sightedness of governments and faculties of medicine. Afterwards he might give a memorable performance on the glass harmonica; and few, perhaps, of those who listened, as night drew in, to the plangent sounds so skilfully drawn forth by those famous hands, did not wonder whether, after all, the old man's claims did not conceal some elements of truth.

A few weeks of Mesmer's company overwhelmed Wolfart, who returned to Berlin fired with enthusiasm and bearing the aforementioned manuscript. (He had, in fact, already published two pieces by Mesmer in $A\Sigma KAH\Pi IEION$.[64]) The new manuscript was in French, which Wolfart turned into German.[65] Publication was delayed, owing to the outbreak of war (during which Wolfart practised animal magnetism in a military hospital[66]). But in 1814 the book appeared at last as *Mesmerismus. Oder System der Wechselwirkungen*.[67] It somewhat resembles an expanded version of the *Aphorismes*, and contains Mesmer's views on a large number of subjects, from physics and physiology to politics and pedagogy. The sections on animal magnetism are disappointing. A long passage on magnetic apparatus and the magnetization of trees is lifted from the *Aphorismes*, whilst the excursus on somnambulism is taken straight from the *Mémoire* of 1799.

Next year, 1815, Wolfart published his own observations on Mesmer's views, *Erläuterungen zum Mesmerismus*. By this time his practice had expanded into a "great magnetic polyclinic".[68] Two large *baquets* of his own design[69] were the centrepieces of a large, tastefully furnished, and dimly lit, salon, which became, says Artelt,[70] not just a showpiece for visitors to Berlin, but a weighty factor in Berlin social and intellectual life. Fichte visited it. Wolfart treated Schleiermacher, Wilhelm and Caroline von Humboldt, and the wife and brother of State Chancellor Hardenberg. And he had the support of Hardenberg's personal physician, D. F. Koreff (1783–1851), a dedicated advocate of magnetism and nature-philosophy.[71] In February 1817, through Hardenberg's influence, Wolfart was appointed a full professor, and a state-subsidized clinic for the magnetic treatment of the poor was set up under his direction.[72] At the same time a prize (never awarded) for the best dissertation on animal magnetism was offered through the (largely hostile) Academy of Sciences.[73]

Wolfart's clinic was now the most notable centre of animal magnetism in Europe, and many interested medical men and scientists, both sympathetic and hostile, came there to observe.[74] It is greatly to his credit that he seems to have admitted all these visitors freely, whatever category they fell into. Some were favourably impressed both by the man and by his work. He was particularly noted for his kindness to and success with child patients.[75] Others felt, perhaps, that someone so socially adept, so agreeable in voice and manner, so popular with the ladies, could not be completely honest. What is probably the truest estimate of him comes from C. I. Lorinser, who acted as his assistant for almost two years:

> Most [Berlin] doctors were hostile to Wolfart, or eyed him askance. Only a few understood him, and these were (like myself) members of the mesmeric association which he had founded. His weakness and good nature were often abused ... Even so he remained a tireless doctor, doing good to the poor, friendly, mild and forgiving towards colleagues, and, when needed, capable

of the greatest dedication. He was one of the few who devote their lives to an
idea, and defend it with noble enthusiasm and unbroken courage despite all
contradictions, insults and enmity ... [76]

Intellectually, Wolfart was no giant. His university lectures were notorious for
their absence of solid content. But he stood for something. Despite the mysticism
which was beginning to infiltrate the magnetic scene, and his own interest in
nature-philosophy, he presented magnetism as before everything else a method
of treatment to be assessed as such. He started a new periodical, the *Jahrbücher
für den Lebens-Magnetismus oder neues Askläpieion* (1818–1823), in which he
published case records and statistics. He contributed largely to keeping animal
magnetism in a position where respectable medical men could at any rate
seriously discuss it.[77] By this time the standing of mesmerism in Berlin was
slipping badly, owing in part to an unfortunate lawsuit involving a young girl
who attended Wolfart's clinic.[78] But other persons in other places kept the ball
in play for some time to come.[79]

And what of the numerous cures performed in Wolfart's clinic, about which
visitors heard much, and of which readers of the *Jahrbücher* could learn details
and statistics? Writing in 1816, Koreff noted that it was not especially in
nervous illnesses that Wolfart obtained beneficial results. He succeeded with
ailments ranging from scrofula, ankylosis, and eye problems to haemorrhoids
and bleeding of the womb. In some cases ordinary remedies had failed, and no
result was anticipated.[80]

INFLUENCE OF WOLFART ABROAD

Despite the doubtful status of animal magnetism in Berlin, there can be no doubt
that around 1820 many doctors in the German-speaking states were taking it
very seriously – a glance at the contents of Wolfart's *Jahrbücher* or of the *Archiv
für den thierischen Magnetismus* is quite convincing on the matter – and that
Wolfart's clinic and publications were important factors in its spread. In Austria,
however, animal magnetism was frowned on by the authorities and officially
banned, though it was practised for all that, and was demonstrated during the
Congress of Vienna in 1815.[81] Soon afterwards the Viennese court sent G.
Malfatti (1775–1859), personal physician to Erherzogin Beatrix of Este, to study
animal magnetism at Wolfart's clinic.[82] None the less, the Austrian government
remained for some decades implacably hostile to animal magnetism.

Curiously enough, the only leading magnetist besides Wolfart who occupied
a university chair was Austrian by birth. Joseph Ennemoser (1787–1854) had
interrupted his medical studies at Innsbruck to join the patriotic rebellion of
Andreas Hofer, and thereafter went to Berlin to complete his medical education.
He was already interested in animal magnetism, and became friendly with

Wolfart. In 1819, he published his substantial *Der Magnetismus*, and on Koreff's recommendation was appointed professor of medicine at the University of Bonn (he later migrated to Innsbruck and to Munich). Over the next forty years he quietly produced a series of substantial treatises on animal magnetism and related topics.[83] His speciality was the interpretation in magnetic terms of the history of magic and religion. There is, he holds, a latent somnambulic element in all of us, which in most cases emerges only in dreams, but which in some people may occasion religious visions and fantasies. There can also be an actual influence of soul upon soul independently of space and time, an influence which underlies true magic and the communion of spirits. He avoids naturalistic explanations of New Testament miracles and revelations. Ennemoser seems to have been more respected than influential.[84]

The country in which German animal magnetism caught on most extensively was Russia. According to Ludmila Zielinski, "there must have been a great many cases of use and abuse of animal magnetism in Russia even at the beginning of the nineteenth century, since in 1816 it was found desirable to confine its use to physicians."[85] But little was published prior to G. F. von Parrot's *Coup d'oeil sur le magnétisme animal* (1816). Parrot (1767–1852), a German, was professor of physics at the university in St. Petersburg. He maintained that the magnetic fluid is probably akin to electricity, and that physicists alone are qualified to investigate it. This opinion was, not unnaturally, opposed by J. R. Lichtenstaedt (1792–1849), a German medical man practising in St. Petersburg, whose *Untersuchungen über den thierischen Magnetismus* also appeared in 1816. Lichtenstaedt does not doubt the therapeutic effects of animal magnetism, but in his book, and in a subsequent article,[86] throws doubt on the existence of the fluid, and supports the views of F. Hufeland.[87] He also praises Wolfart, who in 1819 contributed a foreword to his *Erfahrungen im Gebiete des Lebensmagnetismus*.

After 1816, there was a little flurry of interest in animal magnetism among medical men working in Russia, mostly German ones to judge by their names.[88] In 1817, the Czar sent the Czarina's physician, Stoffregen, to visit Wolfart,[89] and a commission of doctors was set up to which medical men using animal magnetism were required to send weekly reports. Other expatriate German doctors also visited Wolfart.[90] Some odd findings were reported. Dr. Loewenthal of Moscow had a somnambule so sensitive that, with his back turned, he could be magnetized by thought alone, and would fall senseless to the earth.[91] Professor Reuss of Moscow discovered by a series of experiments that once the magnetic influence had been communicated to one of the substances known to store it, almost no treatment – neither boiling, nor burning, nor heating to red heat, nor calcining, nor dissolving in acid – would expunge it again.[92]

There were also some signs of native Russian interest in animal magnetism. In 1818, Count N. P. Panin (1770–1837), a former Russian ambassador to

Prussia, published a case report,[93] and in the same year came possibly the first Russian book on the subject, D. Velianski's *Zhivotniy Magnetizm*.[94] Velianski was professor of physiology and pathology at the Imperial Academy of St. Petersburg. The first two parts of his book are a translation of Kluge's 1811 work, with the references omitted, and the third contains marked traces of the influence of Schelling and the German nature philosophers.

Animal magnetism in Russia seems to have lapsed into comparative obscurity by the mid-1820s.

In the Scandinavian countries, interest in animal magnetism was much less, and again largely derivative from Germany.[95] In Norway during the period under consideration, interest was never more than slight. In Sweden, interest was somewhat greater, partly, perhaps, because of the efforts of Count C. Löwenhjelm, Swedish ambassador to Russia. Löwenhjelm published various case reports (in one of which, incidentally, he mentions stimulating the subject's recollection by touching the organ of memory as defined by Gall, this becoming perhaps the first person to unite the ideas of phrenology with those of mesmerism[96]), and possessed sufficient influence to have a young (and later quite distinguished) Swedish doctor, P. G. Cederschjöld (1782–1848) sent by the government in 1816 to study animal magnetism under Wolfart. On his return, Cederschjöld, who had been interested in animal magnetism for some years, published in the *Journal för Animal Magnetism* (a periodical which he had apparently started himself) an extensive account of his journey, and of the cases that he subsequently treated by magnetic methods.[97] His reports aroused great interest and much press comment, though they do not seem to have convinced any of the country's leading doctors and scientists, with the possible exception of C. A. Agardh (1785–1859), the algologist and follower of Schelling.

A very unusual event was the presentation by a Swedish doctor, Otto C. Ehman, in December 1818, of a thesis for the degree of doctor of medicine, much of which concerned itself with the remarkable phenomena presented in the state of magnetic somnambulism by a nineteen-year-old servant girl, Anna Nilsson.[98] But after this Swedish interest in animal magnetism seems to have declined pretty quickly. In the 1840s the great Swedish chemist, J. J. Berzelius (1779–1848), who had been interested in, though somewhat hostile to, the earlier phase of Swedish animal magnetism, evinced some interest[99] in the claims of Baron K. von Reichenbach, to whom we shall come in due course.

Denmark, too, underwent a minor upsurge of interest in animal magnetism at about the same time as did the other Scandinavian countries. It probably began soon after the turn of the century – in 1811, H. C. Ørsted (1791–1851), the eminent physicist, gave Cederschjöld, who was visiting Copenhagen, practical instruction in animal magnetism[100] – but according to Erik Bjelfvenstam the earliest document in the Danish literature of the subject is an 1817 translation of J. Weber's *Der thierische Magnetismus* (1816), a work heavily tinged with

German nature philosophy. The most interesting Danish work on animal magnetism (if one can properly call it Danish) is the *Ueber psychische Heilmittel und Magnetismus* (1818) of J. D. Brandis (1762–1845), a distinguished German doctor who was court physician to the Queen of Denmark. Brandis had been reluctantly converted, through his own practical experience, to a belief in the beneficial effect of magnetic methods of treatment. He was not, however, a convert to belief in the magnetic fluid. He argues that all the phenomena supposed to prove the existence of the fluid are due to expectation and self-deception on the part of the subject, and to the close relationship between the subject and the magnetizer. Indeed, all the psychological phenomena of the somnambulic state can be obtained without the intervention of any occult influence.[101] In many ways Brandis looks back to the Royal Commissions of 1784 and to Rahn; but he also foreshadows Faria and Bertrand, especially the latter.

Despite Brandis's reputation, and level-headed approach, he did not succeed in arousing serious interest in mesmerism among his Danish medical colleagues,[102] and little more was heard of the subject there.

How central Wolfart was in the spread of animal magnetism in Russia and the Scandinavian countries will not have escaped notice. Interested persons travelled great distances to visit his clinic, which was generally regarded as the best place to learn the principles of magnetic practice. Although Wolfart had his detractors, he kept open house to serious inquirers, and he made, as we shall see, what were perhaps the first systematic efforts to collect statistics concerning the success rate of magnetic treatment. For these things, and whatever one's view of mesmerism, he deserves every credit.

In western Europe, the influence of German mesmerism was less. According to Zorab, from 1815 to 1820 practically complete silence reigned in Belgium on the subject of mesmerism.[103] Holland, on the other hand, enjoyed from 1814 to 1818 what he calls the Dutch Golden Age of animal magnetism.[104] The most influential figure was a lawyer, P. G. van Ghert (1782–1852), who published several detailed case reports somewhat in the manner of Tardy de Montravel,[105] and achieved a great reputation as a powerful magnetizer and promoter of clairvoyance in his subjects.[106] His principal writings were translated into German,[107] and became influential in German magnetism.

An important event in Dutch animal magnetism was the publication in 1814 and 1818 of a two-volume treatise by three medical men, the senior author, C. Bakker, being a professor at the University of Groningen.[108] It was principally cited in the later literature as a source of anecdotes concerning the alleged magnetization of animals.[109] As always, the distinction between an animal's being magnetized and its simply falling asleep remains obscure.

According to Zorab, the first volume of the book by Bakker and his two colleagues "gave rise to an avalanche of books, pamphlets and articles for and against the subject".[110] But the excitement cooled well before 1820, and without becoming totally extinct, Dutch animal magnetism passed more and more into the hands of "fringe" practitioners. Even so, Wesermann in 1822 could still cite eleven medical practitioners in five Dutch cities who were actively engaged in animal magnetism.[111]

NOTES

1 On these rulers see Fauchier-Magnin (1958).
2 Fauchier-Magnin (1958), pp. 114–115; Benz (1968), pp. 80–81.
3 Viatte (1928), Vol. 1, pp. 181–183.
4 Ramm (1967), p. 11.
5 Baas (1971), p. 594.
6 On Nicolai see Lange (1982), pp. 14–16.
7 I have not come across much evidence that animal magnetism was practised to any extent in Germany between Mesmer's departure for Paris in 1778 and the summer of 1785. However, occasional reports of Mesmer's activities were published in the German press, and some of his writings were translated into German (see Grässe 1843, p. 41).
8 On Lavater and animal magnetism see Lavater (1785, 1787, 1821); Puységur (1807), pp. 241–254; Milt (1953).
9 Rahn (1787); Scherb (1787).
10 Rahn (1790).
11 Milt (1953), p. 101.
12 Presumably it is from this society that there emanated the mystical *Extrait du journal d'une cure magnétique* (1787).
13 This account of animal magnetism in Baden is taken mainly from Funck (1894).
14 Olbers is best known as an astronomer and the propounder of Olbers' paradox, the paradox of why, if the universe is of infinite extent, all parts of the sky are not equally illuminated by the infinity of stars to be found in any direction.
15 Bicker (1787a, 1787b, 1788); Olbers (1787a, 1787b, 1788); Wienholt (1788a, 1788b).
16 Wienholt (1787); cf. Wienholt (1802–6), Vol. 3, Part 2.
17 Nicolai (1787); this J.D. Nicolai should not be confused with the C.F. Nicolai mentioned above.
18 Funck (1894), pp. 16–19.
19 Wienholt (1802–6), Vol. 1, p.3.
20 Heineken (1800); cf. Treviranus' contributions to Wienholt (1802–6).
21 Gmelin (1789), p. 276.
22 Ueber den Magnetismus am Oberrhein (1787); presumably this is J.F.C. Pichler (1754–1807), author of Pichler (1787).
23 Funck (1894), pp. 34–35.
24 Selle's account, preceded by remarks of Biester, who was present at some of the experiments, is in the *Berlinische Monatsschrift*, 1789, 14, pp. 469–475.
25 On all this see Artelt (1965).
26 Hufeland (1794), pp. 49–50.
27 Brazier (1984), p. 214.
28 Humboldt (1797), pp. 213–218. A line is the twelfth part of an inch, or roughly 2 mm.
29 One of the first animal magnetists to take up these findings was certainly Wienholt's friend Heineken. See Heineken (1800), pp. 9–11.

30 Brazier (1984), p. 192. Later experiments of this kind were sometimes cited by animal magnetists in support of their position. See e.g. Léonard (1834), pp. 88–89. Aldini also treated mental patients, apparently with some success, by giving them electric shocks across the head.

31 Reil (1796); cf. Tepperberg (1936).

32 See White (1916); Lòpez Piñero (1983), pp. 26–30; W. Fischer (1984).

33 Reil (1796), p. 30.

34 Tepperberg (1936), p. 23.

35 Reil (1807).

36 Pezold (1797).

37 Also known after a certain time as the *Journal der practischen Heilkunde*, under which title it has different volume numbers.

38 *Archiv für den thierischen Magnetismus*, Jena, 1804. It had to fill many of its pages with translations of Petetin and Tardy de Montravel.

39 Wienholt (1802–6), Vol. 1, pp. 31–32, refers both to medical interest, and to a more favourable attitude by the press.

40 Chenevix (1829), p. 220.

41 Two of them are contributed by G. R. Treviranus. In several cases persons other than Wienholt acted as magnetizers under his supervision.

42 Wienholt (1802–6), Vol. 2, pp. 439–635.

43 Cf. Wienholt (1802–6). Vol. 2, pp. 616–617.

44 Cf. Wienholt (1802–6), Vol. 2, pp. 618–619.

45 Wienholt (1802–6), Vol. 3, pp. 163–304.

46 The reference is of course to the findings of the Abbé L. Spallanzani (1729–1799), which purported to show the existence of a "sixth sense" in bats (really, of course, due to their powers of echolocation).

47 There is an English translation by J. C. Colquhoun (Wienholt, 1845).

48 See Wienholt (1845), pp. 137–138.

49 Wienholt (1845), p. 141.

50 Wienholt (1845), p. 144.

51 See especially Artelt (1965); Erman (1925) covers some of the same ground, but is highly prejudiced against the magnetists.

52 Lentin (1800).

53 Hufeland (1809).

54 Not F. W. J. Schelling, the philosopher, but K. E. Schelling, the physician.

55 Kluge (1811). There were further editions in 1815 and 1819, and it was translated into several foreign languages.

56 Artelt (1965), p. 81, says that in later life Kluge expressed regret that he had written it.

57 Zugenbühler (1809). On Zugenbühler see Milt (1953), pp. 114n–115n.

58 Zugenbühler (1809), pp. 124–125.

59 Artelt (1965), p. 417.

60 Artelt (1965), pp. 417–418.

61 Verordnung über die Ausübung des Magnetismus (1812).

62 On Wolfart see Artelt (1965), *passim*; Landauer (1981); *Allgemeine deutsche Biographie*, s.n. Wolfart, K. C.

63 The most interesting of these is that by J. H. Egg, a medical practitioner who visited Mesmer in 1804. See Milt (1953), pp. 108–112.

64 Mesmer (1812c) (the original is Mesmer, 1800), and Mesmer (1812a), much of which is lifted from Mesmer (1799). But occasionally Mesmer expands on the French text. For

instance, where the latter says (Mesmer, 1971, p. 304) "il est prouvé par la raison et constaté par l'expérience continuelle, que ce feu peut être concentré et conservé; que l'eau, les animaux, les arbres et tous les végétaux, ainsi que les minéraux, sont susceptibles d'en être chargés", the German text goes on (Mesmer 1812b, p. 44), "und, woruber man sich noch mehr verwundern kann und wird, dass selbst die Sonne, der Mond und andere Gestirne es empfangen, verstärken und zuruckwerfen können." Now Dr. Egg, who visited Mesmer in 1804, reports him as claiming to have magnetized the sun, a claim which has attracted acid comment from critics. Until I noticed the extra phrase in the German text, I had supposed that Egg had simply misunderstood an example of Mesmer's rather ponderous humour. But it is clear that Mesmer was serious. A German translation of the *Mémoire* was published at Jena in 1808, but I have not had an opportunity to compare this with the translation given in Mesmer (1812b).

65 Not very well according to Tischner (1928), p. 101.

66 Artelt (1965), p. 429.

67 Mesmer (1814). Wolfart created much subsequent confusion by erroneously giving Mesmer's Christian name as Friedrich in the title of this volume.

68 Artelt (1965), p. 429.

69 Brosse (1819).

70 Artelt (1965), p. 443.

71 A letter of Koreff's, describing his experiences as a magnetizer, is printed anonymously in Deleuze (1825), pp. 393–467. Koreff is highly credulous.

72 For varying accounts of this clinic and of Wolfart's activities see Friedlander (1817); Oppert (1817); Brosse (1818); Muck (1818); Meissner (1819).

73 The matter of the prize is treated in great detail by Erman (1925).

74 See Artelt (1965), pp. 433–439.

75 Cf. Brosse (1818).

76 Quoted in Artelt (1965), p. 468.

77 See e.g. the article in Pierer and Choulant (1823), pp. 178–203.

78 Erman (1925), pp. 108–111.

79 E.g. Reichel (1829); Hanak (1833); Wetzler (1833); Siemers (1835).

80 Quoted in Artelt (1965), p. 456.

81 Oppert (1817), p. 192.

82 Oppert (1817), p. 196.

83 E.g. Ennemoser (1842, 1844, 1854).

84 On Ennemoser see Artelt (1965), pp. 465–466; Kiesewetter (1909), pp. 453–469.

85 Zielinski (1968), p. 4.

86 I have not seen the German original of this article. It is translated into French as Lichtenstaedt (1819a).

87 See below p. 143.

88 For a list of names and places see Wesermann (1822), p. 270.

89 Oppert (1817), p. 192. Stoffregen reappears in Kohlschütter and Bähr (1843), pp. 69–70,107.

90 E.g. Dr. P.T. Brosse of Riga and Dr. J.F. Weisse of Dorpat (Brosse, 1818; Weisse, 1819).

91 Hamel (1818).

92 Wesermann (1822), pp. 264–266.

93 Panin (1818).

94 Described by Zielinski (1968), pp. 9–13, and reviewed by J.F. Weisse, *Archiv für den thierischen Magnetismus*, 1819, 5(3), 129–135.

95 I rely here a good deal on the review by Bjelfvenstam (1967).

96 Löwenhjelm (1819), pp. 139–140.
97 Cederschjöld (1815).
98 Ehman (1819).
99 Bjelfvenstam (1967), pp. 215–217.
100 Bjelfvenstam (1967), p. 204.
101 See especially Brandis (1818), pp. 149–162.
102 Bjelfvenstam (1967), p. 242.
103 Zorab (1967), p. 6.
104 Zorab (1967), p. 55.
105 Ghert (1814,1815).
106 Zorab (1967), pp. 55–57.
107 Ghert (1817, 1818).
108 Bakker *et al.* (1814–1818, 1818).
109 Bakker *et al.* (1818), pp. 90–98.
110 Zorab (1967), p. 63.
111 Wesermann (1822), p. 270.

CHAPTER 5

Kluge

In the last chapter, I described C. A. F. Kluge's *Versuch einer Darstellung des animalischen Magnetismus als Heilmittel* (1811) as one of the most useful books in the whole history of animal magnetism. It systematized and encapsulated the literature of the subject – and by no means only the German literature – at a time when, in Germany at least, the magnetic movement was trying to achieve some degree of scientific and medical respectability. It will be convenient, therefore, to look in some detail at the contents of Kluge's book. For here, if anywhere, we find a summary of what animal magnetism had to offer, what its theories were and how they cohered with the science of the time, and what the evidence was for the phenomena on which the theories were based.

Versuch einer Darstellung des animalischen Magnetismus consists of two main parts, "Theoretical Survey", and "Practical Survey". I shall take the latter first. It begins by outlining the requirements of successful magnetization. The magnetizer should be between 25 and 50 years of age, in good health, and possessing a surplus of energy. He must have a strong soul, a mind full of elevated notions, a lively and imperturbable belief in what he is doing and in his own capacity to do it, a firm will, and a zeal to do good. Ideally, treatment should be carried out daily at the same hour. The patient should be in bed, with the magnetizer beside him, or in a comfortable chair, with the magnetizer seated facing him. Rapport should be established by holding the patient's shoulders, or lightly stroking his arms. Treatment should begin with generalized stroking from the head downwards and towards the extremities. Passes may be made with or without actual contact (those with contact being more efficacious), and with the back of the hand, the side of the hand or the flat of the hand. Passes with the back of the hand have no effect, and may be used when returning the hand to its starting point (reversed passes would undo the benefits of the downward passes or even produce nausea, convulsions, etc.). Passes with the edge of the hand block the effects of other passes.[1] Passes with the flat of the hand ("volar manipulations") have most therapeutic value. They may be subdivided into palmar manipulations and digital manipulations, the latter being the more powerful. The strongest effects are obtained by "emballing" the two hands and carrying out digital manipulations with the two projected and touching thumbs.

General manipulations may be followed by localized ones. Should manipulations not fully succeed, various supplementary forms of treatment may be tried. These include breathing on afflicted parts of the body, fixing the patient with the eyes, directing one's thoughts upon him, and using such "magnetic substitutes" as *baquets*, magnetized trees, magnetized glass and magnetized water (excellent for indigestion and constipation). All these were thought of as ways in which magnetic fluid may be transferred to the patient.

The final section of the practical survey purports to tell us in which cases the use of animal magnetism is indicated. What one would most like to know is Kluge's grounds for deciding when animal magnetism is the treatment of preference. But all he gives us is perfunctory references to a few case histories from Gmelin, Heineken, Wienholt, etc. Animal magnetism, he tells us, usually takes effect only on persons who are ill, but even among these some are insensitive, whilst others are sensitive to the influence even when fit. Children are more sensitive than adults, women (especially blondes) than men. An uneven distribution of life-force, as during times of adolescence or premenstrual tension, also makes for sensitivity. It does not matter whether patients are cooperative or uncooperative, or even totally disbelieving. Since animal magnetism works first and foremost on the nervous system, it is especially useful in nervous ailments. Where there is a lack of life-force (nervous force) leading to exhaustion, powerlessness, tremors, etc., animal magnetism can help by bringing about a transfer of life-force from magnetizer to patient. It can also cause a redistribution of the patient's own life-force. So in conditions which are due to an abnormal distribution of life-force, bringing about disharmonious nerve activity (epilepsy, catalepsy, St. Vitus' dance), motor or sensory weakness or hypersensitivity, or localized congestion (gout, dropsy), animal magnetism may be of great service. Its value in acute illness is, however, debateable.

The "theoretical" survey is of more interest than the "practical" one. It contains, perhaps, nothing radically new, but it synthesizes current data and ideas more systematically and more comprehensively than any previous work, and its voluminous references make it an invaluable guide to the literature of its time. It has three sections. The first is largely historical; the second is a survey of the phenomena of animal magnetism; and the third is an attempt to elucidate those phenomena. It is these last two sections which are of importance.

Kluge's section on the phenomena of animal magnetism begins with a brief account (derived especially from Wienholt, Gmelin and Tardy) of the effects of magnetizing on the magnetizer. If the magnetizer has the energy and life-force to magnetize successfully, he will feel whilst operating a warmth and an outflowing from his hand, and may afterwards suffer a corresponding weakness and loss of force.

The effects on the magnetized person are of two broad kinds. First, there are generalized effects, which may endure between sessions: for instance, a general arousal and strengthening of all systems; a gentle stimulation of the whole surface of the body, soothing out disharmony and restoring equilibrium; reduction and removal of symptoms brought on by a disordered nervous system.

Then there are effects which last only as long as the magnetic session. Kluge tries to arrange these effects into six grades or stages, each "higher" or more "advanced" than those preceding. He was not the first person to attempt to define such stages, nor yet the last,[2] and he emphasizes that there are many individual variations, but his proposals formed the starting point for many subsequent discussions, as did the theoretical speculations with which they are entwined.

The first three of Kluge's stages are marked by such phenomena as localized feelings of heat, cold, etc., feelings as of the magnetic influence "streaming through" the body, reddening of the skin, sweating, feverish movements, twitchings, drowsiness, heaviness of the eyelids, difficulty in moving or speaking, etc. The third stage is that of magnetic sleep. The outer senses are shut off. The patient has no memory of events during the sleep, but finds that his condition has improved.

Stages four to six are the grades of waking in sleep (sleep-waking), or somnambulism. Normally, there is complete amnesia for events in the somnambulism, but memory of the events of one somnambulic episode is carried over into the next and later episodes, and may sometimes be recovered in dreams.[3] The transition from stage three to stage four constitutes "inner awakening", and develops gradually over many sessions.[4] It involves the progressive recovery of awareness of objects and events in the surrounding world – an awareness, however, which does not seem to be mediated through ordinary sense-perception – together with a slowly developing ability to speak and to respond to questions without emerging from the somnambulic state. The "rapport" existing between magnetizer and subject is such that the subject can hear through the ordinary hearing apparatus only words spoken or sounds made by his magnetizer, or by a person touched by his magnetizer. Usually the eyes remain closed, but if they are opened the pupils will be wide and unresponsive to light. However, loss of ordinary vision may be compensated for by the development of "stomach-seeing". When objects are laid upon or brought near to the pit of the stomach, subjects can perceive not just their outline but their colour and other details. They can, for instance, tell the position of the hands of a watch, read writing, and recognize the colour, suit and value of playing cards. In this latter connection Kluge, cites some experiments of Gmelin's with a lady somnambule who could tell the value, colour and suit of playing cards which he laid on her stomach with such precaution that he remained unaware of their identity himself. She could succeed even with her eyes bandaged.[5]

When this capacity is further developed, somnambules can perceive through intervening substances (provided these are not non-conductors). For example,

Gmelin placed a playing card in his silver snuff-box, closed it, and brought it into contact with the somnambule. She could always name the card, just as she could tell what was in his closed hand as soon as she touched it.[6] With further development, the capacity to perceive in visual impressions spreads over the whole outer surface of the body as a common or general sense, and gives the somnambule knowledge of distant objects. Eventually, the somnambule may become able to perceive the glowing substance which streams from the hair, eyes, fingertips and iron wands of magnetizers engaged in magnetizing, and which illuminates flasks of magnetized water – Kluge quite uncritically accepts the early reports of Tardy and the later ones of Nasse upon this phenomenon.[7]

There are alterations, too, in senses other than vision. Sometimes the centre of hearing is removed to the pit of the stomach. Pezold and Petetin,[8] for example, had to address their remarks to the stomachs of certain patients; they would get replies even when they whispered so softly that (as they fondly imagined) no sound could have reached the ears of the somnambules. Sensitive subjects would report that magnetized water had a peculiar smell and taste. Kluge even accepts Petetin's claim that sensitivity to gustatory and olfactory impressions may migrate to the pit of the stomach.[9]

The skin senses too may undergo peculiar and characteristic changes, rendering them sensitive to the presence and nature of objects at some distance from the body. The presence of other persons (other, that is, than the magnetizer) is generally repugnant to the somnambule, causing unpleasant feelings, horror, even convulsions; in instances reported by F. Hufeland, by Nasse, and by K. E. Schelling,[10] somnambules were disturbed by the presence, unknown to them, of certain individuals in a neighbouring room. A touch from a person other than the magnetizer can cause such somnambules great distress.

Somewhat comparable are the effects which the approach of certain metals may have upon magnetic somnambules. These effects range from pricking as if with needles, and small shocks like those from an enfeebled Leyden jar, to fear, powerful shocks, convulsions, localized paralyses and numbness, and eventual unconsciousness. The effects vary with the part approached or touched, the depth of magnetic sleep, the quantity of metal (large lumps may have very powerful effects[11]), and kind of metal (steel, iron and gold being least harmful, gold even beneficial, and silver, brass, pinchbeck and copper being the most harmful). The approach of a mineral magnet has similar, but even more marked effects.

Turning now to the response side: most patients who reach this stage speak, either spontaneously, or on demand. Sometimes their speech improves; sometimes they lose the capacity to pronounce certain letters or words, or they may speak a broken German with extraordinary rapidity. Somnambules who know a foreign language may become more fluent. Changes in speech may be accompanied by alterations of personality. Kluge cites one of Gmelin's patients, a young German lady who would fall spontaneously into a somnambulic state in which she believed herself to have been brought up in a Paris convent. In these phases she spoke French well and German poorly, with a French accent.[12]

Once stage four has been regularly reached, patients can be made to pass into it or out of it very quickly by various means, e.g. by the magnetizer breathing on them, staring at them, coming into their vicinity, etc. Practised subjects may even pass into and out of stage four unaided – a development foreshadowing the emergence of numerous amateur and professional clairvoyants, mediums, etc.

Somnambules who progress to stage five become clairvoyantly aware of their own interiors. This awareness may be confined to diseased or malfunctioning parts, or may extend more widely. Kluge cites in detail the remarks of a somnambule magnetized by Fischer.[13] This somnambule gave, in homely language, many details of the lay-out and appearance of her own nervous system and internal organs. A somnambule of whom Kluge had recently been told by "a worthy doctor" described during magnetic sleep the minutest details of all parts of her body, including the interior of her eye.[14] Kluge is clearly impressed by these stories, though he admits that they have all along been much doubted and denied.

Along with this "inner self-observation" develops an "instinct" for the proper remedies. Kluge quotes one of Heineken's somnambules:

> I see the interior of my body ... closely observe the disorders which afflict one or other part, think attentively of means to relieve them, and then it comes to me, as if someone called out: you must use this or that.[15]

Such somnambules may additionally prescribe the proper dosage, and correctly predict the details and duration of the illness. They may also be able to diagnose and prescribe for persons with whom they have been placed in rapport through contact. They may then feel in their own bodies pains and other sensations corresponding to those felt by the patient. This sympathy is at its strongest between clairvoyant and magnetizer. Kluge cites a number of curious examples. When Gmelin suffered from an attack of diarrhoea, his somnambule underwent a similar discomfort.[16] One day Fischer cut his upper arm, and the next day his subject had a severe pain and a hard swelling at the same spot. He conducted some experiments on this "community of sensation". When he tasted pepper, salt, wine, etc., his somnambule reacted appropriately.[17] This phenomenon continued to be reported well into the era of hypnotism.

A somnambule approaching the borders of the sixth stage may be able to diagnose the ailments of a distant person simply by holding to the pit of her own stomach some object which that person has kept for a while in contact with his own body. An object saturated by similar means with the influence of the magnetizer and given to the waking patient will induce somnambulism.

Somnambules who reach the sixth grade enter a new and higher relationship with the whole of nature. Freed from petty and earthly considerations, they enjoy a new sharpness of clairvoyant perception to which space and time constitute only limited barriers. Kluge cites a somnambule of Wienholt's who became aware at a distance of over a hundred miles of the illness of her brother,[18] and a somnambule of K. E. Schelling's who, during the somnambulic state only, became greatly distressed by the belief, subsequently proved correct, that a letter was on its way announcing the unexpected death of a close relative.[19] He even cites examples of somnambules exhibiting foreknowledge of unforeseeable accidents.

The relationship with the magnetizer now becomes so close that his very thoughts are obeyed. Kluge gives some instances of magnetizers putting subjects into the somnambulic state from a distance of several miles simply by concentrating very hard upon them. Several of these cases suffer from the disadvantage that the attempt was made at the time of day when the patient was ordinarily magnetized. One or two are more interesting. Herr Nadler, a gentleman who used to magnetize under Wienholt's direction, reported a couple of examples. In one, the patient was inside her house, and Nadler inside his a short distance away. Later in the day he learned that at the time of his attempt the patient, who had been engaged in housework, suddenly dropped her broom and fell asleep in a chair.[20]

Another topic that continued to be of interest through the nineteenth century and beyond is that of the moral status of somnambules. Kluge reports that somnambules in the sixth stage show an enhanced moral delicacy. They cannot, while in the somnambulic state, endure the presence of an impure heart. One of Wienholt's subjects was sent into convulsions by the presence of a young doctor, a stranger, who entertained (she supposed) lascivious thoughts about her. He turned out to be the black sheep of his family; however, that his thoughts were more lascivious than those of the average young doctor does not seem to have been established.

The final section of the "theoretical" part contains Kluge's attempts at a theoretical synthesis. He lays the groundwork with an account of relevant aspects of the structure and functioning of the nervous system. He refers to the supposed principle of nervous activity as a fluid ("the mediator between soul and body"), but does not attempt to determine its nature. He had earlier rejected the proposals of Gmelin and F. Hufeland[21] that animal magnetism is electrical in nature – the experiments of Nasse[22] applying electroscopes to magnetized persons appeared to him decisive on that issue. He is certain that an active nerve is surrounded by a circle of influence or sensible atmosphere of the kind which Wienholt had already postulated. Like Wienholt, he relied upon the experiments of F. A. von Humboldt, and others, to prove that neural activity can jump a gap in a severed nerve. Like Wienholt again, he goes on to suggest that such a nerve-atmosphere, generated by the nervous system at large, must surround the whole body. This nerve-atmosphere will be a vehicle both of sensitivity (witness the "sixth sense" of bats and polyps, the effect of metals on certain somnambules and sleepers, abnormal sensitivity to the unseen presence of a cat, the successes of dowsers, etc.) and of influence upon other organisms (witness the rejuvenating effect which contact with the young can have upon the old, instant sympathies and antipathies between new acquaintances, etc.).

The nervous system itself Kluge divides into the cerebral system and the ganglion (or sympathetic) system. The sympathetic system consists of two chains of ganglia, lying in front of the spine, with numerous branches, which unite to form "plexuses" of nerves and ganglia (the solar plexus, the cardiac plexus, etc.) associated with important internal organs. The anatomy of this system had been worked out during the eighteenth century, and it became customary, following Bichat, to regard it as subserving the (largely unconscious) "organic" or "vegetative" life, and the various ganglia of the system were regarded as more or less independent "little brains".

"Animal life", involving such functions as sensation, perception, muscular activity, etc., was held to be mediated by the "cerebral system". The focus of all brain function lies in the brain stem and basal ganglia, for damage here, unlike damage to the forebrain and cerebellum, is immediately fatal. The seat of the soul lies at this focus. However, the focus itself is a "dynamic indifference-point", to be compared to the "magnetic indifference-point" which lies within a group of magnets. Move or remove the magnets, and the indifference-point will move accordingly. Change the balance of neural activity, and the "seat of the soul" will likewise shift. The soul itself is, however, not a mere product of the union of different bodily parts, for it has freedom and the consciousness of freedom (and by implication, rationality).

What are the relations between the cerebral system and the ganglion system? Here Kluge takes up some recent suggestions by Reil.[23] Reil argues on anatomical grounds that the nerves which connect the cerebral and ganglion systems normally have the function of "isolators", but that when the ganglion system contains a superfluity of life-force, or else when it is depleted thereof, these nerves instead become conductors. In these terms, he offers interpretations of various kinds of physical and mental disorders,[24] and also of magnetic somnambulism.[25] His speculations concerning the latter form the basis for Kluge's account.

Kluge does not spend much time on the ordinary therapeutic effects of the magnetic passes. The benefits of being magnetized come from a redistribution of life-force, and a boosting of the patient's life-force by the life-force of the magnetizer. The early phases of magnetic sleep are not essentially different from ordinary sleep; both involve a movement of life-force away from the cerebral system and into the ganglion system. The latter becomes in consequence more active whilst the former becomes quiescent, and the organs of the "vegetable life" can go steadily and uninterruptedly about the business of restoration and repair. Of course, ideation continues in sleep, and takes the form of dreams. Sometimes this ideation attains such vividness that it activates the musculature, and then we have natural somnambulism, which has many affinities with magnetic somnambulism, most notably the occurrence in it of clairvoyance and kindred phenomena.

Kluge's avowedly tentative approach to these phenomena is as follows. There are good grounds for supposing that the nervous "atmosphere" which surrounds the body is principally generated by the ganglion system. Fine branches of the ganglion system extend with the circulatory system right to the periphery of the body; nervous fluid from them may pass out through the skin with the perspiration. The impressions occasioned by the disturbances of the nervous atmosphere are generally indeterminate, which would not be the case if they originated from the cerebral system. Such sensitivity is most marked during sleep, when the predominance of vegetative life is greatest. And all the phenomena of abnormal perception during magnetic sleep, particularly the claims of so many clairvoyants to "see" from the neighbourhood of the solar plexus, suggest that this kind of perception is mediated through the ganglion system.

We can now offer the following account of magnetic clairvoyance. As the ganglion system becomes increasingly "potentiated" by the process of magnetization, the "sensible body-sphere" will become strengthened and enlarged and more responsive to neighbouring objects. At the same time, communication between the ganglion system and the cerebral system will be opened up. The consequent inflow of life-force will set the cerebral system into activity. The effect of this will be two-fold. The musculature, especially the speech-organs, may be set in motion; and stimuli reaching the ganglion system (including those originating from the "sensible body-sphere") may be reflected to the brain as indeterminate sensations.

With still further "potentiation" of the ganglion system, the subject passes into the fifth or sixth grade of somnambulism, and his clairvoyance develops *pari passu* with extraordinary changes of functional relationship within his nervous system. The ganglion system ceases to be a scattered collection of more or less independent "little brains", and develops its own indifference-point, centre, or focus. This new focus – which is of course the solar plexus – constitutes a centre of perception opposed to that of the brain. Meanwhile, the brain ceases to be the absolute central-point of nervous functioning, and becomes merely one ganglion within an enlarged, but now unified, ganglion-system, or rather, within a whole nervous system now changed into a ganglion system. I am not absolutely clear whether Kluge thinks the solar plexus is now *the* focus of conscious perception (some later writers certainly did hold this), or how precisely he regards the relation between the new focus and the old.

It may, he adds, seem a contradiction that there should be a true perception of colour, and also hearing of tones, etc., without appropriate sense organs. However, what the clairvoyant actually obtains from the "sensible body-sphere" and the ganglion system is at root only a general awareness of objects and events. The sensory qualities are supplied, as in dreams, by preexisting ideas; they have in short the status of hallucinations.

The same considerations apply to the capacity of clairvoyants to see inside themselves. Their ability to fix upon appropriate remedies Kluge attributes to a "sense of preservation" which is possessed by all animals, but which in man resides within the ganglion system. The "potentiation" of the ganglion system raises it from a dark instinct to a clear idea.[26] Many other contemporary writers adopted a similar view.[27]

Magnetic "rapport" Kluge explains in terms of a vaguely conceived blending of the two nervous systems mediated by their "atmospheres".[28] When, as was sometimes alleged, a magnetizer proved able to influence his patient over a considerable distance, this explanation seems likely to break down, and explanations of clairvoyance in terms of the "sensible body-sphere" seem likewise implausible when the clairvoyance is ostensibly exercized over a distance of many miles. Kluge makes some token references to the influence of the will upon the nervous fluid, and to the speed and distance which other imponderables, such as light and electricity, can attain.[29] But in the end, he seems by implication to accept the view that these and other "higher" phenomena of somnambulism represent a release through the nervous changes wrought by magnetization of faculties latent in the soul.[30]

From today's standpoint it is very easy to criticize Kluge. His principal weaknesses are two, and are closely interrelated. The first is the very variable quality of the empirical data on which he relies. It is revealing to look up his original sources. They range from casual remarks and incidental one-sentence comments (his claims about the depleting effect which successful magnetization has upon the magnetizer are based mainly upon such) to major treatises (such as Humboldt's two volumes on his electrophysiological experiments). The former are certainly more numerous than the latter; and even reports which are in some respects detailed frequently fail to provide essential information. Thus I mentioned a subject of Wienholt's who became convinced that her absent brother, more than 100 miles away, was dangerously ill. Wienholt details the subject's forebodings, but all he tells us as to their accuracy is half a sentence in a footnote: "Her brother really was in a bad way at that time."[31] Or consider the case, which I quoted above, of the somnambule magnetized at a distance by Herr Nadler. We are simply not given enough details to enable us to assess what rôle the imaginative expectation of the subject might have played in producing the phenomenon. Was she usually magnetized at this hour? Could she have suspected from Nadler's absence, or observed movements, that he was "up to something"? And so on.

Somewhat better are certain experiments of Nasse's, cited by Kluge as evidence concerning somnambules' sensitivity to the near approach of metals and other substances.[32] Although in many instances the somnambule could have perceived the substances by ordinary means, Nasse on some occasions presented different substances in similar containers – a step on the road to a properly controlled double-blind experiment. Somewhat better too are Gmelin's experiments, mentioned above, on the apparent discrimination of playing cards presented to the pit of a somnambule's stomach. Comparison of Kluge's summary with Gmelin's original suggests that Kluge is slightly exaggerating the success rate.[33] However, it is still impressive, and Gmelin was clearly making considerable efforts to ensure that the somnambule could not perceive the card by normal means, even though he placed an entirely unjustified faith in the efficacy of blindfolds. Like Nasse, he was on the right lines when he presented cards concealed in a container. Even so, he does not give us enough detail of the circumstances to assess for ourselves the possibility that the subject was able to obtain ordinary knowledge of the target card. We have simply to assume that Gmelin was intelligent enough to foresee and forestall the possible pitfalls.

The second of Kluge's principal weaknesses follows on from the first. Insofar as he bases his claims and proposals upon reports which contain insufficient details, it becomes very difficult to assess the alternative explanations or interpretations. This problem does not much worry Kluge, for he seems often unaware that there might be other interpretations. To modern eyes it is quite astonishing that he should, for instance, accept the ability of some people, and of all bats, to avoid obstacles in total darkness, or following loss of vision, as evidence for the "sensible body-sphere," whilst not assessing the possibility that other normal senses could be involved; that he should accept instant sympathies or antipathies between individuals as further evidence for the same, without weighing up possible psychological explanations; that he should accept the evidence for "community of sensation" without asking himself what covert sensory cues might have been involved. He has in addition a regrettable propensity to take over other peoples' interpretations of data in some relevant area and build them into the foundations of his own interpretative scheme, regardless of the fact that these interpretations have been propounded by persons – especially Reil – who are often as neglectful of possible alternative explanations as he is himself.

Kluge's weaknesses are the weaknesses of much of the literature of animal magnetism; indeed, they are the weaknesses of much of the fringe literature of science. Perhaps, indeed, they are really the strengths of such literature, because out of such "weaknesses" best-sellers are born. But one must not be misled into supposing that these shortcomings are in all cases sufficient to justify us in totally dismissing the claims of the enthusiasts. Often the enthusiasts are on to something, even if they haven't got it quite right. The animal magnetists were certainly on to *something*, and it is debatable whether the progress we have made in the intervening centuries is such as to justify us in being patronizing towards them.

Kluge's shortcomings are ones which he shares with other writers of his time – say Reil, Sömmering, or Prochaska – who still have a niche in the history of medicine. And his book, as I said before, is at least extraordinarily useful. It encapsulates the literature and findings of contemporary animal magnetism. Many of these findings and ideas passed into received tradition, and Kluge's was a prime influence in transmitting them. Their influence was strongest among German magnetists, including ones of widely different schools. Indeed, some of the teachings were, as we shall see, later reflected back from the other world through the somnambules favoured by mystical magnetists. French and British mesmerists were more pragmatic, more purely interested in the therapeutic benefits of magnetism, less given to speculation and systematization. Most French and British magnetists were, furthermore, ignorant of German magnetic writings. None the less, the tradition fed by Kluge's work gradually filtered through into the French and British literature.[34] Much of the remaining history of animal magnetism can be presented in terms of deviations from and additions to the ideas, practices and alleged phenomena set forth in Kluge's book.

A further reason why Kluge's book is so useful is that it encapsulates not just doctrines and findings, but an important parting of the ways. It is itself, from the point of view of the medicine of its time, relatively respectable, because it was a serious and well-documented work, and it was followed by some other works of the same *genre*. But in it one can discover hints of tendencies soon to manifest themselves in German animal magnetism – a tendency to be more interested in the alleged marvels than the alleged cures, a tendency to dwell upon the supposed religious and philosophical implications of the phenomena rather than to seek scientific explanations, a tendency indeed towards magic and mysticism. From about the date of Kluge's book – 1811 – onwards these tendencies became increasingly prominent. They will require a chapter of their own. But first it is necessary to say a little about developments in France.

NOTES

1   According to Gmelin (1789) they hinder conduction of the life-force.
2   Luetzelbourg (1786), pp. 28–36, is usually credited with being the first – he proposes four stages; C. W. Hufeland (1809), p. 10, from whom Kluge starts, proposes three; Kieser (1826), Vol. 2, pp. 118–119, has six; Lausanne (1818), p. 21, (1819), Vol. 2, p. 300, has twelve.
3   Kluge cites an example from Nasse. Kluge (1811), pp. 187–188.
4   Kluge (1811), pp. 127–129 again illustrates this with a case history from Nasse.
5   Kluge (1811), pp. 131–132.
6   Kluge (1811), pp. 134–135.

7 Kluge (1811), pp. 141–146. On Tardy, see chapter 3 above; cf. Nasse (1809), pp. 246, 300–304.

8 Pezold (1797),

9 Kluge (1811), p. 155.

10 Kluge (1811), pp. 156–159.

11 Kluge (1811), pp. 166–168; Nasse (1809), p. 273 etc.

12 Kluge (1811), pp. 180–181; Gmelin (1791), pp. 3–89.

13 Kluge (1811), pp. 193–194; F. Fischer (1805), pp. 274–275.

14 Kluge (1811) pp. 194–195.

15 Kluge (1811), p. 196; Heineken (1800), pp. 125,128.

16 Kluge (1811), p. 202; Gmelin (1793), pp. 378ff.

17 Kluge (1811), pp. 201–202; F. Fischer (1805), pp. 275–276.

18 Kluge (1811), p. 217; Wienholt (1802–6), Vol. 3(2), pp. 34, 102–107.

19 Kluge (1811), pp. 222–225.

20 Kluge (1811), p. 236; Wienholt (1802–6), Vol. 3(3), p. 384.

21 F. Hufeland (1805).

22 Nasse (1809).

23 Reil (1807).

24 Reil (1807), pp. 242–248, etc.

25 Reil (1807), pp. 232–235.

26 Kluge (1811), p. 348.

27 E.g. Strombeck (1814), pp. 133–134.

28 Kluge (1811), p. 350.

29 Kluge (1811), p. 357.

30 Kluge (1811), pp. 360–372.

31 Wienholt (1802–6), Vol. 3(2), p. 107n.

32 Kluge (1811), pp. 166–168; Nasse (1809).

33 Kluge (1811), pp. 131–132, 134–135; Gmelin (1791), pp. 78–79, 83–84.

34 The second (1815) edition of Kluge was the subject of a critical notice in the *Bibliothèque du magnétisme animal* (Redern, 1818); many ideas similar to Kluge's are expressed in Rostan's famous dictionary article (Rostan, 1825; cf. chapter 7 below).

CHAPTER 6

The revival of magnetism in France

The impact of the French Revolution upon the animal magnetic movement may be gauged from the fact that Dureau's bibliography of animal magnetism lists 212 items for the period 1781–1790, but only six for the period 1791–1800.[1] Although interest had markedly slackened even before 1789, it is surprising that there was not a more evident recovery after 1794 when the confusion and bloodshed began to subside. Deleuze says that the practice of animal magnetism continued in private. It was a matter on which he could speak with authority, and there are other indications that he is correct.[2] But even during the relative stability of the Consulate and the First Empire, exponents of animal magnetism were in no hurry to make themselves known – Dureau has only 15 entries for the decade 1801–1810. Medicine and medical education, like much else in France, were being reformed and centralized, and animal magnetism was "fringe medicine". It also had some links with fringe politics. Napoleon's censors and his police were ubiquitous. Well might mesmerists shrink from attracting too much official attention.

A mesmerist who did not shrink from such attention was Mesmer. In 1798, he returned to Paris in an attempt to regain funds sequestrated during the Revolution. He also took the opportunity to see his latest *Mémoire* through the press, and to address a petition to the minister of the interior offering his services to the republic. He demanded that a jury of savants should examine his doctrines, and that thereafter he himself should be appointed to a clinical chair in a teaching hospital.[3]

Mesmer's ambition was as stubborn as it was out of touch with reality. Neither his book nor his petition made any notable impact. However, after a stay of some four years he obtained the reversion of an annual rent of 3000 florins.

PUYSÉGUR

The gradual revival of animal magnetism in France may perhaps be dated from the publication in 1807 of Puységur's *Du magnétisme animal*. Puységur's story since 1789 had been a truly remarkable one.[4] A liberal in politics, he was at first a supporter of the revolution, and accepted a command in the revolutionary army. But by 1792 things were getting out of hand, and he resigned his commission, retiring to live quietly on his estates. The reign of terror saw him imprisoned for two years. In the difficult times of the Consulate, he was mayor of Soissons. Not until 1805 did he feel free to resign and devote himself to his preferred field of endeavour.

Du magnétisme animal is a miscellany of materials relating mainly to the period before 1789. Its importance was as an icebreaker – Puységur's weight was sufficient to force open a channel through which, before very long, others began to pass. The following year, 1808, saw the publication of a second, and very influential, memoir by Petetin, describing eight further cases of "catalepsy" with stomach vision.[5] In 1809, Puységur brought out a new edition of his *Mémoires*. He was pleased with the fair notices he received in the *Moniteur*, the *Gazette de France* and the *Gazette de santé*. To a hostile article by J. B. Salgues in the *Mercure de France* for 12th February 1810, he replied at some length in his next and perhaps most ambitious work, *Recherches, expériences et observations physiologiques sur l'homme dans l'état de somnambulisme* ... (1811).[6]

The *Recherches*, like all of Puységur's books, is somewhat disorganized. It is as though his materials, and their implications, are too weighty and too extensive for him to manage. Still, from this and other late works we can derive a picture of Puységur's final position, which is indeed not too different from the position he had adopted in the heady days of 1784. A patient's best hope of cure lies in becoming himself or herself a clairvoyant somnambule. (According to Deleuze, only one patient in twenty is a potential somnambule, but I think that Puységur might have put the figure rather higher.[7]) Magnetic somnambulism, Puységur holds, differs from natural somnambulism only in that magnetic somnambules are always more or less under the influence of their magnetizers. Otherwise the two conditions are the same. In support of this assertion, Puységur cites the case, sent to him by M. Donnet of Sisteron, of a young man of 25, a natural somnambule, who could see in the dark, see inside his body, and diagnose and prescribe just like a magnetic somnambule.[8] He gives detailed instructions as to how one should set about magnetizing with a view to producing magnetic somnambulism. The secret seems to lie not in the passes and manipulations employed, but in their being employed with a strong wish to procure somnambulism, and a concentration of attention and will to that end.[9]

The first sign of success is a fluttering of the patient's eyelids. Full somnambulism has three distinctive characteristics:[10] *isolation*, "a patient in this state has communication and rapport only with his magnetizer, hears only him, and retains no relation with external objects"; *concentration*, "a patient in this state should be so preoccupied with himself that he cannot be distracted by anything"; and *magnetic mobility*, "a patient in this state is always more or less responsive simply to the thoughts of his magnetizer". Once a patient has attained somnambulism you can question him as to how long he should remain in that state, and as to the nature of his ailments, the appropriate remedies, etc., always, however, taking care not to put your own ideas into his mind.

In all of this there is little about the magnetic fluid. Puységur certainly thought that the action of a magnetizer on a patient has some kind of physical basis, a basis which he likes to compare to mineral magnetism.[11] But he regarded questions concerning a fluid as of little importance. There are two elements underlying every successful magnetic act: "our force or organic power, always subjugated and secondary; and our will, which, setting this force to work, is always determining, active, and unqualified".[12] Thought and will are always primary: that is all the magnetizer needs to know. In a later publication he says: "Whether electricity is or is not related to animal magnetism, it is not necessary to enquire ... let us not try to explain inconceivable phenomena by imaginary fluids, nor by the logic, so frequently fallacious, of our human reason."[13] He is close to dispensing with "magnetism" altogether, and his position now appears not too distant from that of Barberin and the "spiritualist" magnetizers. Behind animal magnetism there lies the thinking principle or soul; and behind the thinking principle lies God.

After the publication of the *Recherches*, Puységur seems to have engaged in something of a campaign to rearouse interest in animal magnetism. In 1812, he took the hazardous step of publishing two instalments of a somewhat unusual case history before he had completed treatment of the patient.[14] During the winter of 1812–13, he gave in Paris a series of demonstrations of the phenomena of animal magnetism, much as he had done twenty-eight years before. These demonstrations were attended by more than fifty doctors, French and foreign.[15] The magnetic movement was once again stirring into visible life.

In 1813, Puységur completed the case history just referred to, and published all three parts together as his last substantial work on animal magnetism, the *Appel aux savants observateurs*.[16] The case history was that of Alexandre Hébert, aged twelve, the son of a clockmaker of Soissons. About a year before, young Hébert had become subject to nervous attacks. These involved violent headaches, delirium, convulsions, weeping and groaning without cause, outbursts of rage, and prolonged bouts of somnambulism. In his frenzies he would engage in acts of a self-destructive tendency, and become violent and abusive towards others. When he returned to normal he was amnesic for these episodes. He proved an excellent magnetic subject. Puységur was able to control his frenzies immediately by a gesture of the hand and a firm concentration of will. In the somnambulic state young Hébert could recollect all. He attributed his troubles to a surgical operation which he had undergone at the age of four for the removal of a *depôt* on the top of his head. He was convinced that part of his brain had gone. He was pessimistic about the prospects of a cure, and was adamant that during six months or so of necessary treatment he must not be long separated, night or day, from the unfortunate Puységur. It is a remarkable testimony to Puységur's goodness of heart that during this prolonged period he (a grand seigneur of the old aristocracy) allowed this very disturbed child[17] of humble parentage to sleep either in a bed at the end of his own or in a neighbouring chamber with the communicating door open.

In the end, young Hébert was more or less cured. His memory, however, remained so weak that he should not (he said) be forced to study Latin. On the other hand, his memory in the somnambulic state was excellent, and it is obvious that some persons in his entourage suspected him of shamming. The local schoolmaster, whom he had struck in a fit of rage, was particularly bitter, and probably thought that a sound switching would work a wonderful cure.[18] Without doubt young Alexandre profited a great deal from his illness. Not merely did he miss lessons; he received attention, pocket money, carriage drives, a trip to Paris, a visit to the zoo, etc., from the indulgent Puységur, and it was noticeable that when he was enjoying himself predicted attacks failed to materialize.

Puységur believed the case of great interest for the theory of mental illness.[19] In magnetic sleep, Alexandre could recall what had happened during his attacks; he could also recall his dreams and predict future dreams. Dreams, Puységur thought, are a transient disturbance of the brain, a brief nocturnal madness; madness, on the other hand, may be regarded as a dream prolonged through waking life. Now dreams, states of madness (such as young Hébert's), nocturnal somnambulism, and artificial somnambulism are in some sense all of a piece – they are bound into a common chain of memory. Natural som- nambulism is something like the acting out of a dream; in artificial som- nambulism the dream and the acting out are properly regulated by the magnetizer; but in madness both are wild and uncontrolled. Madmen may thus in a way be regarded as disordered somnambules. Similar ideas reemerged frequently in the later history of animal magnetism.

Some modern writers have hailed Puységur's treatment of young Hébert as a pioneering venture in psychotherapy.[20] But in his exchanges with Alexandre, Puységur was not consciously playing the psychotherapist; he was merely being a sympathetic human being. Many forms of psychotherapy may be fore- shadowed or hinted at in such interactions, but that no more makes the sympathizer a pioneer of some psychotherapeutic method than having delivered a few successful kicks makes a bar-room brawler into a pioneer of karate.

The *Appel aux savants observateurs* was Puységur's last major contribution to the movement he had done so much to originate and to revive. Thereafter, others came to the fore, and to them I shall come in a moment. Puységur himself survived an order of Napoleon's to have him summarily shot (he was wrongly suspected of collaboration with the invading armies[21]), and continued to magnetize and to write occasional pieces on animal magnetism right up to his

death in 1824. He remained an important presence and influence on the
magnetic scene, though now it was not so much because of what he did or
discovered as because of what he was. By this date, Puységur's total honesty was
hardly questioned by anyone. If he described, or had described, an unusual case,
or if he demonstrated a magnetic somnambule, his readers, or his audience,
could be absolutely sure that there was *something* here that required consideration and explanation, even if the true explanation might not be that
favoured by Puységur. Enemies of magnetism might mock him – "Le doux, le
généraux, l'innocent Puységur"[22] – but they could not denigrate or disparage.

Puységur was of course no scientist, and it did not require an especially
perceptive reader to see that to Puységur animal magnetism was more akin to
religion than to science. Privately, he united religion and magnetism into one
indissoluble belief-system. Writing to Bergasse in 1822 he said:

> I have made a sort of list of my friends from all periods, living or dead, little
> matter. I make an appeal to them before my evening prayer, as to my patron
> saints, guardian angels. Thus reunited in thought, I make a kind of magnetic
> chain, as at our old baquets, to address myself to God, and draw from that
> inexhaustible source the true principle of life for us all.[23]

But it would be wrong to leave the impression that Puységur was *au fond* a
mystic. Before everything else he valued people, of all ranks, categories and ages,
and his values were translated into immediate action whenever action was
called for. Here is how he responded when the father of a little girl of 10 or 12
months, who seemed at death's door with convulsions, begged him to try to save
her:

> Without saying a word to them ... I took little Honorine in my arms ... sat
> down and placed her on my knees. Then, without thinking or taking note of
> anything that was going on around me, I concentrated entirely on touching
> the little child in the desire only to produce on her the effect which would
> benefit her the most ... My profound concentration imposed a silence which
> ... no-one attempted to break, when suddenly there was heard the reassuring
> sound of an abundant evacuation. I expressed my delight, and ... continued
> my magnetic activities with even greater energy. A general relaxation of the
> muscles and the cessation of the state of convulsions were shortly the happy
> result.[24]

The urgent endeavours to help, the total commitment, the unselfconscious joy
in success, these are utterly characteristic of Puységur, a man whom his
follower, Deleuze, described as one of the most beneficent and the truest persons
who has ever existed.[25]

Figure 3. J. P. F. Deleuze. (From the Library of the Society for Psychical Research.)

DELEUZE

The year 1813 marked the beginning of a decisive, though hardly dramatic, upturn in the fortunes of animal magnetism in France. It saw not just Puységur's *Appel aux savants observateurs*, but the publication of a major work by a new author – new so far as animal magnetism was concerned – namely J. P. F. Deleuze (1753–1835). Deleuze, who was to be the leading figure in the magnetic movement for the next couple of decades, was already sixty years old, but he had been actively interested in animal magnetism since 1785,[26] when he and a friend had investigated a somnambule, a country girl of 16, who presented "most of the phenomena observed by M. de Puységur, by M. Tardy and by the members of the Strasbourg Society." Among these phenomena were some that Deleuze could "neither explain nor conceive".[27]

In 1787, Deleuze moved to Paris, and in 1795 he was appointed assistant naturalist at the Jardin des plantes (where he later became librarian). He produced a number of literary and scientific works, and acquired numerous friends among the *savants* of Paris. But all the while he continued to be intensely interested in animal magnetism. The fruits of his twenty-seven years of reading, reflection and observation were gathered into the two volumes of his *Histoire critique du magnétisme animal* (1813). The first volume is mainly devoted to the methods and phenomena of animal magnetism, the second to a review of the leading French works on the subject down to the year 1812. According to Bertrand (1826), a writer very well-informed on the history of this period, the outstanding qualities of the *Histoire critique* brought it a success which all other works on the same subject had hitherto been far from obtaining. It was, he goes on, "not only useful to the cause of magnetism, by procuring it a great number of converts; it served it further by encouraging those who practised magnetism in secret to declare themselves open practitioners."[28] It even achieved favourable notice in the medical press.[29] It is clear, however, that this was exceptional – the editor of the *Annales du magnétisme animal* speaks of "all the public declamations that have been made against animal magnetism during the course of the year 1813".[30]

Deleuze followed his *Histoire critique* with a number of other works.[31] The most important is the *Instruction pratique* (1825), aimed especially at would-be magnetizers. It is easy to understand why, as Puységur took more of a back seat, Deleuze came to be regarded as animal magnetism's principal spokesman. It is not that he produced radically new observations and ideas. He always declares himself a disciple of Puységur, and his differences from Puységur are mainly ones of balance and emphasis. He believes in the magnetic fluid, but prudently avoids speculation as to its nature;[32] he never tires of extolling the sovereign virtues of magnetized water (to magnetize a bottle of water hold it in one hand and pass the other hand downwards over it for two or three minutes[33]); he is rather more critical than Puységur in his attitude towards the deliverances of magnetic somnambules. But in clarity of style and in the organization and presentation of materials and arguments, he is greatly superior to Puységur. His literary gifts are combined with a caution in judgment, a moderateness in tone, and a reasonableness in debate, which brought him respect even from opponents. He does not make exaggerated claims, or engage, like Kluge, in premature over-systematization. "One must abandon", he says, "all theories, and simply see if there are sufficient indubitable facts to establish the reality of the effects of magnetism."[34] He lays down criteria in accordance with which he thinks the factual claims should be judged. These concern the number of witnesses, their character, intelligence, motives, etc., the agreement between them and the probability that they could not have been deceived.[35] Judged by these criteria, the case for the reality of the phenomena of animal magnetism is overwhelming.

His own acquaintance with animal magnetism was vast – by 1819 he had witnessed the phenomena of somnambulism more than a thousand times[36] – and one could wish that he had described some of his own cases at greater length. Though he writes elegantly, he is too often exiguous on detail – Puységur is greatly superior in this respect. None the less, Deleuze's experience is so considerable, and his judgment so balanced, that his statements about the facts of animal magnetism retain some interest even today. Many even of his incidental remarks are strikingly prescient of future findings and developments. He notes[37] that only one subject in twenty will become somnambulic, which is consistent with modern figures for the incidence of "deep trance" in hypnotic subjects; he claims (contrary to the beliefs of some) that women may magnetize as effectively as men;[38] he outlines the therapeutic uses of what would now be called "post-hypnotic suggestions";[39] he discusses how the "insensibility of somnambules" may be turned to good account for the performance of surgical operations;[40] he notes that animal magnetism may be used to calm the pains of childbirth.[41]

During the period from 1813 to about 1833, Deleuze was in effect animal magnetism's recognized spokesman. He was "recognized" in that both supporters and opponents agreed that no-one could put the case for magnetism with greater authority or more persuasively, and he was a spokesman in that he would very commonly respond in print, with his customary restraint and reasonableness, to whatever criticisms of the magnetic movement were currently being propounded. He replied, for instance, to those who disbelieved in the phenomena altogether; to those who regarded them as the work of the spirits of the dead; to those who attributed them to non-human intelligences; to those who put them down to demonic intervention; to those who ascribed them to the imagination; and to those who alleged that animal magnetism was dangerous to health or morals.[42]

His most important arguments were with the imagination-theorists. These arguments began in the *Histoire critique*, where he discusses the "metallic tractors" which an American physician named Elisha Perkins had devised in the late 1790s.[43] The tractors were two pieces of different metals about two and a half inches long, and so shaped that when brought together they formed half a cone, split lengthwise. The pointed end was drawn over the site of disease, and some remarkable cures were attested – cures in many ways similar to those allegedly achieved through animal magnetism. In 1800, a Bath physician, Dr. J. Haygarth, published a book on the curative power of the imagination,[44] in which he reported results similar to Perkins', obtained by the use of imitation metallic tractors made of wood. These findings clearly worried Deleuze. He does not deny that imagination may be therapeutically effective, but he remarks:

> I have often seen patients convinced that if they were able to consult a doctor whose reputation had struck them, they would soon be cured. The doctor has been called ... the patient has been ... given over to hope, and nonetheless he has undergone no improvement.
>
> How does it happen that a magnetizer who promises nothing ... produces in certain cases a greater effect than a famous doctor, awaited with eagerness, even on persons who tried magnetism without having much confidence in it ... one must seek the cause of the cure in an agent foreign to the imagination of the patient.[45]

THE MAGNETIC SOCIETY AND THE MAGNETIC PERIODICALS

The period of French animal magnetism between the publication of Deleuze's *Histoire critique* in 1813, and the early 1820s, may be termed the period of the Magnetic Society and the magnetic periodicals. The Paris Société du magnétisme developed out of a "numerous group of ladies and gentlemen, amateurs of, or curious concerning, animal magnetism",[46] who in 1813 began to meet at the house of M. du Commun, Paris, rue Vantadour 1. The Society was formally constituted on 25th July 1815. Puységur was its first President, Deleuze a Vice-President. A contemporary journal[47] alleged that the new Society shortly attracted the attention of the police because its organization and doctrines could furnish dangerous weapons to charlatanism, to fanaticism and to party spirit. In fact its members, though tending somewhat to uncritical enthusiasm, could hardly have been more respectable.

The foundation of the Society was preceded by the establishment, in July 1814, of the first important French periodical devoted to animal magnetism, the *Annales du magnétisme animal*. This seems to have been a private venture by the mathematician A. A. V. Sarrazin de Montferrier (1792–1868), who wrote under the pseudonym of M. de Lausanne.[48] It ran with various interruptions until 1st January 1817. The standard of the case material which it published was sometimes regrettably low, which worried the more moderate supporters, such as Deleuze. No doubt it was in an attempt to tighten up standards that when the *Annales* were resumed as the *Bibliothèque du magnétisme animal* the new periodical was run by a committee of members of the Society.[49] The *Bibliothèque* appeared from July 1817 to September 1819, with one brief interruption. Sarrazin de Montferrier remained principal editor, assisted by M. du Commun. In July 1818, a single number of the *Journal de la Société du magnétisme* was put out by the Baron E. F. d'Hénin de Cuvillers (1755–1841), secretary of the Society and a military and literary figure of some distinction.[50]

The Société du magnétisme ceased to exist in March 1820. According to Hénin de Cuvillers, its more reasonable members had deserted in droves, partly from fear of participating in the ridicule which was being poured upon the zealots and wonder-workers, and partly out of disgust at the Society's refusal to countenance discussion of the imagination-hypothesis (of which he was a keen supporter).[51] Hénin de Cuvillers then brought out and edited another periodical, the *Archives du magnétisme animal*, of which two volumes appeared in 1820, and a further two in 1822–3. Many of its pages are devoted to the erudite classical and philological ramblings of its somewhat disorganized editor. He has the distinction of having proposed the substitution for the old magnetic terms by a large number of new ones beginning with the prefix "hypno-," including hypnotic, hypnotism and hypnotist.[52] His turn of mind is indicated sufficiently by the fact that he eventually settled for replacing *magnétisme animal* with *phantasiéxoussisme* (or *fantasiéxoussisme*). But his bibliographical knowledge is sometimes useful.[53]

Although the Society was a Paris society, it had active members elsewhere – the various periodicals seem to have circulated quite widely, and to an extent provided a national rather than a parochial forum for the gradually reemerging magnetic movement. We find, for example, lists of cures worked at what were in effect magnetic clinics at such widely separated places as Saint-Quentin, Poitiers, Châtellerault, Nantes and Béziers.[54] And there was some degree of press reaction, mostly hostile.[55] A number of minor pamphlet wars broke out around matters raised in the magnetic periodicals, and helped to swell the number of publications recorded by Dureau from 15 between 1801 and 1810 to 59 between 1811 and 1820.

By and large, the 32 volumes which these three periodicals between them muster contain only a little that is new and nothing that is revolutionary. There are lucid discussions by Deleuze, and notes of minor historical interest by Puységur – in April 1818, for instance, we find Puységur once again in Victor Race's cottage, treating him before the very hearth where it had all begun thirty-four years previously.[56] Articles by the newcomers Lausanne and du Commun sometimes have the character of mere benevolent waffle. The numerous case reports are in general no different from the many that had preceded them. Somnambules were perhaps less given to stomach-seeing than they had once been, but they continued to diagnose and prescribe for their own ailments and those of others. Their clairvoyant observations tended to dwell a great deal upon *depôts* and accumulations of humours, although there was rarely if ever any confirmation that these really existed. Occasionally, a clairvoyant would perceive the luminous "magnetic fluid" streaming from the hands and eyes of her magnetizer.

It is notable that the magnetizers were quite prepared to tackle ailments which were undeniably physical and sometimes grave. A detailed journal by the chevalier J. M. P. A. Brice[57] (a geographer of some distinction) describes his treatment of Madame G., a lady of fifty-eight. Originally, this lady had suffered from an abscess on her left leg, probably tuberculous, and so serious that amputation was discussed. However, she got away with an incision to facilitate suppuration, and was cured. Two months after she left hospital a second abscess developed at the top of her forehead. It exuded blood and pus and, after a while, occasional little splinters of bone. When Brice first saw her she had had this abscess for six years. It was round, about four inches in diameter, raised and very lumpy, with an opening on the right side. It was accompanied by a constant headache with frequent stabbing pains of the most violent nature. The glands round her neck were enlarged, some to the size of small eggs, and so painful that she could not turn her head or raise her arms. Sleep was impossible – for six months she had not been able to lay her head on a pillow – and doctors expressed the most profound pessimism as to her chances of recovery. She was given over to thoughts of suicide.

Brice treated Madame G. by magnetic passes without actual contact. She did not become somnambulic. Her condition rapidly improved. She began to sleep and to rest; her headaches improved, she perspired a good deal, the enlarged glands dissolved. A new hole opened at the left side of the sore, where there was a large lump, and exuded immense quantities of the most unpleasant blood and pus, together with occasional small fragments of carious bone. But as the discharge began to diminish, the ulcer began to heal. After five months she was completely cured.

In another case of undoubtedly physical disease, described at considerable length by the patient's husband, M. Périer,[58] the patient became somnambulic and diagnosed and prescribed for her own ailments. She had been afflicted for eleven years with ulceration, fistulae and constriction of the rectum. Her sufferings from her disorders were exceeded only by those she endured at the hands of her medical advisers. Her magnetic susceptibility was discovered by a family friend, and her husband began to act as her magnetizer. In her somnambulic state she prescribed treatment and a diet for herself, foresaw crises, and perceived pockets of blood, humours etc., inside herself, often in most unlikely places. However, her husband continued in addition with ordinary magnetic treatment (magnetic passes, magnetized water), and in eight months she was perfectly restored.

For the rest, the magnetic periodicals of the time contain a certain amount that is of interest to the historian. There are foreshadowings of developments still to come: the production of blisters by self-suggestion;[59] self-magnetization;[60] surgical practices carried out with the patient in magnetic sleep;[61] speculations about the rôle in certain phenomena of mental processes that are not conscious.[62] There is a good deal about contemporary press reaction (which for a while focussed on the public demonstrations of the Abbé J. C. de Faria, of which the Société du magnétisme heartily disapproved).[63] And there is a good deal of material concerning the progress of animal magnetism outside France. But, as we have seen, the spread of animal magnetism in Europe during this period drew much of its inspiration from Germany rather than from France.

NOTES

1 Dureau (1869).
2 See e.g. Coll (1817); Lettre de Mme. *** (1817); Cheron (1817), pp. 49–51; Guéritaut (1814), cf. Deleuze (1813), Vol. 2, pp. 259–262; Ramsey (1988), p.196.
3 *Moniteur universel* No. 201, 21 Germinal, an VII (i.e. 10th April 1799); Autographe de Mesmer (1849).
4 A useful biographical account of Puységur is given by Foissac (1833), pp. 229–244. Cf. also Ellenberger (1970), pp. 70–74.
5 Petetin (1808). This work is closely analyzed in Dingwall (1967), pp. 21–28. Similar cases are described by Guéritaut (note 2 above) and Renard (1815).
6 A. M. J. de Chastenet de Puységur (1811), pp. 92–101.
7 Deleuze (1813), Vol. 1, p. 138.
8 A. M. J. de Chastenet de Puységur (1811), pp. 78–82.
9 A. M. J. de Chastenet de Puységur (1811), pp. 14–15.
10 A. M. J. de Chastenet de Puységur (1811), pp. 43–45.
11 A. M. J. de Chastenet de Puységur (1813), Vol. 3, pp. 27–40.
12 A. M. J. de Chastenet de Puységur (1811), p. 73.
13 A. M. J. de Chastenet de Puységur (1817), pp. 166, 168. Cf. A. M. J. de Chastenet de Puységur (1813), Vol. 1, pp. 67–68, where Puységur describes how he told Gall that it was for the learned physiologists to discover the nature of the magnetic agent.
14 A. M. J. de Chastenet de Puységur (1812a; 1812b).
15 A. M. J. de Chastenet de Puységur (1813), Vol. 3, pp. 75–76.
16 A. M. J. de Chastenet de Puységur (1813).
17 I almost wrote, little monster.
18 A. M. J. de Chastenet de Puységur (1813), Vol. 1, p. 35; Vol. 2, pp. 27–28.
19 Cf. especially A. M. J. de Chastenet de Puységur (1813), Vol. 2, pp. 39–40.
20 E.g. Conn (1982).
21 Foissac (1833), pp. 291–292.
22 Fabre (1838) quoted by Dingwall (1967), p. 10.
23 Viatte (1928), Vol. 1, p. 228.
24 A. M. J. de Chastenet de Puységur (1811), pp. 71–72.
25 Deleuze (1819), p. 97.
26 On Deleuze see Foissac (1833), pp. 245–255; Rouxel (1892), pp. 72–116; Dingwall (1967), pp. 14–21.
27 Deleuze (1813), Vol. 1, pp. 216–217.
28 Bertrand (1826), p. 256. For Deleuze's own estimate of the reasons for the success of his book see Deleuze (1815), pp. 35–36.
29 E.g. M. A. C. Savary of the Paris Faculty, writing in the *Journal de médecin, chirurgerie, pharmacie*, etc., Vol. 27, August 1813, quoted in *Annales du magnétisme animal*, 1814, Vol. 1, p. 34.
30 *Annales du magnétisme animal*, 1814, Vol. 1, p. 35.
31 Deleuze (1817; 1818a; 1819; 1825; 1826b; 1836).

32 In an interesting exposition of his theoretical position (Deleuze, 1814), Deleuze says (p. 233), "When we call this principle *magnetic fluid*, *vital fluid*, we are using a figurative expression. We know that something emanates from the magnetizer: this something is not a solid, and we call it a *fluid*."

33 Deleuze (1813), Vol. 1, p. 121.

34 Deleuze (1813), Vol. 1, p. 52.

35 Deleuze (1813), Vol. 1, pp. 37–40.

36 Deleuze (1819), p. 59.

37 Deleuze (1813), Vol. 1, p. 138.

38 Deleuze (1813), Vol. 1, p. 133.

39 Deleuze (1825), pp. 136–138.

40 Deleuze (1825), pp. 139–140.

41 Deleuze (1825), p. 246.

42 On these controversies see, in addition to relevant passages in Deleuze (1813) and Deleuze (1825), Deleuze (1818a; 1818b; 1819; 1826a; 1828).

43 Deleuze (1813), Vol. 2, pp. 266–277. Cf. Dingwall (1967), pp. 16–19.

44 Haygarth (1800).

45 Deleuze (1813), Vol. 2, p. 272.

46 Hénin de Cuvillers (1823a), p. 189.

47 *L'Aristarque*, 11th December 1815, quoted in *Annales du magnétisme animal*, 1816, Vol. 5, pp. 43–44. For the rules of the Society see *Annales du magnétisme animal*, 1816, Vol. 6, pp. 234–240.

48 He wrote his books on animal magnetism (Lausanne, 1818; 1819) under this pseudonym. A leading theme of the second of them was the view, originated by Barberin, that the magnetizer can use his own sensations when passing his hands over a patient to diagnose that patient's ailments.

49 *Bibliothèque du magnétisme animal*, 1817, Vol. 1, p. iii.

50 On the history of these periodicals see Hénin de Cuvillers (1820), pp. 2–4. This work is largely taken from volume 1 of the *Archives du magnétisme animal*.

51 Hénin de Cuvillers (1820), pp. 92–93.

52 Hénin de Cuvillers (1820), pp. 131–134; Gravitz and Gerton (1984–5).

53 Hénin de Cuvillers (1823a; 1823b; 1823c).

54 Drouault (1816); Tanton (1818); Segretier (1819); Traitemens magnétiques (1818).

55 The more important articles are listed in Hénin de Cuvillers (1823c). Two articles by M. Hoffman in the *Journal des débats* 24th June and 10th July 1816 are reprinted in the *Archives du magnétisme animal*, 1822, Vol. 6, pp. 239–248 and 263–273. Hoffman adopts the curious position that the phenomena of animal magnetism are not due to the imagination, but that animal magnetism has no curative efficacy.

56 A. M. J. de Chastenet de Puységur (1818).

57 Brice (1823).

58 Périer (1814–15).

59 Le Lieurre de l'Aubépin (1819).

60 Birot (1814); cf. Rouillier (1817b), pp. 62–64, Kieser (1826), 1, pp. 246–253, and Bursy (1818).

61 S. du M. (1815); Pam (1816).

62 S. du M. (1816a).

63 S. du M. (1816b).

CHAPTER 7

Mesmerism and the medical profession in France 1820–1840

The Paris Société du magnétisme was essentially a society of dedicated amateurs. It had, so far as I can discover, only a few medical members, and the *Annales* and its successors contained only a limited number of contributions from medical men. The most notable were those of M. Cheron, professor and director-in-chief of the hospital at the Val-de-Grâce (the Paris military hospital),[1] and of M. Auguste Rouillier, a physician at Pont-Sainte-Maxence who, in addition to articles in the *Bibliothèque du magnétisme animal*,[2] devoted a book of some substance to the same topics. Rouillier, a product of the vitalist medical school of Montpellier, and a great devotee of Stahl, had been a member of the Strasbourg Society of Harmony in the later 1780s. His book, *Exposition physiologique des phénomènes du magnétisme animal et du somnambulisme* (1817), is one of the more notable works of its period by a supporter of animal magnetism. He regards magnetic somnambulism as not fundamentally different from natural somnambulism, and says:

> Somnambulism is a state, *sui generis*, in which the exercise of the external senses is truly suspended, whilst the faculties of the soul come forward with more energy, as one might express it, following the ideas of Stahl; or, if you like, ... it is physical and moral instinct, further developed, and reaching such a degree of perfection, that it enlightens the patient as to his health, the health of other people, and the proper means of reestablishing it ...[3]

It is easier, Rouillier thinks, to understand these peculiar phenomena if, following Stahl, and his own teacher J. C. M. G. de Grimaud (1750–1789), one distinguishes in the soul

> ... two sorts of ... ideas, one simple, intuitive, anterior to all exercise of the senses; instinctive ideas ... which exist in the soul without our being habitually aware of them, and which cannot be used in any sort of judgment, so long as we remain in the ordinary state; the others, reflexive, and the only ones upon which, in the ordinary state, memory and reason can operate, because the soul owes them to the action of external objects on the senses.[4]

In perfect somnambulism the outer senses are closed off, and the soul acquires consciousness of these "intuitive" ideas. It will not, however, subsequently remember them. It may also recover an awareness of long-forgotten events.[5]

Rouillier thus emerges as one of the first writers to combine a "special state" view of magnetic somnambulism, with the proposal that in this state there may emerge ideas not accessible to the ordinary, everyday consciousness. Yet to professional colleagues he must have seemed an outdated, pre-revolutionary figure, left behind by the many and rapid advances in medical organization and teaching which had come in the wake of the Revolutionary and Napoleonic wars.

Paris (and hence French) medicine at this time centred around the study of pathological anatomy, of which M. P. X. Bichat (1771–1802)[6] was universally acknowledged to be the founder. The so-called "anatomo-clinical" school set out by means of statistics based upon numerous autopsy reports (only the facilities of the newly enlarged and reformed hospitals made the collection of such statistics possible) to establish links between clinically established symptom complexes and anatomically verified lesions. A medical historian who has certain reservations about the value of the approach says, not altogether fairly, that the patient "was treated rather as a living cadaver or a living anatomical preparation, not as a sentient being, endowed with vital force".[7] Among members (in a broad sense) of this school may be mentioned G. L. Bayle (1774–1816), A. L. Bayle (1799–1858), J. Cruveilhier (1791–1874), R. T. H. Laënnec (1781–1821), L. L. Rostan (1790–1866), E. J. Georget (1795–1829), J. N. Corvisart (1755–1821), F. Chomel (1788–1856), P. A. C. Louis (1787–1872) and L. F. Calmeil (1798–1895).

The influence of pathological anatomy was felt in all areas of French medicine. Philippe Pinel (1745–1826), who worked first at the Bîcetre (the Paris hospital for the male insane) and then at the Salpêtrière (the corresponding hospital for females), is remembered now chiefly for his humanity in releasing the patients at these institutions from their chains. But, as Lopez Piñero says, "it is less known that he also forged the link between the Enlightenment and anatomo-clinical pathology".[8] Pinel's successor at the Salpêtrière, another of his pupils, J. E. D. Esquirol (1779–1840) became more nearly a pure psychiatrist, but Georget and Calmeil were his pupils.[9]

To the members of the Société du magnétisme it must have been clear that if they could win over the Paris medical world, or some appreciable part of it, to their cause, the rest of the medical world would soon follow. It must have been equally clear that their chances of so doing were fairly slight. Pinel was not hostile. Before the Revolution he had attended d'Eslon's lectures in Paris;[10] and he received Puységur with courtesy and interest when the latter brought young Hébert to see him.[11] But absence of hostility was, in general, the most that animal magnetists could hope for. Still some doctors had privately shown interest. And one or two aspects of the medical scene worked in the magnetists' favour. Most notable perhaps was simply the kind of philosophy to which some leading younger members of the anatomo-clinical school adhered. These men –

for instance Pinel, Rostan, Corvisart, G. L. Bayle, Laënnec, Chomel and Louis –
were strongly anti-theoretical and admirers only of observation and fact.[12] They
were prepared to consider alleged facts without too much theoretical presup-
position. Leading representatives of Paris medicine in the 1820s had a certain
openness not characteriistic of their pre-revolutionary equivalents. It showed in,
for instance, the interest which several displayed in the claims of Gall, who was
resident in Paris from 1810 to 1819. Among these were Corvisart, Broussais,
Georget, Calmeil and Rostan.[13]

French medicine during this, arguably its greatest, period was encapsulated in
the monumental *Dictionnaire des sciences médicales* (1812–22).[14] Animal
magnetism was thought a topic of sufficient interest to warrant a major article,
and this was supplied in 1818 to the tune of ninety-five pages by J. J. Virey
(1775–1846), a well-known pharmacologist and writer on natural history.[15]
Virey claims that he set out with no prejudices, that he had observed magnetizers
at work, and talked with them, and that, in 1799, he had even heard Mesmer
speak. Certainly, he had read widely in the literature of animal magnetism
(including the German literature, of which most French mesmerists were largely
ignorant), and his historical survey is quite detailed. He offended magnetists by
the use of terms such as folly, charlatanry, credulity, etc., and by implying, for
instance, that a rather high proportion of magnetizers were handsome and virile
men, and a rather high proportion of their best subjects susceptible young
women (a contention which particularly annoyed the somewhat prudish
Deleuze). But his account of magnetic procedures is quite fair, and he does not
deny the phenomena; he does not even deny the cures. He simply attributes
them to ordinary causes. The magnetic passes may cause changes in blood
circulation, and an increase in perspiration, which may by themselves prove
beneficial in certain sorts of disorders. After all, the great Cuvier, reflecting on
the success of magnetic procedures applied to unconscious persons, remarks
that it is impossible to doubt that the proximity of two living bodies may have a
real effect independent of the participation of the imagination.[16] Even some of the
more remarkable phenomena of somnambulism may be perfectly genuine. Why
should not delicate female somnambules, if their attention is directed appropriately, feel what is going on inside them and exhibit an instinct for the right
remedies, such as animals possess? Such sensibility may be a result of the life of
the ganglion system leaking into the brain. What Virey firmly denies is that
there is any evidence for the existence of a magnetic fluid. Prove, he demands of
the magnetists, that there is a veritable *magnetism*. Until they do "we may
justifiably continue to attribute their cures and the other results really obtained,
to neural connections, and to the well-known ways in which tricks and illusions
have perennially worked upon human intelligences."[17]

THE MEDICAL MAGNETISTS

Virey's article seems to have caused some stir. In fact, his acceptance that there were genuine facts in need of an explanation may well have done more good to the cause of magnetism than his rejection of the magnetic fluid did harm. However, the next year (1819) Deleuze published a very long reply,[18] in which he answered Virey point by point. Deleuze writes with his usual lucidity and power of organization, and since he knew a great deal more about the history and practice of animal magnetism than did Virey, he was able to correct him on a good many points. Indeed, it would hardly be an exaggeration to say that Deleuze defeats Virey on almost every issue except the main one. He never satisfactorily answers Virey's challenge to prove that there is a veritable magnetism. He attacks the hypotheses of imagination and imitation, which Virey had applied to some of the more dramatic phenomena. He points out that magnetism works upon persons who have never seen anyone else magnetized, and upon peasants, children, etc., who have no idea what it is supposed to accomplish. He cites the cases of two children aged five and four who, on their first encounter with animal magnetism, quickly became somnambulic. He mentions a girl of eleven whom he magnetized during her ordinary sleep, and quickly put into somnambulism. He talks of doctors who have magnetized patients without their knowing what was being attempted. But he simply does not give enough details for us to assess what was really going on in these cases, still less to convince us that the effects may justifiably be set down to an emanation of some unknown kind from the operator.

1819 saw further developments of interest. A second edition of Deleuze's *Histoire critique* was published; and, for the first time since before the Revolution, a Paris medical man offered a course of public lectures on animal magnetism. The doctor in question, A. J. F. Bertrand (1795-1831),[19] was to become an important figure in the histories both of animal magnetism and of hypnotism. A bent for mathematics[20] led him to begin training for the corps of military engineers in 1815, but political events caused him to switch to medicine. He became interested in animal magnetism in 1818, when he witnessed a demonstration of it in his birthplace, near Nantes, and he offered the course just mentioned in August 1819, shortly after he had received his medical degree. At this period, it seems, he was a fully convinced believer in animal magnetism. His lectures were so well-attended that he put on a second course in January 1820.[21] And soon there began to pass through some of the Paris hospitals not exactly a wave of serious interest in animal magnetism, but at any rate a fairly marked ripple.

The process began at the Hôtel Dieu, where H. M. Husson (1772-1853), physician-in-chief, agreed to a request from certain students to permit a trial of magnetic procedures. The magnetizer for these experiments was J. D. Dupotet

(1796–1881), a young man who later became one of the most famous figures in French animal magnetism. Dupotet came from a family which had some claims to nobility, but had fallen on hard times. He later called himself Baron Dupotet de Sennevoy.[22] At the Hôtel Dieu he magnetized over the next few weeks a young lady of seventeen, Mlle. Samson, who was suffering from severe gastric troubles. She soon became somnambulic, and her ailments grew better under Dupotet's ministrations; however interest mostly centred upon attempts (apparently successful, and attended by quite a number of witnesses, including Bertrand) to defeat the imagination-hypothesis by demonstrating that Dupotet could magnetize her successfully even when hidden.[23] The details given of these experiments are however hardly sufficient to enable us to decide what ordinary cues she might have picked up as to Dupotet's presence and actions.

In November, Husson left the Hôtel Dieu for the Pitié, and his successor, M. Geoffroy, received instructions to suspend the séances. However, Mlle. Samson's health deteriorated so rapidly that he told an interne, M. Robouam, secretly to resume magnetizing her. She was discharged as cured on 20th January, 1821.

One of the witnesses of these experiments had been C. A. Récamier (1774–1856), who became professor of clinical medicine at the Hôtel Dieu in 1821. Récamier invited Robouam to try animal magnetism in the wards of which he (Récamier) had charge. Robouam and Récamier put two patients (one male and one female), who had been rendered somnambulic, through the painful ordeal of a moxa. Neither patient exhibited the least sign of pain, through cries, movements, or changes of the pulse rate.[24]

According to Dupotet,[25] interest in animal magnetism was about this time carried from the Hôtel Dieu to the Salpêtrière by a surgeon, M.Margue. Esquirol was then physician-in-chief at the Salpêtrière, and permitted extensive experiments, although attempts to treat eleven "maniacs" by means of animal magnetism were not successful.[26] The most enthusiastic experimenter at the Salpêtrière was at first E. J. Georget (1795–1829), a brilliant young physiologist who published his preliminary results in his *De la physiologie du système nerveux* (1821).[27] Georget seems to have witnessed most of the standard phenomena of magnetic somnambulism, though he describes them only very briefly. The phenomena which he found most remarkable were certain apparent instances of prevision. "I have seen", he says, "positively seen, a sufficiently large number of times, somnambules announce several hours, several days, twenty days in advance, the hour, even the minute, of epileptic and hysterical attacks, of the arrival of periods; and indicate the duration and intensity of these attacks; matters which were exactly verified."[28] Even a brilliant young physiologist may sometimes be a little naïve.

Georget tells us that he has often produced and seen produced muscular paralyses of fingers, arm, tongue and legs. These were accompanied by prickings, feelings of cold and finally loss of feeling. In some of his somnambules, it was possible to shut down the exterior senses completely, so that they heard nothing, did not show any disturbance when a flask of concentrated ammonia was held under the nose for up to fifteen minutes, and remained absolutely insensible to the burning caused by a moxa, or the irritation produced by immersion in hot water heavily charged with mustard. The after-effects of these treatments were so painful that the victims complained loudly as soon as reawakened.

Georget also experimented with magnetized water, and several times demonstrated, so he assures us, that subjects were able to pick out by taste from among five glasses of water that glass containing water which had been magnetized. But in this matter, as in every other, Georget gives so few details that it is impossible to assess his claims. In effect he relies upon his known abilities and medical qualifications to quell any doubts which his readers might feel.

Georget, indeed, proposed to write further on the subject, and in his first volume tells us in a long footnote[29] that he has recently witnessed phenomena still more remarkable, which he will report in his second volume. He did not report them, but we know from other sources[30] that what had impressed him was the clairvoyance and previsions of an epileptic girl named Pétronille. So impressed was he that he abandoned his materialistic views and turned to dualism and a religious view of life.[31] Pétronille, in somnambulism, one day informed Georget's colleague, M. Londe, that in fifteen days he would be involved in an affair of honour and would be wounded. M. Londe made a careful note of this prediction, which was exactly fulfilled. Returning home, wounded, in a carriage, he read his note to his more fortunate adversary.[32]

Another forthright and courageous supporter of animal magnetism, was Georget's friend and colleague at the Salpêtrière, Léon Rostan (1770–1866), who was noted as a teacher, an authority on hygiene, and a Don Juan.[33] Rostan had been much involved, as witness and experimenter, in the Salpêtrière investigations of animal magnetism, and was invited to contribute an article on the subject to the new *Dictionnaire de médecine* (1821–28). His article appeared in 1825,[34] and created a sensation. Like Georget, Rostan concerns himself primarily with the phenomena of somnambulism, and, like Georget again, he fails in most instances to substantiate his factual assertions with anything more than his own *ipse dixit*. His materials overlap with Georget's, and we need not dwell on them at any length. He emphasizes the exalted powers of memory and imagination found in somnambules, often accompanied by an elevated diction. Some can recite lengthy pieces of verse which they have read only once, or which have long dropped out of their conscious recollections. Especially characteristic of somnambules are amnesia upon waking for all that passed during the somnambulic state, and the recovery of the missing memories upon reentering that state.

Rostan does give us a somewhat more detailed account of experiments conducted by himself and his colleague, G. Ferrus, with a somnambulic lady who seemed able to tell the time to which the hands of a watch, held at the back of her head, and carefully shielded from her, had been set. These experiments started something of a trend.[35]

He goes a little further into the theory of animal magnetism than does Georget. The phenomena, he is sure, belong to the nervous system. He admits the circulation therein of a certain agent, having a marked analogy with the electrical fluid, and able, up to a point, to pass through solid obstacles. This agent, as Reil, Autenrieth and Humboldt have argued, creates a "nervous atmosphere" around the body. It is in terms of the mingling of the atmosphere of the magnetizer with that of the subject that magnetic "rapport" may be understood.[36]

THE MEDICAL CRITICS

By the mid-1820s it had become apparent that the major controversies in French animal magnetism were no longer between gentleman amateurs and medical men, but between different members of the medical profession. In fact, with the exception of Deleuze's *Instruction pratique* (1825), and perhaps of Mialle's *Cures operées* (1826), the contributions of the amateurs to the debate during the 1820s were of little significance. Hénin de Cuvillers' prolix defences of the imagination-hypothesis[37] were too eccentric, and Dalloz's lengthy advocacies of the fluidic hypothesis[38] too uncritically enthusiastic, to have any notable impact. But Georget and Rostan had written concisely and well and to considerable effect; and a position opposed to theirs was soon to be forcefully stated by two other medical writers of distinction.

The first of these was J. A. Dupau, author of divers medical works, whose *Lettres physiologiques et morales sur le magnétisme animal* appeared in January 1826. Dupau is an irritatingly cocksure person, and he does not apply to the evidence against animal magnetism the same critical standards as he applies to the evidence for it.[39] None the less he has some perceptive criticisms of the magnetists. He is well aware of the effect that the patter with which so many magnetizers accompanied their ministrations might have upon sick or sensitive persons.[40] He notes the likelihood that in somnambules whose eyes are closed senses other than vision may play a role in many examples of apparent clairvoyance or thought-transference. He transcribes the utterances of a diagnostic somnambule exercising her faculties upon a lock of hair sent by a sick person, and remarks that "the skill of this false somnambule consisted of repeating, in the manner of those who tell fortunes by the cards, a very large number of circumstances; she would be very unlucky if some one of these did not fit the sick person."[41]

Dupau admits that magnetic cures do take place, but only in cases where the disease is caused by imagination, for the imagination, like the spear of Achilles, has the marvellous property of curing the evils which it has itself occasioned.[42] Any technique which powerfully excites the imagination, can cure diseases due to the imagination, and if, like animal magnetism,it damps down the outer senses, so heightening imaginative activity, its effects will be stronger still.[43] To this advocacy of the imagination hypothesis, Dupau adds[44] a powerful attack upon the logic of the fluidic theory as expounded by Rostan.

More interesting, because informed by a richer practical experience and a wider knowledge of the literature, are the works of Alexandre Bertrand. At the time when he offered his public lecture courses on animal magnetism, he was convinced of the existence of a magnetic fluid. But before long, experience and reflection changed his mind, and in his *Traité du somnambulisme* (1823) he was highly critical of the supposed evidence for the fluid. In the preface to his *Du magnétisme animal en France* (1826), he summarizes the reasons for his reversal of opinion. Originally, certain sorts of phenomena shown by his own subjects had seemed to him to support the fluidic theory. By a strong exercise of will, he could produce in these patients effects quite unlike those generally thought of as simple results of the imagination. Several of his somnambules assured him that they could see streaming from his fingers the fluid by means of which he acted upon them. They could recognize magnetized water by its characteristic taste. Their most violent symptoms could be calmed, and they themselves put into a somnambulic state, by a magnetized handkerchief, glove, coin, or ring. But soon Bertrand was led to reject one by one the proofs on which he had relied.

He began to recognize, for instance, that somnambules' descriptions of the magnetic fluid were simply the result of the ideas upon which they had been dwelling whilst falling asleep. Magnetizers who did not believe in the fluid found that their subjects did not see it.[45] As for experiments done with the aim of discovering whether somnambules can really detect the taste of magnetized water, independently of all prior conviction, "there is not one, of all that I have seen attempted with proper precautions, which gave me positive results." Often somnambules "take for magnetized water water which is not, clear proof that their imagination can produce all the effects attributed to the fluid."[46]

Unfortunately, Bertrand, like so many of his opponents, gives no details of these crucial experiments. He does, however, describe some of his experiments on the effect which magnetized objects might have upon somnambulic subjects even in the absence of their magnetizers. There should be, he points out, no need of such objects, for the patient's own clothes will be saturated with the fluid as a result of the magnetizer's earlier ministrations. But in any case, it is easy to show by experiment that what is important is not whether or not the object has been magnetized, but whether or not the subject believes it has. One of his somnambules could readily be put into the magnetic sleep by a piece of paper he had not even touched, as long as she thought he had magnetized it. On the other hand, magnetized objects laid on her stomach without her knowledge had no such effect.[47]

To test the view of Puységur and others that the power of the will is all-important in magnetization, Bertrand attempted some experiments in which outwardly he expressed a firm will, but inwardly distracted himself with other thoughts; and others in which he willed strongly, but made himself appear distracted. Whether or not he really willed with vigour made no difference. His subjects responded positively so long as they thought he did.

What Bertrand describes as "experiments" are by modern standards rudimentary, and are generally reported with an almost scandalous shortage of detail. None the less he had grasped, more clearly than anyone else of his time, the crucial point that if you are conducting psychological experiments with human beings it is essential so to arrange matters that your subjects either do not know what results are anticipated, or else are misled by their prior expectations on a certain proportion of trials.

Bertrand developed his own approach to the phenomena, but this belongs rather to the history of the hypnotic movement. His books, though spoken of with respect for their scholarship, seem not to have played any great rôle in shaping opinion. He accepted, perhaps, too many of the phenomena of animal magnetism, including the ostensibly paranormal ones, to please the opponents of magnetism, whilst its supporters tended to look upon him as somewhat of a renegade and backslider. Neither party gave careful consideration to his views, except indeed that the veteran Deleuze reviewed Bertrand's volume of 1826 with his accustomed patience and good sense.[48] Deleuze defends belief in the magnetic fluid upon grounds, such as magnetization by a hidden magnetizer, magnetization of children, peasants, animals, that he had advanced many times before. As usual he fails to support his arguments with adequate factual details. He also suggests, not unreasonably, that Bertrand should have given more attention to the simpler, i.e. non-somnambulic, phenomena and cures, which are readily reproducible, and free of the extravagances to which somnambulism sometimes gives rise.

THE LATER COMMISSIONS

Contemporary shades of medical opinion about animal magnetism are interest-ingly revealed in debates which took place at meetings of the Royal Academy of Medicine (founded in 1820 and already of international importance)[49] during 1825 and 1826. In the autumn of 1825, a young Paris physician, P. Foissac (1801–1884?) wrote to the Academy suggesting that the time was ripe for a further examination of animal magnetism. A small committee was set up to report on the matter. Its report, read by Husson, at the session of 13th December 1825, argued that sufficient had changed since 1784 to warrant the appointment of a new commission. There were extensive discussions at the meetings of 10th and 24th January and 14th February 1826,[50] during which it became apparent that a number of quite eminent members of the Academy had thought animal magnetism of sufficient interest and possible importance to have observed and experimented with it themselves. These included Laënnec (who thought it all imposture), Georget (who believed in magnetism, but thought a commission would get nowhere), Virey (who did not believe in magnetism, but favoured a commission), F. Magendie (1783–1855, the celebrated physiologist and pharmacologist), and L. B. Guersant (1777–1848, director of the childrens' hospital). A few members – for instance Baron F. Desgenettes (1762–1837), an authority on military medicine – thought that the commissions of 1784 had said all that needed to be said. But a majority disagreed with him, and the proposal to set up a commission was eventually carried by 35 votes to 25. There was no clear line of cleavage between the more and the less "establishment" members or between the younger and the older. Indeed, the most quintes-sentially "establishment" of all contemporary medical men, M. Orfila (1787–1853), a noted teacher and administrator at the Paris School of Medicine, and an accomplished academic politician, made a strong and reasoned plea for the setting up of a commission – he had been particularly impressed by the testimony of Rostan. Just possibly he was also impressed by the fact that the king, Charles X (d'Eslon's former patron the Comte d'Artois), still retained an interest in the subject. At the end of these discussions, it could well have seemed to many supporters of animal magnetism that at long last, after almost fifty years in the wilderness, animal magnetism was about to be admitted within the portals of orthodox medicine.

The members of the commission were persons of some distinction: M. M. J. Bourdois de la Motte, a fashionable practitioner; Husson and Guersant, both of whom we have already met; J. M. G. Itard (1775–1838), of the institute of deaf-mutes; F. Gueneau de Mussy, of the Hôtel Dieu; P. E. Fouquier (1776–1850), professor of clinical medicine and famed for his eloquence; J. J. Leroux des Tillets (1749–1832), formerly Dean of the medical school; C. C. Marc, noted for work in the field of hygiene; and M. Tillaye. Contemporaries must have supposed that these gentlemen were conducting lengthy and painstaking investigations, for their report was not published until 1831.

Despite this long delay, medical interest in animal magnetism did not subside. One indication of this is the sprinkling of theses on animal magnetism presented by graduating medical students during the earlier 1830s and subsequently published.[51] These probably represented an interest acquired a year or two earlier. Another is the publication by "a society of medical men" of L'Hermès, a new periodical devoted to animal magnetism (four volumes, 1826–9). Some of the contributions are not without interest. It is curious, for instance, to read of a magnetic cure, by Foissac, of a case of chronic pleurisy with effusion in which the patient's chest had been listened to by Laënnec, introducer of the stethoscope.[52] Again, L'Hermès contains some of the earliest accounts of the use of animal magnetism as an anaesthetic. The first of these, published in 1826, is by M. Delatour, a member of the old Society of Magnetism. Delatour describes[53] two minor surgical procedures carried out upon subjects in a state of magnetic somnambulism. The earlier dates from 1819, and involved the application of a painful mustard plaster. The second was that of a tooth extraction performed painlessly in May 1824 on an exceptionally nervous lad of 14. The boy, who had been put into the somnambulic state prior to setting out, did not realize, when awoken, that his tooth had been extracted, and took the dentist's compliments upon his courage for a joke.

Three years later, L'Hermès published an account of a far more severe operation caried out painlessly during magnetic sleep. The patient was Madame Plantin, a lady of sixty-four suffering from an advanced cancer of the right breast, the magnetizer was M. Chapelain, the lady's medical adviser, and the surgeon was Jules Cloquet (1790–1883), a noted anatomist and inventor of surgical instruments. Madame Plantin was a somnambulic subject, though not a specially good one. When an operation became inevitable, Chapelain proposed to Cloquet that it should be carried out with the patient in magnetic sleep. Cloquet agreed. The operation took place on 12th April 1829. It involved the extirpation of the tumour following two immense incisions, and dissection and ablation of the enlarged glands of the arm. During the ten or twelve minutes of the operation Madame Plantin remained calm and immobile, giving no sign of pain, and manifesting no change in pulse or breath rate. When awoken she was astonished to discover that the operation had been carried out. Cloquet reported the case to the Royal Academy of Medicine on 18th April. Madame Plantin died on 28th April, apparently from pneumonia with pericarditis. Her case is the most famous of its kind in the literature.[54]

L'Hermès also gives us some valuable side-lights on the contemporary magnetic scene. For instance, certain court cases reported in its columns[55] show that in Paris, and no doubt elsewhere, there were quite a few professional magnetic somnambules who diagnosed and prescribed for their clients' ailments. Indeed, it is clear that such a class of persons had existed for at least a decade.[56] We learn, too, that the Royal Academy of Medicine was sharply attacked by critics, such as the formidable F. B. H. Hoffman (1760–1828) of the Journal des débats, for so much as lending an ear to the pretentions of animal magnetism,[57] and that at the end of 1826 Bertrand was again offering a course of public lectures on the subject.[58]

To return to the commission: its report, prepared by Husson, was read to the Academy at the meetings of 21st and 28th June 1831.[59] It ends with thirty conclusions, some of them both startling and unequivocal. The phenomena of somnambulism – including clairvoyance, prevision of the onset of symptoms, and anaesthesia – are regarded as established, along with the subsequent amnesia. Magnetic sleep can be produced with magnetizer and subject well separated and without the latter being aware of the endeavours of the former. As for the therapeutic effects of magnetism, the commission did not feel that it had made a sufficient number of experiments to justify any conclusions, but it was clearly impressed by the improvements shown by certain patients, and stated that the Academy ought to encourage further investigations into the subject.

These, and other, positive conclusions were a source of much satisfaction and encouragement to the magnetists, who frequently reprinted them. Less frequently reprinted were the commissioners' accounts of the investigations on which their conclusions had been based. And there is indeed a painful gap between their inadequate and inadequately described investigations, and the revolutionary conclusions they draw from them. The gap is not made less painful by the self-satisfaction with which the commissioners dwell upon the stringency of their experimental conditions.

To be fair, not all the report's shortcomings were the commissioners' fault. A decree of the General Council of Hospitals, dated 19th October 1825, in effect precluded them from using hospital in-patients as subjects. In the end, most of the work was done with just four somnambulic subjects, namely M. Petit, an elementary school teacher aged thirty-two (presented by Dupotet); Paul Villagrand, aged twenty-four, a law student presented by Foissac, and suffering from a paralysis of the left side of the body consequent upon a stroke seventeen months previously; Pierre Cazot, aged twenty, a hatter liable to severe and frequent epilepetic fits (he was a patient of Foissac's); and Mlle. Céline Sauvage, a patient of Foissac's suffering from various nervous complaints.[60]

Such therapeutic effects of animal magnetism as the committee observed were mostly minor and uncertain. Insensibility to external stimuli – loud noises, pinches, pricks with needles, the approach of ammonia bottles to the nose – was apparently demonstrated with several subjects, but the tests do not seem to have been severe. As further evidence for somnambulic insensibility, the report introduces Cloquet's account of the operation upon Madame Plantin. To do so was, the commission claimed, permissible, since the case had already been presented to the Academy.

The commissioners attempted to demonstrate sensitivity to the magnetic influence under conditions where the subjects remained ostensibly unaware of the magnetizer's presence, or else of what he was trying to make them do. In several instances, the experiments showed quite conclusively that the effects were due to the imagination: the subject responded to what he supposed the magnetizer was doing, rather than to what, if anything, he was in fact doing. On two occasions, Cazot was apparently put into the magnetic sleep by Foissac, who was concealed in a neighbouring room. But it is impossible to say what covert sensory cues the subject might have picked up, and what means he might have had for guessing the experimenters' intentions. The experimenters do not seem to have been fully alive to such possibilities, and the details they give are far too sparse for a retrospective assessment.

Experiments on the ability of Petit and Villagrand to read, or to recognize playing cards, with their eyes closed, were totally vitiated by the commissioners' failure adequately to control against normal use of vision. On a number of occasions, Cazot successfully predicted the day and hour of his epileptic attacks; but there are several means by which such predictions could become self-fulfilling.

Three persons were submitted to the diagnostic clairvoyance of Mlle. Céline, and in all cases what she said was at least partially correct. However, in only one case did she vouchsafe details that could not, perhaps, have been guessed at from looking at the patient, and in this case, though the patient died, no autopsy was performed.

By modern standards the experimental methods of the commissioners are hopelessly inadequate, and their report is often scant in detail to the verge of uselessness. It would be unjust to call the commissioners totally uncritical. They were aware of certain possible sources of error, and were probably no less competent than other medical workers of their time. Their fundamental oversight consisted in not appreciating that bizarre and debatable phenomena should be investigated with a far greater than average methodological rigour, and the findings set forth with all the details that a hostile critic might require. The report of 1831 may have pleased committed believers, but it merely hardened the views of sceptics, and it made no impact at all upon the relatively uncommitted. The toehold which magnetism seemed to have gained within "establishment" medicine during the 1820s slipped again.[61] The report was not published by the Academy, owing perhaps to Husson's intransigence in refusing debate, but privately in a small edition by Husson. Eventually, in 1833, it was edited and brought out by Foissac, the young doctor whose letter to the Academy had triggered the whole episode. His extensive annotations are in many ways of more value than the report itself.[62]

It is tempting to say that a real opportunity was lost. The work of the commission attracted considerable interest from fellow members of the medical profession; seventy medical men, apart from members of the commission, signed the protocols of the experiments with Villagrand and Cazot.[63] Had the commission set about its work with greater forethought and care, had it drawn upon the experience and ideas of Bertrand, had it not been effectively barred from the Paris hospitals, some progress might have been made. But if there was such an opportunity, neither the commissioners nor any other members of the Royal Academy of Medicine were capable of seizing it.

The Academy had two further brushes with animal magnetism.[64] In 1837, a young doctor, T.P.G. Hamard, somewhat reluctantly gave the Academy an account of a painless tooth extraction performed upon a somnambulic subject by J.E. Oudet, a dentist. The resulting discussion showed a marked resurgence of hostility towards animal magnetism, several members hinting that the subjects of such "painless" operations, not least Madame Plantin, must have been shamming. This last proposal, however, did not find favour with Cloquet, who was present at the meeting. Following these discussions, another physician, Dr. D.J. Berna, wrote to the Academy offering to demonstrate with his own somnambules phenomena which would establish magnetism conclusively. This offer was accepted. A committee of six persons was set up to undertake the investigation, but after four experimental sessions it concluded that Berna had been deceived.[65] There was so little argument over the acceptance of Berna's proposal that I cannot help suspecting certain factions within the Academy of hoping that a negative report from this second committee would help to eradicate the positive verdict of the first.

Later in the same year, 1837, a member of the Academy, C. Burdin the younger, reasoning that whereas most of the phenomena of magnetism are easy to feign, vision without eyes can be decisively tested, offered a prize of 3000 francs to the first person who could demonstrate his ability to read without the use of his eyes. Two magnetizers brought forward somnambules for testing, J. Pigeaire, a doctor from Montpellier, and Dr. A. Teste of Paris. Both investigations collapsed at the outset.[66] In its report, read on 8th September 1840, the committee recommended that the Academy should not bother itself further with animal magnetism, a proposal which was agreed to with applause.[67] M. Burdin then collaborated with F. Dubois, the perpetual secretary of the Academy, to produce a very substantial *Histoire académique du magnétisme animal* (1841), a work whose great usefulness (it reprints all the reports from 1784 to 1837) is marred by the astonishing amount of petty spite with which its authors remorselessly belabour the proponents of magnetism. So far as "establishment" medicine was concerned, magnetism was dead. But in other, if less respectable, quarters it had remained alive, and was flourishing as never before.

NOTES

1 Cheron (1817).
2 Rouillier (1817a; 1818).
3 Rouillier (1817b), p. 108.
4 Rouillier (1817b), pp. 140-141.
5 Rouillier (1817b), p. 128.
6 According to Bérillon (1884), pp. 5-6, Bichat held that proper mental balance depends upon an equal development of, and functional balance between, the two cerebral hemispheres. But when Bichat's own brain was examined at autopsy, one hemisphere was found to be atrophied and the other greatly overdeveloped.

 7 Baas (1971), Vol. 2, pp. 889–890.

 8 Lopez Piñero (1983), p. 44.

 9 Doerner (1981), p. 127.

10 A. M. J. de Chastenet de Puységur (1813), Vol. 1, pp. 81–83.

11 A. M. J. de Chastenet de Puységur (1813), Vol. 3, pp. 75–76; Deleuze (1819). p. 176.

12 Ackerknecht (1967), pp. 8–10.

13 Ackerknecht (1967), p. 172.

14 Adelon *et al.* (1812–22). Adelon lived to be 114 and occupied his chair of legal medicine until he was 97 (Ackerknecht, 1967, pp. 92–93).

15 Virey (1818).

16 Cuvier (1805), Vol. 2, pp. 117–118; Virey (1818), pp. 529–530.

17 Virey (1818), p. 555.

18 Deleuze (1819).

19 On Bertrand see Foissac (1833), pp. 256–260.

20 His younger son, J. L. F. Bertrand (1822–1903), was a child mathematical prodigy who became a distinguished mathematician.

21 The first lecture of his second course was printed in the *Archives du magnétisme animal*. See Bertrand (1820).

22 There is a brief biographical notice of Dupotet in Dupotet de Sennevoy (1927).

23 These experiments are conveniently described in Foissac (1833), pp. 272–279, and in Dingwall (1967), pp. 54–58. Dupotet's account is in Dupotet de Sennevoy (1821).

24 Foissac (1833), pp. 280–282.

25 Dupotet de Sennevoy (1930).

26 Foissac (1833), p. 521.

27 Georget (1821), Vol. 1, pp. 267–301. On Georget see Lopez Piñero (1983), pp. 49–53; Chardel (1828).

28 Georget (1821), Vol. 1, p. 287.

29 Georget (1821), Vol. 1, pp. 267n–270n.

30 Mialle (1826), Vol. 1, pp. 258–259.

31 Foissac (1833), pp. 289–290; Chardel (1828).

32 Mialle (1826), Vol. 1, p. 258. According to Burdin and Dubois (1841), pp. 264n-265n (a somewhat prejudiced source), Londe shortly became convinced that Georget had been deceived.

33 Ackerknecht (1967), p. 112.

34 Rostan (1825).

35 Dingwall (1967), *passim.*

36 Rostan (1825), pp. 449–450.

37 Hénin de Cuvillers (1820; 1822).

38 Dalloz (1823).

39 Dingwall (1967), pp. 58–73, thinks more highly of Dupau than I do. Dupau is heavily criticized by Bertrand (1826), pp. 533–539.

40 Dupau (1826), pp. 109–110.

41 Dupau (1826), p. 172.

42 Dupau (1826), p. 180.

43 Dupau (1826), p. 83.

44 Dupau (1826), pp. 206–216.

45 Bertrand (1826), p. XII; cf. Bertrand (1823), pp. 320–327, 418–420.

46 Bertrand (1826), pp. XIII-XV.

47 Bertrand (1826), pp. 256n-257n.

48 Deleuze (1826a).

49 Ackerknecht (1967), pp. 116–117.

50 The report of the committee is in Bertrand (1826), Foissac (1833) and Burdin and Dubois (1841), which also give various versions of the discussions. The discussions were also reported in such contemporary periodicals as the *Gazette de santé* and *L'Hermès*.

51 Dureau (1869) lists published theses by A. Fil(l)assier (1832), R. Saura (1834), A. Jozwik (1834), T. P. G. Hamard (1835), D. J. Berna (1835), and H. E. Le Brument (1835).

52 Foissac (1827).

53 Delatour (1826).

54 Rapport fait à l'Académie royale de medicine...(1829); Cloquet (1829). Many further details will be found in Pigault-Lebrun (1829). Bertrand wrote on the case in the *Globe* (Bertrand, 1829).

55 Procès de Mme. Fructus (1826); Procès en police correctionelle (1828); Cour royale de Paris (1828).

56 Cf. Virey (1818), p. 503; Deleuze (1825), pp. 300–302.

57 *Journal des débats*, 24th April 1826, 7th May 1826, 22nd May 1826. See *L'Hermès*, 1826, Vol. 1, pp. 153–156.

58 *L'Hermès*, 1826, Vol. 1, p.372.

59 This report is printed in Husson (1831), Foissac (1833) and Burdin and Dubois (1841). There is an English translation by J. C. Colquhoun (1833), which also forms Appendix No. 1 to his *Isis Revelata* (1836). A German translation is to be be found in J. F. Siemers (1835).

60 Further details of these persons will be found in Foissac (1833), pp. 405–453.

61 A symptom of this is the fact that in volume 18 of the second edition of Adelon *et al.*'s *Dictionnaire de médecine* (30 vols., 1832–1849) Rostan's article on animal magnetism is replaced by a sceptical one by L. F. Calmeil.

62 Husson (1831); Foissac (1833).

63 Foissac (1833), p. 425.

64 For further details of these see Burdin and Dubois (1841), pp. 451–641; Podmore (1909), pp. 110–121; Dingwall (1967), pp. 83–125.

65 Burdin and Dubois (1841), p. 511.

66 See Pigeaire (1839); Frapart (1839); Dingwall (1967), pp. 90–112.

67 Burdin and Dubois (1841), p. 630.

CHAPTER 8

Mystical magnetism: Germany

Most of the early magnetists regarded themselves as primarily healers. Such major authorities as Puységur, Deleuze and Wienholt always emphasized the dangers of exploiting somnambules for the mere satisfaction of curiosity. Some practitioners had, of course, always defied these injunctions, and after about 1810 their number began to increase, especially in Germany. More and more case reports were issued in which it is quite clear that the magnetizer's, or author's, prime interest is not the patient's symptoms, but her clairvoyant gifts, her visions, the information and teachings derived therefrom, and the possible philosophical or religious significance of it all.

MAGNETISM, ROMANTICISM AND NATURE-PHILOSOPHY

It is tempting to relate this development to the burgeoning and growth of the romantic movement in Germany.[1] And certainly romanticism in its varied forms was conducive to an intense interest in the supernatural and the psychologically bizarre. Many writers in the romantic tradition showed an interest in animal magnetism.[2] Even Goethe, who thought that animal magnetism "had too many mouse holes and mouse traps", studied Kluge's book on it.[3]

Linked to romanticism was the "nature-philosophy" developed in the late 1790s by F. W. J. Schelling (1775–1854).[4] Schelling held that reality is at root ideal or spiritual; if reality were not fundamentally akin to the knowing mind, knowledge would be impossible. The reality that is thus known cannot be thought of as a merely mechanical system. It is best looked upon as an organic whole. All parts of it are internally related, and a change in any one part affects the others, and the whole. This organic unity derives from an underlying conceptual unity – the development and unfolding of nature and of natural phenomena are to be explained teleologically, and in part *a priori*, as the development and self-unfolding of an eternal idea. Its unfolding is hierarchical and rule-governed, and the earlier stages are to be understood as occurring to lay the groundwork for the later ones. At each stage there is a conflict of opposing forces which is resolved by means of a new synthesis. The product of the new synthesis is itself the occasion of further conflict, and so on. Nature thus continually exhibits herself in the form of polar opposites, the opposition of light and gravity being resolved in matter, of magnetism and electricity in chemical interaction, of sensibility and irritability in creative instinct, and so on.

It must not be thought that the idea working itself out in the development of Nature is a conscious idea; Nature below man is unconscious "slumbering spirit". The human spirit develops as the organ for Nature's reflective consciousness of herself. And through this consciousness, in certain circumstances, may act the same eternal idea, or will, which acts unconsciously in producing Nature. It is in the symbolic creations of art and of poetry, involving as they do the freely and consciously produced manifestation of the real in the ideal, that the self-development of Nature reaches its highest point.

Some leading magnetists – for instance Reil and Wolfart – fell under the influence of nature-philosophy, and Schelling himself showed a reciprocal interest in animal magnetism.[5] It was, however, certain of Schelling's followers who explored the possible relationships between nature-philosophy and animal magnetism and brought them to more general notice. The earliest and most influential of these was G. H. von Schubert (1780–1860), whose frequently reprinted *Ansichten von der Nachtseite der Naturwissenschaft* first appeared in 1808. Schubert looks back to a time when people were at one with Nature and felt, indeed *lived*, the unity of all its parts with the whole. This primal intuition has largely been lost but is now being recaptured through the endeavours of modern science. Of the findings of modern science Schubert gives a readable resumé, diversified with much curious information and a great deal of colourful metaphor and analogy. His resumé is clearly influenced by Schelling's ideas of hierarchical development of Nature through the resolution, at each stage, of opposite and conflicting tendencies. Furthermore, features which can only be understood prospectively, in terms of what they will develop into at later stages, emerge at each stage. Thus the rudimentary air-breathing lungs of certain fishes and amphibians are of little practical use; but they foreshadow, and lay the groundwork for, the developed lungs which are essential to higher vertebrates. Schubert takes a somewhat similar view of the phenomena of animal magnetism[6] (his account of which is largely derived from Heineken). The clairvoyance, presentiments, self-diagnoses and prescriptions, etc., characteristic of somnambulic and related states, are intimations of a higher stage of being yet to come. And (more generally) the wealth of interesting facts which he has presented are symptoms and manifestations of a higher influence which harmonizes and unifies all things and guides them to good ends.

Schubert was from all accounts a busy, aimiable and much-travelled man, with a wide circle of friends.[7] He wrote, in addition to the *Ansichten*, a number of substantial volumes, several of which touch upon animal magnetism and related topics.[8] His influence on certain contemporary romantics, such as Novalis and E. T. A. Hoffmann,[9] and also on Kleist, was direct and considerable.

And among animal magnetists, the *Ansichten*, a popular and widely read work, without doubt helped to engender a frame of mind in which the medical aspects of a case of magnetic somnambulism took second place to its possible religious and philosophical significance.

Echoes of nature-philosophy can be detected in Kluge (1811), and are even more noticeable in a little book on animal magnetism published in 1812 by E. Bartels (1774–1838), professor of medicine and physiology at Breslau. Indeed, various writers attempted to marry magnetism and nature-philosophy in considerable detail. An early attempt at so doing is that by Friedrich Hufeland (1774–1839), brother of C. W. E. Hufeland, and professor of medicine at Berlin, whose *Ueber Sympathie* first appeared in 1811.[10] According to Friedrich Hufeland, the phenomena of life are to be understood in terms of the tensions within the organism of two opposed poles, those of animal and of vegetative activity (respectively underpinned by the central nervous system and the ganglion system). But these forms of activity only manifest in pure form in, on the one hand, the psychic activity of the brain, and, on the other hand, the vegetative activity of hair or nails or glandular secretions. Mostly the two sorts of activity coexist with a varying preponderance of the one or the other, and they may reciprocally influence each other, use each other's pathways, and so on. When vegetative (constructive, reparative) activity predominates, sleep ensues, when animal, the individual is intellectually and perceptually alert.

The "organic independence" of the individual may, however, be lost as a result of the procedures of animal magnetism. For when a patient, weakened by illness, comes into contact with a magnetizer, it is as if a weaker magnet has come into contact with a stronger, and a new system of polarities is set up, so that the polar centres of activity of the patient's organic life may lie within the organism of the magnetizer. Thus in a sick and sleeping patient we may find that:

> ... the state of the patient, which bordered on the vegetative life of plants, is led back to a higher level, consciousness begins to awake, and the proper sleep changes into a middle state between sleep and waking, which constitutes a transition to the latter, and is usually called somnambulism.[11]

In this peculiar state in which predominance of vegetative activities is combined with the artificial (external) arousal of animal ones (including thought and consciousness), certain instinctive capacities, not usually conscious, may find their way into consciousness, and the patient may become able to articulate them. Hence the capacity, shown by some somnambules, for clairvoyant diagnosis and prescription – animals were widely believed to have an instinctive ability to seek out and recognize remedies for their own ailments. This idea was very popular with other writers[12] – no doubt it harmonized with the romantic tendency to exalt feeling and emotion at the expense of reason. In a more obscure way, perhaps, it harmonized with the nature-philosophers' notion that Nature below man (and especially animate Nature) is "slumbering spirit", and awakens to consciousness – though gradually and fitfully and imperfectly – only through human beings. Ideas are, as it were, latent in Nature, and may in certain unusual circumstances dawn on our consciousness in an intuitive and non-logical manner.

The appeal of animal magnetism to philosophers and scientists steeped in doctrines about polarities was very strong, as was the appeal of such a philosophy to magnetizers. In the next decade or so, such doctrines were worked and reworked until they were at last smothered under the weight of a myriad woolly phrases. C. A. von Eschenmayer (1768–1852), a follower and in part a critic of Schelling, became, though a physician by training, professor of philosophy at Tübingen. His writings on animal magnetism[13] contain much that is derived from Schelling, but also elements from sources as diverse as Paracelsus, Stahl and Reil. Central to Eschenmayer's thinking are the notions of an "organic ether", concentrated especially in the brain and nervous system, and of polarities in the nervous system, the brain being usually positive, the ganglion system negative, and the sphere of indifference somewhere between. The peculiar phenomena of animal magnetism (like those of certain nervous ailments) are brought about, in ways too complex to detail here, by disturbances in these polar relationships.

Speculations about polarities ran riot in the writings of J. Weber (1753–1831),[14] well-known both as a chemist and as a Catholic theologian, and they took a quite baroque turn in the voluminous works of D. G. Kieser (1779–1862), professor of philosophy at Jena.[15] According to Kieser, the magnetic agent is the living activity or inner power of the earth, and to its "telluric" powers are opposed the "antitelluric" or solar powers of the sun. Everything on earth tends towards, or fluctuates between, a preponderance of one or other of these polar opposites. Thus moonlight is telluric, sunlight antitelluric; carbon telluric, nitrogen antitelluric; the south pole of a magnet telluric, the north pole antitelluric; positive electricity telluric, negative antitelluric; acid substances telluric, alkaline ones antitelluric; and so on. Among psychic powers, will and sensory activity are telluric, intellectual activites antitelluric, etc. Yet Kieser was an intelligent and widely-read man, and, as we shall see, a perceptive critic of the spiritualist beliefs to which certain of the mystical magnetists had increasingly become drawn.

MAGNETISM AND MYSTICISM

Into the obscure mix of influences that went to make up German romanticism, flowed streams of mysticism that long antedated German nature-philosophy.

Even before the French Revolution, the ideas of Boehme, and also of Swedenborg, had found an energetic German propagandist in F.C Oetinger (1702–1782). Saint-Martin discussed Boehme whilst at Strasbourg in 1788, and gradually became the centre of a mystical and theosophical school extending into Germany where many masonic lodges were steeped in mystical doctrines.[16] As a result of his friendship with the mystical theologian F. von Baader (1765–1841), Schelling himself fell under the spell of Boehme. Baader had written some small tracts on animal magnetism; but the principal vehicle through which mystical ideas of varying kinds were fed into the magnetic movement was the influential *Theorie der Geisterkunde* (1808) of J.H. Jung-Stilling (1740–1817).[17] Born in Switzerland to poor parents from Nassau, Jung-Stilling had a most remarkable career, being by turns schoolmaster, teacher, novelist, physician, eye-surgeon,[18] professor of "agriculture, technology, commerce and the veterinary art", professor of "economical, financial and statistical science", and private aulic councillor to Duke Charles Frederick of Baden. From youth onwards he had devoured the literature of pietism and mysticism, Paracelsus and Boehme being especial favourites, and later he became much interested in Swedenborg. The *Theorie der Geisterkunde* shows signs of all these influences,[19] especially Swedenborg's, and of Jung's general medical and scientific knowledge. And it picks up and incorporates a good many ideas and observations from the contemporary German magnetic scene – Jung-Stilling had known Gmelin, Wienholt and Böckmann.

From all these sources, Jung-Stilling builds up a comprehensive theory of God, man, space, time, soul and spirit. His central concept is that of the ether, which he identifies both with the magnetic fluid of the mesmerists, and with the luminiferous ether of physics, (the latter is the vehicle not just for light-waves, but for magnetism, electricity and galvanism). The fact that ether can freely penetrate all material bodies makes it a kind of half-way house between immaterial spirit and matter. It is thus ideally adapted to be the medium through which the rational element in man, his "spirit", can control his body. In fact, Jung thinks that man has a duplicate etheric "light-body" interpenetrating the material body and performing just this function. To the combination of an etheric body and a spirit he gives the name "soul". It is the soul, and not just the spirit, which survives death. Persons with unusual sensitivities may at times be able to see the souls of departed spirits. Furthermore, the procedures of mesmerists, which disturb the magnetic fluid, may have the effect of partially freeing the soul from the body. The somnambule then becomes able to exercise clairvoyant faculties appertaining to the soul, and her soul may even visit distant scenes (Jung-Stilling holds in a somewhat Kantian manner that space and time are constructs of our sensory apparatuses). Foreknowledge of future events may likewise, up to a point, be possible.

Many of the ideas subsequently characteristic of mystical magnetism, and of the Spiritualist movement which succeeded it, are already contained in the *Theorie der Geisterkunde*.

OSTENSIBLY PARANORMAL PHENOMENA

As I pointed out earlier, the case reports published after about 1810 were increasingly preoccupied with the patient's supposed paranormal powers. Not all German magnetizers took this exciting if mist-enshrouded pathway. Some even of those who showed an interest in the alleged paranormal phenomena remained primarily physicians. Wolfart, J. C. F. Bährens (b. 1794), W. Arndt and J. F. Siemers (b. 1792) will serve as examples.[20] Nor were all those who prepared and published case reports of this kind directly under the spell of nature-philosophy or the romantic liking for strange phenomena or the writings of Boehme and Swedenborg. But even for those who were not, such factors created an ambience out of which certain ways of handling and interpreting cases readily arose.

Case histories were published at an increasing rate. The five summarized by Dr. J. L. A. Vogel (1771–1840), another medical man influenced by nature-philosophy, in his *Die Wunder des Magnetismus* (1818) are merely representative. The publication of case histories accelerated notably with the appearance from 1817 of the *Archiv für den thierischen Magnetismus* (12 vols., 1817–1824[21]), edited by Eschenmayer, Kieser, and a third nature-philosopher, the physiologist and psychiatrist F. Nasse (1778–1851), professor at Halle, who was eventually replaced by yet another nature-philosopher, the botanist C. G. Nees von Esenbeck (1776–1858), professor at Bonn.

The assessment of these case reports is often extraordinarily difficult, quite apart from the question of the supposed paranormal phenomena. Central to almost all of them was a long-continued and close relationship, involving several hours of daily contact, between magnetizer (usually male) and patient (often younger and female). Sometimes this relationship clearly had strong emotional overtones, and the somnambule's clairvoyance would be especially directed on the magnetizer's doings and welfare. Being a successful somnambule must often have opened up opportunities for manipulating other people, for becoming a focus of attention and respect (sometimes from important persons), for indulging in a favoured diet, and indeed simply for amusing oneself at the expense of others, which would not normally have come the way of the otherwise quite ordinary girls concerned. And of course one can hardly deny that the magnetic manipulations may somehow have had a beneficial effect upon the patient's problems, or a peculiar effect upon her state of mind. With so many layers of complexity in the relationship between patient and magnetizer, and between patient and other persons, there is no hope whatever of any clear or straightforward understanding of what was going on.

Relations of this kind existed between, for instance, Baron F. C. de Strombeck (1771–1848) of Braunschweig and his patient Julie (Julie was a natural rather than a magnetic somnambule, but Strombeck's report[22] was very influential); between W. Arndt, a Prussian legal officer, and his patients Madame S. and Julie D.;[23] between the Dutch lawyer P. G. van Ghert and his subjects the Demoiselle B. and Madame Millet (van Ghert's works were principally influential through the German translations in the *Archiv*[24]); between Dr. Meier of Karlsruhe and Auguste Müller;[25] between Dr. C. C. von Klein (1772–1825) of Stuttgart and Lotte K.;[26] between Dr. J. C. S. Tritschler of Cannstadt and Matheus Schürr;[27] between Dr. F. A. Nick (1780–1832) of Stuttgart and Caroline Krämer (usually called "the Krämerin");[28] between Dr. J. C. Valentin, of Kassel, and Caroline Ramer;[29] between A. Köttgen, a silk manufacturer of Langenberg, and Maria Rübel;[30] between Dr. F. Dürr of Baden and Marie Wilhelmina Koch;[31] between Dr. C. Römer and his daughter;[32] between Dr. Spiritus of Solingen and Anna Maria Joest;[33] between Dr. F. Lehmann of Torgau and Frau K.;[34] between Dr. Bende Bendsen of Odense in Denmark and the widow Petersen;[35] and between Professor Kieser and Anton Arst.[36]

Since examples of supposedly paranormal phenomena figure prominently in many of these case histories, and since they were often taken for gospel by theorists, it may be of interest to enquire whether any of them merits serious consideration.[37] We have certainly a fair range of allegedly paranormal phenomena to choose from. They include instances of most of the sorts of thing reported in the palmy days of Puységur and Tardy de Montravel. For our purposes (which can extend only to a few of the better examples), we can take them under two headings: cases in which the ostensible paranormal cognition came to the somnambule unsought and unwanted (the event cognized being in these instances usually an impending death, danger or disaster), and cases in which paranormal cognition is deliberately sought, whether through systematic experimentation or spur-of-the-moment trials. We may call these two kinds of case "spontaneous" and "experimental" respectively. It may be said at once that very few of these German doctors and philosophers had the slightest idea how a spontaneous case should be recorded, verified and presented, and how, and with what precautions against sensory cues, an experiment on "extra-sensory perception" should be conducted.

In the "spontaneous" cases, the supposed paranormal cognition usually related to a close friend or relative of the somnambule, or else to her magnetizer. Thus Arndt's somnambule, Madame S., a young woman of nineteen, suddenly broke out during a magnetic sleep with exclamations of concern for her distant father, whom she saw covered in blood and apparently dying. In the waking state she showed no recollection of this distressing vision. Several weeks later she received a letter from her father from which she learned that at the very date and time of her vision, as noted down by Arndt, he had suffered an accident from which a serious loss of blood had resulted. Arndt does not, however, put either his contemporary note, or Madame S.'s father's letter, before his readers.[38]

In a curious case narrated by J. C. Valentin of Kassel, the ostensibly paranormal cognition related to a person unknown to both somnambule and magnetizer. On 14th December 1818, the somnambule Caroline Ramer said that in Breitenbach bei Hof, a place four hours from Kassel (the somnambule had never been there), an old Greek of seventy-two had fallen from the barn and had sustained three cuts in his head. He had wanted to get some hay for the goat. Valentin wrote to the minister in Hof, who confirmed the facts, except that the old man had sustained not three cuts but only one. Once again the confirmatory document is not printed.[39]

While some of these "spontaneous case" reports are certainly curious, we would need in each instance a good deal more background information, and much more detailed documentation, than we are given, before we could contemplate taking them very seriously.

The documentation of "experimental cases" is often equally unsatisfactory, the experimenters apparently believing it unnecessary to print details of precautions taken, because all reasonable readers would understand that what neeeded to be done would have been done. Often it is hard to be sure what was done, or indeed that anything was done. The rather numerous experiments on apparent "community of sensation" between magnetizer and somnambule[40] are an example in point. Many magnetic somnambules could readily be induced to taste what their magnetizer was tasting, feel the pins that were stuck into him, etc. – indeed on one occasion van Ghert had three somnambules simultaneously sneezing when he took a pinch of snuff, savouring peppermint when he tasted it himself, and so on[41] – but I have not noticed a single one in which it is clear that adequate precautions were taken against ordinary sensory cues.[42] The small number of cases of experimental "magnetization at a distance" are likewise highly unsatisfactory, because the possibility that the subject might have suspected an imminent attempt is never adequately considered.[43]

Of slightly more interest were certain experiments on "stomach-seeing". For instance, Dr. Tritschler of Cannstadt conducted with his subject Matheus Schürr some experiments on the "stomach-seeing" of playing cards in which considerable efforts were made to select cards at random and lay them on the lad's stomach in darkness, in such a way that neither experimenter nor subject knew in advance what the target card was. Matheus got the two successive target cards right, with odds against chance of 2704 to one.[44]

Köttgen's subject, Maria Rübel (who was subsequently caught in some rather crude cheating) specialized in reading words written on pieces of paper sealed within opaque envelopes or wrappers. In one such experiment a sealed package was given by Herr G. Siebel of Elberfeld to Herr Platzhof of the same town, who brought it to Köttgen's house, where it was placed on the belly of the sleeping

Maria. After a while she named a few letters, remarking that one of them could not be seen properly on account of a red stain. Eventually she gave the first word as "uch" and the later words syllable by syllable as "Mu-si-ka-len-Hand-lung". The package was returned unopened to its sender, who examined the seals and the wrappings. It contained the phrase "Buch- und Musikalienhandlung". The first letter had been partly obscured by sealing-wax.[45]

Perhaps the oddest cases of ostensible paranormal cognition from this period are certain examples of apparent clairvoyance of distant persons and scenes, particularly a number recorded by van Ghert[46] and by Valentin. Valentin's somnambule, Caroline Ramer, specialized in "homing in", so to speak, on the houses and families of strangers brought into her presence. Thus on 8th February 1819, Privy Councillor Gössel asked whether she could see his home (about 140 miles away). She described a big white house with steps in front of it. She saw Frau Gössel – a plump lady, of more than middle size, with a white, healthy, round face, light brown eyes and hair – enter a room on the right of the front door, containing very colourful carpets, and sofas and chairs covered in a dark material. Towards half-past twelve, Frau Gössel moved into the room opposite, which the somnambule described as very beautiful, with a desk to the left of the door and a very fine sofa and chairs with red covers. By the sofa, covered with a woollen cloth, was standing a something which (being unused to the houses of the wealthy) she neither recognized nor knew how to describe. A girl of four or five, with blue eyes, light blond hair and a lively round face, was jumping about in this room. The somnambule also described an older sister, with rather darker hair, and two elderly ladies, one sitting at the stove, and another going to and fro in the living room.

These descriptions were exactly correct. The object under the woollen cloth was a piano. Herr Gössel wrote immediately to his wife for information about the activities of the various persons in the house at the time, and received full confirmation. The "little girl" jumping about was in fact a little boy who happened that day to be wearing a smock frock.[47]

It is curious that Valentin, who has by the standards of the time some notion of evidence, does not seem to realize how much stronger his case would have been if he had published Privy Councillor Gössel's annotations of the somnambule's statements and Frau Privy Councillor Gössel's letter.

JUSTINUS KERNER AND THE SEERESS OF PREVORST

If highly gifted somnambules can have clairvoyant visions of distant scenes, read the contents of opaque packages, and so forth, it is natural to speculate as to the limits of this peculiar sensitivity. Can it reach other planets, other worlds, other realms of being? In this field, such speculations readily turn themselves not, indeed, into fact, but into the semblance of fact. A minor treatise could be written on accounts given by magnetic somnambules of voyages to the moon, the planets, even the sun, and of the inhabitants, buildings, scenery, vegetation, etc., of those localities.[48] The accounts are frequently picturesque, but the interested amateur astronomer will search them in vain for anticipations of modern discoveries. A major treatise would hardly suffice to exhaust somnambules' descriptions of their contacts with the spirits of deceased persons, or with angelic or spiritual beings, and of their visions of or visits to heaven or Hades or limbo (ideas of a "middle state" between heaven and hell being quite common amongst them).[49] Not infrequently the two kinds of voyage – astronomical and eschatological – are combined, and the moon and planets are represented as being the probationary homes of departed spirits. Often the voyages are conducted under the guidance of the somnambule's own particular guardian spirit, who also fulfils the more mundane rôle of medical and moral adviser.

These developments were probably under way by the time Kluge published his textbook (1811),[50] but for whatever reason they did not extensively find their way into print, and this despite the success and influence of Jung-Stilling's *Theorie der Geisterkunde*. Strombeck's report on his spontaneous somnambule Julie (1813) may have been the thin edge of the wedge. In her somnambulic state, Julie paid various visits to heaven, and there held edifying discourse with angels (she narrated these adventures in faultless iambics). From then onwards accounts of somnambulic contacts with spiritual beings appeared with increasing frequency.

In the 1820s and 1830s, these tendencies were powerfully reinforced by the activities, publications and personality of Justinus Kerner (1786–1862), who combined the rôles of writer, poet, magnetizer and psychical investigator with the mundane occupation of district medical officer for Weinsberg in Württemberg. With Kerner and his circle of friends – including Eschenmayer, Schelling, Schubert, Schleiermacher, Baader and Görres – the interplay of literary romanticism, nature philosophy, theology and animal magnetism reached new heights.[51] Kerner has often been dismissed as a credulous mystic, eager to believe in all varieties of supernatural visitation, but this estimate of him is far from just. If he believed somewhat naïvely in certain startling instances of supposedly paranormal phenomena, he had none the less a rather better notion of evidence than most of his contemporaries. Sometimes he even bothered to obtain and print the signed testimony of crucial witnesses, a precaution which, as we have seen, other writers of his time sadly neglected. A respect for documentation is also evident in his pioneering life of Mesmer, for which he obtained materials that might otherwise have been lost.[52]

At the age of twelve, Kerner had been magnetized by the celebrated Eberhardt Gmelin.[53] It was not, however, until 1824, with his *Geschichte zweyer Somnambülen*, that his interest in magnetism found substantial expression. This work, which narrates the case histories of two magnetic somnambules, is not much different from other accounts being published at the time, except perhaps in the fullness and clarity of the records. But in 1829 came something of an altogether different order, Kerner's extremely influential study of Frau Frieder-icke Hauffe (1801–1829), the famous "Seeress of Prevorst".[54]

This lady was the daughter of the district forester of Prevorst in Württemberg. Her upbringing was pious and spartan. From an early age she was subject to prophetic dreams, nocturnal visions and odd experiences. Following an arranged marriage, her life began to turn increasingly inwards. This process accelerated following a severe fever in February 1822. She became subject to violent spasms of the chest, and spent much time in a kind of half-waking state, during which she developed an increasing tendency to see visions. Other ailments ensued, and only animal magnetism seemed to afford her much relief. She began to fall into the somnambulic state spontaneously, and to prescribe for herself. In the somnambulic state she spoke High German, instead of her normal dialect, and also a strange language, which she wrote as well, and called her inner tongue. She wasted away until her life was feared for. At this point (November 1826) Kerner was called in, and had her transported to Weinsberg to be more immediately under his eye. He maintained a detailed journal of her case.

Kerner found that only magnetic procedures could alleviate her sufferings. From then on she passed much of her time in the magnetic state, which Kerner, or she, divided into four degrees, for the last two of which she was subsequently amnesic. The first was what appeared as her ordinary waking state, although her "inner life", i.e. her psychic sensitivity, was already apparent within it. The second was that of the "magnetic dream", in which she dreamed with great vividness, and sometimes prophetically. The third stage was the "half-waking" state. In this she would write and speak the peculiar "inner language" already mentioned. In the fourth, or sleep-waking, state, she might still speak her special language, but she also became clairvoyant, and could diagnose and prescribe both for herself and for others.

In many respects, the Seeress resembled other clairvoyant somnambules of the time. She had a guardian spirit (her lately deceased grandmother); she could see the internal workings of the body, diagnose and prescribe for herself and others, etc.; she developed the faculty of "stomach-seeing"; she was painfully sensitive to the near approach of metals; and so on. But she also exhibited some much less common phenomena. In all of her magnetic stages, including the first, which approximated to an ordinary waking state, she was liable to have visions and to see apparitions. The visions (which were commonly fulfilled) seem often to have resembled instances of traditional Scottish second-sight, and had generally a symbolical character. The apparitions gradually grew more numerous. When she came to Weinsberg, the spooks afflicting persons or places in that locality seem to have been drawn to her like moths to a flame. She did not, it must be emphasized, relish their company, and it does not seem likely that she was doing anything other than honestly report what she thought she saw and heard. Indeed, in several cases the apparitions concerned favoured her with information (subsequently verified) about themselves which it hardly seems likely that she – a stranger in Weinsberg, and an invalid – could have acquired unaided by ordinary means.[55]

In her "sleep-waking" state, Frau Hauffe did more than diagnose, prescribe, exercise powers of clairvoyance, etc. She delivered a whole series of complex psychological, cosmological and theological teachings, which seem greatly to have impressed certain of Kerner's friends who came to visit her, and most notably Eschenmayer. She expounded a complex system of partly symbolical sun-circles and life-circles, with which was connected a good deal of number-magic, and in terms of which she interpreted her life and destiny. And she said a good deal about the relation between soul, spirit, body and nerve-spirit, a topic to which I shall return shortly.

The Seeress was not, of course, the first German somnambule to deliver cosmological and theological teachings;[56] but Kerner's report of them, with commentaries by Eschenmayer, had an unprecedented impact. Kerner and Eschenmayer were already well-known; the teachings themselves were unusually detailed;[57] and her paranormal powers were unusually well-attested.

How Kerner regarded the Seeress's revelations and phenomena is not altogether clear. He says that he accepts none of the theories hitherto advanced. He gives mere facts, and leaves the explanation of them to others[58], especially of course to Eschenmayer. But Kerner certainly thought these matters fascinating, and of the highest possible importance. He published a number of other peculiar case histories, and founded and edited two periodicals – *Blätter aus Prevorst* (12 vols., 1831–1839) and *Magikon* (5 vols., 1840–1853) – in which he inserted not just, or even principally, materials to do with magnetic somnambulism, but all sorts of cases of apparitions, possessions, poltergeists, and other allegedly paranormal happenings.

HEINRICH WERNER AND R. O.

Following publication of Kerner's book on the Seeress in 1829, the didactic, religious and cosmological elements in the published case histories of magnetic somnambules became even more pronounced. Most of the somnambules had, like Frau Hauffe, a Protestant background.[59] However, the somnambule of the Hungarian F. von Szapáry (1804–1875) was Catholic, and Wiener's somnambule, Selma, was Jewish.[60] Some of the investigators of these cases received the sayings and teachings of their favourite somnambules with astonishing reverence and credulity. For instance, J. K. Bähr (1801–1869) and R.

Kohlschütter of Dresden chronicle the prayers, teachings and heavenly excursions of their somnambule, Auguste K., a very clever young lady, in the minutest and most tedious detail.[61] Yet it never occurs to the infatuated pair that they should descend to similar detail over examples of her supposed clairvoyance, the authentication of which might have been taken to support her pretensions as a seeress.

Of all the works produced during this phase of German mystical magnetism, the most interesting, because most characteristic and most comprehensive, is H. Werner's *Die Schutzgeister* (1839).[62] Werner (a great admirer of Eschenmayer) was a doctor of philosophy and pastor of Bickelsberg in Württemberg. He devotes 156 of the 639 pages of this weighty tome to a journal of the case history of R.O., a girl of eighteen, who seems to have been related to him. The remainder is taken up with his attempts to produce, from the literature of animal magnetism, and the teachings of the somnambules, a coherent account of the psychological, physiological, philosophical, theological and cosmological implications of magnetic somnambulism.

R.O. was an intelligent, well-read, and religious girl – she claimed, however, not to have read Kerner's book on the Seeress of Prevorst[63] – who suffered, in the wake of a depression caused by an emotional problem, from convulsions, cramps of the chest, suffocations and troublesome periods. Conventional medicines proving of no avail, she decided to give them up, and shortly began to pass spontaneously into a state which was taken to be that of magnetic somnambulism. She was never subjected to formal procedures of magnetization, but her vital forces were strengthened, and her somnambulic powers enhanced, by having Werner hold her hand, or lay his hand upon her stomach, and she entered into a corresponding "magnetic rapport" with him. She soon became aware of the presence of her guardian spirit, the ineffably virtuous Albert, a former clergyman whom she at first confused with Werner. Albert conducted her on trips to the sun, Venus and the moon, of the inhabitants and flora of which she gave picturesque descriptions.

Ostensibly paranormal phenomena are frequently described in Werner's case report. R.O. in fact exhibited the same sorts of supposedly paranormal faculties as did many other somnambules of that period. In the somnambulic state, her rapport with Werner was particularly marked. Her community of sensation with him could be almost embarrassing, and he could draw her limbs and body after his own in a quasi-magnetic manner (a phenomenon often reported by magnetizers of that period).[64] Several of her clairvoyant visions related to him.

Some instances of her apparent spontaneous clairvoyance were indeed quite odd. On the afternoon of 19th May 1834, while in the magnetic state, she suddenly burst out with anguished exclamations occasioned by an imminent danger to her little sister, Emily. Emily, she said, had been in the upper story of the house, at a window with no balustrade, and had made as if to grasp the rope of a windlass that was hauling up wood outside. Her father (prompted by Albert) had come into the room and saved her in the nick of time. A letter from home, which Werner quotes, shortly confirmed the details of this vision. Her father had felt and surrendered to a powerful urge to return home from his place of work.[65]

In later somnambulic states, R. O. was much troubled by the near approach of a dark and deformed being, from whom she had to be protected by the moral force of Albert. This was the spirit of an evil monk, who had (so Albert informed her) murdered five of his own children, and committed many other shocking crimes; he was moreover a Jesuit. One day, while R. O. was in the somnambulic state with Werner by her side, this degraded creature three times made clearly audible clattering noises on a nearby table.[66] On another occasion he threw down two flower pots in a locked and empty room in a house where Werner was lodging – R. O. apparently became clairvoyantly aware of this incident at the time of its occurrence.[67]

R. O. favoured Werner with much information upon the nature of animal magnetism, the relationship of spirit, soul and body, etc. Some of this came from her own intuitive understanding of these matters, some she transmitted from the invaluable Albert. From these teachings, and those of Frau Hauffe and others, from the speculations of Eschenmayer, and the physiology of Reil, Werner builds up a detailed theory of animal magnetism and its bearings on the nature of the human psyche. He does not see any incongruity in putting together data from sources so diverse. In fact, he holds (for reasons to which we will come) that somnambules who have reached the highest grade are incapable of lying and deception, and that they have deep and direct insights into the life of soul and spirit which it is our duty to utilize.[68]

Werner's views about soul and spirit seem to derive largely from the statements of such somnambules. The spirit is a small emanation from God. It is that within us which causes us to hunger for the divine, and enables us to recognize the good, the beautiful and the true. The soul is not the duplicate etheric body that Jung-Stilling had depicted. Like the spirit, it possesses reason, feeling, fantasy, will, but these are only attributes, not (as with the spirit) its very essence. The soul is in a relation of reciprocal interaction both with the spirit above it and the body below. From the bodily organs it receives sense-impressions which it transmits to the spirit; through it, in turn, the spirit can influence the body.[69]

If all this is a little difficult to follow, it becomes more so when we introduce the notion of the nerve-spirit or *Nervengeist*, with which the Seeress of Prevorst had made great play. In Werner's synthesis, the *Nervengeist*, which is the active principle of life, mediates between body and soul. It may serve, in the nerves, as a material conductor between sense-organ and neural centre, or it may mediate the direct influence of one bodily organ upon another. *Nervengeist*, being "psychophysical" and plastic to the influence of thought, in some way adheres to the soul, which forms it into a subtle body or vehicle around the spirit, a vehicle exactly reflecting the organs of the physical body. This "soul-body" persists after death, and may become visible through binding with a peculiar stuff from the atmosphere – Werner is following the Seeress of Prevorst here.[70]

The *Nervengeist* has its source in the brain and nervous system, and in the normal state the cerebral system is the focus for perception. However, the cerebral system is linked by the *Nervengeist* to the ganglion system; hence the soul, whose seat of operations is generally the brain, may get dark glimpses of the working of human instincts. On its physical side, *Nervengeist* probably possesses polarity like any other imponderable. The brain is normally positive, the peripheral nerves and ganglion system negative. In abnormal states and as the result of magnetic manipulations, this polarity may be reversed. In that case the *Nervengeist* is also loosened or freed from the bodily organs and may extend beyond the boundaries of the body, with the result that the ordinary sense-organs and muscular apparatus are shut down. The "nerve-atmosphere" can now detect objects in the body's vicinity and conduct information back to the "communal sense" in the stomach. At the same time the soul, which is ordinarily linked to the body by the *Nervengeist*, becomes free or freer and can range widely in both space and time. In the highest stage of magnetic somnambulism, the spirit becomes totally free of soul and body and when thus unencumbered may participate in God's immediate knowledge of matters that cannot be reached by reason, including religious matters. It cannot deceive nor yet be deceived.[71] No wonder that these German somnambules spoke and prayed with such heightened religious fervour, and no wonder that their magnetizers hung upon and recorded their every word!

Werner's book, if it does not exactly mark the culmination of German mystical magnetism, may at any rate represent that culmination. From 1840 onwards, somnambules of the kind in question became rarer, or at least less publicized,[72] and after 1850 mystical magnetism was largely absorbed by Spiritualism and related movements.

CRITICS OF MYSTICAL MAGNETISM

It is hard to gauge the extent of hostility to animal magnetism in Germany during this period. There is no doubt that among books and articles published in this area, ones by believers, and not least by believers of mystical inclinations, considerably predominated. But in any fringe field true believers tend to outpublish dedicated disbelievers. It is therefore perhaps more significant that even those writers who were on balance hostile to the claims of the magnetists none the less tended to accept that there was *something* in *some* of these claims. This is true, for instance, of such early critics as Stieglitz (1814), C. W. Hufeland (1816) and Pfaff (1817).[73] All three adopt a similar basic position. The procedures of animal magnetism may have a beneficial effect in certain cases, and this effect is probably due to some sort of physical emanation from the organism of the magnetizer.[74] Beyond this none of them will go. Stieglitz and Pfaff, both admirers of Kant, were resolutely opposed to nature-philosophy, and to acceptance of the more exotic phenomena of magnetic somnambulism which nature-philosophers found so fascinating. Somnambulism proper is a patho-logical nervous disorder, and so-called magnetic somnambulism is simply the arousal by artificial means of this pathological disorder, in persons who are antecedently vulnerable to it. Pfaff's critique of the alleged paranormal phenomena of somnambulism is particularly incisive. He divides the sources of error with regard to such phenomena into two broad categories – those which originate with the magnetizer and those which originate with the subject.

Errors originating with the magnetizer begin with the fact that many magnetizers are medical men. Work in this field requires its own peculiar talents, and a feeling for psychological phenomena that ordinary medical education does not inculcate. Furthermore, even men whose judgment is normally sober may have as great an appetite for wonders as the man in the street, and be as liable to self-deception. If the extraordinary phenomena do not come, they may try in every way to arouse them, and so lead their magnetic subjects into cheating. And even if the phenomena do come, the witnesses are seldom in a position to observe and experiment properly. An observer of magnetic phenomena must also have the will to state the truth, and his credit in the learned world is no guarantee of that. He may not be able to undergo a higher test of love of truth – he may not be able to bring himself to retract publicly statements to which he has publicly committed himself.

Errors originating with the subject have in Pfaff's view generally to do with the fact that the majority of good magnetic somnambules belong to the sex rightly called weaker. Such somnambules can only too easily drift into the habit of speaking half-truths, and from this it is not a big step to outright lying. Magnetizers who prompt their somnambules' fantasies with leading questions share in their guilt; somnambules are faithful mirrors of their magnetizers. It is above all female vanity which leads somnambules astray. If a somnambule is surrounded by admirers and enquirers, she is liable to cultivate and exaggerate her own "gifts" for their benefit, and may well stoop to deliberate deceit. Women can be very skilled at simulating magnetic and somnambulic phenom-ena, and have a particularly nice line in faking the most frightful convulsions.

Stieglitz develops similar views at some length.[75] Women, he says, are very quick at picking things up, and furthermore in nervous illnesses the senses and the memory may become abnormally heightened. There are many ways in which somnambules can acquire the information they pretend has supernatural origins. They may or may not themselves know where it came from. Magnetizers and assistants are not sufficiently cautious in what they say in front of somnambules, and in any case the regularity of many peoples' lives makes it easy to guess what they have been doing.

The possible motives for such deceptions, says Pfaff, can be many and various: amorous intrigues, greed for money, pure pleasure in deceit. Women like to outwit and tease an audience of doctors, theologians or philosophers, and to play the oracle before them. (Despite holding these views, Pfaff was twice married.)

Having set out these sources of error, Pfaff proceeds to seek out illustrative examples from published case reports by thirteen leading magnetizers. He is particularly severe upon Nick and his somnambule, the Krämerin. He dwells somewhat excessively upon the lady's past peccadilloes, and is uncharitably sceptical concerning her later repentance and her newly-acquired religious propensities. We cannot, he maintains, possibly rely upon such a person. She cornered the naïve Nick into agreeing to magnetize her. She knew all about magnetism before she came to Nick, and either consciously intended trickery, or else was led on by an exalted imagination. Nick might reply that cheating by the subject could not explain the impressive clairvoyance and precognition which she sometimes exhibited. But the evidence for this rests largely upon Nick's own testimony, and he is a blind enthusiast upon whose impartiality we cannot rely. For instance, when investigating the Krämerin's claim to possess the faculty of "stomach-seeing", he relied simply upon blindfolding her. He should know from playing blind man's buff in his youth how easily blindfolds can be circumvented. Her apparent rapport with him, as manifested in her knowledge of what he was, had been, or would be doing when away from her house, could be due to ordinary sources of information. Before we can lend credence to stories so extraordinary, we must be sure that Nick is incapable of exaggeration, and we must have the harmonious concurrence of several unimpeachable witnesses. As these conditions are never met, we must set the stories aside. As for the prophecies which were apparently fulfilled, of which the Krämerin provided several examples, they did not fulfil the conditions that a prophecy (to be of any value) must be committed to paper at once, sealed, and safely deposited, and must be opened before impartial witnesses.

Pfaff's treatment of other case reports is in similar vein, and there is no doubt that he makes points of substance. One might, indeed, well suspect that *no* case of alleged clairvoyance or precognition, however well-documented, would ever have been acceptable to him. But he puts his finger on the fundamental shortcoming of most of these early investigations of ostensibly paranormal phenomena when he quotes[76] Daubenton's dictum that if you are attracted by some hypothesis, you should direct all your endeavours towards proving it false, and remarks that in animal magnetism exactly the opposite has been the case.

Later critics of the supposed "higher phenomena" of magnetism were on the whole less full-blooded than Stieglitz and Pfaff; they tended to accept some at least of these phenomena, and to concentrate their fire upon the claims of certain somnambules to have guardian spirits, see angels, transmit religious teachings, etc., and upon certain cases of alleged possession which had become loosely but strongly tied up with such things through the activities of Kerner, Eschenmayer and their friends.[77] The most influential of these critics were D. G. Kieser, whom we have already met, J. U. Wirth (1810–1859), a theologian steeped in Hegel, and Friedrich Fischer, a professor at the University of Basel. First in the field was Kieser, to whom the others were manifestly much indebted.

According to Kieser,[78] there are three sides to the human psyche, Will, Knowledge and Feeling (including fantasy and imagination). In the waking state, Knowledge (cognition) predominates, and Feeling and Will, though active, are subordinated to it. In sleep and the related states of dream and som-nambulism, Feeling rules, and Will and Knowledge are subservient. In dream and somnambulism Knowledge and Will can only find expression through the characteristic activity of Feeling, namely fantasy and imagination. The psyche is not aware of the cognitive and volitional activities, but only of the fantasy pictures into which Feeling transmutes them.

> To the sleeping and dreaming person, and likewise to the clairvoyant somnambule, each feeling, each idea, each impulse, appears in plastic form, as a real shape. He cannot comprehend their origin and essence, for such insight, as a manifestation of cognitive activity, is not available to him ...[79]

These fantasy pictures thus represent or symbolize ideas and wishes of which the dreamer or somnambule is not directly aware. Even the influence of the magnetizer's actions and thoughts on the subject, or the dynamic influence of metals, the baquet, etc., will appear to him as a material picture whose illusoriness he does not recognize. Somnambules' visions of their own interiors are symbolic images of their implicit understanding that their bodies are disturbed; and as for the apparitions of holy persons, demons, good and bad spirits, deceased relations, etc., these are just "symbolic representations ... in which the somnambule has embodied and personified his inner presentiments and apprehensions".[80]

Kieser does not spell out for us in detail the reasoning by which he arrived at these conclusions. They seem to follow quite naturally from a consideration of the rich, but seemingly involuntary, play of fantasy in dreaming and allied states, together with a belief in the polarity, and possible reversal of polarity, of the nervous system. The notion that two, largely insulated, kinds of activity may go on simultaneously in the same mind, one of them being reflected in the other, if at all, only in an indirect and symbolic manner, contains the germs of many subsequent developments in the psychology of the unconscious. Fischer too now and again introduces such ideas into his *Der Somnambulismus* (1839) when criticizing the followers of Kerner, and Wirth works them out a little more systematically in his *Theorie des Somnambulismus* (1836).

A widely held over-simplification of the history of hypnotism has been that the concept of "animal magnetism", a quasi-electrical fluid responsible for certain interesting phenomena, was displaced in the second half of the nineteenth century by the concept of "hypnosis", which was in turn closely linked to newly emerging notions about the unconscious mind. But in fact ideas about the unconscious mind, or dissociated or relatively independent streams of consciousness, initially emerged in the context of animal magnetism – Kieser was not the first to adumbrate them[81] – and were well advanced long before hypnotists took them over. Between 1850 and 1890, a series of German philosopher-psychologists, who represent perhaps the final eddies of romantic nature-philosophy, developed complex doctrines about the unconscious mind, and combined these doctrines with belief in the mesmeric fluid or some descendant or variant thereof. These writers included, for instance, H. B. Schindler, whose *Das magische Geistesleben* (1857) expounds in detail a position very like Kieser's; C. G. Carus, whose *Ueber Lebensmagnetismus* also appeared in 1857; and E. von Hartmann (1842–1906), whose *Philosophie des Unbewussten* (1869) is usually regarded as ancestral to much later speculation about the unconscious. Hartmann also argued for the existence of a "psychic force", externalized from the human body and responsible for some of the odder phenomena of the séance room.[82] For all of these writers the postulated link between animal magnetism or its offshoots and the unconscious is at root similar. They could probably be termed "vitalists" – life, they hold, is the product of a hidden and purposive force, which directs the development of the individual and the evolution of living forms, which has or can have healing properties, and which may be involved in the curative effects of animal magnetic procedures. This force is something cosmic rather than merely earth-bound and parochial, yet it lies within and operates through all of us. Some part of us, though not our ordinary conscious part, is aware of or somehow in touch with it, and in certain abnormal states – natural or mesmeric somnambulism, dreams, drug states – this awareness may find expression in symbolic speech or visions, sometimes conveying information not normally accessible to us. These writers usually fortify their theories with a great variety of facts and anecdotes; but there is often a certain fuzziness as to how precisely the facts are meant to vindicate the theories.

To return for a moment to Fischer and Wirth. Their refusal to take literally the stories of apparitions, guardian spirits, possession, and so forth, with which the literature of animal magnetism was now full to repletion, greatly irritated certain of the more spiritualistically-minded magnetizers. Werner has his say about Fischer and Wirth in the book already cited, whilst U. Gerber's *Das Nachtgebiet der Natur* (1844) – Gerber was much influenced by Kerner – is largely devoted to an attack on them. One remark of Gerber's may be noted. Why, he asks,[83] should Wirth reject the view that in cases of possession the spirit of a deceased person operates the muscles, nerves and speech-apparatus of a living man, when he believes that Dr. Nick could make the Krämerin obey him, sit and stand against the law of gravity, etc., by magnetically activating her nerves? With this argument of Gerber's, the stage has been fully set for the emergence from about 1850 of the modern Spiritualist movement, which took over so many of the ideas and practices of animal magnetism. The "trance medium" – the magnetic somnambule "controlled" not by a living magnetizer but by a discarnate spirit – is, within the framework of thought favoured by the disciples of Kerner, a perfectly reasonable conception.

NOTES

1   On mesmerism and romanticism see Artelt (1951) and Engelhardt (1985). On romanticism and science in general see Cunningham and Jardine (eds.)(1990).
2   Tymms (1955), p. 181.
3   Artelt (1951), p. 9.
4   The principal texts of this phase of Schelling's philosophy are his *Ideen zu eine Philosophie der Natur* (1797), *Von der Weltseele* (1798), *Erster Entwurf eines Systems der Naturphilosophie* (1799) and *System des transcendentalen Idealismus* (1800). The latter has been translated into English by P. Heath (Schelling, 1978).
5   R. F. Brown (1977), p. 231.
6   On Schubert see Straumann (1928), pp. 40–46.
7   Cf. Erdmann (1924), pp. 595–596.
8   Schubert (1833; 1840b).
9   Cf. Müller-Funk (1985).
10 I have used the second edition (1822).
11 F. Hufeland (1822), p. 116.
12 Cf. above pp. 69, 125, and Werner (1839), pp. 230–231. Stieglitz, an opponent of nature-philosophy, also expressed similar views. See Stieglitz (1814), pp. 303–310.
13 Eschenmayer (1816; 1817).
14 Weber (1816; 1817).
15 Kieser's *System des Tellurismus* (1821) is a mine of information about contemporary animal magnetism. I have used the new edition of 1826.
16 Benz (1968), pp. 80–81.
17 Jung-Stilling (1808). There is an English translation by S. Jackson (Jung-Stilling, 1834).
18 According to Wesermann (1822), p. 129, Jung-Stilling gave some 3000 blind people their sight during forty years of practice.
19 A general account of the doctrines of Paracelsus, Boehme, Swedenborg, Jung-Stilling, and many others, will be found in Kiesewetter (1909).
20 Arndt (1816); Bährens (1816); Siemers (1835).
21 It was succeeded by the *Neues Archiv für den thierischen Magnetismus* (2 vols., 1825–6).
22 I have used the French translation of 1814 (Strombeck, 1814).
23 Arndt (1816).
24 Van Ghert (1817; 1818).
25 Meier (1818).
26 Klein (1819).
27 Tritschler (1817).

28 Nick (1817; 1818).
29 Valentin (1820).
30 Köttgen (1819).
31 Dürr (1822).
32 Römer (1821).
33 Spiritus (1819).
34 Lehmann (1819).
35 Bendsen (1821; 1822–3).
36 Kieser (1818b).
37 A contemporary survey of this kind of material is given by Wesermann (1822). Many useful references are given by Fischer (1839) and Werner (1839), pp. 380–398. For modern surveys see Moser (1967), and Podmore (1909), chapter 11.
38 Arndt (1816), pp. 76–79.
39 Valentin (1820), p. 65.
40 The were also, of course, some apparent "spontaneous" examples of this phenomenon. Cf. Fischer (1839), Vol. 2, pp.187–189.
41 Ghert (1818), pp. 4off.
42 See, for instance, Spiritus (1819) – he and his somnambule made rather a speciality of such performances.
43 See e.g. the case described by Arndt (1816), pp.72–73 and Wesermann (1822), p. 231.
44 Tritschler (1817), pp. 86–87.
45 Cf. Moser (1967), pp. 140–141.
46 A number of Ghert's cases can conveniently be studied in Zorab (1967).
47 Valentin (1820), pp. 98–100.
48 Brief reviews of these materials are given by Ennemoser (1819), pp. 141–143, Fischer (1839), Vol. 3, 211–218, and Wesermann (1822), pp. 81–84.
49 On these topics in general see Werner (1839), Part 3.
50 See especially Kluge (1811), pp. 364–372. The extent to which they had progressed in ten years may be judged from Passavant (1821), pp. 194–205.
51 See Straumann (1928) and Grüsser (1987). On Kerner as a parapsychologist see Berger-Fix (1986) and Bauer (1989).
52 Kerner (1856). Extracts from this will be found translated in Watts (1883).
53 On Kerner's youth see Kerner (1819).
54 Kerner (1829). I have used the reprint in Kerner (1904). There is an abridged English translation by Catherine Crowe (Kerner, 1847). Further information about the case will be found in Eschenmayer (1830). A useful account of the Seeress is given by Ellenberger (1970), pp. 79–81.
55 See e.g. Kerner (1904), Vol. 2, pp. 392–443; Kerner (1847), pp. 234–261; Eschenmayer (1830), pp. 58–62.
56 Extrait du journal d'une cure magnétique (1787). Wienholt's respected Bremen ally, J. Heineken, was questioning a somnambule (who was aware of spirits) before the turn of the century on such topics as the relationship of soul and body, the functions of different parts of the brain, etc. His account was published after his death by his son. See P. Heineken (1818).
57 Kerner (1904), Vol. 2, pp. 175, 221; Kerner (1847), pp. 110, 123.
58 Kerner (1904), Vol. 2, p. 52; Kerner (1847), p. 55.
59 One of them gives a most edifying account of Protestant notables grouped around God's throne, a region from which Catholics are noticeably absent. See Reisen in den Mond (1834), p. 71; Fischer (1839), Vol. 3, p. 216.

60 Wiener (1838); Szapáry (1840).
61 Bähr and Kohlschütter (1843).
62 There is an English translation of part of this work (Werner, 1847). The translator, A. E. Ford, is much concerned to establish the compatibility of the somnambules' teachings with those of Swedenborg. Werner also wrote another work in which magnetic somnambulism is discussed (Werner, 1841).
63 Werner (1839), p. 173.
64 Nick was able to draw the Krämerin so far into the air with his thumbs that her toes seemed scarcely to touch the floor! Nick (1817), pp. 101, 111–112.
65 Werner (1839), pp. 88–91; Werner (1847), pp. 68, 70–71.
66 The spooks who so regularly visited the Seeress of Prevorst not infrequently made their presence known by raps and other physical disturbances, and she herself claimed to be able to produce such phenomena during trips out of the body, on one occasion thus disturbing Kerner and his wife. See Kerner (1904), Vol. 2, pp. 308–316; Kerner (1847), p. 194. Such ostensibly paranormal physical happenings (of a generally "poltergeist" character) are only rather rarely reported in connection with these magnetic somnambules. For a similar episode with another somnambule, see Wiener (1838), pp. 135–141.
67 Werner (1839), pp. 190–199; Werner (1847), p. 194 etc.
68 Werner (1839), pp. 211, 213.
69 Werner (1839), pp. 20–28.
70 Werner (1839), pp. 28–32, 216–218, 442–443.
71 Werner (1839), pp. 20–28, 212–214, 224–230.
72 But by no means extinct. See for instance *Commissarische Berichte* (1840); Brendel (1840); F.,R. (1840); Rumpelt (1840); Neuberth (1841); Schmidt (1846); Wideck (1848); Görwitz (1851); Uhlmann (1853).
73 C. H. Pfaff (1773–1852), professor of medicine and chemistry at Kiel.
74 Stieglitz thought of this as a transpiration through the skin, but called it an "excrement", which greatly enraged magnetists such as Wolfart (1816).
75 Stieglitz (1814), pp. 252–285.
76 Pfaff (1817), p. 34.
77 Cf.Kerner (1833; 1834), Eschenmayer (1837), Gehrts (1966).
78 Kieser (1817b; 1820; 1826).
79 Kieser (1820), p. 108.
80 Kieser (1817), p. 109.
81 Cf. above, pp. 125–126, and Stieglitz (1817), pp. 326–343.
82 Hartmann (1891).
83 Gerber (1844), p. 411.

CHAPTER 9

Mystical magnetism: France

During the 1830s, as we have seen, animal magnetism was decisively rejected by the French medical establishment. Its rejection, however, did not lead to any general loss of interest in it, or to any decline in its popularity as a method of treatment. Indeed, if we may judge by the number of books and pamphlets on the subject, interest continued to grow until well after the mid-century. Dureau lists sixty-two French works on animal magnetism published between 1831 and 1840, 129 between 1841 and 1850, and 132 between 1851 and 1860. Many of the authors were in fact medical men; but they were country practitioners rather than eminent Parisian doctors, and often they practised homoeopathy as well as magnetism.[1]

THE TRAVELLING DEMONSTRATORS

It is hard to say why France should at this period have been relatively receptive to the doctrines and practices of animal magnetism. An easier question is that of the channels through which interest spread. It was the age of the public magnetic demonstrator. The most celebrated of the travelling demonstrators were Dupotet, C. Lafontaine, and J. J. A. Ricard of Bordeaux, but there were many others. Dupotet was first in the field. He was a prolific author[2] and lecturer, greatly given to bombast and hyperbole, but none the less a sincere believer in the virtues of animal magnetism and in his mission as its prophet. After establishing his reputation in Paris as a magnetizer and teacher of magnetism, he toured France extensively from 1835 to 1840, staying for prolonged periods in Bordeaux, Montpellier, Béziers and Metz.[3] He gave lectures followed by demonstrations, and after the demonstrations he would open a subscription[4] for classes. He never passed by an opportunity of obtaining publicity, whether by performing cures, interesting the medical profession, or engaging in controversy with local authorities. But his demonstrations were principally devoted to curing the sick, and many of his former pupils continued in later years to address him respectfully as *professeur* or *maître*.

Lafontaine (1803–1892) was introduced to magnetism in Brussels about 1835, by J. B. A. M. Jobard, a Belgian businessman.[5] From about 1840, he travelled widely in France and abroad, giving public demonstrations and courses. His autobiography[6] is very informative about animal magnetism in the mid-nineteenth-century, and is highly entertaining in a somewhat picaresque way. Connoisseurs of the gruesome should not miss his account of the triple execution which he witnessed at Vendôme as a lad.

Lafontaine came from a theatrical family, and his demonstrations were more theatrical than Dupotet's, though he also took "private patients", and would magnetize persons about to undergo minor operations. The centre-piece of his demonstrations seems usually to have involved producing, in a susceptible member of the audience, or in a "good" subject whom he had brought with him, varied phenomena of insensibility – insensibility to a pistol shot beside the ear; to sulphur and concentrated ammonia held near the nose; to pricking with long needles; to powerful shocks from a Leyden jar; to the light of a candle brought close to the eyes. Sometimes he would attempt to demonstrate the "higher" phenomena of magnetism: clairvoyance, community of sensation, etc. His self-confidence must have been considerable: he outfaced rowdy audiences at home; impressed the sceptical British to such an extent that a London footpad, meditating robbery with violence, ran off as soon as he recognized the luxuriant whiskers of the "sleeper"; and braved the official threats that often hung over those who practised or preached animal magnetism in the various states of Italy.

J. J. A. Ricard,[7] self-styled "professor of magnetology", emerged as a demonstrator and popularizer of animal magnetism in the late 1830s.[8] He claims to have demonstrated magnetism in many towns and cities, before audiences of 500, 600 or even 800 persons.[9] His demonstrations seem to have resembled those of Lafontaine, but he relied even more heavily on gifted subjects who travelled with him, and the "higher phenomena" were correspondingly prominent in his exhibitions. His most celebrated somnambule was a young man named Calixte Renaux. Other magnetizers were principally known for their work with just one gifted subject. Such were Auguste Lassaigne[10] (b. 1819), magnetizer of his wife, Prudence Bernard, and J. B. Marcillet, magnetizer of the most celebrated of all French somnambules, Alexis Didier.

The leading members of the old Société du magnétisme had looked upon public and theatrical demonstrations of magnetic phenomena with the utmost disfavour. In 1842, certain persons who regarded themselves as heirs to this tradition founded a new Société du magnétisme de Paris.[11] (Another Paris society, the Société philanthropico-magnétologique, had been in existence since 1841.[12]) The leading lights of the Société du magnétisme were Aubin Gauthier, the erudite but opinionated historian of somnambulism, who made it his business, as a moderate Catholic, to defend magnétism against the hostility of Catholic extremists,[13] Comte Brice de Beauregard, and Comte Louis Lepelletier d'Aunay, Puységur's nephew. Brice and Gauthier fiercely attacked the public demonstrations of Lafontaine and Ricard[14] – the replies of Lafontaine and Ricard are by comparison mild and not unreasonable. The Society did, in fact, hold its own public sessions twice monthly, but no doubt these were discreet and sober affairs.[15]

Despite the objections of traditionalists, public exhibitions of magnetism continued, and the better-known demonstrators drew large crowds. Travelling demonstrators – Dupotet and Lafontaine especially – who remained in a particular town for a few weeks or months would often leave behind a group of active supporters and pupils to form the nucleus of a local society.[16]

Of course, the majority of magnetizers and clairvoyant somnambules did not engage in public demonstrations. They were principally concerned with the cure of disease, and saw clients in private. "Official" medicine remained largely hostile. When Aubin Gauthier by invitation addressed the medical section of the Congrès scientifique de France (Rheims, 1846), he was subjected to bitter attacks. On 10th November 1845, the Congrès medical de Paris voted (with only one dissentient) to accept the report of a committee which recommended that treatment undertaken by non-medical persons (sc. magnetizers) should be deemed illegal even when prescribed by doctors.[17] Many clairvoyant somnambules were prosecuted for the illegal exercise of medicine. To judge by the number of prosecutions, such somnambules must by now have become fairly numerous,[18] and quite often they seem to have managed to exercise their talents, like modern "psychic consultants", without the aid of any magnetizer.[19] Magnetizers were less frequently prosecuted.[20]

Interest in animal magnetism was not at first strong enough to sustain a regular magnetic periodical. Ricard's two ventures – Le révélateur (1837–8) and the Journal du magnétisme animal (1839–42) – soon petered out, and Aubin Gauthier's Revue magnétique(1844–6) had little better success. A number of fringe periodicals in Paris and the provinces gave magnetism, and often also homoeopathy, a hearing.[21] By far the most successful magnetic periodical was the Journal du magnétisme begun in January 1845 with M. Hébert de Garnay as proprietor, and Dupotet as editor.[22] This lasted from 1845 until 1861 (20 vols.) and is a mine of information for the historian of the magnetic movement.[23] It had its financial ups and downs, and in 1848 a Société du Journal du magnétisme was founded to promote and probably finance it. At the annual general meeting of this Society in 1849, M. Hébert told the assembled members that while support was diminishing in France, the Journal had been saved from ruin by an influx of subscribers from abroad: from Germany, Italy, Switzerland, Russia, Sweden, Belgium, England, the republics of Central America, Algeria, Greece, Syria, Turkey, even Spain, and especially the United States.[24]

THE SPREAD OF MAGNETISM OUTSIDE FRANCE

The decline of French subscriptions may have been due to recent political troubles. How real the upsurge of interest was elsewhere is difficult to say. Britain and the United States will be dealt with later. I have little information about the other countries named. In Brazil, there had been some interest in animal magnetism in the 1820s, and it seems to have been maintained long enough to merge after 1850 with the Spiritualist movement, which gained a particularly firm hold there.[25] In Spain and Portugal, conditions were not favourable, though in 1846 Dr. Ramòn Comellas, a Spanish doctor, and member of a Spanish "philanthropico-magnetic" society, published what seems to have been the first serious book on mesmerism printed in Spain.[26] In 1849, P. Meric of Madrid described some cases he had treated by animal magnetism,[27] and during the 1850s various articles on animal magnetism appeared in Spanish medical journals.[28] In Russia, a few magnetizers (one or two of them medically qualified) were active in the period 1830–1850, but our accounts date from many years later.[29] No doubt this was because the practice of animal magnetism by non-medical persons was still banned – a leading magnetizer of the time, Andrey Ivanovitch Pashkov, was sentenced to a long term of imprisonment.[30] The Scandinavian countries have almost nothing to offer from this period,[31] though in 1848 the *Journal du magnétisme* stated, with characteristic over-optimism, that ever since Berzelius had again drawn the interest of the savants of his country to animal magnetism[32] (really to the so-called "Reichenbach phenomena"), the people there had been much occupied with it, and, indeed, experiments had been carried out at the civil hospital in Stockholm.[33] In Holland, the earlier interest in animal magnetism faded very rapidly, though some non-medical and fringe practitioners continued their activities into the 1820s and 1830s.[34] A modest revival of interest began around 1850,[35] associated partly with the activities of certain "electrobiologists", a revival which was shortly to mingle with the burgeoning Spiritualist movement.

Only in Belgium and Italy, among the countries of Europe outside France and Germany, do we find animal magnetism actually advancing during the period 1830–1850. It is to be noted that whereas between 1810 and 1820 magnetism had spread into Europe principally from Germany, the spread into Belgium and Italy, and also into Britain and the United States, came primarily from France.

Zorab says[36] that the decade following 1830 (when Belgium achieved independence from Holland) can be justly styled the Belgian golden age of animal magnetism. The subject began to attract the attention of Belgian philosophers, medical men and laymen. One of the earliest of the medical men was D. Cremmens, co-author of *Le propagateur de magnétisme animal* (1841), who began mesmerizing his patients around 1833. He seems to have relied much upon the deliverances of "gifted" somnambules. He also organized public demonstrations, which were, however, always directed towards medical and therapeutic purposes. In 1840, we are told,[37] one of his somnambules had, by her clairvoyant gifts, converted several distinguished persons to the magnetic cause, including the Deputies MM. Verhaegen and Dumartier.

Another pioneer of Belgian animal magnetism was J. B. A. M. Jobard (1792–1861), the distinguished industrialist, who first introduced Lafontaine to animal magnetism. Jobard's newspaper, *Le courier Belge*, occasionally reported the doings of magnetizers and their somnambules, and in 1839–41 a magnetic periodical, *Le magnétophile*, enjoyed a brief existence. Its founder and editor, V. Idjiez, was a well-known public demonstrator, and, of course, Lafontaine made a number of return visits to Belgium.

In the following decade, 1840–50, the therapeutic use of animal magnetism in Belgium declined, but, says Zorab, "theoretical interest in explaining the uncommon phenomena emerging during the somnambulistic state was growing".[38] Thus in 1842 Professor E. Tandel of the University of Liége published a paper[39] which anticipates many modern controversies. This paper is an attack on the theory, which Tandel attributes to the French philosopher F. P. Maine de Biran (1766–1824), that we each of us possess two independent egos, the ego of the waking state, and the ego which manifests during sleep, dreams and somnambulism. (Maine de Biran was an important influence on Pierre Janet.[40]) The idea that these two egos are wholly independent might be thought supported by waking amnesia for the events of the somnambulic state. However, Tandel cites an observation which he made when watching a demonstration by the Dutchman van Ghert, around 1829. Ghert told a somnambulic subject that after she had awoken she would remember all the events of the somnambulic state the moment he said the word "seven". She did so. Tandel claims this proves that the memories exhibited by the two egos, waking and somnambulic, are essentially one, and that accordingly there exists only one ego.

A notable figure in the Belgian magnetism of this decade was the Comte L. M. G. de Robiano, a very erudite priest.[41] His *Mesmer, Galvani et les théologiens* (1845) went through a number of editions. In part, it is an attempt to defend animal magnetism against members of the Catholic hierarchy, who regarded the more remarkable phenomena of magnetic somnambulism as of possibly diabolic origin. Robiano preferred naturalistic explanations, and held that the magnetic fluid is electrical in nature, the electricity being generated by the nervous system. One of his reasons for believing the fluid to be electrical was the fact that the application of so-called "galvanic rings" (I presume these were rings made of two different metals) could be just as effective as a mesmerist in inducing somnambulism and in curing disease.[42] (Such rings were widely sold at this time as a panacea for all ills.) He conducted many experiments to demonstrate that all mesmeric phenomena originate in the flow of galvanic currents (i.e. bimetallic electricity) through the nervous system. Subjects might be required to hold, for example, a piece of zinc in one hand and a piece of gold in the other. Robiano may be regarded as a precursor of the electrobiologists. He was also convinced that wearing copper and zinc belts round the body would cure all sorts of complaints, and is thus, less directly perhaps, a precursor of "metallotherapy". Zorab regards Robiano's experiments as paradigm cases of "doctrinal compliance"[43] or "experimenter effect",[44] i.e. of the tendency of experimental subjects to give experimenters the results which the latter have inadvertently led them to anticipate.

After about 1850, Belgian interest in animal magnetism went into decline.

In the various kingdoms and duchies that made up Italy at this period, animal magnetism was a plant of very slow growth. In many areas it met with hostility and repression. Wherever ecclesiastical influence was strong, magnetism was not permitted to flourish. To the perplexed bishops who wrote to the Holy See during the early 1840s requesting a ruling on the permissibility of mesmerism, the authorities replied with a certain casuistry. They managed to avoid a definite and total condemnation (which might have exposed the Holy Office to retrospective ridicule) whilst making disapproval plain (thereby satisfying conservatives who feared that only demonic intervention could account for the more remarkable phenomena[45]). Animal magnetism was banned in Naples and in Rome.[46] In Lombardy, which was then under Austrian rule, the imperial decree of 18th October 1845 permitted the practice of magnetism only to medical men (of whom Dr. C. Dugnani of Milan became the best known). The neighbouring Duchies, together with Tuscany, came under the influence of the imperial decree, an exception being Parma. Here the Count Jacopo San Vitale was actively engaged in magnetism by about 1840, when he demonstrated it at a scientific conference in Turin (Piedmont) with great success and scientific and ecclesiastical approval.[47] After 1848 Piedmont, the most forward-looking of the Italian states, was in fact the only one "where magnetism could be studied and applied in an atmosphere of great tolerance".[48]

In a useful survey published at Modena in 1846, M. Sabbatini can find only seven pieces on animal magnetism printed between 1785 and that date.[49] The most substantial were the *Fatti relativi al mesmerismo e cure mesmeriche* of professors A. Cogevina and F. Orioli, published at Corfu in 1842, and the four-volumed *Sulla storia, teoria e practica del magnetismo animale* of L. Verati (G. Pellegrino), published at Florence in 1845–6. The latter is based almost entirely on French sources. In 1847, a medical student at Pavia named Righi published a doctoral dissertation on animal magnetism.[50] In 1849, Count G. Freschi writes[51] that magnetizers and serious investigators are not widespread in Italy, most of them being homoeopathic doctors practising in the North. They include Poeti of Turin, Dugnani of Milan, Count Nani of Venice, Dr. Orioli, Count San Vitale of Parma, Dr. Ferzaghi of Milan, Drs. Granatti and Finella of Turin, and Dr.

Coddè of Mantua. There is a homoeopathic society at Turin which also embraced magnetism. (This association between animal magnetism and homoeopathy is confirmed for example by Poeti's *Saggio sull'azione curative del magnetismo animale* (1848), half of which is devoted to homoeopathy and half to magnetism.) In central and southern Italy, some interest has been aroused by the publication of Verati's book.

Once the revolutionary wars of 1848 had subsided, Italian interest in animal magnetism grew quite quickly, or perhaps became bolder. No doubt it was fanned by the tours of such travelling demonstrators as Lafontaine and Lassaigne.[52] Probably there was also some connection with anti-clerical tendencies in Italian thought. In the 1850s, a number of magnetic societies and periodicals were founded in northern Italy, and the Italian literature of the subject grew quite rapidly.[53] There were some signs of a serious scientific interest, especially in Milan, where, in 1850, two groups of medical men and other savants conducted experiments with Lassaigne's famous clairvoyant subject, Prudence Bernard.[54] Italian interest in magnetism remained strong into the 1860s and, as elsewhere, became partially fused with the nascent Spiritualist movement. In fact Freschi, in the article referred to above, states that he and Dr. L. Coddè of Mantua were even then (1849) doing experiments which involved contact through lucid somnambules with another world and with supernatural beings.

THE PARANORMAL IN FRENCH MAGNETISM

Freschi's and Coddè's experiments almost certainly had French models, to which we shall come. French animal magnetism was a more pragmatic business than German; for most French magnetizers cures came first, and the wonders of clairvoyance, transference of sensation, contact with and revelations from spirits, etc., though naturally of interest, were not the principal object of the exercise.[55] None the less, there was a substantial French literature on these topics, and during the 1830s and 1840s, this side of magnetism became steadily more conspicuous in France. It never attained such predominance there as it had in Germany, but that does not mean that the French were more discriminating than the Germans in the assessment of evidence. The French evidence manifests precisely the faults that we have already noted in the German. I shall therefore disinter only a few of the more curious examples. It will be convenient to tackle them, as I did the German cases, under the headings of "spontaneous" cases and "experimental" cases.

"Spontaneous" cases (cases in which the paranormal cognition comes unsought, as an intrusion upon the destined drift of the somnambule's ideas) are somewhat uncommon, and most are poorly recorded. The most remarkable case of this class from this period is probably a late Dutch one set down by Dr. H. G. Becht, for many years town physician of The Hague.[56] In 1849, Becht and an older medical colleague, Dr. W. H. M., undertook the case of a young lady, Miss M., who was seriously ill with pulmonary tuberculosis. Becht began to magnetize her and she became somnambulic. On 10th April 1850, the patient, being in the somnambulic state, suddenly produced a detailed account of her own forthcoming death-bed and its surroundings (which were quite different from her present surroundings). She dated her death to the 17th January 1851. She also stated that Dr. M. would die in precisely a year. Both deaths occurred as predicted, her own in the circumstances which she had foreseen. According to Zorab both deaths are recorded in the town's register of deaths.[57]

Cases that would by modern standards be regarded as "experimental" are even rarer than "spontaneous" cases. The nearest approximation is perhaps some experiments carried out by Dr. L. J. J. Charpignon (1815–1886) of Orléans, one of the abler and more reasonable of the later French writers on animal magnetism.[58] Charpignon would magnetize one of four empty glass phials, mix the phials up, and present them to his somnambulic subject. She would at once recognize the magnetized phial by the glowing vapour which it contained. To guard against the possibility of thought transference, he sometimes had the phials magnetized and passed to him by other persons, so that he did not himself know which one had been magnetized. Unfortunately, this precaution is nullified by the fact that Charpignon does not tell us whether or not his somnambule was allowed to handle the phials. If she was, the residual warmth in the magnetized phial resulting from the process of magnetization could have provided the necessary clue.

If we extend the scope of the term "experiment" to cover any attempt at paranormal cognition, the number of "experimental" cases at once becomes far greater, though the evidential quality is not correspondingly enhanced. Most celebrated at the time were the performances of the famous somnambules who accompanied various public demonstrators on their rounds – Calixte Renaux (magnetized by J. J. A. Ricard), Prudence Bernard (magnetized by her husband A. Lassaigne), Alexis Didier (magnetized by J. B. Marcillet), and Alphonse Didier (magnetized by a variety of persons, including Lafontaine and Ricard). The acts of these somnambules were so alike that they need not be treated separately, especially since they are the subject of a definitive study by E. J. Dingwall.[59] Staple features were playing card-games while blindfolded; reading (still blindfolded) billets laid on their stomachs or otherwise concealed from them; travelling clairvoyantly to visit places or persons thought of by members of the audience with whom they had been put in rapport; performing medical diagnoses; carrying out actions merely willed by their magnetizers; and "object-reading", i.e. telling something about the history, and personal associations, of objects handed to them. A good part of these performances seems to have been simply mental magic, based upon the evasion of blindfolds, the use of codes, etc., and in most cases the residuum of phenomena that might defeat this hypothesis is not large enough to be worth pursuing. The Didier brothers, however, present more interesting problems,[60] and I shall return to their phenomena later.

Putting aside, then, the set-piece performances of the public somnambules, we are still left with quite a considerable number of instances of apparent paranormal cognition falling under the kinds of headings with which we are already familiar. There were, for instance, occasional cases of apparent transposition of the senses, though none strikes me as having much evidential weight.[61] Cases of "travelling clairvoyance" were frequently reported, but essential details as to the recording and verification of the statements made are generally lacking.[62]

There were some instances of apparently successful magnetization at a distance. The most dramatic is perhaps one in which the magnetizer was Alexandre Dumas, the novelist. Dumas, like Balzac, introduced animal mag-netism into his novels, and fancied himself as a magnetizer.[63] About the beginning of 1848, he was dining at the house of a Deputy, and undertook to summon a somnambule from the other side of Paris. Taking a glass of water, he made cabalistical signs over it, and threw jets of magnetic fluid into it. After an appropriate lapse of time, the lady arrived, accompanied by her husband, who said that his wife had fallen asleep over dinner, and had got up and gone out. He had guessed that Dumas was responsible.[64] It is difficult to know how seriously to take a story dependent upon the sincerity of a person so much larger than life and with so well-developed a sense of the dramatic as Dumas, but he certainly took magnetism seriously enough. He contributed to the *Journal du magnétisme*, and went so far as to obtain signed testimony in support of his account of some remarkable incidents with Alexis Didier.[65]

By this time, the use of magnetic somnambules to detect criminals, find lost or stolen property, locate dead bodies or missing persons, etc., seems to have been fairly common.[66] An example that obtained some publicity concerned the artist A. V. Sixdeniers (1795–1846), who drowned in the Seine on 10th May 1846.[67] On 14th May a friend of the artist handed a somnambule a *portefeuille* which Sixdeniers had carried, and asked, "Can you tell us where the person who owns this *portefeuille* is?" Her reply was: "Empty the Seine and you will find him!" She described the details of his death correctly, but could not locate the body. Another somnambule, handed the *portefeuille*, became afraid and eventually stated that she saw the body between two boats a little above the Pont des Arts. She described Sixdeniers' costume exactly. First thing next day some of his friends and pupils went to the spot and discovered the body.

With regard to such clairvoyant detective work, Charpignon prudently remarks[68] that one should maintain the greatest reserve, because for one fortunate success, there will be ten illusory and vain hallucinations. Things could, indeed, go badly astray, as in the case of Madame Lemoyne (1850),[69] whose suspicions concerning her husband's fidelity led her to consult the somnambule Madame Mongruel. The latter asserted that a certain Mademoiselle G., whose address she supplied, was M. Lemoyne's mistress. Madame Lemoyne accused her husband, who, unable to persuade her of his innocence, took her to confront Mlle. G.. Mlle. G.'s parents were furious, she herself had a fit of hysterics, and her mother wished to kill the scoundrels who had dared to tarnish her daughter's reputation. But Madame Lemoyne kept repeating over and over again, "The somnambule has said so".

To such mundane purposes was the clairvoyance of French somnambules frequently turned. There were, however, some somnambules who aspired to higher things, and claimed to be aware in the somnambulic state of spirits, angels, saints, etc. Such somnambules did not become so numerous as in Germany; but there was none the less a long and probably continuous French tradition of Spiritualistic somnambulism.[70] This did not, however, fully surface until 1839, when Dr. G. P. Billot published his correspondence with Deleuze.[71] The correspondence in question largely emanates from Billot, a prolix and obsessive person, who sent Deleuze long accounts of his encounters, through somnambules, with spirits, angels, the Virgin Mary, the Archangel Michael, etc. At some of these séances occasional peculiar physical phenomena ("apports") took place. While one of Billot's somnambules, Mlle. Laure, was in the magnetic sleep, an angel, at his request, obligingly imprinted the shape of a cross on her right forearm.[72] The general tone of the revelations was pervasively Catholic – Billot strongly desired to reconcile magnetism and Catholicism in despite of such clerical writers as Marne.[73] Also pervasively, if somewhat loosely, Catholic in tone were the phenomena described by A. Possin (1843) with his somnambule Ferdinand, and by J. J. A. Ricard (1841) with three spiritualist somnambules.[74] More heterodox were the revelations vouchsafed by various somnambules to C. M. L. Loisson de Guinaumont (1773–1849), a former Deputy, and J. Olivier, a pupil of Dupotet and President of the Magnetic Society of Toulouse.[75] The somnambules of Drs. Ordinaire and Wiesseké preached reincarnationist doctrines,[76] which afterwards became prominent in French spiritualism. Such views had had some currency among pre-Revolutionary occultists and had no doubt lingered in some quarters. For instance, C. Chardel (at one time a French Deputy) manages in his *Essai de psychologie physiologique* (1831; 3rd edition 1844),[77] to combine belief in a somewhat mystical *magnétisme vital* related to the luminiferous ether, with a large dose of Swedenborgianism, and the conviction that repeated reincarnations are desirable, beneficial and highly likely.

It was in fact an admirer of Swedenborg, Alphonse Cahagnet (1809–1885), who recorded and published the most remarkable somnambulic revelations of this period. Cahagnet was a Parisian cabinet-maker, who became interested in the phenomena of somnambulism around 1845, and began to employ his leisure in recording and studying the utterances of various clairvoyant somnambules.[78] During the magnetic sleep these somnambules encountered numerous deceased persons, who portrayed the next world in strongly Swedenborgian terms. Some of these revelations came from Swedenborg himself, who must have found it highly gratifying to be able to confirm that so many of his ante-mortem visions of the post-mortem world had been correct. Far more interesting, however, were the accounts of the deceased (and sometimes living) friends and relations of persons present which were retailed by a somnambule named Adèle Maginot. These accounts were sometimes strikingly accurate, and Cahagnet published some of them in the first volume of his *Arcanes de la vie future dévoilés* (1848). They were subjected to a good deal of criticism. Cahagnet, whom Podmore describes[79] as "a man of quite unusual sincerity and teachableness", tried to meet these criticisms by preparing a statement of the leading facts communicated at each sitting and getting the persons present or otherwise involved to add comments and sign it, and by seeking in the protocols items of correct information which could not have been conveyed to the somnambule by thought-transference from the sitters. Two further volumes appeared in 1849 and 1854.[80]

In the following case,[81] Cahagnet asked on behalf of a gentleman, M. Fandar (not present), for a description of his late father. The latter seems to have been quite unknown to Cahagnet, thus ruling out thought-transference from the sitter. Adèle said : "I see a man with grey hair, full ruddy countenance, large nose, stern look, and smiling mouth, and this betokens a lively and good disposition; short neck, and breathing with difficulty. I perceive pimples caused by heat of the blood on his face. He is pretty corpulent and of middling stature. I should say that he suffered in his legs. He wears a brown vest, and, I think, coarse grey pantaloons. He is happy, and reunited to his wife, whom he dearly loved on earth, and who was, as he says, very kind to him."

Replying to Cahagnet, M. Fandar describes "the indications given by our somnambulist" as "very exact" and goes on specifically to confirm the breathing problems, the swollen legs, the brown vest and coarse grey pantaloons, the sanguine temperament and pimples, and the combination of sternness and benevolence.

As Cahagnet's cases go, this one is by no means especially striking, and it is hard to deny that Adèle Maginot must have been a very remarkable lady. Her repertoire included "travelling" clairvoyantly to visit distant living persons. After one such excursion, to Mexico, during which she had been much troubled by the tropical sun on her left, she was found to have suffered a severe discoloration of the skin on that side of her face only.[82]

Surprisingly, perhaps, extended and systematic critiques of French mystical magnetism are hard to find. A late critique that made some noise is that by G. Mabru (1858).[83] But this work is merely an immense and totally uncritical collection of every anti-magnetist and anti-spiritualist quotation, passage, snippet or sentiment that its fanatical compiler had clapped eyes on. Altogether different is the knowledgeable and reasoned analysis by A. S. Morin (1860), a book which is essential reading for the student of French animal magnetism. But it lies on the edge of the period with which we are concerned.

MAGNETISM AND MAGIC

I have left until last the most eccentric of all the developments that might be included under the heading of "mystical magnetism". I refer to the attempts by Dupotet[84] to lay bare the secrets underlying various forms of ceremonial magic – the sort of magic that involves ritual invocations, heady perfumes, ritually prepared objects, wands, and above all circles, sigils and other shapes inscribed on the floor of the theatre of operations. Dupotet's explanations of magical phenomena are elliptically expressed and hard to follow, but the general idea seems to be this. Animal magnetism (the quasi-electrical fluid) has, when left to itself, or set in motion by purely mechanical movements of the hands, properties conformable to the laws of physics, and resembling those of ordinary magnetism. It can be stored, transmitted, etc. But when by a deliberate act of the intellect, the soul or spirit of the operator influences, indeed becomes infused into, the magnetism, all kinds of further effects may be wrought, the laws of which we can so far only glimpse. In some way a portion of the magnetic fluid – or at any rate a force – may separate itself from the operator and become attached to, enclosed within, another object, where it will soon begin to affect its surroundings in such a way that a person coming close to it may have alarming or exciting experiences or otherwise feel its influence.

These experiences, however, are not in any way directly controlled by the will of the operator. Dupotet seems to have thought that once the force has become separated from the operator, and installed in some object, it will exhibit a certain purposeful autonomy in the way it develops the idea originally imprinted on it and influences those who come near. Dupotet probably also thought that the force may be implanted directly into another individual.

Dupotet carried out numerous experiments which he thought supported these conceptions. He might, for instance, without speaking draw on the floor in charcoal two wavy parallel lines, representing a path, and at the end put a conventional sign for a precipice. As he drew he would concentrate hard on these ideas. A healthy young man, placed on the "path", might take a few steps, touch the "precipice", hesitate, become giddy, collapse, and need support. Another of Dupotet's favourite procedures was to make what he calls a "magic mirror". This involved drawing on the floor a heavily filled circle of four or five inches in diameter, and fixing upon it the intention that this should be a *magic mirror*. Persons brought close to it would be variously affected, might be unable to look away, might feel shocks, see visions, shake and tremble. Dupotet meanwhile would say nothing, make no suggestions.

Dupotet, though of middle size and rather slender,[85] was generally regarded as one of the most powerful magnetic operators of his time, and there can be no doubt that scenes like these actually occurred at his demonstrations. One should not let the extravagancies of his theory and practice, not to mention his prose style, prevent one from noticing that he was actually doing things of considerable originality and interest. He made no magnetic passes; his subjects were not put into a sleep-like or somnambulic state; he offered (in many instances) no direct verbal suggestions. His procedures were not like those of earlier magnetists, nor yet like those of subsequent hypnotists. Yet clearly he was able to produce effects that later stage hypnotists would have been happy to incorporate into their acts. His approach to his subjects in these demonstrations was not altogether unlike that of some recent workers who have used waking "guided fantasies" instead of traditional hypnotic induction techniques to obtain from their subjects many of the phenomena thought characteristic of hypnosis. Only, instead of providing a pre-packaged standard fantasy, he gave his subjects the materials, and no doubt also intentional or unintentional hints, from which to construct their own fantasies. In some ways, though perhaps almost by accident, Dupotet was more than a century ahead of his time.

NOTES

1 The magnetizer Frapart treated the celebrated Broussais homoeopathically when the latter was in his decline (Ackerknecht, 1967, p.68).
2 See Dupotet de Sennevoy (1834; 1838; 1846; 1930; etc.).
3 He tells the story of his travels in Dupotet de Sennevoy (1840).
4 I have not discovered the amount of the subscription. Mabru (1858), p. 339, states that Dupotet's daily takings exceeded 100 francs; Mabru is, however, a prejudiced source.
5 On Lafontaine's connections with animal magnetism in Belgium, see Zorab (1967), pp. 6–16.
6 Lafontaine (1866); cf. also Lafontaine (1847).
7 On Ricard see esp. Dingwall (1967), pp. 135–144.
8 He wrote a number of popular books: Ricard (1841; 1844; etc.).
9 Ricard (1841), p. 539.
10 Lassaigne's autobiography is Lassaigne (1851).
11 See *Revue magnétique*, 1844–5, Vol. 1, pp. 5–12.
12 *Journal du magnétisme*, 1845, Vol. 1, p. 188.
13 Cf. Gauthier (1844).
14 Brice de Beauregard (1844–5).
15 Morin (1860), pp. 14–15.
16 E.g. Cambrai, Grenoble, Toulouse, Cherbourg, Versailles, Lyon, Rennes, Port-Louis (Mauritius). Cf. *Journal du magnétisme*, 1845, Vol. 1, p. 192; 1846, Vol. 2, pp. 13–19; 1847, Vol. 4, pp. 366–369; 1850, Vol. 9, pp. 276–277.

17 *Journal du magnétisme*, 1846, Vol. 2, p. 54; *Revue du magnétisme*, 1845–6, Vol. 2, pp. 27–37; *Gazette des hôpitaux*, 12th Nov 1845.

18 In the period 1845 to 1850, and beyond, the *Journal du magnétisme* ran what was almost a regular column detailing police court proceedings against these persons. On public somnambules cf. Noizet (1854), pp. 338–349.

19 Cf. Charpignon (1848), pp. 307–310.

20 Ricard was convicted, but his conviction was quashed. Cf. Ricard (1845).

21 The *Revue magnétique*, 1844–5. Vol. 1, pp. 335–336, lists the following: *L'avenir médical* (Paris); *Le Journal du magnétisme* (Paris); *Le somnambule* (Lyon); *L'observateur homoeopathique de la Loire-Inférieure* (Nantes); *Archives de la Société magnétique de Cambrai*; *La mouche de Sâone-et-Loire et de l'Ain* (Mâcon)

22 Dupotet had previously edited *Le propagateur du magnétisme animal* (2 vols., 1827–8).

23 A purported continuation of the *Journal du magnétisme* from vol. 22, 1879, onwards, under the editorship of Hector Durville, is really a quite different periodical.

24 *Journal du magnétisme*, 1849, Vol. 8, pp. 72–78.

25 Cf. Dingwall (1968b), pp. 194–196. A work which I have not seen, *Doctrine de l'école de Rio de Janeiro, et pathogénesie brésilienne* (1849) is reviewed in the *Journal du magnétisme*, 1849, Vol. 8, pp. 156–159.

26 Comellas (1846). I know of this volume only through the reference in Dingwall (1968b), p. 194.

27 *Journal du magnétisme*, 1850, Vol. 9, pp. 43–51.

28 Dingwall (1968b), p. 194.

29 Zielinsky (1968), pp. 13–17.

30 Zielinsky (1968), p. 14.

31 Cf. Bjelfvenstam (1967).

32 Bjelfvenstam (1967), pp. 214–217.

33 *Journal du magnétisme*, 1848, Vol. 6, pp. 57–58.

34 The most notable work of this period is by a pharmacist, B. Meylink (1837).

35 Hoek (1852; 1854); *Het magnetismus* (1858); van der Hart and van der Velden (1986–7).

36 Zorab (1967), p. 20; cf. Van der Hart and Van der Velden (1987).

37 *Le magnétisme en Belgique* (1840).

38 Zorab (1967), p. 22.

39 Tandel (1841–2).

40 Ellenberger (1970), pp. 402–403.

41 Zorab (1967), pp. 23–26.

42 Zorab (1967), p. 25.

43 Zorab (1987), p. 25; Ehrenwald (1957).

44 Rosenthal (1967).

45 Leppo (1968), pp. 140–142; Loubert (1844).

46 But Pius IX told Lafontaine he thought animal magnetism a natural phenomenon. Harte (1902–3), 2, p. 65.

47 Letter of A. Despine, *Journal du magnétisme animal*, 1840, Vol. 2, pp. 11–12.

48 Leppo (1968), p. 140.

49 Sabbatini (1846).

50 Dugnani in *Journal du magnétisme*, 1848, Vol. 6, pp. 33–38.

51 Freschi (1849).

52 Lafontaine, who was the pioneer, seems to have had a fair amount of trouble with the authorities. On one occasion, he was ordered by the Neapolitan police to leave Naples. However, on the intervention of the French ambassador he obtained from King Ferdinand

a written permission to remain, couched in the following terms: "I consent that M. Lafontaine remains in Naples, on condition that he makes no more blind people to see nor deaf ones to hear." Cf. Harte (1902–3), Vol. 2, p. 64.

53 See, amongst quite a few others, Nan͏   ͏350); Tommasi (1851); Guidi (1851, 1854, 1860); *Sulla causa dei fenomeni mesmerici* (1856). Crabtree (1988. pp. 168–169) states that the first Italian periodical devoted to the subject was the *Cronaca del magnetismo animale* (2 vols., 1853–4), edited by G. Terzaghi of Milan. It was shortly followed by *Il mesmerista* (2 vols., 1854–5), published at Turin.

54 Leppo (1968), pp. 144–145, 152–165. Cf. Leppo's account of Beroaldi's experiments at the provincial hospital in Vicenza, pp. 165–168.

55 On the difference of emphasis between French and German magnetism, cf. Fischer (1839), Vol. 1, p. 17.

56 Becht (1876); Zorab (1967), pp. 84–89. Becht had not himself made a contemporary memorandum of the predictions, but his mother-in-law-to-be had!

57 Teste (1843), pp. 101–104, gives, from his own experience, another case of a somnambule predicting her own death-date.

58 Charpignon (1848), pp. 24–25; cf. Dingwall (1967), pp. 214–215.

59 Dingwall (1967–8).

60 Dingwall (1967), p. 205, regards their phenomena as "of a different order from those reported with other somnambules".

61 Cf. for instance Teste (1843), pp. 127–131; Dingwall (1967), pp. 136–137.

62 Cf. the frequently cited case reported by Fillassier (1832), pp. 53–54.

63 See Dumas' *Mémoires d'un médecin: Joseph Balsamo*, and Balzac's *Ursule Mirouet*. Cf. Cailliet (1980), pp. 118–123, etc. On Balzac's activities as a magnetizer see Leppo (1968), p. 142; on Dumas', see Gentil (1848).

64 *Journal du magnétisme*, 1848, Vol. 6, pp. 47–53.

65 *Journal du magnétisme*, 1847, Vol. 5, pp. 146–155.

66 For an earlier German example see Wesermann (1822), p. 214.

67 *Journal du magnétisme*, 1846, Vol. 2, pp. 321–323.

68 Charpignon (1848), pp. 90–91.

69 Morin (1860), p. 339; *Gazette des tribunaux*, 1st and 29th August 1850.

70 Cf. on the Lyons magnetizers pp. 64–67 above; *Sur les faits* (1818); Deleuze (1818b). In 1836 one of Despine's favourite somnambules, Mlle. Estelle, aged 11, enjoyed the friendly protection of a guardian spirit (Despine, 1840).

71 Billot (1839).

72 Billot (1839), Vol. 2, p. 235.

73 Marne (1828). Deleuze's reply is Deleuze (1828).

74 Possin (1843); Ricard (1841), pp. 274–301.

75 Loisson de Guinaumont (1846); Olivier (1849). On Loisson de Guinaumont and Olivier see Rouxel (1892), pp. 155–170.

76 Charpignon (1848), p. 421. In 1845, Dr. Ordinaire, of Mâcon, was the editor of *La mouche de Sâone-et-Loire*. See Dureau (1869), p. 134. He was also the author of *Une somnambule mâconnaise* in Laurent and Ordinaire (1841).

77 Chardel (1844), pp. 353–401. On pp. 166n–167n he offers an interesting explanation, in terms of what would now be called "state-dependent memory", of why most of us do not recall our previous incarnations.

78 On Cahagnet see Podmore (1902), 1, pp. 81–91.

79 Podmore (1902), Vol. 1, p. 83.

80 There is an English translation of the first two volumes as *The Celestial Telegraph* (1850).
    A third volume of the *Arcanes* appeared in 1854.

81 Cahagnet (1850), Vol. 2, pp. 136–137.

82 Cahagnet (1850), Vol. 2, p. 23.

83 Mabru (1858). On the response to Mabru see Dingwall (1967), pp. 228–229.

84 Dupotet de Sennevoy (1852). Much of the material is taken from articles published earlier
    in the *Journal du magnétisme*. There is an abridged English translation (Dupotet de Sennevoy,
    1927).

85 Léger (1846), p. 367.

CHAPTER 10

Mesmerism in the United States

Mesmerism came late to the United States, and to Great Britain, making no significant impact in either country until the middle 1830s. Thereafter it began to attract some degree – in America a considerable degree – of public attention. In both countries, the chief influences on the mesmeric movement came from contemporary France. It is odd, therefore, that so many summary accounts of the history of hypnotism jump straight from Puységur and the 1780s to the English-speaking mesmerists of the 1840s, as though nothing of any moment had gone on in the interim. It is perhaps even odder that they should, as a rule, alight upon British mesmerism in the 1840s, as represented by Elliotson and Esdaile, who published, indeed, some remarkable materials, but had little influence upon succeeding generations, rather than upon American mesmerism of the same period, which was a more important social and cultural movement, and, as I shall argue, had a powerful, though indirect, effect in determining the range of phenomena later thought characteristic of hypnotism.

Once introduced into the United States, mesmerism seems quite quickly to have set the brush ablaze, and it is possible to guess at some factors which made America on the whole more combustible than Britain. In the United States, authority in professional, intellectual, educational and religious matters had not yet become highly institutionalized, centres of advanced learning were scattered, and a habit of independent thinking flourished even among the less well educated. Such a country was bound at a certain level to be more open to new social, religious, intellectual and medical ideas than was Britain with its entrenched professional, intellectual and political establishments. Between 1830 and 1850, the eastern United States was a ferment of new ideas, new movements, and new cults, many of a reformist or utopian character, and was uplifted by an almost euphoric optimism as to the prospects for improving man's lot in this world or assuring his comfort in the next. The literary and intellectual side of these movements met and merged with the New England transcendentalism which grew up along with Unitarianism as the older style of Puritanism began to lose its grip on the American mind.

One might have thought that transcendentalism, with its curious mixture of Plato and Kant, of Carlyle and Swedenborg, of eastern mysticism and the Protestant belief that each individual can find divinity within himself, would have harmonized well with certain aspects of the mesmeric movement. But the *guru* of the transcendentalists, Ralph Waldo Emerson, took, like other eminent literary persons,[1] a poor view of mesmerism and its phenomena. It was not that he altogether disbelieved in the phenomena; rather he looked upon "this monkey of mesmerism" as liable to assume an undeserved and misleading importance, as "Momus playing Jove in the kitchens of Olympus".[2]

Generally speaking, mesmerism in the United States had stronger affinities with the more popular reform and "progressive" movements than with the speculative intellectualism of the transcendentalists and literary brahmins. These movements (which seem to have been at their strongest in the New England states and in Pennsylvania) covered a wide spectrum, and were complexly interconnected. They included movements for the abolition of slavery, the emancipation of women, the limitation of child labour, and the promotion of socialism, Fourierism, "communitarianism", free love, vegetarianism, penal and educational reform, homoeopathy, phrenology and Swedenborgianism.[3] Phrenology and Swedenborgianism developed, as we shall see, particularly close ties with mesmerism. The New York publishing and bookselling firm of Fowler and Wells, which began by issuing works on phrenology, soon became a centre from which literature on many of these topics was disseminated.[4] A branch was opened in Philadelphia.

The history of animal magnetism in the United States may conveniently be tackled by considering three groups of persons whom (in the order of their first appearance) we may label the pioneers, the missionaries and the prophets, though I must add that the distinctions between them are by no means hard and fast.

THE PIONEERS

The pioneers were French, and the first of them was the Joseph du Commun at whose house in Paris the Société du magnétisme first met in 1813. By 1829, du Commun had become a teacher of French at the U.S. Military Academy, West Point. On 26th July, 2nd and 9th August 1829, he lectured on animal magnetism at the Hall of Science, New York. The lectures do not seem to have made a great impression in either their spoken or their published form.[5] Du Commun claims[6] that on his arrival in New York in 1815, he and a few other persons founded an animal magnetic society which had diffused some knowledge of the subject among the public.

A far more effective pioneer was Charles Poyen St. Sauveur, author of *Progress of Animal Magnetism in New England* (1837). Poyen claims that he first became interested in animal magnetism when a medical student in Paris in 1832.

Shortly afterwards he betook himself for the sake of his health[7] to the French West Indies, where he had relatives. He found that animal magnetism, which had been introduced to the French West Indies before the Revolution, was still much in use there, and during his fourteen-month stay he became a proficient practitioner. He next migrated to New England under the impression (surely misguided) that its climate would benefit him. Eventually, he settled in Lowell, Massachusetts, as a teacher of French and drawing. His spare time he occupied in preparing a French translation of Rostan's celebrated Dictionary article. Failing to find a publisher, he turned his materials into four lectures which he delivered at Chauncey Hall, Boston, and thereafter (about March 1836) he began to offer lectures and private instruction in mesmerism. A certain number of physicians and medical students showed an interest. On June 30th 1836, before various doctors, professors and students from the Harvard medical school, one of Poyen's pupils, B. F. Bugard, mesmerized a girl of twelve and a half from whom a carious tooth was painlessly extracted.[8] The operation was reported in the *Boston Medical and Surgical Journal*,[9] but even this did not awaken any great interest, and towards the end of 1836 the discouraged Poyen ended up in Providence, Rhode Island, more or less on his beam ends.

Now at last his fortunes, and those of animal magnetism in the United States, took an upward turn. Providence, and also nearby Pawtucket, proved remarkably receptive to animal magnetism. Why this should have been I cannot pretend to say – both towns were by then principally known as centres of textile manufacture – but it may be that the support quickly given to Poyen by such distinguished persons as Francis Wayland (1796–1865), President of Brown University, and J. C. Brownell (1779–1865), episcopal bishop of Connecticut, whose brother was a doctor in Providence, had a good deal to do with it. A number of local doctors showed interest, and through one of them Poyen discovered a remarkable somnambule, Miss Cynthia Gleason of Pawtucket. Miss Gleason consented to become his regular assistant, and after a trial run in Providence, the two set out for an extended tour of New England. Poyen now lectured to packed halls, and received copious newspaper publicity. There resulted in 1837 and 1838 what may be called the "Poyen phase" of American animal magnetism, a phase during which a considerable proportion of the active magnetic operators owed their inspiration, and often their instruction, to Poyen. Providence itself became for a while the Mecca of American magnetism. In 1837, soon after Poyen's departure on tour, C. F. Durant (1805–1873) – best remembered as an adventurous balloonist – wrote that this "interesting science"

> ...now claims of its supporters a large share of the intelligent men of the country. In the city of Providence alone, there are six professors of the science ...Thomas C. Hartshorn, of Providence, a gentleman eminent in learning... has ...translated and published Deleuze's "Practical Instructions in Animal

Magnetism "... Mr. Hartshorn accompanies the translation with many
remarkable facts produced by the professors of the magnetical science in
Providence.[10]

Durant adds a list, derived from Hartshorn's translation of Deleuze,[11] of twenty-
seven assorted persons from assorted states who have visited Providence to
enquire into animal magnetism.

The appendix to Hartshorn's book is rich both in cures and in cases of
ostensible clairvoyance. Hartshorn has a much better notion of evidence than
the French and German writers whom we considered in the previous two
chapters. He deals in the first-hand signed testimony of eyewitnesses, and often
produces signed supporting evidence. The most remarkable case is that of the
clairvoyant somnambule, Loraina Brackett, of Dudley, Massachusetts, who is in
addition the subject of a pamphlet by Colonel William L. Stone (1792–1844),[12]
editor and part-owner of a leading daily newspaper, the *New York Commercial
Advertizer*,[13] and the author of various historical works. Following a head injury,
Miss Brackett had been totally blind for about eighteen months. She was
magnetized by Dr. George Capron of Providence (another person whose interest
had been aroused by Poyen). On a number of occasions she succeeded in reading
messages in sealed envelopes. In one instance a sentence of thirteen words
which had been slipped between two pieces of sheet lead, and placed in an
envelope, was handed to her by its author. She held it to the side of her head for
about a minute, and then returned it to the writer. Some time after she awoke
she wrote down the sentence, correctly except for the spelling of two words.[14]
Her "travelling clairvoyance" was sometimes even more remarkable. For
instance, Colonel Stone reports in detail on her clairvoyant "visit" to his house
in New York on 26th August 1837.[15] (She had known Colonel Stone only two
days, and had never been to New York.) Much of what she saw related to works
of art in the Colonel's possession. Thus she described, hanging over a settee in
the library, a picture of three indians sitting in a hollow tree. the tree being
"filled with marks". This was a painting by Hoxie which contained the feature
in question. Inside the tree were "hieroglyphics – seldom noticed by visitors". In
the library she also saw a portrait of an Indian chief, with shaven head and
"something on top", and a picture of Christ, wearing the crown of thorns. The
former corresponded to a portrait of Brant [Thayendanega], the Mohawk
warrior, and the latter to a copy of Guido Reni's *Ecce Homo*, which Stone had
only received the preceding week.

Those who believe in the possibility of thought-transference may suppose that
Miss Brackett obtained her information by this means from Colonel Stone
himself. One could certainly find arguments in favour of this. It is quite clear, for
instance, that Miss Brackett's clairvoyant visions, whatever the ultimate source
of the remarkably accurate information which they sometimes contained,
cannot be thought of using the analogy of any form of ordinary perception. In

her visionary excursions, she would believe herself to witness events and to perform actions – e.g. conversing with a person at a distant scene – to which nothing corresponded or could have corresponded. And the mistakes she made when describing places unknown to her sometimes corresponded with what her interlocutor might be supposed to have thought likely.[16] In other cases, however, information was given which was not known to any person present. Many people would find it easier to try to propose ways of dismissing these cases altogether. With regard to the particular case just mentioned, sceptics might suppose that some account of Colonel Stone's art collection had appeared in print and somehow been transmitted to Miss Brackett; but the Colonel does not appear by any means so naïve that this possibility would have escaped him.

After giving these extraordinary demonstrations in Providence for seven months, Miss Brackett was removed by friends (who feared her health was suffering) to the Perkins Institute for the Blind in Boston.[17]

Poyen continued active as a lecturer and demonstrator during 1838. He appears to have returned to the West Indies in the summer of 1840, and to have died in France in 1844.[18] Miss Gleason seems thereafter to have set up as a medical clairvoyant in Boston.[19]

Among the French pioneers of mesmerism in the United States one should also include the small group of French-speaking inhabitants of New Orleans who seem to have begun meeting together informally in the late 1830s to discuss and practice animal magnetism. Out of the group grew the Société du magnétisme de la Nouvelle-Orléans, which was formally constituted in April 1845 under the presidency of M. Joseph Barthet. In 1848 it had seventy-one members. Barthet contributed news items about the Society to Dupotet's *Journal du magnétisme* until the civil war grew imminent.[20]

THE MISSIONARIES

With the 1840s we we move into much the most interesting period of American mesmerism, the period of the "missionaries". "Missionaries" may be defined as individuals who deliberately set out to convert the populace to the gospel of magnetism by giving public lectures and demonstrations, writing popular books on the subject, etc. Poyen himself had passed from his "pioneering" phase into what could be called a missionary phase. Others followed in his wake, and many proved highly successful. Among them were Robert H. Collyer, British by birth (he had studied medicine under Elliotson), a particularly pugnacious lecturer and demonstrator, who in the summer of 1841 was attracting audiences of 500 to 1000, in Boston and in New York, as he tells us with no hint of false modesty in his pamphlet *Psychography* (1843);[21] Samuel Gregory (1813–1872), an energetic advocate of medical education for women, whose *Mesmerism, or Animal Magnetism and its Uses* (1843) is a brief mesmeric *vade mecum*; Charles

P. Johnson, author of *A Treatise on Animal Magnetism* (1844); J. R. Buchanan, M. D. (1814–1899), a medical man, editor of *The Journal of Man*, and author of *Etherology* (1850) and *Outlines of Lectures on the Neurological System of Anthropology*; La Roy Sunderland (1804–1885), until 1833 a renowned Methodist minister, later a prominent abolitionist, who founded and edited *The Magnet* (1842–44); John Bovee Dods (1795–1872),[22] a Universalist preacher, whose *Six Lectures on the Philosophy of Mesmerism* (1843) and *The Philosophy of Electrical Psychology* (1850)[23] achieved exceptional popularity; J. S. Grimes (1807–1903), professor of medical jurisprudence in the Castleton Medical College, "erratic philosopher",[24] and author of *A New System of Phrenology* (1839), and *Etherology* (1845); Dr. H. G. Darling, former professor of physiology in the New England Medical College, Worcester, Mass.; and Dr. Théodore Léger (1799–1853), a pupil of Deleuze, whose *Psycodunamy* (1846), despite its name, is one of the more reasonable works of this period.[25]

These touring "missionaries" often attracted large audiences. Collyer states that when he "publicly espoused the cause", he was heralded with "scoffs, jeers, licentious ribaldry, ridicule, and all the artillery which young scribblers could bring to bear". But that was around 1840, and hostility, whether from public, press or medical profession, soon faded. Many medical men became actively interested in animal magnetism,[26] including ones of some distinction, for example Charles Caldwell (1772–1853), professor at the Louisville Medical Institute, Kentucky, whose *Facts in Mesmerism and Thoughts on its Causes and Uses* (1842) gives an eyewitness account of Elliotson's mesmeric techniques, and describes some of the author's own experiments. Quite a number of surgical operations were carried out using mesmeric anaesthesia[27] (I shall discuss this topic in a later chapter). Indeed, as I have just pointed out, several "missionaries" were themselves medically qualified. The only critical work of any interest by a medical man at this period is Daniel Drake's *Analytical Report of a Series of Experiments in Mesmeric Somniloquism* (1844). Drake (1785–1852), at this time a professor at the Louisville Medical Institute, presents the report of a "committee of gentlemen" on the exhibitions of "mental sympathy" given by a young lady somnambulist. The report, which, within its limits, is an able document, and unique in its time, attributes the clairvoyant's successes in "mental sympathy" to common associations, hints unconsciously given, unconscious leading questions or leading by questions, and the politeness of the experimenters in not misleading her.

Such rational and cautionary voices were unusual, and went almost unheeded in the hubbub of excitement which the missionaries left in their wake. Speaking of the immediate post-Poyen period, Orestes Brownson says:

> Animal Magnetism soon became the fashion, in the principal towns and villages of the Eastern and Middle states. Old men and women, young men and maidens, boys and girls, of all classes and ages, were engaged in studying

the mesmeric phenomena, and mesmerizing or being mesmerized, – some declaring themselves believers, some expressing modestly their doubts, the majority, while half believing, loudly declaring themselves inveterate sceptics.[28]

A class of professional magnetizers soon grew up – according to one con-temporary estimate there were two hundred of these in Boston alone by 1843.[29] On the theoretical side also there was a good deal of activity. Whereas Poyen and his disciples had been mainly followers of Deleuze, many of the missionaries plunged into theoretical speculation with a naive enthusiasm unrestrained by any systematic knowledge of science. Their enthusiasm spilled over into a plethora of new terms: these included neurology and nervaura (Buchanan), etherium, etheropathy and etherology (Grimes), electrical psychology (Dods), electrobiology (Stone and Darling), psycodunamy [sic] (Léger) and pathetism (La Roy Sunderland). A modern reader, faced with such verbiage, might be tempted to suppose that these writers were simply out to milk the gullible while the times were propitious. But this view would, I think, in most instances be mistaken. The naïveté of the missionaries has a quality almost of innocence, and it is impossible to doubt the sincerity of their desire to edify, instruct, alleviate suffering, and (often) to promote religion, even when, like Dods, they ask ten dollars a head for five lessons of two hours each (five dollars to ladies).[30]

For a relatively brief period in the early 1840s, many of the missionaries showed a similar naïve enthusiasm for phrenology. The best-known phreno-logical propagandist, Dr. Johann Spurzheim, had died in New York in 1832 during a lecture tour, but his influence continued, and was energetically spread by Lorenzo and Orson Fowler and their brother-in-law Samuel Wells, proprietors of the publishing company mentioned above. Inevitably, phrenology and mesmerism were brought together to form the new science of phrenomesmerism (or phrenomagnetism). Direct the magnetic fluid through an outstretched finger to a particular phrenological organ (veneration, destructiveness) and the appropriate behaviour will result (prayer, violence). There was much dispute among the magnetists as to who had first thought of this idea. The phenomena of phrenomesmerism were certainly being demonstrated in public by the middle of 1841. Buchanan, Sunderland and Collyer each claimed priority.[31] Sunderland and Collyer very soon developed second thoughts about phrenomesmerism.[32] But for some years mesmerism and phrenology were almost synonymous. Thus when a mesmeric society was formed in Cincinnati in 1842, it called itself "The Phreno-Magnetic Society of Cincinnati",[33] and demonstrations of phreno-mesmeric phenomena became a central part of the public performances of many of the missionaries. I shall come back to phrenomesmeric phenomena.

To return to the theoretical – mainly physiological – speculations of the missionaries: In many cases these of course became entangled with the anatomical postulates of phrenology. The ideas of J. S. Grimes provide an example. Grimes calls the mesmeric fluid etherium, and thinks it kin to the universal fluid of the physicists; the science of animal magnetism he terms etherology, and the magnetic state etheropathy. The process of magnetization is a process of induction. There is a natural tendency for the phrenological organs of the operator to "induct" a similar state in the corresponding organs of a passive subject. Once this has happened, the "insulation" of the subject's brain may be overcome, and currents of etherium may flow directly from the phrenological organs of the operator to those of the subject, so that a movement by the operator may immediately produce a movement by the subject, and so on. Etheropathy may also be induced in a more roundabout manner, by playing upon the subject's organ of Credenciveness. Credenciveness is "a *conforming* social propensity. The whole group to which it belongs have this peculiar character, that they all tend to conform to the wishes, feelings, actions, commands, and assertions of others."[34] If Credenciveness is over-developed or over-stimulated it will induct other organs in the subject's brain and produce appropriate effects, causing other impulses and intellectual faculties to conform to it. Willing by the operator, i.e. the exercise of his organs of Imperativeness, Firmness, and Hopefulness, may induct it; or it may respond to simple assertion, which is its natural stimulus. Indeed, Credenciveness may respond appropriately to an assertion even when the assertor wills otherwise. Thus production of the "etheropathic" state need not necessarily involve any rapport or sympathy between operator and subject, or any transfer of a quasi-magnetic fluid from the former to the latter.

Like Grimes, Dods links the phenomena he is trying to explain to a state – the "electrical-psychological state" – which is distinct from "the mesmeric slumber". A person in the electrical-psychological state continually exerts his will against the operator, is not dependent on the operator's special senses, and afterwards recalls all that has happened.[35].

What, in Dods' view, the electrical-psychological state actually is, is not easy to say, because his ideas on matters electrical derive from an electrical philosophy embracing life, mind, God and the cosmos. Electricity, it seems, is the connecting link between mind and matter, and the agent that mind employs to control the neuro-muscular apparatus. There are two sorts of electricity, positive and negative. Arterial blood contains positive electricity, which it loses in the capillaries, whence it is drawn through the nerves to the brain. Venous blood contains negative electricity (how acquired is obscure). When the electricity of the system is thrown out of balance, disease results. The electricity can be thrown out of balance by mental impressions or by physical impressions. Mental impressions can likewise correct the imbalance and cure the disease. The mind will move the electricity to whatever part of the body is thought of or attended to, thereby correcting the imbalance. Curative mental impressions can be self-generated or can be impressed from the outside by a magnetic operator. If a magnetic operator is to generate the curative impressions, two things are necessary. The operator must have relative to the subject a surfeit of positive electricity, whereas the subject must be negative, and an electrical-psychological communication must be established between the operator's "organ of Individuality" and the subject's. The organ of Individuality appears to be that part of the forebrain from which the mind influences and is influenced by the neuromuscular apparatus. Communication may be established by the operator pressing certain of the subject's voluntary nerves (which lead straight to the organ of Individuality), or by means of a peculiar technique for heightening the subject's sensitivity.

This technique is as follows. The subject holds in the palm of his hand a small disc, part zinc and part silver. He stares fixedly at this disc for twenty minutes or more, so becoming mentally and physically passive to his environment, and hence relatively negative to the positive electricity being thrown off by the disc. Unused electrical fluid accumulates in the brain and a state of torpor results. The coin is removed and the subject is now ripe for the positive impressions with which the operator will assail him. Before proceeding to implant curative impressions, it is desirable to test whether he is in the electrical-psychological state. This may be done by pressing with the thumb upon one of his voluntary nerves, getting him to close his eyes, and saying to him firmly, "You cannot open your eyes!" If he fails this challenge, he is in the electrical state, but you must endeavour to ascertain how far. Accordingly you will proceed to "produce mental impressions by operating upon his mind only", that is, you remove the pressure from the nerve, and influence his mind by verbal statements alone:

> If he is entirely in the state, you can make him see that a cane is a living snake or eel ... You can suddenly show him a boy or girl, and he sees in them the lost father or mother standing before him, and gives the warm embrace. You can change his own personal identity, and make him believe that he is a child two or three years old ... an aged man, or even a woman, or a negro, or some renowned statesman or hero. You can change the taste of water to that of vinegar, wormwood, honey, or of any liquor you please ... When you can produce such mental hallucinations as these on all his senses, we say, for the sake of convenience, that he is in the *psychological state*.[36]

Dods' electrical speculations are perhaps the greatest gobbledy-gook in the whole literature of animal magnetism.[37] Yet historically he and Grimes are important figures, for two interlinked reasons. The first is that they supplied a rationale (of sorts) and more importantly a technique (the metallic or "galvanic" disc or electro-magnetic coin) to a group of magnetic demonstrators who had, perhaps, a wider impact than any others had done before them, namely the itinerant showmen calling themselves "electrobiologists". By the very early 1850s, electrobiologists, with their bags of discs,[38] were stupefying audiences across the United States,[39] and also in Canada, Britain and continental Europe.[40] I shall discuss the electrobiologists in greater detail in the next chapters.

The second may be understood simply by glancing again at Dods' list, quoted above, of the delusions which a subject in the "psychological state" may be induced to suffer and the antics in which he may be made to engage. Grimes gives a similar list:

> Now tell him to open his eyes and put his hands together ... and say, "You cannot get your hands apart," and he cannot ... Now tell him to extend his arm, and ... tell him that he cannot put it down, and he cannot. If he is well inducted, you may tell him that he cannot step, or speak, or see, or hear, or taste, and he cannot ... Tell him that water is rum, or ink, or hot, or cold ... that he cannot lift a feather, or a penny and it will seem so to him. ... Tell him that he is a negro, a female, a dog, a fish, a post, a steam-engine – that his head is a coffee-mill – that he is Richard, Hamlet, or what you please, and he is transformed instantly, and verily believes your assertion to be true. Tell him that he can walk until he gets to such a line, but cannot pass over it, and he cannot.[41]

With the phenomena detailed in these two quotations, the animal magnetic movement has evolved virtually all the phenomena later thought characteristic of the supposed state of "hypnosis". Add the phenomena described in previous chapters, and one can find virtually no residuum of "hypnotic" phenomena not previously produced by animal magnetism. Indeed, Dods and Grimes, unlike many later hypnotists, were well aware that certain susceptible individuals may exhibit such phenomena without being put through an "induction" at all.[42] It was, naturally, the most spectacular of these phenomena that were popularized. Public demonstrators of the next generation – magnetizers like Hansen and Donato, and, before long, stage hypnotists – retained essentially the same elements in their performances. Now and then academic and scientific persons would unbend sufficiently to attend these spectacles, and it was from such visits that they acquired their concepts of what it was that a scientific explanation of hypnosis (or electrobiology) would have to accommodate. I shall return again to the influence of stage performances on the history of hypnotism. Without doubt Dods and Grimes and the electrobiologists helped to determine the shape of the problem of hypnosis as it was envisaged in the later nineteenth century and beyond.

Of course, the phenomena concerned were not new. Even if we set aside the numerous instances of hallucinations and delusions – seeing the magnetic fluid, seeing and talking to angels and spirits, travelling out of the body, etc. – which occurred in a magnetic context but were usually thought of by those involved as not hallucinatory at all, there still remain plenty of instances, particularly from French animal magnetism of the post-Napoleonic period, of the deliberate induction, in somnambulic subjects, of hallucinations, illusions and delusions, and of bizarre actions related to these, that were understood to be such by the magnetizers who occasioned them. Indeed, Puységur's very first experiment with Victor Race was of this kind. Bertrand has some remarkable examples of the production of positive and negative hallucinations, which he says he would not have taken seriously had he not been certain of the honesty of those involved. He regards them as possible instances of thought-transference.[43] Such phenomena were, however, usually held to demonstrate the influence of the operator's will (mediated by the magnetic fluid) on the nervous system of the subject.[44] Thus they were made to seem quite rare and were looked upon as something producible only with the most sensitive subjects. They did not gain a large foothold in public demonstrations.

Why was it that such phenomena had become relatively commonplace by the time that the itinerant electrobiologists set out on their tours? Part at least of the answer must, I think, lie in the brief but considerable notoriety attained in the early 1840s by the demonstrations of the phrenomesmerists. It became a leading feature of such demonstrations that a young lady, in whom the appropriate phrenological organ had been stimulated, might nurse a non-existent baby, become seemingly unaware of her actual surroundings and immersed in unreal ones, act in peculiar and totally inappropriate ways (e.g. by manifesting aggression or passion or generosity towards real or imaginary persons), and so on.[45] By the later 1840s, the vogue of phrenomesmerism had begun to wane, but a tradition of exotic phenomena had been established from which it was impossible for public demonstrators to retreat. The demonstrations of phrenomesmerists opened the way for those of the electrobiologists just as those of the electrobiologists opened the way for those of later stage hypnotists.

Of less historical importance, but greater theoretical interest, than the writings and activities of Dods and Grimes, were those of La Roy Sunderland. Sunderland was a missionary *par excellence*. He took up one cause after another with irrepressible enthusiasm. His command of an audience – whether he preached methodism, abolitionism or mesmerism – was unrivalled. At his magnetic lectures susceptible persons might be affected before he even entered the room. He received very considerable publicity – not least because several times he induced in surgical patients an anaesthesia which enabled an operation to be painlessly carried out – but he does not belong in this chapter. For before long he abandoned not just phrenomesmerism but the mesmeric fluid itself, and all variants, descendants, and synonyms thereof. He belongs in a chapter which I shall later devote to the precursors of hypnotism.

THE PROPHETS

By "prophet" I mean someone who delivers, or is a vehicle for, teachings, commands, pronouncements, predictions and statements of purported fact, all with moral or religious overtones, which he, and at least some others, regard as having some source other than his own everyday intelligence. In this sense of the term many of the German somnambules whom we discussed earlier were "prophets", and it is perhaps surprising, in view of American receptiveness at the time to all kinds of novel cults, that it took so long for "prophets" to appear in the context of American mesmerism. Part of the reason for the delay was no doubt that while American mesmerism was started by its pioneers along orthodox Puységurian and Deleuzean lines, it was quite soon developed by the missionaries in the indigenous and highly unorthodox ways just described. Americans learned little of the "mystical" developments in European magnetism until the middle and later 1840s. Catherine Crowe's English abridgment of Kerner's book on the Seeress of Prevorst was published in New York in 1845. In 1847, came a translation of part of Werner's *Guardian Spirits*, and George Bush's *Mesmer and Swedenborg*. Bush (1796–1854) was professor of Hebrew at New York University and his book seems to have caused some stir. It attempts to prove that the doctrines of Mesmer and of Swedenborg are compatible, and contains a long appendix on the Seeress of Prevorst. Not until about 1850 do we find somnambules in the United States at all comparable to the somnambules of Werner or Cahagnet, and then they are of the kind generally referred to as "spirit mediums". For this development we must seek some further contributory causes.

One of them was certainly Andrew Jackson Davis (1826–1910), the "Seer of Poughkeepsie".[46] In 1843, Davis, a youth of impoverished background and slight education, was living in Poughkeepsie, New York State. After attending some lectures on animal magnetism by J. S. Grimes he discovered that he was himself an excellent magnetic somnambule, and "came out" as a professional clairvoyant, specializing in the diagnosis and treatment of disease. One day in March 1844, while in a state of exaltation, he had a vision of Galen and Swedenborg, who informed him that he would become an uniquely gifted healer and teacher. The following year, 1845, he announced that he would deliver in the magnetic state a series of lectures on philosophy. He had come to believe that his clairvoyance yielded knowledge not just of diseases and of diseased organs, but of the sciences, of society, of theology and eschatology, and of the past, present and future of the universe. With two helpers, Dr. Lyon (his magnetizer) and the Rev. William Fishbough (his official scribe), he took rooms in New York, and dictated his lectures before a small circle of interested persons.

The resultant work, *The Principles of Nature* (1847), contains 786 pages. Those who have read it from cover to cover constitute a select, high-minded, indefatigable and probably defunct company. It is difficult for a modern reader to understand how it achieved its rapid and considerable success. Something was perhaps due to George Bush, who gave it a long and laudatory review in the

New York *Tribune*, no doubt because of its strong echoes of Swedenborg. But more was probably due to the facts noted by Podmore:

> Davis had, in fact, realised something of the orderly progression from the primaeval firemist; something too of the unity in complexity of the monstrous world; something, too, of the social needs of his time and of ours … It was partly because he could appreciate the bigness of the ideas with which he dealt, and in a semi-articulate, barbarous fashion could make other people appreciate them too, that the *Revelations* had such an extraordinary and immediate success. Partly, too, the secret lay in the moral attitude of the author. The whole book is transfused by a vague enthusiasm for the moral regeneration of mankind, like that which in England inspired the Owenite and later the Co-operative movement, which in America expressed itself in phalansteries, in religious revivals, and in abstinence from alcohol, tobacco, or meat, and which in both countries found perhaps its fullest expression for a few years in the movement known as Modern Spiritualism.[47]

Enough people were sufficiently impressed with Davis for a group of religious and social reformers to begin a periodical, the *Univercoelum*, as a vehicle for the "interior and spiritual philosophy" revealed and to be revealed by him, and for contributions on kindred themes from other enthusiasts. It ran from December 1847 to July 1849. By that time, Davis and his supporters had been overtaken by events, and his further career and numerous publications are of no concern to us.[48]

Davis paved the way for modern American Spiritualism in four ways. He accustomed a wide public to the idea that a clairvoyant somnambule might engage not just in medical diagnosis and travelling clairvoyance, but in the transmission of social, religious and cosmological teachings; he propounded neo-Swedenborgian doctrines about the future state and the spirit spheres and about the features and inhabitants of the planets; he propagated the view that some new and stirring revelation was about to rock mankind; and he implied that this revelation would involve a bursting of the barriers that separate our world from the spiritual one.[49] It was, however, not Davis and his revelations that set the Spiritualist ball rolling, but some odd events in a small township in a remote region of western New York State.[50]

The events concerned were the activities of a poltergeist which troubled the cottage of Mr. J. D. Fox, a farmer, or smallholder, in Hydesville, New York State. Fox had moved into the cottage towards the end of 1847, with his wife and two daughters, Margaretta (fifteen and a half) and Kate (almost twelve). For several months, the family was disturbed, mostly at night, by raps and bangs of unknown origin. One night little Kate exclaimed "Now do this just as I do", and clapped her hands so many times. The raps followed suit.[51] Then Mrs. Fox began to communicate with the raps by means of a simple code. Neighbours were called in. The raps continued, and over the next few days a variety of deceased persons, mostly friends and relatives of those present, purported to communicate. In the summer of 1848, Margaretta Fox went to stay with a married sister in Rochester, New York State, and Kate went to friends at Auburn, not far from Rochester. The rappings followed them. Many came to witness the wonders, and to receive communications from the beyond. It seemed that a new revelation was indeed at hand. Furthermore, the rappings spread by a kind of contagion. Visitors would find similar phenomena breaking out in their own homes. The same might happen to those who merely read about the rappings in the newspapers. Within two or three years the rappings, and other phenomena to which we shall come, were to be found over much of the United States, and had acquired numerous enthusiastic protagonists.[52]

At first, Spiritualism spread by means of the supposedly paranormal physical phenomena – the raps, and before long table tippings, etc. – but within a comparatively short time phenomena of divers other kinds had been added. Ladies with clairvoyant sensitivities began to perceive the spirits who made the rappings, which was perhaps inevitable given the growing currency of strange stories about the German somnambules and the widespread publicity accorded to the visions of Andrew Jackson Davis. Others found themselves "possessed" by the spirits and wrote and spoke in the character of deceased persons.

If one remembers that belief in an imminent new revelation was widespread; that the majority of people in the United States still lived in small, scattered and somewhat isolated rural communities; that in such communities the inhabitants have to devise their own evening activities and amusements; that lighting was more often by candle or lamp than by gas; one can perhaps begin to capture something of the feeling of those times, and of the tingling mixture of hope, curiosity and nervousness with which, in so many towns, expectant groups of seekers after truth sat around mediums in dimly lit rooms to await the latest deliverances of the "spiritual telegraph".

The traditional mesmeric framework of thought, with its plethora of quasi-electrical and quasi-magnetic concepts, was tailor-made to accommodate the new phenomena of Spiritualism. Indeed, the spirits seem to have thought so themselves. Robert Hare (1781–1858), emeritus professor of chemistry at the University of Pennsylvania, and a keen advocate of Spiritualism, received the following communication on the subject from the spirit of his late father:

> When we wish to impress the mind of the medium, by the effort of our magic will... we can dispose the magnetic currents of the brain so as to form or fashion them into ideas of our own ... To influence, mechanically, the hand of a medium to write, we direct currents of vitalized spiritual electricity on the particular muscles which we desire to control ...[53]

Even the opponents of Spiritualism found the magnetic fluid a convenient concept with which to explain away Spiritualistic phenomena. It was only necessary to suppose that living persons can in some way unconsciously utilize the fluid to produce rappings, etc., *as if they came from* departed spirits. We need not pursue the heated arguments which resulted.[54]

The new Spiritualism and the old mesmerism accorded so well with each other that a marriage was inevitable, and it was inevitable too that the younger and more exciting partner would be predominant. Mesmeric clairvoyants became Spiritualist mediums, and writers and lecturers on mesmerism turned their attention to the phenomena and philosophy of Spiritualism. La Roy Sunderland was an early convert, but before long grew disillusioned, and abandoned Spiritualism almost as completely as he had phrenomesmerism.[55] John Bovee Dods, on the other hand, joined the ranks of those who explained away Spiritualistic phenomena as resulting from the electrical properties of the nervous system of living persons.[56] He does not appear to have learned any more about electricity in the interim.

Insofar as mesmerism survived in the United States beyond the very early 1850s, it did so largely in association with Spiritualism, or (for a while) as part of the opposition to Spiritualism, though many stage performers, medical cranks, and so forth, retained the term "mesmerism" or some virtual synonym as part of their billing.

MESMERISM AND CHRISTIAN SCIENCE

Modern Spiritualism was one of two major religious movements which emerged directly or indirectly from the mesmeric movement in the United States. The other was Christian Science. The link between mesmerism and Christian Science was Phineas Parkhurst Quimby (1802–1846), a clockmaker of Belfast, Maine. In 1838, Quimby attended one of Poyen's lectures, and was greatly impressed. He found that he was himself a capable operator, and in a youth named Lucius Burkmar he discovered an excellent somnambulic subject. For some years, Quimby and Burkmar toured New England giving mesmeric demonstrations, and offering clairvoyant diagnoses and prescriptions. Anyone who thinks that itinerant mesmeric demonstrators were all conscious charlatans deceiving a credulous public for gain should study *The Quimby Manuscripts*.[57] No more honest or dedicated healer than Quimby ever breathed. After some years, he decided that the therapeutic effects of the prescriptions were due to the positive ideas they generated in the minds of the patients, and not to any intrinsic properties of the medicines themselves. What could be cured by an idea must also be caused by an idea. From about 1847 he devoted himself to mental healing, which he referred to under various names, including "Christian Science", and "Science of Health". Cure comes from the eradication of the false idea, and Quimby seems to have thought that sometimes at least he could achieve this by direct mind-to-mind contact with the sufferer. Quimby's most famous patient, whom he treated in 1862, was, of course, Mrs. Mary Patterson, later Mrs. Mary Baker Eddy (1821–1910), founder of Christian Science. There can be no doubt that Mrs. Eddy (like other contemporary proponents of mind-cure systems) was indebted to Quimby for some of her leading ideas, but (unlike the others) she was unwilling to acknowledge the fact. However, Mrs. Eddy and her works lie outside the scope of the present volume.[58]

Robert Fuller has recently argued[59] that mesmerism has been responsible not just for originating Spiritualism and Christian Science, but for the initiation of "an enduring tendency in American religious thought. The American mesmerists were the first to encourage popular audiences to abandon a scripturally-based theology in favour of psychological principles said to govern the individual's ability to inwardly align himself with a higher spiritual order." Certainly, mesmerism played some part in a loosely definable tradition of American religious thought with the characteristics described, but it is surely a mistake to say that mesmerism *began* this "tradition". The germs of the tradition can probably be found in New England Protestantism with its focus upon conversion experiences, as exemplified, say, in the writings of Jonathan Edwards.

NOTES

1 An exception was Poe, whose views on mesmerism are to be found in his "Mesmeric Revelation" (1844), rather than in the better-known "Facts in the Case of M. Valdemar" (1845).
2 Emerson (1971), p. 339.
3 An interesting picture of these multitudinous cults emerges from Orestes Brownson's autobiographical novel *The Spirit-Rapper* (1854). See e.g. Brownson (1854), p. 46.
4 Russell and Goldfarb (1978), p. 31.
5 Du Commun (1829).
6 Du Commun (1829), p. 19.
7 According to Brownson (1854), p. 4, Poyen had been a leading Saint-Simonian. If so, he may have found it prudent to leave France when that sect was dispersed by the authorities in 1832.
8 Poyen St. Sauveur (1837b), pp. 72–76.
9 West (1836).
10 Durant (1837), p. 47.
11 Deleuze (1837). I have used the revised edition (Deleuze, 1846).
12 Stone (1837).
13 He first described his encounter with Miss Brackett in a letter to the *Commercial Advertiser* of 4th September 1837.
14 Deleuze (1846), p. 345.
15 Stone (1837), pp. 42–44.
16 Cf. the letter from Rev. F. A. Farley to Hartshorn, Deleuze (1846), pp. 257–258. On the thought-transference hypothesis with special reference to Miss Brackett see Poyen St. Sauveur (1837a).
17 On the later history of Miss Brackett see Deleuze (1846), pp. 370–371.

18 Brownson (1854), p. 33; Deleuze (1846), p. iv.
19 Dickerson (1843), p. 18.; Angoff (1968), p. 27.
20 Tomlinson and Perret (1974).
21 Collyer (1843), p. 10.
22 On Dods see Fuller (1982), pp. 83–89.
23 I have used Dods (1876), in which both sets of lectures are reprinted.
24 So the *Dictionary of American Biography* terms him; however, the same source also tells us that at the age of seventy Grimes persuaded an insurance company to change a $4000 life policy into a $400 annuity, and lived another 26 years, so it is clear that being erratic sometimes paid off.
25 On Léger and the invention of the "magnetoscope" see Ashburner (1867), pp. 59–81, and cf. Léger (1852).
26 Cf. Angoff (1968), pp. 16–17, 27–28; the appendix and new appendix to Deleuze (1846); Kiernan (1895); and S. Gregory (1843), who says (p. 10): "Drs. Caldwell and Buchanan of Louisville, Kentucky; Professor Mitchel [sic] of Philadelphia, and Dr. Sherwood of New York, have published works on the subject. Among other mesmerizers and advocates of mesmerism, are Drs. Capron, Brownell, Utley, of Providence; Drs. Lewis, Flint, Dana, Ingalls, Gilbert, Gregerson, Bell, Stedman, Stone, &c., of Boston. Indeed hundreds of physicians in all parts of the country are now zealously engaged in investigating its claims, and proving its utility in practice."
27 A committee of gentlemen (members of a class in animal magnetism) reported to the *Richmond Whig* in December 1845 that a particular operation performed with mesmeric anaesthesia in their presence was the third such operation carried out in the past month in Richmond, Virginia, alone. Deleuze (1846), p. 403.
28 Brownson (1854), p. 23.
29 Fuller (1985), p. 165.
30 Dods (1876), p. V.
31 Sunderland (1847), p. 73n; Deleuze (1846), p. 387; Collyer (1843), p. 10. Cf. *The Zoist*, 1843–4, Vol. 1, pp. 54–55. In fact, the possibility had been mentioned in print at least twenty years earlier. Cf. Loewenhjelm (1819), pp. 139–140.
32 Sunderland (1847), p. 73n; Collyer (1843), p. 9n.
33 Wester (1975–6). This society, or some of its members, seem afterwards to have become involved in the foundation, under the direction of a magnetic clairvoyant, of a cooperative community called the Cincinnati Brotherhood. See Podmore (1902), Vol. 1, p. 175.
34 Dods and Grimes (1851), p. 155.
35 Dods and Grimes (1851), p. 10.
36 Dods (1876), pp. 191–192.
37 Yet they were consistently plagiarized by later writers and lecturers like Stone and Fiske. On the latter's plagiarism, see *Boston Daily Mail*, February 7th 1850, and *The Zoist*, 1852–3, Vol. 10, pp. 63–64.
38 Bimetallic discs were a popular quack remedy of the time. See Dingwall (1968a), pp. 117–118.
39 A number of mesmeric, phrenological and kindred demonstrators seem to have visited Hannibal, Missouri, where Mark Twain lived as a lad, in the period 1847–50, and Twain himself gives a very entertaining account of them. See Wecter (1952), p. 196.
40 For England and Canada, see the appendix to Stone (1850); for Holland, see Zorab (1967), pp. 80–81.
41 Dods and Grimes (1851), pp. 168–169.
42 Dods and Grimes (1851), p. 169; Dods (1876), pp. 188, 192.

43 Bertrand (1823), pp. 246–262.

44 Teste (1843), pp. 215–225, cites cases in which mesmeric passes were made at the spot where the illusions and hallucinations were to appear. Presumably, the rationale was similar to that of Dupotet's magical experiments discussed in the previous chapter.

45 As an example of such a phrenomesmeric performance, see the account of one of his own cited by Léger (1846), pp. 395–397, from the New York *Tribune* 9th April 1844.

46 Davis wrote an autobiography (Davis, 1857). Useful short accounts of Davis are given by Podmore (1902), Vol. 1, pp. 158–176, Brown (1970), pp. 73–97, Daly (1964) and Zwelling (1982).

47 Podmore (1902), Vol. 1, p. 169.

48 A useful digest of *The Principles of Nature* and others of Davis's writings is Davis (1917).

49 Davis (1847), pp. 675, 676; Podmore (1902), Vol. 1, pp. 163, 172.

50 Spiritualism, Mormonism, Adventism all started there.

51 See her mother's statement, Lewis (1848), pp. 6–7.

52 On the history of American Spiritualism in general see Capron (1855); Podmore (1902); S. Brown (1970); Moore (1977).

53 Hare (1856), p. 94; cf. Ballou (1852), p. 85.

54 Mahan (1855); Rogers (1852); Mattison (1853).

55 On all this see Podmore (1902), I, pp. 204–206.

56 Dods (1854).

57 Dresser (1921). On Quimby in general see Fuller (1982), pp. 105–136.

58 On Mrs Eddy and her relation to mesmerism and mind-cure see Podmore (1909).

59 Fuller (1985), p. 164.

CHAPTER 11

Mesmerism in Britain

It is probable that Mesmer's activities in Paris in and after 1778 soon became quite widely known in Britain, but I have not traced any British publication of note concerning animal magnetism before 1784, when a translation of parts of the Report of the Commissioners of the Royal Academy of Sciences was published in *The Gentleman's Magazine*.[1] A full translation was published in 1785,[2] and in the same year we find the first indications that animal magnetism was being practised in Britain. *The Times* of 30th November 1785 announced that a school of animal magnetism was to be opened at the west end of the town under the auspices of a French lady, a disciple of Mesmer. Other "empirical professors of animal magnetism" set up shop, but a year later, on 17th November 1786, *The Times* noted that the supporters of animal magnetism had failed, despite much exertion on their part.

The Times spoke too soon. In 1787, a French pupil of d'Eslon, Dr. de Mainauduc, a "man-midwife",[3] began to attract notice in London. By the following year, animal magnetism was well enough known to be the subject of a farce by Mrs. Elizabeth Inchbald (1753–1821), the actress and playwright. *The Times* remarked of this play that it would perhaps not be fully understood, "as the price at which this visionary science is taught, is such as to exclude the lower order of people from acquiring any knowledge of it."[4]

The "professors" of animal magnetism in Britain in this and the succeeding years were, indeed, unashamedly mercenary. Mainauduc (who had been taken up by certain titled ladies[5]) asked twenty-five guineas for a course of fifteen lectures.[6] Mr. John Holloway required five guineas for a course of three lectures. By July 1790 he claimed to have instructed over 200 persons.[7] Mr. John Parker shortly began an attempt to undercut Holloway, and offered the same service for only two guineas.[8] Those who practised magnetism without teaching it may have been more modest in their demands.[9]

A few small pamphlets on animal magnetism were published in London at this period,[10] but none is of any great interest, and all lean heavily upon the French literature. There were also occasional public debates on the subject.[11] Interest in animal magnetism probably reached its zenith in 1790, when the author of a pamphlet published in May of that year remarked on "the number of illustrious and respectable names which dignify the list of converts to magnetism".[12]

However, on 6th September 1790, *The Times* gave a different assessment: "To the honour of this country, the weak people that are influenced by the folly of Animal Magnetism are but few." Whatever British interest in animal magnetism there may have been died away fairly quickly as the French sources from which it sprang dried up following the Revolution. For nearly four decades we hear almost nothing of animal magnetism in Britain.[13] Now and again we learn of some isolated individual who was practising it or trying to rouse interest in it, but none of these ventures came to anything.[14] The few books that were published were – with one exception – throwbacks to or spin-offs from the the more prosperous days of 1787–90.[15]

The one exception is a strange work entitled *La Prima Musa Clio. Or, the Divine Traveller*, privately printed in London about 1810. The author, or translator, George Baldwin (1743?-1826), a much-travelled person of mystical tendencies, who had begun experimenting with animal magnetism in 1789[16] (probably in the traditions of continental illuminism), was for some while British consul-general in Egypt. In 1795, he encountered there a wandering Italian poet, Cesare Avena di Valdieri, who proved an excellent magnetic subject. *La Prima Musa Clio* is Baldwin's translation, from his earlier Italian version of 1802, of the entranced Cesare's rhapsodical accounts of his highly allegorical visions of other worldly scenes and entities. It has thus the distinction of being the first such magnetic revelation to be actually printed. Its 614 pages of vapid and verbose lunacy give it the further distinction of being the most unreadable book in the whole literature of animal magnetism (not excluding the works of Andrew Jackson Davis).

The German magnetic movement of the early nineteenth century went almost unnoticed in Britain. Of course, Coleridge had wandered much in the mazes of German metaphysics and had reflected long and deeply upon the problems of animal magnetism;[17] but only rarely does this emerge in his writings. The only account that I have noticed of magnetism in its German phase is three anonymous articles – reasonably unprejudiced – in *Blackwood's Edinburgh Magazine* for 1817 and 1818.[18]

French animal magnetism, and the controversies which it aroused in the Paris medical world during the 1820s, eventually began to breach British insularity here and there, though the usual result was simply to heighten the self-satisfaction which that insularity had bred.[19] However, the only extensive practical trials of animal magnetism carried out in Britain at this time derived from French models. The operator, Richard Chenevix (1774–1830), was a distinguished chemist, who had somehow managed to live on the continent of Europe during much of the Napoleonic Wars. In 1816, he had encountered and been impressed by that noted magnetic demonstrator, the Abbé Faria. In May 1828, on a visit to Ireland, he began to try out animal magnetism on assorted Irish peasants, and by January 1829 had "tried the effects of mesmerism upon

164 persons, of whom 98 manifested undeniable effects". Going to England in 1829, he was allowed to try his methods in various hospitals, at all of which, according to the enthusiastic articles which he published in the *London Medical and Physical Journal* for 1829, he cured or alleviated a variety of physical ailments, ranging from dyspepsia to epilepsy.[20] While Chenevix was in London, several distinguished medical men, including John Elliotson (1791–1868), Sir Benjamin Brodie (1783–1862), Henry Holland (1788–1873), William Prout (1785–1850) and Henry Earle (1789–1838), came to witness his demonstrations, and the matter might have been pursued further but for his death the following year.

The French controversies finally surfaced *in extenso* in Britain with the publication in 1833 of J.C. Colquhoun's English translation of the Husson report. Colquhoun (1785–1854), an Edinburgh lawyer, had studied in Germany, and had there acquired an interest in animal magnetism which eventually resulted in his two-volumed treatise, *Isis Revelata* (1836). Colquhoun was well-read in both the French and the German literature of animal magnetism, and his is the first extended English account of them. It was not superseded for over half a century, and is indeed still useful today. The author's somewhat naïve enthusiasm for the alleged paranormal phenomena and the light they shed on the spiritual nature of man is easily allowed for.

Neither Colquhoun's translation of the Husson report nor his *Isis Revelata* seem at first to have had any great impact,[21] but the latter was well-timed in that its publication immediately preceded the first serious signs in Britain of a rising interest in mesmerism. A third edition was called for in 1844.

JOHN ELLIOTSON: EARLY CONTROVERSIES 1837–8

It was the intrepid French lecturer and demonstrator Baron Dupotet who in effect started the mesmeric movement in Britain. Undeterred by his ignorance of the English language, he arrived in England in June 1837, and set out his stall with his customary lack of inhibition.[22] He remained in England for some twenty months. No doubt he came because he scented cash, but he had also a sincere desire to spread the benefits of magnetism. His most important influence on the development of animal magnetism in Britain was, however, indirect; it was he who reawakened John Elliotson's interest in the subject.

Elliotson was professor of the practice of medicine at London University, and senior physician to University College Hospital (which he had had a large share in founding). He was one of the most eminent clinicians of his time, with a large private practice, and a high reputation as a teacher and demonstrator. A man of immense and restless energy, he was noted in the profession for his impatience with conventional methods of treatment and his continued experimentation with new ones.[23] Medical students loved him; but some established practitioners resented his innovations and his assertiveness. And there was another reason why certain professional bretheren fought shy of him. He was an advocate of phrenology and President of the London Phrenological Society. Phrenology, as expounded by Elliotson, embraced the view that mind is a property of the living brain.[24] Although he also argued that immortality remains possible through bodily resurrection by means of a miracle of the Almighty, he was widely suspected of atheism.[25]

Roger Cooter tries to account for Elliotson's "compulsive turning to unorthodox knowledge" by speculating that as the son of a prosperous London druggist, he may have "shared the characteristic ambivalence of the socially and intellectually upwardly mobile", seeking "on the one hand ... through the mantle of learning to bury a past connected with crass petty commercialism", and on the other "to defy the smug superiority of those born into traditional power and privilege". He was, moreover, scarcely five feet tall.[26] Since this kind of explanation could account for exaggerated conformity as readily as for exaggerated non-conformity, it seems more sensible to point out that Elliotson was intelligent, well-read, active, ambitious, self-confident, and courageous, beyond the majority of his colleagues. In pressing innovations upon a conservative profession he was, quite simply, often right. And, far from wishing to defy the smug superiority of the well-connected, he seems to have taken a rather marked pleasure in their society.[27]

Dupotet's attempts to rouse interest in animal magnetism met at first with little success. He was advised to approach Elliotson, who received him courteously, and invited him to mesmerize several patients in University College Hospital. In the autumn of 1837, Dupotet took advantage of Elliotson's temporary absence to expand his activities, which were then halted by the Hospital Committee. Elliotson thereupon took over the mesmeric treatment of certain patients himself,[28] while Dupotet hired a house and began to offer daily lectures and demonstrations which achieved some success,[29] though it appears that next year he got into pecuniary difficulties and had to return to France.[30]

Elliotson was a well-known figure, and the fact that he had begun to practice mesmerism caused a wave of interest which soon spread outside medical circles. He charts its progress as follows:

> Some excellent cures were effected, and such striking phenomena produced that the students regularly attended; then students from other schools; then they all requested to bring their friends, so that for convenience [Elliotson] was obliged to mesmerize no longer in the wards but in the theatre ... Requests were poured in upon him from all classes of medical men and from others to be permitted to witness the cases; the highest nobility and even royalty attended.[31]

The cures included "a severe case of periodic insanity"; "a child who had laboured under paraplegia and incontinence of urine during nine months"; a

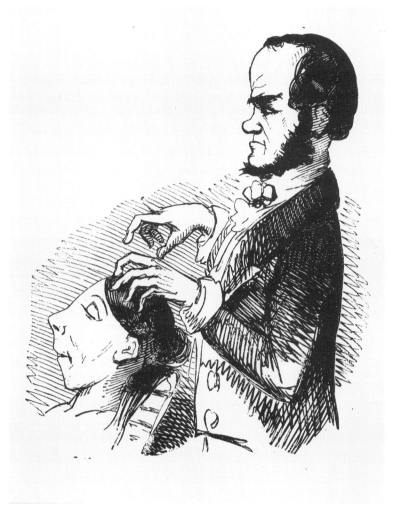

Figure 4. John Elliotson mesmerizing. (From the Mary Evans Picture Library.)

severe case of epilepsy (*grand mal*); a case of delirium "in a young woman subject to hysteria"; and a case of St. Vitus' dance.[32] The "striking phenomena" were mainly though not exclusively those exhibited by two epileptic sisters Elizabeth and Jane Okey, aged seventeen and sixteen respectively. By continental standards the phenomena were not especially remarkable, or remarkable only for the sauciness of the repartee in which the girls engaged while in the state of "delirium", and for the rapidity with which they could be sent by vertical passes from wakefulness to delirium to coma, and by transverse ones to wakefulness again.[33]

While in the magnetic state, the sisters could be "pulled" by the mesmerist's gestures, or induced to move their hands or limbs in sympathy with him, despite the intervention of a screen. There were occasional apparent instances of clairvoyance, transposition of the senses, and foreknowledge of impending ailments. Prominent in several of Elliotson's demonstrations were experiments supposed to show that the mesmeric influence could be stored by various substances. Drinking magnetized water or touching or even approaching magnetized gold would send either sister into a state of sleep or catalepsy. The effects occurred, or (where appropriate) failed to occur, even when precautions were taken to prevent the girls learning which objects had been magnetized and which had not. For instance, gold sovereigns were put in warm water so that residual warmth from the magnetizer's hand should not provide a clue. It was, however, experiments of this kind which ultimately led to the girls' downfall, and had indirectly a profound effect on the future of animal magnetism in Britain.

Accounts of Elliotson's demonstrations appeared in *The Lancet* throughout the summer of 1838.[34] In August, its founder and editor, Thomas Wakley (1795–1862), M.P. for Finsbury and coroner for West Middlesex, undertook personal investigations of magnetism.[35] Elliotson brought the Okey sisters to Wakley's house, and there, in the presence of various medical men:

> ... experiments were made under Wakley's sole direction. On the first day the violent contortions and muscular cramp, which were the characteristic result of contact with mesmerised nickel, were produced when the nickel – unknown to Elliotson and most of the company – was safe in the waistcoat pocket of one of the spectators. It was shown in a further series of experiments that unmesmerised water could produce sleep, whilst water which had been carefully mesmerised had no effect; and that whilst three or four mesmerised sovereigns could be handled with impunity, well-marked catalepsy was produced when Jane Okey stooped to pick up a sovereign which had merely been warmed in hot water, without human contact at all.[36]

Wakley (along with many subsequent opponents of magnetism) regarded these experiments as providing conclusive evidence of deliberate imposture, and as definitively exploding mesmeric doctrines about the magnetic fluid. He had, however, done no more than Bertrand, or the commissioners of 1784. His findings did not prove deliberate dishonesty by the Okeys, who were hysterical girls of the kind so common in the history of hypnotism,[37] and they told against a quasi-physical influence only in so far as this was supposed capable of being stored in certain kinds of substance. Yet he shortly launched himself into a campaign of bitter and often quite disgraceful invective against the mesmerists, a campaign which was to sputter on for much of the next two decades.

Whether (as Elliotson insinuates[38]) because of the jealousy of his colleagues and their fear of Wakley's onslaught, or because of Elliotson's own headstrong disdain for their doubts and palterings, the Hospital Committee, which had already denied him the use of the Hospital's theatre for his demonstrations, ended in December 1838 by discharging Elizabeth Okey, despite his contrary recommendation, and forbidding the practice of magnetism on the wards. Elliotson immediately resigned from both College and Hospital, and never again crossed the threshold of either institution.[39]

The events of 1838 had, however, been sufficient to secure mesmerism a toehold, and something more than a toehold, in Britain. Elliotson continued to demonstrate mesmeric phenomena in his own home to all interested and respectable persons; he claims that by this means he made hundreds of converts, who in turn converted others.[40] And there began a small but increasing trickle of British publications on the subject,[41] though an attempt by Colquhoun, in 1839, to establish a mesmeric periodical, *The Journal of Zoomagnetism*, quickly foundered.

The most influential of these early works was the Rev. Chauncy Hare Townshend's *Facts in Mesmerism* (1840). Townshend (1798–1868), was a wealthy clergyman who held no living and, for reasons of health, spent much of his time abroad. He had achieved early recognition as a poet; later, Dickens was to dedicate *Great Expectations* to him. His experiences with magnetism began at Antwerp in 1836. He conducted his own experiments at Cambridge in 1837, and thereafter more extensively on the continent, where he discovered several gifted subjects. His book deals in part with the treatment of disease, but principally with the more remarkable phenomena of what (following Elliotson and various German writers) he calls the "sleep-waking" state. He speculates extensively as to the psychological and philosophical bearings of the phenomena, and, like Colquhoun and numerous continental writers, postulates an all-pervasive electric ether as the half-way house between soul and matter, and the vehicle of clairvoyance, mesmeric rapport, and so forth. *Facts in Mesmerism* is the intellectual Odyssey of a much-travelled, literate, reflective, unassuming and totally honest man, already well-known among people who mattered. It attracted notice, and came out in a new and enlarged edition in 1844. Townshend was the first of several Church of England clergymen whose books on mesmerism seem to have carried considerable weight with the public.[42]

LAFONTAINE AND S.T. HALL

Further impetus was given to the mesmeric movement in Britain by the arrival in June 1841 of Charles Lafontaine, the French magnetic demonstrator.[43] Lafontaine took rooms in Hanover Square, London, where he gave regular demonstrations. The principal phenomenon upon which he relied to carry conviction to his audience was that of insensitivity to stimulation. *The Times* of 20th July 1841 describes a demonstration in which a young man had pins stuck in his cheeks, ammonia held under his nose, a pistol fired near his head, matches burned beneath his nose, and strong electric shocks sent through his body, without giving any signs of feeling. A gentleman from the audience who tried out the electrical machine on himself "gave a shout that we shall never forget as long as we live". On 24th July 1841, *The Times* called for an impartial inquiry into the merits of mesmerism.

In October 1841, Lafontaine set out for a lengthy tour of the provinces, visiting Birmingham, Manchester, Leeds, Sheffield, Nottingham, Leicester, Liverpool, Ireland and Scotland.[44] He drew large audiences and attracted a good deal of newspaper publicity.[45] He claims that in Sheffield, about January 1842, he acted at the request of Dr. Holland (G. C. Holland, 1801–1865) as magnetizer for the painless amputation of a leg.[46] Towards the end of 1842, he ran into financial difficulties and had to return to France.[47]

Lafontaine's importance for mesmerism in Britain lay only partly in the audiences he attracted and the publicity he received. His chief influence sprang from the imitators he inspired, persons who thought that what an itinerant Frenchman could do, they could do equally well, and moreover just as profitably. Quite a few such imitators were in the field before Lafontaine had left Britain, and by the middle of 1843 they were numerous and widespread. There were even lecturers, such as Duncan and J. P. Catlow (d. 1867) who devoted themselves to showing that mesmeric ideas about the "fluid" were mistaken. Duncan was a follower of Braid (to whom we shall come), and Catlow attributed the phenomena to "suggestive dreaming" (apparently a special state in which unintended suggestions arouse a preternaturally quick association of ideas).[48] After 1843, platform demonstrators became less numerous, but this did not reflect any general decline of interest in mesmerism. Probably the general increase during the 1840s in the number of established local mesmeric practitioners left less scope for the itinerant lecturers.

Most of the demonstrators[49] are, individually, of no special importance to the history of mesmerism, but it will be of some value to glance briefly at the activities of perhaps the most successful of them, Spencer T. Hall (1812–1885). Hall, the son of a Nottinghamshire cobbler, had become known as a versatile journalist and poet under the name of "The Sherwood Forester". His interest in mesmerism was aroused when he attended one of Lafontaine's demonstrations at Sheffield in 1842. He began to experiment for himself, and gave his first public demonstration in the summer of 1842 at Castleton in Derbyshire, as the result of a chance conversation in an inn there. His career rapidly took off. He held meetings at most of the leading Midland cities, and his audiences numbered hundreds; indeed he claims[50] that on 5th December 1842 he lectured to 3000 persons in the amphitheatre, Sheffield. In 1843, he started *The Phreno-Magnet* (which lasted a year), and continued to lecture with even greater success. By now – and in some cases, no doubt, because of Hall's activities – many of the cities he visited (for instance Liverpool, Leicester, Coventry, Nottingham,

Burslem, Halifax and Northampton) had phrenomesmeric societies or committees, composed (naturally) of leading citizens.[51]

The next year, 1844, Hall helped to initiate what was shortly to become a *cause célèbre*, namely the mesmeric cure of the well-known authoress, Harriet Martineau (1802–1876). Miss Martineau had been suffering since 1839 from retroversion of the uterus and uterine polypi, which had caused her intense suffering and had rendered her more or less an invalid. Her surgeon, Mr. T. M. Greenhow (her brother-in-law) thought it might be worth trying mesmerism, and on 22nd June 1844 brought Hall, who was on tour in the North of England, to Newcastle, where Miss Martineau was then living. She felt benefitted by his ministrations, and arranged for a friend, and then for her own maid, to continue the treatment. By September, she was much improved, and by December she was totally free of pain. In November and December, with considerable courage, she published an account of the whole affair,[52] causing a sensation. Her account (somewhat oddly) prompted a reply from Mr. Greenhow, who attributed her cure to the pills of iodide of iron which she had been taking for several years. Miss Martineau (who had taken up the practice of mesmerism herself), replied that the improvement in her condition had begun with the beginning of mesmeric treatment and had continued after she had ceased to take the iodide of iron.[53]

To return to Hall: In the autumn of 1844, he was demonstrating at Edinburgh, where he remained until 1845, apparently by invitation. He seems to have aroused some interest among the professors of the University there. Especially impressed was William Gregory, the professor of chemistry, a pupil of Liebig (who was present at some striking demonstrations). Later, Hall took up other forms of fringe medicine, notably homoeopathy and hydropathy, but he never lost his belief in mesmerism.

MESMERISM AND PHRENOLOGY: *THE ZOIST*

As the title of his journal, *The Phreno-Magnet*, indicates, Hall was a phrenomesmerist, as were most of the other lecturers and demonstrators of the time, including E. T. Craig (1804–1894), the editor of another short-lived periodical, *The Mesmerist*.[54] Who first introduced phrenomagnetic phenomena to Britain is obscure. Mr. H. G. Atkinson, a British gentleman amateur,[55] claims to have hit upon them at about the same time (the middle of 1841) as Collyer in America and Mr. Mansfield in Britain.[56] Both mesmerism and phrenology gained greatly from the union. A successful demonstration of phrenomesmerism was so much more immediate, compelling and dramatic than a demonstration of either phrenology or mesmerism separately.[57] Only a few British mesmerists – for instance W. Newnham and J. C. Colquhoun[58] – were strongly opposed to phrenology.

In allying itself with phrenology, mesmerism acquired a partner that already had some established bases. Phrenology began to gain momentum in Britain in the early 1820s, and a national Phrenological Association was started in 1838 with the deliberate aim of improving the status of the subject. (The Association split and then foundered because of the militant materialism preached by Elliotson's friend, Dr. W. C. Engledue (1813–1858) of Southampton, in his Presidential Address of 1842.[59]) Not, however, that phrenology in Britain lacked eminent supporters, at least in its early days. Many of its supporters were medical men,[60] and it gained adherents in most classes of society, including the highest. Prince Albert was on good terms with Elliotson and had his childrens' heads examined by George Combe.[61] But throughout the 1830s, and with increasing rapidity after 1840, phrenology, and for a time, therefore, phreno-mesmerism, became a movement whose greatest strength was among the artisans and radical working class. It soon came to permeate many of the progressive social movements of the time,[62] to which it appealed for a variety of reasons. It had implications for education, for law and order, for the treatment of the insane, for the adaptation of social and political institutions to the proper expression of human nature. G. J. Holyoake (1817–1906), an Owenite, and later a secularist, found phrenology "a natural corollary of Owenism".[63]

Elliotson, who was a leader of what might be called the intellectual wings of both the phrenological and the mesmeric movements, was by no means an Owenite or a secularist, and disapproved of most popular demonstrators,[64] but he none the less supported many "progressive" causes. The Zoist, which he founded in 1843, and edited jointly with Engledue, is subtitled A Journal of Cerebral Physiology and Mesmerism, and their Applications to Human Welfare. It ran until 1856, and is the most important single source of information about British mesmerism of the period. It did not, however, neglect the "applications to human welfare". If phrenology is correct, it has important applications to education (each person should be educated in accordance with the potentialities revealed by the development of his various phrenological organs), to morality and crime (in youth different organs may be enlarged or diminished by appropriate treatment), to punishment (it cannot be right to punish, rather than to restrain or re-educate, a criminal who is what he is through no fault of his own, but because of the state of his cerebral organs). For Elliotson, these applications to human welfare were not just rational and utilitarian turnings of natural laws to practical purposes. He was moved to passionate indignation at the thought of suffering and injustice brought about by human obstinacy, stupidity or irrationality, and therefore avoidable. To him it was monstrous that congenital criminals or lunatics should be executed for crimes they could not help committing; monstrous that convicts should be subjected to the "disgusting, disgraceful barbarities" of flogging;[65] monstrous that surgeons should torture their shrieking patients without troubling to inform themselves of the possibilities of mesmeric anaesthesia. Capital punishment concerned him particularly. He regularly filled pages of *The Zoist* with analyses of the heads of executed murderers,[66] analyses which demonstrated both the truth of phrenology and the futility of capital punishment.

Elliotson – humane, sensitive, committed, deeply concerned – was in many ways ahead of his time. In these parts of his writings we can begin to discern the man whom Dickens and Thackeray so much admired as a doctor and a human being,[67] the man who, whatever his faults, was generally acknowledged to be "incapable of dishonesty or falsehood."[68] Unfortunately, there was another side to him. His warmth of sympathy for suffering mankind[69] easily turned to a warmth of anger against those who were too stupid to see things as he did. That he should have given his vehement and inveterate medical critics, especially Wakley, as good as he got is understandable, though his occasional sneers at Wakley's lack of a classical education could have been dispensed with. But even persons who had merely passed an incidental anti-mesmeric remark were liable to find themselves pilloried at length in *The Zoist*. And Elliotson's attitude to those who, like Braid and Colquhoun, admitted the phenomena, or many of them, but differed from him over their interpretation, is peculiar. Braid is rarely mentioned in *The Zoist*,[70] and when he is, it is very briefly; the same applies to Colquhoun who is the "bigotted [*sic*] and irascible"[71] Mr. Colquhoun, probably because he had published a pamphlet against phrenomesmerism. Both Braid and Colquhoun had courage; they had been prepared to voice beliefs which were liable to bring them into ridicule and contumely; and it is a pity that Elliotson, who had so much courage himself, did not respond to it when its possessors proved to be not quite on his side.

It was round *The Zoist*, and hence of course round Elliotson, that British mesmerism centred during its period of most active expansion, from 1843 until the early 1850s. Popular practitioners looked up to it, even though they perhaps resented its lofty tone.[72] More serious and more educated adherents subscribed, contributed and sent in cases; interested outsiders turned to it to find out more. Despite Elliotson's unrestrained denunciations of the bigotry of his opponents, his own views on mesmerism are in most respects notably moderate. He avoids committing himself to specific doctrines about a mesmeric fluid, maintaining only that "an occult soporific power over others exists clearly in the animal force";[73] it was long before he became convinced that certain somnambules might manifest a clairvoyant perception of distant scenes; though he does not doubt the silent power of the will upon mesmerized persons, he is opposed to those writers who maintain that all mesmeric phenomena result from the will only;[74] he is well aware of the power of imagination and suggestion upon mesmerized persons, and often used what would now be called verbal post-hypnotic suggestions on certain of his patients. He does not spend much time theorizing about the phenomena; his orientation is pragmatic – the proven usefulness of mesmerism in alleviating disease and suffering is what matters most. *The Zoist* of course reflects this bias. Setting aside articles on phrenology, mesmeric cures of disease fill the greatest percentage of its pages, followed by cases of surgical operations performed with mesmeric anaesthesia. These will be treated in the next chapter. A long way behind come mesmeric and related phenomena of other categories; the theory of mesmeric phenomena[75] hardly figures at all, except incidentally as part of some discussion under another heading.

THE LITERATURE OF BRITISH MESMERISM, 1843–55

Much of the British mesmeric literature of the period shares this pragmatic and practical orientation. A good deal consists of works, brief, unpretentious and middle-of-the-road, intended as guides for the practitioner, usually the un-qualified practitioner, such as the head of a household. No purpose would be served by describing this fairly extensive literature in any detail.[76] It reached a peak in the early 1850s, which I am inclined to think marks the peak of British interest in mesmerism. Systematic treatises, covering the history, phenomena, theories and implications in some detail, were by contrast somewhat scarce, and did not grow more frequent; though by 1845 many of the classical French works were being offered for sale by H. Baillière, the publisher and bookseller, at 219 Regent Street, London. The works by Colquhoun (1836) and Townshend (1840) have already been mentioned. William Newnham's *Human Magnetism: Its Claims to Dispassionate Inquiry* (1845) originated when its author was commissioned to build a case against magnetism. Newnham (1790–1865) was a general practitioner at Farnham, in Surrey, and a writer on medical subjects, and his book is quite substantial. It is, however, heavily dependent on Ricard and Foissac, and is chiefly remarkable as containing an attack on phrenomesmerism. A far more useful book is the Rev. George Sandby's *Mesmerism and its Opponents* (1844; second edition enlarged, 1848). Sandby (1799–1881) was vicar of Flixton in Suffolk, and though not himself a prominent practitioner of mesmerism, knew a good deal about the contemporary mesmeric scene in Britain.[77]

The most substantial British treatise on animal magnetism is the *Letters to a Candid Inquirer on Animal Magnetism* (1851) of William Gregory, M.D., (1803–1858), professor of chemistry at the University of Edinburgh. *The Dictionary of National Biography* describes Gregory as "simple, earnest and amiable", and these qualities[78] are amply reflected in the 528 pages of his book. It is rich in accounts of extraordinary phenomena, some from his own experience, but short in critical spirit and detailed references. Gregory was, I think, himself so totally honest that he found it difficult to doubt, still less to scrutinize minutely, any happening for which any other person had pledged his word.[79] At the same time there can be no doubt that he had heard of from reliable persons, or himself witnessed,[80] various remarkable phenomena (to some of which I shall return), and that he deserves every credit for trying to draw the attention of the learned world to them.

The most interesting feature of Gregory's book is his detailed descriptions of the performances of two visiting electrobiologists, Dr. Darling and Mr. Lewis, who had lately been making a considerable stir in Scotland. Gregory recognized, more clearly perhaps than any writer whom one could properly label an animal magnetist, that it was often possible to produce in persons not entranced or in the somnambulic state, and through fixation of gaze and verbal suggestion only, a whole range of startling phenomena (hallucinations, visions, delusions, role-enactment, responsiveness to post-hypnotic suggestions) once thought to be characteristic only of the deeper stages of somnambulism. This did not, however, lead him to abandon the theoretical presuppositions of the magnetists. He held that magnetic somnambulism involves a redistribution of magnetic fluid (or in his terminology "odyle"), brought about by the external influence of the magnetizer, and having as its main effects a shutting down of the ordinary sensory pathways and a heightening of sensitivity and activity in the "inner senses". Darling and Lewis (and also Braid) act on their subjects by inducing in them a similar redistribution of fluid, which is however not occasioned by any direct influx of force from the magnetizer. The result is the production of a peculiar state (auto-magnetism), which may or may not pass into magnetic sleep, but in which the above-mentioned phenomena can be brought about. In highly susceptible persons (presumably those whose odyllic force tends naturally to the appropriate distribution), such phenomena may even be brought about by straightforward suggestion, without any preliminary process of fixation of gaze, etc.

Gregory casts his net widely and shows an interest in the revelations of Cahagnet's somnambules and kindred topics. But this interest springs from a general curiosity and open-mindedness – he neither accepts nor rejects the revelations, but simply thinks such matters worth inquiring into. There is, indeed, little extremism of any variety in the British mesmeric literature. Spiritual revelations, such as those of the German somnambules, are in very short supply. J. W. Haddock's *Somnolism and Psycheism* (1849) describes clairvoyant visions not unlike those of Cahagnet's subjects. Hill H. Hardy's *Analytic Researches in Spirit-Manifestations* (1852) presents a Swedenborgian viewpoint. A small crop of pamphlets by evangelicals ascribed mesmeric phenomena to diabolical agency[81] – one of them describes the blue lights so often seen streaming from the operator's fingers as "sulphureous flames, perhaps".[82] Radically sceptical works on animal magnetism are likewise notably lacking, which is surprising when one remembers how ready the medical press was to brand mesmerists "humbugs", "charlatans", and the like. Perhaps medical critics were reluctant to hazard a book-length justification of such violent assertions.

The general tone, then, of British mesmeric literature, is pragmatic, putting cures and phenomena before theory and philosophical implications. What of the actual mesmeric practitioners? Who were they? And what was their background? There is a helpful list in the second (1848) edition of Sandby's *Mesmerism and its Opponents*[83] of locally eminent persons who have successfully used mesmerism in the treatment of disease. These 81 persons can be divided up into the following categories:

Medical men (physicians and surgeons)	32
Clergymen	4
Professional men of other categories and persons of independent means	16
Mesmeric lecturers and demonstrators	8
Status unknown	21

This list is heavily biassed towards persons who contributed to or are mentioned in *The Zoist*. It passes over all but a few of the more reputable itinerant demonstrators, and takes no account of the mesmeric experimentation, partly serious, partly born of curiosity, and partly done for amusement, that went on in many middle and upper class households at this period.[84] But it fairly strongly suggests that there existed in Britain a modest but quite solid nucleus of perfectly respectable and non-fanatical individuals who regularly made use of mesmerism in the attempt to cure disease and alleviate pain.

MESMERISM AND THE MEDICAL PROFESSION IN BRITAIN

When one considers the obvious sanity, sincerity, respectability and good intentions of so many of the British mesmerists, and the fact that not a few of them were medically qualified, the violent, almost hysterical, hostility of so much of the medical establishment and the medical press is not easy to understand. That hostility is, however, readily documented, for *The Zoist* made a regular point of collecting and publishing the more outrageous examples of medical invective.

In the pages of *The Lancet*, Wakley was perennial and unchanging. "Are the delusions, caused by the tricks of women, to be called a Science", he asked on 8th September 1838; on 24th December 1853, after years of abusing mesmerists, he was yet more vehement: "Coffinites and herbalists, nostrum-mongers and syphilitic doctors, cancer-curers or mesmerists, professors of biology and a host of other childish nonsense and iniquitous folly, – all these, without exception, are ignorant imposters, whose sole object is to cheat and defraud the public, and gain an easy, because dishonest and disreputable livelihood." Wakley, of course, was rather against than for the medical establishment, but plenty of choice examples can be culled from journals other than *The Lancet*. For instance, *The Medico-Chirurgical Review* for 22nd November 1842 says: "The mesmero-mania has nearly dwindled, in the metropolis, into anile fatuity, but lingers in some of the provinces, with the *gobe-mouches* and chaw-bacons, who, after gulping down a pound of fat pork, would, with well-greased gullets, swallow such a lot of mesmeric mummery as would choke an alligator or a boa constrictor."[85]

From time to time, optimistic supporters of mesmerism would fancy that they could perceive signs of a milder tone in the medical press's attitude towards them. For instance, when John (later Sir John) Forbes (1787–1861), editor of *The British and Foreign Medical Review*, wrote in April 1845 and October 1846 two articles on mesmerism which admitted that "mesmerism has hardly received fair play at the hands of many of our professional brethren, or in the pages of our contemporaries",[86] several mesmeric writers began to feel that the tide might be turning in their favour.[87] *The Medical Times* received praise from mesmerists as the only medical journal which admitted articles in their favour.[88] However, I very much doubt whether during the whole of the period with which we are concerned there was on balance any marked change in the attitude of the British medical press towards mesmerism. Indeed in 1850 (16th February), we find even *The Medical Times* writing, "Time was when Dr. Elliotson was respected, and honoured, and esteemed... But he swerved from the straight-forward course which was before him. He was tempted, with gipsy-like credulity, to wander into the paths of darkness, which so sadly obscured his vision, that he mistook even his duty as a professor...".[89] And so on.

The excesses of British medical men were not confined to the printed word. In the early 1840s, it was not unknown for them to attempt to break up mesmeric demonstrations by conduct bordering on violence. For instance, about the end of June 1843 a demonstration at Northampton by S. T. Hall was interrupted by a group of local doctors, alarmed by the success of a previous meeting. With "Dr. Robertson, the senior physician of the town, at their head, they came to the theatre in a gang, and by clamour, clapping, yells, and hisses, interrupted the lecture for more than two hours."[90]

All in all, it is hard not to agree with an editorial footnote in *The Zoist* for 1851 that it was "curious to observe the very low state of moral feeling, the want of dignity and self-respect, in so many of the medical profession".[91] No doubt the polemical literature of that time was more robust than would be acceptable today; no doubt some mesmerists were deliberate deceivers, and many others were deceiving themselves; but the vehemence of the medical attacks seems enormously out of proportion to the well-intentioned, and probably quite harmless, activities of the more respectable magnetic practitioners. Why such hatred? Part of the answer, I think, is fear, or at any rate consternation. As T. M. Parsinnen has argued,[92] mesmerism in Britain threatened the social and political aspirations of medical men during a critical period in their history. The profession was widely held in low repute. Many common methods of treatment were unpleasant, and of doubtful benefit; no less than nineteen bodies could license persons to practice medicine, and standards were very variable; there were unseemly differences of opinion between reformers like Wakley and Forbes, and certain sections of the medical establishment. That people should so often have turned to fringe practitioners reflected upon the medical profession and hindered attempts at reform and standardization; moreover, the growing number of fringe practitioners meant a loss of fees to the orthodox. Where would the process end? It was not just the poor, who could not afford fees, that turned to the mesmerists. Prince Albert was known to be interested, and to be critical of the medical men of the country for refusing to investigate.[93] When the London Mesmeric Infirmary was set up in 1849, it had a number of distinguished supporters, one of whom was the second Earl of Ducie (1802–1853), a former Lord-in-Waiting to Queen Victoria.[94]

It must have been particularly galling to persons who had gone through a toilsome medical education to find their livelihoods encroached upon, and even their wealthier customers tempted away, by practitioners of a form of healing the essentials of which could be learned in a few hours.

THE SPREAD AND DECLINE OF MESMERISM

With the dust and din of so much combat obscuring the scene, it is not easy to be completely certain how much the practice of mesmerism was spreading. The editorial introduction to the seventh volume of *The Zoist* (1849) claims that at the present time "there is not a town in the United Kingdom which does not furnish a medical man who has experimented for himself, and who is prepared to give an opinion in a positive manner, supported by facts of his own collection, and by views the result of serious conviction. Six years have produced this great change in the medical world ... ".[95] It goes on to remark that literary journals have likewise changed their attitude since 1843. Most of them have now referred to it, "and in a manner calculated to excite attention and calm investigation rather than the reverse". Further on in the same volume occurs the following passage:

> ... nine, or even five years ago, the various mesmeric lecturers in London and the provinces met with the most vulgar rudeness, coarseness, abuse, and even rioting, and especially from the medical portion of the audience. But all this has subsided ... The various English works on mesmerism are in every house; our *Zoist* is read extensively in Europe and America, though

childishly excluded from the public medical libraries of this country. An
ignorant and pert scoffer of mesmerism in a party is sure to meet with a
rebuke.[96]

All this might be dismissed as excessive optimism. But it agrees well with a range
of other pointers: the increase in mesmeric publications; the unremitting stream
of hostile comment in *The Lancet* and other medical periodicals; even the
remarks of a country surgeon, Mr. Tubbs, at the first Annual General Meeting of
the London Mesmeric Infirmary in May 1850. Mr. Tubbs said that when he first
took up mesmerism, his practice at Upwell in Cambridgeshire had diminished
because of local prejudice. But now things had improved, he was mesmerizing
again, and was converting many who were formerly hostile.[97]

Of course, the existence of the London Mesmeric Infirmary was itself a sign of
the spread of mesmerism. The first steps towards founding it were taken at the
Earl of Ducie's house in 1846. A committee was formed, and subscriptions were
solicited. Elliotson was a driving force. In 1849, the time was deemed ripe for
finding premises, and in January 1850 the infirmary came into operation in
premises at 9 Bedford Street, Bedford Square, with Mr. Buckland as secretary
and principal magnetizer.[99] Lord Ducie was President, Earl Stanhope, the
Archbishop of Dublin, and Augustus de Morgan, the mathematician, were Vice-
Presidents. Lord Adare and several M.P.s[100] were subscribers. The following
year, Mr. T. Capern of Tiverton (called "the modern Greatrakes") became
secretary and principal mesmerizer. In 1855, the infirmary moved to 36
Weymouth Street and Mr. Gardiner became secretary and principal magnetizer.
Annual Medical Reports were produced. At the sixth Annual General Meeting,
for instance, with the Earl of Dunraven in the chair, those present learned that
247 individuals had been treated in the previous year, most of them being out-
patients.[101]

A Bristol Mesmeric Institute was founded in 1849, again with Lord Ducie as
President,[102] and a Dublin Mesmeric Dispensary in 1852.[103] Both of these
institutions seem to have failed quite quickly,[104] but this did not prevent the
formation of a Scottish Curative Mesmeric Association in 1854,[105] with Gregory
as its President, and General Sir Thomas Makdougall-Brisbane (1773–1860), a
distinguished astronomer and President of the Royal Society of Edinburgh,
Colquhoun, and Esdaile, as Vice-Presidents. The London Mesmeric Infirmary
continued to flourish for some while, and to issue Annual Reports down to
1866.

It is hard to say to what the rather marked increase in the practice of
mesmerism by and upon respectable persons during the period from 1843 to the
early 1850s may be attributed. Partly it was due to the forceful personality of
Elliotson, and to *The Zoist*, partly, perhaps, to the shortcomings of contemporary
medicine. But the most potent factor was almost certainly the steadily rising
number of surgical operations conducted under mesmeric anaesthesia (these

will be considered in the next chapter). In particular, the reports of Esdaile from 1846 onwards made a considerable impression, and many people now found it impossible to deny that mesmerism could produce effects at the same time extraordinary and beneficial to humanity.

Conversely, the decline in mesmerism in Britain after the mid-1850s was to an extent due to the growing use of chemical anaesthetics.[106] Elliotson's retirement in 1856 and his withdrawal into the home of Dr. G. S. Syme, a former pupil, lost mesmerism its most energetic proponent and its most successful periodical. The decline in the mesmeric movement in Britain had, however, already begun before 1856, and, as in America, it was partly due to the advent of modern Spiritualism. The arrival in 1852 of the first Spiritualist mediums from America led to a widespread craze for raps and table-tilting (a craze which *The Zoist* did its best to stem). Spiritualism stole not a little of mesmerism's thunder. The initial period of Spiritualistic excitement in 1853–5 was succeeded by a period of relative calm. However, from about 1859, when the celebrated medium D. D. Home arrived in Britain (Home, incidentally, converted Elliotson, the life-long sceptic, to Spiritualism[107]), until at any rate the mid-1870s, Spiritualism in Britain grew steadily. It made, as in America and elsewhere, ready use of mesmeric concepts,[108] and for the most part Spiritualism and mesmerism remained in harmony.[109] Mesmerism, however, was now much the less exciting partner and, though one or two new mesmeric works of note were published,[110] it faded more and more from the public eye.[111]

NOTES

1 *The Gentleman's Magazine*, 1784, Vol. 54, Part 1, 944–946.
2 The translation, by William Godwin, was entitled *Report by Dr.Benjamin Franklin and other Commissioners ...*, and is reprinted in Tinterow (1970), pp. 82–128.
3 *Wonders and Mysteries of Animal Magnetism Displayed ...* (1791), p. 7.
4 *The Times*,16th April 1788. The play was first performed at Covent Garden on 26th May 1788. On 29th April 1788, *The Times* reviewed four French books on animal magnetism, one of which, Delandine's *De la philosophie corpusculaire*, it regards as Mrs. Inchbald's chief source.
5 See *The Times*, 4th August 1789; 19th September 1789; 17th June 1790. The ladies were Anne, second wife of George Townshend (1724–1807), first Marquis Townshend, and Lady Wright, wife of Sir James Wright.
6 *The Times*, 4th September 1790.
7 *The Times*, 22nd July 1790.
8 *The Times*, 4th October 1790.
9 In the latter part of 1788, *The Times* contained various satirical accounts of Dr. Freeman, a Jamaican horse-doctor who had turned to animal magnetism. See *The Times*, 7th August 1788; 20th August 1788; 2nd September 1788; etc. On 13th July 1789, *The Times* carried an advertisement for "The many wonderful cures that are daily performed ... by Mr. YELDALL'S CONCENTRATED MAGNETIC EFFLUVIA" Six applications of his magnetical apparatus cost one guinea.

10 Pearson (1790); *Wonders and Mysteries* (1791); *A Practical Display* (1790); Martin (1790); Bell (1792). Edwards (1789) has little to do with animal magnetism.

11 On the debates see *The Times*, 10th, 17th and 24th September 1789; 20th October 1789. These were linked to the cures supposedly wrought by the painter P.J. Loutherbourg (1745–1812) and his wife, which were, however, more religious than mesmeric. On the Loutherbourgs see *The Times*, 16th, 17th and 24th September 1789; *Wonders and Mysteries* (1791), pp. 9–10; Pratt (1789).

12 Pearson (1790); Lee (1866), pp. 34–35. John Pearson (1758–1826) was a London surgeon.

13 Kaplan (1974), p. 695, attributes this in part to traditional British hostility to all things French.

14 For instance, we hear of animal magnetism at Lutterworth, Leicestershire in 1813 (The *Phreno-Magnet*, 1843, Vol. I, p.75); London in 1816 (Corbaux, 1816); and London again in 1817 (Friedlander, 1817, p. 169). In 1816 F. Corbaux made an unsuccessful attempt to establish a journal, the *Magnetiser's Magazine and Annals of Animal Magnetism*.

15 E.g. Mainauduc (1791); Winter (1801).

16 See Baldwin (1818–19).

17 See e.g. Lowes (1927), pp. 254, 546–7.

18 Observations on animal magnetism (1817); On the present state of animal magnetism in Germany (1817–18); The German somnambulist and Miss M'Avoy (1817–18).

19 Cf. the review of Dupau (1826) in *The London Medical and Physical Journal*, 1826, NS1, 463–471; Animal magnetism (1830b); Animal magnetism (1830a). The latter is better informed and more reasonable than the others.

20 Chenevix (1829). For a critical paper by an eyewitness of Chenevix's Irish activities see Cotter (1829).

21 It received a long, and not wholly unfavourable, review in the *Quarterly Review*, 1838, Vol. 61, pp. 273–301.

22 Dupotet's account of his London venture is in Dupotet de Sennevoy (1840), pp. 186–262.

23 On Elliotson see J.H. Hardy (1952), pp. 13–80; Bramwell (1930), pp. 4–14; Quen (1976); Rosen (1936); Miller (1983); and Kaplan's introduction to Kaplan ed. (1982). He was the first or one of the first medical men in England to use a stethoscope; to use dilute prussic acid to allay vomiting; to use iodine in the treatment of goitre; to try acupuncture for the relief of rheumatism; to link hay fever with the state of the atmosphere; to note that glanders is communicable to man.

24 Elliotson (1840), Vol. I, pp. 32–48.

25 Cf. Elliotson (1840), Vol. I, pp.41–48. Elliotson's views seem really to have been exceedingly hostile to Christianity. See the letter from Elliotson to George Combe printed by Miller (1983).

26 Cooter (1984), pp. 52–53.

27 See the letter in Lang (1843), pp. 190–193, by a gentleman who had attended a demonstration in Elliotson's own drawing-room, at which a number of "the fair ones of the British Court" were present.

28 S.I.L.E. (1843–4), pp. 89–90. I presume that S.I.L.E. was Elliotson.

29 See e.g. *The Times*, 18th October 1836; 1st November 1837.

30 S.I.L.E. (1843–4), p. 90.

31 S.I.L.E. (1843–4), pp. 90–91.

32 *The Lancet*, 26th May 1838, p. 283. Most of these cases were later described in detail in *The Zoist*.

33 Elliotson (1843–4d), p. 171.

34 The principal accounts are *The Lancet* 26th May 1838, 282–288; 9th June 1838,
377–383; 16th June 1838, 400–403; 14th July 1838, 546–549; 21st July 1838,
585–590; 28th July 1838, 615–620. The best short account is Podmore (1909), pp.
126–134. Elliotson's own account is Elliotson (1840), Vol. 2, pp. 660–694. Elliotson's
clinical notebooks are in the Library of University College, London.

35 These were reported in *The Lancet*, 1st September 1838, and were followed by editorials on
8th and 15th September.

36 Podmore (1909), pp. 132–133.

37 Elliotson remained in touch with the Okeys, and always continued to defend them against
the charge of imposture. See e.g. Elliotson (1843), pp. 44–45, according to which they were
both cured of their violent fits. "One is already respectably married to a young man in her
own station of life, and a mother; the other lives with her parents and supports herself by
doing needlework for a neighbouring establishment."

38 S. I. L. E. (1843–4), pp. 91–93; Elliotson (1843–4d), pp. 171–172.

39 S. I. L. E. (1843–4), p. 91.

40 S. I. L. E. (1843–4), p. 93.

41 E.g. "Physician, A" (1838); Lee (1838; 1843); Wilson (1839); Lang (1843);

42 Sandby (1844; 1848); Pyne (1844).

43 Lafontaine's luxuriant beard and moustaches (things not often seen in Britain) brought him
much publicity. Women in Regent Street would cover their faces and cry out when they saw
him – Lafontaine (1866), Vol. 1, p. 272. On Lafontaine's appearance see S. T. Hall (1845),
p. 1.

44 From Lang (1843), p. 9, it appears that Lafontaine was not the first to demonstrate
mesmerism in Scotland, that honour belonging, apparently, to P. E. Dove (1815–1873),
whose experiments at Edinburgh in 1839 were witnessed by Sir William Hamilton and J.
Y. Simpson, the discoverer of chloroform.

45 E.g. *Manchester Times*, 13th and 20th November 1841; *Manchester Guardian*, 13th and
17th November 1841.

46 Lafontaine (1866), Vol. 1, pp. 320–321.

47 S. I. L. E. (1843–4), p. 93.

48 On Catlow see *The Phreno-Magnet*, Vol. 1, 1843, pp. 86–87, and Hall (1873), p. 180; On
Duncan, *The Times*, 22nd December 1841, 1st January 1842.

49 Many of these lecturers are listed by Cooter (1984), pp. 272–300.

50 *The Phreno-Magnet*, 1843, Vol. 1, p. 193; Hall (1845), p. 72.

51 See *The Phreno-Magnet, passim*.

52 *The Athenaeum* (1844), pp. 1070–1072, 1093–1094, 1117–1118, 1144–1145, 1173–
1174; afterwards published as Martineau (1845), the first edition of which sold out in four
days. For Hall's account see Hall (1845), pp. 63–75. Further information about Harriet
Martineau and mesmerism can be gleaned from her correspondence with Lord Houghton
among the Houghton papers in the Library of Trinity College, Cambridge.

53 Greenhow (1845); L. U. G. E. (1845–6).

54 On Craig see Cooter (1984), p. 279.

55 Co-author with Harriet Martineau of Atkinson and Martineau (1851).

56 *The Phreno-Magnet*, 1843, Vol. 1, pp. 221–222. Cf. Elliotson (1843–4a), pp. 236–246.
Collyer seemingly hit on the phenomena in November 1839, but did not publicly
demonstrate them until about May 1841.

57 In his Presidential Address to the Phrenological Association in 1843, Elliotson said: " ...no
circumstance, no book of any of Gall's disciples or disciples of his disciples, no lectures, no
societies, have ever given the impulse to the reception of phrenology which has been given

by mesmerism. When I formerly converted one, I have since by means of mesmerism, converted a hundred to the truth of phrenology." Elliotson (1843–4a), p. 230.

58 Newnham (1845), pp. 374–411; Colquhoun (1843).

59 Cooter (1984), p. 94.

60 Cooter (1984), pp. 28–32.

61 On Prince Albert and phrenology see De Giustino (1975), pp. 219–223.

62 On mesmerism's appeal to the Owenites, see Cooter (1985).

63 Oppenheim (1985), p. 226.

64 Exceptions to Elliotson's disapproval included William Davey (d.1858) and J. W. Jackson (1809–1871) who lectured jointly in the West of England, Ireland and Scotland (The Zoist, 1849–50, Vol. 7, p. 328). See Davey (1854) and Jackson (1851).

65 Elliotson (1843), pp. 16–17.

66 Many of these are accompanied by plates depicting plaster casts of the shaven heads of the executed criminals. There seems to have been some trade in such casts at the time.

67 Cf. Russell and Goldfarb (1978), pp. 69–70.

68 Review of Elliotson (1843) in The Examiner, 1st July 1843.

69 He was also an opponent of vivisection; see The Zoist, 1844–5, 2, p. 189.

70 Podmore (1909), p. 147.

71 The Zoist, 1849–50, Vol. 7, p. 327.

72 Cf. The Phreno-Magnet, 1843, Vol. 1, pp. 99–100.

73 Elliotson (1844–5a), p. 235; cf. Elliotson (1846–7e), p. 123.

74 In Thompson (1847–8), pp. 253–254.

75 There are rather more papers on what might be called theoretical issues in phrenology.

76 As examples may be mentioned: Pyne (1844); Kiste (1845); Pasley (1848); Tweedie (1857); Saunders (1852); Buckland (1850); Jackson (1851); Capern (1851); C.M. Friedlander (1852); Miles (1854); Barth (1853); Hastings (1854); Davey (1854); Saunders (1855); Townshend (1854); Neilson (1855); Mesmerism and Media (1855); Mesmerism Solved (1853).

77 In writing about animal magnetism in Britain I have been able to consult a volume of Tracts and Extracts collected and annotated by Sandby, now in the Society for Psychical Research Library.

78 There is a vivid portrayal of Gregory in Hall (1873), pp. 176–177.

79 He was even prepared to take the notorious "snail telegraph" seriously – Gregory (1851), pp. 244–245. Cf. Benoit and Biat (1850).

80 Gregory's book is rich in his own observations and those of his friends. Reviewing the book for The Zoist, W. C. Engledue (1851–2), accuses him of insufficiently acknowledging the work of earlier British pioneers.

81 M'Neile (1842); Mesmerism Considered (1852); A Christian (1852); Corfe (1848). M'Neile's sermon attracted some attention and drew a reply from Sandby (Sandby 1843), which was the nucleus out of which Sandby (1844) was derived. M'Neile also drew forth Braid's first work on mesmerism, Braid (1842; repr. Tinterow, 1970, pp. 317–330).

82 Corfe (1848).

83 Sandby (1848), pp. 143–162.

84 Townshend (1844) provides a good few examples, and I think that many would be found in letters and diaries of this period. For instance, the family papers of F. W. H. Myers, an important figure in late nineteenth century hypnotism, which are now in the library of Trinity College, Cambridge, show that his father, the Rev. F. Myers, of Keswick, and his circle of friends, experimented with mesmerism a good deal during the 1840s.

85 E. W. C. N. (1844–5), p.278.

86 April 1845, p. 430.

87 Sandby (1848), pp. 33–37.

88 E. W. C. N. (1844–5), p. 285.

89 Quoted in *The Zoist*, 1850–1, Vol. 8, pp. 31–32.

90 *The Phreno-Magnet*, 1843, Vol. 1, p. 163. For an extraordinary attempt by a Leamington doctor to expose Hall see *The Cheltenham Journal*, 8th December 1845.

91 *The Zoist*, 1851–2, Vol. 9, p. 31n.

92 Parsinnen (1979).

93 L. U. G. E. (1845–6), p. 88n.

94 Ducie (a well-known agriculturalist) had been himself a mesmeric patient, a fact which led to some offensive remarks in *The Medical Gazette*, 6th July 1849. See Podmore (1909), pp. 141–142.

95 *The Zoist*, 1849–50, Vol. 7, p. 3.

96 *The Zoist*, 1849–50, Vol. 7, p. 328.

97 *The Zoist*, 1850–1, Vol. 8, p. 208.

98 Author of Buckland (1850).

99 See the Report of the First Annual General Meeting of the London Mesmeric Infirmary. *The Zoist*, 1850–1, 8, 203–211. Subsequent reports of AGM's will be found in *The Zoist*, 1851–2, 9, 124–145; 1852–3, 10, 194–218; 1854–5, 12, 180–195; 1855–6, 13, 171–198.

100 *The Zoist*, 1848–9, 6, p. 237n lists fourteen members of the House of Commons who are supporters of mesmerism.

101 *The Zoist*, 1855–6, pp. 171–198.

102 *The Zoist*, 1849–50, Vol. 7, pp. 152–164; 1850–1, Vol. 8, pp. 211–220.

103 *The Zoist*, 1852–3, Vol. 10, pp. 218–222. An Exeter mesmeric institute of which one also reads in *The Zoist* seems to have been a private venture by a local surgeon, J. B. Parker.

104 *The Zoist*, 1854–5, Vol. 12, p. 110.

105 *The Zoist*, 1854–5, Vol. 12, pp. 108–111; 1855–6, Vol. 13, pp. 199–202.

106 Some mesmerists fought a rearguard action by pointing to the high mortality rate with chemical anaesthesia. For instance, Elliotson does so in Elliotson (1854–5a).

107 Oppenheim (1985), pp. 221–222.

108 See for instance Jones (1861) and cf. Barrow (1986), pp. 94–95. Jones calls the mesmeric fluid "mesmerine."

109 Zerffi (1871) attempts to explain away all spiritualist phenomena in mesmeric terms.

110 E.g. Lee (1866); Ashburner (1867).

111 In H. Baillière's *List of English, American, and French Works, on Spiritualism, Magnetism & Mesmerism* for April 1867, works on Spiritualism greatly predominate, and much of the material on magnetism and mesmerism antedates 1860.

CHAPTER 12

Topics from *The Zoist*

Though *The Zoist* is extensively permeated with the prejudices of its editors, it is very informative about the mesmeric scene of the 1840s and 1850s. In particular, it either itself published original accounts of, or else comprehensively reported, many of the more interesting mesmeric phenomena then being obtained both in Britain and abroad. These accounts will form a convenient basis for brief discussions of certain of these phenomena, though there will be no need to limit discussion to instances appearing in *The Zoist*. The phenomena in question are:

1. Mesmeric analgesia.
2. Phrenomesmerism and hemicerebral mesmerism.
3. The so-called "Reichenbach" phenomena and related matters.
4. Electrobiology and related matters.
5. Alleged instances of extrasensory perception (ESP) occurring in a mesmeric context.

MESMERIC ANALGESIA

From its very earliest days, mesmerism was held to be effective in calming otherwise intractable pain, and some striking case histories could be cited, not the least remarkable of which involve the permanent relief of agonizing *tic douloureux*.[1] However, this section will be concerned exclusively with analgesia for the pain occasioned by surgical operations.

So far as I know, the first published reference to the possible uses of mesmeric analgesia in surgery was by Gmelin (1789).[2] Visiting Karlsruhe about 1787, Gmelin had been impressed by the paralyses and analgesia produced in the limbs of some of Böckmann's subjects by repeated localized downward passes. The analgesia was demonstrated by inserting needles as far as the bone, and furthermore bleeding of the resultant wound was inhibited. However when, eventually, mesmeric analgesia came to be used in surgery, it was in general obtained not by localized passes, but by the customary procedures for putting the patient into the "mesmeric sleep" or into the state of mesmeric somnambulism.

It is often implied that the first operation carried out with "mesmeric anaesthesia" was that performed on Madame Plantin in 1829 by Jules Cloquet (see chapter 7 above). This may have been the first major "mesmeric" operation, but it was certainly preceded by a number of minor ones, some of which have already been mentioned.[3] For example, in 1819 Dr. Spiritus of Solingen painlessly extirpated a ganglion from the wrist of his somnambulic subject, Anna Maria Joest.[4]

The Plantin case did not lead to any sudden rush of comparable operations. Between 1830 and 1842, there were, so far as I can discover, only a few cases of (relatively minor[5]) surgical operations with mesmeric analgesia, though one must not forget how very painful some of the dental examples would ordinarily have been.[6] However, 1842 saw the beginnings of a marked increase. This began with what is generally regarded as the first[7] important British case, that of James Wombell, a Nottinghamshire labourer aged forty-two, who underwent amputation of the left leg above the knee because of a carious knee-joint. He was put into a "mesmeric sleep" by William Topham, Esq., of the Middle Temple (later Colonel Sir William Topham), and the operation, which was an unusually protracted one, was performed by W. Squire Ward, a local surgeon. During the operation Wombell moaned a little as if in a troubled dream, but gave no other sign of pain. The moaning did not increase when Ward thrice touched the divided end of the sciatic nerve.[8] After the operation Wombell blessed the Lord to find that it was all over, and stated that he had felt no pain. He made a good recovery and survived for more than thirty years.[9]

Topham and Ward read papers on the case to the Royal Medical and Chirurgical Society of London. These were received with some hostility.[10] But their pamphlet describing the case,[11] and Elliotson's booklet *Numerous Cases of Surgical Operations without Pain in the Mesmeric State* (1843), in which it also figures, brought the whole issue to public notice.

During the next few years, 1843–6, a number of surgical operations of some severity were carried out in Britain, France and the United States with the patients under mesmeric analgesia.[12] Minor operations became quite common.[13] I can mention as examples only a few of the more serious cases.

In the United States, the case which attained the most celebrity[14] was that of the amputation of a breast by Professor L. A. Dugas of the Medical College of Georgia on 20th January 1845. The patient, Mrs. Clark, was mesmerized by Mr. B. F. Kendrick, and the operation, which began with two elliptical incisions about eight inches in length, was witnessed by five other medical men. All present concurred in stating "that neither the placid countenance of the patient, nor the peculiar natural blush of the cheeks, experienced any change whatever during the whole process; that she continued in the same profound and quiet sleep ... and that, had they not been aware of what was being done, they would not have suspected it from any indications furnished by the patient's condition."

The pulse rate remained unchanged throughout.[15] The unfortunate lady had to undergo two further operations when the tumour recurred, and in both exhibited under mesmerism an apparently perfect insensibility.[16]

In France, a good deal of interest was aroused by a sequence of twelve operations with mesmeric analgesia carried out by Dr. Loysel of Cherbourg between 2nd October 1845 and 4th June 1847.[17] For example, on 22nd May 1846, Loysel operated on M. Baysset, junior, a young man of eighteen, with enlarged (probably tuberculous) glands underneath both sides of his jaw. The mesmerizer was M. Delente. The first cut made "a large opening extending from the left and back part of the lower jaw bone to the centre of the chin. The operator then carefully dissected away a considerable mass, which presented seven united glands, the largest of which had the form and size of a hen's egg." Some spectators were so disturbed by the sight of the enormous wound that they had to leave the room. The same process was repeated for the extirpation of two glands on the right side. The total amount of time occupied in actual cutting was nineteen minutes. The patient remained completely tranquil and immobile throughout, and his pulse and breath rates did not alter (the pulse beat at 84). Afterwards he declared that he had no knowledge of what had happened, and that he did not suffer anything. The *procès-verbal* was signed by twenty-three persons, of whom five (including the operator) were medical men.[18]

The use of mesmeric analgesia, indeed anaesthesia, for surgical purposes reached its culmination in, and obtained its greatest celebrity from, the work in India of James Esdaile (1808–1859), a Scottish surgeon. In 1845, Esdaile, an employee of the East India Company, was in charge of a small hospital at Hooghly, in Bengal, about twenty-five miles north of Calcutta. He was also, apparently, in charge of the jail hospital there. On 4th April 1845, he was treating a convict afflicted with double hydrocoele. The drainage and injection of one side of the scrotum caused the patient such pain that Esdaile determined to try mesmerism upon him for the second operation. Although Esdaile had never witnessed mesmeric procedures, he was successful in rendering the convict analgesic, and at once began to experiment with mesmerism both as a means of producing analgesia in surgical cases, and as a method of treatment for medical ones. Only the former aspect of his work will be touched on here.

Before long Esdaile, whose own health was not good, began to delegate a large part of the tiring mesmerization[19] – he would, if necessary, have a patient magnetized for hours each day for ten or twelve days by native assistants, saving his own strength for the performance of surgery. He was content to have his patients in a condition of "mesmeric sleep"; only rarely did he attempt to induce the "sleep-waking" or "somnambulic" state. He soon acquired a great reputation in the vicinity for his numerous painless operations in cases of hypertrophy of the scrotum (endemic in Bengal due to filariasis transmitted by mosquitoes). Some of these scrotal "tumours" were of immense size, one having a weight of about 112 lbs.[20] Victims of this disease flocked to him, but he carried out many other kinds of operation too. His success rate for inducing analgesia among the natives seems to have been remarkably high – he attributes this to widespread disease, debility and undernourishment, and to a submissive and uncomprehending attitude of mind.[21] It was not uncommon for the mesmerized patients undergoing severe operations to moan and stir as if in a "troubled dream", or to show signs of waking up before the end; but even these patients, when fully restored to consciousness, would firmly assert that they had felt no pain, had no recollection of the operation, etc., and would bless Esdaile with pathetic gratitude. Many other patients exhibited no signs of pain whatever, despite undergoing operations of equal severity.[22] It should be noted that we do not have to rely on Esdaile's word for all this. Various reputable persons (including medical men) came to witness the operations, and gave their own accounts of them.[23]

Esdaile reported his early results in the *India Journal of Medical and Physical Science*. His articles were reprinted as a small pamphlet, *Mesmeric Facts*, published at Hooghly about August 1845.[24] In January 1846, he sent a report on seventy-five cases to the Medical Board at Calcutta. Some of his cases were very remarkable, and he had reputable witnesses, including medical ones, but the Board did not even trouble to send him an acknowledgment. About the same time he completed his first and best-known book, *Mesmerism in India and its Practical Application in Surgery and Medicine*, which was published in London later in the year. Esdaile's knowledge of the literature of mesmerism is slight. The most interesting parts of his book are those which are in effect extracts from the journal of a working surgeon. The case histories vary a good deal in the amount of detail given; otherwise they are straightforward, repetitive, unquestionably honest and, in mass, rather impressive even to a modern reader.

About April 1846, by which time his tally of operations had grown to 102, Esdaile wrote a report which he sent directly to the Government. This time he obtained a response. The Deputy Governor of Bengal, Sir Herbert Maddock, appointed a committee of seven persons (of whom four were medically qualified, and three were sceptics concerning mesmerism) to investigate Esdaile's claims. Esdaile was invited to demonstrate in Calcutta, and ten surgical cases were provided for him from the wards of the Native Hospital. Seven of these, after having been mesmerized by Esdaile's assistants, underwent operations of varying severity before members of the committee. In four cases (one of them minor) the patient gave no sign of pain, though in two of them there was an accelerated pulse rate; in three the patient writhed, moaned, and pulled faces; in all, the patient afterwards declared that he had felt no pain. In its report,[25] published on 9th October 1846, the committee somewhat grudgingly agreed that some operations had been painless, and favoured an extension of Esdaile's work. A small "experimental hospital" in Calcutta was thereupon put at his disposal (November 1846). The hospital was to be open to the public, and medical visitors were appointed.

Over the next thirteen months, Esdaile's case records were published regularly by order of the Government,[26] as were the reports of the official visitors. Esdaile's Calcutta case records are very similar to his Hooghly ones, with operations for hypertrophy of the scrotum particularly common. By this time it had become Esdaile's practice to test the readiness of his patients for operation by dropping a hot coal on the inside of a leg, or by giving them the strongest electrical shock which his machine could produce, or (most often) by powerfully squeezing their testicles. Only if the patient did not respond in any way did Esdaile feel able to proceed to surgery. Otherwise magnetization was continued.

His patients, it is worth noting, were by no means all peasants and convicts, as is sometimes implied. Many of them were comfortably off. Esdaile was irritated by the palterings of certain of the medical visitors, and in January 1848 wrote in the *India Register of Medical Science* a reply to their reports, which was afterwards reprinted as *A Review of my Reviewers* (1848).[27] In the same month Lord Dalhousie, the Governor-General of India, who had taken a keen interest in Esdaile's work, promoted him to the post of Presidency Surgeon. No hospital attached to this post, but in September 1848 a mesmeric hospital was opened in Calcutta by public subscription so that Esdaile could continue his work. He issued two half-yearly reports on his cases,[28] but after some eighteen months the Mesmeric Hospital was closed when the Deputy Governor, Sir John Littler, appointed Esdaile superintendent of one of the Government Hospitals, especially to introduce mesmerism into regular hospital practice. In 1850, he was appointed to the position of Marine Surgeon, which was potentially very lucrative. He preferred, however, to devote himself to his mesmeric work.

From about 1848, the new chemical anaesthetics, ether and chloroform, came into use in Indian medical practice, and Esdaile took an interest in this development. On 6th September 1850, he published in *The London Medical Gazette* an article "On the operation for the removal of scrotal tumours".[29] He argued, on the basis of 161 mesmeric operations for the relief of scrotal hypertrophy, with only a 5% mortality rate, that chloroform would be of questionable value in the case of the larger tumours, for it "has a tendency to paralyze the heart, lungs and brain. With tumours over 40 lbs the loss of blood is usually so profuse that the pulse is usually extinguished on the spot. Persons fainting from such a haemorrhage would probably never recover if the vitality of the heart, lungs and brain was lowered by any additional influence whatever."[30] Esdaile also believed that in mesmeric operations haemorrhage was less, and the course of recovery more favourable, than in operations with chloroform or operations with no anaesthesia at all. Elliotson likewise contrasted the dangers of chemical anaesthesia unfavourably with the safety and benign after-effects of mesmerism.

In June 1851, Esdaile completed his twenty years of service with the East India Company, and retired at once, afterwards remarking that he had detested the climate, the country and all its ways, from the moment he first set foot in it. He had carried out altogether 261 operations using mesmeric analgesia.[31] He settled in Perth, Scotland, where he undertook no further medical practice, and only a little mesmerism (chiefly in an advisory capacity).[32] He wrote, however, two small volumes reviewing his mesmeric activities in India.[33] Later, the Scottish climate proving too bracing for the weak lungs which had sent him to India in the first place, he removed to Sydenham, where he died in 1859. His mesmeric work in Calcutta was continued for a short while by Dr. Allan Webb, Professor of Demonstrative Anatomy in the Medical College of Calcutta, who obtained results very similar to Esdaile's, but also experimented with chloroform.[34] Webb, however, seems to have left in 1852, and I have heard no more of mesmeric analgesia being used in Indian hospitals after that date.

We cannot leave Esdaile – the most famous figure in the whole history of mesmeric and hypnotic analgesia – without quoting some examples of his "painless operations". The two following are from his *Mesmerism in India*, and are necessarily short. The first is a commonplace operation for hypertrophy of the scrotum. It illustrates Esdaile's belief that mesmeric analgesia lessens operative shock.

> If patients are fortunate enough to sleep some time after the operation, they not only feel no pain on waking, but none subsequently even. The following is an extraordinary instance of the absence of pain, from first to last.
> Sept. 1st [1845]. – Teg Ali Khan, a tall strong-looking man; has a hypertrophied scrotum ... The tumour is perfectly round, and as big as a man's head. He was mesmerised in two hours, on the first trial, and, in the presence of Drs. Ross and Sissmore, I dissected out all the parts; which was very tedious, from the testes having contracted adhesions all round them; and it was about half an hour before the organs were covered up again by stitching flaps over them. Not a quiver of the flesh was visible all this time, and at the end, his body was as stiff as a log, from head to foot, and his separated legs could be with difficulty put together again. He woke in half an hour after the operation, and felt no pain.[35]

In the following instance, the patient, a man of forty, had suffered for two years from a tumour in the maxillary sinus. This was pushing upon the orbit of the eye, and had filled the nose and entered the throat. The operation lasted half an hour – Esdaile describes it as "one of the most severe and protracted in surgery". The patient, who had initially proved very difficult to mesmerize, afterwards declared that he had felt no pain whatsoever.

> I put a long knife in at the corner of his mouth, and brought the point out over the cheek-bone, dividing the parts between; from this, I pushed it through the skin at the inner corner of the eye, and dissected the cheek back

to the nose. The presence of the tumour had caused the absorbtion of the anterior wall of the antrum [sinus], and on pressing my fingers between it and the bones, it burst, and a shocking gush of blood, and brain-like matter, followed. The tumour extended as far as my finger could reach under the orbit and cheek-bone, and passed into the gullet – having destroyed the bones and partition of the nose. No one touched the man, and I turned his head into any position I desired, without resistance, and there it remained till I wished to move it again: when the blood accumulated, I bent his head forward, and it ran from his mouth as if from a leaden spout. The man never moved, or showed any signs of life, except an occasional indistinct moan; but when I threw back his head, and passed my fingers into his throat to detach the mass in that direction, the stream of blood was directed into his wind-pipe, and some instinctive effort became necessary for existence; he therefore coughed, and leaned forward, to get rid of the blood; and I supposed that he then awoke. The operation was by this time finished, and he was laid on the floor to have his face sewed up, and while this was doing, he for the first time opened his eyes.[36]

More than anything it was these numerous accounts of painless operations, Esdaile's and others, that convinced not a few members of the literate public that there might be something in mesmerism. Members of the medical profession were much less willing to be convinced. The luckless Wombell, who had moaned a little while his leg was being taken off, was, like Madame Plantin, more than once accused of having suppressed the external indications of pain, and this despite his published declaration (witnessed by the vicar of his parish) to the contrary.[37] In support of this contention, surgeons dwelt on instances of persons who had endured amputations with extraordinary fortitude.[38] No doubt there were such persons, such as the Marquis of Anglesey and an unnamed sailor,[39] but they do not seem to have been more numerous than, say, the select few hardy enough to take the cat o'nine tails in stubborn silence.[40] And, as Elliotson pointed out,[41] persons steeling themselves to undergo such suffering without writhing or outcry none the less betray their agony by their posture, their muscular tension, or their accelerated pulse and breath rates, whereas in many cases of mesmeric operations these signs were not manifested. Even granted the ability of so many patients to endure torture, it is quite incomprehensible that they should most of them have maintained afterwards that they felt no pain, thus sacrificing all credit for their bravery in order to boost the false claims of the triumphant mesmerist. That persons, the great majority of whom there is no reason to regard as dishonest, should adopt such a policy would itself be an argument that they had fallen under some powerful and mind-warping spell. Above all, it was the sheer number of painless operations upon persons in the mesmeric state that made the theory of imposture seem so implausible. In the second edition of his *Mesmerism and its Opponents* (1848), Sandby lists[42] from the period 1841–1847 312 examples of surgical operations without pain under mesmeric influence, 178 of them being tooth extractions (of which, however, one should not forget the potential for pain). The figure of 312 is certainly an underestimate, but even so it hardly seems credible that within six years 312 seemingly ordinary persons should have been at once heroic enough to suffer terrible pain without fuss, and pliant enough to insist for no clear motive that they had felt nothing. And it is wholly incredible that (as more than once happened) the same person should have several times voluntarily repeated this pointless and agonizing charade.

PHRENOMESMERISM AND HEMICEREBRAL MESMERISM

Phrenomesmerism rose to prominence in Britain at much the same time, and with much the same velocity, as it did in the United States. However, in Britain it remained prominent rather longer; it was a background assumption of much that appeared in *The Zoist*, those contributors who did not share Elliotson's enthusiasm for it perhaps choosing not to air their absence of conviction; it remained a central feature in public demonstrations of mesmerism, and by the 1850s was regularly billed as a side-show attraction.[43] Given the pervasiveness of phrenomesmeric assumptions and phenomena on the British mesmeric scene, it is perhaps surprising that there were not a larger number of detailed and careful investigations of the alleged phenomena (which were supposedly manifested only by subjects already in the mesmeric state). In the first volume of *The Zoist*, Elliotson lays down some of the basic requirements of such investigations – that not a word should be uttered or any thing done which could by the slightest possibility suggest to the subject the expectation of the operator; that there should not even be contact over any organ, unless it is known with absolute certainty that the subject is totally unacquainted with phrenology; and that the operator should will nothing, if possible not knowing to what cerebral organ he is pointing.[44]

Elliotson claims that he has seen these admirable conditions fulfilled "times innumerable", but *The Zoist* contains relatively few detailed accounts of his own or other peoples' experiments. Elliotson's experiments with a young lady "totally ignorant of phrenology" are described in a letter to Engledue dated 1st September 1842, and printed by the latter as an appendix to his notorious 1842 Presidential Address to the Phrenological Association:[45]

> On placing the point of a finger on the right organ of attachment, she strongly squeezed my fingers of the other hand, placed in her right hand, and fancied I was her favourite sister; on removing it to the organ of self-esteem, she let go my fingers which were in her right hand, repelled my hand, mistook me for some one she disliked, and talked in the haughtiest manner. On replacing the point of my finger on attachment, she squeezed my fingers

of the other hand again, and spoke affectionately. I removed the point of my
finger to destructiveness, and she let go my fingers again, repelled my hand,
mistook me for some one she disliked, and fell into a passion. The finger upon
benevolence silenced her instantly; and made her amiable, though not
attached. I could thus alter her mood, and her conception of my person at
pleasure, and play upon her as upon a piano.[46]

Elliotson found he could produce the same phenomena, though more slowly, by
just pointing his finger at the organ concerned, and could do so with the young
lady's eyes "stopped up with a handful of handkerchiefs".

It was (in terms of Elliotson's simile) possible in such cases for the mesmeric
operator to play upon the subject's head not just single notes but chords, i.e. to
activate two or more organs together, the result being a blending of the
functions normally mediated by the organs in question.[47] One can readily see
what mileage could be extracted from phrenomesmerism at public demonstrations.[48] In Britain, mesmerists who did not accept phrenomesmerism were
few. Newnham and Colquhoun were the most notable.[49] Colquhoun, who is
sceptical concerning phrenology in general, attributes the phenomena of
phrenomesmerism to the existence of a mesmeric rapport between operator and
subject. When such rapport is established, the thoughts, actions and utterances
of the subject are apt to be determined by the will and the expectations of his
mesmerizer. Elliotson denies that the will of the operator plays any part in the
matter. Phenomena may be produced by a sceptic, or by an accidental
misdirection of a finger, or by someone thinking of quite another matter, and he
personally has never obtained them by mere willing, however hard he tried.[50]

It is certainly curious that James Braid, who did not believe in the mesmeric
fluid, and was besides a man of unemotional common-sense, should have
obtained phrenomesmeric phenomena quite as readily as did Elliotson. Braid,
who became interested in phrenomesmerism in 1842, devotes most of chapter
6 of his *Neurypnology* (1843) to it. Many of his cases fulfil the criteria laid down
by Elliotson, and his reports are very similar to the latter's. By stimulating the
appropriate cerebral organs, he induced a lady who was a strict Methodist, and
had never before been mesmerized, to steal two handkerchiefs and a ring, and to
cut a very good figure at waltzing. She esteemed dancing a sin, and it was
perhaps fortunate that on awakening she could not remember what had
happened.

Braid's specious attempts to give a non-mesmeric physiological explanation of
phrenomesmeric phenomena need not detain us. He seems to have abandoned
all belief in phrenomesmerism after 1844, but why he did so is not apparent. To
a modern reader the only question of interest concerning phrenomesmerism is
this: Through what covert channels did the expectations of the experimenters
convey themselves to the subjects, when the former were trying (as did Elliotson
and Braid) to give no hints, and the latter were purportedly ignorant of

phrenology? Of course we have no way of retrospectively discovering the answer, though perhaps recent work on "experimenter effect" may provide some hints.[51]

Related to phrenomesmerism was the equally odd alleged phenomenon of hemicerebral mesmerism (the mesmerization of each hemisphere of the brain separately), which Elliotson seems to have regarded as his own special discovery – he first observed it on 10th March 1839.[52] It led to some extraordinary phenomena – for instance one person might mesmerize one half of the brain and another the other,[53] or the two halves of the subject's body might exhibit contrary emotions or attitudes. It was to flourish again briefly as hemicerebral hypnotism. Somewhat oddly, mesmerization of one hemisphere of the brain had its supposed effect upon the limbs of the same side;[54] whereas by the mid-nineteenth century it was well-known to physiologists (including Elliotson[55]) that damage to one hemisphere paralyzes the limbs of the opposite side. Indeed this had been known to Aretaeus of Cappadocia (fl. 70–80), and was amply confirmed by the great eighteenth century pathological anatomist G.B. Morgagni (1682–1771).[56]

"REICHENBACH" PHENOMENA AND RELATED MATTERS

In the late 1840s and early 1850s, a good deal of interest was aroused in mesmeric circles, and even beyond, by the researches and theories of a well-known Austrian chemist, Baron K. von Reichenbach (1786–1869). These were first published as a supplement to Liebig and Wöhler's *Annalen der Chemie und Pharmacie* in 1845, and then as *Physikalisch-physiologische Untersuchungen* (2 vols., 1849). The first extended account of them in English is the *Abstract* (1846) by W. Gregory, another chemist. Two full English translations were published in 1850 and 1850–1 by Gregory, and by John Ashburner, a London physician and friend of Elliotson. Reichenbach's later works on the same theme need not concern us.

It had been, as we have already noted, a commonplace amongst students of animal magnetism that magnetized subjects were liable to feel peculiar skin sensations, and sometimes undergo muscular reactions, at the approach of the magnetizer's hand, or a magnet, or numerous other kinds of substances, particularly metals. They were also, of course, liable to see luminous emanations from the hands or eyes of the magnetizer, and a luminous glow around animally magnetized objects or water. The reports of the various subjects were far from always consistent with each other. Reichenbach's endeavours were directed towards imposing order and system upon such findings, placing them within a wider context of similar observations, and deriving a preliminary theoretical framework from the whole. It should be emphasized, however, that Reichenbach's own experiments were generally not carried out with "magnetized" subjects, but with "sensitives", many of whom were persons whose nervous systems were in one way or another disordered.

Reichenbach found that such subjects were very responsive to the approach of a magnet to their skins, feeling sensations of prickling and of warmth and chill, the positive pole being warm and the negative cool. He also found that if such subjects were kept in the dark for upwards of an hour they would begin to see lights or flames issuing from the poles. The shapes and properties of these flames varied somewhat from percipient to percipient, and a good deal with the shape of the magnet. Reichenbach says that the light is red and affects a daguerrotype, Gregory that the flame of the negative pole is blue and that of the positive red. Reichenbach also believed himself to have discovered by similar experiments that crystals of all kinds, natural and artificial, likewise give out flames and produce sensations of warmth and cold, from opposite ends (or "poles") of the axis of the crystal. From this he concluded that what his subjects were sensitive to was not magnetism as exemplified by iron magnets, but some other force or imponderable agent altogether. He gave this new force the name of "Od" (changed by Gregory into "Odyle"). By further experiments (he states that he has experimented upon upwards of 100 persons) he discovered that od radiates outwards, though in very different degrees, from all substances, and from the sun, moon and stars (the sun being negative and the moon positive). It is also generated by heat, light, electricity, friction, and all kinds of chemical change (including that involved in the decomposition of dead bodies, a fact which accounts for sundry strange appearances in graveyards). Like all forms of animal and vegetable life, the living human body is a source of od, the right hand being negative and the left positive, the right hemisphere of the brain negative, and the left positive. From this, and the facts that od force can be conducted through, or collected in, suitable objects, and that it attracts the hands of cataleptic patients, we can readily infer that the "magnetic fluid" of the mesmerists is merely one kind of manifestation of the pervasive od force.

Reichenbach's discoveries were taken up with enthusiasm by some British writers on mesmerism, for instance Gregory, Mayo, Elliotson and Haddock.[57] The greatest enthusiast was Gregory,[58] who claimed to have confirmed Reichenbach's findings himself, and proceeded to reinterpret standard mesmeric doctrines about the magnetic fluid in terms of the polarity and flow of od. For instance, he relates the somnambulic or sleep-waking state to Reichenbach's claim that the od of a sleeping person is redistributed away from the head. Perhaps a similar redistribution takes place in somnambules, the effect of which is to cut off the external senses from their brain centres, whilst leaving the intellect awake. When the somnambule's brain is in this state, it may become sensitive to faint odic radiations from distant objects, radiations which would otherwise be drowned by the inflow of ordinary stimuli. Along these lines we may perhaps find an explanation of mesmeric clairvoyance.

The most interesting response to Reichenbach's claims was James Braid's scarce pamphlet, *The Power of the Mind over the Body* (1846).[59] Indeed, from the viewpoint of today this pamphlet is of greater interest than the work which it criticizes. It occurred to Braid, as it would to any modern reader of Reichenbach did he have one that in designing his experiments the latter had taken no account of "the important influence of the mental part of the process, which is in active operation with patients during such experiments". Using subjects, most of whom were apparently not in a hypnotic state during the experiments, and some of whom had never been mesmerized or hypnotized, Braid was able to show that susceptible subjects could be induced, by manipulating their expectations, to feel all the skin sensations reported by Reichenbach's subjects, and to do so in the total absence of a magnet or crystal, and, in darkness, to fancy they could see flames shooting from the place of a magnet, even when the magnet had been surreptitiously removed. On the other hand, they would not see flames coming from a magnet whose presence they had no reason to expect.

These experiments were not perhaps decisive against Reichenbach – Braid's suggestions to his subjects seem often to have been quite direct, and more direct than any Reichenbach can reasonably be supposed to have made – but they certainly required a reply from Reichenbach, a reply that was, so far as I can discover, never forthcoming.[60] Braid – and this is the most impressive part of his pamphlet – makes it quite plain what form this reply should have taken. He points out that Reichenbach's sensitive subjects may have picked up and inadvertently responded to auditory cues from the experimenter's movements and tone of voice. Experiments should be conducted in which all such cues have been eliminated. There should be "no regular order in the experiments", i.e. in whether the electromagnet is on or off; and sometimes the subject should be misled into believing that a change has been made when it has not and not been made when it has.[61]

Braid had in fact a much firmer grasp than anyone who had preceded him, or than anyone who was to succeed him in the rest of the century, of the readiness with which psychological experiments, especially ones involving mesmerism or hypnosis, may be contaminated by "experimenter effect" or "doctrinal compliance",[62] and of the channels through which that contamination may run. But he wrote in vain. Reichenbach's claims were still being taken seriously in the 1850s and 1860s, and even later, by writers who seem never to have heard of Braid's criticisms.[63]

Equally persistent was the notion that certain substances, particularly metals, possess intrinsic virtues or properties capable of producing sensations, muscular spasms, etc., in magnetized persons, or persons afflicted with nervous disorders.[64] In the late 1840s, a Paris physician, Dr. V. Burq (1823–1884), convinced himself that these properties of metals could, if metal, patient, and disease were properly matched, be turned to therapeutic purposes in the treatment not just of nervous problems, but also of cholera and other physical ailments. His works on the subject[65] are decorated with illustrations of curiously shaped pieces of metal which his patients wore like body armour (very uncomfortable they must have been). Burq's metallotherapy is of little importance in the history of mesmerism, but it, and he, survived long enough to be sucked into the maelstrom of superstition and self-deception which in the 1870s and 1880s dizzied the heads of eminent hypnotists in the Salpêtrière and other Paris hospitals.

ELECTROBIOLOGY AND RELATED MATTERS

Electrobiology seems to have reached Britain from America in the latter part of 1850, and during 1851 achieved a fair notoriety there. It was propagated principally by demonstrators who had learned their art in the United States – Dr. H. G. Darling, Mr. H. E. Lewis, Mr. G. W. Stone, the Reverend Theophilus Fiske, and Mr. Warren – but native practitioners also arose. These demonstrators put on shows, often well-attended, in most of the major cities in the United Kingdom.[66] Electrobiology continued to have a vogue through much of 1852, but was thereafter largely eclipsed by Spiritualism and the table-tipping craze. In fact it was Stone who, in October 1852, first brought a medium (Mrs. W. R. Hayden) over from the United States.[67]

I have already said something about the performances of the electrobiologists in America, and it will not be necessary to go into much detail about the phenomena with which they regaled their British audiences. The most interesting accounts are probably those given by Gregory in his *Letters on Animal Magnetism* (1851)[68] of the methods and the results obtained by Darling and Lewis, who seem to have enjoyed considerable success in Edinburgh and Glasgow, not least among the academic and intellectual communities. According to Gregory, both men were open about their methods and fully answered all enquiries, a matter in which they contrasted favourably with Fiske and Warren, who followed Dods' practice of demanding money for initiation.[69]

Gregory had, it seems, ample opportunity during 1851 to observe both Darling and Lewis perform in public and before private parties, to question them as to their methods, to induce them to experiment, and to experiment himself as subject and as operator. He gives the following account of Darling's methods:

> The process followed by Dr. Darling ... is to cause a certain number of persons ... to gaze for ten or fifteen minutes steadily at a small coin, or double convex mass of zinc with a small centre of copper, placed in the palm of the left hand. The other conditions are, perfect stillness, entire concentration ... on the object, and a perfectly passive will, or state of mind. Dr. Darling does not profess to affect those who sit down with an active determination to resist ...
>
> He ascertains ... which of them have been affected, by desiring them singly, to close the eyes, when he touches the forehead with his finger, makes a few

passes over the eyes, or rather presses the eyelids down with a rapid sideward motion, and then tells them that they cannot open their eyes. If, in spite of him, they can do so, he generally takes hold of one hand, and desires them to gaze at him intently for a moment, he also gazing at them, and then repeats the trial. It it fail, he tries no further at that time, but goes on to the next case.

... Those thus discovered to be susceptible are requested to remain and to keep their eyes shut, the others are dismissed.[70]

Gregory proceeds to summarize the effects which Darling could obtain with subjects thus found to be susceptible – the summary, I should remark, is amply backed up with observations of individual cases presented later in the book. The simpler effects were such matters as inability to open the mouth, to separate the hands when pressed palm to palm, to remove the hands from the top of the head, all of which could be induced by verbal suggestion and removed by simply saying, "Now you can", or "All right!" More complex effects could be similarly induced:

In the same way, Dr. Darling proceeds to shew his power over the sensations of his subject. For example, he deprives one hand, or one arm, of all feeling, and renders it utterly insensible to the most acute pain; or he makes his subject feel a cold pencil-case burning hot, or himself freeze with cold, or taste water as milk, brandy or any other liquid...

In like manner, he controls the will, so that the subject is either compelled to perform a certain act, to fall asleep in a minute, or to whistle, &c. &c., or is rendered unable to perform any act...

Dr. Darling further controls the memory. He causes the subject to forget his own name, or that of any other individual; or to be unable to name a single letter of the alphabet, &c. &c.

Moreover, he causes him to take any object to be what Dr. Darling says it is, a watch for a snuff-box, a chair for a dog, &c. &c., or to see an object named, where nothing really is, as a book in Dr. Darling's empty hand, or a bird in the room, where none is. The illusion is often absolutely perfect.

Again, he will cause the subject to imagine himself another person, such as Dr. Darling, Father Matthew, Prince Albert, or the Duke of Wellington, and to act the character to the life... Lastly, Dr. Darling can control, perfectly, the emotions...

Every one of these forms of influencing the subject I have seen, varied in a hundred details. The effect is usually ... instantly produced, and as instantly removed, by the operator's simple word. And there is no mystery, no secret, nothing supernatural in it. It is a perfectly natural phenomenon, and any one who tries, may do it...Lord Eglinton, Col. Gore Browne, and other gentlemen, as well as myself, have found no difficulty, when we lighted on a susceptible subject.[71]

Lewis (a negro) produced similar effects "by gazing for five minutes only, with extreme earnestness and concentration, at the subject, while the latter gazes either at him, or at an object in the same direction." He added various gestures and passes, "all of which are most deeply imbued with that energetic concentration of will, which I have never seen so strongly developed, nor so beautifuly exhibited in the natural language, as in Mr. Lewis."[72] Otherwise Lewis proceeded as did Darling.

There is hardly an effect figuring in the repertoire of modern stage hypnotists that was not already produced by Darling and Lewis. It should be noted, however, that unlike later hypnotists, and earlier mesmerists, Darling and Lewis did not, generally speaking, put their subjects in a sleep-like or "somnambulic" state before inducing the phenomena. The subjects remained conscious, could comment on their sensations, and had full recollection afterwards. Yet it remains a widely prevalent superstition that the fact that "hypnotic" phenomena can be obtained from individuals in the waking state is a relatively recent discovery.

The whole business struck contemporaries as most extraordinary. One of the more curious upshots was a series of pamphlets by medical men attempting to explain the phenomena.[73] The explanations offered ranged from the trite (Bennett) to the ludicrous (Buchanan). What is interesting is that medical men in general clearly accepted that studying and explaining the performances of itinerant electrobiologists was a perfectly proper activity; whereas those same medical men had for the most part vehemently repudiated any truck with the very similar happenings which took place in mesmeric contexts, i.e. contexts linked to healing. Investigation of the phenomena was acceptable only so long as those phenomena did not constitute a threat to the profession.

The reaction of mesmerists to the "electrobiological" phenomena was equivocal. The good news was that phenomena, not a few of which had long been familiar to mesmerizers and their audiences, were now being not just widely demonstrated, but accepted by members of the scientific and medical establishments. The bad news was that it was concentration by the subject combined with verbal suggestion by the operator, rather than procedures distinctively mesmeric, that seemed primarily responsible for those phenomena. No serious mesmerist thought that the little bimetallic discs were sources of the mesmeric influence. Several persons, e.g. Mr. G. S. Nottage and Mr. Froy,[74] demonstrated the phenomena using small pellets of paper instead of metallic discs, and Lewis used neither discs nor pellets. Elliotson endeavoured to show[75] "that the phenomena resulted from imagination, excited by suggestion in a slight degree of mesmerism", which he termed "submesmerism".[76]

Elliotson certainly cannot be accused of becoming conveniently wise after the event concerning the power of the imagination, and the way in which imagination can contaminate the results of mesmeric procedures. He had himself witnessed, indeed brought about, mesmeric phenomena quite as odd as those which formed the highlights of electrobiological performances. He had induced anaesthesias, paralyses, positive and negative hallucinations, changes of personality, etc., many of them in response to verbal suggestions. The subjects were, of course, persons who (unlike most of the subjects in electrobiological demonstrations) either were, or had previously been, in "the mesmeric state", but Elliotson is quite clear that the effects are usually due to the imagination of the subjects,[77] and not to the will of the operator, or to any transference of mesmeric fluid from operator to subject. Of one such patient he had written in the very first volume of *The Zoist*:

> ... a large number of the realities which I produced in this case, were, I feel satisfied, the result of an impression only that they would occur. Thus metals had various effects, just as I led her to expect them. A glass of water would send her to sleep for hours, if she said it would, provided it was mesmerised, when I did not mesmerise it at all: and yet the sleep was, I fully satisfied myself, perfectly real.
>   I could almost believe that the stigmata on the hands and feet of certain Roman Catholic sleep-waking females might not be artificial, but the result of a strong imagination that they would have these marks.[78]

Gregory's view of the electrobiological phenomena was, as we have seen, rather different. He holds that the "electrobiological" state, and the state produced by animal magnetism, are not identical, though in both "suggestion has the force of fact". The difference is that the electrobiological state (like the hypnotic state produced by Braid's methods) is one of "auto-magnetism" in which a redistribution of od force (or odyle as Gregory calls it) in the brain (a redistribution which causes damping down of the external senses and vivifying of the "inner" ones) is brought about without any influx from the magnetizer. In effect, Gregory (though he does not say so) is explaining electrobiological and hypnotic phenomena in non-mesmeric terms, for only where there is postulated what he calls "a foreign influence thrown into the brain" is it strictly correct to call the theory a mesmeric one. Of course, Gregory's "hypnotic" theory is vague in the extreme, though it is not a great deal woollier than other early theories of hypnotism, or, indeed, than some that have been propounded rather more recently.

ALLEGED INSTANCES OF EXTRASENSORY PERCEPTION

Instances of alleged extrasensory perception are fairly common in the pages of *The Zoist* and of other contemporary mesmeric publications. Most are quite similar to instances discussed earlier, and most suffer from the same short-comings. Thus there are not a few reports of experiments in community of sensation, experiments in which it seems that almost no precautions were taken to ensure that auditory and olfactory cues of an ordinary kind had been properly excluded;[79] reports of "stomach-seeing", concerning which similar problems arise;[80] reports of apparent "magnetization at a distance" in which an initial success or two has not been followed up by systematic experiments, so that it is impossible satisfactorily to rule out chance coincidence, or correct surmise by the subject;[81] reports of cases of "travelling clairvoyance", some remarkable, but very few based upon adequate contemporary notes; and so on. Space does not permit any detailed examination of these debatable materials, and such an examination is in any case rendered superfluous by the comprehensive survey which we owe to the late Dr. E. J. Dingwall.[82] Indeed in 1856, Elliotson himself went so far as to say " ... We well know that gross imposition is hourly practised ... both by professional clairvoyants and private individuals considered to be trustworthy but influenced by vanity and wickedness ... even in the genuine the power is uncertain."[83] I shall confine myself therefore to a few brief remarks upon one of the most remarkable "clairvoyantly gifted" subjects of all time, the Frenchman Alexis Didier.

Alexis's clairvoyant abilities first attracted public attention when he was a very young man in Paris in the early 1840s. I think it would be fair to say that if no case can be made for the genuineness of Alexis's powers, *a fortiori* no case can be made out for the powers of any lesser fry. The literature concerning Alexis is very considerable, but much of it consists of obscure French newspaper reports which it would be very difficult to disinter. However, he paid extended visits to Britain in 1844 and 1849, and was in turn visited in Paris by various British magnetists. As a result there is enough material in *The Zoist* and other British publications,[84] as well as in the *Journal du magnétisme*, to construct a reasonably comprehensive picture of his procedures and ostensible gifts.

Alexis's career began when he attended a public demonstration of magnetism and was found to be an excellent subject. Before long he was himself acting as clairvoyant subject in demonstrations of magnetism. His magnetizer for most of his career was J. B. Marcillet, a former cavalry officer, who managed the Paris road haulage firm in which Alexis worked as a clerk. In 1844, Alexis's charge for performing before a private party was 30 francs – about 24 shillings in contemporary English money.[85] He was a modest and polite young man, and soon became almost as well-known in "good" society as did the medium D. D. Home in the ensuing decades.

Alexis's performances tended, with minor variations, to follow a set pattern.[86] He was first of all blindfolded, and in that condition would play a few hands of cards, or read the titles of books held before him. Next, without the blindfold, he would read a few words or lines from a page ten or twenty below the page at which a book had been opened. Then he would attempt to read words written on pieces of paper or cards concealed inside opaque envelopes or packages, or to describe, and give information concerning, objects concealed inside closed boxes or other containers. Finally, he would attempt "travelling clairvoyance" to places thought of or designated by persons with whom he had been placed in rapport (usually by holding their hands). He would also (especially in private consultations) attempt to locate lost property or persons, and to detect thieves.

Much of Alexis's routine is similar to, and probably derived from, the routines of other magnetic performers of the time, for instance Prudence Bernard[87] and Calixte Renaux,[88] whose powers there is little reason to regard as genuine.[89] His feats of blindfold card-playing and book reading sound very much like ordinary mental magic, of which there were numerous stage practitioners in France and Britain by the mid-nineteenth century.[90] Some people regarded Alexis as just another exponent of these arts. John Forbes, for instance, later Sir John, editor of *The British and Foreign Medical Review*, who watched Alexis perform in London in 1844, thought that the latter could see under the bandages which made up the blindfold, and noted that he repeatedly touched and fiddled with them. Forbes also thought that Alexis managed unostentatiously to flick over the pages of the books from which he had to read lines or words several pages below a currently opened page.[91]

Whether Forbes, whose tone was temperate though his eyesight appears to have been poor,[92] was correct, or was merely seeing what he expected to see, is not of much importance, since no experiments dependent upon the effectiveness of blindfolds can possibly be accorded weight.[93] (In fairness to Alexis it should perhaps be pointed out that the leading conjurer of the day, J. E. Robert-Houdin (1805–1871), having himself bandaged Alexis's eyes, could find no ordinary explanation for his knowledge of cards taken from newly opened packs or for others of his apparent successes[94]). Far more curious are certain examples of Alexis's ability to identify, and to discover salient facts concerning, objects or messages concealed in opaque boxes, or otherwise hidden, and of his feats of "travelling clairvoyance". Even with material of these classes, it is unusual to find instances in which the records reach anything like a satisfactory standard; instances, that is, in which the testimony is first-hand, comes from intelligent witnesses, is properly attested, based upon contemporary notes, etc. And when one does find it, it is often noticeable that Alexis had more failures and made more mistakes than are generally reported by persons writing about him from memory or at second-hand. Still, it would be going too far to dismiss all of the testimony in favour of Alexis's reputed paranormal powers on the grounds that a good deal of it does not measure up to criteria which at that time many people would have thought unnecessarily severe. Some cases, set down in apparent good faith by capable persons, are certainly quite striking.

A somewhat remarkable instance of Alexis's ability to divine the contents of an opaque box is described by the Reverend George Sandby in a letter dated 8th July 1844 and published in *The Zoist* the following year. The incident concerned took place at a "private reunion" in Welbeck Street, London, on 2nd July 1844. A certain Colonel Llewellyn, a veteran officer who had been severely wounded at Waterloo, handed Alexis a morocco case, eight inches long, and an inch and a half thick, looking like a surgical instrument case, or a small jewel case. According to Sandby no one in the room was aware of its contents; only two or three persons present were acquainted with the Colonel; and he and Alexis had never before met. Alexis held the case to his stomach for a short time in silence, and then gradually and slowly gave the following description:

> The object within the case is a hard substance.
> It is folded in an envelope.
> The envelope is whiter than the thing itself. (The envelope was a piece of silver paper.)
> It is a kind of ivory.
> It has a point ... at one end (which is the case).
> It is a bone.
> Taken from a body – from a human body – from your body.
> The bone has been separated and cut, so as to leave a *flat side*. (This was true; the bone, which was a piece of the colonel's leg, and sawed off after the wound, is *flat* towards the part that enclosed the marrow.)
> [Here, Alexis removed the piece of bone from the case, and placed his finger on a part, and said that the ball struck *here*.] (True.)
> It was an extraordinary ball, as to its effect. You received *three* separate injuries at the same moment. (Which was the case, for the ball broke or burst into three pieces, and injured the colonel in three places in the same leg.)
> You were wounded in the *early part* of the day, whilst charging the enemy. (Which was the fact.)[95]

A second account of the same incident, taken from notes, and from Colonel Llewellyn's statement made after the séance, was furnished by Lord Adare.[96] It should be observed – and this is quite characteristic[97] – that Alexis did not just *describe* the bone; he furnished all sorts of information about matters connected with the bone. His "travelling clairvoyance" likewise occasionally yielded information that (whatever its source) could hardly have been obtained from an immediate quasi-visual inspection of the designated place or person.

Turning now to Alexis's feats of travelling clairvoyance. Dr. Edwin Lee, of Brighton, author of a volume, already mentioned, on animal magnetism, describes several instances of travelling clairvoyance in his published notes of séances which he attended while Alexis was touring the south coast of England in January 1849. The following one is extracted from Lee's notes of a séance at a private house in Montpelier Road, Brighton, on 24th January:

> Mr. W. [who had probably been put in rapport with the entranced Alexis by contact] asked for the description of a house. Alexis said, "it was sixty leagues to the right of London, about a league from a railroad; the sea on one side, and sands along the shore; the house very old; of stone; an inscription engraved on it in stone; Latin; five words; five letters in the second word.["]
> At length, after some efforts, Alexis – having been correct in the former

particulars – wrote the words, "Non Nobis Domine," in characters similar to the inscription. He tried hard at the other words, but seemed confused, which was accounted for on the words being stated. The two first were repeated thus: "*Non Nobis Domine, Non Nobis.*" He further said, "that the house was two storeys high; that one portion was much newer than the other; that there was a servant living in the stables about forty years old, not good-looking (he is much marked with the small-pox); a large dining-room in the house with three windows, they look out on trees on either side; there are two wells in the grounds; the oldest well contains good water; the newer one is dry, or has at times brackish, or rain water. In the park, near the centre, a pillar or column, with something on the top; a transverse cross-bar," of which he drew a representation. (The object was, as I understood, a high post with a frame to hold a slate for marking the points at archery shooting.) "There was no game in the park." All correct.[98]

Alexis was, allegedly, capable of clairvoyantly homing in upon objects and persons as well as upon places. Hence he acquired some reputation in the location of missing persons and objects, and in the detection of crime.[99] His most celebrated feat of discovering, or assisting the discovery of, a missing person, involved a French officer named Bonfilh who had been saved during the Peninsular War by Colonel John Gurwood (1790–1845), later Wellington's private secretary. The case is unfortunately too long and complex to be described here; it is analyzed in detail by Dingwall.[100]

Alexis was only one of the numerous mesmeric clairvoyants whose feats are reported in *The Zoist* and other publications of the time. Nearly as celebrated was Alexis's younger brother Adolphe.[101] Among British clairvoyants of the time, the most interesting were Emma, the subject of Dr. J.W. Haddock, of Bolton (1800–1861), who added to mundane clairvoyance extended clairvoyance of the next world, and Ellen Dawson, the subject of Dr. J. Hands, a London physician.[102] There were many others,[103] but Alexis was without doubt the most remarkable – a whole volume might be written about him – and he may therefore stand for them all. By this time, too, standards of evidence, though still generally far too slack,[104] had improved just a little, so that it is no longer exceptional to find oneself presented with the near-contemporary verbatim testimony of relevant eyewitnesses. In a sense, therefore, Alexis may represent all mesmeric clairvoyants when we come to the inevitable question – is there any evidence at all that any among the so-called mesmeric clairvoyants really did possess gifts of extrasensory perception, the alleged ability to acquire factual information other than through the channels of the ordinary senses?

Now so far as Alexis's feats of blindfold card-playing, etc., are concerned, there appears to be nothing that goes beyond the reach of ordinary conjuring, and, so far as I am aware, no-one ever attempted any properly controlled experiments with him. His successes at discerning the contents of closed packets or boxes, and at "travelling clairvoyance" are another matter, and present problems of considerable complexity. None of the more obvious and everyday explanations can make much headway with these problems. It is, for instance, no good supposing that the successes were just occasional lucky shots which found their way into print whilst all the failures were passed over. We have a sufficient number of moment-by-moment reports of Alexis's séances to make it abundantly clear that successes were far too frequent for this explanation to be even remotely plausible. For similar reasons we cannot ascribe all or even many of Alexis's "hits" to errors of memory and wishful thinking by the witnesses. Nor is it any use attributing all or most of the "hits" to intelligent guesswork based on the appearance, social status, age, etc., of the sitters, together with judicious "fishing", or such simple manoeuvres as shaking the box containing the target object. The details given were quite often at once so trivial and so precise as to render these proposals altogether insufficient. Townshend, for instance, who introduced himself to Marcillet only as a friend of Elliotson, was correctly told, *inter alia*, that in a picture of a stable (by Morland) which hung in his house at Lausanne, there were a man with a wheelbarrow, and a grey horse lying down, which had wounds on its flanks.[105] The extraordinarily detailed knowledge which Alexis so often exhibited of matters with which he certainly had no prior personal acquaintance could have come only from exhaustive research in works of reference combined with the bribery of servants and the employment of a widespread network of agents. Some agent must have learned and transmitted the story of Colonel Llewellyn's bone, and Alexis must have memorized it on the assumption that the Colonel would one day come to a sitting and would be likely to offer the bone as a test; Alexis must have had in his mind a mental card index of English country houses and their owners, so that he knew immediately what Latin phrase was written over the door of which house; he must have known all about Colonel Gurwood's search for Bonfilh, and the reasons therefore, and acquired information Gurwood had not yet got wind of; then he must have got Marcillet to entice the Colonel to a sitting; and so on and so on for all the very numerous other cases. And all this for the very modest fee of thirty francs for attendance at a private party.

One can only say that while for a given individual case such an explanation might be made to work, for the *ensemble* of recorded cases the idea is impossible. There is no evidence whatever that Alexis bribed servants and paid agents in the numbers that would have been necessary, or indeed that he bribed any servant or paid any agent. And the scale of the operations required would have been so large, the expense so great, the risk of being betrayed so acute, and the difficulty of retaining all the information and producing individual items of it for the right persons at the right times so enormous, that the whole enterprise would have collapsed before it had been properly started. I find it impossible to believe that Alexis could have maintained an intelligence network of the requisite scope and penetration. Consequently I cannot help suspecting that he, and certain other magnetic somnambules, did sometimes acquire and transmit information which they could not have come by in any of the ordinarily recognized ways.

NOTES

1 For instance, nine cases altogether are reported in vol. 5 of *The Zoist* (1847–8).

2 Gmelin (1789), pp. 276–277, 278–279, 341–342.

3 On Mesmer's painless delivery of a child cf. above p. 62. Another such case is somewhat obscurely reported by Wienholt (1802–6), Vol. 3, Pt. 3, pp. 385–386. Deleuze (1825), pp. 139–140, 246–247, writes as though this were an established practice. Early examples of surgical procedures carried on mesmerized subjects are given by Pam (1816) and Delatour (1826).

4 Spiritus (1819), p. 87.

5 Chapelain, the magnetizer of Madame Plantin, painlessly delivered a baby in 1829, and the following year painlessly removed a tumour from the same lady's neck. See Fillassier (1832), pp. 65–66, 66–67.

6 Cf. Elliotson (1843), pp. 67–78.

7 If Lafontaine is to be trusted this case was not the first. He claims (1866, Vol. 1, pp. 320–321) that in Sheffield, about January 1842, he acted as mesmerist, at the suggestion of Dr. Holland, for the painless amputation of the leg of a young man injured in an accident. Dr. G. C. Holland (1801–1865) was a well-known Sheffield physician, and author of, *inter alia*, *The Philosophy of Animated Nature* (1848) which contains passages about animal magnetism. In an account of Lafontaine's second conversazione at Sheffield in *The Sheffield and Rotheram Independent* for 5th February 1842, the interpreter, Dr. Bartolomeo of Sheffield, is reported as saying that he, "in company with several of the medical gentlemen in the town, had seen experiments performed by M. Lafontaine in private much more extraordinary than any he had produced in public, which, were he to narrate, they would call him a fool or a knave. Indeed, so extraordinary were some of the phenomena, that he scarcely dare give them credence."

8 Elliotson (1843), p. 9, remarks of this: "The pain which such an experiment would occasion to a person in his waking state must be equated to a strong dart of tic douloureux; and I defy any human being, in his ordinary condition, to be subjected to such an experiment without, not to say an increase of low moaning if he were already moaning, but without giving some other more decided sign of anguish."

9 See the statement by Captain John James, *The Spiritualist*, 27th June 1879.

10 See Elliotson (1843), pp. 22–58.

11 Topham and Ward (1842).

12 See the lists given by Sandby (1848), pp. 51–52 and by Elliotson, *The Zoist*, 1853–4, Vol. 11, pp. 216–217.

13 Many of these, as well as some major ones, are reported in Elliotson (1844–5b).

14 For other American cases see Deleuze (1846), pp. 397, 402–403; Leger (1846), pp. 392–394; Elliotson (1845–6e), pp. 384–385 (from *The Boston Medical and Surgical Journal* – the mesmerist was P. P. Quimby); Elliotson (1846–7c), pp. 204–205; and Gravitz (1987–8).

15 Elliotson (1845–6e), pp. 380–383; Dugas (1845).

16 Elliotson (1846–7a), pp. 1–12.

17 The first of these cases, that of Marie d'Albanel, was often cited. In it the patient "conversed several times, smiling, with her magnetizer, even during the most painful moments of the operation", which was the amputation of a leg. Cf. Durand (1845); Elliotson (1845–6d).

18 *Journal du magnétisme*, 1846, Vol. 2, pp. 295–299; *Journal de Cherbourg*, 1st May 1846; Elliotson (1846–7c), pp. 199–203.

19 On Esdaile's somewhat unusual methods of mesmerizing, see below p. 257.

20 According to Esdaile (*The Zoist*, 1846–7, Vol. 4, p. 567) this tumour was "voted to Elliotson by acclamation", and was then in rum awaiting the latter's acceptance. Elliotson adds that as soon as the mass arrives he will have great pleasure in showing it to any gentleman who may call at his house in Conduit Street. Even this immense tumour was by no means a record one. Gould and Pyle (1962), pp. 800–803 present, with a photograph, the case of such a tumour weighing 120lbs.

21 Esdaile (1846), pp. 14–15.

22 For some figures on this question see below p. 255.

23 A good many appeared in the Indian press. Elliotson reproduces quite a few of these in *The Zoist*.

24 I know this material only from Elliotson's accounts of it; Elliotson (1845–6d) and (1846–7a), pp. 21–50.

25 *Report of the Committee . . .* (1846).

26 Esdaile (1847–8); Esdaile (1847); cf. Elliotson (1848–9d); Esdaile (1848); cf. Elliotson (1848–9a).

27 Elliotson (1848–9b).

28 Esdaile (1850a); cf. Elliotson (1849–50b; 1849–50c).

29 Esdaile (1850b).

30 Esdaile (1850b), p. 453.

31 Esdaile (1856), pp. 26–27.

32 Esdaile (1854–5); Elliotson (1852–3b; 1854–5c).

33 Esdaile (1852a; 1852b; 1856).

34 Elliotson (1852–3a).

35 Esdaile (1846), p. 151.

36 Esdaile (1846), pp. 148–149; cf. illustration in Esdaile (1856).

37 *The Zoist*, 1843–4, Vol. 1, pp.208–210.

38 Elliotson (1843), pp. 12–15, 36–37.

39 Elliotson (1843), p. 15.

40 "Once the operation started, even the bravest screamed and struggled to be free of the agonizing pain." (MacQuitty , 1971, p. 67, quoted by Perry and Laurence, 1983, p. 357). On flogging in prisons see Priestley (1985).

41 Elliotson (1843), pp. 15–18.

42 Sandby (1848), pp. 50–54.

43 Cooter (1984), p. 183.

44 Elliotson (1843–4a), p. 241.

45 Engledue (1842). See Lang (1843), pp. 186–189. Lang's chapter 7 is the most useful single account of phrenomesmerism in Britain.

46 Lang (1843), p. 187.

47 For an entertaining example of "chords" see Baldock (1845–6), pp. 337–8.

48 S. T. Hall had a subject, a young man named Wilkinson, who "in his somniloquence composed the most beautiful poetry, the theme of which might be changed, or modified in an instant, as I moved my finger from one part of his head to another." S. T. Hall (1845), pp. 6–7.

49 Newnham (1845); Colquhoun (1843).

50 Lang (1843), p. 205.

51 Rosenthal (1966).

52 Elliotson (1843–4c), p. 327. For a detailed case see Elliotson (1844–5a).
53 Elliotson (1851–2a), pp. 176–180.
54 Cf. Lang (1843), pp. 187–188.
55 Elliotson (1846), p. 748.
56 The first person to demonstrate anatomically the decussation of motor nerve fibres in the pyramidal tracts was Gall.
57 Elliotson (1846–7e); Gregory (1851); Haddock (1851); Mayo (1851).
58 Gregory (1846); Gregory (1851), pp. 247–319.
59 See Tinterow (1970), pp. 331–364. This material originally appeared in *The Medical Times*, vol. 14.
60 Gregory replied in the *Phrenological Journal*, October 1846 and January 1847. He met Braid and found his experiments very interesting.
61 Tinterow (1970), p. 333n.
62 See above, p. 168.
63 See e.g. Reichenbach (1853–4; 1855–6); Ellis (1854–5); Lee (1866); Ashburner (1867). The matter was still considered of sufficient interest in 1882 when the Society for Psychical Research was founded for that body to set up a committee to investigate "Reichenbach" phenomena. See Barrett *et al.* (1882–3); Barrett (1884); Jastrow and Nuttall (1886)
64 A.M.J. de Chastenet de Puységur (1817); Lamy-Sémart (1817). On different effects produced by some metals on different subjects see Kluge (1811), pp. 162–169; Werner (1839), pp. 237–248; Siemers (1835), pp. 150, 152, 163.
65 Burq (1852–3; 1853).
66 Cf. *The Zoist* in the relevant period.
67 Podmore (1902), Vol. 2, p. 4.
68 Gregory (1851), pp. 186–210, and Part II, *passim*. He was criticized by some writers in *The Zoist* for the attention he paid to electrobiology. See e.g. Acland (1852–3), p. 54n.
69 Acland (1852–3), pp. 62–63.
70 Gregory (1851), pp. 190–191.
71 Gregory (1851), pp. 192–193.
72 Gregory (1851), pp. 154–155.
73 Cf. Podmore (1909), p. 149; A. Wood (1851); Bennett (1851); Buchanan (1851). For Braid's views on electrobiology see Braid (1851).
74 Acland (1852–3); Dingwall (1968a), p. 118.
75 Elliotson (1851–2b).
76 Elliotson (1851–2b), p. 425.
77 Yet T.M. Parsinnen (1979), p. 115, claims that Elliotson and his followers rejected imagination theories of mesmerism as not being medically respectable!
78 Elliotson (1843–4b), p. 441. Kieser (1826), Vol. 1, p. 261, attributes stigmatization to unconscious self-magnetization (one's own will can magnetize oneself, he thinks). It was common ground among magnetizers that the vascular system could come under special control of the magnetizer. Cf. Ennemoser (1842), pp. 251–252.
79 For sample cases see e.g. Scoresby (1849), pp. 24–32, 78–80, 102–105; Lang (1843), pp. 98, 102–103; Engledue (1844–5); Elliotson (1847–8a), pp. 242–246.
80 Townshend (1844), p. 50; Parsons (1849–50).
81 Thompson (1845–6; 1848–9); Ashburner (1847–8); Townshend (1854), pp. 73–86; Adams (1849–50); Gregory (1851), pp. 107–108.
82 Dingwall (ed.) (1967–8).
83 Elliotson (1855–6).

84 Elliotson (1844–5d); Lee (1866), pp. 255–280; Dingwall (1967), pp. 158–201, 205–206; Dingwall (1968a), pp. 92–96. Alexis's autobiography is Alexis Didier (1856).

85 Elliotson (1844–5d), p. 486. Elliotson's informant was not charged for attending a demonstration by Alexis at Marcillet's premises.

86 A useful and unusually full account of one of Alexis's performances, taken from contemporary notes, is given in Elliotson (1844–5d), pp. 521–529. It is by John Auldjo, FRS, (1805–1886), at whose house the session was held.

87 On Prudence's blindfold card-playing see Braid (1852), pp. 111–118.

88 It seems that J. J. Ricard, the magnetizer of Calixte, played some part in developing Alexis. See Dingwall (1967), p. 159.

89 Cf. Dingwall (1967), pp. 125–157.

90 Lee (1866), pp. 123–125; Dingwall (1967), pp. 198–199; Gandon (1849).

91 Forbes (1845).

92 *The Zoist*, 1844–5, Vol. 2, pp. 399–400.

93 See the interesting experiments of Hodgson (1884–5); cf. Braid's researches on blindfolds in connection with Prudence Bernard, Braid (1852), pp. 114–115.

94 Mirville (1854), pp. 18–32.

95 Elliotson (1844–5d), p. 509, slightly repunctuated.

96 1812–1871, later 3rd Earl of Dunraven, at this period M. P. for Glamorgan, known for his interests in astronomy and archaeology.

97 Cf. the details given to Townshend by Alexis concerning the writer of a letter which the former had presented to him in an opaque envelope, Townshend (1851–2), pp. 408–409.

98 Lee (1866), pp. 268–269.

99 This talent seems now and again to have backfired. In 1851, he was taken to court for making a false accusation of theft, and it was revealed that thirty other similar complaints had been made against him. Dingwall (1967), pp. 188–190.

100 Dingwall (1967), pp. 186–188. Gurwood, who committed suicide on Christmas Day 1845, figures as a post-mortem "communicator" in another curious "psychic" incident. See F. W. H. Myers (1903), Vol. 2, pp. 162–167.

101 On Adolphe Didier see Dingwall (1967), pp. 201–206, and his autobiography, Adolphe Didier (1860).

102 On these see Dingwall (1968a), pp. 96–99, 120.

103 The whole subject is treated in detail in the classic volumes edited by Dingwall (1967–8).

104 To take just one example. Gregory reports a case of vision at a distance, of which he evidently thought highly, which he can only date as having taken place "about the end of last April, or early in May", and though he says he immediately wrote all the details down and later received confirmation of them, he prints neither his own notes nor the confirmatory letter he received from an unnamed gentleman (Gregory, 1851–2).

105 Townshend (1851–2), p. 407.

CHAPTER 13

Animal magnetism: Retrospect and reflections

If one were to look only at a small segment or a limited aspect of the history of animal magnetism one might be tempted to regard the magnetic movement as at root a manifestation of prevailing social and intellectual forces. This sort of treatment has been accorded to other "fringe" movements, for example phrenology, Spiritualism and parapsychology.[1] And certainly, if one limits one's perspective, it is easy enough to represent the magnetic movement as riding upon political undercurrents in France in the 1780s; or as a "secularization of the relation that had existed between exorcist and demonically possessed";[2] or as evolving from and with the dark complexities of German nature-philosophy; or as carried onwards beneath a common banner with other mundane reform movements among the educated working class in mid-nineteenth-century Britain.

Now the phenomena of mesmerism were of course influenced by their cultural settings, and by the views entertained by particular mesmerists or local groups of mesmerists. But the magnetic movement, and the magnetic phenomena (real or illusory), had a momentum of their own, which transcended any merely local influences. Many phenomena emerged in very similar forms in a fair variety of appreciably different cultural and geographical settings, from New York to Calcutta and from the Shetlands to Brazil.[3] An obvious inference is that some at least of the phenomena are not the direct or indirect products of culturally determined beliefs, but rather reflect features of man's psychological con-stitution, which function independently, not perhaps of all his belief-systems, but of his beliefs as to his own constitution. This conclusion was regularly drawn by the mesmerists themselves, many of whom, searching extensively through historical works and traveller's journals, thought that they had unearthed numerous examples of mesmeric phenomena in cultures temporally or geographically remote from their own. In this way was built up a kind of mesmeric perspective on history, and (to an extent) on what would now be called social anthropology.

There were many contributors to this literature, some of whom deployed a great deal of ingenuity and scholarship.[4] The same topics recur again and again.

Any past healer whose technique involved anything remotely resembling the mesmeric passes was very likely to be represented as a pioneer or precursor of animal magnetism, Valentine Greatrakes (1629–1683), the Irish "stroker," being a particular favourite. If the techniques led to the induction of a state that could, however vaguely, be assimilated to that of somnambulism, the mesmeric writers were even better pleased. Often cited was the curative and vision-haunted "temple-sleep" which sick persons might enjoy within the bounds of certain temples during classical times. Mesmerists tended to assume (with little or no substantive evidence) that the sleep-procuring practices of the officiating priests at these healing temples *must* have resembled those of modern mesmerizers.[5] The state of artificial or induced "somnambulism" was also thought to shed light upon states of religious ecstasy less tranquil than that of temple-sleep. Thus the gyrations and ecstasies, the visions and wild speeches, of participants in methodist or revivalist meetings, of the tremblers of the Cevennes, of the convulsionaries of St. Médard, of deranged prophets and ancient oracles, of Siberian shamans and Lapland seers, of whirling dervishes and possessed demoniacs, were explained by supposing that excitement, or narcotics, or rhythmic drummings, or the magnetic power of such charismatic personalities as Wesley or Whitfield, had disturbed the fluidic balance of nervous systems that were naturally unstable, and produced a form of somnambulism. Where the nervous systems concerned, though still in a sense unstable, were of a high order, and the lapses into or intrusions of the somnambulism or sleep-waking state were spontaneous or even self-controlled, rather than precipitated by environmental disturbances, the result might be such extraordinary careers as those of Mohammed, Socrates, St. Hildegard, Swedenborg, or Joan of Arc. Even the occasional clairvoyance of second-sighted persons, ghost-seers, etc., was regarded as indicating what might be called a somnambulistic diathesis, with the form of the hallucinations, and their symbolism, reflecting the folk-beliefs of the percipient's time and culture.[6]

When – as was not infrequently the case – a mesmeric perspective on healing was combined with certain characteristic views as to the relation between, or identity of, the nervous and magnetic fluids, and of these with the universal ether of physics,[7] we end up with quite grandiose schemes which might be called mesmeric "world-views". The details, of course, vary somewhat from one writer to another, the spiritualistic Werner, say, differing on many points from his opponent Wirth, or from Kieser. But by some writers the enterprise was carried through with much learning and persistence, and often with a surprising amount of insight, and the upshot, however misguided, was sometimes fairly impressive.

To return, however, to the phenomena which presented themselves in the context of the mesmeric movement: These phenomena invite certain questions which, though still impossible to answer decisively, merit some discussion. It will be easiest if we tackle these issues by considering, first, the cures of disease allegedly worked by mesmeric methods, and, second, the various other kinds of unusual psychological and physiological phenomena which those methods apparently engendered. And it will be convenient if we take, with respect to each of these two areas, first, those phenomena produced by basic mesmeric procedures (passes, drinking or application of magnetized water, etc.) without the subject falling into a somnambulic or sleep-waking (as distinct from sleep-) state, and, second, those phenomena which seemed (at least until the advent of the electrobiologists) to follow only upon the induction of a somnambulic state.

THE CURES

So first of all the alleged cures: By drawing a distinction between the cures ostensibly wrought by the "passes" and other simple contact or near-contact methods, and cases in which the patient became somnambulic and clairvoyantly diagnosed and prescribed for her own (or other peoples') ailments, I am in effect raising the question as to whether the clairvoyant diagnoses and prescriptions ever contributed to the supposed cures. There were very few cures which can safely be categorized as purely somnambulic, for in the great majority of cases somnambulism was induced by passes and allied methods, even though as the case developed induction might take a progressively shorter time, and a few practised subjects might in the end be entranced or awoken by verbal command alone.[8] And the somnambulic prescriptions often included liberal doses of animal magnetism. I shall begin therefore by considering the cures in general, and shall afterwards say a little about the somnambulic contribution (if any).

An obvious preliminary issue is, of course, what were the kinds of ailments for which mesmeric treatment was ostensibly successful? And here, at the risk of being repetitive, I must again emphasize that those writers who have presented the mesmerists as ancestral psychiatrists, specializing in the treatment of mental disturbances, are egregiously wrong. Thus in Mialle's survey of rather over 600 cures described in the early literature of animal magnetism, I can find only eight cases of possible insanity, and five of hysteria,[9] while of 4232 cases treated at Wolfart's clinic in 1819–21, only 64 (1.5%) were cases of mental problems.[10] It is true that in a certain percentage of further cases, the patient may well have been a hypochondriac or *malade imaginaire*, and that the treatment may have worked for psychological rather than medical reasons. Some of these patients, unfortunately, attracted the compassion of well-meaning and long-winded magnetizers who wrote up the cures in great detail. But such cases form a small minority. The majority of mesmeric patients undoubtedly had something wrong with them, and the mesmerists claimed successes with a wide variety of ailments.

As for the question of how well different kinds of ailment responded to

248 RETROSPECT AND REFLECTIONS

Table 1. Statistics from Wolfart's clinic

Category of ailment	N (% of total)	Recovered (%)	Improved (%)	Uncertain (%)	Worse (%)	Died (%)
Mental Illnesses	64 (1·5)	21·9	45·3	32·8	0	0
Fits	315 (7·4)	46·7	39·7	11·1	2·5	0
Fevers	386 (9·1)	64·2	32·6	2·1	0	1·0
Inflammations	156 (3·5)	73·7	19·2	7·0	0	0
Paralyses and apoplexies	145 (3·4)	15·9	54·5	20·7	4·8	4·1
Wasting diseases	380 (9·0)	19·5	55·0	16·3	4·2	5·0
Discharges of blood and mucus	270 (6·4)	27·4	57·4	12·2	1·9	1·1
Cachexia, scrofula and rickets	474 (11·2)	45·6	43·7	7·4	2·1	1·3
Rheumatism and gout	709 (16·7)	41·3	45·4	10·7	2·5	0
Suppressed excretions	161 (3·8)	55·3	39·1	5·0	1·0	0
Organic changes	134 (3·2)	21·6	35·1	39·6	3·0	0·7
External ailments and chronic skin afflictions	109 (2·6)	67·0	23·9	7·3	0	1·8
Eye complaints	734 (17·3)	46·3	44·6	8·4	0·7	0
Hearing problems (two years only)	195 (4·6)	30·3	42·0	27·7	0	0

mesmeric treatment, the most extensive and the most systematic data appear to be those of Wolfart, who published a statistical breakdown of the cases treated in his clinic in the period 1814–1821.[11] By combining his figures for each of the years 1819, 1820 and 1821 we may construct the table shown, in Wolfart's own categories, of the 4232 cases treated both in his private practice and in the public mesmeric clinic of which he was in charge.

It must be borne in mind here that many of Wolfart's patients were treated round the baquet, and that they were probably also receiving mild conventional medication. Unfortunately, too, his classification of ailments obscures many distinctions of possible relevance, and, though like Wienholt he deserves credit for including all patients, not just those who were cured, neither his data, nor the lists of cures furnished by, for instance, Mialle and Elliotson,[12] help us very much with the central problem, that of whether there were categories of disease for which mesmeric methods were more successful than conventional treatments, or than simply leaving the patient be. What would be required to tackle this problem is something we do not have – proper background information as to the frequency at the relevant periods of different categories of disease, and as to the success or otherwise of the then customary treatments.

In the absence of such information, the best we can perhaps do is to look at the opinions of persons who were both experienced magnetizers and scientifically or medically trained. For instance, in his *Instruction pratique* (1825), Deleuze[13] lists the kinds of malady which he thinks may be beneficially treated by animal magnetism, and indicated for each the magnetic methods which he regarded as appropriate.[14] In cases of acute illness, magnetism should only be used as a supplement or ancillary to conventional treatments. It may be of value through calming spasms, and bouts of pain, assisting laboured breathing, regularizing the pulse, reducing fever, procuring sleep and strengthening the *vis medicatrix naturae*. Among chronic ailments, migraine, colics, spasms, menstrual disorders, asthma, and hysteria are particularly amenable to magnetic treatment. Mental illnesses, however, will not yield unless recent and accidental; hereditary and long-standing cases will prove refractory. Rheumatism is the most responsive of all disorders to magnetic treatment. Many eye disorders can be helped or cured (eyes were generally bathed in magnetized water) and so can some cases of deafness.[15] Cases of paralysis can often be improved – Deleuze knows of sixty cures of such cases in France, though he does not think magnetism will cure paralyses due to brain damage. With regard to epilepsy, Deleuze claims that magnetism can both calm attacks and, with persistence, cure the disease. He makes the following strong statement: "...it is certain that, out of the large number of epileptics who have resorted to magnetic treatment, many more complete cures have been obtained than would have been achieved by medicine."[16]

In general, other sources, such as Mialle, Rouillier, Elliotson, Teste and Foissac,[17] are in agreement with Deleuze over the special amenability to magnetic treatment of such disorders as rheumatism, dropsy, digestive troubles, constipation, dysmenorrhea, epilepsy, catalepsy, St. Vitus' dance, headaches and neuralgia, paralyses and ophthalmia. With regard to such commonly progressive and fatal diseases as consumption and cancer, most reponsible magnetists were very cautious. Deleuze thinks that something may perhaps be done if the malady has not progressed too far. Otherwise, magnetism may for a while produce the appearance of great benefit by relieving accessory conditions, but the hopes thus raised will prove empty. Foissac warns that certain supposed cures of malignant disease may rest upon misdiagnoses.[18] Some writers warn that magnetism may bring dangers as well as benefits.[19]

I must not create the impression that all, or even a majority, of the alleged magnetic cures were of the rheumatisms, headaches, menstrual disorders, etc., just mentioned – kinds of ailments, some of which are, for whatever reason, notoriously amenable to the ministrations of "fringe" healers. We also find alleged examples of (for instance) the accelerated healing of burns, wounds, ulcers and fractures, the correction of wry neck and curvature of the spine, the rapid and permanent relief of tic douloureux, the quieting of mania, the cure of scrofula, etc., etc.[20] There are even some alleged examples, very difficult to assess, of the cure of malignant disease.[21]

The most celebrated of these last is the case described by Elliotson in his article "Cure of a true cancer of the female breast with mesmerism", which was published in *The Zoist* for October 1848, and also separately as a pamphlet.[22] Elliotson commenced treatment of this patient (Miss Barber, an unmarried dressmaker of forty-two) in March 1843. He was well aware of the possibility of misdiagnosis, and took care that his diagnosis of a true cancer was amply confirmed by independent authorities.[23] He persisted with regular mesmeric treatment for more than five years; and by the end of 1848 all sign of cancer had disappeared. Miss Barber died the following year of inflammation of the lungs. A post-mortem revealed no trace of cancer anywhere in the system.[24]

What is so impressive about so many mesmeric case histories is not just that the patient got better – most patients get better anyway given time – but that so often the victims of long-standing and burdensome complaints began to improve as soon as mesmeric treatment was commenced. I have already remarked on this point, but a couple of further brief examples selected on no better principle than that of readiness to hand will make it more concrete.

Both were contributed to *The Zoist* by Dr. Storer of Bath.[25] The first is a case of severe epilepsy of three or four years standing. The patient, a woman of twenty-three, suffered from five or six attacks a week, and sometimes as many as five or six a day. They were so violent that she required two or three persons to hold her, and she had frequently injured herself by falling suddenly against the wall, down the stairs or into the fireplace. She had been treated by various medical men, and the late Dr. Barlow, of Bath, had pronounced her case incurable. Storer mesmerized her regularly for about six weeks, producing simple (i.e. non-somnambulic) sleep. By the end of a month, the fits had completely ceased and at the time of writing had been in abeyance for more than eighteen months. It may be remarked that despite the hazards of misdiagnosis this case has all the hallmarks of *grand mal* epilepsy. Though in some instances epileptic attacks may be precipitated by psychological factors,[26] and some sufferers may successfully adopt psychological and other techniques for averting attacks, true epilepsies are only doubtfully responsive to psychotherapeutic procedures,[27] and many neurologists would expect most patients to be radically helped only by drugs or other treatments which directly influence neurotransmitter systems in the brain.

The second is a "most violent case of tic douloureux" (trigeminal neuralgia, a complaint which was then often, and is still sometimes, most refractory[28]). The patient, Mrs. West, aged fifty, "states, that she has suffered severely for the last three or four years, that sometimes the pain is so severe as to cause her to bite her lips, that she has frequently been without sleep for two or three weeks together, her eyes are constantly suffused with tears, and her mouth drawn aside by the pain. She has had several teeth removed with the hope of relief, but all to no avail. She has been under several medical men, and her case has excited much commiseration." A fortnight's regular mesmerization induced a remission of the more severe symptoms, and she slept better. After three months of treatment, with three weeks' interruption during which she lost ground, she was "so much improved that some who met her did not know her for the same person". Even during a bad bout of rheumatism the tic "scarcely returned". "In such an extreme case", says Storer," with a disposition to rheumatism, I can hardly hope for an entire cessation; but the good already effected has gone far beyond anything yet accomplished by medicine."

To any accumulation of such cases – cases, that is, in which there seems to have been a direct and obvious relationship between the commencement of magnetic treatment and an improvement in the patient's condition – it might be objected that only those cases in which a remission of symptoms chanced to coincide with the start of treatment would have found their way into print. Against these cases must be set an unknown, but probably much larger, number of unpublished failures, consideration of which would greatly lessen the tendency to imagine a link between mesmerization and cure. But one has only to look at what we know of the more reputable magnetic clinics to realize how little the theory has to recommend it. There is nothing to indicate that those clinics (such as the Strasbourg Society) which published long lists of cures were selecting the cases from a much larger number of patients most of whom were not cured. On the contrary, it seems probable that the majority of those who came to the clinics ended up classified as "cures", or as having benefitted from the treatment. Even though Wolfart's figures, quoted above, which purport to include all those who attended his clinic, may be unduly optimistic (probably most of those classified as "uncertain" did not benefit), they certainly indicate a relatively small proportion of clear and complete failures. The Annual Reports of the London Mesmeric Infirmary confirm this general picture. Thus at its sixth Annual Meeting (8th June 1855), the audience was told that 247 patients had been treated during the previous year. Of these, 69 had been cured; 9 nearly cured; 49 had improved; 41 were still under treatment; and 71 did not continue attendance after the first one or two visits. (This leaves 8 unaccounted for.) If for "41 still under treatment" we read "41 showed no improvement", we shall probably be not too far from the truth.[29] It really is quite difficult to avoid believing that in an appreciable proportion of cases there was not some direct relationship between regular magnetization and an improvement in the patient's condition.

What this relationship might consist of will shortly be discussed. But we must briefly consider the possibility that those patients who passed into the "somnambulic state" may have enjoyed extra benefits compared with those who did not.

This question can be subdivided into three parts. The first issue is that of whether or not the somnambulic state (whatever can be meant by this term) is somehow in and of itself specially conducive to the amelioration of illness; the second is that of whether the supposed clairvoyant self-diagnosis, or diagnosis of the ills of others, which was so characteristic of magnetic somnambules, ever went usefully beyond what the somnambule concerned could have learned or inferred by ordinary means; the third is that of whether or not the prescriptions with which those same somnambules so often accompanied their diagnoses commonly, or even now and again, benefitted the patient more than the remedies they would otherwise have received.

The first of these issues is tied to some particularly difficult and far-reaching problems, and I shall confine myself to making one tentative point. It is quite clear that the majority of mesmeric cures (including apparent cures of quite severe ailments) were performed upon subjects who did *not* pass into a somnambulic state, but (at most) fell asleep, or into a relaxed and sleep-like state, or indeed even remained awake throughout all the mesmeric procedure. Magnetizers of course tended to think it a great step forward if the patients passed into the "sleep-waking" state. However, though the cures performed upon somnambulic patients tended to be protracted and to undergo many and sometimes dramatic vagaries, no doubt because the patients were in effect managing their own cases, I cannot see on balance any grounds for supposing that patients who became somnambulic stood a better chance of being cured than those who did not.

Likewise, I can see very little reason for supposing that the clairvoyant diagnoses and prescriptions added anything, except by chance, to the patient's prospects of getting better. Often the diagnoses were anatomically impossible and physiologically preposterous.[30] Often they could only have been verified by post-mortem, and the patients disobligingly survived. Sometimes, when the diagnosis was given by a clairvoyant somnambule other than the patient, its general accuracy could be quite impressive, and I have noted several cases in which fairly detailed confirmation was received upon autopsy (in one the clairvoyant diagnosed, as was not uncommon, simply from a lock of the patient's hair).[31] The trouble is that this small collection of confirmed successes has to be evaluated against an unknown number of unreported clear failures, and an immense background of dubious material produced by the very numerous professional (or non-professional) somnambules who enjoyed a considerable vogue in the mid-century (particularly in France), and who had no doubt picked up a good deal of useful knowledge of medicine and anatomy. A. S. Morin (1807–1888), a French lawyer keenly interested in the magnetic movement, and not a total disbeliever in diagnostic clairvoyance, gives an amusing account of the fast footwork by means of which these consultants would retrieve their own errors, and he suggests that no conclusions can usefully be drawn about the accuracy of such diagnoses without a full stenographic record of the proceedings which would reveal the clairvoyant's mistakes as well as her hits, and the help she was inadvertently given by the client.[32] Against such a background a small number of apparently striking successes can very probably be discounted.

As for the medicines and treatments commonly prescribed by somnambules possessing the "instinct for remedies", many were homely nostrums which could have come straight out of great-grandmother's commonplace-book or some volume of domestic medicine.[33] Mediums and clairvoyants are apt to recommend such remedies to this day, and their value, in most instances, could only have been as placebos. Occasionally, however, forms of treatment were recommended which could have been positively dangerous, and it is perhaps surprising that there were not more fatalities.[34]

In sum then:

1. It seems by no means unlikely that quite a few patients benefitted from magnetic treatment.
2. The treatments were apparently effective with a fair variety of different kinds of disorder.
3. There does not seem much reason for supposing that magnetic patients who became "somnambulic" benefitted more than patients who did not.

If, as I have just suggested, traditional mesmeric treatments (those involving "passes", and other manipulations, but not somnambulism) may sometimes have genuinely benefitted patients, it becomes a matter of some interest to enquire through what means the benefit may have been conveyed. Five hypotheses have been commonly advanced, of which the first three may be quickly dismissed.

The first is the well-tried imagination-hypothesis with its latter-day descendant the placebo hypothesis. And there is of course no doubt that patients given dummy treatments or inert drugs under appropriate double-blind conditions often show substantial improvements, improvements which are presumably due to the imaginings, expectancies and false hopes engendered by the treatments. Now had mesmerism been commonly the treatment of first resort a comparable explanation of its effectiveness might be tried out; but more often than not it was the treatment of last resort, turned to when orthodox medicine and orthodox practitioners had failed. Why were the patients' imaginations and placebo-propensities stirred into life by mesmeric methods, rather than by medicines backed with the weight of established authority? Of course a dozen specious answers could readily be invented, but none, I think, that would be more than a piece of evasive ad hoccery.

The second is the hypnotism hypothesis, the hypothesis that since we can now with hindsight perceive that "magnetization" was merely a method of hypnotizing the patient, we can explain mesmeric cures in the same terms as hypnotic ones. This hypothesis raises more problems than it resolves, particularly as to the nature of hypnosis. However, it can be immediately disposed of, for most mesmeric patients were ostensibly benefitted without passing into a state resembling hypnosis. Many stayed awake, whilst others simply dozed. If it is replied that suggestion can benefit even patients who have not passed into hypnosis, the same problems arise as confronted the first hypothesis.

The third hypothesis is that the orthodox medical practices of the time were so barbaric that their termination at the start of magnetic treatment could not but benefit the patient. And certainly for much of the period we are concerned with patients were liable to find themselves bled, cupped, blistered, burned, setoned, shocked and purged, all in the cause of promoting health. As for the medicines that were commonly administered, Oliver Wendell Holmes remarked: "I firmly believe that if the whole materia medica, as now used, could be sent to the bottom of the sea, it would be all the better for mankind – and all the worse for the fishes."[35]

This hypothesis cannot be wholly dismissed, but strong reservations are appropriate. Orthodoxy did not always preach the same doctrine or prescribe the same medicines. At the height of the French medical revolution of the post-Napoleonic period, heroic treatments went out of fashion, and it was said, "The British kill their patients; the French let them die."[36] Among the quite numerous mesmeric case histories which I have studied I have found only a few in which there are clear indications that what started the patient on the road to recovery might have been the cessation of misguided conventional treatments. It is of course probable that these conventional treatments sometimes added appreciably to the sufferer's miseries, and that the cessation of such miseries may now and again have aided recovery.[37] And that is as far as we can safely go.

The fourth and fifth hypotheses are much harder to evaluate. The fourth relates the mesmeric cures and alleviations to some unknown or little understood kind of force or energy that emanates from the organisms of magnetic healers, perhaps from the organisms of all healthy persons. This hypothesis would be regarded with profound suspicion by most modern workers, though of course the mesmerists thought they had evidence for their version of it.[38] A certain amount of possibly relevant material is to be found here and there in the current literature on "mental healing" and kindred topics.[39] The most interesting studies are those conducted upon animals and other non-human life forms[40] or organic systems which may be presumed immune to the effects of "suggestion". The best-known are those of Grad, most of which have used a single reputed healer, Oskar Estebany.[41] In two experiments with mice, Grad found that the approach of Estebany's hands reduced goitres and promoted healing of small skin lesions significantly more than did various kinds of control treatments or than simply leaving the animals to themselves. Grad is also one of a number of persons who have demonstrated in controlled experiments an apparent ability by certain healers to influence the growth of seedlings, usually through "treating" the water or saline solution with which the experimental group of seedlings is to be watered.[42] The procedures in these experiments are somewhat reminiscent of the activities of mesmerists with magnetized water.

Recently, D.P. Wirth[43] has described some experiments on the treatment of small skin wounds in humans by the near approach of a healer's hands. By an ingenious arrangement, subjects were kept unaware of the presence of the healer, and the experiment was run double-blind. The mean wound surface areas of the treated group, measured on the 8th and 16th days, were smaller than those of the untreated control group, the difference being significant with a p of less than .001. Regrettably, controls were not instituted for possible heat effects from the healer's hands.

An extended review of this literature would not at present be profitable. None the less, future developments in the area should be watched, because it is perhaps just possible that clues to some of the mesmerists' successes may emerge from it.

The fifth and final hypothesis would in practice be exceedingly difficult to distinguish from the fourth. It is that direct stimulation of the patient's skin – mechanical, thermal, even electrical – such as may be occasioned by the contact and near contact involved in mesmeric manipulations may unleash beneficial secondary effects. To assess this possibility we need to return to what was said in the previous chapter concerning major surgical operations for which the patient was rendered mesmerically analgesic. Some recent writers[44] have attempted to discount these cases, or rather to lessen the theoretical problems which they present, by suggesting (a) that the patients concerned actually suffered some degree of pain (shown by groans, muscular rigidity, etc.) but chose, for somewhat obscure reasons to conceal the fact, and (b) that the painfulness of major surgery has been greatly exaggerated – once the skin has been cut, severance of muscles, bone, etc., causes little additional anguish.[45] Proposal (a) was (as we have already noted) often made by medical sceptics in the mid-nineteenth century and contemporary mesmerists were well aware of it. Elliotson, for example, presents some relevant statistics derived from Esdaile's cases. He notes[46] that of forty-nine consecutive mesmeric operations carried out by Esdaile in 1847, seventeen took place "with the patient like a corpse", fourteen "with no difference from the appearance of a corpse except slight contractions of brow, fingers, or toes", thirteen "with considerable indications of suffering, but subsequent denial of it", and five "with the patients becoming sentient before the operation was completed". He also himself witnessed a major surgical operation (the removal of a breast) carried out with mesmeric anaesthesia at the London Mesmeric Infirmary in April 1851, and he looked especially for signs of pain. The patient, he says, "sat perfectly still, silent, and relaxed, like anyone in the sweetest sleep ... her lips were relaxed and motionless; and, in order further to show that she exerted no effort to restrain herself while the gashes were making, I moved the ends of her fingers backwards and forwards, in complete relaxation with the tip of one of my fingers. There was no holding or catching of her breath; all was the tranquillity of complete repose."[47]

Suggestion (b) raises a number of complex issues, but here it is perhaps sufficient to note that mid-nineteenth-century surgeons were not prone to economize on skin incisions. In one of the mesmeric cases, an incision of 33 inches was made.[48] We are left with the fact that the 1840s saw a remarkable upsurge of surgical operations – some of them very severe – performed with mesmeric analgesia, and that nothing quite comparable is to be found in the subsequent history of the hypnotic movement. It is not unreasonable to ask – though with no strong hopes of obtaining an answer – whether some additional factor (additional that is to the workings of hypnotic suggestion) may have contributed to the sometimes remarkable effectiveness of mesmeric analgesia. And an obvious possibility is this. In the last couple of decades we have acquired a considerable, though still very incomplete, knowledge of the body's previously unsuspected ability to buffer itself against pain, shock and stress by the release from neurosecretory brain cells of opium-like chemicals, the so-called en-dogenous opioids.[49] There is some evidence that acupuncture brings about analgesia in part by stimulating the production of endogenous opioids, whilst hypnotic analgesia is not mediated by these substances.[50] So it must be reckoned not inconceivable that the procedures of protracted mesmerization, involving various kinds of bodily contact or near-contact, might in some persons procure a release of endogenous opioids with similarly analgesic effect. Certainly if (as appears to be the case) tying a loop of thread round a horse's upper lip – a procedure which causes the release of endogenous opioids – will calm the animal sufficiently to induce it to submit to a tooth extraction without fuss,[51] it does not seem impossible that the peculiar activities of mesmerists might facilitate comparable analgesia in the human case.

Now analgesia is not the only effect of potential medical usefulness that endogenous opioids may produce. Different opioids may work upon different neural systems to induce sedation or euphoria, and some are known to interact with immune system functioning in a complex way. If mesmeric manipulations do indeed procure the release of endogenous opioids, certain of their therapeutic benefits may become easier to understand. We should at least remain open to these possibilities, and to the further implication that the almost universal contemporary tendency to regard mesmerization as simply a variety or subspecies of hypnotization may be misguided.

As a remote analogy we may note that in monkeys the contact and skin stimulation involved in mutual grooming seems to cause the release of endogenous opioids,[52] no doubt with a resultant feeling of enhanced well-being. It may be that in humans the practice of "therapeutic touch", which has been canvassed in recent years, and which is alleged, though upon less than compelling evidence, to result in (for example) reduced blood pressure, improved immune functioning and increased blood haemoglobin, owes some of its effectiveness to similar factors.[53]

In principle, the proposal that mesmeric methods might have physiological results not usually occasioned by hypnotic induction procedures could be submitted to experimental test. In practice, there would be considerable difficulties, and an additional complication would be that different mesmerizers adopted different techniques of mesmerizing. Among earlier mesmerizers, it was common practice to hold the hands or thumbs of the patient (who was usually seated opposite the mesmerizer) for protracted periods, or to lay a hand for a while on his or her shoulders or stomach. The "passes" of the hands from head downwards or shoulders downward would often involve contact with the body (or rather the clothes).[54] Mesmerizers nearer the mid-century (like Dupotet or Teste) seem to have thought contact not quite seemly, and often kept their hands a little above the body.[55] Esdaile's method was to make the patient lie down in a dark room, wearing only a loin cloth, and to repeatedly pass the hands in the shape of claws, slowly over the body, within one inch of the surface, from the back of the head to the pit of the stomach, breathing gently on the head and eyes all the time.[56] He seems to have sat behind the patient, leaning over him almost head to head and to have laid his right hand for extended periods on the pit of the stomach.[57] Elliotson commonly began by extending two fingers of his right hand near to and pointing towards the subject's eyes. Afterwards he might resort to downward passes.[58] Elliotson and Esdaile regarded the process as purely mechanical. Dupotet, following the tradition of Puységur, emphasized the importance of the operator's will. Obviously, if the endogenous opioids are in any way involved in mesmeric analgesia and other benefits, these differences in technique could be of importance.

THE PHENOMENA

We come now to the various odd psychological phenomena traditionally associated with mesmerism. And, as with the alleged mesmeric cures, we can consider them under two subheadings, viz., the phenomena brought about by mesmeric passes (with or without contact) and other mesmeric manipulations, in subjects who had not yet passed into the so-called state of somnambulism, and the phenomena occurring with somnambulic subjects. Concerning the former there is not a lot to say. The accounts of their experiences given by persons who had been successfully magnetized, though varying from individual to individual, are selections from a range of phenomena which remained pretty constant throughout the whole period of the mesmeric movement. The following summary, from Charpignon (1848), of the effects of magnetism preceding any transition to somnambulism is absolutely typical, and may be compared with numerous other accounts:[59]

> In susceptible individuals the sensations and changes felt are very variable ...The eyes shed tears, the skin becomes hot, dry or moist; sometimes perspiration is abundant; yawns follow each other; there manifest themselves a general restlessness, tinglings at the extremities, twitchings of the limbs, wishes to sleep or to evacuate; the pulse quickens, more rarely slows down; the eyes grow heavy, the eyelids stick together, a beneficial calm spreads over the patient; at other times cold shivers run down the spine; they follow the hand of the magnetizer. Sometimes there occur general or partial convulsions, or else the breathing appears stifled; there is a sort of delirium. Sometimes, by contrast, one sees certain individuals fall into a kind of lethargy; ...they can neither move nor speak; sometimes they hear without being able to understand; if one puts their limbs in a certain position, it is retained. Shocks like those given off by an electrified body are felt at the approach of the magnetizer's finger.[60]

The magnetists, of course, interpreted the peculiar sensations and other phenomena as signs of the working of the magnetic fluid. It is not unknown even today for persons who believe in the healing power of the human hand to interpret in an analogous way the sensations felt by patients undergoing healing.[61] On the other hand, as we have seen, the commissioners of 1784 regarded these sensations as most likely the result either of imagination or of commonplace natural causes, and (more to the point) supported this view with some simple controlled experiments. In the absence of further and better experiments their opinion must provisionally hold the field.

Throughout much of the history of the animal magnetic movement, its supporters were quite clear as to the leading characteristics of the supposed state of mesmeric somnambulism or sleep-waking. It constitutes a phase of existence, or state of consciousness, quite different from that of ordinary wakefulness or (to an extent) of sleep, and its various recurrences are united by a chain of memories not accessible in waking life (though the memories of waking life may be available to the somnambulic consciousness). It is characterized by the divers remarkable phenomena on which we have already touched so often, phenomena which do not occur, or occur much more rarely, in waking life (this assumption ran into heavy weather with the arrival of the electrobiologists). It presents, however, many analogies with "natural" sleepwalking and sleeptalking, and with certain other pathological nervous conditions, and is probably linked to changes in nervous system functioning similar to those which underlie these conditions. It may, indeed, exhibit a chain of memory in common with them.

The proposition that mesmeric somnambulism or sleep-waking constitutes a phase of existence or state of consciousness quite different from that of ordinary waking consciousness raises a variety of issues, some of which are conceptual and to be avoided at this point. The parallel which the mesmerists themselves frequently drew between "natural" somnambulism or sleepwalking and mesmeric somnambulism or sleep-waking, would nowadays be generally regarded as quite misconceived; but in the context of the times it was by no means absurd or without plausibility. Somnambulism is fairly common among children and adolescents, affecting boys more often than girls. In adults it is rarer and more often of pathological significance, but it also more often presents greater complexities.[62] Sometimes episodes, or series of episodes, are precipitated by fever or stress.[63] Sleepwalkers generally have their eyes open and staring, but can move around in their own home and avoid obstacles even in the dark. They seem often to have a restricted or selective, but sometimes acute, awareness of their surroundings. They may carry out routine activities or tasks, may reply, usually briefly, to questions, and in some cases may be responsive to verbal suggestions. Sometimes, on returning to a normal state, sleepwalkers may have a confused memory of the events of the somnambulistic episode; but more usually there is complete amnesia.

In rather rare cases, the phenomena develop to a point where they are quite strongly reminiscent of "artificial", i.e. mesmeric, somnambulism. The subject may talk intelligibly, or even fluently, may exhibit abilities or skills exceeding those of his ordinary life, and, despite darkness or closed eyes, show an awareness of persons and objects that seems almost preternatural, though curiously limited. In very rare instances[64] there may be signs of what would nowadays be called "state-dependent memory". In ordinary waking life, the patient is amnesic for previous somnambulic episodes, but when somnambulic can apparently remember events from such past episodes, as well as some at least of the events of waking life. Recollection thus appears tied in some way to the psychological or physiological state of the patient, a series of like states (of which ordinary waking consciousness may be just one) being united by a common thread of memory not available in other states.[65] This kind of phenomenon has been investigated experimentally, and cases are on record in which (as in that classic Victorian detective story, Wilkie Collins' The Moonstone) events which took place when a person was drugged or drunk become available to him again in memory when he is drugged or drunk once more.[66] When this kind of state-dependence of memory is combined with a somnambulic phase in which the patient exhibits capacities different from those of his or her ordinary waking consciousness – e.g. an enhanced fluency or vivacity – we have some-thing that begins to resemble a case of multiple personality. Among such cases those of Mary Reynolds and Jane C. Rider[67] are often referred to by the later magnetic writers (who sometimes talk of dual consciousness and dual personality). Some of these writers present us with collections of cases of "natural somnambulism" which they think have strong affinities with cases of mesmeric somnambulism and sleep-waking – Colquhoun has a particularly large collection[68] – and although quite a few of these cases are almost cetainly not ones of somnambulism proper, but of hysterical dissociation, fugue, psychomotor epilepsy, and so on, this does not perhaps matter a great deal for present purposes, because the distinctions between adult somnambulism, fugue, and even epilepsy are by no means completely clear.[69]

Here is a case from Colquhoun's collection. He takes it from Abercrombie's *On the Intellectual Powers*, the original case account being by Dr. Dyce of Aberdeen:

> The patient was a servant girl, and the affection began with fits of somnolency, which came upon her suddenly during the day, and from which she could at first be roused by shaking or by being thrown out into the open air. She soon began to talk a great deal during the attacks, regarding things which seemed to be passing before her as a dream; and she was not, at this time, sensible of any thing that was said to her... In her subsequent paroxysms, she began to understand what was said to her, and to answer with a considerable degree of consistency... She also became capable of following her usual employments during the paroxysm; at one time she laid out the table correctly for breakfast, and repeatedly dressed herself and the children of the family, her eyes remaining shut the whole time. The remarkable circumstance was now discovered, that, during the paroxysm, she had a distinct recollection of what took place in former paroxysms, though she had no remembrance of it during the intervals. At one time she was taken to church while under the attack, and there behaved with propriety, evidently attending to the preacher; and she was at one time so much affected that she shed tears. In the interval, she had no recollection of having been at church; but, in the next paroxysm, she gave a most distinct account of the sermon, and mentioned particularly the part of it by which she had been so much affected... During the attack, her eye-lids were generally half-shut, her eyes sometimes resembled those of a person affected with amaurosis, that is, with a dilated and insensible state of the pupil; but sometimes they were quite natural... At one time during the attack, she read distinctly a portion of a book which was presented to her: and she often sung, both sacred and common pieces, incomparably better, Dr. Dyce affirms, than she could do in the waking state.[70]

In a footnote, Colquhoun adds (probably on information from Dyce himself) that in one of the paroxysms "this girl was afterwards abused, in the most brutal and treacherous manner. On awaking she had no consciousness whatever of the outrage; but in a subsequent paroxysm, *some days afterwards, it recurred to her recollection, and she then related to her mother all the revolting particulars*. This case presents a very striking instance of the phenomenon of *double personality*."[71]

One does not often find cases as complex as these in the modern literature. However, complex and peculiar cases still occur. Some of the oddest are cases of sleeptalking, or somniloquism, a phenomenon related to sleepwalking, and presenting even readier analogies to reports of mesmeric somnmabulism. According to Arkin,[72] sleep utterances occur in both REM and NREM sleep, though favouring the latter. If mentation is reported, the sleep utterance may or may not correspond to it. Sleep speech may develop to the point at which conversation with an interlocutor becomes possible. In a case described by Arkin[73] from an original report by Burrell (1904), the patient, a thirty-eight-year-old farmer, who had had two psychotic episodes, would, during sleep, repeatedly speak as if conducting a prayer meeting, with hymns and quotations from scripture; correctly carry out complex activities; tell humourous anecdotes; and engage in dialogue with others in the room. On one occasion, witnessed by Burrell, he conversed with a visiting brother who was also asleep. In the waking state, the patient was amnesic for all his sleep speech, but during sleep speech episodes he would, in conversation with observers, recall speeches from previous episodes.[74] In another case, without any apparent pathology, the subject's sleep utterances covered an even wider range of topics.[75] A few writers have reported patients who exhibit during sleep speech accomplishments superior to those of their waking state. Arkin describes[76] from personal observation the case of a lady who had visited Russia at the age of eleven, and had acquired some fluency in the Russian language. While in Russia she had been badly treated by her father. Later she forgot her Russian almost completely. However, in adult life she began during sleep both to utter Russian words and to write in Russian, and what she said and wrote related unmistakably to her father's treatment of her so many years before. Sometimes (as with the examples cited by mesmerists) the distinction between sleeptalking and fugue seems obscure. For instance in a case described by Rice and Fisher,[77] the patient, a man of forty-six, underwent episodes of fugue which revolved around the recent death of his father, and his nighttime sleeptalking exhibited many similar characteristics, so that the authors conclude that the latter was a nocturnal equivalent of the former. They note that in many fugue patients the fugue states are preceded by drowsiness or sleep.

On the face of it, then, "natural somnambulism" presents many analogies to the "artificial" or mesmeric kind in respect of the activities carried on (talking, quasi-purposeful actions), the dense waking amnesia for events of the somnambulic state, and the state-dependent recollection of past somnambulic episodes exhibited in a current episode. Whatever the ultimate value of the parallel between "natural" somnambulism and mesmeric somnambulism, the mesmerists are not to be blamed or thought simple-minded for drawing it. The similarities, though not complete, are obvious. And it was inevitable that the mesmerists should try to fit both categories of phenomena into the same kind of theoretical framework. Of course when that framework was confined, as it often was, to vague talk about a progressive shutting down of the ordinary channels of sense, and a redirecting of the mesmeric fluid, or vital force, to other parts of the nervous system, it must indeed be reckoned simplistic to the point of vacuousness.[78] But when developed by, for example Kluge, who fortified it with all the latest neurophysiological speculation and discoveries,[79] or by Kieser and Wirth, whose proposals so clearly foreshadowed modern doctrines about the unconscious origins of mental abnormalities,[80] the framework, whether or not misconceived, was, in the context of its times, in no way simplistic, but in some respects sophisticated and forward-looking. For these writers the somnambulic consciousness is a state of being very different, and quite separate, from the waking consciousness. Fischer, in 1839, used the metaphor of the individual (still unified) passing from one region to another, totally strange one:

> The soul wanders in a region hitherto closed to consciousness, and in it begins a completely new course of life, one separated from the day-life, and possessing its own peculiar circle of ideas, its own closed memory and history ... In this somnambulic region there rules an unquestioned and accomplished plastic power, a creative fantasy and penetrating intelligence, such as the often very moderately endowed day-individual does not possess; at the same time these plastic powers have ... in their willing and creating a rigid, instinct-like necessity, so that their productions, like the fixed ideas of madmen, obtrude themselves as corporeal realities.[81]

This kind of view may be compared with that of modern writers who regard hypnosis as a "special state" of "loss of generalized reality orientation", in which "primary process" thinking comes to the fore.[82] On the other hand, Elliotson, as might be anticipated from his rigidly materialist views, does not hold with speculations about a "night-side" of the mind, different from our waking consciousness; but he does believe in what might, in a broad sense, be called "unconscious cerebration", including unconscious perceiving, unconscious willing and unconscious reasoning, all conceived just as phases of brain activity unaccompanied by consciousness.

> The truth is, that sleep-wakers are more or less abstracted, and abstracted in regard to some things and not to others: that they, through unconscious and involuntary inclinations, may not be consciously percipient of many things, which however their brains really perceive, and which they in the mesmeric state, and sometimes in the ordinary only, may never afterwards know they have perceived; and their brains may have various internal feelings, and will many things, quite unconsciously; and afterwards they may act upon and be influenced by the knowledge thus unconsciously received, without ever suspecting that they had received it ... No point in cerebral physiology is more curious than our unconscious reception of sensations or unconscious prevention of consciousness of them, and the influence of unconscious knowledge and feeling over our actions. Materialism only can explain this. The brain acts in all these wonderful ways: one part doing what another is ignorant of.[83]

Elliotson here emerges as a "dissociation" theorist with respect to some at least of the phenomena of the sleep-waking state. He is, in effect, a precursor of those more recent writers who have sought to explain the phenomena of hypnosis as due to the operation of brain systems or subsystems that have begun to function independently of the system which underpins the main stream of everyday consciousness. In these terms, it is easy to handle, for instance, those curious cases in which the subject exhibits more than one distinctively different somnambulic phase, each with its own separate or partially separate state-dependent memory – both Elliotson and Gregory claim to have observed cases of triple consciousness.[84] It is also easy to handle instances of the fulfilment of what later came to be called post-hypnotic suggestions. Elliotson himself carried out various experiments with such suggestions. He would get subjects in the somnambulic state to "promise" to see or not to see certain people or things (positive and negative hallucinations) in the waking state or to perform certain actions; and often they would comply. However, one young lady who had thus promised to whistle on next entering Elliotson's library, and had no waking recollection of her promise, failed to do so. Sent to sleep, she said she did not comply because the butler was in the room. Elliotson's comment is as follows:

> The circumstance of a desire occurring in the *brain*, and being repressed by
> an unconscious reason, is a striking fact, but in harmony with the facts
> previously mentioned.[85]

It is the extraordinary combination of subtlety with simple-mindedness, and of enlightenment with prejudice, that makes the character of John Elliotson so fascinating.

To return to the question of the supposed analogy between "mesmeric" and "natural" somnambulism: I have tried to establish that the parallels between them so often drawn by the mesmerists were by no means foolish, and to show also that some of the related theoretical speculations in which certain mesmerists engaged were, if one disregards nonsense about the "magnetic fluid", a good deal more advanced than is generally realized. Indeed, the parallel with natural somnambulism remained commonplace into the early phases of the hypnotic movement. Why is it nowadays so generally rejected? There are, as it were, two layers of reasons.

The first is as follows. We may assert with confidence that whatever the "state" of mesmeric somnambulism was, it was identical with the "state" of hypnosis. For all the phenomena thought characteristic of mesmeric som-nambulism can be reproduced exactly with deeply hypnotized subjects. Now we know that natural somnambulism is a phenomenon of stages 3 and 4 of NREM sleep. But "hypnotized" subjects are not in these sleep stages; the EEGs and other physiological measures of deeply hypnotized persons are not essentially different from those of waking persons. Hence the state of hypnosis, and *a fortiori*

that of mesmeric somnambulism, is in no way like that of natural somnambulism.[86]

If one is prepared to agree that there is such a "state" as the state of
"hypnosis", this line of argument is acceptable so far as it goes; but a few
reservations would be appropriate. The assertion that somnambulic episodes
take place in, or out of, stages 3 and 4 NREM sleep is based largely upon work
with children, and matters are not always so clear with adult sleepwalkers and
sleeptalkers, some of whom seem to exhibit EEGs with a preponderance of alpha
rhythms resembling EEGs found in the immediate pre-sleep stage.[87] Some
workers have likewise claimed that the EEG of deeply hypnotized subjects
resembles that of persons in the immediate pre-sleep stage. And again if it turned
out that some additional factor (additional, that is, to those involved in hypnosis)
were involved in the production of mesmeric analgesia or of mesmeric cures –
and I suggested above that this is at any rate just conceivable – then it would of
course be worth considering the possibility that such a factor might also have
some influence in producing the state of mesmeric somnambulism, which would
then not necessarily be altogether the same as the state of hypnosis. We have for
instance some evidence that endogenous opioids may promote catalepsy, a
condition not infrequently observed in mesmerized subjects.[88]

The second layer of argument goes as follows. There is no such state as the
state of hypnosis. It is a myth handed down through our culture, a myth which,
like other myths, has none the less a compelling influence upon our experiences,
beliefs, and behaviour. The same is of course therefore true of the state of
mesmeric somnambulism. A fortiori neither of these two states can be identical
with or closely related to the state of natural somnambulism which is in some
sense genuine and not factitious. At most they might be looked upon as in part
originating from beliefs about natural somnambulism. This view of hypnosis, and
the grounds on which it is based, will be discussed later on. Of course if reasons
(such as I suggested in the preceding paragraph) came to light for supposing that
mesmeric somnambulism is not wholly to be equated with hypnosis, we might
explore the possibility that the imaginary state of hypnosis sprang from the
dissemination of knowledge about the genuine state of mesmeric somnambulism.
But speculation along these lines would be premature.

THE DECLINE OF MESMERISM

I have brought my account of the mesmeric, or animal magnetic, movement
down to the year 1850, or a little beyond. At that period the movement was still
flourishing in many areas without obvious signs of decay. I have said little about
its decline, except to indicate that in certain countries its impetus, ideas and
personnel were to a fair extent transferred to the nascent Spiritualist movement.

Nor shall I now say much about that decline, for two very good reasons. The first is that the decline is of little interest, and of virtually no interest for the purposes of the remainder of this book. The second is that the decline was not a sudden or marked one at all. Histories of psychology often give the impression that mesmerism was killed off in the 1840s by the compelling counter-demonstrations of Braid and Wakley; but this is quite untrue. The mesmerists had little difficulty in accounting to their own satisfaction for the findings of Wakley, or Braid, or the electrobiologists. The animal magnetic movement went ahead with scarcely a halt in its stride. In fact, as we have seen, some quite weighty manuals of mesmerism were published after 1850. When we talk of mesmerism's decline in the third and fourth quarters of the nineteenth century, we must be clear what we mean. Mesmeric ideas and practices did not disappear in the later nineteenth century – the magnetists held a conference in Paris in 1889 at the same time as the first International Congress of Hypnotism – they simply diversified and fragmented and changed context. The ideas and some of the practices and quite a few of the personnel were absorbed into Spiritualism; the ideas and practices also turned up in theosophy;[89] they are pervasive in the writings and activities of occult groups, especially French ones, certain of whom renovated the *Journal du magnétisme* in 1879 under the editorship of Hector Durville; they crop up with disconcerting regularity in the early hypnotic movement, especially in the form of a belief, apparently verified by experiment, in the direct influence of metals and other substances on the nervous system;[90] they were put to the test and supposedly confirmed by the designers of assorted little instruments that were held to register mysterious emanations from the human hand or eye (this too was very much a French preoccupation[91]); in 1905 the participants at the Fifth International Congress of Psychology at Rome heard a paper by L. Favre on the influence of animal magnetic procedures on the germination of seeds and the development of bacterial cultures, and one by J. Courtier on the effect of magnetic passes on an ebonite plate wired to an electrometer.[92] Indeed, magnetic ideas and practices are in a sense still with us, though I will not attempt to relate the views of practitioners of "radionics" and "psychotronics", or those of modern Russian believers in "bioplasma", directly to the ideas of the mesmerists.[93]

Though mesmeric ideas did not disappear, it became increasingly difficult for educated and rational men to take them seriously. Of course, mesmeric ideas never had commanded the general assent of educated persons; but at various times there had been a substantial sprinkling of such persons who were keenly interested in them. And mesmeric notions about a "nervous fluid" were for a long time not too far removed from orthodox beliefs about how the nervous system worked. But from about 1850, scientific knowledge concerning these things moved rapidly, and in directions alien to mesmeric ideas. Understanding of the laws of electricity and magnetism came with a rush and was quickly turned to practical use. In 1831, Faraday discovered electromagnetic induction; by the 1880s there were electric tramways. It gradually became apparent that the nervous system is not powered by some quasi-electrical energy or fluid. In 1850, Helmholtz showed that nerve impulses (unlike electric currents passing along a wire) travel hardly faster than a steam train. Yet many magnetists in the 1850s still had the same vague ideas about the magnetic fluid, the ether, the ganglion system, and so forth, as their predecessors had had near the beginning of the century. In fact Kluge, writing in 1811, gives a more detailed and original, and certainly more scholarly, presentation of the theory than they do. It soon became impossible for anyone with a smattering of scientific knowledge to lend any credence to mesmeric notions. The "magnetists" who continued to speculate and practice in the later part of the nineteenth century were for the most part either persons involved in occultisms of one sort or another, or unorthodox healers catering for a rather simple-minded clientèle. It has to be added, however, that a few persons continued to hold that certain of the results of the mesmerists required an explanation that was both non-occultist and non-psychological (i.e. non-hypnotic), and that a hard and fast line between occultism and science would not at this period always have been easy to draw.

In their heyday, or rather heydays (for there were several), the mesmerists were a group of people for whom it is impossible not to feel a certain admiration. Though some of them were without doubt charlatans and confidence-tricksters, others were persons of the highest and most unselfish ideals. They made what they thought were discoveries of great importance, and they had the courage to proclaim them in the teeth of hostility and derision. And without doubt they were on to something, though even today we are unable to say with exactitude just what that something was. Meanwhile, we can at least try our best to give an accurate and (so far as possible) unprejudiced account of them. The history of the mesmeric movement has very often been misrepresented, the same mistakes and misunderstandings being passed from one author to another. Among the errors thus perpetuated, either by direct statement, or by implication, are: that nothing of interest took place between, roughly, 1784 (Puységur) and 1843 (Elliotson and Braid); that the mesmerists' patients were for the most part suffering from mental disturbances of various kinds, so that the magnetizers were really playing the role of psychotherapists; that Braid and/or Wakley finally undermined the whole movement; that the mesmerists were one and all simple-minded men who explained all the phenomena in terms of "magnetic fluid" without exhibiting any awareness of the psychological complexities involved; that there is no evidence whatever for mesmeric clairvoyance; that all mesmeric phenomena are quite obviously at root the same as hypnotic phenomena (whatever that may involve). The mesmerists are not quite so easily dismissed or assimilated to certain stereotypes as has commonly been supposed. Their most conspicuous failing was one that they shared with a very large proportion of the hypnotists who followed in their footsteps in the second half of the nineteenth century – an almost complete failure to appreciate the powerful workings of "experimenter effect" and "doctrinal compliance"[94] upon mesmerized or hypnotized subjects, and to grasp the methods of controlled experimentation which are necessary to offset these dangers. Even those mesmerists who, like Elliotson, showed now and again some awareness of the issue, and of the ways in which it might be tackled, did not act on their own prescriptions. But they can hardly be blamed for not coming to terms with a problem which was not even adequately spelled out until well into the present century.

NOTES

1 Cooter (1984); Oppenheim (1985).
2 Spanos and Gottlieb (1979).
3 Gregory (1851), pp. 459–466.
4 E.g. Kieser (1826); Wirth (1836); Gauthier (1842); Thouret (1784); Bertrand (1823; 1826); Colquhoun (1836; 1851); Ennemoser (1842; 1844; 1854).
5 Cf. the valuable paper by Stam and Spanos (1982).
6 Ennemoser (1842), pp. 105–106.
7 Among British writers, Colquhoun and Townshend engage extensively in such speculations.
8 See Fischer (1839), Vol. 2, pp. 38–39; Kieser (1826), Vol. 1, pp. 233–234. Kieser suggests that verbal commands are effective because the magnetizer's will operates more intensively when he is giving commands.
9 Mialle (1826). It is hard to be sure of the exact number, since some cases seem to be mentioned twice under different headings.
10 Wolfart (1822–3).
11 Wolfart (1822–3).
12 Elliotson (1853–4).
13 Deleuze (1825), pp. 187–264. Cf. Koreff's list, in Deleuze (1825), pp. 451–454.
14 The curious will find a list of kinds of malady and preferred methods of magnetic treatment in Gauthier (1845), pp. 369–428.
15 Reports of improvements in the condition of deaf-mutes were not uncommon, e.g.: Morin (1860), pp.295–298; Braid (1960), pp. 240–256; Ricard (1840); Rapport de l'Académie des sciences (1840); Lafontaine (1866), Vol. 1, pp. 246–247, 2, pp. 140–143. These apparent ameliorations were probably due to the fact that many deaf-mutes have a slight degree of residual hearing to which the treatment encouraged them to pay attention.
16 Deleuze (1825), p. 224.Cf. Koreff in Deleuze (1825), pp. 451–454.
17 Elliotson (1853–4); Rouillier (1817), pp. 187–234; Teste (1843), pp. 278–279; Kluge (1811), pp. 520–533; Foissac (1833), pp. 499–549; Mialle (1826).
18 Foissac (1833), p. 541.
19 Cf. Deleuze (1825), pp. 265–324. P.I. Hensler (1775–1861), professor of physiology at Würzburg, devoted himself to documenting the dangers of magnetism. He believed that there are different sub-varieties of animal magnetism, and that unless magnetizer and patient are properly matched danger may ensue. See Hensler (1833; 1837).
20 For ample illustrations see Elliotson (1853–4).
21 Cf. above, p. 15.

22 Elliotson (1848).

23 For certificates see Elliotson (1848–9c).

24 Elliotson (1849–50a).

25 Storer (1846–7), pp. 448–449.

26 In a famous case described by Mitchell, Falconer and Hill (1954), attacks were precipitated by staring at a safety pin.

27 Lishman (1987), pp. 270–271, holds that psychotherapy is rarely indicated in cases of epilepsy, but discusses various psychological "manoeuvres" that may avert impending attacks. Kroger (1977), p. 264, claims, with supporting cases, that hypnotherapy can avert attacks by reducing tension.

28 Storer (1847–8), pp. 18–19.

29 *The Zoist*, 1855–6, Vol. 13, pp. 172–173.

30 See above pp. 45–46. Cf. Morin (1860), p. 195.

31 Pigault-Lebrun (1829); Foissac (1833), pp. 445–447; Deleuze (1846), pp. 221–222; Sloman and Mayhew (1851–2; 1852–3).

32 Morin (1860), pp. 151–154.

33 It is only fair to point out that in an extensive and erudite account of these somnambulic prescriptions, Du Prel (1889), Vol. I, pp. 280–332, takes a more positive view. He has some interesting remarks on the forms, including symbolic forms, in which the prescriptions present themselves to the somnambule.

34 See above, p. 63; and cf. Morin (1860), pp. 331–335; Teste (1843), pp. 286–298; and Charpignon (1848), pp. 216–221.

35 Quoted in Ackerknecht (1967), p. 129.

36 Ackerknecht (1967), p. 129.

37 One of Lausanne's patients, who very likely had appendicitis, tells us that over a three-month period his previous doctors had ordained for him two hundred baths, a hundred enemas, thirty leeches, three hundred cataplasms, nine or ten pints of whey, and well over a hundred pints of assorted (probably harmless) herbal infusions, syrups and distillations. See Razy (1815).

38 The principal categories of evidence they deployed consisted of examples of the successful magnetization of very young children (Cf. the list extracted from Mialle, 1826, in *L'Hermès*, 1826, Vol. I, pp. 224–225); of the magnetization of sleeping persons (Deleuze, 1819, p. 153; Townshend, 1844, p. 86; Elliotson, 1845–6b, p. 48); of the magnetization of idiots and persons ignorant of magnetism; of the magnetization of animals (Deleuze, 1825, pp. 262–264; Wesermann, 1822, p. 229; Martineau, 1850–1; Saunders, 1852–3; Elliotson, 1850–1, 1851–2c; Bakker *et al.*, 1814–18, Vol. 2, pp. 90–98; Wilson, 1839); and of the effects of magnetization on plant-growth (Hervier, 1784, pp. 9–10; Passavant, 1821, pp. 47–48; Ennemoser, 1842, pp. 210–211; Kieser, 1826, Vol. I, pp. 63–64; Charpignon, 1848, pp. 52–54; and cf. Morin, 1860, pp. 93, 98–99.)

39 A valuable review of this literature is Solfvin (1984).

40 See the reviews by Solfvin (1984), and by Rush in Edge *et al.* (1986), pp. 248–251.

41 Grad (1965); Grad *et al.* (1961). A review by Grad of his own work is Grad (1989).

42 Grad *et al.* (1961). For other comparable experiments see Solfvin (1984), p. 41. On the claim that the infra-red absorbtion spectrum of water thus treated is changed see Dean and Brame (1972); Dean (1983); Grad and Dean (1984). Fenwick and Hopkins (1986) fail to replicate, but they do not follow Dean's methods precisely.

43 D. P. Wirth (1990).

44 E.g. Chaves and Barber (1976); Wagstaff (1981), chapter 8; Spanos (1986).

45 I have discussed these issues in slightly more detail in Gauld (1988).

46 Elliotson (1848–9b), p. 170.

47 Elliotson (1854–5a), p. 116. For examples of the agonies of operations without anaesthetics see e.g. Elliotson (1847–8c), p. 199 and Burney (1975), p. 610.

48 See *The Zoist*, 1844–5, Vol. 2, pp. 390–393, and Elliotson (1844–5c).

49 For reviews see Akil *et al* (1984); Basbaum and Fields (1984); Olson *et al.* (1990). The first to suggest a link to mesmeric analgesia was Gibson (1982), ch. 5.

50 The administration of naloxone, which reverses opiate analgesia, also reverses acupuncture analgesia (Mayer *et al.*, 1977; Pomeranz, 1982). On the other hand it is alleged not to reverse hypnotic analgesia (Spiegel and Albert, 1983; Goldstein and Hilgard, 1975) which suggests that the latter is not mediated by endogenous opioids (but see Frid and Singer, 1979, Levin *et al.*, 1978). Moret *et al.* (1991) fail to find any reversal by naloxone of either hypnotic or acupuncture analgesia.

51 Lagerweij *et al.* (1984).

52 *New Scientist*, 20th January 1990, p. 32.

53 Krieger (1976); Levitan and Johnson (1985–6); Quinn (1987).

54 A. M. J. de Chastenet de Puységur (1811), pp. 14–15; Deleuze (1825), chapter 2; Kluge (1811), pp. 403–405, 415–422.

55 Teste (1843), p. 152; Dupotet de Sennevoy (1930), pp. 424–429; cf. Scaresby (1849), p. 15.

56 Esdaile (1846), pp. 145–146.

57 This is how his native assistants proceeded (Elliotson, 1847–8b, p. 52).

58 Caldwell (1842), p. 67; S. T. Hall (1845), pp. 79–80.

59 Kluge (1811), pp. 101–104; Bertrand (1823), pp. 218–219; Teste (1843), pp. 48–51; Fischer (1839), Vol. 2, p. 112; Deleuze (1825), pp. 48–49; Deleuze (1826b), p. 17.

60 Charpignon (1848), pp. 44–45.

61 Benor (1984), pp. 21–22; Grad (1989), pp. 108–109.

62 Luchins *et al.* (1978); A. Kales *et al.* (1980); E. Hartmann (1983); Oswald and Evans (1985).

63 Huapaya (1979); J. D. Kales *et al.* (1979).

64 Cf. the case cited in Carpenter (1881), p. 596. Arkin (1981), p. 111, cites some instances in connection with sleeptalking.

65 This notion of state-dependent memory is spelled out in some detail by Du Prel (1889), Vol. 2, pp. 56–61 and following pages.

66 Mello (1972); Overton (1972); Eich (1977; 1980); Ho *et al.* (1977); Goodwin et al. (1969), cited in Lishman (1987), p. 510. Many of Goodwin *et al.*'s 64 subjects had had the experience of hiding money or alcohol when drinking, forgetting it when sober, and finding it again during a subsequent drinking bout.

67 Belden (1834); Kenny (1986), pp. 25–61.

68 Colquhoun (1836), Vol. 1, pp. 299–395.

69 Lishman (1987), pp. 259–265; Rice and Fisher (1976).

70 Colquhoun (1836), Vol. 1, pp. 311–313.

71 Colquhoun (1836), Vol. 1, pp. 347–348. A longer account of this case, together with several other comparable cases, will be found in Elliotson (1846–7b).

72 Arkin (1981).

73 Arkin (1981), pp. 100–101.

74 Arkin (1981), p. 111.

75 McGregor (1964).

76 Arkin (1981), pp. 380–383.

77 Rice and Fisher (1976).

78 The theorizing of Gregory (1851) will serve as an example.
79 See above, pp. 104–107.
80 See above, pp. 158–159.
81 Fischer (1839), Vol. 1, pp. 7–8.
82 Cf. Shor (1979).
83 Elliotson (1845–6a), pp. 361–362.
84 Gregory (1851), p. 177; Elliotson (1843–4b), p. 436.
85 Elliotson (1845–6a), p. 362.
86 E.g. Sarbin and Slagle (1979).
87 Arkin (1981), p. 185.
88 Olson *et al.* (1990), p. 1291.
89 Sinnett (1892).
90 Cf. Harrington (1988).
91 The equivalents of these machines in the earlier part of the century were various devices centring on pendulums, starting with Gerboin (1808). The idea was that a pendulum suspended from the finger might oscillate via the magnetic fluid in directions determined by the will of the operator. As suspicions grew that the oscillations might really be due to unconscious muscular movements by the operator, the pendulums came to be attached to various kinds of frame and not suspended directly from the finger of the operator. But the hand of the operator was always somewhere where it seems possible it could have disturbed the apparatus. On these early devices see Dingwall (1967), pp. 215–216, 220–221, 239–240; Louvier (1844), pp. 263–273; Morin (1860), pp. 101–113; Mayo (1851), pp. 197–235; Rutter (1851; 1854); Léger (1852). Later in the century and into the twentieth century somewhat more sophisticated devices became popular, and delicately balanced objects under glass covers were made to flutter or rotate at the approach of a hand as though under the influence of mysterious emanations. See e.g. Amadou (1953); Pascal (1936); Montadan (1927); Carrington [1939]; Bonnaymé (1908); Baraduc (1904); Baréty (1882; 1887); Clement-Martin (1926); Menager (1926). Advances in photography opened up new possibilities for recording the supposed emanations. See Baraduc (1896; 1897).
92 Courtier (1905); Favre (1905).
93 On the persistence of mesmeric ideas, see Glowatzki (1983).
94 See p. 168 above.

PART II

The heyday of hypnotism

CHAPTER 14

Precursors of the hypnotic movement

In this chapter I shall consider writers with the following in common:

1. They were keenly interested in animal magnetism.
2. They came to reject the magnetists' own explanations of these phenomena.
3. They developed their own explanations of the phenomena, explanations which had unmistakable points of resemblance to the explanations offered by adherents of the hypnotic movement in the last quarter of the nineteenth century.
4. They wrote before this movement had got under way.
5. They derived their ideas, at least in part, from their own practical investigations.

The writers in question are: the Abbé Faria; Alexandre Bertrand; James Braid and the "braidists"; and the Americans La Roy Sunderland, W. B. Fahnestock and J. K. Mitchell.

THE ABBÉ FARIA

The Abbé José-Custodio de Faria (1756–1819) was born in Goa, his father being an Indo-Portuguese who later became a priest.[1] As a boy he accompanied his father to Portugal[2] and then to Rome, where he studied for the priesthood and was ordained in 1780. An attempt to establish himself in Lisbon as a preacher was unsuccessful, and in the spring of 1788 he came to Paris, where he seems to have lived on the fringes of "Society", and to have shown a regrettable liking for the gaming tables. He survived various risky political involvements to emerge in 1811 as professor of philosophy at the lycée of Marseille. His efforts as a teacher proving disastrous, he returned to Paris, and on 11th August 1813 opened "un cours libre et payant" on magnetism at 49, rue de Clichy.

Faria's concern with animal magnetism was not new. Chateaubriand, who met him at a dinner-party in 1802, speaks as though his magnetic gifts were even then well-known in private circles. Faria (who had, according to Chateaubriand, links with the Swedenborgians[3]) undertook to kill a canary by the power of magnetism; however the canary gained the upper hand, and Faria felt it prudent to depart.[4] It seems likely that he then believed in some sort of magnetic fluid or emanation.[5]

Faria's public demonstrations were, in several respects, a complete departure from anything that had gone before. According to General F. J. Noizet (1792–1885), a distinguished soldier who attended them regularly, Faria "did not fail to declare loudly that he possessed no secret, no extraordinary power, and finally that he obtained nothing save through the will of the persons on whom he operated."[6]

His procedures are thus described by his close friend Alexandre Bertrand:

> He placed the person who wished to be influenced in an armchair, and urged him to close his eyes and meditate; then, suddenly, he uttered in a strong and commanding voice the word "Sleep," which usually made a keen enough impression on the patient to produce in him a slight jolt of the whole body, warmth, perspiration, and sometimes somnambulism. If the first attempt did not succeed, Faria submitted the patient to a second, then to a third and even a fourth, after which he declared him incapable of entering into lucid sleep.[7]

Faria would sometimes try other methods on refractory subjects.[8] But verbal command remained his method of choice. He claimed to have put more than 5000 persons into the state of "lucid sleep" (somnambulism).[9] "Perhaps", says Bertrand, "there was some exaggeration in this claim, but it is incontestable that he succeeded very often."[10] Many of his subjects were well-known persons who can hardly be suspected of having shammed. Indeed, Noizet himself gives an interesting account of his sensations while under Faria's influence.[11] It is likely that Faria's imperfect command of French, his dark complexion, and his reputed Brahman ancestry, conveyed subtle intimations of arcane knowledge and strange powers, and wrought wonderfully upon the minds of his subjects.

With his "lucid somnambules" Faria could demonstrate a wide range of remarkable phenomena. According to Noizet, he could induce such persons, during their sleep, to undergo powerful illusions of taste and smell, and to feel sensations of cold, heat, indeed of every kind. He would order a somnambule to see some absent person to whom he was attached, and would command him to fix the image of that person in his memory and continue to see him even when woken up.

> I remember having seen this experiment made on a Russian officer who wished to see his wife who had remained in his homeland. The illusion gave rise to the most touching spectacle. The officer shed bitter tears because he thought he saw the object of his devotion refuse his embraces; but soon both the spell and the sorrow were dispersed.[12]

In fact Faria's repertoire seems to have fallen little short of that of modern public demonstrators of hypnotism. By verbal suggestion he would occasion in his somnambules not just illusions and hallucinations of sense, but (for instance) particular actions, or, conversely, paralyses of particular limbs;[13] responsiveness to varying kinds of what would now be called post-hypnotic suggestion, even including suggestions of negative hallucination;[14] anaesthesia sufficient to permit the performance of surgical operations;[15] and the cure or amelioration of various diseases. In this last connection, Faria remarks[16] that the effects of the illusions he has induced are by no means illusory. The illusions are so vivid that they produce the same bodily effects as the actual objects would do. Thus a glass of water may act as an intoxicant, as a purgative, or as an emetic, depending upon the illusion as to its nature under which a somnambule is currently labouring. All these effects may be achieved not just with subjects in the state of "lucid sleep", but with waking subjects, provided that they have been put into that sleep at least once.[17] Indeed, these effects may be obtained, though not so strikingly, with certain exceptional subjects who have never been *endormi* at all.[18]

Faria's public courses seem to have done reasonably well until about the time of the restoration of the French monarchy (1815). Thereafter, he found himself increasingly under attack. Clergymen denounced him as a sorcerer;[19] he was insulted in the press, denounced by the Société du magnétisme,[20] and caricatured on stage;[21] his demonstrations were noisily interrupted, and he was more than once fooled by persons who had merely pretended to be under his influence.[22] In 1816, he closed his salon, and sought a position as almoner to a convent for young ladies. His leisure he devoted to the composition of a full answer to his critics and traducers. This work, *De la cause du sommeil lucide*, was to have been in three volumes, but owing to Faria's sudden death in 1819, only one was published.[23]

Probably no-one who studies this volume seriously would dismiss Faria as a mere charlatan. Indeed, though he was no doubt in some sense an adventurer, living much by his wits, persons who knew him reasonably well were prepared to testify to his fundamental sincerity.[24] As a piece of exposition, however, *De la cause du sommeil lucide* leaves much to be desired. It is rambling and disorganized, and the many striking insights which it contains have to be disentangled from a background of exceedingly obscure philosophical, theological, psychological and even physiological speculations. To the limited extent that I can make sense of Faria's scattered and frequently opaque theoretical utterances, the gist of his opinions seems to be as follows:

The ideas of the animal magnetists concerning the "magnetic fluid" are quite preposterous.[25] The notion of such a fluid is incoherent, and the evidence upon which belief in it has been based is flawed. "Magnetized" trees, and "magnetized" water (as Faria has shown by his own experiments) have an effect upon subjects only insofar as the latter believe them to have been magnetized, and without regard to whether they have in fact been magnetized.[26] Those who stress the influence of the magnetizer's will in producing the phenomena are likewise at fault; one can put subjects into the magnetic sleep with an exercize of the will, without one, or even with a contrary but unexpressed will.[27] Nor is the imagination hypothesis in better case. One does not lose the memory of one's imaginings, whereas somnambules have no recollection of the events of the somnambulic state.[28]

The theory of the magnetists being so misconceived, we need a fresh terminology. Faria proposes[29] to call magnetic somnambules *époptes*;[30] the magnetic state *sommeil lucide*; the process of magnetization *concentration*; the magnetizer the *concentrateur*; and "magnetize" *concentrer*. By *concentration* he apparently means a sort of turning of the mind inwards, away from the immediate deliverances of the senses.

Every human being has a body and a soul, intimately bonded. In its pure state, the soul possesses intuitive knowledge of eternal verities, can transcend space and time, etc. But when it is bound up with a human organism these faculties are suppressed, and the everyday consciousness knows nothing of them.[31] At the same time the soul assumes control of the "necessary activity" of the body (heartbeat, breathing, digestion, etc.). Again, the everyday consciousness is unaware of the soul's exercise of this function.

The soul is linked to the body through the blood, with which it is somehow interfused.[32] There are, however, certain persons, usually unhealthy or anaemic, whose blood is uncommonly liquid. In such persons the soul may at times partially free itself from the body. These persons are liable to various remarkable experiences, especially at times (sleep, profound abstraction) when their faculties are not wholly absorbed by the demands of everyday existence. These experiences may convey, through sensory imagery, something of the intuitions of the soul. Thus dreams "often express hidden verities, sometimes clearly, sometimes in obscure shapes."[33] The imagery may be misinterpreted, and the subject is liable to attribute it to some outside influence.[34]

"Lucid sleep" is simply the dreaming sleep had by these *époptes*. The dreams may result in external actions, especially if the dreamers are natural sleepwalkers and sleeptalkers.[35] The wiles of a *concentrateur* may promote lucid sleep – but, generally speaking, only in those individuals who are constitutionally disposed to it. "Lucid sleep" is not fundamentally different from ordinary sleep. It is sleep had by a certain kind of person.

The influence of a *concentrateur* over an *épopte* arises from the trust (*confiance*) which the latter has come to repose in the former.[36] *Époptes* are disposed to be trusting because of their impressibility, which in turn derives from the liquidity of their blood. The statements of the *concentrateur* accordingly arouse in the *épopte* what Faria calls an "inmost conviction" (*conviction intime*). Now "inmost conviction" is a "modification of the soul";[37] and since the soul can act anywhere in the body where the blood is sufficiently liquid, inmost conviction may influence the workings of most parts of the body. Thus once the *concentrateur* has put his subject into a sufficiently abstracted state, shutting out all countervailing influences, his commands or statements, however ridiculous, may profoundly affect his subject's mind, senses, and "necessary activities". The *épopte*, meanwhile, does not realize that these bizarre happenings are at root manifestations of his own faculties; for so long as his soul is (partially) embodied, its inner workings remain hidden from him. Hence he attributes the phenomena, which are really his own doing, to the power which the *concentrateur* has over him.

Though aspects of Faria's doctrine and practices foreshadow both those of the "Nancy School" and those of early "dissociationists", his direct influence upon later writers was probably slight. It can be traced here and there in Noizet's *Mémoire sur le somnambulisme et le magnétisme animal* (1854).[38] Noizet attributes many though not all of the phenomena to the "conviction" of the subjects. Towards the end of the nineteenth century, Faria was much praised by Liébeault, Bernheim and other members of the Nancy School, and also by Gilles de La Tourette from their opponents of the Salpêtrière,[39] but their praise hardly amounts to an acknowledgement of indebtedness. To a writer whose works did influence Liébeault, Faria's views and methods were certainly known. I refer to Noizet's close friend, Alexandre Bertrand.

ALEXANDRE BERTRAND

Bertrand's career, and his criticisms of the animal magnetists, were outlined in Part I of this book.[40] It remains for us to consider his own views of the nature of "magnetic" phenomena. His writings exhibit qualities – a clarity of style, breadth of information, moderation of judgement, a refusal to rest content with merely conventional solutions – which should still commend him to modern readers. His approach is primarily historical. The phenomena of "magnetic somnambulism" are, he holds, closely paralleled not just by the phenomena of "natural somnambulism", but by the phenomena presented in certain "epidemics of ecstasy".[41] Of these epidemics he gives a fairly full account in his *Traité du somnambulisme* (1823) and a somewhat more succinct one in his *Du magnétisme animal en France* (1826). They include: the possession of the nuns of Loudun (1633–4 and after); the strange fits of exaltation which, after 1685, broke out among the persecuted Protestants known as the "tremblers of the Cevennes"; the "miracles" worked on and by the Jansenist "convulsionaries of Saint-Médard" (1727–32); and the effects produced on certain sick persons by the activities (c. 1770–1777) of Father J. J. Gassner, the exorcist.

Bertrand was by no means the first sceptic to use accounts of these, and similar, epidemics to throw light on the phenomena of animal magnetism. But he differs from his predecessors in significant ways. He is at pains to unearth the best available sources for his historical information; he is prepared to admit that some of the happenings recorded defy any ordinary explanation;[42] and he does not do what he is widely said to have done, namely ascribe all the phenomena of animal magnetism straightforwardly to "the power of the imagination". What he does do is claim that in the above "epidemics", and also in cases of magnetic somnambulism, the patients or sufferers exhibit such similar phenomena that we may justifiably suppose their organizations have undergone "a singular modification, which gives birth to physical or intellectual phenomena which together constitute a particular state"[43] which may appropriately be termed *ecstasy*. Magnetic somnambulism represents one form which the state of ecstasy can take, or one context in which it can manifest itself.

The phenomena characterizing the state of ecstasy are as follows:

1. Subsequent amnesia for the events of the ecstatic state.
2. Heightened appreciation of the passage of time.
3. Bodily insensibility.
4. Exaltation of the imagination.
5. Development of the intellectual faculties (including memory).
6. The instinct for remedies.
7. Prevision.
8. Moral inertia.
9. The communication of the symptoms of ailments to the ecstatic.
10. The communication of thoughts.
11. Seeing without the use of the eyes.
12. A peculiar influence exercised by the somnambule on his own organism (including, particularly, control of certain processes normally automatic).

These phenomena need not all be present in any given case. As to the presumed physiological nature, or underpinning, of this state of "ecstasy", Bertrand is understandably reticent.[44] He does, however, give some account of "the causes, as much organic as mental, which predispose either to the state of ecstasy, or to the production of the other effects attributed to magnetism."[45]

With regard to the organic causes, Bertrand begins by combatting the view that magnetizers obtain their effects only with ladies of nervous temperament, in whom the slightest disturbance produces twitchings and the vapours. On the contrary: in such organizations "magnetism confines itself to the reproduction of these trivial misadventures".[46] The best subjects are often uncultured but healthy peasants, whose brains have not lost the ability to react upon their organisms. As for those nervous conditions which predispose to ecstasy, at the head

> ...one must place hysterical affections, those which manifest through attacks of convulsions, returning at certain periods, regular or not, but usually coinciding with the time at which menstruation should appear. Between whiles, the patients enjoy an apparent good health, or at any rate

suffer only from more or less isolated afflictions, such as the momentary loss
of a sense, or the transient paralysis of a limb ... [47]

Turning next to the mental predisposing causes, Bertrand once again begins by
correcting certain widespread misconceptions. Firm belief in the power of
magnetism, or a strong desire to be influenced, do not favour the production of
ecstasy. What does promote it, quite independently of reasoned conviction,
belief or will, is a powerful imagination, especially if activated by fear.

> That which goes on in disbelievers liable to fall into somnambulism is wholly
> analogous to what happens to certain persons whose education does not
> allow them to believe in ghosts, apparitions and presentiments, but who
> cannot hear stories designed to strike the imagination vividly without being
> profoundly affected; in some of these persons, despite their beliefs, such tales
> even suffice to bring about veritable visions relating to the fantastic objects
> that are filling their minds. [48]

Bertrand thus emerges as an early proponent of a "special state" theory of the
phenomena of magnetic somnambulism. He claims to have shown that the same
constellation of peculiar phenomena crops up in a variety of different contexts in
subjects who are at any rate not in a normal state of mind. We may therefore
legitimately suppose that despite the differences of setting the individuals
concerned are in the same underlying abnormal psychophysiological state. The
nature of that state, and the reasons why certain antecedent conditions
predispose to it, remain as problems to be elucidated.

   Bertrand is often represented as a precursor of the Nancy School, or as the
"real founder" of the theory of suggestion.[49] These proposals are misleading.
The founders of the Nancy School did not regard hypnosis as "a special state
completely different from all those which have hitherto been recognized";[50]
Liébeault thought it identical with ordinary sleep, and Bernheim came to
disbelieve in the existence of any state of hypnosis at all. Furthermore, Bertrand
lays no stress upon verbal command as a method of inducing magnetic
somnambulism,[51] and he makes no use of the term "suggestion". In certain
respects – his presenting magnetic somnambulism as an abnormal state with
links to hystero-epilepsy – he is more nearly a forerunner of the school of the
Salpêtrière.

JAMES BRAID AND HIS FOLLOWERS

With James Braid[52] (1795–1860) we reach one of the most famous names in the
history of hypnotism. Like Esdaile, he was born in Scotland and became a
surgeon; like Esdaile and Elliotson he was educated at Edinburgh University. For
some while he practised in Scotland, but then removed to Manchester where he
resided for the remainder of his life. He seems to have been well-regarded in his
profession, but his present fame, of course, rests entirely on his publications in

Figure 5. James Braid. (From the Library of the Society for Psychical Research.)

the field of hypnotism.[53] These publications were, without doubt, many years ahead of their time, but their prescience is somewhat obscured by their resolutely pedestrian style. Braid never pretended to be other than he was – a competent and fairly well-educated surgeon in general practice, who wrote from no hankerings after literary achievement, but rather from a sense that it was his duty to communicate discoveries which he believed to be of the highest importance. Robust common-sense, and a certain reassuring solidity of mind and person were his leading characteristics as a man and a doctor, and they are everywhere apparent in his writings. From the vanity and intolerance which so disfigured Elliotson's character he was totally free. He shared, however, the latter's "kindness to humble sufferers", and in private life was of a "warm-hearted and cheerful" disposition.

Braid's interest in mesmeric phenomena was first aroused by Lafontaine, who visited Manchester in November 1841. After observing Lafontaine, and experimenting for himself, Braid came to believe that he had discovered the key to both phenomena and cures. Feeling it his duty to communicate his findings, he began to give lectures and demonstrations – the first on 27th December 1841.[54] During 1842, he gave many private demonstrations, lectured widely, and published his first pamphlet on the subject.[55] Though he never repaid spite with spite, he was quick to reply to critics, firm in asserting his priority,[56] sometimes incensed by the injustice of attackers, and quite prepared to interrupt the performances of magnetic demonstrators with a statement of his own views. In 1843, he published his *Neurypnology; or, the Rationale of Nervous Sleep*,[57] his first systematic, and only book-length, exposition of his views and findings.

The title of this book, *Neurypnology*, (originally neurohypnology, "the science of nervous sleep"), is part of a scheme of terminological revision with which Braid had been experimenting for some time. The rest of this scheme ("hypnotism", "hypnotize", "hypnotist", "dehypnotize", etc.) has been incorporated into our ordinary language, but, mercifully, "neurypnology" has not. However the state of "nervous sleep" – Braid's "hypnotism" – is now usually called "hypnosis", and "hypnotism" has acquired the meaning which he gave to "neurypnology".

Braid regarded the "nervous sleep" produced by mesmerism and hypnotism[58] as differing in significant respects from ordinary sleep.[59] The most effective way to induce it is by visual fixation of a small bright object held above the eyes and about eight to fifteen inches away, "at such position ... as may be necessary to produce the greatest possible strain upon the eyes and eyelids." If, after a while, "the fore and middle fingers of the right hand, extended and a little separated, are carried from the object towards the eyes, most probably the eyelids will close involuntarily, with a vibratory motion."[60] If the eyelids do not close, the process may be repeated and the subject told to close them, but to keep his eyeballs in the same position and "the mind riveted to the one idea of the object held above the eyes".[61] The first signs of "artificial hypnotism" are likely to be found soon thereafter. Subjects who have been frequently hypnotized may become more susceptible, and may even be affected purely through the imagination.[62]

As to the physiological condition underlying nervous sleep – a condition brought on by straining the attention, over-exercising the eye muscles, and suppression of respiration – Braid thinks it premature to speculate, though he suspects that many of the phenomena are "attributable to the altered state of the circulation in the brain and spinal cord".[64] Theories about the "mesmeric fluid" he totally rejects on the ground that "any one can hypnotize himself by attending strictly to the simple rules that I lay down".[65]

The nervous sleep itself has two phases or stages, transition from one to another of which comes about either spontaneously or through the application of simple physical stimuli.[66] The first stage may be demonstrated in a susceptible subject ten or fifteen seconds after eye closure by "gently elevating the arms and legs". It "will be found that the patient has a disposition to retain them in the situation in which they have been placed", and also that "all the organs of special sense, excepting sight", and likewise certain mental faculties, "are *at first* prodigiously *exalted*, such as happens with regard to the primary effects of opium, wine, and spirits". During this stage, subjects may be fully conscious, and may therefore suppose that they have not really been "hypnotized".

The first stage passes into the second. The extended limbs become rigid and involuntarily fixed, and the organs of special sense pass into a state of "the most profound torpor". The subject may now "be pricked, or pinched, or maimed, without evincing the slightest symptom of pain or sensibility." But any sense-organ, or any group of muscles or individual limb may at once be returned to its stage one condition of "extreme mobility and exalted sensibility" by blowing on it, pressing on it or stroking it. And the patient may be returned to the waking state by "very gentle pressure over the eye-balls".

Part II of *Neurypnology* is devoted to a description and discussion of some of the cures, ameliorations or other benefits which Braid had effected by his "hypnotic" methods. These sixty-nine[67] cures and ameliorations (thirty-four males and thirty-four females and one of sex not given) may be divided up as follows:

"Affections of the eyes"	8
Deafness with dumbness	5
Impaired sense of smell	I
Tic douloureux	3
Paralysis with or without anaesthesia	14
Rheumatic disorders	10
Muscle pains following injury	3
"Nervous headache"	5
"Spinal irritation"	2
Epilepsy	4
Backward curvature of the spine	I
"Neuralgic pain in the heart and palpitations"	5
Analgesia for minor operations	3
Skin disorders	2
Tonic muscular spasm	3

It is quite clear that Braid regarded himself as treating largely physical ailments by a largely physical method. And indeed it does not seem likely that the bulk of his patients were hysterics or hypochondriacs. The "affections of the eyes", for instance, were cases of weak sight rather than of hysterical blindness;

in one, a long-standing opacity of the cornea was quickly dispersed following the start of hypnotic treatment. The cases of paralysis included seven in which the condition was clearly occasioned by a stroke; several of these cases were long-standing and were accompanied by speech problems.

Unfortunately, Braid made no attempt to compare the effectiveness of hypnotic and of non-hypnotic treatments of the same disorders. There is, however, no doubt of his professional competence. If he thought that hypnotic treatment produced results which ordinary treatment would not have done, this opinion is not to be altogether dismissed. It does not seem in the least likely that he was simply forgetting about numerous failures. Furthermore, in many instances long-standing and intractable complaints responded to hypnotism with startling rapidity. Thus several patients in the agony of tic douloureux were relieved almost immediately, and future attacks cut down or abolished. It is very difficult not to believe that through Braid's hypnotic endeavours patients recovered or improved who would not otherwise have done so. But were the cures wrought by the methods or by the man? I have little doubt that like other healers, and like Mesmer himself, Braid possessed not "charisma" exactly, but what may, for therapeutic purposes, be yet more effective – the sort of powerful but kindly "presence" that by itself helps to lift the lethargy and depression which envelop so many of the chronically sick.

Assessment of Braid's case reports is hindered by his failure in most instances to give proper details of his procedures. We can, however, derive some plausible guesses from the general rationale for treatment with which he prefaces Part II of his book.[68] He holds that in the second stage of the artificial sleep "the rigidity of the limbs, and consequent obstruction to the free circulation through them, is the chief cause" of the "determination" of blood to the brain, spinal cord, ganglion system, and "all other parts not compressed by rigid muscles", and also of an increase in heart rate and general quickening of the circulation. This quickening and "re-determination" of the circulation may itself be beneficial in certain cases by "subjecting the brain and spinal cord, and whole ganglion system, to a high rate of excitement". If we want to redirect the circulation to just one limb or organ, which it is desirable to stimulate, we free that limb alone from the cataleptic state, or that organ from its torpor, in the ways already indicated. Braid emphasizes that his "rationale" is highly speculative, and indeed to a modern eye it is scarcely more convincing than the proposals of the more moderate mesmerists.

It is on Braid's later writings,[69] and the very high estimate of them by Milne Bramwell in his well-known text-book of 1903,[70] that his reputation as a precursor of the Nancy School largely rests. Unfortunately, these writings (a mixture of pamphlets and articles and pamphlets derived from articles) are for the most part simply "occasional" pieces called forth by contemporary publications. Nowhere does one come across a really sustained and systematic exposition of his revised views. These views have to be pieced together from scattered passages, and I am not sure that a totally coherent picture can be assembled. The difficulty is increased by the fact that Braid tends to set out his views *tout court*, without directly relating them to his observations; it is often unclear what observations he regards as licensing him to take up the position he has adopted, and why. However, one can at any rate fairly easily discern certain general similarities and differences between the views and practices set forth in these works, and those to be found in *Neurypnology*.

His preferred method of inducing the hypnotic sleep remained the same.[71] But he was now prepared to make greater use of mesmeric passes, because he had come to see that they "produce their effects from what is essentially the same exciting cause as that which induces hypnotic phenomena".[72] He does not seem to have deliberately employed verbal suggestion as part of the induction process, but since the patient was commonly "requested to engage his attention ... on the simple act of looking at the object, and yield to the tendency to sleep which will steal over him",[73] there can be little doubt that verbal suggestion was at work.

Braid continued to regard the hypnotic sleep as having two distinguishable stages, the second of which is only attained by a limited number of patients. These corresponded roughly to the stages of excitement, and of torpor with rigid catalepsy, which he had proposed in 1843. In 1852, he refers to them as "the sub-hypnotic or partial state of the nervous sleep",[74] and the "full or double conscious stage of the sleep".[75] Both are to be explained in much the same terms. To understand Braid's approach we must look at some of the ideas about "the influence of the mind over the body" which he expressed as early as 1846 following his extensive experiments on the Reichenbach phenomena.[76] "I believe", he says

> ... this was the first systematic and *extended* course of experiments published, in which satisfactory proof had been adduced that a fixed dominant expectant idea in the mind of the patient might either *excite* or *depress*, or temporarily suspend the function of *any* one or of *all* of the special senses, or even the function of any organ or part of the human body; and that an audible suggestion from another person, and the dominant expectant idea excited thereby in the mind of the subject would become so engrossing with some subjects, that they could not avoid realizing the *expected* result as regarded *all* the organs of special sense, and some of the mental faculties ...[77]

Likewise, if the mind becomes wholly concentrated upon the idea of a certain bodily movement, that movement is likely to follow, for

> ... a current of nervous force is sent into the muscles which produce a corresponding motion, not only *without* any conscious effort of volition, but even in opposition to volition in many instances.[78]

To return to Braid's proposed "stages of hypnotism", the "subhypnotic state" is simply a condition of "monoideism" in which the subject is peculiarly liable to find his thoughts, perceptions and actions controlled by dominant ideas in the way just described.

> ... by engendering a state of mental concentration ... upon some unexciting and empty thing ... the faculties of the minds of *some* patients are ... thrown out of gear ... so that the higher faculties ... become dethroned from their supremacy, and give place and power to imagination ... , easy credulity, and docility or passive obedience; so that, even whilst apparently wide awake ... they become susceptible of being influenced and controlled entirely by the suggestions of others upon whom their attention is fixed ... they *see* and *feel* AS REAL, and they consider themselves *irresistibly* or *involuntarily fixed, or spell-bound, or impelled to perform whatever may be said or signified by the other party upon whom their attention has become involuntarily and vividly rivetted ...* [79]

It is clear that what Braid has particularly in mind here are the "electro-biological" phenomena which had lately come to Britain, and which Elliotson had christened "submesmerism". The "suggestions" to which Braid refers can be administered verbally, or by gesture, or touch, or "agitation of the air", or in any other way.

In the "subhypnotic stage" the subject is apparently awake, and afterwards has full recollection of what took place (Braid emphasizes that this is no barrier to the effectiveness of therapeutic suggestions). In the stage of "double consciousness" into which the subhypnotic stage may pass, the subject enters a sleep-like state (which does not, however, unless very deep, preclude a responsiveness to suggestion), and is amnesic on waking. He can, however, recall the events of the trance when again hypnotized. This stage is only a development of the preceding one, and is to be understood in the same terms, the patients' "attention being still more concentrated, their imaginations and expectant ideas more vigorous, and the counteracting influence of reason being more in abeyance."[80] Much – indeed almost everything of importance – here remains unelucidated.

Braid's new, more psychological theory, had certain therapeutic implications, but much of what he says in this connection differs little from the views which he presented in *Neurypnology*.[81] His main therapeutic aim is still to increase or diminish blood supply either generally or to particular organs or parts of the body. This is to be accomplished by the methods previously expounded. He also sets great store by whether the blood supply in question is highly or poorly "arterialised" (i.e. oxygenated), which depends, of course, on the patient's respiration. Blood of the former sort produces excitation and of the latter sort torpor.

In addition to these purely physical influences on the bodies of his hypnotized subjects, Braid now feels it possible to employ mental ones. "...through influencing the mind by audible language ... or by definite physical impressions, we fix certain ideas, strongly and involuntarily on the mind of the patient, which thereby act as stimulants, or as sedatives, [and influence] the current of thought in the mind of the patient, either drawing it to, or withdrawing it from, particular organs or functions."[82] Physical and mental influences may be exercised together, old-fashioned mesmeric passes, for example, both change the balance of functions in the system by stimulating the skin, and call the patient's attention to some particular organ or function.[83]

Braid seems to have found this "mixed" way of operating especially effective in cases of constipation and menstrual disorder, for both of which he reports some startlingly rapid results.[84] His account of the *modus operandi* of hypnotic analgesia is another interesting example of his thinking on the matter:

> During the nervous sleep, by inducing the low circulation and suppressed respiration ... the blood, from being thus insufficiently arterialised, acts as a narcotic ... and if the attention has also been fixed in some particular train of thought, every other function becomes deadened in an extraordinary degree, so that severe operations and inflictions may be done in that state without the patient evincing any apparent consciousness of pain.[85]

If the hypnotic state is indeed one in which the mind is liable to be occupied by one idea to the exclusion of others, this would suggest a certain affinity between this state and the condition of those suffering from various forms of mental disturbance. It would also suggest the possibility of treating such cases by displacing a noxious dominant idea with a series of harmless ones. This is in fact what Braid claims to have done in cases of "monomania and delirium tremens",[86] but also, and most particularly, in cases of hysterical paralysis occasioned by a dominant idea:

> In such cases ... by breaking down the previous idea, and substituting a salutary idea of vigour and self-confidence ... (which can be done by audible suggestions ...), on being aroused ... with such dominant idea in their minds, ... the patients are found to have acquired vigour and voluntary power over their hitherto paralysed limbs ...[87]

This foreshadows (though it probably did not influence) much that was to come in the hypnotic and psychotherapeutic movements of the late nineteenth century.

Braid's later writings are exceedingly difficult to assess. There is so much in them that is so sensible, and so much that is incoherent or sometimes almost incomprehensible. It would perhaps be fairest simply to point out that Braid always hoped to bring out a fully revised version of his *Neurypnology*.[88] Had he had leisure to complete this difficult undertaking, much that is obscure in his later writings would no doubt have been partially clarified. Meanwhile, we may look upon those writings as the incomplete and probably rather hurried attempts of a busy man to work, within the limits of the medical concepts available to him, towards views which, modernized and fully expounded, would have anticipated much that was to be found in the vanguard of advanced thinking in these areas thirty or forty years later.

Braid has acquired a very considerable posthumous reputation; in his own time he achieved a real but more moderate celebrity. At first, he was attacked both by mesmerists and by members of the medical profession (including the egregious Wakley). The hostility of Elliotson and his closest supporters was largely covert – mention of Braid was almost excluded from *The Zoist*, and Elliotson went so far as to use his influence to prevent Braid obtaining a certain public appointment.[89] Other prominent mesmerists – Gregory, Esdaile, Colquhoun, Mayo, Haddock – recognized Braid's merits, and were friendlier towards him, and his later more frequent use of mesmeric passes made it tempting for them to set him down as a mesmerist despite himself. The attitude of the medical profession gradually improved,[90] especially after the "electro-biological" demonstrators who travelled the country in 1851[91] had brought it forcefully home that some of the alleged phenomena really did require an explanation. Braid's influence is apparent on such attempts to provide an explanation as those of J. H. Bennett (1851), W. B. Carpenter (1853) and Henry Holland (1852).[92]

After Braid's death in 1860, his ideas ceased to command much attention in Britain, because the mesmeric phenomena to which they were a response were passing more and more from public view. However, as interest waned in Britain, so, for a while, it grew in France. Around 1858, C. M. E. E. Azam (1822–1899), deputy professor at the School of Medicine in Bordeaux, read an encyclopaedia article on hypnotism. He was struck by the resemblance between the state of artificial somnambulism described by Braid and the pathological state entered by one of his own patients. He obtained a copy of *Neurypnology* and began to experiment for himself. He did not succeed with so large a proportion of his subjects as had Braid, but was able to repeat many of the latter's results. He published his findings in January 1860.[93]

Azam had been a fellow-student of Paul Broca (1824–1880), after whom "Broca's aphasia" is named. On a visit to Paris in November 1859, Azam told Broca (who was a physician at the Necker Hospital) of his hypnotic experiments. Broca at once decided to verify the findings. He chose a female patient aged twenty-four who was suffering from an exceptionally painful abscess, and proceeded to hypnotize her by the method of Braid and Azam. When the abscess was lanced she gave only a light cry. On being awoken she remembered nothing and was astonished to find that she had been operated upon.[94] Broca immediately reported this result to the Academy of Sciences, and it was communicated to the meeting of 7th December 1859 by A. A. L. M. Velpeau (1795–1867), one of the most noted surgeons of the time. His authority lent such weight to Broca's observation that numerous French doctors hastened to experiment for themselves.[95] Dr. Guérineau of Poitiers reported a hypnotic leg amputation. The patient said on awakening that he had felt no pain, and was so grateful that he attempted to kiss the surgeon's hand.[96] Dr. J. P. Durand de Gros (1826–1900) published in Paris under the name of A. J. P. Philips a *Cours théorique et pratique de Braidisme* (1860). Various pamphlets of lesser note also appeared.[97]

This little burst of interest in "braidism" had largely evaporated by the end of 1860, perhaps, as Figuier suggests,[98] because "braidism" did not prove a consistently effective anaesthetic. However, publications on the subject continued to appear sporadically throughout the decade,[99] and "braidism" became quite widely known to the French public from the account of it given in the successive editions (1860 onwards) of Louis Figuier's *Histoire du merveilleux*. In Germany, so far as I can discover, it attracted little attention before Preyer's "discovery" of and tribute to Braid in 1881.[100]

Many accounts of the history of hypnotism make Braid a forerunner of the "Nancy School". Yet insofar as Braid had an influence on the history of hypnotism, that influence was almost certainly exercised upon Charcot and the School of the Salpêtrière rather than upon the School of Nancy. Liébeault, the "father" of the Nancy School, cites Bertrand in his first book (1866), a good deal more than he does Braid. And I can find no evidence that Bernheim, Liébeault's most prominent follower, knew at the outset anything at all about Braid.[101] On the other hand, Charles Richet, whose experiments may well have aroused Charcot's interest in hypnotism, knew of the French braidists, as Charcot certainly did himself.[102] Now "braidism" was based almost entirely on the Braid of 1843, the Braid whose subjects exhibited "stages" of hypnotism, characterized in terms of excitement, torpor, catalepsy, anaesthesia, hyperaesthesia, muscular rigidity, etc., produced and terminated by mechanical stimulation, and understood in terms of altered cerebral circulation. All the elements of Charcot's "grand hypnotism" are here. Braid may well have helped set the stage for Charcot; he did not prepare the way for the Nancy School.

THE UNITED STATES: SUNDERLAND, FAHNESTOCK AND J. K. MITCHELL

The most interesting of the early American mesmerists was undoubtedly the Rev. La Roy Sunderland (1804–1885) whom we have already met. At the age of nineteen, in 1823, Sunderland had embarked on a career as a revivalist preacher in New England, and enjoyed from the outset an extraordinary success.[103] His audiences were often profoundly affected. They might jerk, twitch, shake, laugh, weep, pray, groan, jump, roll on the ground, hear voices, have visions and become entranced. By about 1835, Sunderland had become convinced that these phenomena were of human rather than divine origin, and he thereupon gave up his ministry. For a while he devoted himself to the abolitionist cause.

In 1839, he first came in contact with the mesmeric phenomena as manifested by a lady who some fifteen years earlier had fallen into a trance at one of his revivalist meetings. The analogy between the revivalist phenomena and the mesmeric ones struck him very forcibly, and he soon became involved in the mesmeric movement as lecturer, demonstrator and writer. He quickly developed his own system of ideas to which he gave the name "pathetism".[104] The early versions of this theory[105] need not detain us; they are permeated by phrenological doctrines, afterwards abandoned.[106] Later versions[107] are best introduced by saying something about Sunderland's methods.[108]

The most interesting thing about Sunderland's methods of "pathetizing" is that at first sight he did not appear to have any. Though he was small in stature, and troubled with physical infirmities, his name, eloquence and demeanour would by themselves often suffice to overwhelm his hearers. He would enter the hall and begin his lecture, sometimes with, and sometimes without, a preliminary hint that he expected that members of the audience might be "affected". Sure enough, after he had talked for a while, he would break off his discourse to announce that certain persons were now in a state of trance (the "pathematic" state). These subjects would then advance, apparently asleep, to the platform, often without Sunderland needing to say or do anything further. (On one occasion he found a number of subjects already entranced on the platform when he entered the room.[109]) He would then induce in them, by visual or verbal instruction, most of the more entertaining phenomena that had been the stock-in-trade of mesmeric and electrobiological demonstrators. His subjects might be made to exhibit catalepsy or anaesthesia, sing, dance, have visions of deceased friends, religious figures, pass into a state of ecstasy, etc. It was by no means always necessary for subjects to be entranced before they manifested these phenomena, but with entranced subjects clairvoyance and thought-reading, as well as ecstasy and hallucination, became more likely.

At some of Sunderland's meetings, minor operations, especially tooth extractions, were performed upon entranced persons whom he had rendered anaesthetic; he seems to have done this by inducing a state of cataleptic rigidity whilst sending them upon interesting imaginary journeys. The cure of ailments was also frequently reported following "pathetization" of the sufferer; in one case a breast tumour larger than a hen's egg, and suspected of being cancerous, was found to have disappeared during the seventeen days in which Sunderland had been pathetizing the sufferer to produce anaesthesia for its extirpation.[110] However, as Harte remarks,[111] Sunderland, though willing enough to publicize these cases,[112] does not seem to have posed as a healer, and regarded the cures as incidental to the trance state.

Sunderland's theoretical stance has been represented by some writers, for instance Podmore and Harte, as in many ways ahead of his time. And it is indeed true that he clearly and repeatedly stated certain points which completely differentiate him from contemporary mesmerists. He rejects the mesmeric fluid and all kindred and correlated notions; he regards the mesmeric, or pathematic, phenomena, as set in train simply by an idea implanted in the mind of the subject; and the process by which that idea is translated into effect he describes as one of "self-induction" by the subject himself. Having said so much, it is difficult to know what else to say about Sunderland's theory, for it is expressed in such idiosyncratic terminology, or in ordinary terms used so idiosyncratically, and is so tied in with expositions of popular psychology and physiology, that its details are exceedingly elusive.

The distinguishing idea of pathetism, says Sunderland, is that

> ... when a relation is once established between an operator ... and his patient, corresponding changes may be induced in the nervous system of the latter (awake or entranced), by mere volition, and by suggestions addressed to either of the external senses.[113]

What precisely this relation may be remains obscure. It seems to be a relation in which the operator is agreed by both parties to be positive and active, and the patient passive and negative, and it can only be established if there is a certain sympathy or love or concordance of purpose between them. To the extent to which such a relation is formed, the subject's own will executes, in his own nervous system, the dictates of the operator, provided that these have been conveyed (by verbal suggestion or otherwise) in such a way as to compel the subject's whole attention. The relation is capable of degrees, or stages, which Sunderland calls sensuous, mental and spiritual. In the sensuous stage, the operator's will reaches the subject through the ordinary channels of sense, and the phenomena produced are those commonly observed at electrobiological demonstrations; in the mental stage, which is inducted by suspension of the subject's external senses, the subject's mind is directly controlled by that of the operator, so that he wills only what the operator wills, perceives only through his senses, etc.; in the spiritual stage the subject knows the operator's spirit and exercises spiritual and extrasensory faculties.

In some ways similar to Sunderland's were the ideas of Dr. William Baker Fahnestock, of Lancaster, Pennsylvania. Though his principal work, *Statuvolism; or Artificial Somnambulism*, was not published until 1869, he had been active in the subject for some thirty years, and had published what seems to have been the first account of painless mesmeric or hypnotic births in the United States.[114] According to Fahnestock, statuvolism (state produced by the will), though like mesmerism a state of artificial somnambulism, differs from mesmerism in that it is brought about by the subject's own will and not by that of the operator. Thus the operator is not really necessary. But he may help subjects into statuvolism by seating them comfortably, telling them to close their eyes, and taking them on a sort of guided fantasy tour in which they "throw their minds to some familiar place ... This must be persevered in for some time, and when they tire of one thing, or see nothing, they must be directed to others successively ... until clairvoyance is achieved."[115] Of clairvoyance, Fahnestock gives some quite curious examples from his own experience. However, his theory is so shot through with eccentric physiological notions that there is no point in exploring it.

The ablest and most interesting of these American "precursors of hypnotism" is also, regrettably, the least known. I refer to the eminent Philadelphia physician, J.K. Mitchell (1798–1858), father of the well-known neurologist Silas Weir Mitchell (1829–1914). Mitchell seems to have begun investigating mesmerism during the 1840s, but his essay on the subject was published posthumously in his *Five Essays* (1859), a volume which Garrison rightly says "reveals an originality of mind far above the average".[116] Mitchell's experiments, whatever their shortcomings by modern standards, were more systematic, and conducted with a clearer awareness of the likely sources of experimental error, than those of almost anyone who had preceded him, and of most of those who were to succeed him during the rest of the century.

He quite quickly convinced himself "that there is a mesmeric state, and that it may be artificially produced by means which are independent of the volition or imagination, or any other known moral power of the subjects of it." He bases this conclusion on experiments with young and ignorant persons who knew (or so he believed) nothing about the supposed effects of mesmerism. To make assurance doubly sure he endeavoured by disguising or concealing his actions to prevent his subjects from knowing what he was about. None the less, not a few of his subjects passed into the "mesmeric state". Conversely, subjects already in that state could be quickly awoken by reverse or transverse passes even though a thick, doubled shawl through which Mitchell himself could not "see the slightest ray of light" was thrown over their heads, and every precaution was taken to minimize cues from sound or air movement. By reversed passes he also managed to arouse from post-epileptic coma two patients whom other means had failed to waken.

Mitchell tried to determine some of the characteristics of the mesmeric state and of the susceptible subject. Mesmerized persons (unlike sleeping persons) show an increased pulse rate, and have cold damp hands and flushed faces. Out of 118 individuals mesmerized by him, some one in seven proved susceptible. The mean time taken to reach the state of artificial sleep was ten to eleven minutes. Neither women nor invalids were especially susceptible; the very young and the very old were less susceptible than other age groups.

Certain claims of the magnetists Mitchell was able to confirm. He performed a number of quite gruesome tooth extractions using mesmeric analgesia. One lady who had undergone a series of extractions unmoved responded with "a loud cry and a projection of her body backward"[117] when a stump as yet unextracted was merely touched while she was in the waking state. Mitchell notes[118] that the sense of touch can remain though pain sensations for the same area have been abolished, and concludes that different nervous pathways must be involved. Subjects put into artificial sleep may be made to dream, and to describe their dreams, on the command of the magnetizer, or of any other person (he also obtained the same effect with a natural somnambule), and will exhibit amnesia on waking.

However, there were many magnetic phenomena which Mitchell could not obtain under satisfactory conditions. These included rapport, phrenomesmeric phenomena, and clairvoyance. Above all, Mitchell could find no evidence to support the idea of a mesmeric fluid. Mesmerizing a series of subjects one after another does not exhaust the mesmerist in the way this theory would predict. It cannot be the case that metals and other objects may be charged with the fluid, for "If metals, especially those of the currency, were so susceptible of being imbued with mesmeric force, the business of a money-changer, shopkeeper or banker, would be fraught with inconveniences."[119]

Braid's theory that the state of artificial somnambulism is produced by fixity of gaze may be ruled out because the blind, and people with closed eyes, may be successfully mesmerized. Imagination theories Mitchell has already dismissed. He proposes instead that the first step in the production of artificial somnambulism is that the mesmerizer unconsciously acts on and produces a change in his own nervous system. This change correspondingly influences the subject's nervous system by a process of what Mitchell, citing copious parallels from the physical science of his day, calls "induction". Manipulation is not essential for such induction. The essentials are relative position, contact at some point, and "a fixed idea of slow and laboured motion".

It is curious, and a little disappointing, to find that after so much rational experimentation and so many perceptive observations, Mitchell ends up propounding views so vague and simplistic.

NOTES

1 Materials on his life are scanty. A useful account is that by F. D'Amat in the *Dictionnaire de biographie française*. Cf. Dalgado (1906).
2 A Brazilian travel diary of 1774–5 has been attributed to him. See Faria (1876).
3 It would not surprise me to learn that Faria was connected with one of the occult fraternities of the time.
4 Chateaubriand (1947), pp. 59–60.
5 And, indeed, he seems still to have been teaching such doctrines much later. Cf.Dalloz (1823), pp. 23–24; Chenevix (1829), pp. 220–221.

6 Noizet (1854), p. 88.
7 Bertrand (1826), p. 247.
8 Faria (1906), pp. 152–153.
9 Faria (1906), p. 27.
10 Bertrand (1826), p. 248.
11 Noizet (1854), pp. 90–91.
12 Noizet (1854), pp. 111–112.
13 Noizet (1854), p. 113.
14 Bertrand (1823), pp. 256–257.
15 Faria (1906), p. 190. I have not found any account of an operation, but Bertrand says (p. 386): "A woman of whom he made use for his experiments has assured me that, in her somnambulic state, her sensitivity was so completely suspended that one could thrust a nail in her arm up to its head without her feeling anything, and even without causing the loss of a single drop of blood."
16 Faria (1906), p. 61.
17 Faria (1906), p. 62.
18 Faria (1906), p. 44.
19 Faria (1906), pp. 18–23.
20 S. du M. (1816).
21 He thought himself satirized as Soporito in J. Vernet's *Le magnétismomanie* (1816). See Dingwall (1967), pp. 35–36.
22 Faria (1906), pp. 11–12; Noizet (1854), p. 89; Barrucand (1967), pp. 69–70.
23 I have used the reprint of 1906. The *Nouvelle biographie générale*, vol. 17, (1856), p. 115, seems to imply that the later volumes were then still in manuscript.
24 E.g. Chenevix (1829), pp. 220–221.
25 Faria believed in some kinds of emanations from the human body.
26 Faria (1906), p. 350.
27 Faria (1906), p. 315.
28 Faria (1906), p. 286.
29 Faria (1906), p. 32.
30 From the Greek ἐπόπτης, watcher, a somewhat inappropriate word which, however, perhaps not unintentionally, also means "initiate into the greater mysteries".
31 Faria (1906), p. 218.
32 Faria (1906), p. 56.
33 Faria (1906), p. 56.
34 Faria (1906), pp. 51, 56.
35 Faria (1906), pp. 184–185.
36 Faria (1906), p. 230. Trust is "the surrender of the mind to the discretion of another".
37 Faria (1906), p. 65.
38 Like Bertrand (1823), it was written as an entry for the Berlin Prize; but it arrived too late.
39 Dalgado, Preface to Faria (1906), pp. LIII-LX. Modern commentators have often regarded Faria as a forerunner of the Nancy School. Cf. Barrucand (1967), pp. 69–79; Perry (1976).
40 Pp. 128, 132–133 above.
41 Bertrand (1826), p. 311.
42 Especially the monumental assemblage of documents concerning the *convulsionnaires* of St. Médard in Carré de Montgeron (1745-7). On the churchyard of St. Médard see Hillairet (1958), pp. 84–86.
43 Bertrand (1826), p. 309.

44 In 1823 he talked somewhat vaguely of a heightened awareness of interior sensations and a heightened activity of the brain. See especially Bertrand (1823), pp. 482 ff.

45 Bertrand (1826), p. 403.

46 Bertrand (1826), p. 403.

47 Bertand (1826), pp. 404–405.

48 Bertrand (1826), p. 407.

49 Janet (1925), I, p. 156.

50 Bertrand (1826), p. 473.

51 Though he acknowledges its possibilities. Bertrand (1826), p. 248.

52 On Braid see Bramwell (1896–7a); Bramwell (1930), 21–29, 40–41, 278–294, 460–467; Kravis (1988); A. E. Waite's introduction to Braid (1899).

53 On these see A. E. Waite's bibliography, Appendix V to Braid (1899), and Bramwell (1896–7a), the materials from which are incorporated in Bramwell (1930).

54 Bramwell (1896–7a), p. 129.

55 Reproduced in Tinterow (1970), pp. 318–330. It was a reply to a printed sermon (M'Neile, 1842) by a strongly evangelical Liverpool clergyman, the Rev. H. M'Neile (1795–1879), afterwards Dean of Ripon. M'Neile (formerly an Irvingite) supposed that diabolical influences were at work in mesmerism.

56 He always denied that he had been influenced by Bertrand or Faria. Cf. Braid (1899), pp. 88–90.

57 Braid (1843). I have used the reprint of 1899, edited by A. E. Waite.

58 At one point (Braid, 1899, pp. 102–104), Braid expresses a slight reservation as to whether the states produced by hypnotism and by mesmerism are completely identical. He does so because of the somewhat numerous reports of mesmeric clairvoyance. He was unable to obtain such phenomena himself. But I think this reservation is merely a token one.

59 According to Braid nervous sleep, unlike ordinary sleep, is characterized by pupillary dilation, followed later by contraction with insensitivity to light; by a firmness of grasp on objects held in the hand; by an enhanced sense of balance; and by the extraordinary phenomena which it promotes.

60 Braid (1899), pp. 109, 109–110.

61 Braid (1899), p. 110.

62 Braid (1899), pp. 116–117.

63 Braid (1899), p. 126.

64 Braid (1899), p. 220.

65 Braid (1899), p. 113.

66 See especially Braid (1899), pp. 110–112 and 136n–139n.

67 Braid's numbers run from I to LXVI, but V, X and XXVII occur twice.

68 Braid (1899), pp. 225–227, and cf. 220–222.

69 The most important of his later writings are Braid (1846; 1850; 1851; 1852; 1853; 1855).

70 Bramwell (1903; 1930). For fuller bibliographies see Bramwell (1896–7a; 1930) and A. E. Waite's Appendix V to Braid (1899). Cf. also Kravis (1988). Braid (1846) and Braid (1855) are reprinted by Tinterow (1970), pp. 331–364 and 365–389.

71 Braid (1852), p. 56.

72 Braid (1852), p. 58.

73 Braid (1852), p. 58.

74 Braid (1852), p. 67.

75 Braid (1852), p. 65. Later – Braid (1855), pp. 9–10; Tinterow (1970), p. 371–2; cf. Braid, 1853, p. 38 – he proposed to use the term "hypnotism" for the latter stage, and to call the

deepest stage of all "hypnotic coma". He also toyed with a set of terms centring on "monoideism", "the condition resulting from the mind being possessed by a dominant idea".

76 See above, p. 230; cf. Tinterow (1970), pp. 360–362.
77 Braid (1853), p. 7.
78 Braid (1855), p. 4; Tinterow (1970), p. 367.
79 Braid (1852), pp. 65–66.
80 Braid (1853), p. 8.
81 See especially Braid (1853), pp. 8–14.
82 Braid (1853), p. 8.
83 Braid (1853), p. 11.
84 See especially Braid (1852), pp. 94–97 and (1853), pp. 24–31.
85 Braid (1853), p. 9.
86 Braid (1853), pp. 19–20.
87 Braid (1853), p. 20.
88 Cf. Bramwell (1896–7a), pp. 154–155.
89 At least, I assume it was Elliotson to whom Braid refers, Braid (1852), p. 25n.
90 Braid gives examples of favourable medical opinion (1852), pp. 25ff.
91 See above, pp. 231–234.
92 Holland (1852), pp. 89n, 92; Carpenter (1853); Bennett (1851). Henry Holland (1788–1873), shortly to be Sir Henry, was one of the best-known physicians of his time; J. H. Bennett (1812–1875), professor of the theory of physic and clinical medicine at Edinburgh, is best remembered for his advocacy of the virtues of cod liver oil; W. B. Carpenter we shall meet again.
93 Azam (1860); Azam (1887), pp. 15–16; cf. Figuier (1881), pp. 380–382.
94 Broca (1859).
95 Figuier (1881), p. 385.
96 Guérineau (1860); Figuier (1881), p. 386.
97 E.g. those by Demarquay and Giraud-Teulon (1860); Dunand (1860); and Desmartis (1860). Desmartis contends that the discoverer of hypnotism was Dr. Piorry of Bordeaux, who had lectured on it there in 1846–8.
98 Figuier (1881), pp. 386–387.
99 Hébert (1861); Lallart (1864); Laségue (1865); Broca (1869). Laségue found that in some cases merely closing the patient's eyes by hand caused various degrees of sleep from drowsiness to catalepsy.
100 Preyer (1881).
101 Of course he knew about him later, though he certainly did not "popularize" him as Kravis says (1988, p. 1201). The point is that it was not the study of Braid which launched Bernheim into his hypnotic career or gave him his ideas about hypnotism.
102 Charcot acknowledged an indebtedness to Braid (Kravis, 1988, p. 1201); Kravis, who exaggerates Braid's role in the history of "dynamic psychiatry", also points out that Charcot possessed copies of the French and German translations of Braid's *Neurypnology*. But these translations considerably postdated Charcot's involvement with hypnotism.
103 On Sunderland's career see especially Sunderland (1868).
104 Sunderland (1847), p. 13. On p. 40n of this book will be found what must be the first account of a human "split brain" patient. The patient operated on himself, accidentally, by raising his head unthinkingly whilst operating a steam-driven circular saw.
105 As expounded in *The Magnet* and in Sunderland (1843), pp. 99–116.
106 On his abandonment of phrenomesmerism, see Sunderland (1847), p. 73n.

107 E.g. Sunderland (1847) and (1853).

108 I base this generalized account of Sunderland's procedures mainly on contemporary newspaper reports, of which he quotes not a few in the works cited above.

109 Sunderland (1847), p. 147.

110 Sunderland (1853), pp. 105–106; Gravitz, 1985.

111 Harte (1902–3), Vol. 2, pp. 214–215.

112 Sunderland (1847), pp. 147–150; (1853), pp. 102–113.

113 Sunderland (1853), p. 51, italics omitted.

114 Fahnestock (1846).

115 Fahnestock (1871), pp. 69–70.

116 Garrison (1929), p. 439.

117 Mitchell (1859), p. 178.

118 Mitchell (1859), p. 180.

119 Mitchell (1859), p. 249.

CHAPTER 15

Hypnotism and scientific orthodoxy
1875–1885

With the last quarter of the nineteenth century, we at last reach a period in which hypnotism began rather suddenly to engage the serious attention of a fair number of respectable scientists and medical men. This change of attitude was not at first accompanied by any softening of medical scepticism concerning the more extravagant claims of the mesmerists; their claims, for instance, as to clairvoyant diagnosis, thought-transference and so forth. In fact, the 1870s marked the apogee of a certain kind of rather brash medical and scientific materialism which had risen to the ascendant around the mid-century as the last remnants of romantic nature-philosophy sputtered fitfully towards the horizon. Such views soon came to permeate the medical profession, and particularly its vanguard. And this was as true of psychiatry, insofar as there was a psychiatry, and of neurology, as of any other branches of medicine. Whereas in the aftermath of the romantic movement there had been occasional doctors, especially in Germany, who recognized psychological as well as physical methods of treatment for mental disturbances, and devised psychological methods of treatment shaped to the needs of each patient considered as an individual, by the mid-century European psychiatry was taking a marked turn towards a purely organic approach, with a strong emphasis upon neuroanatomy and neurophysiology, and upon the hospital, the clinic, and even the autopsy table, as the arenas within which alone significant advances were likely to be made. The psyche was almost squeezed out of psychiatry. The leading figures in this development were (in France) B. A. Morel (1809–1873), L. F. E. Renaudin (1808–1865), J. M. Charcot (1825–1893) and V. Magnan (1835–1916), and (in Germany) W. Griesinger (1817–1868). The process attained its culmination, and perhaps first overreached itself, in Freud's teacher at Vienna, T. Meynert (1833–1892).

The organicist, or somatological, approach to psychiatry, which often went with a tendency to suppose that many kinds of mental abnormality were forms of hereditary degeneration, centred on France and Germany, but was reflected elsewhere. In Italy, it was represented by, for instance, S. Biffi (1822–1899) and C. Lombroso (1836–1909), and in Britain by H. Maudsley (1835–1918).

Another leading British figure, D. Hack Tuke (1827–1895), was perhaps too eclectic to be readily categorized – in his great *Dictionary of Psychological Medicine* (1892) all points of view are represented – but the ideas of the noted British neurologist, J. Hughlings Jackson (1834–1911), who propounded a doctrine of evolutionary "levels" in the nervous system, came, as we shall see, to be influential in psychopathology, including continental and American psychopathology.

It was against the background of this heavily organic psychiatry that the hypnotic movement of the last quarter of the nineteenth century got under way. Academic psychology played only a small and rather peripheral role in it. Not a few of its leading figures were in fact neurologists, or else belonged to that impressive group of individuals of this period who seemed equally at home as clinicians and anatomists, neurologists and experimenters, psychiatrists and histologists.[1] Obvious examples are Charcot and his pupils Babinski and Gilles de La Tourette; Heidenhain of Breslau; Forel of Zürich, distinguished alike for his studies of ants, his discoveries in brain anatomy and his work as a psychiatrist; Bekhterev of St. Petersburg; and Tamburini and Seppilli of Modena.

The term psychotherapy – the treatment of disease, especially psychological disorders, by psychological methods – came into currency during the later 1880s.[2] That the concept of psychotherapy and of the psychological causation of mental disturbance gained ground between 1880 and 1900 was probably due to two closely linked factors: a growing interest among medical men and educated laymen in sexual psychology and pathology (a field in which such noted hypnotists as Moll, Krafft-Ebing and Schrenck-Notzing were prominent)[3]; and the hypnotic movement itself.

CHARLES RICHET

If one were asked to name the date at which and the person with whom the modern hypnotic movement began, as good a choice as any would be 1875 and Charles Richet. Richet (1850–1935), later professor of physiology at the University of Paris, and a Nobel Prize winner, was at this time a house physician in the Paris hospitals. A man of wide interests[4] and great personal charm,[5] his curiosity and lively intelligence often swept him into regions which orthodoxy avoids.[6] His interest in hypnotic phenomena (he does not often use the term "hypnotism") sprang from a casual attendance at a magnetic demonstration. He seems then to have experimented himself for several years before publishing his preliminary results in 1875.[7] Thereafter, despite a heavy programme of physiological researches, he found time to write a series of further papers[8] on the topic, a series which culminated and was recapitulated in his book *L'homme et*

Figure 6. Charles Richet.

l'intelligence (1884). From an historical point of view, the earlier of these
publications were the most important, for they almost certainly helped turn the
attention of Charcot and his school to the problems of hypnotism.

Though Richet was aware of Braid's eye-fixation technique, he preferred to
use the traditional mesmeric passes, with the subject facing him on an armchair
or sofa, preferably in silence and semi-obscurity. He found that susceptible
subjects first exhibit a kind of torpor, with sensations of warmth, cold or tingling,
and perhaps twitchings of the muscles. Then come heaviness and fluttering of
the eyelids. Finally, the eyes close and the subject is asleep. The state that
supervenes is not, however, that of ordinary sleep, for at first the subject is aware
of her condition, and can converse about it. The reports thus obtained suggest
that magnetization produces an effect like that of hashish or of opium, with a

general anaesthesia or analgesia, and a feeling of detachment and happiness.[9] Unlike Braid's subjects, Richet's never exhibited a marked catalepsy. Later, perhaps under the influence of Charcot and his school, Richet emphasized the muscular hyperexcitability which somnambules may exhibit when in the early stages of somnambulism.

Once the state of *somnambulisme provoqué* has been achieved, the hypnotist can enter into conversation with his subject, and can obtain further effects by direct verbal suggestion and interplay. In 1875, Richet emphasized[10] four groups of phenomena thus obtainable, hallucinations, hyperideation, automatism, and abolition of memory. Hallucinations can be perfectly realistic, and in any sense-modality. By hyperideation, Richet meant the fact that in certain somnambules intelligence may become exceptionally lively, language elevated, and emotions readily aroused. Despite this liveliness of intellect, somnambules are subject to the will of persons around them. One can force them, by verbal command or by gesture, to stand up, sing, remain upright, keep silent, join the hands, etc., and even to do so repeatedly. This is what Richet refers to as automatism. The abolition of memory (post-hypnotic amnesia) is "absolutely characteristic", Richet says, and he goes on: "I have not seen it lacking a single time, but, and this is very strange, the events of the sleep do not disappear completely, but the return of the somnambulic state brings back the memory of them." He does not mention the possibility that these memories may, by suggestion or other techniques, be recovered in the waking state.

In later writings, Richet distinguishes between "active" and "passive" memory. By active memory he means the capacity to commit facts and words to memory as they strike the senses; by passive memory he means the capacity to recall items that have been committed to memory in the past. He holds that in somnambules active memory is abolished or impeded, whilst passive memory is exalted.[11] He also interested himself in the relationship between memory and the phenomenon which he christened "the objectification of types"[12] (a term which gained some currency). By appropriate suggestions, somnambules can be induced to develop "partial amnesias", e.g. for their names, for proper names, for the names of places. An amnesia for one's personality, for who and what one is, would be a form of partial amnesia. Two female somnambules studied by Richet – ladies of very different social station but (he felt) of unquestionable veracity – exhibited the capacity for this kind of partial amnesia to a very striking degree. In addition, each could give various different forms to her self, believe herself to be a soldier, a priest, a little girl, a rabbit, and so forth. These characterizations were what Richet calls "the objectification of types" – amnesia for the real personality combined with a new, and assumed, personality. The *self* is preserved, but conforms to the various temporary personalities suggested to it. Richet transcribes some of his subjects' performances, and they undoubtedly exhibit a certain dramatic zest. This example is from an eminently respectable and highly religious matron:

> *As an actress.* Her face, instead of seeming harsh and angry as it had done a moment ago, takes on a smiling expression. "You see my skirt? Well, it was my director who had it made longer. They are tiresome, these directors. I find the shorter a skirt is the better. They are always too long. A simple vine-leaf, my God that's enough! You think there's no need for anything more than a vine-leaf, don't you?...Now then, my love (she begins to laugh). You are a bit timid with women; you're wrong. Come and see me sometimes. You know I'm always at home at three o'clock. Come and pay me a visit and bring me something."[13]

With the curative aspects of hypnotism, Richet was little concerned. He thought that with certain hysterical or neurotic subjects there was a notable remission of symptoms following "artificial sleep". Patients slept better, were less subject to mental agitation and erratic pains, etc.[14] But his interests lay in the psychological and physiological peculiarities that went with the state of *somnambulisme provoqué* rather than with any therapeutic benefits that it might confer. Concerning these peculiarities, he developed the outlines of a theoretical stance[15] which has some affinities with that of the later Braid, though I doubt that there was any direct influence.

Richet begins by remarking that normally higher nervous centres have a moderating influence upon automatic nervous actions. If this moderating influence is removed, the result is an animal showing an exaggerated reflex excitability, and a tendency to persevere pointlessly with behaviour once set in motion. Such animals really are automata.

In "artificial" somnambules it is *intelligence* that has become automatic. In the mind of a normal individual there are at any one time a large number of coexisting ideas which balance and mutually complement each other. Out of this balance results the apparent spontaneity of our minds. In somnambules, by some inhibitory process which, Richet admits, we do not as yet understand, these multitudinous ideas have been stilled, put to sleep. Thus when one idea gains a foothold – say the idea of *snake*, insinuated by the operator – there will be no competing ideas to constrain it or hinder its development. The subject may become afraid, speak and act as though a snake were present, even see a snake, and so on.

An issue to which Richet repeatedly returns[16] is that of whether he could conceivably have been misled throughout by subjects who have merely simulated the phenomena of *somnambulisme provoqué*. The theory of simulation crops up again and again in the history of hypnotism, and it is therefore of interest to note what so acute an observer as Richet has to say about it. Richet points out that the most compelling of all arguments against the simulation theory is to be put into the artificial sleep oneself, before witnesses whose testimony one cannot doubt; and he claims to have vanquished more than one sceptic by this method. His remaining arguments can be put under three heads, as follows:

1. Wicked though the world may be, it is not so full of liars as the theory of simulation would require. This theory, says Richet, "appears to me absolutely ridiculous. Take, for instance, three of my good friends, educated and enlightened young people, in whom I have complete confidence. Must I admit that they have deceived and made mock of me?...Moreover I would then have to believe that the fifty persons (or nearly) of every age and condition, whom I have been able to send to sleep, were all, without exception, *without a single exception*, rascals and liars. That is in truth something quite inadmissible."[17]

2. The theory of simulation cannot account for the coincidence of the phenomena described by different investigators in different times and places, sometimes with subjects who were previously quite ignorant about magnetism, hypnotism, etc.

3. The theory of simulation does not explain why certain subjects are able to perform feats or endure unpleasantness which it would be beyond their waking powers to perform or endure, e.g. breathing the fumes of ammonia with apparent pleasure.

GERMANY: THE HANSEN PHASE

Richet's interest in *somnambulisme provoqué* had, as I noted above, much more to do with the peculiar psychological and physiological phenomena supposedly characteristic of it than with its alleged therapeutic efficacy. Perhaps this was because his interest in these matters was first aroused by a magnetic demonstrator bent upon startling audiences rather than upon curing them. If we turn from France to Germany, we find there too the beginnings of scientific and medical interest in hypnotism; an interest which was likewise primarily in the psychological and physiological effects rather than in the therapeutic possibilities. And this interest was largely unleashed by the activities of a highly successful, itinerant magnetic demonstrator, to wit the Dane Carl Hansen (1833–1897).

Hansen's name crops up repeatedly in the hypnotic literature of this period.[18] We find him moving from city to city in, for instance, Sweden, Denmark, Finland, Russia, France, Germany, Austria and Britain, always attracting a good deal of notice in the press, and occasionally arousing serious interest in the scientific community. His impact was greatest in German-speaking countries around 1879–80, where he initiated what might be called the "Hansen phase" – roughly 1879–1884 – of German hypnotism. In Chemnitz he stirred the interest of Weinhold; in Breslau of Heidenhain and Berger; in Berlin of Preyer and Eulenberg; in Würzburg of Rieger; in Leipzig of Möbius and Wundt; in Vienna of Krafft-Ebing and Benedikt.

To magnetize a subject, Hansen would make him stare for a while at a glittering piece of glass, perform a few passes without contact over his face, and lightly close his eyes and mouth, at the same time gently stroking his cheeks. A few more passes over the forehead would see him in a sleep-like state in which, like a will-less automaton, he would at command perform the most ridiculous actions.[19]

Among the ridiculous actions which Hansen would compel his subjects to perform were the assumption of cataleptic postures or of muscular rigidity sufficient for the performance of the well-known "human plank" feat; the inability to move or speak; the striking of absurd attitudes; standing up and singing; engaging in preposterous pantomimes; eating a raw potato believing it to be a pear; and drinking imaginary champagne from real glasses and thereafter behaving as if intoxicated. Sometimes Hansen would succeed in producing positive and negative hallucinations, anaesthesia, or the fulfilment of post-hypnotic suggestions.

Hansen had many imitators and successors, and it is an interesting question how far subsequent, and indeed current, notions of what phenomena are most readily, or most characteristically, elicitable from hypnotized subjects have been shaped by the resourcefulness of public performers. Many items in current scales of hypnotic susceptibility are recognizably drawn from the same range of phenomena as those with which Hansen entertained spectators over a hundred years ago, and one could go further back to the performances of the electrobiologists, or to Faria. Unlike Faria, however, and to a greater extent than the electrobiologists, Hansen had an immediate and enduring effect on the notions of scientists and medical men concerning hypnotic phenomena, and some of his imitators (whom he was always pleased to help) were individuals already known in the world of science, and anxious to experiment for themselves.

The first of these orthodox scientists to experiment along neo-Hansen lines seems to have been Professor A. F. Weinhold (1841–1917), a physicist of Chemnitz, whose pamphlet, *Hypnotische Versuche*, first appeared in 1879.[20] He succeeded in obtaining some of the more elementary of Hansen's phenomena, and came further to suspect that a mild electric current applied to the forehead may have a direct effect on certain people. He develops no theory of his own, confining himself to the statement that we are not dealing with any magnetic influence, but only with "Braid's hypnotism". However, he does not seem to have read Braid himself.

The same is true of Professor R. Heidenhain (1834–1897) of Breslau, a physiologist of some repute, and author of the most influential hypnotic work of this period, *Der sogenannte thierische Magnetismus* (1880). Using as subjects principally medical men and medical students, Heidenhain soon succeeded in reproducing some of Hansen's phenomena. In particular, he caused some of his subjects to assume absurd postures and carry out ridiculous actions either by cutaneous stimulation, or by manipulating their limbs into cataleptic attitudes, or by himself performing or hinting at the required action in front of them, a procedure which would trigger an imitative response. Certain sensitive subjects (he discovered), would exhibit catalepsy of the right limbs and side of the face, together with inability to speak, if the left forehead and temple were stroked; right-sided stroking produced left-sided catalepsy but not speech problems.[21]

Verbal suggestions Heidenhain regarded as superfluous. Later he seems to have changed his mind on this issue, following a communication from his colleague Professor O. Berger, who had found that a hypnotized individual became responsive to verbal commands as soon as the operator's hand was laid on his head. Heidenhain at once tried the experiment on one of his best subjects, his brother A. Heidenhain, described by an American visitor, G. S. Hall, as "a tall, athletic, duelling medical student".[22]

> A glass containing ink was given him, with the request to drink some beer. Without the least hesitation he began to drink the ink. He also, on being told to do so, thrust his hand into a burning light, and with scissors so unmercifully cut off his whiskers, which he had assiduously cultivated for a year, that on awakening he was greatly enraged.[23]

Heidenhain also found that in certain hypnotized subjects it was possible to induce dreams, hallucinations and illusions, by verbal suggestions.

Among others at Breslau who interested themselves in hypnotism were Professor O. Berger (1844–1885), a neurologist, and P. Grützner (1847–1919), a physiologist, later professor at Tübingen. Heidenhain soon became almost as influential as Hansen in promoting interest in hypnotism. Professor H. Senator (1834–1911) of Berlin paid him a visit and on his return reported to the Berlin Medical Society (11th February 1880) on what he had seen.[24] Other distinguished Berlin *Gelehrten*, for instance Professor A. Eulenberg (1840–1917), a neurologist, and P. Börner (1829–1885), took up the subject. Heidenhain's own pamphlet reached a fourth edition by March 1880, and influenced other writers, for instance Dr. Ernst von den Steinen of Heidelberg, whose *Über den natürlichen Somnambulismus* (1881), is much in the Heidenhain tradition, and C. Bäumler, professor of medicine at Freiburg, author of *Der sogenannte animalische Magnetismus oder Hypnotismus*, a very useful guide to the literature and ideas of the Hansen phase.

The writers of the "Hansen phase" varied somewhat in their theoretical ideas. Most were physiologically oriented, but in a series of articles on "Hypnotische Zustände und ihre Genese" in the *Breslauer aerztliche Zeitschrift* for 1880,[25] Berger of Breslau argued that physical manipulations owe their effects entirely to their canalization of the subject's thought and imaginative power. G. H. Schneider, a zoologist, later assistant to Haeckel at Jena, presented a psychophysiological theory somewhat resembling Richet's.[26] Sensations and perceptions all have a tendency to transform themselves into action. However, the tendencies to motor expression are normally modified or inhibited by the multitude of competing impressions and action-tendencies. Causes (especially narrowing of attention) which restrict or cut off these inhibitory factors can result in a state of "abnormally one-sided concentration of consciousness"[27] leading to hypnotic automatism in which the subject will, for instance, unreflectingly initiate actions which he has seen carried out before him.

Heidenhain, whose experience of hypnotism at the time when he wrote his book was very limited, had the extraordinary idea that a deeply hypnotized subject is actually unconscious. He bases this conclusion on the post-hypnotic amnesia which most of his subjects apparently displayed.[28] The unconsciousness of hypnotized subjects is due to "the inhibition of the activity of the ganglion-cells of the cerebral cortex...the inhibition being brought about by gentle prolonged stimulation of the sensory nerves of the face, or of the auditory or optic nerve."[29] Once the higher cortical centres have been thus inhibited, stimulation from the sensory centres will bypass them, and go directly to the most nearly associated motor centres. These will naturally be the centres which produce the same movements or utterances as those just seen or heard. Hence the imitative automatisms which Heidenhain regards as the central phenomenon of the hypnotic state. He is left, of course, in some difficulty over hypnotic dreams and hallucinations, with the accompanying spontaneous conversational interplay between hypnotist and subject.[30]

The most popular kind of physiological explanation – one of which Braid gave a foretaste, and which many later workers were to exploit in greater detail – was that offered by Krafft-Ebing's pupil, H. Kaan of Graz,[31] following suggestions by Carpenter, W. Preyer, D. Hack Tuke, T. Rumpf of Düsseldorf, H. Brock of Berlin, G. S. Hall, and Heidenhain himself.[32] Kaan argued, from his own (somewhat unconvincing) plethysmographic experiments with human subjects, and from the results of investigations of cerebral circulation in animals, that the physiological basis of the hypnotic state is changes in cerebral blood supply. It is curious to note how very different are these somewhat crudely materialistic explanations of the phenomena from the deep inklings which the last nature-philosophers had tried to articulate upon the subject only a generation or so before.

The writers of the "Hansen phase" of German hypnotism were, as I remarked before, generally speaking more interested in the academic and theoretical problems raised by the phenomena than in the therapeutic application of hypnotic methods. When in 1884 Dr. A. Wiebe of Freiburg im Breisgau reported his hypnotic treatment of three hysterical patients to the Berlin Medical Society,[33] he was able to claim that in Germany only Berger had preceded him in trying such methods. Wiebe seems to think that the hypnotic state is by itself therapeutic. He used passes to treat patients, and also occasional verbal suggestions.

CHARCOT AND HIS SCHOOL: HYSTERIA

Though the braidists, Richet, and the German writers of the "Hansen phase", all roused some degree of passing interest among their scientific and medical contemporaries, there can be no doubt that the final establishment of hypnotic phenomena as a legitimate subject of scientific inquiry was largely due to the intellectual and personal force of a single clinician and scientist of international reputation. It is a sad irony that that reputation, so boldly and confidently risked, and for a while enhanced, by involvement with hypnotism, was through that involvement almost destroyed, and for a long while undeservedly abased. The individual to whom I refer is of course the French neurologist J. M. Charcot (1825–1893).[34]

Though Charcot contributed to many aspects of medicine,[35] neurology was his overriding interest, and it was almost certainly this interest which led to his becoming, in 1862, physician to the Salpêtrière, the great Paris asylum for women. Behind its imposing façade, the Salpêtrière was an immense complex of large and run-down buildings, almost a town in its own right, but inhabited by the disturbed, the destitute, the insane, the senile, and the victims of all kinds of nervous diseases, to the total of 5000 persons – an unrivalled pool of subjects for the aspiring neurologist. Charcot's growing private practice had won him some influential friends, and through these, and his own determination, he raised funds to enlarge the Salpêtrière's clinical facilities.[36] He reorganized the wards. By acute observation combined with careful anatomo-pathological investigations,[37] he was able to produce important studies on a series of neurological topics. His reputation became first national and then international. In 1872, he was appointed professor of pathological anatomy in the Paris Faculty of Medicine, in 1882, a chair of diseases of the nervous system was created specially for him, and in 1883, he was elected to the Academy of Sciences. Among his pupils were many of the most celebrated French neurologists of the rising generation – including Pierre Marie (1853–1940), G. Gilles de La Tourette (1857–1904), Paul Richer (1849–1933), A. Pitres (1848–1928), F. Raymond (1844–1910), C. S. Féré (1852–1907), D. M. Bourneville (1840–1909), and J. F. F. Babinski (1857–1932), who came from Poland. Numerous foreign students paid him shorter visits, the most famous being Sigmund Freud (1856–1939) and V. M. Bekhterev (1857–1927).

Figure 7. Charcot in case conference.

From this it is apparent that as well as being a great clinician and an energetic administrator, Charcot was an inspiring and highly successful teacher. His efforts were not directed just towards his pupils and assistants, but also towards a much wider audience of medical students, practising physicians, and other interested persons. This aspect of his teaching activities crystallized in later years into two series of weekly classes, formal lectures on Fridays, and less formal clinical demonstrations on Tuesdays.[38] The Friday lectures were most impressive set pieces, prepared in advance down to the last detail, and so clear in organization and expression that they could be published as they stood. They were delivered in a miniature theatre, and the theatrical impression was heightened by stage lighting, the use of photographic slides, and the somewhat excitable patients who were occasionally put through their paces.[39]

All this, together with the fact that the lectures were often attended by interested laymen, including writers, journalists and society figures, led some to accuse Charcot of an excessive love of publicity, and of playing the great man, the "Napoleon of neuroses". But in fact Charcot seems to have been reserved and shy by nature. Lecturing made him nervous, and his motives for doing it certainly included a sense of the obligations which his position laid upon him to propagate medical and scientific knowledge.

Privileged visitors might also be invited to the Tuesday case demonstrations.

At these, Charcot would examine patients, including ones new to him, and offer diagnoses and reflections, freely admitting when he was baffled or had been in error. Discussion was open. "Never", says Freud, did Charcot "appear greater to his students than on these occasions, when he thus did his best to lessen the distance between teacher and pupils by giving them a complete and faithful account of his own train of thought by stating his doubts and misgivings with the utmost frankness."[40]

It was against this background of semi-public and increasingly publicized case demonstrations that Charcot and his school developed their singular approach to the problems of hysteria and hypnotism.

The modern approach to hysteria[41] is sometimes (though at the cost of great over-simplification) said to have begun with the publication by P. Briquet (1796–1881) of his *Traité de l'hystérie* (1859). Briquet's ideas were based on a close study of 430 patients at the Charité hospital in Paris. He concluded[42] that "hysteria is an illness consisting of a neurosis of the part of the brain destined to receive the affective impressions and the sensations". A strong element of hereditary disposition underlies it. It affects women much more than men, and the lower classes more than the middle. It is precipitated by emotional shocks and stresses, including disappointment in love, but *not* sexual frustration – prostitutes are, whilst nuns are not, much subject to hysteria. Its symptoms are distortions or disturbances of the ordinary manifestations of emotion.

Briquet's work leads on to that of Charcot and Richer. Many of the same leading ideas are detectable in both.[43] However, in the Charcot version the whole clinical picture has become more definite, and more scientific, but at the same time almost totally artificial, and one can, in hindsight, readily see how this came about. In 1872, Charcot had assumed charge of a ward containing large numbers of both epileptics and hysterics. The hysterics had become well acquainted with epileptic seizures; hence epileptoid attacks were unusually prominent in their symptomatology. Now in studying hysteria and hypnosis, Charcot was disposed to take an approach which had proved successful in his other neurological studies: to look for the most fully developed form of a disease, of which other forms might be regarded as partial manifestations. This kind of approach, brought to bear upon case material so metamorphosed and so malleable, soon created the fully developed hysterical and hypnotic symptom-sets which it had set out to find.[44]

The view of hysteria developed by Charcot and his pupil Richer in the late 1870s and early 1880s[45] may be summarized as follows: hysteria is a peculiar mode of feeling and reacting, not confined to women, and not specially linked to frustrated sexual urges. It is known only through its symptoms, but there is a marked hereditary predisposition, no doubt involving some kind of neural abnormality. Its most characteristic symptom is the hysterical attack, or *grande hystérie*, in which (as in epileptic attacks) a number of stages can be detected.[46]

Often there are prodromi – hallucinations, organic disturbances, disorders of sensibility. These are followed by an epileptoid attack, with tonic and clonic phases, passing into a sterterous sleep, the whole lasting from five to ten minutes. After this, the patient, still lying down, but usually now conscious, enters a period of disordered movements, frequently of a gymnastic kind ("clownism") and beginning with a tetanic *arc de cercle* or opisthotonos.[47] After five to ten minutes this stage passes into a stage, termed the stage of *attitudes passionelles* or *poses plastiques*, in which the patient mimes or acts out some emotion or series of emotions – erotic, fearful, sentimental, ecstatic – perhaps related to a pathogenic episode in his or her own past. This stage may in turn merge into a period of disorientation and delirium.

From this idealized picture, there can be many deviations. In some patients one phase becomes predominant – thus Charcot calls a prolonged stage of disordered movements a "demoniacal attack" because the patient's convulsions and cavortings resemble those of hysterical *possedées*.[48] In other patients, the attack is greatly simplified or abbreviated, and in others again there may be an admixture of lethargy, catalepsy or somnambulism[49] (an observation which it is hard to suppose uncontaminated by Charcot's views about hypnotism and its relation to hysteria).

Hysterics, whether or not liable to these acute attacks, generally suffer from various chronic, though frequently labile, symptoms, which include disturbances of sensibility, and development of hypersensitive "hysterogenic points", often on the midline of the body or bilaterally symmetrical about it (e.g. on the ovaries). Prolonged stimulation of these points may induce a grand attack, or sometimes arrest one. There may also be various kinds of motor derangements, particularly paralyses and contractures and tremors of the skeletal muscles. Such paralyses tend to be accompanied by anaesthesias. It is to be noted that the anaesthesias, being of a hand, a limb, etc., do not coincide with the areas innervated by particular sensory nerves. In other words they are determined by the patient's anatomical ideas rather than by his anatomical structure.

Other kinds of troubles which may afflict hysterics include respiratory, digestive, reproductive, vocal and circulatory disorders. There may be bouts of prolonged apparent sleep, and attacks of ambulatory automatism in which the patient wanders off quite oblivious to what he has been or should be doing. And there is a characteristic hysterical personality. Most authors, say Charcot and Marie, "mention as peculiar to hysteria a certain mental condition – excessive psychical re-action, desire to do something remarkable, love of everything brilliant and extraordinary, a tendency to lying, exaggeration and even simulation, absence of will, irritability, and frequently more or less absolute loss of moral sense."[50]

As for the causes of hysteria, the view of the Charcot school might be described as catholic: a large number of different kinds of diseases, infections, poisonings, intoxications, shocks, traumas, and misfortunes, can activate an hysterical neurosis in a predisposed constitution. Later, Charcot came to lay increasing emphasis upon psychological factors, especially in "traumatic neuroses" following accidents.[51]

CHARCOT AND HIS SCHOOL: HYPNOTISM

We know from Charcot's own statement[52] that he and his pupils began serious investigation of hypnotism in 1878. What, beyond his general knowledge of the work of Braid and the braidists, and of Richet,[53] turned his mind in that direction at that time is not completely clear. Moritz Benedikt (1835–1920), the Viennese neurologist, implies[54] that it was through his advice that Charcot finally ventured into this uncertain terrain, but Benedikt was an individual whose exaggerated opinion of himself was only matched by his disapprobation of those who failed to share it, and his claim must be treated with a corresponding reserve. A factor that almost certainly was of importance was Charcot's involvement, from 1876 onwards, with metallotherapy.[55] Metallotherapy was being energetically promoted in Paris by Dr. V. Burq. Burq claimed, amongst other things, that contact of the skin with metals could relieve hysterical anaesthesias, and also the paralyses and contractures which he regarded as secondary to the anaesthesia.[56] Charcot became convinced that there was something in Burq's claims, and in consequence a commission of the Societé de Biologie was set up in 1877 to investigate the matter. It consisted of Charcot, V. A. A. Dumontpallier (1826–1899) of the Pitié hospital and J. Luys (1828–1897) of the Charité. Its report was favourable.[57] Meanwhile, Charcot had continued his own experiments, with the special assistance of R. Vigouroux. They found that the "cures" brought about by contact with metals tended to be evanescent, and that sometimes a symptom removed on one side of the body would immediately reemerge in the corresponding part of the other side, a phenomenon which was called "transference". They also discovered other agencies (magnets, tuning forks, electricity) that would produce similar effects. These substances were termed "aesthesiogens" because of their effects in restoring sensation. Charcot's interest in aesthesiogens seems to have diminished quite quickly, but they continued to be studied with mounting gullibility by others, and most notoriously by Professor Luys, whose credulous involvements in this and kindred areas enlarged the frontiers of folly in several directions simultaneously.

A long tradition linked magnetic somnambulism with a sensitivity to the presence of metals, and it seems likely that Charcot's involvement with metallotherapy encouraged him to tackle hypnotism, which he investigated intensively between 1878 and 1882.[58] His work had that curious semi-public character which we have already mentioned, and naturally received wide publicity.[59] The mere fact that Charcot, a scientist of international reputation, was known to be busying himself with hypnotism, and was apparently analyzing its principal phenomena in terms acceptable to physiologists, gave the subject a sudden surge in respectability which culminated on 13th February 1882 with his reading to the Academy of Sciences a paper summarizing his results to date. This was an event which had, as Janet points out, very important results. "It was as if [Charcot] had broken down a dam behind which a vast head of water had been accumulating .... Everywhere, 'hypnosis redivivus'... gave rise to numberless books and articles. An enumeration of the authors of these would include the names of most of the neurologists of that day, both in France and in other lands, for there were few who failed to be influenced at this epoch by the teachings which emanated from the Salpêtrière."[60]

Of these teachings, so far as they related to hypnotism, the most systematic and thorough account is probably that contained in P. Richer's *Études cliniques sur la grande hystérie* (2nd edition 1885),[61] with which may be compared A. Binet and C. Féré's more popular *Le magnétisme animal* (1887). From these and other sources[62] I shall endeavour to construct a brief summary.

The Salpêtrière's doctrines (it is important to note) were largely founded upon the performances of not more than a dozen star subjects, all of whom were hysterical female patients. Charcot was once again pursuing his principle that diseases – and hypnosis was in his view undoubtedly a morbid state, closely linked to hysteria, and nurtured by the same hereditary predisposition – are best studied in their most developed form. The most developed form of hypnosis he called *grand hypnotisme*, less developed or partial forms *petit hypnotisme*. He emphasized that his nosography of hypnotism applied only to *grand hypnotisme*, in which he thought that there are three stages.

Each stage has its own distinctive physiological characteristics, most notably reflexes, which Charcot, always alive to the possibility of simulation, thought could not be counterfeited by anatomically uneducated subjects. The stages are most conveniently induced in the following order:

1. *Stage of Catalepsy*. Catalepsy may be induced gradually by having the subject look at a bright object close to her eyes, or immediately by an unexpected loud sound, or by lifting the eyelids of a patient in the next stage (lethargy). The subject is immobilized like a statue. The limbs, however, are not stiff, and will maintain for long periods any position they are pushed into. The pulse and heart rates are reduced, skin sensibility is usually absent, but vision, hearing and muscular sense are at least partially active.

Patients in a state of catalepsy are peculiarly receptive to certain kinds of suggestion. Put the limbs or hands or fingers of such a patient into an attitude suggestive of such and such an emotion or state of mind and she will at once assume the whole posture and expression of someone in that state of mind. Close her fists, for example, and she will assume an attitude of anger and aggression. Some observers found the poses thus struck exceedingly realistic, not to say dramatic.[63] Similar effects are produced if the facial muscles are electrically stimulated so as to mimic the expression of an emotion. At once the patient assumes a corresponding posture.

2. *Stage of Lethargy.* The stage of lethargy is reached either by prolonging fixation of a bright object beyond the period that produces catalepsy, or by pressing shut the eyelids of a patient already cataleptic. The subject appears profoundly asleep. The skin and mucosa are completely insensitive, though the senses may show some vestigial functioning. There is no responsiveness to suggestion. The most characteristic feature of this stage is "neuromuscular hyperexcitability". Pressure on a muscle produces the contraction proper to that muscle. Likewise, compression of a superficial nerve immediately causes contraction of the muscles it supplies. For example, by pressing the ulnar nerve at the elbow a contorsion of the hand termed *griffe cubitale* is occasioned. Muscular contractions so produced are exceptionally powerful, but are relieved immediately by pressure or friction on the antagonist muscle.

3. *Stage of Somnambulism.* The somnambulic stage is most readily obtained by gently rubbing the top of the head of a patient who is already in the stage of catalepsy or the stage of lethargy. Somnambulic subjects are anaesthetic in skin and mucosa, and generally, but not always, have their eyes closed. Senses other than cutaneous ones may be exalted, and hyperacuity of vision and hearing may be found. Like cataleptics, somnambules are highly suggestible, but whereas the cataleptic responds to suggestion like a mere automaton, somnambules will interpret suggestions intelligently and even question them, and will respond to suggestions made verbally, gesturally, or in any other way.

"Neuromuscular hyperexcitability" is to be found in the somnambulistic stage also, but now it is produced just by light cutaneous stimulation, which makes the underlying muscles contract. Even a draught of air will suffice.

Subjects may be awakened by blowing on their eyes; and they are usually amnesic for the events of the preceding hypnotic states.

So much by way of a synopsis of the Charcot-Richer view of hypnotism. Two obvious questions now arise. The first is that of the relationship between *grand hypnotisme* and hysteria. What grounds do Charcot and his school have for their frequent claim that these two states have a deep underlying affinity? The second, and more difficult, anticipates the upshot of the next chapter. It is that of how men of such intelligence as Charcot and his collaborators could have misled themselves over hypnotism, or have been misled, so egregiously.

The final Salpêtrière position with regard to the first question was set forth by Babinski in a series of parallels between the phenomena characteristic of hysteria and those characteristic of hypnosis.[64] For example:

1. The somatic characteristics – anaesthesias, flaccid paralyses, contractures, catalepsy – which one always finds, at least in part, in hypnotic subjects, constitute the most common manfestations of hysteria.

2. The heightening of suggestibility, which is the fundamental characteristic of hypnotism, also belongs to hysteria. This was always emphasized by Charcot, especially in his later years when he became particularly interested in "hystero-traumatic paralyses", i.e. the paralyses which, in hysterically disposed individuals, may follow certain sorts of accidents.[65] Charcot and Marie outline the putative process as follows:

> It seems probable that hystero-traumatic paralysis is, amongst others, formed by the following process: A man predisposed to hysteria has received a blow on the shoulder. This slight traumatism ... has sufficed to produce in the nervous individual a sense of numbness extending over the whole of the limb and a slight indication of paralysis; in consequence ... the idea arises in the patient's mind that he might become paralyzed; in one word, through auto-suggestion, the rudimentary process becomes real.[66]

3. Hypnotic somnambules, like hysterical ones, are sometimes plunged into "second mental states", i.e. states of somniloquism and somnambulism for which there is amnesia on awaking, but which exhibit amongst themselves a continuity of memory.

4. The therapeutic influence of hypnotism is chiefly effective with troubles that depend upon hysteria.

5. An hypnotic "attack" may sometimes mingle with an hysterical one.

Babinski concludes: "For all these reasons, it seems to me impossible not to see a close relationship between hypnotism and hysteria, and one would almost be justified in maintaining that hypnotism is a manifestation of hysteria."[67] This idea, whether or not well-founded, by no means withered away with the death of Charcot and the dissolution of his school.

With regard to the second question, that of how Charcot came to be so grossly in error, three sorts of explanation have been offered, none wholly satisfactory.

The first lays stress upon the personal shortcomings of Charcot. He is said to have been domineering, egotistical, hungry for publicity, resentful of all criticism, and surrounded by place-seeking sycophants eager to find and train hysterical and hypnotic subjects to jump through the expected hoops at his public demonstrations.[68]

Now Charcot had his enemies, professional, political, and religious, and his public demeanour tended towards the formal and unbending. He was often referred to as the "Napoleon" or the "Caesar" of the Salpêtrière (no doubt because his shape and stance suggested the former, and his profile the latter). Individuals who wasted his time with doubts or questions that he considered foolish were probably made to regret their temerity. But those who knew Charcot best – his pupils and associates – though they may have stood in some awe of him, were far from regarding their *patron* as a petty dictator or a deluded egomaniac. Charcot was a dedicated and inspiring teacher, devoted to his chosen pupils. A reserved and shy man, his private face was much less constrained and severe than he public one. Kindness, humour, and the capacity for intellectual excitement were all notable elements in his character. He loved animals and had a horror of vivisection. He was widely read in several languages, a bibliophile,[69] a connoisseur of art, a gifted artist prone to execute amusing caricatures of his colleagues, and a keen traveller who always read up in advance the history and archaeology of the places he visited.[70] These qualities are not those one associates with petty dictators. And Charcot's neurological writings remain tremendously impressive even today. His lectures, in particular, present the current state of play on any topic with great clarity and impartiality, indicate without a trace of either vainglory or false modesty the direction of his own ideas, and are as concerned to bring out what we do not know as what we do. Of Napoleonic aggrandisement and self-aggrandisement there is no hint.

A somewhat milder theory is propounded by Pierre Janet,[71] whom Charcot appointed in 1890 as director of a newly-established psychological laboratory at the Salpêtrière. Janet's proposals have three prongs. The first is that at his demonstrations and clinics Charcot would treat hypnotic and hysterical patients much as he did neurological ones. He would without sending them away freely expound their cases to the assembled company, taking it for granted that the patients themselves were lethargic and therefore insensible, or else too uneducated to comprehend his medical discourse. But many of the patients were neither insensible nor idiotic, and were well able to pick up hints and suggestions. The second prong is that the subjects thus put through their paces had often been trained beforehand by Charcot's assistants, who had looked for, and naturally obtained, certain patterns of reaction. In his demonstrations and clinical sessions, Charcot merely took over these pre-trained subjects. The third is that Charcot's "stages" of hypnotism were derived from "stages" previously "discovered" and standardized by animal magnetists. Janet bases this assertion both on his own experiences with a subject ("Léonie") who had previously been in the hands of a magnetizer, and on his own reading of the magnetic literature. Furthermore, some of Charcot's pupils had actually enlisted the help of a well-known amateur magnetizer, the Marquis de Puyfontaine,[72] to teach them the secrets of the art. But Janet exaggerates the extent to which the Salpêtrière "stages" were foreshadowed in the mesmeric movement and underestimates the possible influence of Braid.

A final theory, that of deliberate simulation by the subjects, is not altogether incompatible with the preceding one, and indeed, almost forces itself upon any modern reader who studies some of the pictures in the *Iconographie photographique de la Salpêtrière*.[73] But the direct evidence bearing on the question is slight and far from conclusive.[74]

A related, but subtler and more interesting hypothesis, is suggested by a remark of J. Delboeuf, a perceptive Belgian professor of psychology, made after watching the performance of one of Charcot's somnambules: "There is not simulation, but there is something approaching it. It is an excess of obligingness (*complaisance*) on the part of the subject; she might be able to speak, but it becomes a duty to remain silent."[75] As Ellenberger implies,[76] if we only knew much more than we do, or are likely to do, about the offstage social relationships and social pressures that formed and shifted between patient and patient and between staff and patients within the strange, hothouse subculture that was Charcot's Salpêtrière, much that is now obscure might be illuminated.

NOTES

1   Cf. Hécaen and Lanteri-Laura (1977), p. 130.
2   See below, pp. 341, 359 n. 125.
3   See especially Ellenberger (1970), pp. 290–301.
4   His literary output included novels.
5   He is said to have spoken French so beautifully that even persons who could not understand a word of it were enthralled. I have myself met several persons who could remember him.
6   On his career in psychical research see Richet (1923).
7   Richet (1875).
8   Richet (1880a; 1880b; 1880c; 1880d; 1881; 1884a; 1884b; 1884c).
9   Richet (1874), p. 349.
10 Richet (1875), p. 370.
11 Richet (1884a), p. 194.
12 Richet (1884a), pp. 235–244.
13 Richet (1884a), p. 238.
14 Richet (1875), p. 363.
15 See especially Richet (1884a), pp. 226–232.
16 Richet (1875), pp. 363–370; Richet (1881); Richet (1884a), pp. 152–170.
17 Richet (1884a), pp. 155–156.
18 On Hansen see Bjelfvenstam (1967), pp. 242–244. Contemporary accounts of him will be found in, for instance, Benedikt (1880); Morand (1889), pp. 424–428; Gilles de La Tourette (1887), pp. 440–450; Marin (1890), pp. 217–219; Tuke (1884); Weinhold (1880); Heidenhain (1880; 1888); Bottey (1886). Freud, when a student, attended a demonstration by Hansen (Chertok and Saussure, 1979, p. 115). Many of the above references also contain accounts of another famous demonstrator of the time, Donato (i.e. A. E. d'Hont, 1845–1900, a Belgian), on whom see also Morselli (1886), pp. 270–299; Delboeuf (1890b), pp. 96–98; Moll (1890), pp. 64–65; Senso (1881).
19 See, e.g. Heidenhain (1888), pp. 4–5; Bottey (1886), who prints one of Hansen's programmes; and *Journal of the Society for Psychical Research*, 1889–90, Vol. 4, pp. 85–86, 99–100.
20 I have used the 3rd edition of 1880.
21 On which see Heidenhain (1888), pp. 84–89.
22 G. S. Hall (1881), p. 99.
23 Heidenhain (1888), p. 63. (I have used this, the English translation of Heidenhain, 1880, throughout.)

24 *Berliner klinische Wochenschrift*, 1880, Vol. 17, 277–278.

25 Berger (1880b); cf. Berger (1880a).

26 Schneider (1880).

27 Schneider (1880), p. 32.

28 Heidenhain (1888), pp. 15n–16n. He has therefore to give a somewhat tortuous account of the recall which certain stratagems might unleash. He supposes that the sensory events leave changes in the lower centres which are only "propagated" to the cortex when the subject emerges from hypnosis.

29 Heidenhain (1888), p. 46.

30 For an extended criticism of Heidenhain's ideas, and also the related ones expressed by P. Despine (1880), see Gurney (1884a).

31 Kaan (1885).

32 Carpenter (1881); Preyer (1881); Tuke (1881); G. S. Hall (1883); Brock (1880); Rumpf (1880).

33 Wiebe (1884).

34 The first notable biography of Charcot is Guillain (1955; Eng. tr. 1959). Cf. Owen (1971), and see also Ellenberger (1970), pp. 93–101, and Chertok (1984).

35 Cf. Garrison (1929), p. 640.

36 Guillain (1959), p. 53.

37 He once said "A physician is only as good as a clinician as he is as a pathologist". Guillain (1959), p. 15.

38 Guillain (1959), pp. 56–57; Owen (1971), pp. 45–48.

39 Garrison (1929), p. 759.

40 Freud (1924), p. 18.

41 For a useful introduction to this topic see Meares *et al.* (1985).

42 Briquet (1859), p. 601.

43 Charcot and Richer think that Briquet's observations confirm their own views as to the sequence of stages in hysterical ailments. See e.g. the references to Briquet and other mid-nineteenth-century writers on hysteria in Richer (1885), pp. 154–161, and Charcot (1877), pp. 247, 283, 302.

44 I do not want to imply that Charcot's account of *grande hystérie* lacks *any* support from cases outside the Salpêtrière. E.g. some of the symptoms were apparently shown by two lads of seventeen and thirteen sent to the Salpêtrière from Moscow; but it is unfortunately not wholly clear what symptoms were occurring prior to their arrival in Paris (Charcot, 1889, Lecture VI).

45 See especially Richer (1885). Charcot's views are to be found in Charcot (1877), especially Lectures IX–XIII, and Charcot (1889), Lectures III, VI–VIII, XVII, XVIII–XXVI. The French originals are in Charcot (1886) and Charcot (1887). A useful summary article is Charcot and Marie (1892). A later work is Gilles de La Tourette (1891–5).

46 These are brilliantly captured in Richer's line drawings of hysterical patients.

47 I.e. resting on head and heels with the back arched.

48 Cf. Charcot and Richer (1885), and Richer (1885), pp. 195–201, 914–956. The way in which Charcot and his School brought such phenomena within the domain of psycho-pathology did not endear them to the church. Cf. Goldstein (1982).

49 Richer (1885),pp. 253–323.

50 Charcot and Marie (1892), p. 638.

51 See Charcot (1889), Lectures VII, VIII, XVIII–XXIII. Of course, he had recognized such factors from the beginning. E.g. in 1877 he ascribed the prolonged hysteria of one of his classic cases, a woman of forty-eight named Ler – , to her having been frightened by a mad

dog at the age of eleven, and to her having been frightened by the sight of a corpse and by being robbed at the age of sixteen. See Charcot (1877), p. 279. Despite this he accords only a limited value to psychological methods of treatment. He advocates exercize, good hygiene, firm handling, and above all separation from over-solicitous parents. See Charcot (1889), Lecture XVII.

52 Guillain (1959), p. 167.
53 Charcot (1890b), p. 297n.
54 Benedikt (1894), pp. 21–23.
55 Charcot's papers on metallotherapy are in Charcot (1890b). The best short account of metalloscopy and metallotherapy is that by Pierre Janet, who took a special interest in the matter. See Janet (1925), Vol. 2, pp. 788–857.
56 Burq (1882).
57 Charcot, Dumontpallier and Luys (1879).
58 Charcot's articles on hypnotism, and those of Charcot and Richer, are to be found in Charcot (1890b). Several of the articles therein reprinted originally appeared in periodicals in instalments. See the references in Dessoir (1888).
59 On this see Figuier (1881), p. 412.
60 Pierre Janet (1925), Vol. 1, pp. 170–171.
61 See especially pp. 505–795.
62 A very useful article summarizing Charcot's mature views is Charcot and Gilles de La Tourette (1892).
63 E.g. J. Delboeuf (1886b), p. 8 (despite his strong reservations about the goings-on at the Salpêtrière).
64 Babinski (1891), pp. 14–17.
65 See especially Charcot (1889), Lectures XVIII-XXIV.
66 Charcot and Marie (1892), p. 633.
67 Babinski (1891), p. 17.
68 Cf. the opinions of Léon Daudet and the Goncourts, Ellenberger (1970), p. 92.
69 The reprints of rare witchcraft books, edited by D.M. Bourneville as the *Bibliothèque diabolique* to illustrate hysteria and kindred phenomena in earlier times, are said to have come from volumes in Charcot's collection.
70 I construct this more favourable picture of his character from Guillain and Owen, and also from Freud (1924), a particularly interesting account.
71 Janet (1925), Vol. 1, pp. 186–192.
72 Dingwall (1967), p. 257, refers to Puyfontaine as the Comte.
73 Bourneville and Regnard (1877–80).
74 In later life Babinski, who had repudiated Charcot's doctrines, questioned some of the old Salpêtrière subjects, but could not obtain admissions of fraud (Guillain, 1959, p. 171). Dingwall (1967. pp. 256–257) cites a confession of fraud by one of Charcot's star subjects, but his source does not seem to be totally trustworthy.
75 Delboeuf (1886b), p. 31.
76 Ellenberger (1970), p. 98.

CHAPTER 16

The Nancy school 1882–1892

Though the mesmeric and hypnotic movements had their share of charlatans and opportunists, both were fortunate in the moral quality of certain of their most prominent advocates. Thus, the honesty, philanthropy, and resolution in the face of difficulty and disbelief exhibited by the Marquis de Puységur, the aristocrat and great landowner who more than anyone set the mesmeric movement on its future path, are worthy of the highest praise. And the same praise must be bestowed for similar reasons upon the country doctor of peasant stock who was not the founder exactly, but the cause of the founding, of the hypnotic movement as we know it today. I refer of course to A. A. Liébeault (1823–1904).[1]

Liébeault qualified in medicine at Strasbourg in 1850, and spent most of the remainder of his life in practice in the vicinity of Nancy. In his student days, he had acquired an intense interest in animal magnetism, and wished to practice it; but the French peasants who made up the bulk of his patients were not receptive. He therefore worked tirelessly to make himself financially independent. By about 1860 he had done so, and felt able to devote himself to his chosen area of investigation. To procure subjects he "profited from the legendary parsimonious tendencies of the French peasant",[2] and offered magnetic treatment at no charge.

After a good deal of experimentation with different methods of inducing artificial sleep,[3] he adopted what he termed a "mixed method". Making his subjects look him in the eye for a minute or two, he would emphatically tell them to go to sleep. If their eyes did not close, he would push down their eyelids, meanwhile informing them over and over again that they needed sleep, that their eyelids felt heavy, that they were sleepy, that their senses were becoming dulled, and so on.[4] Later accounts reveal that during the phase of eye contact he would usually stand with his right hand on the patient's forehead.

Once his patient was asleep, or at any rate somnolent or in the state which Liébeault calls "spell-bound" (*charme*), he would proceed with "affirmations" of amelioration, symptom remission, progress towards recovery, etc. When there was disease or pain in a part, it might be stroked or rubbed to accompany the

Figure 8. A. A. Liébeault in his clinic.

suggestion.[5] He seems to have enjoyed at least as much success as he would have done by use of the drugs then in vogue. In 1864, he partially retired from practice to reflect upon and write up his materials. The resultant book, *Du sommeil et des états analogues* (1864), made no great impact on the medical or any other world. Yet it is, in many ways, a remarkable work. Its 534 pages present not just case histories in the therapeutic applications of artificial sleep, but a whole psychology, indeed philosophy, of man. One wonders how an exceptionally busy country doctor could have found time for the reading, and the sustained thought, that went into it.

It is in three parts. The first presents Liébeault's theoretical framework; the second applies this framework in the explanation of what might be called mental epidemics (table-turning, Spiritism, possession, apparitions of religious figures, etc.); the third has to do with the mental causes and mental cure of maladies.

With regard to his cures, Liébeault's claims are exceedingly modest. He warns readers of his inexperience, the shortage of time, the treatments left unfinished, the cases in which he administered both suggestions and medicines, the limited range of kinds of diseases he has tackled, the preponderance amongst them of nervous ailments.[6] This last point is undoubtedly true, though from later accounts of his clinic it is apparent that he was prepared to tackle almost any sort of illness.[7] Unfortunately, his case reports are very brief, and he gives almost no details of the actual suggestions.

Liébeault's chief concern is with the cure of diseases. But it is also apparent (though he gives us no details) that he had experimented widely in the production of the more exotic kinds of psychological phenomena favoured by electrobiologists and other magnetic demonstrators. He tells us, for instance, that by simple verbal affirmation he has caused very impressionable waking subjects to see the Virgin Mary, and the ghosts of deceased persons. Another of his somnambules, profiting from the faculty which he had acquired of evoking spectres, occasionally summoned to his bed the simulacrum of the woman that most attracted him, and demonstrated his passion for her in a highly satisfactory manner.[8]

Despite the failure of his book, Liébeault continued his work patiently. After he was "discovered" by Bernheim in 1882, he became quite quickly famous, and was adopted as a kind of father figure to the whole hypnotic movement. That an obscure country doctor should for so long have followed his own star, unconcerned with fame and fortune, and should towards the end of his life have found such recognition, does great credit both to the man himself, and to those distinguished persons who took pains to accord him the praise which they felt was his due. His book was republished in two volumes (1889 and 1891).[9] When in May 1891 a banquet was given in his honour at Nancy, *savants* from thirteen countries attended.[10]

Delboeuf, who visited Liébeault's clinic in 1888, describes him as follows:

> ...a little man, of brisk carriage, with a forehead deeply grooved   by horizontal lines, crossed by others which fan out from the root of his nose. The brown complexion of a countryman; a brilliant and lively eye; speech resonant and hurried; an open caste of   countenance, a mixture of seriousness and simplicity, of authority and gentleness; the gaiety of a child; something of the priest.[11]

A. W. van Renterghem, a Dutch psychiatrist, attended a session at Liébeault's clinic in 1887. It lasted from 7.00 a.m. to 12.00 noon, and 26 patients were treated. Most, but not all, suffered from functional nervous troubles. In the *salle des consultations* (which, according to a British visitor, C. Lloyd Tuckey, was a sort of bungalow, situated in a garden[12]) there was a constant coming and going of patients, often accompanied by their friends and relatives, who talked freely and sometimes loudly amongst themselves. Each new patient was treated in front of the others. Liébeault would begin with routine sleep-suggestions and tests of the patient's responsiveness to suggestion (arm rigidity, rotation of the hands round each other). From these he would proceed to suggestions designed to benefit the health in general, and then to ones devoted to the patient's specific ailment.[13] As an example of the latter, here is how van Renterghem describes Liébeault's treatment of a little boy of ten suffering from enuresis nocturna:[14]

> ...L. said to him. "You will go...on sleeping well at night, but you will no longer sleep that very heavy sleep! You will wake up of your own accord when the need makes itself felt!" Then, placing his hand at the level of the boy's pelvis, Liébeault continued: "Feel this heaviness, you won't be able to

.

piss now. Try. Just make an effort. You can't, impossible!" I saw the lad try
as hard as he could. "Ah, you didn't manage it! That's how it will always
be from now on while you are asleep. You will feel as though a stopper is
closing the orifice. You won't be able to piss until you have got up and left
your bed, wide awake." The doctor repeated this theme with variations
several times more.[15]

Some of Liébeault's subjects apparently passed into a state of artificial
somnambulism. Even where subjects were not asleep, but merely somnolent or
"spell-bound", Liébeault would still make his therapeutic suggestions, apparently with success. Indeed, he would try them on persons in the waking
state.[16] Delboeuf felt that some of Liébeault's subjects, like some of Charcot's,
were merely compliant to the good doctor's wishes, and remarked, "...this is
inevitable. M. Liébeault's burning faith in the excellence of hypnotism and of his
method, a faith which he transmits to the simple people who are his ordinary
patients, is bound to develop among them into unconscious fraud."[17]

To account for the phenomena of hypnotism and for hypnotic cures Liébeault
develops an extensive and systematically worked-out theoretical framework. His
approach is through the concept of attention, which he expounds at the start of
his book. His ideas on this subject are peculiar. Attention is to be identified with
"nervous force", which proceeds from the brain, and diverges into two broad
currents. One of these flows to the cerebro-spinal division of the nervous system,
and then yields conscious awareness of sense-impressions, the laying down of
memories, etc. The other flows into the nutritive (autonomic) branch of the
nervous system, and is not usually accompanied by consciousness,[18] though it
can under some circumstances reciprocally influence consciousness, through
the intermediary of the "grand sympathetic nerve". The flow of attention can be
influenced by a stimulus or by thought or by emotional disturbance, in such a
way that it will "move upon a faculty of the brain or one of the sense organs"
at the expense of others, or home in upon the nutritive functions. The nervous
functions concerned will be correspondingly enhanced or depressed. It is easy to
see how various functional nervous disorders may be understood in these terms.
An excess of attention in certain organs or parts of the nervous system may lead,
for instance, to epilepsy, idées fixes, neuralgia or digestive disturbances; a
shortfall to paralyses and all kinds of nervous debilities. Since this nervous force,
or attention, emanates from the highest nervous centres, which are those upon
which hypnotism acts in the first instance, cures or ameliorations may be
effected by redirecting the patient's nervous force, or attention, through hypnotic
suggestion.

There is one obvious snag in all this. The aim of most magnetic and hypnotic
procedures is to produce sleep (Liébeault believes "artificial sleep" is identical
with natural sleep[19]), and most people regard sleep as a state of greatly
diminished capacity for attention. Liébeault replies that in sleep, attention,

though withdrawn from the sense organs, is concentrated on a memory-idea, usually in the first instance the idea of sleep. Now if the subject has been hypnotized, and while falling asleep has retained in his mind the idea of the hypnotizer, this idea too may be carried over into the sleep state, and the subject may then put his senses and his accumulated attention at its disposal.[20] The subject is now in rapport with his hypnotizer, and the latter, by suggesting new ideas to the former's pent-up attention, can alter it and turn it towards any sense organ or part of the nervous system. The sleeper then becomes an automaton controlled by the hypnotist.[21] Somewhat similar results can be obtained without the induction of sleep by producing a state of inertia of the attention, discovered by the electrobiologists, which Liébeault calls *charme*.[22]

In these sorts of terms, Liébeault can very readily give an account of the principal phenomena of hypnotism. Enhancement of memory, the capacity to hallucinate, heightening of intellectual function, are all due to the concentration of attention in the parts of the brain concerned. Post-hypnotic amnesia is explained by a sudden expansion and attenuation of the attention that prior to awakening had been concentrated on a single idea. The imitative propensities so characteristic of subjects in a state of *charme* come about when a state is reached in which the inactive attention abandons itself to impulses coming from without, giving birth to involuntary ideas which are copies of events taking place in the vicinity. Such ideas, unconstrained by any other influences, tend to realize themselves in action. The effects of suggestion on the sympathetic system (manifested in suggested skin reddening, digestive changes, etc.[23]) Liébeault seems to wish to explain by proposing that there may be unconscious thoughts linked to the functioning of that system. In the state of artificial sleep, these thoughts, and the nervous force at their command, may for reasons which are unclear fall under the influence of conscious attention.[24]

Liébeault's views about "nervous force" are only a whisker away from the doctrines of certain animal magnetists about the mesmeric fluid.[25] The main difference is that Liébeault did not believe the fluid to be directly transmissible from operator to patient, whereas the magnetists did. Even this difference seemed to be in danger of vanishing in the early 1880s when Liébeault managed to convince himself that there might after all be something like the transmission of a fluid.[26] He was led to this view by his own success in curing very young children by simply laying hands on the afflicted region. None of the infants showed any symptom of hypnosis, and several were asleep when treated. He published his findings in a brief pamphlet, *Étude sur le zoomagnétisme* (1883). Bernheim argued against him that an infant observes and understands more than one might think, and can realize that the laying on of hands is done to help him. He is cured by a sort of indirect suggestion.

Liébeault seems not to have accepted these arguments, and in 1887 he began to try, with apparent success, the effect of doses of "magnetized" water on very

young children. Bernheim proposed to him that he should test for a possible
suggestive effect by instituting a series of control experiments with unmagnetized
water. Unmagnetized water proved a remarkably successful remedy and
Liébeault capitulated.

BERNHEIM: THE FOUNDER OF THE SCHOOL

It is highly likely that Liébeault would have lived out the last twenty years of his
life, as he had the first sixty, in modest obscurity, had not Hippolyte Bernheim
(1840–1919),[27] professor in the faculty of medicine at Nancy, and a man of high
reputation as a clinician of the anatomo-pathological school,[28] been induced in
1882 to visit Liébeault's clinic.[29] Bernheim was at first sceptical, but was soon
interested enough to try hypnotism for himself in the wards of Nancy's very
modern hospital, and subsequently in his own private practice.[30] He acquired
enthusiastic supporters in J. Liégeois (1823–1908), professor of law at Nancy,
who inquired into the bearings of hypnotism on crime, and H. E. Beaunis
(1830–1921), holder of the chair in physiology, who undertook a series of
investigations into the psychophysiological aspects of hypnosis.[31] Both were
inspired by Bernheim's ideas, as Bernheim had been by Liébeault's, and it is
these four persons, together perhaps with Bernheim's pupil J. Brullard,[32] who
(though not wholly in agreement on all significant matters) properly speaking
constituted the "School of Nancy". The term is sometimes loosely used to
include individuals who merely visited Nancy and carried Nancy opinions away
with them; and still more loosely, and quite inappropriately, to cover almost
anyone who came to reject the doctrines of the Salpêtrière.

The views of the Nancy School first became known to a wider audience in
1884 through the publication of Bernheim's *De la suggestion dans l'état hypnotique
et dans l'état de veille*. This brief work was ultimately to have a greater influence
than anything which had appeared in the hundred years since Puységur's
Mémoires. Its publication made, however, no great noise, only a quiet, though
audible, detonation, as if a demolition expert had fired a small but strategically
placed charge to cause the slow collapse of an imposing edifice. The unsound
structure that shortly began to collapse was of course Charcot's.

Bernheim begins by describing his methods of hypnotizing, which are clearly
derived from those of Liébeault. He likes to allow new subjects to witness the
hypnotization of others, which dispels mystery, and no doubt undermines
scepticism and resistance. He then tells the patient to look at him and fix his
mind upon the idea of sleep, and proceeds with reiterated suggestions of
tiredness, heaviness of the limbs and eyelids, closing of the eyes, sleep, etc.
Sometimes he adds gestures, passing his hands downward before the subject's
eyes; and he says "sleep" in a commanding tone.[33]

Figure 9. H. Bernheim. (From the Wellcome Institute Library, London.)

If these suggestions do not shortly result in eye closure, he pushes the patient's eyelids gradually down and holds them closed, with suggestions that the lids are stuck together and of sleep. It is, he says, "very seldom that more than three minutes pass before sleep or some degree of hypnotic influence is obtained".[34]

Bernheim has often been represented as an authoritarian hypnotizer on the lines of Faria, and certainly he was prepared to use authority. But it is apparent from what he says, and from the accounts of eyewitnesses, that he would vary his attack from patient to patient or from occasion to occasion in a guileful and persistent manner. His subjects were mostly men, chosen to controvert the belief that the subjects affected are "all weak-nerved, weak-brained, hysterical, or women". Following Liébeault, Bernheim recognizes six degrees (later nine) of hypnotic influence.[35]

The first degree is that of mere somnolence, and the second degree of what Liébeault called *charme*. The third and fourth degrees are characterized by increasing drowsiness and loss of relationship with the outer world, and by the performance of suggested or imposed automatic movements, e.g. rotation of the arms about each other. The fifth and sixth degrees constitute light and deep somnambulism respectively, the chief distinction between them being the amnesia that follows the latter (though the missing memories can always be revived artificially). In these stages we may obtain such "classic" phenomena as anaesthesia, hallucinations, the execution of bizarre actions, rôle-playing, spontaneous dreams, the fulfilment of post-hypnotic suggestions, etc. But, as if to undo all that he has just said, Bernheim adds that "docility to suggestion and the ease with which diverse phenomena are provoked, are not always in proportion to the depth of the hypnotic sleep."[36]

Much of the rest of the pamphlet is devoted to accounts of Bernheim's own successes, and occasional failures, in inducing phenomena of these various kinds. A topic that particularly interested him was that of hypnotic illusions and hallucinations, especially the latter. With deep somnambulic subjects he would often attempt to induce hallucinations post-hypnotically. Sometimes the resultant images were indistinct or obscure or incomplete; sometimes the subject on awakening would simply remember what he was supposed to see or hear without seeing or hearing it; but sometimes the hallucinatory percept would be as clear and vivid as a real one. A deep somnambulic patient was told that on awakening in the hospital ward he would see a big dog in each bed instead of a patient, and was astonished to find himself apparently in a hospital for dogs.[37] Bernheim also experimented a good deal with negative hallucinations, and has been credited with inventing the term.[38]

With certain subjects in a state of deep somnambulism, Bernheim was able to induce whole hallucinatory scenes accompanied by actions; these subjects would as it were act out little dramas, for instance the commission of a crime, or involvement in a battle, or would assume different roles at command (Richet's "objectification of types").

Like Liébeault, Bernheim gives various examples of previously hypnotized subjects who now manifested a responsiveness to similar suggestions even in the waking state.[39] He does not, he emphasizes, have to speak in a deep authoritative voice to obtain these phenomena, or select only docile and obliging, or highly suggestible, patients.[40]

In a penultimate chapter, Bernheim develops his own theoretical interpretation of the phenomena of suggestion. The interpretation is neither crystal clear nor very convincing. Bernheim himself says that it is only a formula; he does not pretend to advance a theory. Although Bernheim and the Nancy School were often referred to as having a "psychological" approach to the problems of hypnotism, as opposed to the "physiological" approach of the Salpêtrière, this distinction is somewhat misleading, except insofar as the Salpêtrière attempted to establish physiological criteria of the "stages of hypnotism." Bernheim's early speculations as to the nature of the hypnotic state are largely an exercise in *a priori* brain physiology. What characterizes the hypnotized subject is a peculiar aptitude for translating ideas (received no matter how) into action, sensation or movement. The translation is brought about so rapidly and so actively that intellectual or voluntary inhibition have no time to act. A suggestible person is one "whose paths of intra-cerebral reflectivity are more distinct and more easily traced, and in whom the condition of consciousness which moderates the reflex automatism, is at the same time weakened."[41] Subjects who have been frequently hypnotized may contract an increase of this ideo-reflexive excitability through frequent activation of the pathways involved.

Like Liébeault, Bernheim, at this stage of his thinking, regarded hypnotic sleep as not essentially different from actual sleep. The chief difference between hypnotic and natural sleep is that in the former, the subject's mind retains the memory of the hypnotizer, who thus acquires the power of influencing his mental activity. Sleep favours the production of suggestive phenomena, by suppressing or weakening the moderating influence of reason and will on the "paths of intra-cerebral reflectivity", but it is not essential to the production of such phenomena.

THE WAR WITH THE SALPÊTRIÈRE

Bernheim's pamphlet makes no frontal assault upon the Salpêtrière, though the implications of what he says, and also of what he does not say, are strongly anti-Salpêtrière. All, or almost all, hypnotic phenomena up to and including somnambulism can be obtained in mentally normal individuals. The stages of *grand hypnotisme* are not to be found at Nancy even among hystericals.

The Nancy School was not alone in failing to confirm the Salpêtrière's claims. For instance, the school of V. A. A. Dumontpallier (1828–1899) of the Pitié Hospital, Paris, later of the Hôtel-Dieu, differed from the Salpêtrière in minor respects.[42] Dumontpallier had become a keen devotee of metallotherapy,[43] and held that the deep stages of hypnosis may be occasioned not just as described by Charcot, but by temperature changes, atmospheric vibrations, metals and magnets.[44] He was the founder, in 1886, of the *Revue de l'hypnotisme*, and in 1891 of the Société d'hypnologie et psychologie. His pupil, P. Magnin (b.1854) gave in 1884 the most systematic exposition of the Pitié's stance.[45] Although he attaches much importance to metallotherapy and related matters, his general position resembles that of the Salpêtrière. Hypnotism is closely related to hysteria and has physiologically definable stages. He recognizes, however, mixed or intermediate stages between the three classical Salpêtrière states, and tells us that he and Dumontpallier have managed to produce all three states by means of just one kind of stimulation, for instance rubbing the top of the head.[46]

Another individual who deviated slightly from the Salpêtrière was A. Pitres (1848–1928), of Bordeaux.[47] Pitres was a former pupil of Charcot, and modelled his teaching methods very much on the latter's, as can be seen from a glance at his two volumes of *Leçons cliniques sur l'hystérie et l'hypnotisme* (1891). His most notable contribution was to embroider upon the proposal, current at the Salpêtrière, that there may be found on the surface of subjects' bodies hysterogenic and hypnogenic points or zones the touching or stimulating of which may provoke attacks of hysteria or hypnosis.[48] He accepts the Salpêtrière view that hypnosis is nearly allied to hysteria, and agrees that the three classical states of hypnotism occur and may be defined physiologically. However, he holds that each has various subdivisions – cataleptoid states, lethargoid states, several sorts of somnambulic state – not to be found in the Salpêtrière's scheme of things.

In deeper conflict with the Salpêtrière were the findings of Dr. P. Brémaud (1846–1905) of Brest.[49] Brémaud set out to interview and experiment with young persons who had been successfully magnetized by the famous travelling demonstrator Donato.[50] Donato practised "fascination", which involved prolonged fixation of a bright object or of the eyes of the operator. Subjects did not pass into a sleep-like state, but after a while would begin to follow the operator, to imitate his movements, etc. They might be "fixed" in a certain spot by a look or gesture, or rendered anaesthetic, made to follow orders, suffer from illusions, see hallucinations and visions, etc., in response to verbal command. Brémaud succeeded in repeating many of these results for himself. After talking extensively to his subjects, he was prepared to reject absolutely the hypothesis that they were not telling the truth about their experiences. And he could find nothing whatever to suggest that they should be diagnosed as hysterics. Now many of these subjects, as well as being susceptible to fascination, could be made by means of the standard manipulations to exhibit the states of catalepsy, lethargy and somnambulism as defined by the Salpêtrière. Brémaud therefore concludes (a) that there are four states of hypnosis, viz. fascination, catalepsy, lethargy and somnambulism, and (b) that all of these may be induced in non-hysterical subjects. This is in fundamental conflict with the Salpêtrière position.

Of course the views and findings of the School of Nancy were in even more fundamental conflict with that position, but it seems to have been some while before the Salpêtrière realized that these small clouds on the horizon heralded a serious storm. The initial Salpêtrière reaction was probably that of Richer, who devotes just three sentences of his 1885 historical review[51] to Liébeault, Bernheim and Liégeois. This no doubt represents the rather lofty attitude of one who is an acolyte in the high temples of science towards those who serve at some distant provincial shrine. But one can see Richer's point. The assertions of the Salpêtrière School are backed with detailed, sometimes minute-by-minute, case reports, with the minutiae of clinical observations, with drawings and photographs and anatomical diagrams, with the tracings from pneumographs, myographs and sphygmographs, with control tracings from unhypnotized subjects. Compared with all this the claims and case-reports of Liébeault and Bernheim seem exiguous, almost casual.

The first important salvoes fired against Nancy from a Salpêtrière position came not from a member of the Salpêtrière School, but from a philosopher who had some links with that school, namely Paul Janet (1823–1899), a professor at the Sorbonne, and uncle of Pierre Janet, the psychologist. In a series of articles (starting July 1884)[52] on the psychology of suggestion in the hypnotic state, he successively attacked Liébeault, Bernheim, Brémaud, and Liégeois.[53] Liébeault's attention theory of hypnotic phenomena has confused theory with facts, and his therapeutic successes may well depend as much upon the imagination of the doctor as upon that of the patient. Bernheim's claim that his subjects are not hysterics is not supported by his own case reports, which are either imprecise and insufficiently detailed, or else relate to disorders which, where not hysterical, could be supposed likely to have occasioned neuropathic troubles. Liégeois presents his experiments on the production of criminal acts through hypnotic suggestion "without any consideration for the rigorous demands of scientific method and medical observation". Throughout Janet's articles the magisterial findings of Charcot and his School are taken as the standard against which all else is to be judged.

Bernheim replied to Paul Janet later the same year in a little pamphlet[54] which constitutes Nancy's definitive declaration of war on the Salpêtrière. Bernheim professes his strong indebtedness to Charcot, but adds that the reason why he has not taken the three stages as his point of departure is that he has not been able to confirm their existence by his own observation. He has only been able to produce them by suggestion. As for the allegation that his subjects are covertly neuropathic or hysterical, this is nonsense, a point which he proceeds to hammer home with detailed examples. In his experience the best subjects are common people, soldiers, artisans, those accustomed to passive obedience, or with docile dispositions.

After this exchange, Charcot's pupils and supporters were drawn into the fray,[55] though the *maître* himself always remained aloof. Their point of view was especially championed by Alfred Binet (1857–1911), a psychologist who had worked at the Salpêtrière. The climax came with the publication in 1886 of Bernheim's *De la suggestion et de ses applications à la thérapeutique*, and the subsequent passages at arms between Bernheim and Binet.

Bernheim's book consists of two parts. The first is in essence a reprint of *De la suggestion dans l'état hypnotique* with additional material, much of it a criticism of the School of Charcot. The second part concerns the applications of suggestion to therapeutics. This topic is not our immediate concern, but it will be convenient to begin by devoting a few words to it.

Bernheim points out that hypnotic suggestion can cause both inhibitory phenomena (anaesthesias, paralyses) and excitatory or dynamogenic ones (hallucinations, actions, contractures).[56] If simple hypnotic suggestion can produce inhibitory and dynamogenic effects in healthy persons, why should not such effects be turned against pathological symptoms? Nearly all of Bernheim's hypnotherapeutic endeavours have the straightforward aim of directly neutralizing the symptoms, whether these are positive, as in choreas or contractures, or negative, as in anaesthesias or paralyses. He was prepared to tackle a wide variety of ailments including not just hysteria, but organic and functional diseases of the nervous system, gastro-intestinal problems, rheumatism and neuralgia and disorders of menstruation. His success rate over the 105 cases he cites is quite high, and he claims that they have not been specially selected. Not a few of them appear to have been refractory to previous treatments. Unfortunately, he gives in his otherwise quite full case reports few details of the suggestions given.

To return now to the conflict between the two schools in 1886–7. This may be taken under three headings: first, Bernheim's attack upon the Salpêtrière's central claims – that hypnosis is an abnormal state allied to hysteria, and that it has three physiologically distinguishable states or stages; second, the controversy between Bernheim on the one hand and Binet and Féré on the other concerning certain of the latter's claimed experimental findings; and third, the extravagant claims of certain associates of the Salpêtrière concerning various highly implausible phenomena which might broadly be described as the offspring of metallotherapy. A fourth area of conflict – that concerning the possibility of crimes committed on, or by, hypnotized persons – will be taken up in a later chapter. I shall now look at the first three areas in greater detail.

1. *Hypnotism and hysteria; the three stages of hypnotism*

Bernheim's view on these closely linked issues was still that which he had expressed in 1884. The stages of *grand hypnotisme*, or rather something resembling them, can be obtained in normal and in hysterical subjects alike; but only as a result of direct or implicit suggestion. "Only once", says Bernheim, "did I see a subject who exhibited perfectly the three periods of lethargy, catalepsy and somnambulism. It was a young girl who had been at the Salpêtrière for three years ... the case was no longer one of natural hypnotism, but a product of false training, a true suggestive hypnotic neurosis."[57] And even if there is such a thing as *grand hypnotisme*, only a dozen cases have, by the

Salpêtrière's own admission, been seen there. Should these cases, as opposed to thousands of others, serve as the basis for our theoretical conceptions of hypnotism?

Binet's answers to Bernheim may most conveniently be gleaned from his review of Bernheim's 1886 book.[58] Some parts of this review are so full of evasions and non-sequiturs that they make embarrassing reading even after the lapse of a hundred years. Binet does not even acknowledge the main problem confronting the Charcot School – the fact that the three states of *grand hypnotisme*, with their unmistakable physiological concomitants, could be found almost nowhere beyond the confines of the Salpêtrière. Even such allies as Dumontpallier, Magnin and Pitres emphasized, as we have seen, how many mixed, partial and intermediate phases there are, a point brought out by Pierre Janet in a paper of 1886.[59] What independent confirmation there was for the occurrence of the three phases of *grand hypnotisme* came mainly[60] from Italy, where Charcot had exercised considerable influence in the early 1880s.[61] In 1881 and 1882, two Italian alienists, A. Tamburini (1848–1919) and G. Seppilli, both of Modena, published two physiological investigations, supporting certain Salpêtrière claims, of a patient showing the three classic phases.[62] In 1885, F. Vizioli of Naples described what he believed to be a case of spontaneous hypnotism presenting the Salpêtrière stages,[63] and cites a comparable case from V. J. Drosdow of St. Petersburg. It was not exactly a rich haul, and Binet was no doubt wise not to lay emphasis upon it.

Binet in fact offers only one comment of any note on Bernheim's criticisms of the supposed stages. It is that if everything in the hypnotic state is due to suggestion, hypnotism itself has only suggested characteristics; there is nothing interesting about it except the very fact of suggestibility. Furthermore, it will always be impossible to prove that the subject has not been simulating. It was, he felt, a merit of Beaunis to have looked for physiological indices of *petit hypnotisme*, for these constitute its necessary criteria.[64] Binet here raises an exceedingly important issue, one that is still with us. Can we, in any given instance, be sure that we are dealing with a "hypnotized" subject, when there is no index of his supposed state other than the suggestibility which led us to postulate it? Much of the material under our next heading arises from the attempts of Binet and his frequent collaborator C. Féré to find objective indices of the hypnotic state which cannot be set down to suggestion or simulation.

2. *The experimental findings of Binet and Féré*

It will be convenient to consider these findings, and Bernheim's ripostes, under two subheadings, viz. experiments on perceptual processes, and experiments on "transfer".

1. *Experiments on perceptual processes.*[65] Binet and Féré conducted a number

of experiments of the following kind. A hypnotized subject would be presented with a blank white card and made to see an hallucinatory image, e.g. a portrait, on it. The card would then be placed in a pack of similar cards, and the pack shuffled. The subject presented with the cards, would still see the portrait on the correct card and on that one only. Binet and Féré argued that the hallucinatory portrait must have become associated with reference points (*points de repère*) on the surface of the card. If the card with its supposed reference points were then magnified, made more distant, doubled, inverted, etc., etc., by optical appara-tuses, the hallucinatory pictures changed correspondingly. Binet and Féré seem to hold that since the subjects could not have known what to expect, imagination and ongoing suggestions played no role in these phenomena. Once established, the link between the reference points and the hallucinations operates in a quasi-mechanical fashion.

Binet and Féré also investigated the chromatic properties of hypnotic hallucinations. They claim that monochromatic hypnotic hallucinations, so to speak projected onto appropriate cards, will occasion contrast colours and after-images of the correct hues. Such experiments, performed upon ignorant hysterics, are a peremptory reply to believers in simulation. Binet and Féré also claim to have demonstrated correct colour mixture with hallucinatory colours. A simple optical device superimposed the images of white cards to which hallucinatory colours had been assigned. The resultant mix could be compared with the mix yielded by the corresponding "real" colours.[66] Since the subjects were ignorant of the laws of colour mixture, we may conclude that the mixture of hallucinatory colours to give a correct resultant goes on continually and automatically and is not due to imagination or simulation.

Bernheim attacks the experiments on doubling of hallucinations and on colour mixture by means of a long series of experiments of his own which purport to demonstrate that all the findings in these areas are due to "suggestion".[67] But his attack misfires almost completely. His experiments are crucially different from those of Binet and Féré. Binet and Féré are not claiming that looking through optical apparatuses doubles (and so forth) the hallucin-ations or causes the colours of the hallucinations to blend; they are saying that the apparatuses have various optical effects on the reference points, effects which then automatically set in train changes in the hallucinations to which the reference points are linked. Most of Bernheim's experiments are such as to eliminate the reference points altogether. To experiment on the doubling, inversion, etc., of visual hallucinations, he causes his subjects to see hal-lucinatory lights so to speak in mid-air, and to experiment on colour mixture he uses rapidly rotating discs, segments of which have previously been assigned different hallucinatory colours. The fact that in these situations subjects' reports were strongly influenced by suggestion by no means proves that they must have been so influenced in the situations employed by Binet and Féré.

2. *Experiments on hypnotic transfer.* It had been noted by Burq, and by Charcot, Dumontpallier and Luys (the members of the commission on metallotherapy appointed by the Society of Biology in 1877), that an hysterical symptom, e.g. an anaesthesia, removed from one side of the body by the application of a metallic plaque, tended to reappear on the other. This phenomenon could be produced by all kinds of "aesthesiogenic" substances, and particularly by magnets.[68] It was a natural step to attempt likewise to transfer effects produced by hypnotism. Charcot and Richer at the Salpêtrière and Dumontpallier at the Pitié conducted such experiments with apparent success, and the "classic" study was that of Binet and Féré (1885).[69]

The phenomenon of transfer naturally allied itself with that of hemicerebral hypnotism, which had evolved from hemicerebral mesmerism,[70] and was first investigated at the Salpêtrière in 1878.[71] It was also studied by, for instance, Ladame[72] in Geneva, and (especially) Dumontpallier and his pupil Bérillon[73] at the Pitié. Hemilethargy and hemicatalepsy could be induced in cases of *grand hypnotisme* by the usual procedures of hypnotization applied only to one eye; hemisomnambulism by touching or rubbing one side of the top of the head. Sometimes the bodily effects were ipsilateral to the eye manipulations or friction of the scalp, sometimes contralateral.[74] Some subjects could be made, for example, somnambulic on one side of the body and lethargic on the other. The transfer of hemicerebral hypnosis from one half to the other of the body could be effected by the approach of a magnet to the hypnotized side.

Binet and Féré's own experiments on the production of transfer with magnets in hypnotized hysterics were conducted, they assure us, with careful precautions to prevent subjects from guessing anticipated effects, and with deliberate attempts to deceive them.[75] The genuineness of the phenomenon was none the less confirmed.

Binet and Féré laid especial stress upon two kinds of transfer phenomenon to the establishment of which they clearly felt they had themselves made significant contributions. The first was the transfer through the approach of a magnet to one side of the body or head of unilateral phenomena brought about by verbal suggestion – unilateral actions, paralyses, etc., and above all unilateral hallucinations, that is hallucinations that can be seen only with one or other eye.[76] The second was the "pain of transfer" which subjects felt shooting through their brains at the moment of transfer, in locations which Binet and Féré supposed anatomically appropriate for the effect (hallucination of a given sense, muscular contraction) being transferred.[77]

Bernheim's attack on Binet and Féré's transfer experiments[78] is direct and straightforward. He maintains, on the basis of his own experiments conducted with somnambulic subjects in the presence of several of his colleagues, that the transfer of hypnotic catalepsies, contractures, paralyses, etc., through the application of a magnet (or anything else) never occurs without suggestion, but may be readily induced by even a slight hint or suggestion, or by witnessing the phenomenon in others. The same holds for the pain of transfer. Somnambules are very quick in guessing the hypnotist's intentions. In fact it can easily be shown that subjects hear and note everything in all phases of hypnotism, including lethargy. It is not safe to utter a single word that may give them a clue.

Binet's reply[79] is certainly feeble. Bernheim's conclusion, he says, amounts to a rejection of the great fact of transfer as demonstrated by the experiments of Charcot, Dumontpallier, etc. To deny the action of the magnet on the organism is to deny the action of electricity. Would Bernheim go as far as this? Bernheim has not experimented with victims of *grand hypnotisme*. His experiments can therefore not be compared with their own. As for Bernheim's claim that subjects in the lethargic and cataleptic phases are well aware of what is going on around them, according to Bernheim these persons are merely the victims of suggestion; but in that case the all-powerful suggestion in which Bernheim believes would have made them blind and deaf anyway.

3. *Associated extravagances*

The Salpêtrière was responsible, as we have seen, for starting a number of strange hares, many of them the offspring of metallotherapy. When pursued by other, and even less critical, investigators, some of these hares became positively crazy, and begat still crazier offspring of their own. The pursuers were, for the most part, persons who, though not members of the Salpêtrière School, shared some of its assumptions, and the pursuit, which was noisy and prolonged, did little to enhance the Salpêtrière's standing. The prime pursuer was Dr. J. B. Luys (1828–1897), a neurologist, of the Charité Hospital in Paris.[80]

From all accounts Luys was the soul of courtesy, sincerity and kindness. Along with, and perhaps arising from, these amiable qualities went an inexhaustible and incorrigible credulity. At the Charité he maintained a corps of semi-professional subjects. From the numerous photographs of these little baggages published in his books, it is hard to believe that anyone, even poor Luys, could have been taken in by them for a single moment.[81] When I add that his *chef de clinique*, and frequent collaborator, was Gérard Encausse (1865–1916), better known in occult circles as "Papus",[82] perhaps all has been said that needs to be said.

Though not wholly in accord with the Salpêtrière, Luys continued to believe in Charcot's three stages and in hypnogenic zones at a period when most people had abandoned them. And experiments on transfer reached heights of credulity not previously attained. Yet even here the Salpêtrière led the way. In 1886, Babinski began to publish accounts of a new development in the phenomena of transfer,[83] the transfer of some hysterical symptom (paralyses, contractures,

mutism) from one somnambulic subject to another. The subjects would be
separated by a screen and the transfer induced by placing the second subject in
contact with a magnet.[84]

Luys pushed absurdity a stage further. He adopted this kind of transfer as a
form of therapy. He would transfer the symptoms of a waking patient to a
hypnotized subject in a state of lethargy. This could be done by having the two
hold hands, and passing the north pole of a large magnet down the arms of the
patient and onto those of the subject. The subject would assume not just the
symptoms but (upon being put into the somnambulic stage) the personality of
the patient. She would then be woken, and disease and personality would alike
disappear. With any luck, the personality would return to its original owner,
whilst the symptoms would vanish into limbo, or at any rate diminish in
intensity.[85] Later, Luys achieved similar effects by having the patient wear a
circular magnetic "crown", which was then removed to the head of the
hypnotized subject.[86]

Best known of all Luys' extravagances were his experiments using drugs
sealed in test-tubes. These were a development of previous studies with
aesthesiogens, and again they did not originate with Luys. The pioneers were
M.J.H. Bourru (1840–1914) and P. Burot, professors in the medical school
at Rochefort. About the spring of 1885, whilst investigating the effects of
aesthesiogens on two hysterical subjects,[87] they tried the effects of drugs held a
few inches behind the subjects' heads. Solid drugs were wrapped in paper, liquid
ones were contained in bottles. After a period of general disturbance, symptoms
would become apparent which corresponded to the normal action of the drug
concerned. The experiments were repeated with other subjects, and attempts
were made to rule out the transmission of information by ordinary means.[88]

Luys took up this line of work using hypnotized hysterical subjects and drugs
in sealed glass tubes.[89] Some idea of his standards of experimentation may be
gained from the following account by Dr. G.C. Kingsbury, of Blackpool:

> The subject selected was one of the regular habituées of the laboratory. As
> soon as the lethargy was induced, a tube containing brandy was slipped in
> at the neck of the girl's dress, and Dr. Luys, who stood at one side of the
> subject, while I was at the other, remarked to me in a loud whisper, "*Eau de
> vie.*" In a few seconds, the subject remarked that she felt drunk, began
> singing excitedly, and finally left her seat, staggered, and fell on the
> floor.[90]

Luys reported his results to the Society of Biology and the Academy of Medicine,
with the result that a commission of the Academy of Medicine was appointed to
look into the phenomena. Its report[91] was entirely negative, as were the attempts
of J. Voisin and Bernheim to repeat the experiments.[92] Luys, of course, disputed
these findings.

Into Luys' further extravagances[93] there is no point in entering. Bernheim and his colleagues did not need to engage in extensive experimentation to demolish them, and Charcot might well have reflected in the later 1880s that with members as uncritical as Babinski, and fellow-travellers as credulous as Luys and Dumontpallier, his School needed no enemies. Yet – and here is a real puzzle – we find Charcot himself, a man of great ability, well-read, enlightened, and a dedicated teacher of medical science, prepared to demonstrate, without any serious critical precautions, and before sceptical foreign visitors, the direct transfer of symptoms from one hypnotized subject to another by the sole agency of an ordinary magnet.[94]

SPREAD OF THE NANCY VIEWS

From 1887 onwards, it became increasingly apparent that the Salpêtrière was rapidly losing ground in its disputes with Nancy. In the Salpêtrière, perhaps, the fact was not openly acknowledged, but Binet and Féré quietly withdrew from the struggle,[95] leaving the defence mainly in the hands of Babinski and Gilles de La Tourette. No effective answer was found to Bernheim's arguments, and the somewhat numerous foreign and other visitors to the Salpêtrière could hardly have been impressed by the demonstrations of transfer and other absurdities which we have just been considering. Foreign visitors interested in hypnotism began to show a tendency to visit Nancy as well as – and before long instead of – the Salpêtrière. Among them were, from Belgium, J. Delboeuf; from Germany, A. Moll; from Switzerland, A. Forel; from Holland, A. W. Van Renterghem; from England, A. T. Myers, F. W. H. Myers, Edmund Gurney, C. Lloyd Tuckey, H. Rolleston, H. Wingfield and J. Milne Bramwell; from Austria, S. Freud (who translated Bernheim's 1886 and 1891 books into German[96]); from the United States, Hamilton Osgood, J. M. Baldwin and Morton Prince; and from Sweden, O. Wetterstrand.

   Delboeuf was much impressed both with the spaciousness and cheerfulness of the hospital at Nancy, and with Bernheim himself. In appearance, says Delboeuf, Bernheim was "small, somewhat plump, a fine regular face, nose slightly aquiline, compact grey hair, bluish eyes, lively and gentle, moustaches and beard pointed in military fashion, voice caressing, piercing and penetrating."[97] "A very genial gentleman," says Hamilton Osgood, "a very able clinician, with an unusual degree of ability as a lecturer by the bedside and in the class-room ..."[98]

   Bernheim seems to have been a very effective hypnotist, and if baulked would vary his line of attack again and again. Here is how, according to Delboeuf, he dealt with a patient, a "solid day-labourer", who was suffering acutely from erysipelas. This man had had no prior experience of hypnotism or animal magnetism.

> "Now, my friend, I am going to relieve you of all your pains; I am going to send you to sleep...Here, I put my finger on this spot on your forehead. Do you feel the sleep coming?" "I don't know." "Oh! but you do. Already you can't keep your eyes open (M. Bernheim closes the man's eyes). Heaviness overwhelms all your limbs; you can't move your arm (he lifts up the patient's arm); you can't lower it. And if I make it turn (he sets it moving), you can't stop it. Even better – the more you try to stop it, the faster it will go (this happened). Let's see, where does it hurt?" "My head." "The pain in your head is going to go. It's away! It's gone! You have no more pain!" "No." "You are asleep?" "I don't think so." "You are asleep! You will not remember anything when you wake up. You don't feel anything (he is pricked). When you wake up you will drink half a glass of water."
> All this took scarcely the time that I need to write it. The man was deeply asleep... He was visibly relieved of his pains [and could remember nothing of the events of the hypnosis until told to do so].[99]

Bernheim was prepared to treat a very wide range of diseases by hypnotism. A. T. Myers, an English physician who visited Bernheim in 1885, saw him in the course of several mornings tackle hypnotically, and with apparent benefit, cases of chorea, paralysis agitans, heart and kidney diseases, rheumatoid arthritis, sciatica and pneumonia. In the latter case, hypnosis was used simply to combat pain and procure sleep.[100]

The "hypnotherapeutic" or "psychotherapeutic"[101] movement which now, under the influence of Nancy, began to spread across Europe, conformed at first very much to Bernheim's ideas and practices, and was accordingly by no means a movement exclusively for the therapy of ailments that would nowadays be thought of as psychological. Different practitioners had different views on the value of hypnotherapy in acute or organic diseases. As the century neared its end, hypnotherapy was increasingly thought of by many as appropriate only for functional nervous disorders, but there always remained more venturesome individuals who would draw an hypnotic bow at almost any disease.

It was outside France that Nancy doctrines spread in the clearest and most decisive way. In France itself, their progress, though rapid, is not always easy to chart, mainly because the great respect in which Charcot was widely held made many people reluctant to reject his doctrines completely. There were, for instance, some individuals of an older generation, with a long-standing interest in magnetism or braidism or somnambulism, who took up hypnotism again in the 1880s, and, though to a greater or lesser extent influenced by Nancy views, continued to believe that Charcot's three "stages" of hypnotism were not always artifactual. Such were C. M. E. E. Azam of Bordeaux, whom we have already met as a pioneer of braidism; E. Mesnet (1825–1898) of the Hôtel-Dieu, well-known for his studies of natural somnambulism;[102] and A. Voisin (1825–1898), physician to the Salpêtrière, who was especially known for successfully hypnotizing maniacal and frenzied patients. Durand de Gros had ploughed his own furrow through this difficult field from even earlier times, and continued to do so, a venerated figure, until the end of the century. Some members of the younger generation – for instance Professor J. Grasset (1849–1918) of Montpellier, and Professor G. Ballet (1853–1916) of Paris – though in the main followers of Nancy, maintained at this period a qualified belief in Charcot's "phases".[103] Some who had been followers of or fellow-travellers with the Salpêtrière School – for instance Binet and Dumontpallier – moved quietly nearer to Nancy on many, though not all, issues.[104] Some who professed strong adherence to the Nancy School, such as J. Fontan and C. Ségard of Toulon, made assertions which Bernheim would never have countenanced (in the case of Fontan and Ségard these had to do with the hypnotic production of organic changes).[105]

Still, Nancy doctrines, whether in a pure or a somewhat adulterated form, steadily gained ground, and attracted support from persons as differently situated as J. Dejérine (1849–1917), of the Bicêtre, Genevan by birth, who was to become one of the most famous neurologists of the age,[106] and Dr. Bourdon of Méru (Oise) who describes himself as a "humble country doctor".[107] The encroachment of Nancy is not altogether obvious from the popular and semi-popular surveys of the time,[108] which still tended to devote more space to the illustrious Charcot and his School than to Liébeault and Bernheim. But if we turn to the more professional sources, such as the *Revue de l'hypnotisme* (founded by Dumontpallier in 1886, and edited by his pupil Bérillon), that encroachment is unmistakable. In 1889, came the First International Congress of Hypnotism, held in Paris under the Presidency of Dumontpallier,[109] and one has to search the printed papers and reports of discussions somewhat carefully before finding traces of the dispute between Nancy and the Salpêtrière.[110] In the same year, also in Paris, came the International Congress of Physiological Psychology,[111] and the story was repeated.

A principal agent in spreading information about the Nancy School to French-speaking Belgium was J. Delboeuf (1831–1896), of the University of Liège. Delboeuf had achieved some celebrity for his work in the field of psychophysics; but to judge from his more popular writings he could certainly have made his living as a journalist. He is an incisive writer, sometimes perceptive, often combative. He was involved in a number of controversies, the most important of which was with Liégeois and others who believed that hypnotism might be turned to criminal purposes.[112] Other figures important in Belgian hypnotism were also laymen. These incuded A. Bonjean, a lawyer of Verviers, whose book *L'hypnotisme* (1890) is much influenced by the writers of the Nancy School, and H. Nizet, a journalist of Brussels, whose *L'hypnotisme: étude critique* ([1892]) also in general follows the Nancy position. Among Belgian medical men interested in hypnotism were Dr. F. Semal of Mons, whose pamphlet *De l'utilité et des dangers de l'hypnotisme* (1888) is strongly pro-Nancy, Dr. E. Masoin (b. 1844), professor at the University of Louvain, who brought the subject to the attention of the Royal Academy of Medicine in 1888, and Dr. J. Crocq, fils, (b. 1868), of Brussels, author of a substantial work, *L'hypnotisme scientifique* (1896; enlarged edition 1900), a keen propagandist on behalf of the introduction of hypnotism into medical education, who somehow managed to combine belief in hypnogenic zones with a general support of the Nancy position.

Hypnotism in Italy remained a good deal under French influence. The earlier phase,[113] associated especially with the names of Tamburini and Seppilli, in which the ideas of Charcot had been predominant, passed and was replaced by an increasing openness to the ideas of Nancy.[114] The leading figures were both from Turin, namely C. Lombroso (1836–1909), the criminologist, and E. Morselli (1852–1929), one of Italy's most noted psychiatrists.[115] Lombroso distanced himself only a certain amount from the School of the Salpêtrière. He continued to suspect that deep hypnotic subjects have some underlying nervous abnormality or are subject to an almost pathological credulity. He carried out many experiments on psychological functioning in hypnotized subjects, both in his *Studi sull'ipnotismo* (1886), and, with S. Ottolenghi, in *Nuovi studi sull'ipnotismo e sulla credulità* (1889). In the latter work, the authors repeated some of Binet and Féré's more controversial experiments – those on the effect of optical instruments on visual hallucinations, and on the chromatic properties of visual hallucinations – with some degree of apparent confirmation. The experiments, however, were not conducted with hysterical subjects of the Salpêtrière ilk.

It was above all Morselli who led Italian hypnotists away from the ideas of Charcot. Morselli had, he tells us,[116] been interested in the subject since 1879, recognizing the truth of Braid's teachings. He was the first in Italy (he believes) to derive the cause of hypnotic phenomena from suggestion. In 1886, he published his influential *Il magnetismo animale: la fascinazioni e gli stati ipnotici*. Like Brémaud he had been very impressed by the performances of the magnetic demonstrator Donato whose speciality was "fascination";[117] like Brémaud he claims that hypnotic susceptibility, even advanced susceptibility, is not necessarily a sign of a morbid state of the nervous system;[118] but unlike Brémaud he accords no special status to the three phases of *grand hypnotisme* and the somatic signs by which they were supposed to be characterized.

Another eminent Turin psychiatrist with an interest in hypnotism was E. Tanzi (1856–1934). In 1890, he collaborated with Morselli in some studies of the circulatory and respiratory changes which occur in hypnosis.[119] Such changes can readily be demonstrated, the authors conclude, but are due to suggestion rather than to intrinsic properties of the hypnotic state.

A fair number of articles on hypnotism and suggestion appeared in Italian medical and scientific periodicals at this time (almost as many, in fact, as appeared in German ones[120]); but most Italian books rely heavily on French sources. Thus Dr. G. Magini, whose *Le meraviglie dell'ipnotismo* (1887) is an elementary but quite reasonable survey, is principally a disciple of Richet. Dr. G. Belfiore's *L'ipnotismo e gli stati affini* (1888), which has a preface by Lombroso, is very like the semi-popular French surveys of the time. So are A. Veronesi's *L'ipnotismo e magnetismo davanti alla scienza* (1887) and G. Stucchi's *I problemi dell'ipnotismo* (1892), which is much concerned with hypnotism and crime. If anything, these works are less favourable to the Salpêtrière and more favourable to Nancy than are their French equivalents.

In Holland, hypnotism made only rather slight progress, but Amsterdam enjoyed the distinction of possessing what was perhaps Europe's most celebrated hypnotic clinic. This clinic was established in 1887 by the curious partnership of A. W. van Renterghem (1845–1939) and F. van Eeden (1860–1932).[121] Van Renterghem, a former military doctor, and author of a standard compendium on dosimetric medicine, was in practice in Goes (Zealand), when he chanced to read Beaunis's *Le somnambulisme provoqué* (1886). This led him to read further works by members of the Nancy School, and, in 1887, to visit Nancy himself. He was particularly impressed by the personality and achievements of Liébeault,[122] and on his return to Goes set up a polyclinic modelled on Liébeault's, in which he treated all manner of ailments by hypnotic suggestion.

Van Eeden, a young doctor who was later to achieve fame as a poet, novelist and social reformer, had already become interested in psychological medicine, and had visited Charcot in Paris. Learning of van Renterghem's achievements in Goes, he visited him there and was most impressed. Van Renterghem's healing methods harmonized well with his own vitalistic ideas. He proposed to van Renterghem that they should set up a joint clinic in Amsterdam. Van Renterghem agreed, and the clinic was opened in August 1887. Its success was so considerable that they soon required larger premises. From van Renterghem's account,[123] it is apparent that their Amsterdam clientèle was drawn from a more prosperous section of society than his patients at Goes had been. Hysterics and neurasthenics preponderated, and the doctors soon found it wisest to see patients individually rather than to treat them in front of each other, as Liébeault did. There were certain differences between the partners. Van Eeden preferred not to plunge his patients into the deep or somnambulic state, which he thought hindered achievement of the principal aim of therapy, to enlarge the sphere of effectiveness of the patient's conscious will. He developed instead techniques of waking suggestion, or of treating patients while they were in a state of "daze" or "passive lying still with closed eyes".[124] He thought it best to avoid the words "hypnotism" and "hypnosis", which have come to have such overtones of the pathological and the theatrical, and preferred the term "psychotherapy".[125] Van Renterghem, though prepared to put patients into the somnambulic state, insists that it is not enough merely to hypnotize them and negate their symptoms. One must locate the "accidental cause" of the "psychic lesion" and try to eliminate it. "Accidental causes" (as opposed to essential causes, such as hereditary disposition) might include "worry, care, fear, or indeed relate to previous organic troubles, to physical or intellectual overwork, to weakness following childbirth or lactation, to the abuse of alcohol, morphine, cocaine, tobacco."[126] With patients capable of understanding, one can go on to explain the rationale of the treatment.

Van Renterghem and van Eeden became well-known among contemporary hypnotists for the detailed statistics which they published of their therapeutic successes and failures.[127]

In 1893, van Eeden left the partnership to devote himself to literary work, but they remained on good terms. Van Renterghem continued alone, and in 1900 moved his clinic to a building in Van Breestraat, Amsterdam, specially designed for the purpose. He called it the *Institut Liébeault*.

Elsewhere in Holland, hypnotism found only limited favour with medical men. The best-known practitioners were Drs. Arie de Jong and S. Reeling Brouwer of the Hague, whilst two professors of psychiatry – Winkler at Utrecht and later Amsterdam, and Jelgersma, director of the sanatorium of Velp, and later professor at Leyden – taught and practised suggestive therapy in accordance with the doctrines of Nancy.[128]

In the German-speaking countries, the Nancy view of hypnotism spread at first somewhat slowly.[129] The individuals who did most to introduce Nancy doctrines and practices to the German-speaking world were A. Moll (1862–1939) of Berlin and A. Forel (1848–1931) of Zürich. Moll[130] is perhaps better known today for his later studies on sexual psychopathology, but his interest in hypnotism seems to have begun even earlier. Soon after qualifying he went, towards the end of 1886, to visit the Salpêtrière, and thence journeyed to Nancy, where he became a firm convert. On his return, he began to apply Nancy methods in his own practice, and on 26th October 1887, he described some of his results to the Berlin Medical Society. His reception was somewhat discouraging, but he persevered, and a second presentation, on 10th April 1889, was rather better received. In the same year, he brought out his *Der Hypnotismus*, a very useful[131] general survey whose immediate success[132] was at once a sign and a cause of a growing German interest in hypnotism.

Moll was at this time a young and relatively unknown psychiatrist, and until the publication of his book his influence in promoting hypnotism in Germany was limited. A greater immediate impact was made by the publication in German-language periodicals of Forel's first excursions into hypnotism.[133] Forel was professor of psychiatry at the University of Zürich, and director of the Burghölzli asylum.[134] A man of great energy, Forel is still remembered for his studies of the behaviour of ants, a hobby which he pursued for the best part of his life. He trained, however, as a medical man, and became known for the studies on brain anatomy[135] which led to his appointments in Zürich. A materialist in philosophy, he tended at first towards somaticism in psychiatry; none the less, when he learned of the exciting new hypnotherapeutic ventures at Nancy, he travelled there in 1887, and upon his return as a convert immediately began to practice the Nancy methods and to publish his results. (The pioneers of hypnotism in Switzerland, Ladame and Yung[136] from French-speaking Geneva, had been much under the influence of Charcot.) His little pamphlet of 1889, *Der Hypnotismus: seine Bedeutung und seine Handhabung*, illustrated with his own cases, is intended as an introduction to hypnotism for medical men. It was enlarged through successive editions into a fairly substantial textbook.[137] Another German-Swiss work of some note (with a preface by Forel) is G. Ringier's *Erfolge des therapeutischen Hypnotismus in der Landpraxis* (1891), which contains some useful tabulations of successes and failures in the hypnotherapy of different kinds of ailments.

It was during 1888 and 1889, and not a little through the influence of Forel and Moll, that the hypnotic movement in Germany really began to gain momentum. 1888 saw the first publications of A. von Schrenck-Notzing (1862–1929) of Munich,[138] a Nancy supporter who soon became known for his work on the hypnotherapy of sexual disorders, and for his studies of hypnotism and crime. Another pamphlet of 1888 showing Nancy influence was by J. G. Sallis, a physician of Baden-Baden, *Ueber hypnotische Suggestionen*. Rather similar was a pamphlet by E. Baierlacher of Nürnberg, *Die Suggestions-Therapie und ihre Technik* (1889), a work in which the customary recital of therapeutic successes is rather refreshingly leavened with accounts of some dozen failures. In Austria, an early work was that of H. Obersteiner (1847–1922), professor of anatomy and nervous pathology at Vienna University, whose pamphlet *Der Hypnotismus* shows an acquaintance with the leading writers of the Nancy School and with the therapeutic uses of hypnotism, though he does not present cases of his own. Also in Vienna was R. von Krafft-Ebing (1840–1902), professor of psychiatry, and already well-known for his *Psychopathia Sexualis* (1886). His hypnotic experiments with an hysterical subject, Ilma Szandor, are described in his *Eine experimentelle Studie auf dem Gebiete des Hypnotismus* (1889). The findings – especially those involving hypnotically produced skin markings, and regression to childhood – were so bizarre that the pamphlet occasioned a good deal of notice and some controversy. Associated with Krafft-Ebing in this case was E. Jendrássik of Budapest, whose interest in hypnotism had roots in the Heidenhain era.[139]

By 1890, in the second edition of his book, Moll can cite a fair number of German-speaking medical men or psychiatrists who "have made experiments in medical treatment by hypnotism",[140] and the number continued to grow for several more years. Most of these persons, being medical men, were chiefly interested in the therapeutic uses of hypnotism. A mainly academic interest in the phenomena of hypnotism for their own sake is hard to find in the literature of this period. Perhaps it would be appropriate to mention here a well-known Danish psychologist, A. Lehmann (1858–1921), whose *Hypnosen og de dermed beslaegtede normale Tistande* (1890), given as lectures at the University of Copenhagen in 1889, was more influential in its German translation *Die Hypnose und die damit verwandten normalen Zustände* (1890). Lehmann's starting point is the work of the Nancy School. He develops, with an impressive amount of detail, an attention-theory of hypnosis linked to psychophysiological speculations about changes in cerebral blood flow. He proposes that in sleep the flow of blood through the brain is evenly distributed and probably diminished. A hypnotized person falls asleep with his mind still directed on the activities of the hypnotist. He remains responsive to stimuli originating from this source when his brain has become numb to everything else. A hypnotist who intervenes at this stage, before the subject is fully asleep, can keep him in a state of partial wakefulness by continually suggesting ideas or actions which keep him in a sort of dreaming state in which the dreams centre round the hypnotist. This canalization of attention and mental activity has as its physiological basis vasomotor changes as a result of which only part of the brain is fully active; this part may indeed become hyperactive owing to the greater availability of blood there. On this basis, Lehmann is easily able to give some sort of explanation of many of the leading phenomena of hypnotism – susceptibility to hallucinations, enhanced memory, responsiveness to action-suggestions, etc.

Very different from Lehmann's approach is that of another academic worker, Dr. Phil. H. Schmidkunz (1863–1934), Privat-Dozent in philosophy at the University of Munich, author of *Psychologie der Suggestion* (1892). Mental phenomena, far from being constituted out of inert elements passively obeying the laws of association, possess their own intrinsic energies, which urge them, so to speak, towards actualization, and continually pull them into groupings determined by closeness of relationship. When some sensory input so disturbs the flux of mutually attractive or repulsive ideas as to give some transient grouping undue prominence and an excessive tendency towards self-realization, suggestion is at work. Schmidkunz's large tome is not notable for its clarity and it is hard to know whether to regard him as a believer in mystical occult sympathies between ideas (he seems to hold that telepathy is due to dynamical relationships between the ideas in different minds) or as a precursor of some aspects of Gestalt psychology.

Academic studies of a more practical and even experimental kind were carried out by members of the Society for Scientific Psychology of Munich, founded in 1886 with Schrenck-Notzing as its General Secretary, and of the Berlin Society for Experimental Psychology, founded, apparently the following year, with Max

Dessoir (1867–1947) as General Secretary. (Dessoir, the two volumes of whose *Bibliographie des modernen Hypnotismus*[141] are invaluable to the student of this period, was a keen convert to Nancy doctrines, but developed views which will more appropriately be discussed in a later chapter.) These two societies (which seem to have been inspired, in part, by the Society for Psychical Research in Britain) united in November 1890 to form the Society for Psychological Research (Gesellschaft für psychologische Forschung). The first issue of the *Schriften* of this society contains a paper by Schrenck-Notzing on the use of narcotic agents in promoting or producing hypnosis, and an account by Forel of his examination of the trance states of a mediumistic medical clairvoyant;[142] the second, a long report by Moll of the experiments he and Dessoir had carried out on the nature of the alleged "rapport" between hypnotist and subject.[143]

Moll was one of the few medical hypnotists of his time who had any idea at all how experiments ought to be carried out. None the less, the activities of the Society (some of whose members had carried out experiments on telepathy) did not appeal to the leading academic psychologist of the time, W. Wundt (1832–1920) of Leipzig, who, in 1892, devoted a small book to hypnotism and related matters.[144] His stance is the slightly pained one of an authority figure who feels it a tedious duty to correct certain absurd aberrations, which would be best passed over in silence were it not that they have become dangerously popular. This stance is all the odder because, in the midst of his pontifications, Wundt makes almost a virtue of the fact that he has neither done practical work on the subject himself, nor permitted any to be carried out in his laboratory. He strongly opposes the notion that hypnotism may reveal phenomena not satisfactorily demonstrable in the psychological laboratory, and attempts to smother all investigation of telepathy, etc., under the blanket label "occultism", a term of abuse which he does not pause to clarify.

Wundt develops his own theoretical position concerning hypnotism, but there is nothing in it that is particularly new.[145] He proposes a principle of functional balance that is at work in dreams and hypnosis. When one part of the brain, acted upon by inhibitory neural influences, is in a state of latency, the excitability of the part still functioning is correspondingly increased, but the associative pathways available to it are restricted because of the inhibition of other cerebral areas. From this kind of starting point many of the phenomena of hypnosis can be derived in ways which we have already dwelt upon *paene ad nauseam*.

German hypnotism, then, like hypnotism in most other places at this period, remained largely medical and therapeutic in aim. Most approaches to hypnotism were, however, represented in a new and impressive German periodical devoted to the subject, the *Zeitschrift für Hypnotismus*, begun by a Berlin psychiatrist, Dr. J. Grossmann, in 1892, and later edited by Forel and Oskar Vogt. The general standard of the contributions (in four languages) is far higher than that of the

Revue de l'hypnotisme, and it soon became apparent that the centre of serious work in and thought about hypnotism had shifted decisively from France to Germany.

Further east, in Russia, interest in animal magnetism had to some extent survived, despite legal prohibitions on its practice, and received a powerful boost in the 1870s and early 1880s from the spreading wave of interest in Spiritualism and related paranormal phenomena which passed across Russia at this time, so much so that Ludmilla Zielinski, in her valuable survey of nineteenth century Russian hypnotism,[146] calls the period 1880–90 the "Golden Age" of animal magnetism in Russia. The leading propagandist for both magnetism and Spiritualism was A. N. Aksakov (1832–1903), who published extensively on these subjects in both German and Russian. The first person to bring hypnotism to the serious attention of Russian scientists and medical men seems to have been Professor N. P. Wagner (d.1907), a zoologist of the University of St. Petersburg, who wrote and lectured about it from 1882, showing a good deal of interest in the alleged paranormal phenomena. By the end of the 1880s, despite legal obstacles in the way of using and investigating hypnotism, there was a modest flow of publications on the topic by Russians,[147] and at the 1889 Paris Congress on Hypnotism there were sixteen[148] Russian members, a number which Germany, Britain and the United States could only muster between them. How far the views of Bernheim prevailed among the Russian members it is impossible to say; certainly, undiluted Salpêtrière views did not. Dr. L. Stembo, of Vilno, tells us that he began to study and use hypnotism in 1885 following the precepts of the Paris school, but soon switched to those of Nancy.[149] Dr. Drzewiecki of St. Petersburg, who seems to have believed in the "stages" of catalepsy, lethargy and somnambulism, described Russia as "the land of suggestion and hypnotism *par excellence*", and went on to describe his own experiments with a very good subject who presented no symptoms of hysteria.[150] The general tenor of Russian medical hypnotism was physiological. The first Russian work on hypnotism to become widely known in the West was I. R. Tarkhanov's *Hypnotisme, suggestion et lecture des pensées* (1886, French tr. Tarchanoff [1891]). Tarkhanov, a well-known physiologist, offers a physiological theory which is only two removes from Heidenhain's, though he is also a great admirer of Richet. Russian interest in hypnotism and its therapeutic possibilities seems to have continued growing well into the 1890s.

In strife-torn Poland, little was heard of hypnotism, and what was heard had little to do with the School of Nancy. The leading students of the subject were Dr. N. N. Cybulski, a follower of Charcot,[151] and J. Ochorowicz (1850–1917),[152] who still inclined to views more characteristic of the animal magnetists than of hypnotists. Ochorowicz was most widely known for his participation in the investigation of various mediums and clairvoyants,[153] and for his invention of the hypnoscope,[154] a device for revealing an individual's susceptibility to hypnosis. This consisted of a hollow cylindrical magnet into which the subject thrust a finger. Hypnotizable persons were alleged thereupon to feel certain characteristic sensations. Unfortunately, other workers who tried out the hypnoscope were unable to replicate Ochorowicz's results.[155]

In the Scandinavian countries, interest in hypnotism among medical men does not seem to have been great. Setting aside Lehmann, the Danish psychologist already mentioned, and a few Danish medical men,[156] the most notable contributions came from Sweden. In 1885, Frederik Björnström (1833–1889), professor of psychiatry at Stockholm, drew attention of the Swedish Medical Society to "the mischief caused by an itinerant Danish mesmerist" named Sixtus. After some discussion, the Society took no action.[157] Björnström, however, went on gathering materials on the subject, and in 1887 came out with his *Hypnotismen: dess utvecklung och nuvarande ståndpunkt*, a broad if somewhat uncritical survey, which was translated into German and English. The following year, 1888, M. Huss (1807–1890), the doyen of Swedish doctors, whose experience went back to the age of animal magnetism, was stirred by the new interest in hypnotism to publish his *Om hypnotismen*.

The great name in Swedish hypnotism, however, and a considerable name in European hypnotism, was that of O. G. Wetterstrand (b. 1845) of Stockholm. Wetterstrand's interest in hypnotism was aroused by Bernheim's 1886 book. He tried out Bernheim's methods on his own patients, with such success that from the spring of 1888 he devoted himself entirely to hypnotherapy, which he applied to a wide variety of ailments. He set forth his early results in a long article in 1888; it appeared in book form later the same year.[158] He describes successful or beneficial treatment of cases of, *inter alia*, paralyses following cerebral haemorrhage; epilepsy and chorea; homosexuality; melancholia; alcoholism and other forms of drug addiction; asthma; nervous cough; functional heart disorders; diarrhoea; vomiting; and menstrual troubles. He seems to have been a remarkably successful hypnotist – only ninety-seven persons, he tells us, of the 3148 he has hypnotized since January 1887, failed to respond.[159] He was also remarkably patient, succeeding with one male subject only after seventy trials.[160] Unfortunately, however, he gives us almost no details of the way in which he approached each individual case and of the suggestions which he actually made.

Wetterstrand's reputation as a wonder healer spread rapidly, and patients flocked to him from all over Sweden, indeed from abroad. Van Renterghem, who visited him in June 1892, was astonished to find many of the numerous patients in his clinic speaking German. Wetterstrand explained that this was because a considerable proportion of his patients came from Russia and the Baltic countries.[161]

According to van Renterghem,[162] Wetterstrand was a man of middle size, forty to forty-five years of age, blond turning to grey, with blue eyes. The oval of his face crossed with blond moustaches, and a dimple on his chin, gave an impression of both gentleness and energy. He had the *tic* of half closing his eyes like a short-sighted person. His voice was soft and persuasive. By this time he had established his clinic on the second floor of a large house in the aristocratic quarter of Stockholm.

The English physician, J. Milne Bramwell, who visited this clinic in August 1894, describes Wetterstrand's procedures as follows:

> He hypnotized his patients in three rooms communicating with each other; these were darkened, and great silence was observed. The patients who had previously been hypnotized were first put to sleep, and while they rested quietly the fresh cases were dealt with. In addition to making verbal suggestions, Wetterstrand employed passes more largely than is usually done by members of the Nancy school; these were made over the eyes, face, and arms, and always with contact. Sometimes he would touch the forehead with one hand, while with the other he pressed heavily over the region of the heart. Before commencing the passes, he generally requested the patient to look at his eyes, but this was never continued long. The patients rested quietly for about an hour, while Wetterstrand passed from one to the other whispering suggestions audible only to the person addressed.[163]

Wetterstrand was by no means alone among the medical hypnotists of this period in continuing to make use of mesmeric passes. He soon became known for his successes with cases of alcoholism and drug addiction,[164] and for his use in treating these conditions and others of the technique of "prolonged sleep"[165] – hypnotic sleep prolonged for days or even weeks – which had been pioneered by Voisin when dealing with highly disturbed patients. It may also have owed something to Weir Mitchell's rest and rich diet treatment for neurasthenia; both this disease and the cure had become fashionable in the Eastern United States among those who could afford them. Wetterstrand set up a special establishment for his sleep treatment. The patients were closely supervized by trustworthy overseers, many of whom had themselves previously been treated by hypnotherapy. His successes with morphine addicts were so striking that some cast doubts on them.[166]

Another Swedish hypnotherapist with considerable experience was Dr. Velander of Jönköping, who told the 1889 Paris Congress on Hypnotism[167] that in a year and a half he had treated more than 600 persons by hypnotic suggestion. His experience was in accord with that of Bernheim, Voisin, Bérillon, and other distinguished French doctors.

Turning to the other end of Europe, we find in Spain and Portugal only a rather limited interest in hypnotism and hypnotherapy, partly, perhaps, as Dingwall suggests,[168] because the strong association between Spiritualism and mesmerism-hypnotism rendered the Catholic authorities there suspicious of and hostile to the whole subject. The most notable Spanish contributor to hypnotic literature was Dr. A. Sanchez Herrero, professor at Valladolid and later at Madrid, whose *El hipnotismo y la sugestiòn* was first published in 1889. In Portuguese-speaking Brazil, on the other hand, there was some tradition of interest in animal magnetism, and a growing obsession with Spiritualism. Hypnotic ideas and practices took readier root. Commenting on the history of hypnotism in Brazil, Dr. D. Jaguaribe, of Sao Paolo, states that "the natural suggestibility of the inhabitants of Brazil constitutes a vast field for the application of hypnotism and psychotherapy", and goes on to mention a considerable number of medical men who have become hypnotic practitioners.[169] Interest was not confined to medical men. In 1887, the Emperor of Brazil, Dom Pedro, attended Luys' demonstrations at the Charité as he had done those of Charcot at the Salpêtrière (the Emperor, it appears, was a frequent visitor to Charcot's home, and used to play billiards with him).[170] The best-known Brazilian hypnotic practitioners seem to have been Professor E. Coelho, of Rio de Janeiro, and Dr. F. Fajardo (1864–1906),[171] also of Rio, author of *Hypnotismo* (1889), a wide-ranging survey which favours the views of Bernheim as against those of Charcot.

In Great Britain, interest in hypnotism was slow to get under way. Braid was almost forgotten, and would perhaps have been entirely so, but for W. B. Carpenter's *Mesmerism, Spiritualism, &c. Historically and Scientifically Considered* (1877) and his frequently reprinted *Principles of Mental Physiology* (1874). Carpenter holds that all the phenomena of natural and induced somnambulism may be reduced to two general principles: (1) "The *entire engrossment* of the mind with whatever may be for a time the object of its attention", and (2) "The *passive receptivity* of the Mind ... for any notion that may be suggested to it".[172] Early papers (1878, 1880) on hypnotism by G. J. Romanes (1848–1894), the naturalist, are reviews of, respectively, Preyer's work on "animal hypnosis" and Heidenhain's *Der sogenannte thierische Magnetismus*.[173] In the early 1880s, some faint echoes of the "Hansen phase" of German hypnotism could be detected in Britain. Thus in November 1880, Professor Stirling gave in Aberdeen a lecture on hypnotism in which he expressed the opinion that Braid's facts had been rediscovered by Weinhold and Heidenhain.[174] After this, occasional short notices about hypnotism, including the activities of Charcot and his school, began to appear in the medical press.[175] In October 1883, G. F. Yeo, professor of physiology at King's College, London, gave a paper to the Science Society there on "The nervous mechanisms of hypnotism", which was afterwards printed as a pamphlet, and in March 1884, J. N. Langley (1852–1925), later professor of physiology at Cambridge, lectured to the Royal Institution on "The physiological aspect of mesmerism". Both authors develop notions of cerebral inhibition somewhat resembling Heidenhain's; Langley draws a good deal on current work on "animal hypnosis". The most interesting work of this period is, however, D. Hack Tuke's *Sleep-Walking and Hypnotism* (1884). Tuke (1827–

1885) was the co-author of a standard textbook of psychological medicine,[176] and author of *Illustrations of the Influence of the Mind on the Body* (1872). Later he edited the monumental *Dictionary of Psychological Medicine* (2 vols, 1892), a work which amply reflects his own eclecticism and breadth of learning. *Sleep-Walking and Hypnotism* is much influenced by Charcot and Richer (Tuke had recently visited the Salpêtrière) but the author's candour and moderation keep him from all excesses. He draws some interesting parallels between the mental state of hypnotized persons and that of certain of the more unusual cases of sleepwalking, and in the course of so doing gives long accounts of the sensations undergone by various educated subjects hypnotized by Hansen during a recent visit to London.

Neither the claims of the Salpêtrière nor those of Nancy attracted at first much interest in Britain. So far as I know, the first British visitors to the Nancy School were certain members of the Society for Psychical Research (the "SPR") who went there in 1885.[177] The publications of the SPR were a principal forum for hypnotism in Britain, but the Society and the persons involved in it will more appropriately be discussed in a later chapter. A minor curiosity which appeared in 1887 was a pamphlet by Dr. M. Roth,[178] a homoeopath, who had, in his day, seen Dupotet and visited the London Mesmeric Infirmary, and could now compare these with what he had recently read about the Nancy School. In fact for an English author he shows a remarkable knowledge of contemporary French writers on hypnotism, combined with an almost equally remarkable inability to spell their names.

The first English medical man to adopt the Nancy form of treatment seems to have been C. Lloyd Tuckey (1855–1925) who first visited "dear old Dr. Liébeault," and then Bernheim, Bérillon and van Renterghem, in the autumn of 1888.[179] He became especially known for his work on the hypnotic treatment of alcoholism,[180] but he had an extensive general hypnotic practice, and his text-book *Psycho-Therapeutics: or Treatment by Sleep and Suggestion* (1889) went through a series of progressively enlarged editions (after the third it was called *Treatment by Hypnotism and Suggestion; or Psycho-Therapeutics*), reaching a seventh in 1921. He was shortly followed to Nancy by J. Milne Bramwell of Leeds, H. Rolleston of St. Bartholomew's Hospital, H. E. Wingfield of Cambridge, R. W. Felkin of Edinburgh, G. C. Kingsbury of Blackpool, and Sir Francis Cruise of Dublin, each of whom wrote a book or paper(s) on hypnotism. Milne Bramwell (1852–1925) was the most influential. His interest in hypnotism, he tells us,[181] was aroused as a student at Edinburgh by Professor J. H. Bennett, whose pamphlet on *The Mesmeric Mania of 1851* we have already mentioned. In May 1889, he first tried hypnotic methods on a patient. At this time he knew only Braid's *Neurypnology* (he always remained a great admirer of Braid); only afterwards did he visit continental clinics and learn more. On 28th March 1890, he gave a demonstration of hypnotic anaesthesia to a large gathering of medical men at Leeds, which was reported in *The Lancet* and *The British Medical Journal*.[182] Various dental and minor surgical operations were performed. In July of the same year, the whole question of hypnotism was discussed at the British Medical Association's Annual Meeting at Birmingham.[183] Whereas, says Kingsbury,[184] the Association had received with incredulity a paper on hypnotism read by Voisin the previous year, a single year had sufficed to bring about a remarkable change in its attitude; so much so that a committee, with a number of distinguished members, was appointed to investigate the nature and therapeutic value of hypnotism. Two years later, this committee reported back[185] lending powerful support to the genuineness of the hypnotic state and to its therapeutic potentialities. Meanwhile, Bramwell had had so many patients referred to him that he gave up general practice and came to London in November 1892 as an hypnotic specialist. Also a number of English books and pamphlets on hypnosis were published, mostly by medical men.[186] Interest in and practice of hypnotism continued to grow in Britain for several years, but it never wholly, or even largely, overcame the strong conservatism of the British medical profession.

In the United States, too, the climate of medical opinion was far from favourable to the employment of hypnotism. In psychiatry, what N. G. Hale[187] calls "the somatic style", typified by S. Weir Mitchell (1829–1914) of Philadelphia and J. J. Putnam (1846–1918) of Boston in his earlier days, prevailed from the 1870s until after the turn of the century. Psychiatry, it was held, was destined to become a branch of neurology, and the younger neurologists, many of whom had studied in Europe and acquired the latest knowledge about brain mechanisms and disease from German and French laboratories and clinics, were cheerfully confident that their young and growing science would soon be able to handle such "psychological" disorders as hypochondria, hysteria, mania, melancholia, paranoia and neurasthenia. Underneath these disorders was presumed to lie a "predisposing cause" in the shape of some inherited defect of the nervous system, though this defect might be so subtle as to escape detection by existing technology. "Precipitating causes" might be found in the stresses and strains of modern living. Prevention could up to a point be achieved by the avoidance of stress, and by proper "training" and hygiene of the susceptible nervous system. Treatment in some cases might consist in fostering by various means – rest, massage and a fortifying diet, as in the celebrated Weir Mitchell treatment mentioned above, sometimes coupled with electricity, exercise and tonics – a supposed "nervous energy" of which defective nervous systems possess inadequate supplies.[188]

Against the somaticist tide, hypnotherapy was understandably slow to make headway, and progress was not rendered easier by the prevalence, in New England and elsewhere, of mind-cure cults to which members of the medical profession were not unnaturally hostile. And there were features of the general ethos of upper and middle class America (an ethos often shared by the neurologists we are considering) to which hypnotic treatment was uncongenial – a belief in the power of self-help and self-determination, a feeling that the individual should make his way under his own steam.

Among the early American neurologists, the only one who showed a significant interest in hypnotism was G. M. Beard (1839–1883), remembered today mainly for his advocacy of "neurasthenia" as a nosological entity.[189] His interest in hypnotism was connected with a wider interest in trance states in general, a topic on which he wrote a good deal in the late 1870s and early 1880s.[190] For Beard, trance, including hypnotic trance, is a phenomenon of concentration, there being "suspension, more or less complete, of nervous activity in all directions except one."[191] There is no special technique for inducing trance. "If a thousand persons sit down or stand up, with their eyes closed, or even opened, with the expectation that they are to go into trance, quite a number of them will go into that state, even though they have no faith in the matter, and there be no operator or experimenter on the same planet with [them]."[192] Beard is a curious mixture of naïveté and conceit. His ideas are very like those of the later Braid, whose unpublished manuscripts came into his possession, and were passed by him to Preyer, who published them. Yet his account of Braid is such as to make one doubt that he had ever read him,[193] and he regards himself as a pioneer for propounding ideas most of which Braid had anticipated and handled more impressively.

After 1880, Charcot's *grand hypnotisme* received some attention in America[194] – more, perhaps, than it did in Britain – and as the 1880s progressed, a growing, though still small, number of American doctors (mostly ordinary physicians rather than neurologists) began to experiment with the Nancy methods. The chief centre of these activities was the New England states[195] – perhaps something of the transcendentalist tradition still lingered there. In New York, interest was less, perhaps because of the extravagances of Dr. W. A. Hammond who proposed in 1887 to call hypnotism syggignoscism (the agreeing of one mind with another) and supported his view that it is due to the automatic functioning of lower nervous centres with a number of gruesome stories about the apparently purposive behaviour of decapitated animals.[196] A New York physician, R. Osgood Mason, states that when in 1888 he reported to the New York Academy of Medicine on some cases he had treated by hypnotism, the President's comment was, "Well, doctor, you taxed our credulity."[197] All in all these activities did not amount to very much. Speaking to the Annual Meeting of the Medical-Legal Society in December 1889, Clark Bell, a New York lawyer keenly interested in hypnotism, remarked: "The medical profession in American do not give this subject the attention its importance deserves. We know of no medical man of prominence in America who has publicly identified himself with the investigation of this science as some of the most eminent men in foreign countries have recently done."[198] He adds that although *The American Journal of Psychology* (founded in 1887) has devoted more space to hypnotism than any other American scientific journal, it has carried no original article from an American author.

In 1890, two events added a distinct momentum to hypnotism's hesitating progress in the U. S. A. William James published his very influential *Principles of Psychology*, with its celebrated chapter on hypnotism supporting the Nancy viewpoint; and Hamilton Osgood (b. 1839), a Boston physician who had recently visited Bernheim, reported in *The Boston Medical and Surgical Journal* on thirty-five cases he had himself treated by hypnotism.[199] Rheumatic and chest complaints are prominent amongst them, and the depth of hypnosis attained did not appear to have much relation to the degree of therapeutic success. Osgood's report attracted a good deal of attention and the following year he published another paper on the same theme. As the 1890s advanced, it became apparent that interest in hypnotism was increasing on a number of different levels, and that its use by medical practitioners was becoming more frequent. Popular books and articles on the subject and on its legal ramifications became increasingly common, as did what might be described as semi-popular ones by medical men. It was taken up, in a modest way, by some quite prominent neurologists in New York State,[200] and in Boston by a group of persons, to be discussed in a later chapter, all influenced to some extent by William James, whose concerns were not just with its therapeutic applications, but with the light that it might shed on psychological issues of deep theoretical importance. Such theoretical concerns were to become very characteristic of the hypnotic movement in the United States.

From this brief, and necessarily incomplete, regional survey of the progress of hypnotism in the late 1880s and early 1890s, it is apparent that the medical use of hypnotism had gained some currency over much of Europe and America, and a good deal of currency in some areas. The terms "hypnotherapy" and "psychotherapy," and their cognates, were used more or less interchangeably, and hypnotherapeutics was becoming widely recognized as a respectable branch of medicine, though it is probable that somatic approaches to psychological disorders everywhere predominated. The Nancy School had blown the pretensions of the Salpêtrière to the winds, and by the early 1890s it made little sense to contrast these two "schools" at all. Insofar as adherence to Nancy simply involved regarding the symptoms of *grand hypnotisme* as the products of suggestion, almost everyone[201] was a supporter of Nancy. The important divisions were now between persons all of whom were in this broad sense "supporters" of Nancy.

The high point of the progress of hypnotherapy is perhaps marked by a little book edited in 1894 by J. Grossmann (the editor of the *Zeitschrift für Hypnotismus*). This work, *Die Bedeutung der hypnotischen Suggestion als Heilmittel*, was prepared to help Russian doctors to persuade their government that the long-standing legal restrictions on the practice of hypnotism should be modified. It consists of expert opinions by twenty-seven medical men and three lawyers as to the therapeutic value and alleged dangers of hypnotism. The majority of the contributors came from France and Germany, but there are also one or more from Switzerland, Russia, Belgium, Holland, Austria, Italy, Sweden and Great Britain. What is of interest, however, about this volume, is not just the range of countries represented, but the fact that the Nancy position in its broader sense is so completely taken for granted that few contributors even bother to mention Charcot and the Salpêtrière. Those that do mention them do so only to assure us in the next breath that they had never adhered to Charcot's views, or else that they had dropped them like a hot potato as soon as they learned of the work being done at Nancy.

DISINTEGRATION OF THE SCHOOLS

While the influence of the Nancy School was thus spreading across Europe and into America, the original members of the school were moving farther apart, both literally and metaphorically. The last major statement of the Nancy position was Liégeois' 758 page *De la suggestion et du somnambulisme dans leurs rapports avec le jurisprudence et la médecine legale* (1889), which, though concerned especially with the legal implications of hypnotic phenomena, extensively surveys the historical development and contemporary position of the whole subject. Liébeault's *Du sommeil et des états analogues* was reprinted in two volumes, with additions and minor alterations, in 1889 and 1891.[202] He may be regarded as largely restating his original position. Beaunis published little more in the field. Bernheim, who had been the main driving force of the school, developed views which diverged more and more from Liébeault's. These are to be found in various of his articles,[203] and most notably in a new large book, *Hypnotisme, suggestion, psychothérapie: Études nouvelles* (1891). Like his 1886 book, this volume is divided into two parts, the first primarily theoretical and expository, the second primarily a recital of case histories. The case histories cover a wide range, from multiple sclerosis to dysmenorrhoea, and from hysterical contractures to phlebitis, with neurasthenic and hysterical afflictions predominating. Although Bernheim remains adamant that suggestion cannot directly produce organic changes, it can, he thinks, produce functional changes that may promote organic changes, and since he also holds that suggested ideas can have effects upon the workings of any innervated organ, the scope for its influence is considerable. The changes in Bernheim's theoretical stance, though not dramatic, are significant, and appreciably widen the distance between

himself and Liébeault. Bernheim's revised theory will be a topic for discussion in
a later chapter. Here it will be sufficient simply to note the extent and grounds
of his divergences from Liébeault. Liébeault continued to hold that hypnotic
sleep is identical with ordinary sleep and that its induction greatly furthers the
efficacy of therapeutic suggestions. Bernheim, on the other hand, maintained
that therapeutic success, and success in obtaining the "phenomena" of
hypnotism, were not directly correlated with the depth of "hypnotic sleep"; that
though a few subjects will pass into true (i.e. dreamless) sleep, most remain or
become partially conscious, indeed have to do so if suggestions are to take effect;
and that all the phenomena of hypnotism can be obtained in the waking state
with subjects who have never been put into the "hypnotic sleep" or seen
anyone else enter that condition. Sleep, or more often the appearance of sleep,
or a partial sleep, is one phenomenon amongst others which can be produced by
suggestion. It would be best to drop the term "hypnotism" altogether and
replace it by "state of suggestion."[204] In his final statement of his position,
published not long before his death, Bernheim went so far as to say:

> One could have discovered these phenomena directly in the waking state,
> without passing through the unnecessary intermediary of induced sleep;
> and then the word hypnotism would not have been invented. The idea of a
> special induced magnetic or hypnotic state provoked by special manoeuvres
> would not have been attached to these phenomena. Suggestion has been
> born of the old hypnotism, as chemistry was born of alchemy.[205]

Few even of the keenest Nancy supporters would have upheld so radical a
position. Indeed, as we shall see, by the end of the nineteenth century Bernheim
had become in effect the founder of a second school, with tenets very different
from those originally propagated in Nancy.

And what, finally, of the Salpêtrière School? Paul Richer had moved more and
more into the field of anatomical illustration. Binet, worsted in various battles
with Bernheim, turned discreetly away from the Salpêtrière position. Charcot's
most persistent supporters, Babinski and Gilles de La Tourette, particularly the
former, continued to pound away at Bernheim, but the only effect was to bruise
their own fists. The gist of what Babinski said against Bernheim in a series of
articles published between 1889 and 1892[206] may be compressed into the
following. That Charcot's proposed phases of *grand hypnotisme* are not artifacts
is proved by the fact that patients quite ignorant of medical matters will, while
in the lethargic state, show the characteristic muscular contractions in response
to pressure though no hint or suggestion has been given them. One patient, R...,
newly arrived at the Salpêtrière, showed the three phases in developed form
although she had never heard of them before.[207] The phases, or something very
like them, have been observed by workers – Tamburini, Seppilli, Ladame, Yung,
Vizioli[208] – quite outside the Salpêtrière, and therefore cannot be due to

surreptitious or unintentional training given there. Bernheim, who can find in his patients no such physiologically indexable stages, is left with a problem. How, in the absence of such objective physical signs, can he be sure that his patients are really hypnotized, that they are not shamming? And how can he be sure that all the phenomena characteristic of the three phases can be produced by suggestion? Bernheim is wrong, too, in insisting that there is not a special relationship between hypnotism and hysteria. The existence of this relationship is proved by the numerous parallels between the two conditions and by their interdependence.[209] Many of the therapeutic successes reported by Bernheim involve patients who are probably hysterics, though Bernheim does not realize this. Bernheim's own definition of hypnosis is so vague and general that he could not distinguish hypnosis from the waking state, let alone from hysteria.

Some of these arguments are not without cogency, but proffering them was a mere shouting into the wind. There was no bilking the fact that cases of *grand hypnotisme*, if they were ever found in true independence of the Salpêtrière, were at best so very rare that no useful account of hypnotism could be founded upon them, or the fact that even the most bizarre and extraordinary hypnotic phenomena had apparently often been obtained with subjects who could only be thought hysterical, or else deceivers, by persons with a paranoid preconception that they must be the one or the other.

Charcot himself never became directly involved in any of these controversies, though in 1892 he again quietly restated his position.[210] Did he "set up" his pupils to take a more aggressive stance behind which he could hide, or did they take such a stance voluntarily because they were totally convinced by the phenomena they had observed at the Salpêtrière? The latter is perhaps more probable. Charcot, though he clearly disliked controversy, seems always to have answered the doubts and queries of visitors to the Salpêtrière with the utmost candour.[211] There might, however, be some room for doubt as to the motives of some of his pupils.

Though Charcot had misled himself and others for so long, in the end his intellectual integrity asserted itself. Some hints of germinating change can perhaps be detected in the interesting and well-known paper on "the faith-cure" which he published at the beginning of 1893.[212] Like most of Charcot's writings it is notably moderate in tone and not in the least authoritarian or assertive. In it he seems to assign a greater and more general role to suggestion as an agent in the removal of hysterical symptoms than he had done hitherto.[213] In the few weeks before his death (which took place in August 1893) he was, as he told his medical secretary, Georges Guinon,[214] preparing a complete revision of his views on hysteria and related topics. A stack of papers on his desk testified to the seriousness of this endeavour, though what they may have contained, or what the drift of Charcot's new thinking may have been, we do not know. It is just possible, however, that he may have been influenced by the ideas of a recent arrival at the Salpêtrière, to whose book on the mental state of hystericals[215] he had contributed a preface, namely Pierre Janet (1859–1947).

NOTES

1 On Liébeault see Delboeuf (1889); Renterghem (1895–6; 1896–7; 1897a; 1897b; 1898a; 1898b; 1898c); Walser (1964); Chertok (1966); Kissel and Barrucand (1964); Barrucand (1967), pp. 90–99.

2 Renterghem (1895–6), p. 335.

3 See his "Confessions d'un médecin hypnotiseur", Liébeault (1886–7a); (1891), pp. 290–305.

4 Liébeault (1891), p. 293.

5 Rolleston (1889), p. 123.

6 Liébeault (1866), p. 355.

7 Cf. Rolleston (1889), pp. 123–124. Among the diseases which Rolleston saw Liébeault treat were chlorosis, pathological disease of the spine and cystitis.

8 Liébeault (1866), pp. 280, 281–282.

9 Liébeault (1889; 1891).

10 Renterghem (1895–6), pp. 340–341.

11 Delboeuf (1889), p. 30.

12 Tuckey (1909), p. 5.

13 Renterghem (1895–6), p. 350.

14 The treatment of this was rather a speciality of Liébeault's. See Liébeault (1886–7b).

15 Renterghem (1895–6), pp. 352–353.

16 E.g. Liébeault (1866), p. 452.

17 Delboeuf (1889), p. 40.

18 Liébeault sometimes talks as though the autonomic system were the seat of a sort of second, though inferior, consciousness, separate from the main one. See e.g. Liébeault (1866), p. 135.

19 See Liébeault (1866), pp. 15–32. He holds that both natural and artificial sleep involve relative sensory isolation, concentration on the idea of sleep, need for repose, and mechanical means for immobilizing attention.

20 Liébeault (1866), p. 52.

21 Liébeault (1866), p. 76.

22 Liébeault (1866), pp. 17, 18.

23 Liébeault (1866), p. 140.

24 Liébeault (1866), pp. 149–150.

25 Cf. pp. 105–106 above.

26 On this episode see Chertok (1966), pp. 872–877; Renterghem (1895–6), pp. 366–369; Liébeault (1891), pp. 246–268; F. W. H. Myers (1903), Vol. I, pp. 443–444.

27 On Bernheim see especially Barrucand (1978). His year of birth is sometimes given as 1837. I have followed the *Dictionnaire de biographie française* in giving 1840.

28 On Bernheim's contributions to the pathology of infectious diseases see Théodorides (1978).

29 By M. Dumont, chef des travaux physiques of the Nancy Faculty. Cf. Bernheim (1884a), pp. 3–4.

30 Bernheim (1883a; 1883b).

31 Liégeois (1884); Beaunis (1886).

32 Author of Brullard (1886), a doctoral dissertation.

33 Bernheim (1884a), p. 5.

34 Bernheim (1884a), pp. 5–6.

35 Bernheim (1884a), pp. 7–9.

36 Bernheim (1884a), p. 9.

37 Bernheim (1884a), p. 24.

38 Bernheim (1884a), pp. 26–27.

39 Bernheim was interested in this early on. See Bernheim (1883a; 1883b).

40 Bernheim (1884a), p. 48.

41 Bernheim (1884a), p. 87.

42 Barrucand (1967), pp. 154–157.

43 Like Charcot he had been a member of the Society of Biology's Committee on metallotherapy. See above, p. 310. Dumontpallier gives his own account in his Presidential Address to the First International Congress of Hypnotism (Bérillon, ed. 1889, pp. 21–26).

44 In Bérillon, ed. (1889), p. 23.

45 Magnin (1884).

46 Magnin (1883).

47 On Pitres see Barrucand (1967), pp. 144–148.

48 Cf. Crocq (1892–3); Crocq (1896), pp. 33–37; Pitres (1885). Pitres ended up with a long list of exotic zones – spasmogenic zones, spasmofrenetic zones, hypnogenic zones, hypnofrenetic zones, ideogenic zones, ideo-ecmnesic zones, impulsive zones, zones of ecstasy, chatter, laughter, etc. Chambard (1881) and Féré (1883) added erogenous zones.

49 Brémaud (1884).

50 I.e. A.E. d'Hont (1845–1900). See above p. 315 n. 18.

51 Richer (1885), p. 509.

52 Paul Janet (1884).

53 Liégeois had just published a pamphlet on the legal and criminal aspects of the subject (Liégeois, 1884).

54 Bernheim (1884b). This comes from the *Revue medicale de l'est*, and is incorporated in Part I., chapter IX, of Bernheim (1886).

55 The most interesting passage at arms was Binet and Féré (1885a) and Bernheim (1885). See below, pp. 333–334.

56 Bernheim (1889), p. 203. I have used the English translation (1889b; 1897b) of Bernheim (1886).

57 Bernheim (1889b), pp. 90–91.

58 *Revue philosophique*, 1886, 22, 557–563. Bernheim's reply is Bernheim (1887).

59 Pierre Janet (1886d).

60 Ladame of Geneva supported Janet. See Ladame (1881).

61 Charcot's works had been translated into Italian.

62 Tamburini and Seppilli (1881; 1882)

63 Vizioli (1885).

64 See pp. 440–441 of his review of Beaunis (1886), *Revue philosophique*, 1886, Vol. 22, pp. 439–442.

65 Binet (1884a; 1884b); Binet and Féré (1885c). There is a useful summary in chapters IX and XI of Binet and Féré (1887a).

66 Binet and Féré (1887a), p. 255.

67 Bernheim (1889b), pp. 95–104.

68 A useful resumé of experiments on aesthesiogens will be found in F. W. H. Myers (1885–7c), pp. 149–154.

69 Binet and Féré (1885a).

70 P. 228 above. Braid and Heidenhain had produced the phenomenon.

71 Richer (1885), p. 790. Cf. Descourtis (1882).

72 Ladame (1881).

73 Bérillon (1884).

74 There does not seem to have been any consistency on this point. Bérillon (1884), pp. 148–152.

75 For an independent eyewitness account of some of these experiments see F. W. H. Myers in *Journal of the Society for Psychical Research*, 1885–6, Vol. 2, pp. 443–446. Myers thinks suggestion was not excluded.

76 Not, as would have been more logical in view of the anatomical arrangements of the visual system, in one or other visual hemi-field.

77 Binet and Féré (1887), pp. 264–265.

78 Bernheim (1885).

79 Pp. 560–562 of his review of Bernheim (1886), *Revue philosophique*, 1886, Vol. 22, pp. 557–563.

80 On Luys see Kingsbury (1891), pp. 141–157; Hart (1893), *passim*; Foveau de Courmelles (1891), pp. 47–60. His own most general account of his position is Luys (1890c). A useful summary in English is Luys (1890b).

81 One of Luys' subjects (who turned out to be a young actress) told Camille Flammarion, the astronomer, that the whole performance was fraudulent and that she occasionally used confederates. Dingwall (1967), p. 259.

82 In 1891 he became leader of the French Martinist Order – a curious link between nineteenth-century hypnotism and animal magnetism as practised by Barberin and Saint-Martin in the 1780s.

83 Babinski (1886).

84 For a highly critical eyewitness account of some of Babinski's experiments, and some recommendations as to proper procedure, see F. W. H. Myers, *Journal of the Society for Psychical Research*, 1885–6, Vol. 2, pp. 444–450.

85 Cf. Luys (1890b), pp. 180–183.

86 Luys and Encausse (1891).

87 One was Louis Vivé, famous in the annals of multiple personality, the other was a female *grande hystérique* who had been a patient at the Salpêtrière.

88 An interesting account of Bourru and Burot's early experiments is A. T. Myers (1885–6). For their work in general see Bourru and Burot (1887) and Berjon (1886).

89 Luys (1887; 1890a).

90 Kingsbury (1891), pp. 155–156.

91 Dujardin-Beaumetz (1888).

92 J. Voisin (1887–8a); Bernheim (1887–8).

93 E.g. his claims that north and south poles of magnets produced opposite effects upon hypnotized subjects, that such subjects can see lights emanating from the poles of magnets, that their skin sensitivity can be transferred into a glass of water. On the latter cf. *Journal of the Society for Psychical Research*, 1898–9, Vol. 5, p.327. If the water was drunk, the patient swooned!

94 F. W. H. Myers in *Journal of the Society for Psychical Research*, 1885–6, Vol. 2, p. 448.

95 On Binet's later work on suggestion see Carroy-Thirard (1980).

96 Bernheim (1888; 1892b).

97 Delboeuf (1889), p. 50.

98 Osgood (1890), p. 412.

99   Delboeuf (1889), pp. 69–70.
100 A. T. Myers (1890), pp. 200–201. On Bernheim's clinic at this period cf. Baldwin (1892).
101 For a while these two terms were almost synonymous; but then non-hypnotic forms of psychotherapy came to the fore.
102 Mesnet (1860).
103 Grasset (1888–9), p. 334, claims to have had a female patient in whom the fixed somatic symptoms were anterior to all suggestion.
104 Cf. Binet (1897), p. ix.
105 Fontan and Ségard (1887); Fontan (1889).
106 Dejérine (1890–1).
107 Bourdon (1889).
108 E.g. Morand (1889); Cullerre (1887); Foveau de Courmelles (1891); Bottey (1886); Marin [1890].
109 Charcot was a President of Honour, but does not seem to have attended.
110 Only Gilles de La Tourette spoke decisively and at length on the Salpêtrière side. Bérillon, ed. (1889), pp. 99–100. Magnin's paper (Magnin, 1889) in effect summarizes differences between the School of Dumontpallier and that of the Salpêtrière.
111 See *Congrès Internationale* ... 1890. There is a useful summary of the activities of this Congress by A. T. Myers (1889–90).
112 Delboeuf also engaged in intermittent warfare with the medical profession over the question of whether or not the practice of hypnotism should be limited to the medical profession. As a non-medical person who was also a very effective hypnotist, Delboeuf was strongly opposed to such a limitation, which was, however, legally enforced in Belgium by the law of 30th March 1892. He had a violent controversy with Ladame of Geneva on this issue at the end of the 1889 International Congress of Hypnotism.
113 For a useful account of Italian hypnotic literature at the time of Charcot's influence see Belfiore (1888), pp. 71–76.
114 Tamburini seems to have remained faithful to some at least of the Salpêtrière ideas while Seppilli shifted further away from them. For Tamburini see p. 166 of Bérillon and Farez, eds. (1902), and Tamburini (1891). On Seppilli see Seppilli (1891).
115 On Morselli see Guarnieri (1988) and Rossi (1984).
116 See his *Gutachten* in Grossmann, ed. (1894), pp. 80–84.
117 Morselli (1886), pp. 270–300.
118 Morselli (1886), pp. 388–401.
119 Morselli and Tanzi (1890). Tanzi (1887) is an attack on some of Binet and Féré's experiments.
120 In the two volumes (1888 and 1890) of Dessoir's bibliography of modern hypnotism there are altogether 503 French language items, 172 German ones, 148 English ones and 120 Italian ones. In the period up to 1888 the Italian items exceed the German ones by 88 to 69.
121 On this partnership see Verkroost (1980) and Renterghem (1897; 1898b).
122 Hence his *Liébeault en zijne school* (1898).
123 Renterghem (1898b), p. 67.
124 Renterghem (1897), p. 41.
125 Eeden (1892–3), pp. 53–54. Eeden claimed to have introduced the term (Ellenberger, 1970, p. 330). But he only took up hypnotic practice in 1887, and we find Morselli (1886, p. 374) talking in the previous year of the "efficacia psico-terapica" of hypnotism.
126 Renterghem (1898b), p. 72.
127 Renterghem and Eeden (1889; 1894); Renterghem (1899).

128 Renterghem (1902).
129 But, as we have seen, Berger of Breslau had taken a psychological rather than a physiological view of hypnotism. So had Hückel of Tübingen (Hückel, 1885).
130 On Moll see Moll (1936); Winkelmann (1965).
131 It would be much more useful if it had proper references.
132 A second and enlarged edition was published the following year, and a third in 1895. An English translation, which I have used throughout, appeared in 1890, and had gone through five editions by 1901.
133 E.g. Forel (1887; 1889a).
134 On Forel see Ellenberger (1970), pp. 285–286; Forel (1935); Wettley (1953).
135 Later he became one of the select band to survive a cerebral vascular accident and write about his symptoms "from the inside". See Gardner (1977), pp. 400–401.
136 Ladame (1881); Yung (1883).
137 The third edition of 1895 was enlarged with notes by Oskar Vogt. The English translation of 1906 is from the fourth German edition of 1902.
138 Schrenck-Notzing (1888).
139 See Jendrássik (1886).
140 Moll (1890), p. 18. They include F. Maack of Kiel; von Corval of Baden-Baden; Schuster of Aix; A. Barth and A. Sperling of Berlin; W. Brügelmann of Paderborn; O. Binswanger of Jena; J. Hess, J. Michael and M. Nonne of Hamburg; R. von Hosslin of Neuwittelsbach; P. J. Möbius of Leipzig; and Hirt of Breslau. In Austria, he mentions Frey, A. Schitzler and S. Freud of Vienna; and F. Müller of Graz; in Hungary, Laufenauer of Budapest; in Switzerland, K. Bleuler of Zürich.
141 Dessoir (1888; 1890b).
142 Schrenck-Notzing (1891); Forel (1891).
143 Moll (1892). On these societies, their personnel and background, see Bauer (1991), pp. 25–26.
144 Wundt (1892; French tr. 1893). Cf. the review by F. W. H. Myers (1893). A long analysis of Wundt's ideas is given by Sarlo (1893).
145 See especially Wundt (1893), pp. 95–96.
146 Zielinsky (1968b).
147 Titles of these are given in German by Dessoir (1890b).
148 Fifteen are listed, but Dr. Tokarsky, of Moscow, took part in discussions.
149 See p. 116 of his *Gutachten* in Grossmann, ed. (1894), pp. 116–124.
150 Bérillon, ed. (1889), pp. 188, 289.
151 See Cybulski (1887).
152 On Ochorowicz see Zielinsky (1968a).
153 Cf. Dingwall (1967), pp. 266–270, Zielinsky (1968a) and Ochorowicz (1891).
154 On the hypnoscope see Ochorowicz (1884–5; 1890).
155 Moll (1890), p. 38; Loewenfeld (1901), p. 92.
156 E.g. Fraenkel (1889).
157 Bjelfvenstam (1968), pp. 226–227.
158 Wetterstrand (1888a; 1888b; 1891; 1897a).
159 Wetterstrand (1897a), p. 1.
160 Wetterstrand (1897a), pp. 3, 24–26.
161 Renterghem (1896–7), pp. 53–54.
162 Renterghem (1896–7), p. 52.
163 Bramwell (1930), pp. 42–43; cf. Forel (1906), pp. 220–222.
164 Wetterstrand (1895–6a).

165 Wetterstrand (1892–3; 1897b).

166 E.g. by Dr. R. Binswanger, against whom Wetterstrand was defended by Dr. S. Landgren
     (Landgren, 1893–4).

167 Velander (1889), p. 323.

168 Dingwall (1968b), pp. 192, 198.

169 Jaguaribe (1902), p. 258.

170 Owen (1971), p. 230.

171 On Fajardo see Rezendre de Castre Monteire (1976).

172 Carpenter (1881), p. 608.

173 Romanes (1878; 1880).

174 Felkin (1890), p. 9.

175 A useful list of references will be found in Felkin (1890), pp. 7–11.

176 Bucknill and Tuke (1858 and many subsequent editions).

177 The MS diaries of F. W. H. Myers, now in the Library of Trinity College, Cambridge, show
     that he and his brother A. T. Myers, and Edmund Gurney, visited the Salpêtrière and Nancy
     at the end of August and beginning of Septermber 1885.

178 Roth (1887).

179 Tuckey (1907; 1909).

180 E.g. Tuckey (1892; 1904).

181 Bramwell (1909), pp. 9–10.

182 *British Medical Journal*, 1890, i, pp. 801–802; *The Lancet*, 1890, i, p. 771.

183 *British Medical Journal*, 1890, ii, pp. 442–446.

184 Kingsbury (1891), p. 15.

185 Bramwell (1930), pp. 36–37. *British Medical Journal*, 1893, ii, p. 277.

186 A. Nicoll (1890); R. W. Felkin (1890); G. C. Kingsbury (1891); Sir F. Cruise (1891); R. W.
     Vincent (1893); T. Crisfield (1893). Crisfield appears to have been a masseur – his book is
     a discreet piece of self-advertisement.

187 Hale (1971).

188 Hale (1971), p. 49; Ellenberger (1970), p. 244.

189 Ellenberger (1970), pp. 242–244.

190 For a list see Beard (1881), p. 31.

191 Beard (1881), p. 5.

192 Beard (1881), p. 7.

193 Beard (1881), p. 8.

194 See, for example, the able and extended review by C. K. Mills (1882) of various hypnotic
     works.

195 See e.g. Herter (1888a; 1888b); Cory (1888); and cf. the discussion between Herter and
     others in *The Boston Medical and Surgical Journal*, 1888, Vol. 119, pp. 483–485. C. A. Herter
     (1865–1910), of Glenville, Connecticut, wrote widely on nervous diseases and on the
     biochemistry of disease. He translated Bernheim (1886) into English (Bernheim, 1889b).

196 Hammond (1887).

197 Mason (1901), pp. 298–299.

198 Bell (1889), p. 363.

199 The title of his paper refers to thirty-four cases, but there appear to be thirty-five.

200 C. L. Dana of Cornell Medical College; M. A. Starr of Columbia; also Joseph Collins, Mary
     Putnam Jacobi and A. L. Hamilton. Hale (1971), p. 123.

201 The pull-out table in Crocq (1896) still represents a number of persons as believing in the
     Charcot stages; but in some cases, e.g. Pitres, Crocq is out of date.

202 Liébeault (1889; 1891).

203 E.g. Bernheim (1891–2; 1892b; 1892–3).
204 E.g. Bernheim (1892), p. 1214.
205 Bernheim (1917), p. 47.
206 Babinski (1889; 1890; 1891).
207 Babinski (1889), p. 260.
208 See above p. 331, and cf. Babinski (1889), pp. 262–264.
209 See above pp. 312–313.
210 Charcot and Gilles de La Tourette (1892).
211 One gets this impression e.g. from his replies to Delboeuf (1886), who must have been a fairly persistent and incisive questioner.
212 Charcot (1893; 1897).
213 Hypnotism, and likewise suggestion, say Charcot and Marie (1892), p. 640, may be of some service in the treatment of hysteria, "but not so much as one might *à priori* expect."
214 Owen (1971), p. 122.
215 Janet (1892a).

CHAPTER 17

Pierre Janet and his influence

In dismissing the idea that there is a peculiar state of "hypnosis", akin in its deeper stages to sleep or to natural somnambulism, Bernheim was decisively severing himself from a tradition already more than a hundred years old. The animal magnetists had very commonly supposed that the psychophysiological state of a "magnetic somnambule" was distinctively and importantly different from that of a normal, wakeful human being, and that it was furthermore by no means wholly identical with that of a sleeping person. They found support for this view, as we have seen, in certain marked similarities which they thought they could detect between the state of "magnetic" somnambulism and the state of "natural" somnambulism, not perhaps as exhibited by commonplace sleepwalkers, but rather as manifested in a small number of complex and more decidedly pathological cases. I argued in chapter 13 that the apparent analogies were striking enough to make the magnetists' assimilation or near-assimilation of the two states by no means a foolish or implausible move. The amnesia, the state-dependent memory, the insensitivity to normally painful and disturbing phenomena, even the ostensible clairvoyance and other paranormal phenomena though characteristic of mesmeric somnambulism were also apparently to be found in certain examples of "natural" somnambulism. When the amnesia and state-dependent memory were well-developed it became customary to talk of "dual" or "double" consciousness. In some rare but remarkable cases both of natural and of mesmeric somnambulism, the subject developed personality and intellectual qualities different from and sometimes superior to those of the waking life, and to these cases terms such as "double personality" were sometimes applied.[1] There were even cases of both "natural" and "magnetic" somnambulism in which the subjects, when somnambulic, gave themselves new names or spoke of their normal selves as alien personalities.[2]

In the heyday of the animal magnetic movement, hardly any writer who had given these matters serious thought doubted that magnetic somnambulism and the more developed cases of natural somnambulism were fundamentally similar and perhaps identical. This conclusion was no doubt reinforced by the fact that natural somnambules often proved highly responsive to magnetic manipulations. For example: In 1859, J.K. Mitchell described the case of a girl aged thirteen, and completely ignorant of mesmerism, who was subject to attacks of somnambulism lasting two or three hours, and beginning soon after she fell asleep. The likeness to a waking person was – if we set aside a susceptibility to certain delusions – complete, even to the fruitlessness of all the usual efforts to awaken her. But "half a minute's mesmeric action brought her to her senses, and she had no recollection of what had passed in her sleep."[3]

Mitchell adds the following in a footnote:

> It was curious to mark the sudden transitions from one state of consciousness to another; and the unvaried character of each state; while there existed as little resemblance between the two, as is found in the character of the most opposite persons. The ideas, sentiments, passions, forms of expression and gesticulations – even the temperaments, were those of two contrasted individuals. She was slow, indolent, and querulous when awake; quick, energetic, and vivaciously witty, when asleep.[4]

No wonder that Mitchell – an exceptionally capable man and a good physician – concluded from this and other cases that there is a mesmeric state which may properly be termed an artificial somnambulism and that it may be produced independently of the subject's will or imagination. He noted too that the peculiarities of memory associated with mesmeric somnambulism can also be found in "cases of insanity and double consciousness".[5] All these conditions are, to his way of thinking, closely allied, and although mesmeric somnambulism is not *per se* a pathological state, it can become one, and is not lightly to be experimented with.

Cases of complex natural somnambulism, and double consciousness and dual personality, long preceded the animal magnetic movement,[6] and, though never common, continued to turn up after the hypnotic movement had got under way. Just as cases of natural somnambulism or dual consciousness had often proved amenable to the manipulations of mesmerists, so these new cases of dual consciousness often appeared highly susceptible to the procedures of hypnotists; so much so that certain workers, whilst not following the neatly circumscribed pathways laid down by the Salpêtrière, could not but think there must be a close and highly significant relationship between the deepest (or somnambulic) stage of hypnosis and the pathological "dual consciousness" apparently manifested by certain natural somnambules (or "hysterical somnambules" as they now tended to be called to distinguish them from more commonplace sleepwalkers). It will be useful at this point to look at some examples of these cases of "dual consciousness". A convenient selection is provided by Professor E. Azam of Bordeaux in the article on "double consciousness" which he contributed to Hack Tuke's *Dictionary of Psychological Medicine* (1892). Azam cites seven cases, which he divides into cases of "somnambulism" or "ambulatory automatism" and cases of "dual consciousness"; the grounds for the distinction are far from clear. One, a case of episodic automatism consequent upon brain injury, can be disregarded for present purposes. Two[7] are cases of young women who, during the daytime, were liable to attacks of "spontaneous somnambulism" in which they appeared more vivacious and capable than normal, and had full memory of their "waking" life, but for which they were subsequentlly amnesic. A fourth case is that of Tissié's patient, Albert D.,[8] aged thirty, who suffered from irregular visitations of a morbid state, sometimes lasting fifteen or twenty days, the leading characteristic of which was an impulse to walk. This impulse arose out of ambulatory dreams, which were of two kinds. In the first he would simply dream of visiting a town, and (still in bed) would move his legs as if walking. In the second, he would get up, still dreaming, and actually depart. In this state he was "more intelligent, more volatile" than in his ordinary state, and could remember all the phases of his life. None the less, he got into all sorts of scrapes, and would often return to his normal state while travelling, or in prison, not knowing where he was or how he came there. Under hypnosis, he would recover the memory of his extraordinary journeys, one of which took him to Russia, where he was in danger of being hanged as a nihilist.

The remaining three cases are all ones that have subsequently been classified as multiple personality. The earliest in time is that of Mary Reynolds (1816)[9] of Pennsylvania. At the age of twenty-six, she suddenly fell into a profound slumber from which she emerged having forgotten all she knew and in need of reeducation on all fronts. This was rapidly achieved, but then, after a further sudden sleep, she reverted to her old self with amnesia for the intervening period. These two phases of personality alternated until the second one, which was altogether livelier and more fun-loving than the first, took over permanently.[10]

The other cases, those of Louis Vivé and Félida X., achieved some celebrity, and were a background influence on a number of the theoretical ventures of the time. Louis Vivé (b. 1863)[11] was a juvenile delinquent who at the age of eight had been sentenced to ten years of reform school. At the age of fourteen he had a bad fright from a viper which occasioned hystero-epilepsy and paralysis of the legs. He was sent to the asylum of Bonneval where, following a prolonged hystero-epileptic attack, he was relieved of his paralysis, but also of the memory of all events subsequent to his fright from the viper, and became, in contrast to his previous docile personality, insubordinate, belligerent, and given to indulgence in alcohol. He was studied for a while (1880–1) by Dr. L. Camuset, but escaped and, following a spell at the Bicêtre asylum in Paris under J. Voisin (1883–5), he ended up at the asylum of Rochefort in the character of a private of marines, convicted of theft, but considered to be of unsound mind. There he fell into the care of Professors H. Bourru and P. Burot, and Dr. Mabille. He was now suffering from right hemiplegia and hemianaesthesia, combined with speech difficulties. The latter did not prevent him from abusing his physicians and inflicting long and impassioned diatribes on political and religious subjects upon anyone who could not get away from him. His memory for his past life was fragmentary.

Bourru and Burot had been experimenting with transfer.[12] Experimenting with Louis Vivé they found that contact with steel on the right arm transferred the paralysis and insensibility to the left. At the same time there was a great improvement in the patient's character and speech.[13] And his memory now extended only to brief periods when his paralysis was supposedly likewise on the left, and his character irreproachable. By various metallic, magnetic and electrical manipulations, the Rochefort doctors uncovered or created in Louis a further four states, defined by variations in his analgesias and paralyses and in the scope of his memory. Despite this meddling he left Rochefort in 1887 much improved.[14]

Félida X., Azam's own patient, was, with Louis Vivé, the best-known case of double consciousness of this period. Azam wrote and speculated extensively about the case for over thirty years.[15] I quote the following condensation of his accounts:

> Up to the age of fourteen, Félida was a quick, industrious, somewhat silent child, remarkable chiefly for a varied assortment of pains and ailments [including convulsions] of hysterical origin. One day, when engaged in her regular occupation of sewing, she suddenly dropped off to sleep for a few minutes, and awoke a new creature. Her hysterical aches and ailments had disappeared, she had changed from gloom to gaiety, from morose silence to cheerful loquacity. Presently Félida slept again, and awoke to her usual taciturnity. Asked by a companion to repeat a song which she had just been singing, Félida stared in amaze – she had sung no song. In brief, all the incidents of that short hour between a sleep and a sleep were for her as though they had never been. In a day or two the same sequence was repeated, and so on day by day, until her friends learned to look for and welcome the change; and her lover grew accustomed to court her in the second state, when her somewhat gloomy stolidity had given place to brightness and gaiety. In due course she married [an event hastened by the kindness which, unbeknown to her normal self, she had shown to her lover during her livelier second state]; and as time went on the second state came to usurp more and more of her conscious life – until, in her prime, she would spend months together in that state, with only short intervals of recurrence to her normal condition. The characteristics of the two states remained unchanged; in her first, or what we must now call her *normal*, state, she retained the remembrance of those things only which had come to her knowledge when in the normal state, but the memory of the second, or abnormal, state embraced her whole conscious life. Thus in her later life an occasional relapse to her primary state was attended with very serious inconvenience, for with it the memory of large tracts of her life would disappear .... She would not know the whereabouts of her husband and her

children; she would not recognise the dog which played at her feet, nor the acquaintance of yesterday. She knew nothing of her household require-ments, her business undertakings, her social engagements. Once the relapse came during her return from a funeral, and she had to sit silent and learn gradually from the conversation around her whose obsequies she had been attending. Her gloom and despair during these brief intervals of interrupted fragmentary life are so great as almost to impel her towards suicide ... Finally, Félida occasionally relapses into a third state, characterised by terrifying hallucinations and hyperaesthesia of the skin. In this state she recognises her husband only; and behaves as if stricken with madness.[16]

The strength of the amnesic barrier between the normal and the second state is shown by the fact that in 1878 Félida became convinced in her second, and possibly more perceptive, state that her husband had a mistress. She burst out in threats against the suspected lady, and tried to hang herself. However, in her "normal" state she knew nothing of the matter, and was very friendly towards the woman.[17]

Félida was very prone to haemorrhages and circulatory disturbances, which led Azam to speculate that her two states might be due to inhibition of one or other hemisphere by changes in cerebral circulation.[18] She was an excellent hypnotic subject, the first on whom he experimented. On being hypnotized she always passed into her "normal" state, and showed no knowledge of her second state. Azam was unable to use her hypnotic susceptibility to ameliorate her condition.

Azam develops a theoretical framework for these cases of double con-sciousness. They are an exaggerated form of ordinary sleepwalking. The number of somnambulists is considerable, especially among children, "and from the simple case in which the somnambulist accomplishes a limited act, to the extraordinary somnambulist who seems to have an existence which is independent of that of the waking state, there are a large number of stages." The sleeper's senses may awaken to various degrees. "But generally, the principal sense, that of *sight*, is absent or incomplete; moreover, the ideas of the somnambulist, having lost their equilibrium and proper arrangement (co-ordination), may be directed at random, the senses not acting, or acting imperfectly, and the patient thus being able to have only a false or incomplete idea of the outside world." If, however, the patient's sense of sight is added, this "gives the just notion of the external world, and consequently rectifies the ideas and aids to a right arrangement of them." The somnambulist, "thus perfect, strongly resembles an ordinary person ... This is precisely what happens in the cases of dual consciousness, of which we have cited observations."[19]

Azam's argument is by no means unreasonable. Though (as we have already noted) many would dismiss his assimilation of double consciousness to somnambulism because recent researches have shown that somnambulic episodes generally take place during NREM (non-dreaming) sleep, the position is a good deal more complex than this counter-argument (based mainly on work with children) allows. There are indeed cases, rare but not vanishingly so, which fall, as it were, between ordinary somnambulism and instances of double consciousness. What the EEG of these individuals may be during their somnambulic episodes is far from clear; it is unlikely to be that of NREM sleep.[20] Early collections of such cases are plentiful,[21] and the footings of a bridge between cases of ordinary somnambulism and cases of double consciousness could be constructed from works of Azam's own day, most notably, perhaps, from Hack Tuke's *Sleep-Walking and Hypnotism* (1884).[22] It would not be easy to find defensible criteria for establishing a division anywhere along the continuum between cases of commonplace natural somnamblism and cases of double consciousness, and in not a few instances of "double consciousness" the patient seems to have been a childhood sleepwalker.[23] Furthermore, in certain cases of sleepwalking we find, as it were in embryo, the leading characteristics of cases of double consciousness (or dual personality): amnesia for actions and events of the somnambulic state, even when the actions concerned have been quite complex and apparently intelligent; the possibility of breaching the amnesia by hypnotism; signs of a state-dependent memory linking one sleepwalking episode to another;[24] even hints of personality changes during somnambulism.[25] To take an instance which unites the second and third features: Dufay describes the case of a servant girl whose mistress had accused her of stealing certain items. In consequence, though protesting her innocence and ignorance of the matter, she was imprisoned pending trial. Dufay, the prison physician, recognized the girl as having been previously in the employ of a professional colleague, who had found her a good hypnotic subject. He discovered from the prison nurse that the girl regularly walked in her sleep, though when awake knew nothing of having done so. Accordingly he hypnotized her. She at once remembered where the valuables were. During a previous somnambulistic attack she had concealed them in a place of greater safety. They were shortly discovered there, and the girl was released.[26]

As may be supposed, Azam holds that spontaneous and induced som-nambulism (hypnosis) are essentially identical. In support of this he would no doubt have cited the hypnotic susceptibility of many somnambulic patients, and the considerable similarities between natural and induced somnambulism when manifested by the same individual.[27] It thus appears that the state of deep hypnosis (induced somnambulism), like that of natural somnambulism, differs from that of double or second consciousness, if at all, chiefly in degree of development. In fact, deep hypnosis *is* a state of double or second consciousness, and it is thus not surprising that events from states of pathological second consciousness have so often proved recoverable by hypnosis; nor is it surprising that in certain individuals, perhaps with a constitutional nervous irritability, repeated hypnotization has led to the development of pathological double consciousness (secondary personality).[28]

In all these cases of "double consciousness", however, we have only what Azam calls "the appearance of leading a double life". The patient exhibits two (or more) phases of personality, which may overlap hardly at all, or may be so related that one so to speak includes the other as a portion of itself. But in neither case is it suggested that two developed streams of consciousness continuously coexist in connection with the same nervous system. The idea that more than one stream of consciousness might simultaneously exist in connection with the same nervous system (an idea which came to have considerable influence on turn-of-the-century theories of hypnosis) was brought somewhat dramatically to the notice of psychologists and philosophers through certain very curious case studies published by Pierre Janet in the later 1880s.

SECONDARY STREAMS OF CONSCIOUSNESS

Pierre Janet (1859–1947), who had graduated in 1882 from the école normale supérieure in Paris, was at this period professor of philosophy at the Lycée of Le Havre. It was, however, already apparent that his interests lay as much, or more, in the direction of abnormal psychology as in that of philosophy. Casting around for a suitable topic for a doctoral dissertation in philosophy, he was provided with facilities for studying the bizarre behaviour of several highly hypnotizable hysterical subjects by two local physicians, Drs. J. H. A. Gibert and Powilevicz. Janet's resulting thesis was published in 1889 as L'automatisme psychologique, but his previous accounts of certain of these cases in the Revue philosophique and elsewhere had already established his reputation.

In 1889, Janet moved to Paris, where from 1890 to 1897 he taught philosophy at the Collège Rollin. In 1898, he was appointed to teach psychology at the Sorbonne, and in 1902 became professor of experimental psychology at the Collège de France. He managed to combine teaching at the Collège Rollin with obtaining a medical degree, which he did in 1893. During this period of medical training, Charcot (who was an old friend of his uncle, Paul Janet, the philosopher) gave him access to wards at the Salpêtrière, so that he was able to spend time working with neurological and psychiatric patients. After qualifying, he built up a private practice, but his connection with the Salpêtrière continued under Charcot's successor, F. Raymond (1844–1910), and lasted until 1910 when Raymond died and was replaced by J. Dejérine.

Janet's clinical work set new standards in the detailed recording of case histories and in the close analysis of the mental states of neurotic persons. It resulted in a series of books in which, building upon the ideas expressed in his L'automatisme psychologique, he developed and elaborated his own highly original approach to the theory and therapy of the neuroses.[29] He was a

Figure 10. Pierre Janet. (Courtesy of Clark University Press.)

reserved, shy, scholarly and proud man, who made no attempt to found a school, and did not need constant reassurance from a chorus of admiring converts. None the less, his indirect influence was at first considerable, and though it waned later, it is now again increasing.[30]

In *L'automatisme psychologique*, Janet states that his aim is to show that there are forms of human activity that can properly be termed "psychological automatisms". Their characteristics are as follows: 1. They exhibit certain regularities and may be regarded as rigorously determined. 2. None the less, they have something spontaneous about them, and are triggered by external circumstances, rather than enforced by them at each step. 3. There is a continuum between the simplest and the most complex ones, and since the latter must be thought of as accompanied by consciousness and intelligence, the former must be so thought of also.

Janet divides the psychological automatisms which he had observed in his hysterical subjects into two broad classes, total automatisms and partial automatisms. Total automatisms are "a disposition, a mode of mental being, of the whole subject. The persons under study were completely either in the waking state, or in somnambulism, or in delirium, but they were never half in one state or half in another; so their consciousness, extended or narrowed, whatever its nature, included all the psychological phenomena of the subject."[31] Total automatisms as thus defined range from catalepsies, through hysterical somnambulism and fugue states, to alternating personalities.

Sometimes, however, psychological automatisms, "instead of ... governing all conscious thought, can be partial, and govern a small group of phenomena separated from the others, cut off from the total consciousness of the individual which continues to develop ... in a different manner."[32] Partial automatisms include, for instance, partial catalepsy (catalepsy involving one limb only), hysterical paralyses and contractures, limb anaesthesias in which, for example, an anaesthetic hand adjusts itself correctly to an unseen object presented to its grasp, water-divining, pendulum swinging, planchette writing, and other forms of automatic writing and speaking.

It was the "automatic" writing and speaking of certain of Janet's hysterical subjects that provided him with some of his most striking examples of partial automatisms. The activities concerned are intelligent, and presumably conscious – we have the same reasons for attributing consciousness to them as to intelligent activities carried on by persons in a normal state[33] – but the consciousness involved is largely separated from and inaccessible to the ordinary waking consciousness. There is little awareness of or memory for what is written and the hand which writes is likely to be one which is already anaesthetic. In Janet's terminology, the activity of writing, thus carried on outside the main consciousness, is "subconscious".

Janet developed a number of ingenious ways of inducing automatic writing and speaking in his hysterical subjects. The simplest was what he called the "method of distraction". He would arrange for the attention of the patient to be totally distracted, for instance by having a third person engage her in interesting conversation. He would then approach her from behind, speak to her in a low voice, and give her trivial commands, for instance to raise an arm after he had clapped his hands so many times. Often the subject would obey the command without realizing that she had received it or (so long as the conversation continued) that she had obeyed it. Such acts, says Janet, demonstrate a partial somnambulism, in which it is easy to recognize the workings of intelligence.

Thus with some subjects, for example the celebrated Madame B. (known as

Léonie), Janet would surreptitiously put a pencil in their right hands and ask such questions as "How old are you? In what town are we here? etc.", with the result that

> ...the hand stirs and writes the answer on the paper, without, during this time, Léonie ceasing to speak of other things. In this way I have made her carry out arithmetical operations which were sufficiently correct; I have made her write quite long answers which clearly showed a fairly well developed intelligence.[34]

Automatic writing might also be unleashed in a waking subject by post-hypnotic suggestion. A "subconscious" intelligence so well developed must, Janet thought, be supposed to enjoy continual existence outside the experimental periods. Confirmation of this view came from certain patients who in addition to manifesting these partial automatisms were also subject to attacks of total automatism of various kinds. Léonie in particular became celebrated as a case of multiple personality.[35] When Janet first encountered her she was already middle-aged. She had been a somnambulist from the age of three, and had been frequently mesmerized by an assortment of operators. In her ordinary state, as Léonie (Léonie 1), she was a rather dull peasant woman. In the mesmeric state, she became an altogether livelier (and incidentally much less suggestible) personality, who called herself Léontine (Léonie 2) and refused to identify with her waking self, of whose existence she was aware. By renewed mesmeric passes on Léonie 2, Janet obtained a third state, Léonore (Léonie 3) whose memories included the memories of the other two but whose own memories were not shared by them. Now when Léonie 1 was dominant, Léonie 2 could be contacted by Janet's usual techniques of distraction and induced to perform actions of which Léonie 1 remained unaware. But as Léonie 2, Madame B. could remember and describe these actions. And similarly, instructions given to Léonie 3 would be carried out, if desired, during the dominance of Léonie 2, without the knowledge of the latter. Léonie 3, however, would remember what she had done. Sometimes, also, she would subsequently claim responsibility for an hallucinatory voice heard by Léonie 2. These three personalities thus gave every indication of coexisting as autonomous streams of consciousness,[36] even though only one of them would be dominant at any given time. And in other cases too, for instance that of Lucie,[37] who also exhibited three phases of personality, a similar state of affairs seemed to obtain.

Now if in these cases of dual or multiple personality we are forced towards postulating dual or multiple systems of psychological phenomena, coexistent and indeed coactive, why not in other cases too? What more general implications might such a move have for normal and abnormal psychology? In 1889, Janet wrote: "We have insisted on these developments of a new psychological existence, no longer alternating with the normal existence of the subject, but absolutely simultaneous. Recognition of this fact, indeed, is indispensable for understanding the behaviour of neuropaths and of aliénés."[38] Around this notion he developed over several decades his own highly original approach to the understanding of hysteria and other neurotic disorders.

Janet's theoretical stance is perhaps best approached through his concepts of *shrinkage of the field of consciousness* and *subconscious fixed idea*.

According to Janet, the "field of consciousness" is "the largest number of simple, or relatively simple, phenomena, which might be simultaneously connected with our personality in one and the same personal perception."[39] In hysterics, due to a "general cerebral exhaustion" or "psychological impoverishment", the field of consciousness has shrunk and become very small. Hence the characteristic absent-mindedness of hystericals, and their weathercock lability to each passing external impression. Hence too the anaesthesias, contractions of the visual field, paralyses, etc., from which they are so liable to suffer. (Janet calls these the "stigmata", as opposed to the "accidents", of hysteria because they are essential to the diagnosis.) The field of consciousness is "filled with one relatively simple sensation, one remembrance, a small group of motor images, and cannot contain others at the same time."[40] The excluded sensations, images, etc., continue, however, to exist and to exert an influence on behaviour, as is shown by the adaptive responses that can be elicited by various stratagems from anaesthetic or paralysed limbs, etc. Janet talks of these sensations etc. as "dissociated", but he also talks of mental "disintegration" (*désagrégation*), which more correctly expresses what he has in mind. For we are not dealing with a mere splitting apart of associated elements; rather we have a failure of synthesis of certain limited psychological phenomena into the larger unity which may be termed a personality. What exactly this synthesis consists of remains obscure.

The obsessions, aboulias, etc., characteristic of what Janet calls "psychasthenia",[41] are also to be understood as manifestations of a certain cerebral exhaustion, but in psychasthenia the exhaustion results not in the shrinkage of the field of consciousness, but in a general attenuation of all psychological functions, causing feelings of indecision, depersonalization and inadequacy.

In hysteria, the cerebral exhaustion is frequently due to constitutional factors, but there are often precipitating circumstances, usually of an emotive and stress-inducing kind, which finally snap the fragile hold which the subject's mind has upon the elements it has so far managed to synthetize. Some of these elements go astray. The "stigmata" of hysterics – anaesthesias, paralyses, diminution of sensory fields – manifest or become more prominent. Among the "accidents" of hysteria the most frequent is a loss of the memories of the shock or period of stress which produced the initial cleavage. These memories retain their internal coherence but break away from the main personality or stream of consciousness, which can no longer incorporate them into a synthesis. To this nucleus other ideas and other reminiscences may accrete, and the whole complex, now unrestrained because of its relative isolation, may grow beyond measure, forming a "subconscious fixed idea"or system of ideas, which may exercise a most baneful effect upon the principal consciousness, or temporarily displace it, giving rise to repeated episodes of hysterical "somnambulism" in which the events of the initial shock are acted out.[42] These episodes can thus be direct representations of an original trauma, as can other "accidents"; the stigmata, however, are not representative of traumata, but are simply precipitated by them.

Take the case of Irène,[43] a girl of twenty who fell ill because of the despair caused by the death of her mother. The mother lived alone with her daughter in a poor garret, and there died a lingering death of consumption. The girl fought desperately against the inevitable, nursing her mother during sixty nights, struggling meanwhile to earn a few sous from her sewing machine. After her mother's death she tried to revive the corpse, to make it breathe. The corpse fell out of bed and she had the greatest difficulty in lifting it back in again.

Some while after the funeral she began to have attacks of somnambulism, for which she was afterwards amnesic, in which she would give a detailed rendition of these harrowing scenes, with a realism that no actress could have imitated. She would then continue the same series of ideas by planning and (in her imagination) committing suicide through placing herself before an oncoming railway engine. Her waking amnesia included not just the events surrounding her mother's death, and the subsequent reenactments of them, but all kinds of matters to do simply with the fact that her mother was dead. This fact she now seemed able to contemplate without emotion, and she engaged in behaviour inappropriate for a person still in mourning. Thus it was not just the specific memories that had sunk out of sight; they had dragged down with them a whole complex of related ideas, affecting the whole tenor of the principal consciousness.

There are also cases of more complex hysterical somnambulism in which the somnambulic episodes are not, as it were, all performed with the same script. Hysterical fugues (which in turn are not sharply separable from cases of multiple personality) would be an example. Here what has become submerged is not one distinct idea or closely linked system of ideas, but a vaguer set of thoughts or tendencies bound up with some feeling or emotion (e.g. the feeling of curiosity for distant lands, the feeling of bondage towards a hated master), the whole possessing, however, a degree of mental unity.

Janet's position with regard to "subconscious fixed ideas" has certain implications for therapeutic practice. If hysterical somnambulism, and various kinds of pathological happenings intruding into the principal consciousness, can be due to the continued activity of dissociated traumatic or stressful memory-systems, then obvious therapeutic stratagems are either 1. to restore the dissociated memories to the principal consciousness, which may need to be fortified so as to be capable of bearing them, or 2. to eradicate the pathogenic memories altogether, or 3. to transform the pathogenic memories so as to render them innocuous.

In the application of stratagem 3. Janet showed himself at his most ingenious. A striking case was that of Justine,[44] who had nursed patients dying from cholera, and had suffered for twenty years from hysterical crises in which she cried out "Cholera...it's taking me" and had hideous visions of corpses. Redintegration of the memories did not effect a cure, but by means of hypnosis Janet was able to transform them. Thus a particularly unpleasant naked and greenish putrefying corpse was clad and turned into a rather comic Chinese general whom Justine had seen at the Universal Exposition. Finally, he eliminated the idea of cholera by suggesting that the Chinese general was called "Cho Lé Ra".[45]

JANET ON HYPNOTISM AND SUGGESTION

In these therapeutic endeavours, Janet made frequent use of hypnotism and suggestion, and his ideas about them are closely linked to his views on hysteria and hysterical "somnambulism". He is often hailed as a forerunner of the current "neodissociationist" approach to hypnotism, but in fact today's neodissociationists are closer to Grasset or Sidis or indeed Gurney than to Janet.

I shall begin with Janet's views on suggestion. A question to which he frequently returns is that of how suggestion and suggestibility are to be defined. Bernheim's approach to this issue, according to which almost any idea which penetrates the mind and produces an effect qualifies as a suggestion, is far too broad, and fails to capture the essence of the phenomenon. About true suggestion there is something abnormal and pathological, as suggestible subjects are well aware:

> Their minds are not filled all day long with suggestions. They know very well how to distinguish what is suggestion in them from what is not. A patient has sometimes answered me in a vulgar but quite characteristic way: "Sir, I do not know the reason, but the thing did not take."
> "What do you mean? You did not understand what I said?"
> "Yes. I understood quite well."
> "Then you do not wish to do that, you do not accept?"
> "I accept all you please. I am quite ready to obey you, and I will do it if you choose, only I tell you that the thing did not take."[46]

But why should an idea work itself out in this aberrant and automatic way? The answer lies in the diminished capacity for personal synthesis which characterizes the hysterical patient – it is in hysterics above all that suggestion is liable to "take". "Why is it that in our normal state former ideas, called up by some means or another, do not develop completely? It is because we connect them from their beginning with the enormous mass of other recollections, of other images, which constitute our personality. They have their place; they play their part in the great system, but they are not isolated and independent, and their development is restrained by the development of all the other thoughts."[47] In the diminished field of consciousness of the hysteric, ideas, once inserted, may develop in relative isolation without any of the normal restraints; hence the automatic working out of suggestions, and the characteristic susceptibility of the hysteric to both hetero- and autosuggestion.

"Monoideism" as an explanation of suggestion was not new with Janet – we have already encountered Braid's and Richet's versions of it – but his treatment of the topic possesses novel features, and is greatly expanded in *Les médications psychologiques*, a work which, though first published in 1919, belongs essentially to the years immediately preceding the First World War. The starting points are the notions of *impulse* and of *assent*. Both notions, it has to be said, are left fairly obscure. Assent is approached through an analysis of the activity of speaking. Originally, speech was intimately bound up with the action of which it formed part.[48] But gradually speech and action became more and more separated, until men "learned to play with language, to enjoy the excitement of language for its own sake."[49] This evolution has gradually led to the emergence of the *idea*, in which the action-tendency associated with the word, and constituting its meaning, has become vanishingly weak. In an attempt to restore the stability of language, men have "striven to make their own words become once more commands unto themselves; hence arise such things as *promises*, *pacts*, *affirmations*, and various kinds of *assent*, which are at the root of *belief*."[50]

Assent is a response to a verbal question or statement, it matters not whether propounded by oneself or another. Thus *will* is a response of immediate assent to a proposed action. "Will you go for a walk with me?" "With pleasure." Action ensues at once. *Belief* is an assent to a question, request or proposition to which immediate action is not required and will in any case depend upon the future fulfilment of certain conditions. If I assent to "It is raining", and state that I believe it is raining, I make up my mind to take an umbrella if I go out, and to behave in certain ways should it be raining. Assent can be negative or affirmative. The choice negation or affirmation may take place immediately and unreflectingly, or it can be done reflectingly, with a slowing down of assent. During this period, the idea that has been evolved is compared with other ideas carrying other response tendencies. Eventually a resolution or decision is reached.

The labour of "reflective assent" is long and difficult and it can go wrong. Thus certain patients, "abulics, and doubters, perform the action of ideation, and begin interrogation and discussion, but can get no further. They never reach the stage of reflective affirmation or negation; they never get beyond ideas and imaginations, never achieve resolution or knowledge."[51] Another disorder of reflective assent takes the form of *impulse*. Impulse is a kind of automatism that supervenes upon a process of reflection. The subject "will suddenly give up deliberating or reasoning, and will assent more or less fully to one or other of the conflicting ideas, according as either may chance at the moment to preponderate in his mind."[52] Suggestion arises when the play of ideas is so influenced or manipulated as to bring about the induction of a certain impulse in place of the transformation of an idea into reflective will or reflective belief. It is above all hysterics who are liable to impulse and hence to suggestion. The hysteric suffers from a general "lowering of psychological tension" which Janet calls "depression" (an idea not too distant from the "cerebral exhaustion" which he had talked about in earlier years). The effect is greatest upon higher mental functions which "can no longer undergo complete activation, and remain in one of the initial stages of activation, or enter the field of activity slowly and late."[53] Whereas in the psychasthenic, reflection is never completely arrested, the hysteric "ceases to question himself, ceases to call up new pros and cons, ceases to surround the evoked idea by other ideas. If the circumstances or the words of an individual who is on the watch continue to evoke and strengthen this idea, it remains alone, isolated, and powerful. It is then transformed by automatic assent, so that it becomes a suggestion. The curtailment of the mental field, monoideism, results from the disappearance of all other ideas and from the survival of the primary idea just as it existed before reflection began – the idea which has now become impulsive."[54]

I turn now to Janet's ideas about the nature of the hypnotic state, ideas which, in essence, changed very little over the years.[55] For Janet, hypnosis is not a special state of heightened suggestibility, itself produced by suggestion, for there are many examples of subjects who, though highly suggestible in the waking state, are not susceptible to suggestion when hypnotized;[56] nor is it a state of, or akin to, ordinary sleep, for it has neither the physiological nor the psychological characteristics of the latter.[57] Hypnosis is simply the artificial production of a state of hysterical somnambulism in a person liable to such attacks. About methods of achieving this, including his own, Janet says very little, though it is apparent that at times he used verbal suggestion and at times mesmeric passes (he was a great admirer of the old mesmerists). His initial approach often seems to have been to introduce himself into the subject's somnambulism by speaking of ideas currently dominating it. Thereafter, having become himself so to speak a part of the somnambulic dream, he could gradually transform it, call it up at will through a designated signal, and so forth.

According to Janet, there is no reason for regarding hypnosis as in any way distinct from hysterical somnambulism. The hypnotic state never exhibits any characteristics which cannot be found in "natural" hysterical somnambulism; hypnotizable persons (that is, ones who can be put into the deep, or somnambulic stage) are all either hysterical patients or possessed of an undoubted hysterical

diathesis; patients suffering from other diseases than hysteria are not hypnotizable; like natural somnambulism, susceptibility to hypnosis diminishes and
disappears as the patient improves and recovers; it is possible to transform
natural into artificial somnambulism (hypnosis), whilst a hypnotized person, left
to himself, will drift into a state of hysterical somnambulism; memories of past
somnambulic states (though inaccessible to waking recall) may be recovered
during hypnosis, and vice versa, suggesting that the same "subconscious" or
"dissociated" stream of consciousness comes to the fore in both.[58]

This account of hypnosis leaves three very obvious clusters of unanswered
problems, each of which Janet tackles, none altogether happily. They are as
follows:

1. If hypnosis is simply the reinstatement of a pathological state, how are we
to account for the therapeutic usefulness of hypnotic methods? Janet does not
even propose that suggestion can produce physiological or mental effects that
the subject could not ordinarily achieve. Instead, he points out[59] that the
automatic performance or blocking of an action may have helpful physiological
consequences for persons suffering from disorders of the will, amnesia, hysterical
paralyses, tics, constipation, indigestion, etc. Furthermore, the automatic
exercise of a function may suffice to prevent deleterious effects of its prolonged
suppression, and the mere fact that the subject knows he is capable of exercising
the function will help to remove from his mind the idea that he cannot be cured,
and hence restore his confidence. Indeed, automatic functioning will gradually
restore the integrity of depressed function, until "at length we are surprised by
finding that the will has readopted them and can direct them once more."[60]
Hypnosis *per se* may bring certain benefits. It may be useful as a means of repose.
Inasmuch as it differs from the waking state, it may promote the disappearance
of certain harmful symptoms, for example paralysis, aphagia, mutism. It may
restore memories, so helping us to understand the genesis of symptoms. It may
enable us to revive the patient's memory of the symptoms of other secondary
states related to it, especially the somnambulistic state, and then help us to
modify or transform them. Above all, it is a state favourable to the influence of
suggestion.

2. Why are deeply hypnotized persons suggestible? Part of the answer is that
in hypnosis a secondary stream of consciousness has come to the fore, and is *ex
hypothesi* likely to be a restricted and impoverished consciousness, detached from
the principal consciousness, and lacking in all higher, i.e. critical and reflective,
functions. In such a stream of consciousness, the monoideism which either is, or
is a precondition of, the automatic realization of a suggestion, will be more
readily obtained. Occasionally, however, a "secondary" stream of consciousness
may be superior to, more highly endowed with critical faculties than, the
"normal" consciousness, and then it may be less rather than more susceptible
to suggestion than the latter.

3. Lastly, and most importantly, comes the following question. Janet is heavily committed to the view that suggestion and hypnosis are pathological phenomena, part and parcel of the symptomatology of hysteria. Yet Liébeault and Bernheim, and following them numerous others, claimed, with a great deal of circumstantial detail, that they had produced these phenomena in many perfectly normal individuals. How can Janet accommodate these claims without actually denying them, which it would have been very difficult for him to do? His moves in the matter[61] are uneasy to the verge of paltering and equivocation. At times he hints strongly, apparently on the basis of his own experience, that many such subjects are not "normal" at all; fuller investigation of them would reveal many hysterical symptoms. Another of his tactics was to point out (with supporting evidence[62]) that the "depression" (or "lowering of tension" or "cerebral exhaustion") which underlies hysteria, and promotes suggestibility and the dissociation of consciousness, though often a matter of hereditary weakness or predispostion, may also be brought about or intensified by other factors, for instance fatigue, strong emotion, or various kinds of intoxicants or anaesthetics.

Now that such factors should promote suggestibility is one thing, but that they should occasion artificial somnambulism is quite another. If these states acted alone, so Janet thinks, "somnambulism would only occur from time to time. Such states of exhaustion only leave the field free for the invasion of all kinds of inferior tendencies, and these inferior tendencies, developing without guidance, could give rise to simple fits of hysteria or to other states which in no way resemble somnambulism."[63] However, given the right kind of "guidance" from an experienced hypnotist, who can seize this period of depression to minimize the normal personal consciousness, and appeal to and call forth other and more elementary tendencies, hypnosis may be brought about. But this still seems to require "a disposition in certain tendencies to become automatically active",[64] and in any case it could not seriously be maintained that the numerous "normal" individuals hypnotized by the members of the Nancy School were all in a state of "depression" occasioned by fatigue, emotion or intoxicants.

Janet's position of last resort is that psychological depression, though mainly the outcome of exhaustion, emotion, or intoxication, can also to some extent be voluntarily induced in response to commands (e.g. to "sleep"). "We can, to a degree at least, spontaneously relax ourselves; just as in other cases we can deliberately render ourselves tense, we can 'buck up.'"[65]

Since "depression", which is a condition favouring suggestibility and the development of secondary states such as natural and artificial somnambulism, can thus be induced voluntarily and in persons who are presumably normal, it appears that Janet has, albeit late and reluctantly, conceded that susceptibility to hypnosis and suggestion need not always be a sign of underlying pathology.

INFLUENCE OF JANET

As I remarked earlier, Janet founded no school. He always displayed what Ellenberger calls "an unrelenting independence",[66] deferring to no master and expecting no pupil to defer to him. Though he would vigorously defend his views, and his claims to priority, proselytization was quite alien to him, and the not inconsiderable rôle which he played in the history of psychiatry and psychology in general, and of hypnotism in particular, arose not from the direct propagation of his own views but from the influence which he had in shaping the views of others. Jung, Bleuler, Adler and the early Freud were all more or less indebted to him, as were others to whom we shall come.[67] His influence was perhaps greatest in the United States, being overshadowed by that of psychoanalysis shortly before the First World War.[68] In France, he continued to command respect, but he also had enemies, notably Babinski and Dejérine, and his affinities with Charcot made him to many eyes an anachronism as the latter's star posthumously faded. L. Laurent's *Des états seconds* (1892) is almost pure Janet; the author differs from Janet principally in being prepared to suppose that there may be subconscious streams of consciousness even in normal individuals. But for our purposes, the two most notable French works strongly influenced by Janet are Binet's *Les altérations de la personnalité* (1892), and Grasset's *L'hypnotisme et la suggestion* (1903), with each of which belong certain earlier articles by each author.[69] Grasset will be discussed in a later chapter, but the new developments in Binet's thinking – he left the Salpêtrière in 1889 to become (in association with Beaunis) director of the laboratory of physiological psychology at the Sorbonne – merit some notice at this point.

These developments – which began independently of Janet, but soon converged upon his ideas and findings – resulted mainly from work with hysterical patients at the Salpêtrière, and only now and again involved hypnotism. Fundamental to them were observations and experiments (similar to Janet's) upon patients suffering from hysterical anaesthesias. Binet was immensely interested (as was Janet) by the apparently intelligent behaviour which an ostensibly anaesthetic hand might engage in without its owner's knowledge. It was not necessary to hypnotize the subject. "The only preparation for the experiments consists in hiding his anaesthetic arm from sight, either by putting it behind his back or by using a screen. Things being so disposed, it is easy – at least in some cases – to bring out, without the patient's knowledge, intelligent movements in his insensible member. We witness the awakening of an unconscious intelligence; we can even communicate with it and direct it, hold a coherent conversation with it, measure the extent of its memory and the acuteness of its perception."[70] In some subjects, automatic writing could be induced. Binet adopted Janet's usage of referring to the unconscious intelligence as "subconscious".

Binet became convinced that in certain of his hysterical patients these experiments revealed a fundamental division of consciousness, so fundamental that he is quite prepared to talk of there being two egos. None the less, he had also to recognize that a kind of communication or traffic went on between the two consciousnesses, a traffic of which the patient was ordinarily quite unaware. For example, if the patient (pencil in hand) was told to think of a word or a number, her screened hand would, unknown to her, write that word or number, or otherwise indicate knowledge of it. If, *per contra*, the information were given first to the anaesthetic hand (e.g. by making it form words), it would find its way into the main consciousness as an intrusive thought, image or even hallucination. Binet refers to these phenomena as "suggestions from unconscious indications",[71] and thinks they show that unperceived stimulation may arouse association of ideas which are "preserved notwithstanding the mental disintegration", and serve as "a connecting link between separate consciousnesses that have no longer any knowledge of each other."

A sign of the extent to which by 1891 Binet had distanced himself from the Salpêtrière school, stranded as it now was in a mirage-haunted desert of its own making, is that he devoted a chapter of *Les altérations de la personnalité* to "plurality of consciousness in healthy subjects". His evidence for this phenomenon depended mainly upon utilizing Janet's technique of distraction, whilst the subject's hand was led, by delicate initial guidance, to write or draw protracted sequences of simple shapes (rings, pothooks, etc.). Subjects apparently lost awareness of the hand's activities., which could then be influenced by suggestions conveyed by different kinds of light pressure.[72] According to Binet, these facts, "seem to me to be amply sufficient to demonstrate the possibility of rousing an unconscious personality in healthy persons or those who are very nearly so."[73] This was a possibility which Janet could never bring himself to admit, but outside France there were others who had already reached, acknowedged and embraced it.

NOTES

1 See e.g. Colquhoun (1836), Vol. 1, pp. 345–6, 384.
2 Deleuze (1813), Vol. 1, pp.176n-177n; Lang (1843), pp. 105–106. Some writers, following Wigan (1844), talk of duality of consciousness when an individual claims to have been aware of two separate streams of consciousness in his own mind, e.g. two quite different "inner voices" talking at cross-purposes. However, this is not a duality of consciousness properly speaking, but a certain sort of duality in the contents of consciousness.
3 Mitchell (1859), pp. 163–164.
4 Mitchell (1859), p. 164n.
5 Mitchell (1859), p. 186.
6 Some writers have tried to maintain – but I think mistakenly – that cases of spontaneous double consciousness were not known prior to the "discovery" of mesmeric somnambulism. See Gauld (1992).

7   The cases of Mlle. R. L. (Dufay, 1876) and Mlle. X. (Bonamaison, 1889–90). See e.g. Azam (1887). Cf. Binet (1896), pp. 6–20.

8   Tissié (1890).

9   One of the most famous early cases, often referred to as "Macnish's case" because of a second-hand account of it published in R. Macnish's *The Philosophy of Sleep* (1830). See S. L. Mitchell (1816); S. W. Mitchell (1889); Plummer (1887); Kenny (1986).

10 Goodwin (1987) has tried to relate Mary's disorder to traumatic events experienced in childhood.

11 The literature on Louis Vivé is very considerable. See Camuset (1882); Bourru and Burot (1885); J. Voisin (1885); Berjon (1885); A. T. Myers (1886); Bourru and Burot (1887; 1888); Binet (1896), pp. 25–32, 262–266; F. W. H. Myers (1903), Vol. 1, pp. 338–343.

12 Cf. above pp. 333–335.

13 This led some writers to suppse that there had been a sudden recovery of left hemisphere function and depression of right. See e.g. *Journal of the Society for Psychical Research*, 1885–6, Vol. 2, pp. 225–228; Cf. Azam (1887), pp. 180–183. Ireland (1893), pp. 320–376, is of interest in connection with this topic in general.

14 F. W. H. Myers (1903), Vol. 1, p. 339.

15 See e.g. Azam (1887). Cf. Binet (1896), pp. 6–20.

16 Podmore (1897), pp. 401–402.

17 Azam (1887), pp. 174–175.

18 Azam (1887), pp. 180–183.

19 Azam (1892), p. 406.

20 Cf. above, pp. 261, 264; and below, pp. 620–621.

21 E.g. Horstius (1593), and cf. the references in Franzolini (1882).

22 Cf. Carpenter (1881), pp. 591–601.

23 Cf. Dyce's case, above p. 260, and the case of Madame B., below, p. 371.

24 The earliest case of the latter kind that I have come across is a sixteenth-century one from Paracelsus, cited by Volgyesi (1963). Carpenter (1881) has two somewhat similar cases, pp. 596–598.

25 Extreme examples of this are the commission of violent crimes during somnambulic episodes by persons who were not violent during their waking state. Cf. Tuke (1884, pp. 12–19), and below pp. 620.

26 Cited by Myers (1885–7a), pp. 228–229. Podmore (1897), p. 392, cites a somewhat similar case from his own knowledge. Here it was her own money that the somnambulist hid.

27 Cf. the case cited by Tuke (1884), pp. 37–38, 51–60.

28 As in the case described by Forel (1906), pp. 298–303.

29 The most important are: Janet (1893–4; 1898; 1901; 1903; 1907; 1909; 1919; 1925); Janet and Raymond (1898); Janet and Raymond (1903).

30 Boosted especially by the long and sympathetic account of him in Ellenberger (1970). See in addition van der Hart and Horst (1989), and van der Kolk and van der Hart (1989).

31 Janet (1889), p. 223.

32 Janet (1889), p. 224.

33 Janet (1889), p. 22.

34 Janet (1889), p. 244.

35 For a short account of the case of Madame B. (Léonie) see Podmore (1897), pp. 389–390. This has been condensed from Janet (1889),pp. 128–133. See also Janet (1886d; 1888); F. W. H. Myers (1888–9b) and F. W. H. Myers (1903), Vol. 1, pp. 322–326. Photographs of Madame B. in different states will be found in Stead [1891], pp. 18–19.

36 The metaphor of "stream of consciousness" no doubt gained currency from James' famous

chapter on the "stream of thought" (James, 1890). But before that Janet was talking of the conscious life of a subject such as Lucie being composed of "three parallel currents" (Janet, 1889, p. 335).

37 The case of Lucie is described by Janet (1886a; 1887; 1889). Cf. F. W. H. Myers (1903), Vol. I, pp. 326–330.
38 Janet (1889), p. 323.
39 Janet (1901), p. 501.
40 Janet (1901), p. 503.
41 On psychasthenia see Janet and Raymond (1903).
42 On this whole question of traumatic memories see Janet (1925), Vol. I, pp. 589–600.
43 Janet (1936, pp. 6–8; 1907, pp. 29–31, 37–38).
44 Janet (1907), p. 65.
45 Janet (1898), pp. 156–212. In another case, Janet cured a young woman disappointed in love by transforming her lover's head into that of a pig – Janet and Raymond (1898), Vol. 2, p. 135.
46 Janet (1907), pp. 284–285.
47 Janet (1901), p. 274.
48 Janet (1925), Vol. I, p. 233.
49 Janet (1925), Vol. I. p. 233.
50 Janet (1925), Vol. I, p. 234.
51 Janet (1925), Vol. I. p. 240.
52 Janet (1925), Vol. I, p. 241.
53 Janet (1925), Vol. I, p. 269.
54 Janet (1925), Vol. I, p. 273.
55 I follow here mainly the summary of his ideas, Janet (1907), pp. 110–116, which also occurs in Janet (1936), pp. 270–276.
56 Janet (1925), Vol. I, pp. 277–282.
57 Janet (1925), pp. 282–287.
58 Janet (1925), pp. 288–289, 324.
59 Janet (1925), pp. 318–325.
60 Janet (1925), Vol. I, p. 322.
61 Especially Janet (1925), Vol. I, pp. 294–307.
62 Janet cites (1925, Vol. I, pp. 269, 298) Lagrange and Tissié's case of enhanced suggestibility in a bicycle racer in a state of fatigue. This may be compared with the more recent experimental study by Banyai and Hilgard (1976) of an "active-alert" hypnotic induction utilizing an exercise bicycle.
63 Janet (1925), Vol. I, p. 302.
64 Janet (1924), p. 135.
65 Janet (1925), Vol. I, pp. 305–306.
66 Ellenberger (1970), p. 407.
67 On Janet's influence see Ellenberger (1970), pp. 406–409.
68 Cf. Hale (1971).
69 Binet (1889), with which cf. Myers (1889–90); Grasset (1902–4).
70 Binet (1896), p. 97.
71 Binet (1896), p. 207.
72 Binet (1896), p. 243.
73 Binet (1896), p. 245.

CHAPTER 18

Dipsychists and polypsychists: Dessoir, Gurney, Myers and James

One of the earliest, and not the least interesting, of those who came to hold that there is "an unconscious personality" even in healthy persons, was Baron Carl Du Prel (1839–1899), of Munich,[1] a former regular soldier who became a prolific writer in that debatable land where philosophy, psychology, mysticism and mesmerism overlap and cause confusion. His *Philosophie der Mystik* (1885)[2] is an attempt to prove, on empirical as well as logical grounds, the existence of a transcendental ego, which he believed would manifest in such phenomena as duplication of consciousness, alternations of these states of consciousness (each with its own state-dependent memory and functions of knowing and willing), and clairvoyance transcending ordinary limitations of space and time.

That such phenomena occur, Du Prel demonstrates to his own entire satisfaction, drawing, curiously enough, not upon the hypnotic and psycho-pathological investigations that were attracting so much notice at the time, but upon much earlier mesmeric and related literature, of which his knowledge was profound. The student of that literature can still study him with profit, but upon contemporary hypnotists he had little influence.

One of the few hypnotists who was acquainted with Du Prel and his works was another prolific writer, the Berlin psychologist Max Dessoir (1867–1947).[3] Whether Du Prel had any serious influence on Dessoir's thinking it is hard to say, but both are, to adopt Ellenberger's terminology,[4] *dipsychists* (in Dessoir's case perhaps a *polypsychist*). Both held, that is, that linked to the organism of every normal and healthy person (as well as of every hysteric) are two (or more) simultaneously active streams of consciousness. Both were also much interested in parapsychological phenomena (Dessoir is credited with introducing the term "parapsychology"[5]). Dessoir's pamphlet, *Das Doppel-Ich*, which expounds a dipsychist position, was first published in 1889,[6] at about the same time as Janet's *L'automatisme psychologique*, and soon attracted considerable notice. Dessoir's aims in this pamphlet are to show that human personality in reality "consists of two clearly separable spheres..., each of which is held together through its own chain of memories",[7] and (linked to the preceding) to reach a new definition of hypnosis.

In support of the first of these aims, Dessoir cites three categories of material. The first is that of certain "automatic" actions of everyday life, which seem to require conscious regulation, but go on outside the ordinary consciousness – such actions as dressing oneself or following a known route. Some of these actions, Dessoir thinks, require not only an unconscious intelligence, but an unconscious memory. He cites, for instance, the claims of Mr. Thomas Barkworth, a member of the Society for Psychical Research, who could engage in, and give his whole attention to, a lively conversation, and at the same time quickly and correctly add up large numbers. Since consciousness and memory are the two essential elements of personality, we can assert that every human being has within himself the germs of a second personality. Dessoir talks of our everyday consciousness as the "overconsciousness" and the hidden or secondary consciousness as the "underconsciousness".[8] He rests, it seems to me, rather a lot on Mr. Barkworth's somewhat casual self-observations.[9]

Dessoir's second category of material comes from psychopathology, and ranges from the abnormal states of consciousness to be found in dreams and epilepsy, to cases of dual consciousness and dual personality such as we have already extensively discussed. He is much indebted to Janet and to Binet.

His third category is that of the phenomena of deep hypnosis. Hypnosis, he holds, consists of "an artificially induced preponderance of the secondary self",[10] the same secondary self that is active in the automatisms of everyday life, in somnambulisms, hysterical fugues, etc. The well-known techniques of hypnotization have the effect "of reducing the sphere of influence of the ruling synthesis" as much as possible, until some element of the hidden personality emerges, which will then pull other associated elements after it.[11] In deep hypnosis, there may emerge memory-chains embracing past somnambulic episodes inaccessible to the ordinary consciousness. These episodes are not necessarily ones of major hysterical somnambulism. Dessoir quotes[12] a case of Moll's[13] in which a person who revealed his dreams through sleeptalking, but was afterwards amnesic for them, recaptured the dream-images under hypnosis. Is there likewise evidence that hypnotized subjects can recall some of the "automatic" actions of everyday life, thereby demonstrating the continuity of the hypnotic with the "automatic" consciousness? Dessoir has only one ostensibly relevant case.[14] It involved a Herr W. who was reading whilst other persons present chatted. One of them mentioned a name which interested him, upon which he became attentive but had no idea what had been said prior to that moment. He was immediately hypnotized and proved able to remember the whole course of the conversation. As further evidence for the continuity of hypnotic- and automatism-consciousness, he cites[15] observations by Janet and Binet that hypnotic states may evolve out of automatic behaviour (such as automatic writing developed in waking individuals by Janet's method of distraction).

Certain hypnotic phenomena, Dessoir believes, indicate that our personality may be multiplex rather than duplex, that there may be tertiary or even quaternary streams of consciousness. He cites[16] an experiment, apparently by a committee of the Berlin Society for Experimental Psychology, in which one of four playing cards was made invisible to a hypnotized subject. The subject could name only the three remaining cards. However, when a pencil was put into his hand with the instruction to write the names of all the cards, he could do so. This indicates the existence of a consciousness as it were below the hypnotic consciousness still able to perceive the "invisible" card. Such findings, together with cases of multiple personality, could lead one, Dessoir says, to a kind of "onion-theory" of the structure of the mind; penetrate one layer and another will be found below it.[17]

The relation between under- (or sub-) consciousness and overconsciousness can vary in different cases. This is well exemplified by instances of the fulfilment of post-hypnotic suggestions. Sometimes the subject falls into a new hypnosis; here the secondary consciousness overwhelms and displaces the primary one. Sometimes the subject feels an inexplicable impulse to act; here the impulsive idea intrudes just sufficiently to occasion the deed. Sometimes the subject is in the ordinary waking state, and carries out the action automatically without knowing what he is about; here "the territory ruled by the hypnotic self is very sharply separated from that ruled by the waking self." From these examples we see that the secondary self "can blend in a weaker or stronger fashion with the overconsciousness."[18]

Since the underconsciousness is not a pathological entity à la Janet, but a feature of the mental structure of all ordinary human beings, it cannot be supposed to have split off from a normal waking consciousness through some abnormality of development or constitution, or the effects of psychological stress. Dessoir therefore tackles the question of the origin of the secondary consciousness in exactly the opposite way. He asks how, in the course of evolution and individual development, our present mental structure, with its apparently unitary everyday consciousness, could have been built up. And the answer he arrives at is something like this.[19] In lower animals, the various parts of the nervous system (and hence the corresponding functions), possess a great deal of local autonomy. Evolution of the nervous system, and correspondingly of mental functions, has been along the route from republic to monarchy, with many intervening stages. Separate elements or functions gradually assemble "under a chief", and there are chiefs over chiefs, the upshot being a kind of Jacksonian hierarchy (though Dessoir does not spell this out in his first edition[20]) in which evolutionarily most recent ("highest") brain centres modulate or control older ("lower") ones. Consciousness is associated with neural activity that has not become merely automatic, that encounters checks and difficulties, and is therefore in inverse proportion to the ease and speed of neural transmission. It is thus particularly linked to the operations of the "highest" centres, the ones which have to grapple with the continually varying problems of everyday life. But there may still be a kind of consciousness linked to centres below the highest. During the life-history of each individual some functions are prone, with increasing practice and automatization, to sink below the "highest" or "normal" sphere of consciousness, but are maintained by a lower "stratum" of consciousness, from which they may or may not be retrievable by an effort of will, or by the artificial (hypnotic) uncovering of the lower "stratum" concerned. We end up with the following "scale" of psychoneural activities: wholly unconscious, underconscious (relatively conscious, dream-conscious, half-conscious), and overconscious (fully conscious, simply conscious).[21]

Among the matters which may be relegated to the underconsciousness are the memories of past events. Such memories are never lost, Dessoir thinks, but persist in the underconsciousness even when unavailable to the overconsciousness. Their continued existence is shown by the fact that they may recur with hallucinatory vividness under certain circumstances, for example at the approach of death or during fever.

The underconsciousness is not, of course, just a repository for the detritus of the overconsciousness. It is a simpler kind of consciousness than the overconsciousness, but has characteristics of its own. In particular, it is characterized by its suggestibility and by a tendency to find expression through sensory hallucination. Hallucination is the natural form in which ideas express themselves when unconstrained by competing ideas, perceptions and memories. The underconsciousness is also the sphere in which instinctual impulses and states of feeling arise. The underconsciousness is in fact the unknown and unacknowledged source of much that goes on in the overconsciousness. It projects itself outwards, so to speak, into the overconsciousness, and in some circumstances – those turnings away from the outer world which promote fantasy, imagination, artistic inspiration – may "externalize" its contents with hallucinatory vividness.

I am far from sure that Dessoir's views are consistent or coherent. To take just one point: According to Dessoir's scheme of things, cases of deep hypnosis, hysterical somnambulism, fugue, multiple personality, etc., represent the displacement of the normal stream of consciousness – the overconsciousness – by an underconsciousness with different characteristics, and this displacement corresponds to a transient predominance of evolutionarily earlier neural centres. But as F. W. H. Myers says, in an interesting review of *Das Doppel-Ich*

> We can conceive of our nervous systems [as Hughlings Jackson has shown] as consisting of three strata, or three levels of evolution, and we can trace in dissolutive processes the results of the cessation of the activity of one stratum after another. But this is not the kind of cleavage which will make a fresh personality [or, one might add, any sort of personality]. For that purpose the

cleavage must not be horizontal, but to some extent at least vertical; that is to say, that each personality must include a certain amount of work done by the highest centres of all; – as well as much work done by the middle centres, and *all* the work done by the lowest centres, – as heart-action and vegetative processes.[22]

Dessoir's *Bibliographie des modernen Hypnotismus* (the first volume of which he published at the age of twenty-one) had made him well-known in the hypnotic movement; and *Das Doppel-Ich* was noticed and was often referred to by other writers on hypnotism; more frequently, however, by those who took him as a representative of a point of view with which they disagreed than by those who largely accepted his contentions. Moll, with whom he had worked in Berlin, was not wholly unsympathetic to his ideas;[23] but the writers whose views accorded best with his were the members of a small but impressive Anglo-American[24] group to which I now turn.

THE SOCIETY FOR PSYCHICAL RESEARCH: EDMUND GURNEY

It is hard to characterize this group succinctly, and its members were linked by bonds and affinities of very varying strengths. Perhaps it is simplest just to say that its central figures were (in Britain) F. W. H. Myers (1843–1901) and (in the United States) William James (1842–1910), and that it had strong connections with the British and the American Societies for Psychical Research (the 'SPR' and the 'ASPR'). Closely associated with Myers in Britain were his brother A. T. Myers (1851–1893), Edmund Gurney (1847–1888), Henry Sidgwick (1838–1900), and the latter's wife Eleanor Mildred Sidgwick (1845–1936). Other associates were Richet, and Frank Podmore (1855–1910), whose histories of mesmerism and Spiritualism I have frequently mentioned. Gurney and F. W. H. Myers became first correspondents,[25] and then close friends, of William James, and James in turn had a varying but often considerable influence upon the ideas of many American philosophers, psychologists, psychiatrists and neurologists. Amongst these one might, in the present context, particularly mention Morton Prince (1854–1929), Boris Sidis (1867–1923), Richard Hodgson (1855–1905) and W. R. Newbold (1865–1926). Several of these persons were at least as able as any who have ever investigated the problems of mesmerism and hypnotism. Their speculations and activities have a corresponding interest, the more so as the speculations accord hypnotic phenomena a key role in helping us to understand many aspects of normal and abnormal psychology and even (in the case of one or two of these writers) in illuminating man's relationship with the universe at large.

Our starting-point must be the foundation, in 1882, of the Society for Psychical Research, an institution very much of its time, yet concerned with perennial problems. Though never a large society, it quickly made an impression on contemporary intellectual life both in Britain and abroad, an impression that was sustained for several decades by the energy, ability and dedication of its principal workers – most notably Gurney, F. W. H. Myers and the Sidgwicks.[26] Its influence was not lessened by certain "establishment" connections – Sidgwick, for instance, who was professor of moral philosophy at Cambridge, and a figure greatly respected in his own right, enjoyed the extraordinary distinction of being the brother-in-law of an archbishop of Canterbury (E. W. Benson) and of a future prime minister (Arthur Balfour).

A similar society was founded in the United States in 1884, with headquarters in Boston. Among those who lent it support was William James. It did not, however, make as much progress as the British Society, and in 1889 amalgamated with it.[27]

The stated purpose of the SPR was "to investigate that large group of debatable phenomena designated by such terms as mesmeric, psychical, and Spiritualistic", and to do so "without prejudice or prepossession of any kind, and in the same spirit of exact and unimpassioned inquiry which has enabled Science to solve so many problems, once not less obscure nor less hotly debated."[28] One of the working committees formed at the Society's inception was devoted to "The study of hypnotism, and the forms of so-called mesmeric trance". Between that time and the end of the century the *Proceedings* and *Journal* of the SPR contained a good deal of material on hypnotism and related topics, and were perhaps the most useful English-language source of detailed information about continental publications in that field.[29] Among the members of the "committee on mesmerism" those whose work requires special mention were A. T. Myers, Edmund Gurney and F. W. H. Myers.

The least known of the three is Arthur Myers, youngest brother of the more celebrated Frederic. Arthur Myers was a classical scholar who became a physician with a special interest in nervous problems.[30] His interest in the problems of hypnotism antedated the foundation of the SPR[31] – he had visited Charcot in Paris in 1881 – and it was no doubt his contacts in France that expedited the visits which he and his brother and Edmund Gurney paid to the Salpêtrière and to Nancy in 1885. His own articles on hypnotic matters were few and relatively brief, but they are of great clarity, and show a considerable knowledge of the continental literature.[32] I must acknowledge a particular indebtedness to him, for he was largely responsible for assembling an extensive collection of mesmeric and hypnotic books and pamphlets (continued by his brother) on which I have frequently drawn in the preparation of this work.

Edmund Gurney, though little remembered today, was one of the most remarkable men of his time, and made a great impression upon many of his contemporaries – George Eliot is said to have based the high-minded Daniel Deronda in part upon him, and Jane Harrison, the classical scholar (a lady little given to gush), described him as "perhaps the most lovable and beautiful human being I ever met". To these personal qualitites was added an enviable assortment of intellectual gifts. He achieved high classical honours, and a Fellowship, at Trinity College, Cambridge, almost (as F. W. H. Myers put it) in the intervals of practising on the piano.[33] Music was, indeed, his great passion, and for a while he cherished hopes of excellence as a composer or performer. These hopes being disappointed, he turned first to medicine, and then to the law; but his heart proving too sensitive for the former, and his intellect too lively for the latter, he ended up from about 1880 living on his moderate private means, and devoting himself to certain neglected but interesting psychological issues. In 1880, he published *The Power of Sound*, a monumental study of the psychology of music, and in 1882 became joint honorary secretary of the SPR, in which position he worked tirelessly until his death in 1888. His analysis of a large number of first-hand accounts of apparent telepathy resulted in the two volumes of *Phantasms of the Living* (1886), and his experiments on hypnotism in a series of papers in the *Proceedings of the Society for Psychical Research* and in *Mind*.

Gurney's hypnotic experiments were unusual in that he did not himself act as hypnotist, but rather as observer and notetaker. As hypnotist he employed various other persons, most notably his secretary Mr. G. A. Smith, a young man from Brighton who had formerly acted as a stage hypnotist. The subjects were mostly young Brighton men, located by Smith and presumably recompensed for their time. Under these circumstances, the issue of deceit might well be raised.[34] However, Gurney was fully alert to this possibility and found that the subjects could not break the spell even when offered substantial cash rewards for so doing. It is worth noting that Smith seems generally to have used mesmeric passes as his principal method of induction.

Gurney's experiments may be taken under two heads, 1. experiments on certain post-hypnotic phenomena, and 2. experiments on "stages" of hypnotic memory.

1. *Experiments on certain post-hypnotic phenomena.*[35] These experiments made use of a planchette board, which was first demonstrated to the subjects. Subjects were then given some inconsequential piece of information and told that when awoken they would be able to write it with the planchette. Immediately on being woken they would be offered a guinea if they could say it, but none ever succeeded, or claimed to recall it. However, if one of their hands were placed on the planchette (which was hidden from them by a screen) they would write it without difficulty, though unaware of what had been written. A variant of this experiment involved setting the hypnotized subjects arithmetical problems or other simple puzzles the (approximate) answers to which (sometimes the working also) would be written by the planchette. A further variant involved giving the subjects a post-hypnotic suggestion to be fulfilled a designated number of minutes after awakening. Waking subjects remained unaware of the suggestion but the planchette could be induced to write the number of minutes

remaining to the designated time and the number that had already elapsed. When rehypnotized, subjects would recall the sense of what had been written but not the muscular sensations occasioned by moving the planchette.

2. *Experiments on "stages" of hypnotic memory.*[36] Gurney held that one can distinguish between an "alert" stage of hypnosis, in which the subject is apparently awake, and a "deep" stage outwardly resembling ordinary sleep, and that the two may be shown to possess distinct chains of state-dependent memory, neither accessible to the waking consciousness. Thus a subject told something (call it A) in the alert stage, and then thrown into the deep stage, and told something else (call it B), would be able to remember A but not B in the alert stage, and B but not A in the deep stage, and neither – despite the offer of a £10 reward – when normally awake. "The phenomena," Gurney remarks, "are singularly constant. I have obtained them with a large number of 'subjects', and with three operators in three different parts of England, two of whom certainly had no idea what I was expecting."

These experiments of Gurney's, demonstrating as they apparently did the existence of secondary streams of consciousness or subconscious "personalities" in normal persons, considerably influenced F. W. H. Myers and James, and also Binet and Dessoir. But I do not know of any independent worker who was able to repeat them satisfactorily.[37]

Gurney was by far the most original and most dedicated person experimenting on hypnotism in the English-speaking world in the mid-1880s. He knew nothing of the cognate researches of Janet until he was informed of them by F. W. H. Myers,[38] and his early death was undoubtedly a great loss to the subject. His various passages on the theoretical problems of hypnosis are also of some interest.[39] Though he never developed a detailed theoretical system of his own, indeed scarcely had time to do so, he has many pertinent remarks to make on theoretical matters – even his footnotes are worth reading.

On theoretical issues, Gurney was in some respects out of accord with the tendency of his time. He was inclined to believe that mesmeric passes, and other physical methods of inducing hypnosis, had some direct effect not attributable to suggestion. The effect of passes could not, he felt, be set down, as was customary, to "monotonous stimulation", for so far from passes "being explained by being called a form of monotonous stimulation, the burden of supporting the credit of monotonous stimulation, as a hypnogenetic agency, seems to fall almost entirely upon them."[40] In fact, he believed he had experimental evidence for a localized mesmeric influence. In the experiments concerned,[41] subjects passed their hands, with fingers separated, through a screen, behind which the hypnotist directed the supposed influence, by pointing, at an arbitrarily selected finger, which would shortly become anaesthetic, paralyzed, etc. Indeed, he alleged that hypnosis "seems, in the first instance, always to require some distinct *physical* stimulation; though, after it has been once induced, the mere idea of it, associated with that of the original hypnotiser – *e.g.*, if he gives the command '*Dormez!*' – may be enough to cause its recurrence."[42] However, the physical stimulation concerned might be a strain on the eye muscles (as in Braid's method) or a sudden loud sound or bright light (the Salpêtrière's way of producing catalepsy), as well as mesmeric passes.

Gurney's own approach to the nature of the state so induced is through what he calls "psychical reflex action"[43] (a term which he uses to distinguish his position from that of theorists like Heidenhain and Despine who try to explain hypnotic phenomena in terms of "automatic cerebration" or kindred concepts).

> That is to say, I should confine the term "hypnotic trance" to a state in which (or in some stage of which) inhibition reaches the higher inhibitory and co-ordinating faculties; and particular ideas, or groups of ideas, readily dissociating themselves from their normal relation to other groups and to general controlling conceptions, and throwing off the restraint proper to elements in a sane scheme, respond with abnormal vigour and certainty to any excitations that may be addressed to them. Such response may be shown (1) in the inhibition, by command, of ordinary muscular movements or control of movements; (2) in the ease with which the "subject's" mind can be steered, so to speak, in the course of conversation or narration; but chiefly (3) in the ready imposition, by external suggestion, of sensory hallucinations, or (4) of abnormal lines of conduct.[44]

As an account of the commonplace phenomena of Gurney's supposed "alert" stage of hypnosis this is all very well. But it helps us little when we come to consider Gurney's own results which seemed to show the continued existence of an "hypnotic" stream of consciousness during waking life. Indeed, Gurney apparently regards the formation and emergence of such streams of consciousness not as a phenomenon of the hypnotic trance *per se*, but as a phenomenon to which hypnosis may secondarily give rise, for he says

> Hypnotism assumes a wholly new significance when it leads … to results *beyond itself* – when it appears as the ready means for establishing a secondary train of consciousness, to which when established … there is no ground for attributing any special hypnotic character.[45]

Gurney was, we may incidentally note, one of the few writers of his time who had a clear grasp of the philosophical difficulties raised by the postulation of secondary "selves" or secondary streams of consciousness, and of the philosophical issues upon which it might have a bearing.[46]

THE SOCIETY FOR PSYCHICAL RESEARCH: MYERS AND JAMES

A collaborator of Gurney in his hypnotic and other work, and a close friend, was F. W. H. Myers, who succeeded him as Honorary Secretary of the SPR. Like Gurney, Myers had been a classical fellow of Trinity College, Cambridge, and early on established some reputation as a poet and essayist. Afterwards, he became, like Matthew Arnold, an inspector of schools, but substantial private means enabled him to devote much of his time to the unusual interests which he pursued with an almost evangelical fervour. Among these interests were hypnotism and related phenomena. Whereas Gurney was principally an experimenter in hypnotism, Myers was primarily a theorist. His approach, however, was not to attempt a theory of hypnotism *per se*, but to place hypnotism in a much wider context of phenomena, a context against which alone we can begin to make sense of it. This context included not just abnormal and pathological phenomena of the kinds that figured in Janet's and Binet's discussions, but normal and paranormal phenomena of various varieties, up to and including religious and mystical experiences, the whole yielding a picture of the human psyche very different from that presented by the received conventionalities of medical materialism. Myers's most ardent desire was to prove – but by empirical and properly scientific means – that there is more to human personality than medical materialism admits. This desire now and again found vent in somewhat orotund passages of quasi-religious uplift which for many readers seriously damage his claims to scientific credibility. He is not, however, quite so easily dismissed. Though he permitted himself, and perhaps could not restrain, an occasional rhapsody, he had rigorously disciplined his will to believe, and his critical standards are greatly superior to those of many continental, particularly French, workers of that time.

Myers is best known today for the two large volumes of his *Human Personality and its Survival of Bodily Death* (1903). The 193 pages of this work which he devotes to hypnotism are an invaluable source of accurate references to nineteenth-century hypnotic literature, particularly the continental literature, which he probably knew better than any other British or American writer of the time (he was the first person in England to write appreciatively and at length about the work of Janet and that of Breuer and Freud[47]). However, his theoretical stance is perhaps best approached through the five long articles on "the subliminal consciousness" which he published in the *Proceedings of the Society for Psychical Research*, between 1892 and 1895.[48] Substantial chunks of these articles are incorporated in his *Human Personality* (1903). But the most straightforward approach to his ideas is through the first article (containing two chapters).[49]

Myers criticizes as piecemeal and insufficiently comprehensive the explanations so far given of hypnosis, automatisms and so forth. "I hold", he says, "that both that group of facts which the scientific world has now learnt to accept (as the hypnotic trance, automatic writing, alternations of personality, and the like), and that group of facts for which ... we are still endeavouring to win scientific acceptance (as telepathy and clairvoyance), ought to be considered in close alliance and correlation, and must be explained, if explicable at all, by some hypothesis which does not need constant stretching to meet the exigencies of each fresh case."[50]

In keeping with these assertions, Myers presents us with a large framework of ideas which – despite a shortage of detail – may, he thinks, have the requisite degree of generality:

> I suggest ... that the stream of consciousness in which we habitually live is not the only consciousness which exists in connection with our organism. Our habitual or empirical consciousness may consist of a mere selection from a multitude of thoughts and sensations, of which some at least are equally conscious with those that we empirically know. I accord no primacy to my ordinary waking self, except that among my potential selves this one has shown itself the fittest to meet the needs of common life. I hold that it has established no farther claim, and that it is perfectly possible that other thoughts, feelings, and memories, either isolated or in continuous connection, may now be actively conscious, as we say, "within me," – in some kind of co-ordination with my organism, and forming some part of my total individuality. I conceive it possible that at some future time, and under changed conditions, I may recollect all; I may assume these various personalities under one single consciousness, in which ultimate and complete consciousness the empirical consciousness which at this moment directs my hand may be one element out of many.[51]

That such hidden streams of consciousness exist even in perfectly normal persons is clearly shown, Myers believes, by (for instance) the fact that hypnotized subjects will recollect events of previous hypnotic states (sometimes even of other periods of altered consciousness) inaccessible to the waking mind, by the fact that such subjects will afterwards execute post-hypnotic suggestions in complete unconsciousness of having received a command, by Gurney's experiments on the "stages of hypnotic memory" (referred to above) which appear to demonstrate that *more than one* hidden stream of consciousness may simultaneously exist in the same individual, and by numerous examples of automatic writing in which a waking subject finds his hand writing intelligent sentences of whose tenor he is unaware. Such streams of consciousness he designates "subliminal", holding that "unconscious" or even "subconscious" would be directly misleading.[52] The term "subliminal" is, however, not altogether apposite, for the notion of a threshold (limen), suggesting as it does a passage from outer darkness into a well-lit domain still has overtones of a "searchlight" view of consciousness,[53] in terms of which material not caught in the beam would indeed be unconscious. His favourite of all metaphors – that of the visible part of the spectrum, flanked by the invisible infra-red and ultra-violet – carries this connotation even more strongly. He introduces it to point up the possibility (or probability as he holds) that there may be very different regions, or streams, in the subliminal. At the lower end, so to speak, of the subliminal consciousness, may be an awareness of the automatic regulatory processes of the body; at the superior or psychical end a grasp of "an unknown category of impressions which the supraliminal consciousness is incapable of receiving in any direct fashion, and which it must cognise, if at all, in the shape of messages from the subliminal consciousness."[54]

That there may be traffic between the different streams of consciousness which make up a man's total personality, or Self, is central to Myers's position. The various strata of the Self, he says, combining two of his favourite metaphors, "are strata (so to say) not of immovable rock, but of imperfectly miscible fluids of various densities, and subject to currents and ebullitions which often bring to the surface a stream or a bubble from a stratum far below."[55]

Myers adopts the term "message" for any intrusion of material from one stream of consciousness into another, whether that intrusion be accidental or deliberate. Thus information acquired telepathically or clairvoyantly by a "superior" subliminal stream of consciousness may pass, or be sent, into the supraliminal consciousness in the shape of visions, inexplicable impulses to act, monitory voices, apparitions, remembered dreams, automatic writing, etc. Also from the higher or "psychic" levels of the subliminal come the "subliminal uprushes" of ideas and images characteristic of genius. "Messages" may also go in the reverse direction, from supraliminal to subliminal, as when an idea, or set of ideas, of unusual intensity, affects physiological processes normally overseen by "lower" regions of the subliminal – stigmatization or hypnotically induced blistering would be extreme examples.

In some individuals, material may cross more readily from the subliminal to the supraliminal consciousness, or vice versa, than in others. Myers talks of this in terms of the not very helpful metaphor of "permeability of the psychic diaphragm". Hysterics, and other psychoneurotics, suffer from an excessive permeability of this putative diaphragm. Unwanted matter from "below" may intrude into the everyday stream of consciousness; and ideas in the latter (suggestions, autosuggestions) may sink "downwards" and take hold in the stratum of consciousness which still exercises a regulatory influence, lost to the waking mind, on basic physiological processes. Not just particular mental contents, but whole faculties or portions thereof may likewise pass from supraliminal control into that of subliminal strata, and move back again. Hence the anaesthesias, automatisms, and so forth, so characteristic of hysteria. Finally, it is possible that a subliminal stream of consciousness may assume complete control of the organism, displacing the ordinary waking consciousness. Then we get somnambulism, fugue, multiple personality, possession states, and so forth.

We can now turn to Myers's treatment of hypnotism and suggestion, which can usefully be broken down under four headings:

Hypnotism and hysteria

Myers holds that the analogies between hypnotic phenomena and hysterical ones are so striking and obvious as to indicate an unmistakable underlying relationship. It is, he says,

> ... a striking characteristic of the hypnotic self that it can exercise over the nervous, the vaso-motor, the circulatory systems a degree of control unparalleled in waking life. Told to hold out his arm, the hypnotised subject will hold it out for an indefinite time in a state of painless contracture, and with no disturbance of pulse or respiration. Told that the ammonia which is held to his nose is otto of roses, he will inhale it with unwinking, unwatering eyes. Told that he has burnt himself in a given spot, his skin will redden or even form a blister. Told to sleep, he will sleep profoundly until told to wake. I do not here attempt to explain *why* that obedience is rendered. But the *power* of obeying such commands – that is the function, the prerogative, the secret of the hypnotic self.
>
> Are we aware in practice of any malady or group of maladies in which these functions, these capacities, are the subject of special disturbance? Are there cases of prolonged and apparently causeless contracture? Are there anaesthesias appearing, shifting, and disappearing as rapidly as the suggested anaesthesiae of hypnotism? Are there anomalous vaso-motor disturbances which seem to follow the patient's mere caprice?
>
> The reader will answer with the word *hysteria*.[56]

Hysteria may be regarded as a disease of the "hypnotic stratum", i.e. the stratum of consciousness into which the faculties submerged in hysteria sink; or rather, perhaps, as due to an excessive permeability "of the psychical diaphragm which separates ordinary consciousness from the deeps below." This or that group of sensory or motor abilities drops out of waking knowledge or out of control of waking will. In hypnotism, on the other hand, instead of losing control over the supraliminal stratum, we gain control over the hypnotic stratum. "We purposely increase the permeability of the psychical diaphragm in such a way as to push down beneath it various forms of pain and annoyance which we are anxious to get rid of from our waking consciousness, while, on the other hand, we stimulate in the depth of our being many sanative and recuperative operations whose results rise persistently into the perception of our waking life."[57]

Suggestion and self-suggestion

According to Myers, all suggestion is at root self-suggestion, a conclusion which (if we exclude mesmeric or telepathic effects) is at a certain level necessarily true. However, we know no better with self-suggestion than with suggestion why, even with the same individual, suggestions sometimes work and sometimes do not. An effective suggestion, Myers thinks, is a "successful appeal to the subliminal self".[58] Hypnotism is just "a name for a group of empirical practices by means of which we can manage to get hold of subliminal faculty."[59] Why the curious practices concerned work, at least part of the time, is as yet obscure. But once subliminal faculty is engaged, it can be turned to therapeutic ends. "Beneath the threshold of waking consciousness there lies, not merely an unconscious complex of organic processes, but an intelligent vital control. To incorporate that profound control with our waking will is the great evolutionary end which hypnotism, by its group of empirical artifices, is beginning to help us to attain."[60]

Hypnotism and sleep

Traditionally, the state of hypnosis has been regarded as a form of trance or sleep. According to Myers, natural sleep is a phase of being in which supraliminal activity is stilled, and subliminal faculties can guide the recuperative forces of the body undisturbed. There are also signs of other kinds of heightened subliminal activity during sleep. Thus dreams, which, if recollected, represent flowing of subliminal into supraliminal streams of consciousness, may manifest heightened powers of imagination and incorporate material acquired through telepathy and clairvoyance. Furthermore, the memory of dreams may be recovered under hypnosis and vice versa, suggesting that the dream stratum and the hypnotic stratum are the same, or closely related. Indeed, there is at least one case on record in which an hysterical patient apparently had a vivid dream that two letters (W and B) had been written under her breast, and afterwards developed a corresponding inflammation of the skin.[61] The parallel with hypnotically induced skin markings is obvious, and again implies a close affinity of the dream with the hypnotic-hysterical stratum of the personality.

Now if sleep be a phase of personality specially consecrated to subliminal activity, it follows that any successful appeal to the subliminal self will be likely to induce some form of sleep. But manifestly it is not ordinary sleep. "For my part", says Myers,

> ... I shall abandon the attempt to force all the varied trances, lethargies, sleep-waking states, to which hypnotism introduces us into the similitude of ordinary sleep. Rather I shall say that in these states we see the subliminal self coming to the surface in ways already familiar to us, and displacing just so much of the supraliminal as may from time to time be needful for the performance of its own work.[62]

Obedience to suggestion

"I have tried to suggest", says Myers, "the source of these subliminal powers; but what explanation can I give of this subliminal *obedience?* The explanation that he in fact offers is ingenious and not uninteresting. Sleep and waking life are so to speak adjacent strata. There are faculties common to both states; and each state has its faculties which the other does not share.

> But under certain conditions – and I am not now assuming any state profounder than common sleep – the sleeper will begin to exercise in a random way some of the powers which belong properly to waking life. He will strike out with his limbs, or spring to the floor, under the impulsion of a disturbing dream. Meantime the waking self, – the proper ruler of the faculties of progression and balance, – stands aside, and for the moment these faculties obey the random commands of the sleeping usurper. But if these faculties are too rashly drawn upon the sleeper will awake; and the waking self will resume its sway before serious harm is done.
>
> Now I would suggest that somewhat similarly, in these cases of blind obedience to hypnotic suggestion, we see a certain knowledge, consciousness, memory, which are properly subject to the will which commands one stratum of the personality, falling temporarily into dreamlike compliance with impulses which reach them from a stratum not their own. For the moment the subliminal will stands aside; but just as the waking will resumes sway when the dreamer becomes too violent, so also does the subliminal will intervene if the obedience to hypnotic suggestion is in danger of being pushed too far.[63]

Myers's proposals about the nature of hypnotism and hysteria cannot be disentangled from the rest of his theory of the subliminal self and held up for separate assessment. The *tout ensemble*, however, though at times profoundly impressive, is often most elusive where clarity is most essential. This did not stop it from achieving – perhaps it helped it to achieve – some measure of respect and acceptance in both Europe and the United States.[64] But to many, and especially to the self-consciously scientific, it appeared, as it still appears, a mere expression of the mystical yearnings of a mind shaped as much by Plotinus and Wordsworth as by modern neurology and psychiatry. Certainly if one looks from a modest distance at the rather grandiose, but somewhat indefinite, main lines of the theory, one is apt to receive such an impression. But a nearer look reveals that the structure is filled with a massive accumulation of factual materials which, however one regards them, could hardly be termed indefinite or lacking in detail. The obvious stumbling-block is over the issue of justification: How far, and through what logical steps, does the data base justify the broad framework of thought that has been founded upon it? The broad framework is not one that can be used to derive the details of the phenomena that are used to support it. It may "make sense" of the phenomena, but it does not enable us unequivocally to predict any particular phenomenon. This situation obtains commonly enough in psychology, but it would generally be thought undesirable in the "hard" sciences and by philosophers of science. A partial parallel, however, is provided by the Darwinian theory of evolution. Here too we have a broad and abstract hypothesis which "makes sense" of a great mass of observations; yet it would be hard to maintain that the details of the data can be directly derived from the theory. Of course since Darwin's time certain paths have been established which fill some of the space between the theory and particular features of the phenomena. Nothing similar has been accomplished in respect of Myers's theory of the subliminal self. If it had been, Myers would perhaps now be as famous as Darwin.

A comparison between Myers and Darwin was in fact drawn several times by William James who, as the 1890s progressed, became increasingly Myers's friend and admirer – he was present at Myers's death-bed in 1901.[65] James observes, for instance – and no-one who has looked at Myers's *Human Personality* can fail to see the aptness of the remark – that Myers "shows...a genius not unlike that of Charles Darwin for discovering shadings and transitions, and grading down discontinuities in his argument."[66] After the publication in 1892 of Myers's first article on "the subliminal consciousness", James began to use the notion himself with increasing frequency. His unpublished Lowell Lectures of 1896, so interestingly reconstructed by Eugene Taylor,[67] are shot through with ideas acknowledgedly culled from Myers, which James uses to illuminate such topics as dreams, hypnotism, automatic writing, hysteria, multiple personality, mediumship, possession and genius. In a letter to Myers of 19th January 1897, he calls Myers *cher maître* (Myers reciprocated on 3rd February by saying "I wish that there were more people like you, and that I were one of them!").[68] Soon James took up other aspects of Myers's thinking. Myers held that our normal consciousness is only a portion of our nature, one specially adapted to "terrene" conditions. The "supernormal" faculties of the subliminal can hardly be vestiges, degenerations of something which our ancestors once possessed. Rather we should regard them as germs of something still evolving, something carrying us into a wider cosmic environment, an environment which, for Myers, assumed more and more the character of a "spiritual world".[69] The concluding portion of James's *Varieties of Religious Experience* (1903) is heavily influenced by these ideas.

NOTES

1 On Du Prel see Kiesewetter (1909), pp. 840–901.
2 English translation Du Prel (1889b).
3 For Dessoir's reminiscences, see Dessoir (1946).
4 Ellenberger (1970), p. 145.
5 Hövelmann (1987).

6 Dessoir (1889). Much enlarged second edition Dessoir (1896).

7 Dessoir (1889), p. 1.

8 *Unterbewusst* is usually translated as "subconscious", but for consistency with *Oberbewusst* I translate it as "underconscious(ness)" when discussing Dessoir.

9 Barkworth (1889–90), p. 90.

10 Dessoir (1889), p. 27.

11 Dessoir (1896), p. 51.

12 Dessoir (1889), pp. 15–16.

13 Moll (1890), p. 127.

14 Dessoir (1889), p. 26.

15 Dessoir (1889), pp. 26–27.

16 Dessoir (1889), p. 23.

17 Dessoir (1889), p. 26.

18 Dessoir (1889), p. 16.

19 See especially Dessoir (1889), pp. 33–39.

20 But cf. Dessoir (1896), pp. 43–46.

21 Dessoir (1889), p. 34.

22 F. W. H. Myers (1889–90b).

23 Moll (1890), pp. 239–241.

24 In view of Richet's place on the fringes, and Janet's on the further fringes, of this group, perhaps I should say "Franglo-American".

25 Correspondence between Gurney and James and between Myers and James is in the Houghton Library, Harvard University.

26 I have written about this group of persons in Gauld (1968).

27 It was reestablished as a separate society in 1907.

28 *Proceedings of the Society for Psychical Research*, 1882–3, Vol. 1, pp. 3–4.

29 Among corresponding members of the SPR in 1900 were Beaunis, Bernheim, Dessoir, Féré, Flournoy, Janet, Liébeault, Liégeois, Lombroso, Ribot, Richet, Schrenck-Notzing, and Wetterstrand; among its American members were F. H. Gerrish, William James, Morton Prince, J. J. Putnam, and G. Stanley Hall.

30 He became himself subject to attacks of epilepsy, and has a place in medical history as the anonymous physician on whose case Hughlings Jackson founded the modern concept of temporal lobe epilepsy.

31 According to family papers, now in the Library of Trinity College, Cambridge, his father, the Rev. F. Myers, perpetual curate of St. John's, Keswick, was experimenting with mesmerism in the 1840s, along with various friends.

32 I have particularly in mind A. T. Myers (1890; 1892), and his article on Esdaile in the DNB.

33 F. W. H. Myers (1888–9c), p. 360.

34 And has been raised by T. H. Hall (1964). Hall's claims have, however, been subjected to severe criticism, e.g. by Gauld (1965–6) and Nicol (1966).

35 Gurney (1885–7a).

36 Gurney (1884c; 1885–7b)

37 As many as eight different stages of memory were obtained in some experiments carried out by Mrs. E. M. Sidgwick and Miss A. Johnson in July 1891, but the hypnotist was again Mr. G. A. Smith. F. W. H. Myers (1903), Vol. 1, pp. 452–455.

38 F. W. H. Myers (1888–9c), p. 369.

39 Esp. Gurney (1884b).

40 Gurney (1888–9a), p. 251.

41 Barrett *et al.* (1882–3); Gurney (1884a); Gurney (1885); Gurney (1888–9b), pp. 14–17; cf. E. M. Sidgwick and Johnson (1892), pp. 577–593.

42 Gurney (1888–9a), p. 218.

43 On this see especially Gurney (1884b).

44 Gurney (1888–9a), p. 217.

45 Gurney (1885–7a), p. 323.

46 See Gurney, Myers and Podmore (1886), Vol. 1, pp. 69–70.

47 The recently published papers of Breuer and Freud on the psychic mechanism of hysteria are discussed on pp. 12–15 of Myers (1893–4).

48 An earlier series of long articles is largely devoted to automatic writing. See F. W. H. Myers (1884; 1885; 1885–7a; 1888–9a).

49 On hypnotism and suggestion cf. F. W. H. Myers (1898–9).

50 F. W. H. Myers (1891–2b), p. 299.

51 F. W. H. Myers (1891–2b), p. 301.

52 F. W. H. Myers (1891–2b), p. 305.

53 Searchlight views of consciousness were widespread among early exponents of the unconscious. Schopenhauer uses the analogy of objects on the screen of a magic lantern, and also that of objects passing across the field of view of a telescope. See Schopenhauer (1909), pp. 330, 334–4, and cf. pp. 327–328, where the analogy (somewhat reminiscent of Myers) is of ideas struggling up from the depths to the (illuminated) surface of an ocean. In *The Interpretation of Dreams*, Freud says, "We see the process of a thing becoming conscious as a specific psychical act, distinct from and independent of the process of the formation of a presentation or idea; and we regard consciousness as a sense organ which perceives data that arise elsewhere." (Freud, 1954, p. 144.) Myers followed James in rejecting the idea of mental events that can move in and out of the inner searchlight, or theatre, and when out are literally unconscious. See F. W. H. Myers (1891–2a), p. 117.

54 F. W. H. Myers (1891–2b), p. 306.

55 F. W. H. Myers (1891–2b), p. 307.

56 F. W. H. Myers (1891–2b), pp. 308–309.

57 F. W. H. Myers (1898–9), p. 102.

58 F. W. H. Myers (1903), Vol. 1, p. 169.

59 F. W. H. Myers (1891–2b), p. 336.

60 F. W. H. Myers (1898–9), p. 107.

61 F. W. H. Myers (1903), Vol. 1, pp. 127–128; Krafft-Ebing (1889), p.91.

62 F. W. H. Myers (1903), Vol. 1, p. 170.

63 F. W. H. Myers (1891–2b), pp. 353–354.

64 Among those (other than James) influenced by Myers's speculations about the subliminal self were Lloyd Tuckey, H. E. Wingfield, T. W. Mitchell, R. Osgood Mason, Milne Bramwell, F. Podmore and William McDougall.

65 The account of this given by Axel Munthe in *The Story of San Michele* loses nothing in dramatic effect.

66 James (1986), p. 212; cf. p. 101.

67 E. Taylor (1984). On James's views on hypnosis cf. Kihlstrom and McConkey (1990).

68 Letters in the Houghton Library, Harvard University.

69 Cf. James (1986), p. 207.

CHAPTER 19

Hypnotism and multiple personality in the United States: Sidis and Prince

Some biographers of William James have seemed to wish to minimize the influence of F. W. H. Myers upon him;[1] in this I think they are wrong, but for present purposes the issue is not important. The point is that James was an exceptionally well-read, intellectually active and influential person. He had read and assimilated Janet; he had read and assimilated Myers and Gurney; he had read and assimilated Dessoir. These influences, and many other cognate ones, informed his thinking, but were transmuted and clarified and directed to new ends. His *Principles of Psychology* (1890), which incorporates them, at once became a classic. He wrote and lectured a good deal, and above all he knew and in turn influenced many prominent American psychologists, psychiatrists and philosophers. He was a friend and colleague of Josiah Royce and Hugo Münsterberg; G. S. Hall, W. R. Newbold, J. R. Angell and Boris Sidis were his pupils; J. M. Baldwin, Morton Prince, J. M. Cattell, J. J. Putnam and Joseph Jastrow were his friends. James's ideas upon hypnotism, hysteria, multiple personality and so forth passed into circulation and reinforced tendencies already current. And as a background to it all could be felt the first stirrings, themselves owing much to James, of the psychological school (or, better, attitude of mind) that came to be called "American functionalism". Functionalists asked of any aspect or feature of mental life not (as had the "structuralists"): What are its elements? What are the laws of their interaction? but: What is its function? Why has the mind come to act in this way? Why has the function evolved or developed? Myers was always asking such questions, and so were others of more materialistic beliefs. If mental functions are in the process of evolution, it is natural to enquire also into their dissolution, for the lines of dissolution will surely tell us something of the hierarchy and relatedness of functions in the normal mind. We end up with a psychology which attaches considerable significance to the study of abnormal and unusual psychological phenomena, including hypnotism and kindred matters.[2]

Hypnotism itself, now linked with forms of psychotherapy that recognized the causal rôle of "buried" experiences, enjoyed both a medical and a popular vogue during the middle and later 1890s.[3] It merged at the fringes with the "mental healing" cults still prominent in New England. Practitioners such as the flamboyant and appropriately named Dr. J. D. Quackenbos of New York occupied an intermediate place.[4] Interest in hypnotism and hypnotherapy was by no means confined to the New England States, at least if we may judge from the American delegates who attended the Second International Congress of Hypnotism in Paris in 1900.[5] After 1900, interest fell somewhat, but there was a further resurgence both medical and popular in and after 1909, due to the influence of the Rev. Elwood Worcester's Boston-based Emmanuel Movement, a curious blend of liberal Christianity, mental healing, social reform, and speculations about the subliminal.[6] Meanwhile, the work of Pierre Janet had begun to attract a good deal of interest. He was twice invited to America to lecture, once at the Lowell Institute (1904), and once at Harvard (1906), and his influence was probably greater than that of any indigenous psychiatrist.

Two American psychiatrists of this period are of some importance in the history of hypnotism. Both were considerably influenced by Janet and by James, and even to some extent by Myers. I shall begin with the younger of the two.

BORIS SIDIS

Boris Sidis (1867–1923) emigrated from Russia to America as a young man, and after struggling to support himself while studying in his spare time eventually graduated from Harvard University, where he had been a pupil of William James. His interest in hypnotism and suggestion was encouraged by James, and he also became a friend of Morton Prince. He undertook a fair amount of psychotherapeutic work, although he did not qualify in medicine until 1908. His first book, *The Psychology of Suggestion* (1898), is both innovative and idiosyncratic, and, like most of Sidis's writings, it contains many unsubstantiated and dogmatic assertions.[7]

Sidis begins by working towards a definition of suggestion. He rejects Bernheim's loose usage according to which "suggestion" denotes any occasion on which an idea brings another idea in its train. From an analysis of several examples of what most people would regard as suggestions, he reaches his own definitions of "suggestion" and "suggestibility":

> By suggestion is meant the intrusion into the mind of an idea, met with more or less opposition by the person; accepted uncritically at last; and realized unreflectively, almost automatically.
> By suggestibility is meant that peculiar state of mind which is favourable to suggestion.[8]

Suggestibility can be demonstrated in perfectly normal people, and Sidis describes a series of his own experiments designed to reveal the conditions which favour it. Subjects were briefly presented with a column of nine letters or nine digits, and were then required immediately to write down "anything that came into their head at that particular moment – letters, numerals, words, phrases, etc."[9] The aim in each trial was by varying the frequency or order of the letters or digits to induce the subject to think of a particular letter or digit. Success in so doing would constitute a successful "suggestion". By varying the factors of frequency, repetition, last impression (i.e. being at the bottom of the column), and coexistence (letters or digits repeated three times side by side), Sidis arrives at an order of effectiveness of the "modes of suggestion". a combination of frequency and last impression (success rate 75.2%) being much the most effective.

From suggestion of ideas, Sidis turned to suggestion of acts. Subjects had to fix their attention for five seconds on six small squares of different colours placed in varying orders in a line before them. At the end of five seconds, "the subject had immediately to take one of the coloured squares, whichever he liked."[10] On every trial, one of the squares was marked out or different in some way. Abnormality of position or shape constituted the most effective "suggestions" (response rates 47.8% and 43% respectively).

Sidis thinks that these experiments "establish the fact of normal suggestibility on a firm and unshakable basis."[11] From incidental observations made during his experiments he infers that conditions which promote suggestibility in normal subjects are fixation of attention on the stimuli presented followed by distraction of attention; monotony of the stimulus conditions; limitation of voluntary movements; limitation of the field of consciousness; inhibition of competing ideas, etc.; and immediate, unreflecting choice, action or decision. From the fact that in his second series of experiments direct verbal suggestion was much less effective than the indirect suggestion constituted by abnormality of position or shape, Sidis reaches his fundamental law of normal suggestibility – normal suggestibility varies directly with indirect suggestion and inversely with direct suggestion.[12]

From "normal suggestibility" Sidis passes to "abnormal suggestibility," as instanced by hypnosis, and from a brief examination of the findings and pronouncements of such leading authorities as Moll, Lehmann, Bernheim and Braid, he concludes that the conditions which promote it are the same as those which promote normal suggestibility, with the exception of distraction of attention and immediate execution. A "close examination of the nature of hypnotic suggestion" reveals that the more direct and unambiguous an hypnotic suggestion, the greater the chance of its being realized, whereas the more indirect or oblique it is, the less its likely efficacy. Hence we may formulate the law of abnormal suggestibility: – Abnormal suggestibility varies directly with direct suggestion, and inversely with indirect suggestion.[13]

In all this, everything hinges on Sidis's curious little experiments on suggestibility. Many would deny that these experiments can properly be called experiments on suggestion or suggestibility. They do not even fulfil Sidis's own definition of suggestion, according to which a suggested idea has to be met with "more or less opposition" by the subject. For the moment, suffice it to say that his experiments, which insinuate certain ideas or images, however transiently, into a subject's mind, and then require him to act without any time for critical reflection, do seem to present us in miniature with some of the features of more traditional suggestion situations. At any rate, Sidis deserves credit for attempting something that was in his day largely novel[14] – a systematic experimental investigation of the factors which make waking suggestions effective and of the conditions which promote waking suggestibility.

As for the nature of abnormal suggestibility, Sidis takes the following position. Hypnotization produces a "deep cleft in the mind of the subject, a cleft by which the waking, controlling consciousness is separated from the great stream of conscious life."[15] When this cleft is less deep we have "the different slight hypnotic states, but as the cleft becomes deeper and deeper the hypnosis grows more profound, and when the controlling consciousness is fully cut off from the rest of conscious life, we have a state of full hypnosis which is commonly called somnambulism."[16] That is why the introspective reports of certain hypnotic subjects who were on the verge of somnambulism reveal "that during hypnosis they were indifferent to the actions of their body – the latter acted by itself, that they were mere spectators of all the experiments performed on them ... [17]

Upon all this Sidis puts a physiological gloss. The principal function of the higher cortical centres of the brain is inhibitory. Hypnosis is an inhibition of these inhibitory centres:

> In the normal condition of man the superior and the inferior centres work in perfect harmony; the upper and the lower consciousness are for all practical purposes blended into a unity forming one conscious personality. In hypnosis the two systems of nervous centres are dissociated, the superior centres and the upper consciousness are inhibited, or, better, cut off, split off from the rest of the nervous system with its organic consciousness, which is thus laid bare, open to the influence of external stimuli or suggestion ... [18]

In the concluding chapter of Part I of his book, Sidis returns to a consideration of waking suggestibility. The near-sameness of the conditions that promote normal and abnormal suggestibility indicates that both flow from one common source, and this can only be a disaggregation of consciousness. However, in normal suggestibility "the cleft is not so deep, not so lasting as it is in hypnosis, or in the state of abnormal suggestibility; the split is here but momentary, evanescent, fleeting, disappearing at the very moment of its appearance."[19] This is why normal suggestibility requires immediate execution. In both cases, however, we have a dissociation of the waking from the subwaking, reflex[20] consciousness, with suggestion acting only through the latter. "It is the subwaking, the reflex, not the waking, the controlling, consciousness that is suggestible. Suggestibility is the attribute, the very essence of the subwaking, reflex consciousness."[21]

Part II of Sidis's book is largely concerned with the subconscious self. It is tempting to say of this that what is comprehensible is not original, and that what is original is not comprehensible. Indeed, it is far from clear what he means by "subconscious self". He uses "subwaking" more or less as a synonym for "subconscious", and in a later article tells us that by "self" he designates "not personal consciousness, but mere consciousness."[22] He argues at length against the doctrine of unconscious cerebration,[23] maintaining that one cannot admit intelligent subconscious cerebral activities without allowing that a conscious event must exist in connection with them, and yet in a later book he develops a doctrine of "thresholds" which leads him at times to express something like a "searchlight" view of consciousness.[24]

At any rate, Sidis believes that there can be intelligent subconscious processes in normal as well as abnormal individuals; and in support of this claim he relies heavily upon the findings of Binet and Gurney which we ran through in earlier chapters of the present book. He thinks he has evidence for the persistence of the subconscious self during the waking state of ordinary individuals. Thus people who hear, in a waking state, whispers too faint for them to make out, may remember them correctly if immediately hypnotized – the hypnotic self and the subconscious self may be regarded as identical, Sidis thinks, because enhanced suggestibility is a leading characteristic of both. Or again, in some experiments by Sidis himself, subjects were presented with drawings on slips of paper, and required to reproduce them. Each drawing had an inconspicuous digit in the margin as if to number it. Upon completing their reproductions, subjects were then required to choose one digit from a row of eight ("to break up the attention", Sidis told them). Subjects showed a modest tendency to select the digit that had been on the drawing despite the fact that the connection did not register with them. Their subconscious selves, Sidis thinks, had none the less taken the digit on the slip as a suggestion and communicated it to the waking consciousness" as an insistent idea".[25] Like Myers, Sidis talks of communication between the two selves in terms of "messages" passing from the one to the other in the form of automatic writing, intrusive images, hallucinations, etc.

Sidis now begins to give a more detailed theoretical account of the aggregation and disaggregation of consciousness and of the nature and degrees of the possible organization which a subconscious self may achieve. The unit of his analysis is the "moment-consciousness" (plural "moments-consciousness"), a concept which I cannot pretend to have fully grasped. A moment-consciousness seems to be a period of consciousness, however long or short, which maintains a certain direction or tendency, or has ("synthetizes") a constant, central "moment-content", and changes or varies only in inessentials. Moments-consciousness can be linked to each other by associative links, so that they may succeed each other with some regularity, and a moment-consciousness may contain ("synthetize") another as it were past or fading moment-consciousness among its contents, so that they can stand in hierarchical relations to each other, and form progressively higher syntheses. The highest achievement of such synthesis is self-consciousness – a moment-consciousness that knows itself in the process of thinking.

Sidis seems to suppose that the simplest forms of moment-consciousness have as their physiological bases or concomitants the activity of a single nerve cell in the brain. Cells that become linked together or "associated" have as a group "a concomitant mental activity resulting in some form of psychic synthesis",[26] and so on for higher and higher forms of mental synthesis. The more complex such an organization of nerve cells, the more likely it is to suffer a breakdown, a disaggregation. Disaggregations can be due to duration and intensity of stimuli, to weakness in the energy of the synthetizing agency, to shock, to hypnosis, etc., etc., and its physiological basis is the retraction of the dendrites which form the connections between cells (an idea which had some vogue at the time). Moments-consciousness which are not "synthetized" into the main or highest synthesis may fall into subconsciousness and remain there, forming part of a larger, looser, secondary assemblage of moments-consciousness, whose degree of organization may vary in different individuals. Only rarely does this "subconscious self" achieve the self-consciousness that would justify us in calling it a "personality". Lacking a personality of its own, and being servile and suggestible and extraordinarily plastic, it can readily be induced to take on the semblance of all sorts of personalities, as in Richet's "objectification of types".[27]

In the third and final part of *The Psychology of Suggestion*, Sidis turns to the suggestibility of man in society,[28] or, strictly, to man and horses in society, since a number of his examples relate to stampeding horses. He points to the evolutionary advantages to group-living animals of responsiveness to the movements of others, and sees the root of the susceptibility of crowds to suggestion in the constriction of voluntary movement suffered by the individual members – this being one of the "conditions of normal suggestibility" revealed by his own experiments. He then gives us a highly entertaining, if not always entirely accurate, survey of "mental epidemics", such as tarentism, the flagellants, the witch-persecutions, financial crazes, and American religious epidemics, in very few of which could the voluntary movements of the participants be regarded as directly constricted. He even attempts to derive a mathematical expression for the energy of a mob, of number m, in relation to the energy, s, of the "hero" who incites it. On the assumption that the hero awakens half his own energy in each member, and each member awakens half that in his fellow, the total mob-energy is $m^2 s/4 + ms/2$, a fact which will no doubt be of great use to senior riot policemen.

Sidis's ideas are developed further in later works, *Psychopathological Researches* (1902), *Multiple Personality* (1904, with S. P. Goodhart), and *The Foundations of Normal and Abnormal Psychology* (1914), without, however, any notable gain in clarity. *Multiple Personality*, however, is of value for the assortment of cases which it contains.

The simplest cases of multiple personality discussed by Sidis and Goodhart are hardly distinguishable from fugue states. Such was the well-known case of Ansel Bourne, studied by James and Richard Hodgson,[29] and mentioned in the former's *Principles of Psychology*. Here, the memories of the fugue state were recovered by the use of hypnosis, but the two sets of memories (those of the normal life and those of hypnosis and the fugue state) were never amalgamated. Bourne's misadventure seems to have been linked to psychological stress. In a remarkable case described by Osborne in 1894,[30] the subject, who owned a prosperous tinning and plumbing business in Pennsylvania, had apparently no domestic, financial or other worries at all. He left his house for a walk one Sunday morning, and vanished completely. Almost two years later he "came to" working as an employee in a tin-shop in the far South. He seems not to have been hypnotized and the missing memories were not recovered.

Somewhat different are the cases described by Dana (1894)[31] and by Sidis and Goodhart themselves. In both these cases, the patients suffered a physical injury (gas-poisoning and a fall on the head respectively), which resulted in a dense retrograde amnesia involving the loss of nearly all acquired attainments, including speech, which was partly lost in Dana's case and completely so in Sidis's. Both patients had therefore to be largely reeducated, and both proved remarkably apt pupils. Dana's patient was not a good hypnotic subject. His memory returned suddenly and spontaneously, and thereafter he remained amnesic for events of the three months since his accident.

Sidis and Goodhart's patient, the Rev. T. C. Hanna,[32] a Baptist minister, who had suffered a fall from a carriage which rendered him unconscious, was apparently amnesic for his whole life before the accident. However, Sidis and Goodhart discovered that he had dreams in which he saw scenes from this lost period, even though he failed to recognize them. They therefore made use of a method called by Sidis "hypnoidization"[33] (Hanna was not a good subject for orthodox hypnotism). This involved asking the patient to close his eyes and relax, whilst attending to some fairly monotonous form of stimulation. He was then asked to relate the ideas and images that entered his mind. In the case of Hanna, this method brought forth a flood of materials, veridical but still unrecognized, relating to his "missing" life. Further methods of stimulating recall – especially confronting him with once familiar places and scenes, and even with recreated episodes from his past – were adopted, and Hanna began to have phases in which he returned to his original, pre-accident state, and lost all recollection of subsequent events. Finally, almost two months after the accident, and following some hours of intense mental struggle, in which both streams of memory were somehow present before his mind yet still eluded synthesis, he succeeded in welding them into a single continuous history.[34]

Hanna in effect exhibited three personalities or phases of personality – his pre-accident self, his post-accident self (these two alternated for a while) and his final self which united the memories of the other two. Between this case and the classic cases of multiple personality, such as the Beauchamp and Doris Fischer cases, it would be possible to set up a continuum of intermediate cases. For example, Dr. J. A. Gilbert of Portland, Oregon, describes[35] the case of a young man of twenty-two, who had had a very rough upbringing. He was much given to wandering, and for seven years had exhibited three phases of personality, with independent memories and mutual amnesia. These personalities went under the same name, and would sometimes amalgamate for brief periods. However, they differed a good deal. It proved possible to transfer him by hypnosis from one state to another. Eventually, the three personalities were united by hypnotic suggestion accompanied by repeated consecutive narration of the events of his different lives. However, the patient proved liable to relapses. It should be noted that these three personalities could in no way have been created by hypnotism, and the patient's antecedents were such that it is most unlikely that he would himself have read any literature about such things. Of course, had he come at an early stage into the hands of a hypnotist who thought it appropriate to give the three personalities names, the personalities would have been shaped and made more distinct accordingly, and we would no doubt have had something more closely resembling the classic cases of multiple personality.

MORTON PRINCE

Sidis is quite frequently referred to in the hypnotic and psychiatric literature of the time, but he founded no school and seems to have had few followers. No doubt his idiosyncrasies and his idiocies, his dogmatism and the obscurity of his theoretical formulations, prevented him from achieving the degree of influence which the originality of some of his contributions might otherwise have won for him. A more influential (and quintessentially "establishment") figure was Sidis's friend Morton Prince (1854–1929). Prince, a physician from a wealthy Boston family, became an active and prominent member of what Hale calls "the informal Boston School of abnormal psychology and psychotherapy".[36] Widely travelled in Europe, he was well up in the European psychiatric and hypnotic literature, and in 1890 published a paper on hypnotism and related matters in *The Boston Medical and Surgical Journal*[37]. He had read and been influenced by Janet, Gurney, Binet, James and Dessoir, and reported experiments of his own (closely resembling theirs) on double consciousness as evinced by the automatic writing of hypnotized subjects. A number of distinguished persons – including James, Putnam and Royce – took part in the ensuing discussion, and James warned how easy it is in the ordinary hypnotic subject "to suggest during a trance the appearance of a secondary personage with a certain temperament, and that secondary personage will usually give itself a name. One has therefore to be on one's guard in this matter against confounding naturally double persons and persons who are simply temporarily endowed with the belief that they must play the part of being double."[38]

The following year, 1891, Prince presented a paper on hysteria and hypnotism to the American Neurological Association, which was, however, not published until some years later.[39] In it he develops his own "physiologico-anatomical" theory of the phenomena, based upon Hughlings Jackson's proposal that three different levels of evolution may be distinguished in the nervous system, of which the highest, represented principally by the frontal lobes, monitors and regulates the functions of the others, while the second includes the primary motor and some sensory areas, and is capable of some degree of associative linking of the sense impressions of the different modalities, and of such impressions with motor actions.

At times, the highest level and the second level may act more or less independently, "like two distinct but connected brains", as when the highest level converses intelligently, while the second level guides the individual through a crowd, or knits, sews, etc. When sensory impressions enter the second level and do not reach the highest level and the dominant consciousness, we have absent-mindedness, and when the middle level performs actions of which the highest level is unaware we have automatism. The experiences concerned may none the less be conserved, and come out in, for instance, dreams.

With regard to the phenomena of hysteria and hypnosis, Prince proposes the following:

> ... anaesthesia of hysteria is the inhibition or going to sleep of certain limited areas (or centres) of the highest level (frontal lobes), while hypnosis is the more or less (according to the stage) complete inhibition or going to sleep of the frontal lobes as a whole.
>
> In hysteria there is a local suppression of function; in complete hypnosis, a total suppression of function of the highest level.[40]

What exactly this "inhibition" or "suppression of function" could consist in, or why it comes about, Prince is unsure. But he thinks it can provide an intelligible explanation of various of the phenomena of hysteria and hypnosis. Hysterical anaesthesia, for instance, occurs when tactile sense impressions fail to reach the frontal lobes, in which functions have been locally suppressed. Under hypnosis, however, the function of the frontal lobes being now suppressed as a whole, "the states of consciousness which are 'awake' and in rapport with the outer world are those of the middle level. The association of these states constitutes a 'second

personality,' which feels all impressions."[41] This personality will remember the
pricks and so on administered to the anaesthetic hand, and will lack spontaneity,
since middle-level actions tend towards the automatic. It is easy to think up
explanations of automatic writing and of double consciousness along these lines.

Prince does not claim that this sort of explanation can be applied to more
complex cases of spontaneous somnambulism, double personality, etc., and
obviously it would not cover those kinds of hypnotic phenomena, for example
Richet's "objectification of types", which require a great deal of intelligent
inventiveness on the part of the subject.

Prince developed his theoretical position further in a series of articles published
in *The Journal of Abnormal Psychology* in 1909 and 1910,[42] and in his book *The
Unconscious* (1914), which elaborates on the articles. Despite its title, *The
Unconscious*, although it bears signs of having been hastily put together, marks
both the climax and the end of the golden age of the *subconscious*, 1889–1914.
But it contains no further systematic treatment of hypnotic phenomena, and
Prince's importance for the history of hypnosis lies elsewhere. He is, of course,
now chiefly remembered for his involvement in and hypnotic treatment of
multiple personality cases, particularly the case of Miss (or the Misses)
Beauchamp. His book about Miss Beauchamp, *The Dissociation of a Personality*
(1905),[43] though somewhat rambling, was the fullest report of its kind to be
published up to that time.[44] This case received a great deal of publicity (it was
even the basis for a successful play[45]), and more than any other served not
indeed to initiate, but to make definite and give wide currency to certain now
traditional ideas about multiple personality disorder and its relations to hypnosis.
These ideas are:

1. That the different personalities in such cases may call themselves by
   different names.
2. That they may enjoy a continued conscious existence even when not "in
   control" of the organism.
3. That their genesis may be related to unhappy or shocking experiences in
   earlier life.
4. That they may be called forth, sent away, or even created by hypnotism.
5. That hypnotism is the treatment of choice.
6. That from a study of the phenomenology of such cases we may reach
   significant conclusions concerning the structure of human personality,
   hypnosis in relation thereto, and the dynamics of subconscious streams
   of consciousness.

"Miss Beauchamp"'s real name was Clara Norton Fowler,[46] but I shall continue
to use the better-known pseudonym. In April 1898, when she sought treatment
from Prince, she was twenty-five years old and, though from a poor background,
had achieved a fair level of education. Her symptoms – headaches, insomnia,

bodily pains and persistent fatigue – were of a kind then generally classified as "neurasthenic". These symptoms dated from, or had been exacerbated by, an emotional shock received in 1893. Her personality at this time (referred to as "B I" in the case report) was proud, reserved, idealistic, religious and excessively prim. She was treated by hypnotism and readily passed into a state of "somnambulism". The somnambulic personality, known for convenience as B II, shared the memories of B I, whereas B I was amnesic for the B II phase. After some while, another personality spontaneously emerged during hypnosis, and shortly began to manifest during the waking state also. This was B III, also known as "Sally", a personality quite different from B I and B II, who were amnesic for her interventions. She was lively, childish, active, mischievous, and possessed of a marked animus against B I, whom she regarded as hopelessly dull. When she succeeded in ousting B I she would do her best to leave behind all kinds of trouble and embarrassment for the unfortunate and highly sensitive B I Sally was, in Prince's term, a "coconscious" personality. She claimed a continuous existence, independent of B I and B II, going back to Miss Beauchamp's childhood. She was aware of, or could tune into, what was going on in the minds of B I and B II, and thus had access to their past histories even if she did not exactly share their memories. She would intrude upon B I's activities and stream of consciousness, forcing visions upon her and causing her to tell lies and perform socially unacceptable actions.[47]

In 1899, another emotional shock, connected with the first, precipitated the appearance (without hypnotic midwifery) of a fourth personality, B IV. B IV had no memory of the events since the earlier shock in 1893, i.e. no memory of B I's career, and no knowledge of B III (Sally). Her personality was in many ways the reverse of B I's – she was assertive and irritable while B I was meek; irreligious while B I was pious; disliked children and animals while B I loved them; and so forth. B III, though she remained conscious during B IV phases, had no direct knowledge of B IV's thoughts, which helped B IV considerably in the ensuing struggles between them. Eventually, Prince discovered that B I hypnotized and B IV hypnotized were identical with B II (who had now become a more mature and decisive personality), and that B II had the memories of both. After divers complicated hypnotic manoeuvres, Prince was able to coalesce B I and B IV to form the "real" Miss Beauchamp. Sally (B III), who claimed that she had existed prior to the division of "real" Miss Beauchamp into B I, B II and B IV, was not part of this amalgamation and eventually disappeared "back to where she came from"[48] (a limbo whose location is not further specified). Prince, however, came to feel that there were features of the finally unified Miss Beauchamp's character that corresponded to aspects of Sally's personality.

The first issue to arise in reviewing this case is obviously that of whether and to what extent Miss Beauchamp could have deliberately deceived Morton Prince. By 1898, any well-read American, such as Miss Beauchamp, could easily have learned enough about multiple personality and kindred oddities to have a fair shot at playing the rôle(s) of a multiple personality patient. Among easily accessible popular works one might mention W. R. Newbold's articles in the *Popular Science Monthly* (1894–6) and R. Osgood Mason's *Telepathy and the Subliminal Self* (1897). In this book, and in a more professional paper published in 1893,[49] Mason describes a case of multiple personality with coconsciousness, in which one of the personalities had, like Sally Beauchamp, certain child-like qualities. It is also possible, though hardly likely, that Miss Beauchamp, who read French, could have known something of those aspects of *fin-de-siècle* literature in France for which psychopathology, derived from the Salpêtrière, provided somewhat sinister materials.[50]

There is no doubt, too, that in the long run Miss Beauchamp profited from the interest which Prince took in her; indeed much later she ended up married to one of his Boston colleagues, Dr. G. A. Waterman (who had himself studied at least one multiple personality case), and became a noted hostess in "good" society.[51] But one must also also remember how much she had suffered from her affliction, particularly because of the maliciousness of Sally.[52] And Prince assures us that she always denied any acquaintance with psychological works, that he never learned anything to the contrary, and that he had a firm belief, based on much experience, in the truthfulness and honesty of her dealings with him.[53] The hypothesis of deceit by Miss Beauchamp has therefore little to recommend it except the fact that it would make psychological science simpler if it were true.

Another approach to the case is to allege that the various supernumerary personalities were created and made more definite by suggestions consciously or unconsciously conveyed by a psychiatrist who had a penchant for interpreting cases in this way. It is true that Prince was well up in the literature of "double consciousness" and had already studied some less florid cases of this disorder. And he almost certainly made the personalities more definite and encouraged their persistence even while he was trying to unite or dispel them. In fact in another case (the BCA case[54]), it is quite clear that as soon as Prince began hypnotic treatment the three personalities involved became more definite and amnesic barriers developed between them which had not previously been there. But in the Beauchamp case he did not by hypnotism or suggestion directly create B III, whose arrival was so unexpected that for a while he refused to accept her, or B IV who appeared following a shock and quite independently of suggestion or hypnosis. There is evidence that he was on his guard against the possibility of direct suggestion – he had after all heard William James's warning, quoted above.

Now if we are prepared to suppose that the Beauchamp personalities were not suggested or hypnotized into existence by Morton Prince, and were not invented by the well-read and imaginative Miss Beauchamp, some interesting possibilities come into view. For in Prince's opinion to communicate with B II, B III and B IV was to communicate directly with a coconscious, or what others would have called a subconscious or subliminal personality of a considerable degree of sophistication, and from such communication, he thought, important in-formation might be gleaned as to the nature of subconscious streams of consciousness and their manner of interaction with each other and with the waking consciousness. To that end he frequently pressed Sally Beauchamp (B III) for her introspections and persuaded her to write her autobiography. The autobiography, says Prince

> ... is a descriptive history of a dissociated mind, of thoughts, feelings, emotions, and even of a "will", of which the personality whom we call the primary consciousness has no knowledge whatsoever, excepting of course so far as she has learned of it by the revelations of this study.[55]

Sally claimed that she had existed from Miss Beauchamp's early childhood (which had been far from happy), not at first as an independent consciousness, but as a conscious nucleus within or alongside Miss Beauchamp's consciousness, and thinking and feeling independently of it.[56] At first they thought about the same things at the same time, but differently about them. "Now we think about different things at the same time because my life is different from hers and I know things she doesn't know."[57] When both personalities were engaged in activities which Sally enjoyed, for example climbing trees, both would together will the same thing, and Sally would have "a certain feeling of being one with her". But when Miss Beauchamp was engaged in lessons or serious matters, Sally lost interest and the feeling of unity went. Thus Sally could not learn French, and always felt herself inferior to the other personalities on that account.[58]

It is far from clear how Sally gained her knowledge of events in the minds of B I and B II. Sometimes she writes as though she could somehow observe them. Miss Beauchamp when younger had been much prone to daydreams and visions, which totally absorbed her. But Sally always saw these visions projected against reality and was never taken in.[59] The same held true of dreams – Sally did not sleep herself and had not dreamed since the double consciousness became "fixed".[60] But how she knew of Miss Beauchamp's (verbally expressible) thoughts is not made clear. Perhaps she heard the latter's "inner speech", but, *teste Ulysses*, it seems very doubtful that someone's thoughts could be adequately reconstructed from "overhearing" the fragments of inner speech that go on in the person's mind.

Sally also says a little about how she influenced Miss Beauchamp's thoughts and actions. She made her talk by simply talking herself, but how this worked she was not clear.[61] Similarly, to make her do this or that, she "just willed".[62] To force an hallucination, for example of a snake, into the consciousness of B I or B IV, Sally "thought of a snake, and 'willed,' and straightway B I or B IV saw a snake."[63] To replace herself (when in control) by Miss Beauchamp, she "folded herself up" (put herself into a state of abstraction) and fixed her thoughts upon B I.

From the point of view of the other personalities, the visions and hallucinations injected by Sally into their streams of consciousness were intrusions (they just "bubbled up", to use a phrase applied by "Real Doris" in the Doris Fischer case[64]), and the forced actions or inhibitions of action which she inflicted upon them were experienced as totally alien.[65]

In the Beauchamp case, we have what Myers would certainly have regarded as messages "sent from one stratum of the personality to another", and we have an account of how such messages were deliberately sent and of what it was like to receive them. Sometimes the "messages" intruded without being deliberately sent.[66] Whether or not concepts of "dissociation" (or disaggregation) and the "subconscious" (or subliminal or coconscious or subwaking consciousness) will ultimately help us towards an understanding of the phenomena of hypnotism, there is no doubt that in some circumstances these concepts can have exceedingly plausible applications.

NOTES

1 Barzun (1983), p. 230n, makes out that James anticipated Myers's doctrine of the subliminal self by two years, because in 1890 James published a paper on "The hidden self" in *Scribner's Magazine*. But, as we have seen, Myers had been developing his theory for some years prior to that.

2 Compare, for instance, the brief textbooks of James (1892) and Baldwin (n.d.) with those of Wundt (1912) and Titchener (1897).

3 Compare the figures of numbers of articles on hypnotism and suggestion, by countries, which I give below, p. 561.

4 Cf. Quackenbos (1908).

5 Bérillon and Farez, eds. (1902), pp. 15–16.

6 Cf. Parker, ed. (1909).

7 "Woman can not leave long the routine of her life, the beaten track of mediocrity; she can rarely rise above the trite; she is a Philistine by nature." (Sidis, 1898, p. 363). Sidis would not have been well-suited to present day America.

8 Sidis (1898), p. 15.

9 Sidis (1898), p. 29.

10 Sidis (1898), p. 37.

11 Sidis (1898), p. 44.

12 Sidis (1898), p. 65.

13 Sidis (1898), p. 86.

14 The work of Seashore (1895) on certain illusions and hallucinations anticipates Sidis, and also Binet (1900), in certain respects.

15 Sidis (1898), p. 65.

16 Sidis (1898), p. 65.

17 Sidis (1898), p. 65.

18 Sidis (1898), pp. 69–70.

19 Sidis (1898), p. 88.
20 He means reflex in Gurney's sense of "psychic reflex".
21 Sidis (1898), p. 89.
22 Sidis (1912–13).
23 Sidis (1898). pp. 118–128, and cf. Sidis (1913).
24 Sidis and Goodhart (1904).
25 Sidis (1898), p. 174.
26 Sidis (1898), p. 209.
27 Sidis (1898), pp. 252–268.
28 Cf. Sidis (1896).
29 Sidis and Goodhart (1904), pp. 376–382.
30 Sidis and Goodhart (1904), pp. 365–368.; Osborne (1894).
31 Dana (1894).
32 Sidis (1898), pp. 216–227; Sidis and Goodhart (1904), pp. 83–226.
33 For Sidis's later proposal that the "hypnoidal" state which results is intermediate between sleep and hypnosis, and is the primitive rest state from which both have evolved, see Sidis (1910).
34 See the patient's own autobiographical account, Sidis and Goodhart (1904), pp. 224–226.
35 Sidis and Goodhart (1904), pp. 404–419; Gilbert (1902).
36 P. 4 of his Introduction to M. Prince (1975).
37 M. Prince (1890).
38 M. Prince (1975), p. 55.
39 M. Prince (1898–9).
40 M. Prince (1898–9), p. 95.
41 M. Prince (1898–9), p. 95.
42 M. Prince (1909–10; 1910).
43 I have used the second edition of 1908. For an earlier and shorter account of this case see Prince (1900–1), which is reprinted in F. W.H. Myers (1903), Vol. 1, pp. 341–352.
44 At least if we do not count Flournoy's Des Indes à la planète Mars (1900); see Flournoy (1963).
45 "Case of Becky". See Kenny (1986), pp. 157–159.
46 See Kenny (1986); Rosenzweig (1987; 1988).
47 "May 6. Miss Beauchamp reports that she is still telling fibs. For example, she told Miss M. that Mr. X was a great admirer of Swinburne, and had busts of him all about the house; that he had named his baby Algernon Swinburne X.; that this baby was boneless; and that Mr. X. fed him and Mrs. X. on nothing but oatmeal... Miss Beauchamp is much horrified at telling such nonsense, but does not seem to be able to help it ... " M. Prince (1908), p. 60.
48 M. Prince (1908), p. 524.
49 R. O. Mason (1893).
50 Ellenberger (1970). pp. 164–167.
51 Kenny (1986), pp. 155–156.
52 On one occasion, Miss Beauchamp regained consciousness to find that in her "absence" Sally had borrowed a large sum of money, had given $40 to a beggar, and had pulled a valued watch to pieces. Prince (1908), p. 138.
53 M. Prince (1920).
54 M. Prince (1919).
55 M. Prince (1908), p. 368
56 M. Prince (1908), p. 278.
57 M. Prince (1908), p. 374n.

58 M. Prince (1908), pp. 122, 539.
59 M. Prince (1908), p. 375n.
60 M. Prince (1908), p. 376n.
61 M. Prince (1908), p. 62.
62 M. Prince (1908), p. 93.
63 M. Prince (1908), p. 510.
64 W. F. Prince (1915–16), pp. 47, 477, etc.
65 Cf. M. Prince (1908), p. 275.
66 M. Prince (1908), pp. 507–509.

CHAPTER 20

Hypnotism and suggestion at the turn of the century. I – Preliminary considerations

So far, I have dealt with the history of the hypnotic movement largely in terms of the origin and development of certain rather loosely characterized schools of thought – the School of the Salpêtrière, the School of Nancy, the "School" of those who found the key to the mysteries in the concept of the subconscious or some kindred or cognate notion. But of course, not everyone of importance in the movement can be satisfactorily assigned to one or other of these schools, and pursuing the history of the separate schools has hindered me from giving a systematic exposition of the phenomena of hypnotism as these allegedly manifested themselves at this period. I shall therefore attempt to remedy these deficiencies by giving a more or less static picture, or survey, of the hypnotic scene as it was at the turn of the nineteenth and twentieth centuries. I shall attempt to sketch the principal phenomena of hypnotism, the principal applications of hypnotism, and the principal theories of hypnotism, as these might have appeared to a fairly eclectic contemporary writer.

This project amounts to the composition of something like a mini-textbook of hypnotism from the standpoint of (say) the year 1900. It would be very convenient if I could find and merely summarize an actual work of the time, as I did earlier on with Kluge's excellent textbook of animal magnetism. Unfortunately, it is not easy to find a textbook that satisfactorily fulfils the purpose.[1]

On balance the most useful is perhaps L. Loewenfeld's *Der Hypnotismus* (1901), a substantial and commendably eclectic work which covers most aspects of the subject.[2] I shall follow its outlines, reserving the right to depart from them as necessary, and to incorporate, where convenient, certain materials which had not yet appeared at the date when it was published. Basing a survey on a German source is appropriate, because by the end of the century the German-language hypnotic literature, whilst not perhaps more copious than the French, considerably outweighed it from every other point of view. A similar shift had taken place in the early history of animal magnetism. The Germans systematized and intellectualized what the French had pioneered.

HYPNOTISM IN EUROPE

It will be useful to begin by enlarging somewhat on Loewenfeld's survey of the turn-of-the-century hypnotic scene. On hypnotism in the United States I have already touched.[3] Britain continued to lag behind the United States, France and Germany; indeed, interest there was still rising in the years immediately prior to 1914, when it was generally declining elsewhere. Milne Bramwell's *Hypnotism*, published in 1903, became very influential – it reached a third edition in 1913. In 1906, a Medical Society for the Study of Suggestive Therapeutics was founded.[4] Bramwell, Lloyd Tuckey, Felkin, Wingfield and T. W. Mitchell were all involved, and Lloyd Tuckey became the first President.[5] On the more academic side, William McDougall (1871–1938), who was Wilde Reader in Mental Philosophy at Oxford from 1904 to 1920, was considerably interested in hypnotism, and wrote some forceful passages on its theory and implications.[6]

In Italy, medical and scientific interest in hypnotism, which had been not inconsiderable in the 1880s, declined quickly and then became more or less stagnant.[7] The tendency of Italian psychiatry was strongly somaticist, and not friendly to any form of psychotherapy.[8] Such medical interest as there was seems to have been rather backward-looking. For instance, G. Belfiore's *Magnetismo e ipnotismo* (1905) is still heavily concerned with the disputes between Nancy and the Salpêtrière. Popular interest, probably associated with Spiritualism, seems to have been much greater; it received support from at least one noted alienist, Cesare Lombroso, whose *Ricerche sui fenomeni ipnotici e spiritici* appeared in 1909. It could not be said that Lombroso showed in his later, Spiritualist, phase any keener appreciation of the logic of scientific endeavour than he had in his earlier, materialistic one.

Italian Spiritualism, and matters related thereto, had somewhat complex relations with Italian anti-clericalism; it is therefore not altogether surprising to find in Italy certain lingering clerical suspicions of hypnotism. A powerful influence in certain Catholic quarters was G. G. Franco's *L'ipnotismo tornato di moda* (1886). According to Franco, a Jesuit, hypnotism is not merely a form of mental disorder; there is a diabolical element in it. Hypnotism was ably defended against Franco by a French Dominican, Father M. T. Coconnier, in his *L'hypnotisme franc* (1897), but suspicions lingered.[9]

Turning east, to Russia, we find interest in hypnotism still at a high level. Fifteen Russian delegates attended the Second International Congress of Hypnotism in 1900, the same number as had attended the earlier Congress in 1889. It should not be supposed that at this period Russia was an intellectual backwater. The general tone of the Russian medical approach to hypnotism, indeed to psychiatry, was physiological. The leading figure in Russian hypnotism at the beginning of the twentieth century was V. M. Bekhterev (1857–1927), the distinguished physiologist. Bekhterev's interest in hypnotism was originally focussed on the possibility of utilizing hypnosis for his research on reflexes; he was also concerned with its possible applications to therapy in general and to the treatment of alcoholism in particular.[10] Later, he became interested in the theory of hypnotism and suggestion, though his contributions here are somewhat naïve,[11] and wrote extensively on the importance of suggestion in the aetiology of hysterical epidemics and in the understanding of social and crowd phenomena.[12] Many of his principal publications appeared in German, and in 1904 he founded the *Zeitschrift für Psychologie, Kriminal-Anthropologie und Hypnotismus.*

In France, interest remained strong and publications continued to be quite numerous. The quality and originality of those publications, however, were no longer what they had been. Liébeault was growing very elderly (he died in 1904); Bernheim had founded what was virtually a new school; Janet steadily pursued his psychopathological researches with no special, separate interest in hypnotism; Dumontpallier died in 1899 and Auguste Voisin in 1898; Binet was moving into educational psychology. Their successors were not men of comparable stature. The *Revue de l'hypnotisme* was increasingly filled with miscellaneous medical oddities. If one discounts Bernheim's successive re-workings of his earlier books in later editions, the only really substantial French work on hypnotism from the turn of the century is Grasset's *L'hypnotisme et la suggestion* (1903); and Grasset is a survivor from an earlier period.

Belgium produced at least one solid and useful textbook, namely J. Crocq's *L'hypnotisme scientifique* (1896; 2nd. edn., 1900). Belgium also had, for a while, its own hypnotic journal, the *Journal de neurologie et d'hypnologie*. But, Crocq apart, Belgian contributions to the hypnotic literature of this period were not of any great significance.

Without doubt, by the end of the nineteenth century, the intellectual and clinical centre of hypnotism had shifted from France to Germany. Until 1902 the *Zeitschrift für Hypnotismus* continued to provide an impressive forum, and contained frequent contributions from outside Germany. From 1895 onwards it was edited by Vogt and Forel, and published articles on clinical, theoretical, experimental and historical aspects of hypnotism, and reviews of a wide range of related areas. By comparison, the *Revue de l'hypnotisme*, especially in its later years, is a mere fringe magazine.

The most respected figure in German hypnotism at this period was probably Oskar Vogt (1870–1959) of Leipzig and then Berlin, a man of considerable and wide-ranging abilities[13]. He was at home in psychiatry, psychology, neurology, and even (to an extent) philosophy, and merits a fuller study than he has so far received, even if only as a representative of a certain category of German intellectual. He enjoyed the reputation of being one of the most effective hypnotists of his time.

Vogt first made his mark by the lengthy notes which he contributed to the third (1895) edition of Forel's *Der Hypnotismus*. These notes were followed in the later 1890s by a series of substantial articles in the *Zeitschrift für Hypnotismus*. Vogt seems to have been working his way towards a large-scale synthesis of theoretical and clinical hypnotism with an approach through hypnosis to certain problems of experimental psychology. At the turn of the century, while he was in practice in Berlin, he was for a while joined in this endeavour by what was almost a "school" of pupils or associates – van Straaten, Marcinowsky of Paderborn, W. Hilger of Magdeburg, and K. Brodmann (1868–1918), still remembered for the "Brodmann numbers" assigned to different cytoarchitectural areas of the cerebral cortex.   But after 1900 his interests shifted increasingly towards neurology, though we still find him keenly discussing hypnotism at international conferences and gatherings.[14]

The German hypnotic literature, like all the other hypnotic literatures of the time, was predominantly clinical. The *Zeitschrift für Hypnotismus* published some substantial articles on therapeutic methods and results.[15] Various larger collections of cases appeared in book form. Of these, the most frequently mentioned was perhaps *Der Psychotherapeut* (1896) by H. Stadelmann (b. 1865) of Würzburg, which shows the influence of Breuer and Freud. Also influential was Loewenfeld's textbook. Though general in scope, it was principally intended for medical men. Loewenfeld (1847–1924) was a Munich nerve-specialist who had established a reputation with earlier works on neurasthenia and hysteria and on sexual problems. Another textbook for medical men was *Hypnotismus und Suggestivtherapie* (1905), by L. Hirschlaff (b. 1874) of Berlin. This is ostensibly the second edition of an earlier work by Max Hirsch. It is, however, full of Hirschlaff's own highly individual opinions and of his clinical findings.

Theoretical approaches to hypnosis – those of Vogt and of A. Döllken of Marburg[16] were perhaps the most highly regarded – were mainly derived from clincial materials. More academic, but also highly thought of, were the theoretical speculations of Theodore Lipps (1851–1914), Professor of Psychology at Munich.[17] I shall return later to theories of hypnosis.

Of experimental approaches to the problems of hypnosis, as we would understand the term "experiment", there were very few, whether in Germany or elsewhere. Perhaps the resolute hostility displayed by Wundt, the great Cham of experimental psychology, towards all forms of dabbling in hypnotism, had a dampening effect. A less constricted approach to experimental psychology developed in the early years of the twentieth century at Würzburg, under the influence of Oswald Külpe (1862–1915). Narziss Ach (b. 1871), one of Külpe's "Würzburg School", used hypnotism in some of his experiments,[18] and in 1907, an American visitor to Würzburg, Lillien J. Martin (1851–1943) of Stanford University, produced what was arguably the first extensive and systematic laboratory study of hypnotic phenomena[19] to emanate from Europe.[20] In the tradition of the time she used a smallish number of subjects (who were

probably skilled introspectionists in the Würzburg manner), and a fair number
of experimental situations. She investigated the effect of hypnosis upon rapport,
analgesia, involuntary movements and catalepsy, perception, memory, imagination, hallucination, thought, will, attention and feeling, and where
appropriate ran control trials with the subjects in a waking state. Her pioneering
work has certain strengths and certain weaknesses. Its strengths are the care
with which she supplies us with all kinds of relevant background information
about the subjects and the setting (a practice which could with benefit be more
widely followed today), her earnest endeavours to understand the hypnotic
situation from the subject's point of view (to aid in this understanding she was
herself many times hypnotized by Vogt), and the way in which she illuminates
the experimental findings with the introspections of her subjects (surely *these*
subjects were not compliant!). Its weaknesses are wholly insufficient details of
the experimental methods used, and absence of truly quantitative data. A
certain amount of experimental work on hypnotism was also undertaken at
Geneva, under the influence of Edouard Claparède (1873–1940), professor of
psychology there, who had published an *Esquisse d'une théorie biologique du
sommeil* in 1905.[21]

About hypnotism in other European countries, and other parts of the world,
there is little to be said. Participants came to the International Congress of
Hypnotism in Paris in 1900 from as far afield as Brazil, Canada, Ecuador, Egypt,
Cuba, Colombia, Haiti, Greece, Iceland, Mexico, Persia, Poland, Rumania,
Turkey and Venezuela. The movement was without doubt a multi-national one,
and it showed signs of growing.

What inroads hypnotism had made against other methods of treatment it is
hard to say. Writing in 1897, Vogt's pupil Brodmann alleges that there are
hundreds of clinics and doctors using hypnosis, along with general methods of
psychotherapy.[22] In 1901, Loewenfeld was a good deal more pessimistic. He
noted that in Germany, particularly South Germany, nervous illnesses were still
largely treated in institutions, hydros, sanitaria, etc., so that hypnotic treatment,
which would be much more useful for most such ailments, is not considered.[23]

In the end, hypnotism, though for a time widely discussed and even practised
in medical circles, penetrated the medical establishment only to a limited extent.
In 1895, Schrenck-Notzing found it a matter of note that hypnotism had been
favourably presented in two new textbooks of nervous diseases.[24] Soon
afterwards, Forel and Crocq made strong pleas for the teaching of hypnotism in
medical schools.[25] These pleas were only sporadically answered, and in the years
after 1900, favourable answers became less and less likely.

In literature, hypnotism made a much greater impression. The 1890s were
characterized by a reaction against the rationalism, positivism and exaltation of
science which had marked the preceding decades. The ideas of Bergson and of
Nietzsche were widely influential, but the reaction went even further. Disser-

tations could be – no doubt have been – written about the occultist tendencies of *fin-de-siècle* literature in several countries. The psychopathological novel, working hypnotism, secondary personality, etc. into its plot, flourished.[26] But into this fascinating terrain space does not permit us to enter.

DEFINITIONS OF SUGGESTION

Next in Loewenfeld's book come several chapters discussing and analyzing certain key concepts, of which the first is the concept of suggestion. He gives a fairly extensive review of contemporary definitions.[27] I can pick out only a few for brief notice.

Some definitions were so brief and so general as to be quite useless. Such was Bernheim's: "suggestion is an event through which an idea is introduced into the mind and accepted by it."[28] Again there is Bérillon: "Suggestion is the act of utilizing the ability which a subject has to transform an idea, once accepted, into action."[29] Likewise, Edmund Parish says that a suggestion is any sense-perception you like insofar as it arouses ideas, significantly affects existing ideas, in short insofar as it is an influence on the flow of ideas.[30] In terms of these definitions, a great part of mental life would be determined by suggestion, and a theory of suggestion would be virtually all-embracing. Even Bernheim and Bérillon in practice use the word "suggestion" in a much more restricted way when talking about the actual phenomena of suggestion. They implicitly recognize that in "suggestion" there is something bizarre and aberrant about the relationship between the implanted idea and its subsequent effects.

Other writers try to pick out a fairly precise and narrow defining characteristic. Their proposals sometimes highlight important aspects of the phenomena, but never succeed in wholly capturing the concept. Thus Bergmann defines suggestion as "an idea that, because of its intensity, realizes itself with an instinct-like necessity."[31] As Hirschlaff remarks,[32] it is hard to maintain that a fufilled suggestion of, say, anaesthesia, could be looked at in this way. Vogt regards as "suggestions" such "psychophysical phenomena as manifest the abnormally intensive after-effects of purposive ideas." And by a "purposive idea" he understands "the idea of the occurrence of a psychophysical event. The memory-image of this event is contained as content in the purposive idea." There can be unconscious purposive ideas, but the physiological underpinnings to which they correspond are weaker.[33] According to Stoll, suggestion is simply an idea awoken in us in various ways by events in the outer world, which form the starting point for further thought processes without the causal connection between them coming clearly to consciousness.[34] But of course many people who respond to suggestions are well aware of the causal connection between the suggestion and their response to it. Forel makes almost the reverse proposal. Suggestion is the production by another individual of a dynamic alteration in a person's nervous system, or in functions dependent upon it, through the calling forth of the conscious or unconscious idea that this alteration is taking place, has taken place, or will take place.[35] In these terms, dropping or insinuating a purposive idea into someone's mind in such a way as to arouse no reflection upon it would not count as "suggestion", even if the idea took effect.

The more complex definitions offered at this period were often heavily theory-laden, and are unsatisfactory because they would not survive the collapse of the theories in which they are embedded. In effect, they take some unexplicated, perhaps "common-sense", concept of suggestion, which they do not discuss at all, give (to their own satisfaction) a theoretical account of some of the relevant phenomena, and then present us with a redefinition of the concept in terms of this account. The resultant definitions usually invoke some sort of weakening or narrowing of associative activity, so that the suggested idea can take effect with the minimum of interference from ideas of a contrary tendency. Lipps, Schrenck-Notzing and Loewenfeld himself[36] all offer definitions of this kind. I shall look at Loewenfeld's version.

Loewenfeld distinguishes three different aspects of the concept of suggestion. The term "suggestion" may be used: 1. To refer to the event through which the phenomena designated "suggestive" are caused, for example when I verbally announce to someone that he will feel a fly on his nose, my words constitute a "suggestion"; 2. To refer to the event which fulfils the suggestion, for example the tickle that he felt was my "suggestion", that which was suggested; or 3. To refer to the happenings in the subject's mind which bring about the event fulfilling the initial suggestion (sense 1), for example the "associative activity" aroused by my suggestion concerning the fly. In this sense "suggestion" is the working of a certain idea in the subject's mind (a "suggestion" is active therein). Loewenfeld proposes to call "suggestion" (sense 1) *Suggeriren* or *Eingebung*, and to reserve the term *Suggestion* for "suggestion" (sense 3). But I think it will be clearer if we continue to talk of the suggesting event as the "suggestion", and of the effects of a successful suggestion in the subject's mind (whatever these may be) as that suggestion having "taken" (as one of Janet's subjects expressively put it[37]).

Loewenfeld then gives us the following definition of suggestion (sense 3), i.e. of what it is for a suggestion (sense 1) to "take" or be assimilated. The concept of suggestion (sense 3), is that of "... a mental or psychophysical occurrence ... which, in consequence of a narrowing or abolition of associative activity through the bringing about of this occurrence, manifests an extraordinary efficacy."[38]

Now it may or may not sometimes or often be the case that a suggestion "takes" *because of* a certain restriction or narrowing (however brought about) of the range of associations to which it gives rise in the subject's mind. But this is clearly not what is *meant* by a suggestion "taking", any more than what might be meant by a paralysis "setting in" is some one of the various kinds of nervous change that can cause it to set in. And as a matter of fact, some suggestions which regularly "take" seem to need a considerable upsurge of "associative activity" for their fulfilment, rather than requiring a diminution of it as a precondition of their taking – consider, for instance suggestions leading to the "objectification of types". I shall come back to the definition of suggestion. But to return to Loewenfeld: Having given us his account of suggestion in general, he proceeds to differentiate various kinds of suggestion. He seems to regard himself as differentiating ways in which suggestions can "take", but most of his distinctions relate to the ways in which suggestions enter consciousness, and thus in a way straddle the making of a suggestion and the "taking" of a suggestion. He distinguishes first between heterosuggestions and autosuggestions. The former are suggestions aroused externally by a person or object, the latter are suggestions occasioned by the individual's own associative activity without a direct external cause. If the suggestion is given in words we talk of "verbal suggestion", if by an action or event of "material suggestion", and if by an object of "object suggestion".

A further distinction may be made between direct and indirect suggestion. In direct suggestion, the content of the suggestion is directly specified; in indirect suggestion, nothing is spelled out and the suggestion is so to speak "read into" the suggesting events by the patient himself. Such indirect suggestions are a frequent cause of symptoms in hysterical and hypochondriacal persons.

Since we can find some sort of (indirect) outer impulse for most auto-suggestions, an absolute distinction between heterosuggestions and auto-suggestions cannot be maintained. Furthermore, the details of the working out of a direct heterosuggestion are often supplied by the subject's own associative activity, and are thus autosuggestive within the context of a heterosuggestion.

There is also, Loewenfeld thinks, an important distinction to be made between suggestions which run their course ("take") in the subject's consciousness, and those which do not. The latter are "unconscious" or "subconscious" suggestions, and are particularly liable to have pathological results. For instance, an hysterical girl may develop wry-neck just from observing another person similarly afflicted, yet upon being questioned may manifest no idea of the cause of her condition.

Loewenfeld remarks that autosuggestions in general have a notably more compulsive character than heterosuggestions. This may be explained by a number of factors. Autosuggestions develop especially in persons who suffer from neurotic disorders, which favour the development of compulsive ideas. Freud was the first to propose that an idea acquires a compulsive character if it becomes bound up with a transferable affect, and Loewenfeld's own experience has been that autosuggestions often emerge for the first time in states of strong emotional excitement, especially anxiety states, as though the ideas pull something of the affect to themselves, and are correspondingly energized.[39] Autosuggestions can also be extremely difficult to get rid of. This may be partly explained by the readiness with which a patient can convince himself, in a kind of vicious circle, of their past and likely future fulfilments.

SUGGESTIBILITY

Loewenfeld turns next to the topic of suggestibility. He proposes to define suggestibility as "the tendency to form suggestions from inner or outer stimuli"; thus (in the alternative terminology I proposed above) a suggestible person would be one in whom suggestions are peculiarly liable to "take". Someone may be said to exhibit abnormal or pathological suggestibility if suggestions "take" in him which would only "take" in ordinary persons when hypnotized.

Some categories of persons are more suggestible than others. Thus children, because of their restricted judgment and experience and their lively fantasy, are more suggestible than adults. For opposite reasons, the elderly are less suggestible than the middle-aged. Intelligence and level of education are both related to suggestibility. Narrowness and ignorance increase it – hence hysterical epidemics take place among ignorant country people – whilst those accustomed to think critically are relatively immune. However, Vogt (for whom Loewenfeld shows the greatest respect) holds that degree of suggestibility is not always in inverse relation to critical capacity. Rather is it linked to the capacity for attentional concentration and to liveliness of memory-images. So Loewenfeld tones down his claim to the more modest assertion that on average persons of higher intelligence and education will be less suggestible than others.

Most of these assertions are apparently based on personal observations or on the consensus of clinical opinion. We do not find at this period many attempts to develop and apply systematic tests of suggestibility. Two exceptions were the endeavours of Bérillon and Binet to devise suggestibility tests for children.[40] Bérillon's test involved asking a child to look with intense attention at a chair a certain distance away, and telling him he would feel an irresistible need to go over and sit on it. Most children, he says, respond in a minute or two, and those children who respond tend to be the brightest and most educable. Binet, who by the turn of the century had embarked on the work in educational psychology which has left him enduringly famous, made extensive efforts to devise and standardize tests of suggestibility. For instance, a child would be shown a line about 40 mm in length, which he had to match, either from memory or by direct comparison, with one of a considerable number of other lines. At the very moment at which the child chose a line he would be asked, "Are you quite sure it is not the next line?" A change of choice was taken to indicate suggestibility.

Much of the remainder of Loewenfeld's discussion of suggestibility is concerned with the supposed heightened suggestibility of hysterics. Until recently, he says,[41] the essence of hysteria has been thought of as a peculiar "character", involving abnormal sensitivity, capriciousness, instability, egoism, and an inclination to exaggeration and simulation. Modern authorities (Moebius, Ringier, Forel[42]), noting that in hysterics a great variety of symptoms can be called forth through the arousal of the corresponding ideas, have concluded that the central peculiarity of the hysteric is his heightened suggestibility, especially his autosuggestibility. However, Loewenfeld maintains[43] that the abnormal suggestibility of hysterics is not an original phenomenon, or is so only to a small extent, but is tied up with their other mental peculiarities, for example a decline in powers of dispassionate reasoning, inferior development of the will, and emotionality.

There are, indeed, hysterical patients who show very little suggestibility and are not good hypnotic subjects. In fact, according to Breuer and Freud some very strong-willed and critical people (such as Breuer's celebrated patient Berthe von Pappenheim, known as Anna O.) may be hysterics. Breuer uses this point, in an exceptionally interesting theoretical essay, to rebut Janet's contention that the "splitting" of the personality to be found in hysteria is always due to a constitutional weakness in the synthetic function of the mind.[44] According to Breuer and Freud, hysterics suffer mainly from reminiscences of painful and affect-laden events. These events have not been assimilated to the main stream of consciousness, but have become split off, either because the patient has repressed them, or because she was in a "hypnoid" state (daydreaming, autohypnosis, etc.) at the time of the occurrence. The memory of the trauma, with its corresponding affect, "acts like a foreign body which long after its entry must continue to be regarded as an agent that is still at work."[45] The ideational complexes thus formed may influence or at times gain control of the patient's mind, producing symptoms or fresh "hypnoid" states. Treatment consists in the recovery of the pathogenic memories, together with the corresponding affect, and its assimilation to normal consciousness or removal through the physician's suggestions.

A few writers of this period put forward what might be described as theories of the physiological nature of the underlying disposition to suggestibility,[46] but these are hardly to be distinguished from theories of hypnosis, and do not require separate consideration.

HYPNOSIS AND ITS STAGES

Loewenfeld turns next to the definition of hypnosis. By the turn of the century, many people (not all) regarded hypnosis as an artificially aroused, peculiar state, closely related to natural sleep, and characterized by heightened suggestibility. Theories of hypnosis were theories concerning the nature of the supposed state;

or, *per contra*, denials that such a state really exists. Of course, insofar as the notion of "suggestibility" is built into the concept of hypnosis, the latter will naturally suffer from any unclarities and ambiguities afflicting the former. It would be far more satisfactory if we could mark out the state of hypnosis independently of heightened suggestibility, for example in terms of measurable physiological changes, and not a few workers of this period, though in no sense followers of Charcot, tried, as we shall see, to do so.

Finding a satisfactory characterization of hypnosis was made harder by the fact, accepted by almost everyone, that hypnosis admits of degrees or stages. A subject may be more or less "deeply" hypnotized. Different writers recognized varying numbers of grades – Bernheim, nine; Döllken seven; Liébeault six; Max Hirsch four; Fontan and Ségard, Schrenck-Notzing and Forel three; Gurney, Delboeuf, Dessoir, Crocq and Hirschlaff only two.[47] The most widely accepted categorization was probably that of Forel, who recognized three grades of hypnosis as follows:

1. Somnolence. The lightly influenced person can resist the suggestion by the exercise of his energy, and can open his eyes.
2. Light sleep, otherwise called hypotaxis or 'charme.' Here the influenced person can no longer open his eyes, and is obliged to obey a part of the suggestions or all of them, with the exception of amnesia. He does not become amnesic.
3. Deep sleep or somnambulism. This is characterized by amnesia after waking (and by post-hypnotic phenomena).[48]

These proposals as to stages, or grades, of hypnosis are, I suppose, ancestral to the various scales of hypnotic susceptibility which have come to play such an important part in modern experimental investigations of hypnotism. But even at the time, many writers had considerable reservations about them. Milne Bramwell, for example, in his textbook of 1903, argues that the sequence of phenomena, from which "stages" have been derived, is quite artificial, and itself a product of suggestion. Post-hypnotic amnesia, for instance, arises because subjects almost universally believe that hypnosis is identical with sleep, which to them means forgetfulness on awakening.[49] Others noted how much the sequence varies from one individual to another. For example, Schrenck-Notzing points out[50] that degree of suggestibility is often out of phase with the "depth" of hypnosis as indicated by anaesthesia and post-hypnotic amnesia. Loewenfeld makes similar observations.[51] Some subjects in deep hypnosis exhibit the capacity to hallucinate but will not manifest phenomena easily obtained in light hypnosis. Others who are "somnambulic" by the criterion of post-hypnotic amnesia may not hallucinate. Thus for the distinction between lighter and deeper hypnosis neither the presence nor the absence of a particular phenomenon can be taken as authoritative, but only the total state of affairs.

A radical proposal of some interest was put forward by Hirschlaff.[52] Hirschlaff holds that there are two different states passing under the name of hypnosis – superficial hypnosis and deep hypnosis. In superficial hypnosis,[53] suggestions of anaesthesia, diminished capacity to move, and so forth, can succeed, but they do not have a direct, compulsive and irresistible effect. They are often complied with more or less voluntarily, whether out of obedience or obligingness or as a result of the persuasion of the hypnotizer.[54] In other words, we are not dealing here with a special hypnotic state, but with a pseudohypnotic, somnambuloid or hypnoid one. But experience plentifully shows that this "state" can be therapeutically effective. This is because it gives scope to a number of therapeutic influences, for example complete quiet, mental concentration upon curative ideas, belief that the therapy will work.

Deep hypnosis, true hypnosis, is something quite different. It involves a profound alteration of mental life. Suppression of critical faculty enables the realization of hypnotic suggestions quite unrelated to the subject's own motives. But for this very reason, deep hypnosis should be used therapeutically only in exceptional circumstances. We end up with the apparent paradox that therapeutic hypnotism should make no use of true hypnosis or of real suggestions; only the experimental psychologist should have to do with phenomena that are, in a strict sense, hypnotic.

One assumes that Hirschlaff bases these provocative assertions on his own clinical experience; he does not spell out his grounds for making them.

TECHNIQUES OF HYPNOTIZING

Loewenfeld devotes a substantial chapter to the various techniques of hypnotizing. He begins by remarking[55] that although hypnosis can be induced by an extraordinary variety of means, these can be divided into two main groups: 1. sensory stimulation as such, and 2. the direct arousal of ideas of sleep (by verbal suggestion or otherwise).

1. The former methods (if we set aside some of the more bizarre practices of the Salpêtrière School) involve principally the use of monotonous and uniform sensory stimulation. The subject fixates some bright object, listens to the ticking of a watch, is subjected to mesmeric passes, etc. All these produce a certain state of mental emptiness, with concentration on a stimulus that does not stir up further associations. Furthermore, continuation of this uniform stimulation produces feelings of tiredness which readily give rise to the idea of sleep.

The oldest and commonest of these methods (says Loewenfeld) has been eye fixation, with or without strong convergence or upward gaze. All kinds of objects have been used for eye fixation,[56] and some hypnotists have invented special seductively glittering devices to act as fixation points.[57] The most startling claims in this connection (not mentioned by Loewenfeld) were made by Sanchez Herrero, who invented an apparatus for hypnotizing in the manner of Braid. This consisted of a flexible length of articulated metal. One end would be fixed to the bed-head or chair-back. The other end, on which were set two bright jewels on adjacent stalks, could be bent over the subject's head until the jewels were close to and slightly above his eyes. Fixation of the jewels produced convergent strabismus, and this, aided by passes, verbal suggestions, etc., was sufficient to bring about hypnosis *in all cases*. So strong a claim deserves further investigation.[58]

A surprising number of hypnotists still made some use of mesmeric passes. Van Renterghem, for instance, used passes as well as suggestions,[59] and Wetterstrand, one of the most successful hypnotizers of the period, made greater use of mesmeric passes than was usual among members of the Nancy School.[60] J. F. Woods, one of the most sensible and successful of British hypnotists, who seems to have had great success in the treatment of various refractory complaints, would place one hand on his patient's head, and with the other lightly stroke his forehead, at the same time repeatedly suggesting sleep.[61] H. Stadelmann,[62] a widely respected German hypnotherapist, and Loewenfeld himself, both made use of passes from time to time, the latter especially when hypnotizing women or children.[63] He believed that mesmeric passes have some "hypnosigenic" effect in their own right.

Another writer of this period[64] who thought that eye fixation and mesmeric passes may have such an effect was T. W. Mitchell,[65] who cites H. E. Wingfield's experiments on hypnotization without direct suggestion.[66] Wingfield's method was "simply to gaze at the subject from thirty to sixty seconds, generally stroking his temples and forehead at the same time and making strokes or passes over the top of his head. He was then asked to shut his eyes, and generally a few more passes or strokes were made over them." Some subjects reached a stage in which anaesthesia and illusions could be produced. "It would thus seem", says Mitchell, "that the apparently unimportant gazing and passes produce of themselves some mental change which is characterized by an increase of suggestibility."[67] If, as I suggested above, it is conceivable that mesmeric passes may sometimes have effects not wholly reducible to the workings of suggestion, interpretation of the results obtained by these and other late-nineteenth-century practitioners of hypnotism will be greatly complicated.

2. We now come to methods of hypnotizing involving direct, mainly verbal, suggestion, usually of sleep or of phenomena which promote sleep. This was much the commonest method at the time, and the procedures generally adopted, though susceptible of many variations, owed much to Liébeault and to Bernheim.

Loewenfeld considers first certain preliminaries to induction. These involve mainly reassurance, clarification, the allaying of groundless fears and the dispelling of unfounded expectations. Sometimes a preliminary demonstration with another subject will help. A quiet room, moderate lighting, comfortable clothes, a sofa, etc., should all be arranged to suit the patient. Should there be a witness? Some doctors advise it, but patients may not like discussing their problems before a third person. A compromise is to have the patient's chaperone in a neighbouring room with the communicating door open.

Loewenfeld now moves into a discussion of the techniques of verbal hypnotism. He remarks that the Liébeault-Bernheim procedures have been widely adapted and modified. Verbal suggestions (of heaviness of the limbs, tiredness of the eyes, increasing drowsiness, etc.) are commonly supported in two ways: by eye fixation, which concentrates the attention on an indifferent object, so keeping other ideas at bay; and by intoning the suggestions in a quiet and monotonous voice, which strengthens their sleep-generating effect. Loewenfeld has reservations about aspects of these procedures. Tiredness of the eyes is not a suggestion easy to put across at the very beginning of fixation, and if tiredness of eyes and limbs ensues rapidly, a powerful hypnotic influence frequently fails to manifest. Many people devote so much attention to the fixation that they do not feel tired.

Loewenfeld describes his own procedures in some detail. He begins with his subject quietly sitting, or lying, eyes closed, on an armchair or sofa. The subject is told to count from one to a hundred over and over again in a quiet voice. This is of value because the subject is put in a position to self-induce a certain sleepiness. In most cases, Loewenfeld makes some limited use of eye fixation (though with women and children he sometimes employs passes instead). The actual sleep-suggestions are not at first directed at the eyes, but towards a general tranquillity and tiredness. Then come suggestions specially designed to promote tiredness of the eyes, and to these are shortly joined suggestions relating to the psychological changes which immediately precede sleep. "Everything in your head is getting confused, it's all going, all disappearing. You are sinking more and more into sleep, such quiet and gentle sleep [here the hypnotist's voice sinks to a whisper]. You long more and more for sleep, you are becoming more and more stupefied. You have stopped thinking. You don't worry about anything any more, you don't know about anything any more, you don't see anything, you hear only faintly, you feel nothing more, consciousness has vanished completely, you are sinking further and further into sleep, your sleep is getting deeper, sounder and sounder, deeper and deeper still. You are sleeping so calmly and softly and still more calmly and softly."[68]

Loewenfeld does not like the various little "tests" (induction of limb catalepsy, automatic movements, etc.) which have customarily been used to establish how far the subject is responding – these interfere with the process of hypnotization, and their success or otherwise does not really tell us whether or not the subject is in an hypnotic state. On the other hand, we need to know how the hypnotization is progressing. It often lags behind the subject's expectations. The answer is perhaps a technique invented by Vogt and christened by Brodmann "the fractionation method".[69] The patient is hypnotized briefly, awoken briefly, rehypnotized, and so on, for a whole series of brief hypnotic episodes. After each awakening, he is questioned about the effectiveness of the suggestions, his mental state under hypnosis, etc., and the suggestions given during the next hypnosis are adjusted accordingly. By this means, the hypnotic state may be progressively deepened. The method may be combined with fixation, mesmeric passes, etc. Loewenfeld has used it in quite a few cases and thinks that on balance it is more effective than any other method. Bernheim's method is of course simpler, and he obtained remarkable results with it. But this was in a public clinic where the patients obtained an exaggerated idea of the doctor's power. Bernheim himself remarks on the difference between hospital practice and private practice,[70] and it is in the latter that the fractionation method comes into its own.

For subduing really refractory subjects, various tricks have been resorted to; for instance, galvanization of the head, the application of magnets, etc., the subject being assured that these means are infallible. Some workers have used various sedatives and narcotics to induce a state of sleepiness which may be turned into hypnosis. Rifat was the first to try chloroform on difficult cases. Sanchez Herrero, Wetterstrand and Kraft-Ebing also reported favourable results. A. Voisin has found it possible to procure hypnosis in maniacs by first subduing them with a whiff of chloroform.[71] Bernheim and Wetterstrand also make the converse claim – that prior hypnotic suggestion can enable one to use less chloroform when procuring anaesthesia for surgical purposes.[72] Loewenfeld states that he has himself tried chloroform, ether, bromides, morphine, chloral and paraldehyde as aids to hypnosis, but has found them of no avail. He thinks that where success has been achieved it has been due to the suggestive factor, and anyway the fractionation method renders such dodges superfluous. Opinions differed on how far it is possible to hypnotize subjects against their will.[73]

On the other hand, good subjects who have been frequently hypnotized can be sent to sleep by a photograph, an amulet or charm, a letter, a telegram, a promise to hypnotize from a distance, or by just about any device the hypnotist chooses to adopt.[74] It is only necessary to make the subject believe that it will work. Some subjects can acquire the ability to put themselves in the hypnotic state (autohypnotization) by verbally picturing to themselves the hypnotist's proceedings or adopting the position that they generally assume when being hypnotized. It was somewhat rare at this period to hear of individuals who claimed that they had put themselves into a state of autohypnosis without having been previously hypnotized by someone else. Braid, Forel, Coste de Lagrave and Liébeault all claim that they have succeeded in inducing self-hypnosis, or found themselves in such a state and experimented with it.[75] Forel has a case of a lady who, after watching him hypnotize, hypnotized herself by inner repetition of similar commands, and succeeded in extinguishing a toothache.[76]

Dehypnotization generally presents no difficulty. Simple command usually suffices, if necessary supplemented by blowing on the face, shaking the arms, or calling the patient's name. It is a good idea, Loewenfeld thinks, to warn the subject in advance of the impending wakening, and to tell him how good he will feel afterwards.

HYPNOTIC SUSCEPTIBILITY

Loewenfeld's chapter on hypnotizability summarizes conventional contemporary wisdom on the subject; it is based on accumulated clinical judgment, and is of rather limited interest. It is, he says, generally agreed that every normal person possesses some degree of hypnotic susceptibility, though it may not be elicited on each occasion or by every hypnotist. What sort of people make good hypnotic subjects? In general, older persons are difficult to hypnotize, and the young more easy. But views on the hypnotizability of children vary – Loewenfeld cites[77] the somewhat diverging views of Wetterstrand, Liébeault, Moll, Beaunis and Max Hirsch. Sex makes little difference to susceptibility, though women are more easily trained as somnambules than are men. In the past it was often supposed that hysterics are particularly susceptible, but in fact they have very varied degrees of susceptibility.

Of relevance to an individual's hypnotic susceptibility is the manner in which he ordinarily falls into natural sleep. Individuals who fall asleep quickly are more easily hypnotized than those who do not. Again, people without much culture, not given to reflecting, and prone to a certain passive obedience, are better subjects than the more highly educated, who find it hard to restrain critical thoughts, or to put themselves into a state of passivity. But this applies only on the occasion of first hypnotization, because the educated person soon learns how to control his thoughts in the required way, and to concentrate his mind on a certain course of ideas. Conversely, people who cannot control their attention and their thoughts, who suffer from obsessive ideas, who are pathologically anxious, who cannot rid themselves of their worries, make bad subjects. Many hypnotists have found neurasthenics, who are rarely tranquil enough to fix their attention, particularly difficult to hypnotize,[78] though Loewenfeld has not found them so himself. Everyone agrees that the great majority of insane persons cannot be brought into a state of hypnosis. A. Voisin (whose work with highly disturbed patients we shall come to[79]) found only 10% of the insane hypnotizable. Forel says that certain insane patients may be easily hypnotized and cured of minor problems (such as constipation), though the mental disturbances continue refractory. But psychotics of other kinds are impossible to hypnotize.[80]

Certain background factors are conducive to success, natural tiredness for example – hypnosis is easier to achieve in the afternoon and evening, especially with refractory subjects. The trance will grow deeper with repetitions as the individual learns to free himself from cares and worries, and of course post-hypnotic suggestions may be used to facilitate subsequent entries into the hypnotic state. The personality of the hypnotist is highly relevant. Hypnotist A may succeed and hypnotist B may fail with one and the same subject, and the position may be reversed with another subject. The subject's trust in the hypnotist and belief in his hypnotic power are all-important, and trust and belief are influenced by his appearance, voice, etc., and by his skill. Part of his skill consists in appropriately varying his approach to different subjects. Surround-ings have of course also a considerable influence upon the success or otherwise of the hypnotization – quiet is essential, for even small noises can be an impediment. Clinics in which many patients are hypnotized can build up a kind of "suggestive atmosphere" that is highly conducive to success. Nothing helps a patient overcome his reservations more effectively than seeing others willingly enter hypnosis.

Loewenfeld concludes his chapter on hypnotizability with three statistical tables. The most interesting summarizes figures on hypnotic susceptibility assembled from the reports of fourteen hypnotic practitioners by Schrenck-Notzing (1893).[81] The figures for 8705 subjects are as follows: achieved somnolence 2557 (29%); achieved hypotaxis[82] 4316 (50%); achieved som-nambulism 1313 (15%); refractory 519 (6%). These figures are not too different from ones to be found in textbooks of hypnotism today.

Loewenfeld also modifies from Bernheim a table prepared by Beaunis from Liébeault's figures showing hypnotizability in relation to age.[83] From this it appears that, for instance, somnambulism is attained by 55.3% of subjects aged 7–14, 22.6% of those aged 28–35, and 11.8% of those over 63. Figures like these (which may be compared with, for instance the statistics given by Bramwell in 1903[84]) have of course to be received with caution. There were no settled criteria of whether a subject was hypnotized or not, or of what stage he had attained, and no settled and dispassionate procedure for applying such criteria as might have been candidates. Of some interest are statistics relating to 351 of his own patients given by W. Hilger of Magdeburg (1902).[85] Hilger is well aware how subjective hypnotists' judgments of hypnotic depth can be, and is correspondingly cautious. He found male subjects if anything somewhat more prone to reach the somnambulic stage than female, but thinks it not impossible that a lady hypnotist might obtain the reverse result. He gives figures for the hypnotizability of patients suffering from various kinds of ailments, though it could not be said that anything very startling emerges.

More interesting are his statistics concerning the hypnotizability of his patients in relation to their socio-economic class.[86] Of his working class subjects, 42.06% became somnambulic; of his middle class ones, 29.92%; and of his wealthier ones, 15.28%. Hilger attributes the apparent greater susceptibility of the working classes to their great respect for and faith in doctors.

NOTES

1 The well-known books by Bramwell (1903) and Lloyd Tuckey (1900) are too medically oriented. Crocq (1900) is useful, but oriented round stale controversies. Forel (1902) is too short, as is Wingfield's excellent book (1910). F. W. H. Myers (1903) is valuable for its numerous and accurate references, but is too slanted to the author's own theoretical framework. Sanchez Herrero (1905) is the largest work of its time, but is based almost entirely on its author's own experiences and the late-nineteenth-century French literature. Among other textbooks of the time are Hilger (n.d.); Hirschlaff (1905); Mason (1901); Bonnet (1905); Ash (1906); Velsen (1912); Belfiore (1909); Hollander (1910); Trömner (1908); and Munro (1911).

2 Its principal shortcoming is that while the text contains many citations, and there is an extensive guide to the literature at the end, the citations are largely unconnected with the guide. This shortcoming may be partly rectified by reference to Hirschlaff (1899–1900), four very useful articles.

3 See above, pp. 403–404.

4 See *The General Practitioner*, 24th November 1906; *Journal of the Society for Psychical Research*, 1907–8, Vol. 13, pp. 14–15.

5 See his Presidential Address of 30th March 1907, Tuckey (1907).

6 McDougall (1911; 1926; 1967). During the First World War, McDougall became interested in the application of hypnosis to war neuroses (McDougall, 1920–1).

7 Even Morselli's enthusiasm for hypnotherapy cooled somewhat, though it did not disappear. Cf. Morselli (1893–4).

8 See Tagliarini (1985).

9 See, for example, the extraordinary work by Jeanniard du Dot (1913).

10 Bechterew (1895); Zielinsky (1968), p. 97.

11 Bechterew (1904; 1905–6; 1906–7).

12 Bechterew (1899; 1905).

13 On the early Vogt see Farez (1897–8).

14 See below, pp. 567–572. He became head of the Neurobiological Laboratory at the University of Berlin in 1902, Director of the Kaiser Wilhelms Institute for Brain Research in 1919, and Director of the Berlin Neurobiological Institute in 1921.

15 E.g. Corval (1892–3); Brodmann (1897–8); Delius (1896–7, 1898).

16 Döllken (1895–6).

17 Lipps (1897a; 1897b; 1898).

18 Ach (1900; 1905).

19 Martin (1907).

20 I add "from Europe" because Sidis could be a contender. Binet and Féré, Beaunis and Bernheim might also be considered, depending on one's concept of experimentation.

21 E.g. Claparède and Baade (1908–9); Claparède (1911); Chojecki (1911–12; 1912)

22 Brodmann (1897), p. 3.

23 Loewenfeld (1901), p. 343.

24 Schrenck-Notzing (1895a), p. 28.

25 Crocq (1896), pp. VIII-XI; Forel (1895-6).

26 Ellenberger (1970), pp. 165–167; Decker (1986).

27 Loewenfeld (1901), pp. 36–43. This may be compared with other reviews by Hirschlaff (1899–1900), pp. 206–213; Bechterew (1904); Bechterew (1905), pp. 240–246; Bunnemann (1913). And cf. the remarks of Trömner apud Janet (1911), p. 337.

28 E.g., with slight variations, Bernheim (1892b), p. 1213; Bernheim (1980), p. 19.

29 Bérillon (1898), p. 9.

30 Cited in Hirschlaff (1905), p. 241.

31 Bergmann (1894–5), p. 174.

32 Hirschlaff (1905), p. 242.

33 Vogt (1896–7b), pp. 334–335.

34 Stoll (1904), p. 3.

35 Forel (1906), p. 63; Forel (1902), p. 43.

36 Lipps (1897a); Schrenck-Notzing apud Lipps (1897a), pp. 120–124; Loewenfeld (1901), pp. 36–39.

37 See above, p. 375.

38 Loewenfeld (1901), p. 38.

39 Loewenfeld (1901), pp. 50–51.

40 Loewenfeld (1901), pp. 66–67, drawing on Bérillon (1898) and Binet (1900).

41 Loewenfeld (1901), p. 62.

42 Forel (1897). p. 372.

43 Loewenfeld (1901), p. 63.

44 Breuer and Freud (1974), pp. 309–315.

45 Breuer and Freud (1974), pp. 56–57.

46 E.g. Bergmann (1894–5); Döllken (1895–6).

47 Hirsch (1895); Döllken (1895–6).

48 Forel (1906), pp. 98–99, paragraphing inserted.

49 Bramwell (1930), p. 153.

50 Schrenck-Notzing (1891), pp. 7–8.

51 Loewenfeld (1901), p. 129.

52 Hirschlaff (1905), pp. 212–213; Hirschlaff (1899–1900), pp. 215–217.

53 He calls this "epistasis", from the Greek for attention, ἡ ἐπίστασις.

54 Cf. Hirschlaff (1899–1900), p. 333.

55 Loewenfeld (1901), p. 96.

56 E.g. horshoe magnets (Baierlacher, 1889); candle flames (Preyer, 1890, p. 59).

57 The most celebrated were Luys' rotating mirrors. Cf. Luys (1890b); Kingsbury (1891), p. 24. Bramwell (1930), pp. 48–50, devised an improved version.

58 Sanchez Herrero (1905), pp. 85–89.

59 Renterghem (1898b), p. 88.

60 Cf. Bramwell's personal observations, Bramwell (1930), pp. 42–43.

61 Woods (1897), p. 249.

62 Stadelmann (1896), pp. 34–36.

63 Loewenfeld (1901), pp. 101–102.

64 Loewenfeld also cites Moll (Moll, 1890, p. 29, but see p. 68), Tarkhanov, Heidenhain and Berger as supporters of the "hypnosigenic" effect of passes.

65 T. W. Mitchell (1908), pp. 12–13.

66 Wingfield (1908).

67 Mitchell (1908), p. 13.

68 Loewenfeld (1901), pp. 113–114.

69 On Vogt's hypnotic techniques see Brodmann (1897; 1898); Straaten (1900); Marcinow-
    sky (1900).

70 Bernheim (1889c) says that whereas four fifths of hospital inpatients will sleep profoundly,
    only one fifth to one sixth of patients seen in private practice will do so.

71 Rifat (1887–8); Sanchez Herrero (1889b); Wetterstrand (1897), pp. 4, 10; Voisin
    (1891–2a); Loewenfeld (1901), p. 120; cf. F. W. H. Myers (1903), Vol. 1, pp. 440–441.

72 Loewenfeld (1901), p. 120; Wetterstrand (1897), p. 109.

73 Moll (1890), pp. 44–46.

74 On hypnotization by telephone see Liégeois (1889), pp. 729–736.

75 Cf. Bramwell (1930), p. 52; Coste de Lagrave (1889); Tuckey (1900), p. 21 (on Liébeault);
    Forel (1888–9).

76 Forel (1906), p. 119.

77 Loewenfeld (1901), p. 92.

78 Loewenfeld cites Bernheim, Krafft-Ebing, Binswanger and Schrenck-Notzing in support of
    this assertion. Presumably he is referring to Krafft-Ebing (1891) and Binswanger (1892).

79 Cf. pp. 480–481 below.

80 Forel (1906), p. 204.

81 Schrenck-Notzing (1893), p. 30.

82 According to Schrenck-Notzing the principal defining characteristic of "hypotaxis" is the
    inability to resist certain suggestions despite energetic efforts of will. There is diminished
    awareness of the outer world, and there may or may not be partial amnesia on waking.

83 Bernheim (1889b), p. 20.

84 Bramwell (1930), pp. 57–62.

85 Hilger (1902).

86 Hilger (1902), p. 198.

CHAPTER 21

The turn of the century. II – Hypnotic phenomena

I call this chapter "Hypnotic phenomena" rather than "The phenomena of hypnosis" in order to make clear that it is about the phenomena which hypnotic induction procedures have allegedly brought about in "good" hypnotic subjects. I do not wish to prejudge the issue of whether or not these phenomena derive from a special state of "hypnosis" into which such subjects are liable to fall.

A general comment that is often retrospectively directed at nineteenth-century claims concerning the phenomena of hypnotism is that these claims are likely to be unsound because the experiments from which they derive lack proper controls, i.e. no attempt was made to compare the performances of hypnotized persons with their waking performances, or with the performances of matched unhypnotized persons in similar circumstances. It is therefore impossible to say with complete confidence that the phenomena exhibited by hypnotized subjects depended upon hypnosis, and that they could not have been achieved in other ways. Indeed, to call this nineteenth century work "experimental" is to dignify it unduly. Most of these "experiments" were no more than semi-systematic attempts to discover, through repeated trials, what remarkable effects could be wrought by means of hypnotic suggestion upon individual gifted subjects. And experimental hypnotism bulked small in comparison to therapeutic hypnotism.[1]

There is some truth in these rather glib assertions, but the actuality was more complex and much harder to assess. Experimenters as practised as Binet, Beaunis, Vogt, Claparède or G. Stanley Hall were well aware that there was no point in alleging that hypnotism may systematically change reaction times, or muscle tone, or that it may enhance or diminish powers of memory or resistance to muscular fatigue, unless one can compare one's hypnotic data with data obtained by the same or matched subjects under appropriate non-hypnotic conditions. The controls may not always have been satisfactory, but they were there. There were, however, some alleged phenomena for the acceptance of which such controls seemed obviously superfluous. If I tell a hypnotized person there is a dog on the floor when there is not, he may see one there; if I say the same thing to an ordinary, non-hypnotized person he will look round bemusedly, or simply laugh. The difference is patent and does not need to be established by detailed investigations. Of course, it was widely admitted that there are some people susceptible enough to fall prey to such suggestions even in the waking state, and even without prior experience of hypnosis, but they are quite rarely met with, probably mentally abnormal and do not affect the argument. Thus no-one at this time set out by suitable stratagems systematically to compare the susceptibility to hallucinations of waking persons with the susceptibility of deeply hypnotized ones. The same was true, for similar reasons, of other startling alleged phenomena, for example hypnotic blistering. And this renders the assumption that such phenomena are dependent on the induction of hypnosis a fragile one; for it would take only a few well-established counter-instances to undermine it.

No purpose would be served by discussing every kind of phenomenon in equal detail. I shall give more extended treatment to certain key areas, whilst merely indicating the main outlines of others. The phenomena concerned may, following Loewenfeld, be taken under two broad headings:

1.   Phenomena, other than those of suggestion, which are supposed to be enduring, and as it were natural, characteristics of the hypnotic state, or else regular by-products and hence signs of that state.
2.   Phenomena which persons supposedly in the state of hypnosis do not continuously exhibit, but may be induced to exhibit through commands and suggestions, usually those of the hypnotizer.

PERMANENT CHARACTERISTICS OF THE HYPNOTIC STATE

Phenomena supposed to be enduring characteristics of the hypnotic state may be further subdivided into 1. psychological phenomena, and 2. physiological phenomena. Neither subcategory need detain us for very long.

Psychological phenomena

Most of the proposals concerning the enduring psychological characteristics of the state of hypnosis were fairly unsatisfactory. Somewhat surprisingly, in that age of introspectionist psychology, there were not many accounts by serious investigators of what, subjectively, it was like to be hypnotized[2] and those that there are are in general not of much interest. They tend to dwell upon heaviness of the limbs and eyelids, increasing drowsiness and inertia, fragmentariness of subsequent recollection, and so on.[3]

Some authorities, for instance Loewenfeld himself, the school of Vogt, and Forel,[4] maintained that there are features of a hypnotized subject's mental state which make it possible to understand his enhanced suggestibility. Thus Loewenfeld lays emphasis upon a supposed "restriction of associative activity", which he thinks demonstrable in all of a hypnotized subject's intellectual performances. Suggestions take effect in hypnosis because they no longer call up the conflicting ideas which would otherwise have restrained them. The narrowing of intellectual horizons can also enhance remaining intellectual functions.

It is hard not to suspect that these claims represent a contamination of the introspective reports, or of the interpretation of such reports, by theoretical presuppositions derived from contemporary associationism. The data *ought* to be like this. But are they? Bleuler says of his experience of hypnosis, "... my conscious thought was not influenced otherwise than in the waking state; none the less these suggestions were for the most part realized."[5] Indeed, under hypnosis associative activity may seemingly be enhanced, new associative pathways (to talk for a moment in this fashion) be opened out, perhaps as a result of suggestion, for example to imagine oneself Julius Caesar or Abraham Lincoln, perhaps spontaneously, for some hypnotic subjects will dream at command, and fill in the details for themselves, or even dream spontaneously, a topic which especially interested Vogt and his pupils.[6] Of course, in a sense associative activity is restricted in hypnosis because many forms of induction procedure encourage the subject to turn away from the outer world and "look inwards"; but this does not necessarily involve a restriction of associative activity *per se*, but only a redirection of the general current of mental activity.

Rather similar considerations apply to another supposed general characteristic of the hypnotic state, the absence of "will-energy" emphasized by Wundt and by Loewenfeld.[7] Hypnotized subjects lack mental spontaneity, and their thoughts and actions are guided by the hypnotist. They are (Loewenfeld thinks) clearly aware of this weakening of will. But it could hardly be said that absence of will or a weakening of "will-energy" is an essential characteristic of the mental state of hypnotized persons. Some somnambulic subjects, as Loewenfeld himself points out,[8] may exhibit a lively spontaneity, and make decisions as to the details of their actions, or even (rather more rarely) initiate actions and trains of thought for themselves.[9] Others may pursue (or resist) a suggested course of action with greater than usual energy and determination. The postulated "weakening of will" seems likely to reduce to the "heightened suggestibility" which it is supposed to help explain.

This brings us to a third mental characteristic alleged by many to be at any rate quite commonly found among hypnotized subjects, namely the so-called hypnotic "rapport". Some even gave rapport a key rôle in their theoretical speculations. Hypnotic rapport was of course a lineal descendant of the old mesmeric rapport, and it was quite widely observed – Bramwell says that in the somnambulic state it "generally appears".[10] Kingsbury describes it as follows:

> ... the patient is, as it were, only "in touch" with the operator. He hears the
> gentlest whisper of the hypnotist, whereas all other voices and sounds are
> unnoticed. If his arm is raised in a cataleptic position, a stranger may use all
> his force to lower it to no purpose, but it falls at the lightest touch from the
> operator. If the hypnotist tells him he will now hear and feel a third person,
> he is at once "*en rapport*" with such person ... [11]

Psychological explanations of rapport were developed by Braid, Liébeault and
Bernheim,[12] and taken up by others, for instance Loewenfeld.[13] These explanations all hinge upon the supposition that during the process of hypnotization
cortical elements linked to the idea of the hypnotizer will "remain awake" and
will constitute a channel through which the subject can be readily influenced.
An extensive and ingenious series of experiments on rapport was undertaken by
Moll and Dessoir and published by the former in 1892 as *Der Rapport in der
Hypnose*.[14] Moll endeavours to show that even in "isolated rapport" the subject
is by no means completely cut off from stimuli which do not originate from the
hypnotist. For instance, in experiment 100,[15] Y. was hypnotized and answered
only Dessoir. Moll then several times repeated the command, "In two minutes
you will rise from your chair. An irresistible force will impel you to go to the door
and knock on it. Then you will return to your place and go to sleep again." This
happened as commanded, though the subject could give no satisfactory account
of his action. Rapport emerges as an hypnotically generated illusion, not a
fundamental feature of the hypnotic relationship.

The alleged phenomena of enhanced memory-capacity during the hypnotic
state, and of spontaneous post-hypnotic amnesia, both of which were thought
by some writers to be, but by others not to be, regularly produced by deep
hypnosis, will be dealt with later in the chapter.

Physiological phenomena

Concerning physiological phenomena supposed to be general characteristics of
the hypnotic state, there is little to be said, not from lack of materials, but from
lack of agreement between the various investigators.[16] Some writers, for
instance Heidenhain, and Tamburini and Seppilli, claimed that there is an
increase of breath and pulse rates under hypnosis. Preyer, Bernheim and Moll
found that eye fixation alone will have these effects. They do not occur in persons
hypnotized in quiet and unemotional conditions. Some authors (Heidenhain,
Preyer, Demarquay) think that there is increased secretion of sweat during
hypnosis. Loewenfeld thinks this is due to emotion or the strain of eye fixation,
and has never seen it himself. Bekhterev found that simple reaction times
increased in hypnotized hysterics, whilst simple calculations and word-
association trials were performed more quickly. Beaunis found that in hypnosis
auditory and tactile reaction times were usually shortened in the absence of

suggestion, and gustatory ones increased. G. S. Hall found a general shortening of reaction times, William James a somewhat fluctuating increase. It was maintained by some – notably Döllken – that there are certain sensory changes characteristic of the hypnotic state, for instance a diminution of skin sensitivity sometimes amounting to anaesthesia, and all grades of reduction of visual sensitivity up to blindness. Björnström, however, asserts that hyperaesthesia, especially of the skin senses, is characteristic of the somnambulic stage, and cites Berger in his support. Trömner reported enhanced visual sensitivity in hypnotized subjects, and Vogt thinks that he has observed certain sequences of change in muscle tonus characteristic of both sleep and hypnosis. Loewenfeld's strong reservations concerning most of these claims and counter-claims seem amply justified.

THE SUGGESTED PHENOMENA OF HYPNOSIS

These are the phenomena which persons supposedly in the hypnotic state may be brought to exhibit through commands and suggestions, deliberate or otherwise. No distinction will be made between phenomena produced within hypnosis, and phenomena produced "post-hypnotically". However, certain issues connected with post-hypnotic phenomena will require a section to themselves. I shall reserve until the next chapter all discussion of the supposed therapeutic uses and practical applications and implications of hypnotism.

The first thing to be noted about hypnotic phenomena is that they do not of themselves constitute a special class; almost any behavioural or psychological phenomenon may under some circumstances be produced in a subject rendered hypnotically hypersuggestible. It is the hypersuggestibility of the hypnotized subject which is the prime phenomenon of hypnosis, and the principal issue which any theory of hypnosis has to confront.

There is, however, a good deal more to be said about the phenomena of hypnosis than just this. For sometimes a hypnotized subject will allegedly do or experience or achieve things which he does not, presumably cannot, do or experience outside the hypnotic state, or at best does or experiences rather rarely and probably in highly abnormal circumstances. We might refer to such phenomena as "extraordinary" hypnotic phenomena, and it was upon them, rather than upon the more commonplace effects, that much work and discussion centred at this time, and has continued to centre ever since. For manifestly, the more extraordinary the phenomena producible by hypnosis, the harder they are going to be to explain, and the more significant they may possibly be for our understanding of the human psyche. I shall select nine categories of such phenomena for brief discussion, sticking generally, but not exclusively, to materials which were current in the very early years of the twentieth century.

Hallucinations and illusions of sense

The literature on these topics was very considerable, but surprisingly little of it has to do with what is obviously the central question, namely that of the degree of subjective reality which such hallucinations and illusions possess for the experient. Loewenfeld, for instance, devotes just one paragraph to this issue.[17] We cannot, he admits, deny that subjects may simulate in order to please the hypnotist, but adds that this will not suffice as a general explanation of their reports. Simple and uneducated people, who lack the capacity for playing tricks, speak and behave quite naturally when undergoing hypnotic hallucinations, and do so without hesitation or reflexion. Furthermore, suggested sensory deceptions may be accompanied by corresponding bodily effects – the suggestion of taking a cold bath may produce goose-pimples, that of drinking wine a flushing of the face. Loewenfeld's doubts concerning the "reality" of hypnotically induced hallucinations are, like those of most other writers, hardly more than token. He does not hesitate to give us a long list[18] of the simple and complex sensory illusions and hallucinations that may be produced in suitable subjects by hypnotic suggestion. They range from feelings of heat and cold and illusions of taste or smell to complex visual and auditory hallucinations, for example seeing and conversing with hallucinatory persons or hearing hallucinatory music. Bernheim, whose power to produce hallucinations was apparently most remarkable, entertained few doubts as to their subjective "reality"; neither did Liégeois, who in 1889 devoted a substantial chapter to the subject.[19] Yet most of Bernheim's subjects at Nancy were uneducated persons accustomed to fulfil the *professeur's* instructions with unquestioning obedience, and we have already noted how highly trained in the "right" responses were the Salpêtrière's hysterical and hypnotic star subjects. It is somewhat surprising that doubts about hypnotic hallucinations were not expressed on a wider scale, and that subjects were not more often intensively questioned as to the nature of their hallucinatory experiences.

Some respected writers clearly did have certain doubts. After his visit to Nancy, Delboeuf questioned the vividness of most hypnotic hallucinations, which he thought were often pale and lacking in three-dimensionality,[20] and Beaunis reported some experiments on the issue.[21] He had subjects trace the outlines of various suggested hallucinations. The results made him doubt that these hallucinations have the reality and definition of objective images. He suspects that imagination plays a considerable rôle in their production. However, Lehmann and Parish protest that lack of clarity in a sketch does not necessarily imply lack of clarity in the image of which it is a sketch.[22]

In theory at least, Binet and Féré had hit upon an *experimentum crucis* when they demonstrated that subjects, ignorant of optical laws, would experience correctly coloured after-images to coloured hypnotic hallucinations, and the correct resultant colour when hallucinatory colours were mixed. However, their claims received little independent support. Gurney remarked that "these special optical delusions seem as peculiar to the atmosphere of the Salpêtrière as *mirage* to that of the desert."[23] The nearest to an independent confirmation of Binet and Féré's claims was some experiments reported by Ottolenghi and Lombroso in 1889.[24] They trained subjects in the use of a spectroscope and found that in the majority of cases contamination of the image by a glass slide or a blank card to which an hallucinatory colour had been assigned produced similar effects on the perceived spectral image to actual coloured slides or cards. But since Lombroso was almost in the same league of credulity and incompetence as Luys, little credence can be attached to his findings.

The question of the "reality" of hypnotic hallucinations was left, therefore, in a somewhat unsatisfactory state, and I shall touch on it again. Meanwhile, some preliminary points may be made. The investigations of Galton, and of the Society for Psychical Research,[25] had shown clearly that it is by no means unknown, even though not particularly common, for a waking person, not otherwise abnormal, ill or drugged, to undergo an hallucination of one or more sensory modalities, which is subjectively indistinguishable from the perception of an actual object. Curiously enough, Delboeuf, one of the more critical nineteenth-century students of hypnotism, whose somewhat sceptical views on hypnotic hallucinations we have just noted, had himself undergone a spontaneous hallucination. He woke one morning to see his late mother seated at his bed head, looking at him "with eyes of an extraordinary brightness". Her mouth "seemed ready to speak". The figure, which he realized was hallucinatory, was at first opaque, but after a few minutes became transparent and faded away.[26]

Now if it is indeed the case that individuals who are not in other respects greatly abnormal may now and again for no immediately obvious reason experience an hallucinatory percept that seems to them no less vivid than a true percept, it would be rather rash, and an unjustifiable piece of a priorism, to dismiss all comparable reports from hypnotic subjects as necessarily due to exaggeration and compliance. When we find, for example, both Forel and Bernheim reporting on apparently articulate and intelligent subjects who, made post-hypnotically to see, touch and smell an hallucinatory flower, could not tell the hallucinatory flower from a real one,[27] we must at any rate preserve an open mind as to the verisimilitude of the hallucinations.

Negative hallucinations

The term "negative hallucination" – meaning failure to perceive an object or event which the subject would normally have perceived – seems to have been coined by Bernheim, who was particularly interested in this phenomenon.[28]

Loewenfeld notes[29] that, depending upon the suggestion given, hypnotically produced negative hallucinations can be of the whole of an object or part of it, and can involve one, two or three sense modalities (e.g. someone can be rendered invisible to a subject, but continue to be audible, or rendered invisible and inaudible, or invisible, inaudible, and insusceptible of being felt). When a person or object is rendered invisible to a hypnotized subject (or post-hypnotically invisible to a subject who has been hypnotized but is now awake) there is no gap in that subject's field of view. The location of the obliterated object (on a chair, in front of a wall) is filled in by fantasy; in other words (a point often made[30]) a negative hallucination has to be supplemented by a positive one (and of course vice versa).

In the early days of the influence of the Nancy School, some remarkable instances of suggested negative hallucinations were reported.[31] Thus in his account of the hysterical patient Ilma Szandor,[32] Krafft-Ebing describes how he protected a female night-attendant from this patient's lesbian advances by rendering the attendant post-hypnotically invisible to the patient, to the latter's great terror and bafflement.

One might well suppose that in a case such as this all sense-impressions emanating from the designated object have been totally excluded from the subject's main stream of consciousness. She would naturally be alarmed by the apparent outbreak of poltergeist phenomena in her vicinity.[33] That sense-impressions can thus be excluded is on the face of it supported by some famous experiments of Binet's,[34] which were subsequently repeated in variant forms by other investigators. Here is how Moll proceeded (Binet originally used blank cards not matches):

> I took a match and marked its end with a spot of ink. I then suggested that the match was invisible. I took twenty-nine other matches and put the whole thirty on the table in such a manner that X. could see the ink spot. To my question X. replied that there were only twenty-nine matches on the table. I then, while X.'s eyes were turned away, moved the marked match so that X. could not see the spot. He looked at the matches and said there were thirty of them. Thus the marked match was only invisible as long as X. could distinguish it from the others.[35]

This brings out very clearly the latent paradox of negative hallucination – for an object to be excluded from perceptual awareness it must first be perceived and recognized. Some authors, particularly once the euphoric days of the early Nancy School were over, came to believe that the facts are not such as to force this paradox upon us. They thought that in most cases of negative hallucination they could detect signs that the subject was to some degree aware of the excluded object, enough at any rate to avoid bumping into it and perhaps to act appropriately towards it, even though he might not acknowledge this even to himself.[36] The state of mind of a subject undergoing a negative hallucination can certainly be complex and full of conflicting tendencies, as is interestingly evident from the self-observations of one of Martin's hypnotic subjects at Würzburg, a philosophy student named Ernst Bloch. Bloch had been fairly responsive to suggestions of positive hallucinations and illusions. By way of negative hallucination-suggestion, he was first told that he would not see a certain clock. This clock in fact appeared transparent to him, so that he could see the table through it. Then he was instructed first not to see and then not to hear Külpe who was present. His protocol runs in part as follows:

> At first I took the fountain pen as the intended object, and turned my eyes instead in the direction which this indicated – but only so as not to see Professor Külpe and have to name his name. When I was assured he had gone away I at first avoided looking directly towards him and therefore said, with reference to the empty space at which I was looking, that I saw no-one. When I was forced to look at him I had to admit, with great embarrassment, that he was still there. When this should have been repeated, I turned myself, on the command to turn my head, to the other side with intentional misunderstanding ... [37]

The Würzburg School of introspectionist psychology was famous for the scrupulous self-observations of its highly trained subjects, and on reading this subject's protocols one might be forgiven for refusing to take at face value any straightforward account of an apparent negative hallucination. I can only say that, like many other hypnotists, I have myself watched various subjects undergoing negative hallucinations, and their reactions have differed considerably. Some have appeared most realistically frightened and bewildered on seeing an object carried through the air by an "invisible" person. Others have very noticeably kept their eyes averted from the invisible person, looking anywhere but at him or near him. They certainly in some sense knew where he was. What we should say about negative hallucinations remains far from clear even today.

Anaesthesia and analgesia

The topic of hypnotic anaesthesia and analgesia would seem to be closely related to that of negative hallucination. The applications of hypnotic analgesia to surgery will be dealt with in the next chapter. Loewenfeld's scattered remarks on the remaining aspects of these phenomena are not of particular interest.[38] Suggested anaesthesia can be of a whole sense, for example vision or hearing, or of one ear or eye, or of a visual hemifield, or of a limb, and it resembles negative hallucination in that (as with hysterical anaesthesia) the subject seems to remain sufficiently aware at some level of the excluded sensory data to act upon them for any important purpose. For instance, it was often noted that subjects rendered hypnotically deaf would none the less respond to verbal commands to cease being deaf. Suggested analgesia is not necessarily accompanied by suggested anaesthesia, though suggested analgesia of the kind used in public demonstrations and surgical operations seems usually to have involved anaesthesia also.

Worthy of note, according to Loewenfeld, are various phenomena which may accompany suggested somatosensory anaesthesia. Thus anaesthesia of a limb may be accompanied by paralysis of that limb even when suggestions of that tendency have been avoided.[39] Anaesthesia of half the body may lead to complete anaesthesia of the other senses on that side of the body.[40] Likewise, a suggestion of unilateral deafness may weaken the other senses on the same side.[41] Loewenfeld adds that hysterical paralyses, hemianaesthesias, etc., produce the same sort of additional effects as hypnotic ones.[42]

As a general argument against the possibility of shamming in cases of suggested anaesthesia, Bekhterev notes that where hemianaesthesia has been suggested to a good hypnotic subject, there will be no pupillary reaction to a needle-prick on the anaesthetic side, whereas there will be such a reaction to a prick on the other side.[43]

Motor effects – performances of the voluntary musculature

Many of the standard phenomena featured at public demonstrations of hypnotism fell under this heading, and Loewenfeld devotes some pages to the topic.[44] Such movements, he points out, can be induced in hypnotized subjects by verbal suggestion, through the arousal of imitative tendencies, by gestures, even through cues given by eye-movements. What is important is that the subject grasps the hypnotizer's intention. The relation between susceptibility to motor suggestions and depth of hypnosis is by no means consistent. Some subjects will respond readily in light hypnosis, and not in deep, and vice versa.

A favourite simple motor suggestion at this time – often used as a kind of test of hypnotic depth – was to set the subject's hands in motion, for example round and round each other, and leave the movement to continue, which it might do for a longer or shorter period, perhaps sustained or modulated by verbal suggestion. Liébeault and Bernheim call these movements "automatic", and Dessoir and various German writers call them "continued". Other kinds of continued movements included nodding the head, and bending and stretching the forearm or lower leg.

Of course in addition to simple limb movements, hypnotic suggestion can, Loewenfeld observes, produce coughing, laughing, yawning, and more complex kinds of limb movements such as dancing, singing, whistling, etc. But straightforward motor phenomena are not our immediate concern, which is with what I described as the "extraordinary" phenomena of hypnotism, i.e. those which, it is widely assumed, do not happen to, or cannot be produced by, persons who are not hypnotized. It is hard to say what among the "motor" phenomena producible by hypnotism should be classified as "extraordinary" in this sense. Loewenfeld mentions various categories of motor phenomena which might perhaps qualify.

There are, for instance, various kinds of motor inhibitions and paralyses producible by hypnotic suggestion, and it could be argued that to find oneself suddenly unable to carry out some action that would normally present no problem is a phenomenon that could be called "extraordinary" in the sense indicated above. All sorts of parts of the body can be robbed of voluntary motion – not just limbs, but vocal cords, throat, eyes, tongue. It is sometimes possible to produce "systematic paralyses" in which what is inhibited is not the movement of certain groups of muscles, but the performances of certain tasks or actions, such as writing or speech, or even the utterance of one particular word.

A phenomenon which still attracted some attention at this time was that of hypnotic "catalepsy". If the limb of a hypnotized person is put into a certain position it may retain that position when the experimenter's hand is removed. If the disposition to exhibit this phenomenon is very strong, the whole body can be induced to maintain a given attitude, even a very awkward one, though there are likely to be small alterations of posture. Opinions differed as to whether any extraordinary muscular strength or resistance to fatigue was involved. Bernheim held that usually the duration of the catalepsy does not exceed fifteen to twenty minutes,[45] and Charcot showed that a strong man can hold up his arm for pretty nearly the same length of time as a cataleptic subject.[46] But according to Loewenfeld, there have been instances of much longer duration – Berger has a case of a hypnotized girl who maintained a cataleptic posture for seven hours.[47]

The famous "human plank" demonstration was no doubt a showman's development of the phenomenon of rigid catalepsy. This feat was often illustrated in contemporary books on hypnotism, and was at its most noteworthy when executed by a young lady wearing a bustle. Benedikt, however, cites evidence that it can be performed by an athletic person without benefit of hypnosis.[48]

Hypnotism and memory phenomena

Among the "extraordinary" phenomena allegedly found in or obtainable with deeply hypnotized subjects certain alleged alterations in the capacity to lay down, or to retrieve, various kinds of memories were possibly the most frequently reported, and yet their occurrence and interpretation have remained topics of sharp controversy down to the present day. The animal magnetic movement produced quite a few accounts of the apparent enhancement of memory capacity (as of other intellectual capacities) in somnambulic subjects. It was, for instance, from time to time reported that such subjects, while in the somnambulic state, could memorize long passages of prose or verse with a

Figure II. The human plank feat. (From the Library of the Society for Psychical Research.)

rapidity and accuracy greatly exceeding their normal capabilities.[49] It was also sometimes claimed that they might show an extraordinarily minute remembrance of events from previous somnambulic states,[50] or might exhibit an enhanced capacity to speak languages imperfectly known, or else forgotten since childhood.[51]

The belief that somnambulic subjects might exhibit heightened memory capacities passed from the mesmeric movement into the hypnotic mainly, I suspect, through the observations and remarks of Richet and of Liébeault, and was accepted both by Nancy and the Salpêtrière. However, it was generally held that what was enhanced was what Richet had called "passive" memory, i.e. the capacity to retrieve memories (especially from the long-forgotten past) rather than to lay down new memories. This view was supported mainly by anecdotal evidence of a not always very convincing kind.[52]

Reviewing the situation at the turn of the century,[53] Loewenfeld raises two questions: What is the relation of memory to hypnosis without the influence of suggestion, and what differences appear when suggestion is added? Taking the former first: So far as the ability to lay down new memories is concerned, Loewenfeld could find little to support the claims of the earlier magnetizers that magnetic somnambules might show an enhanced ability to memorize poems, etc. Beaunis, for example, could find no difference between the ability of hypnotized and of unhypnotized subjects to memorize lists of figures and of letters of the alphabet.[54] Bernheim concurs on the basis of his own experiments, which are, however, not reported in detail.[55] Dessoir,[56] using deeply hypnotized subjects who were required to learn lists of syllables read out to them, found that, if no suggestions to sharpen memory were given, performance under hypnosis was actually worse than in the waking state. In his thesis on memory in the hypnotic sleep (1887), A. Dichas, a pupil of Pitres at Bordeaux, reports that subjects set to memorize passages of Lafontaine did so as effectively in the waking as in the hypnotic state.[57] Loewenfeld himself found that short series of numbers and spoken sentences in foreign languages were not better retained when presented in the hypnotic than in the waking state.[58] Somewhat similar experiments by Lombroso obtained positive results, but utilized only a single subject.[59]

If we pass a few years beyond the date of Loewenfeld's book, we come to the work in the Psychological Laboratory at Geneva of Claparède and Baade, and of Chojecki,[60] which, though utilizing only small numbers of subjects, was methodologically more sophisticated than anything which had preceded it. With learning tasks involving pairing numbers with nonsense syllables (Claparède and Baade), and lists of single-figure numbers (Chojecki), memory proved to be, if anything, better in the waking state than under hypnosis. At Würzburg, Martin (1907) found indications in hypnotized subjects (in contrast to the same subjects in an unhypnotized condition) of an increase in vividness of memory images, and an increase in ability to recall ordinarily inaccessible material, but she gives inadequate details of her procedures, and is herself in some uncertainty as to how the results should be interpreted; for example, do they reflect increased alertness, heightened attention, lessened tiredness, etc.

Somewhat more impressive results were sometimes reported when the materials to be learned were not mere *ad hoc* series of numbers or syllables, but were connected and meaningful passages of prose and verse. But these claims are again based upon inadequately reported anecdotes.[61]

With regard to the recovery of memories of distant events, Loewenfeld holds that, in the absence of positive suggestions of improvement, this is no better in deep hypnosis (or light) than it is in the waking state. His own observations agree with those of Crocq[62] – hypnotized subjects asked what they were up to on a particular day in the preceding week, or on a certain week many months or a year ago, gave in general no better answers than they could manage in the waking state. However, occasional observations suggest, he thinks, that long-forgotten memories can sometimes emerge spontaneously under hypnosis. Benedikt describes how an English officer, hypnotized by Hansen, began to speak in a language which he had used only in childhood[63] (Felkin has a somewhat similar case[64]). Memories of forgotten dreams can emerge in hypnosis – but, as Loewenfeld notes, such things can happen in waking life, triggered off by chance circumstances.

A few years later, Loewenfeld had changed his position somewhat, and had become convinced that some hypnotized subjects might manifest a striking improvement in their capacity to recover long-forgotten events, even though they had received no special suggestions to that effect. In a paper published in 1910,[65] he described his experiments with four subjects, who were "sent back" under hypnosis to relive events of designated years. Their recollections were stenographically recorded, and contained many details, subsequently verified, that were not accessible to them in the waking state. It is, however, not apparent that much was done to promote or encourage waking recall. One of the subjects, taken back to the age of four, recovered the memory of an event which she came to think underlay a minor pathological symptom which had dogged her during her adult life.[66]

To return to Loewenfeld's textbook: Having dealt with the spontaneous widening of memory under hypnosis (for which he finds little evidence), he goes on to consider the possibility of heightening memory under hypnosis by direct suggestion. He thinks that by this means memory may be so enhanced as to permit the recovery not just of vanished memories, but of usually inaccessible experiences which took place during abnormal states of consciousness, and even of waking impressions which did not reach consciousness at the time of their occurrence. Recovery of such memories, he asserts, can be important for medical practice. He gives, however, no examples.

The most notable modification of memory in hypnosis, Loewenfeld thinks, is not its enhancement, but the fact that deeply hypnotized persons may on awakening be unable to recall what they have just experienced. During subsequent hypnoses, however, the events of previous hypnoses would again become available to them. A simple example of this traditional phenomenon is provided by Pitres' subject Albertine. Pitres gave Albertine a louis d'or while she was awake. She put it in her pocket. She was then hypnotized and rendered lethargic, and another subject was induced to steal the coin. Pitres next woke Albertine and asked her if she had the coin. She was astonished at being unable to find it, and, though pressed with questions, she could throw no further light on the matter. Rehypnotized, she recalled at once that it had been stolen, and by whom, and why.[67]

Authorities differed somewhat as to the generality of this spontaneous post-hypnotic amnesia, and as to its density. Delboeuf remarks[68] that while Bernheim finds amnesia the rule, his own experience has been the reverse. Moll says that often a dim memory persists, like the memory of a dream, and that some cue may revive memory of the whole.[69] Loewenfeld himself observes that there are many exceptions, and the amnesia may be only partial or temporary, or may start to wear off after a while.[70] Post-hypnotic amnesia may be prevented by an appropriate suggestion given during hypnosis, and it may be lifted afterwards by a firm command. The subject himself may prevent it, as Noizet observed long ago, by formulating a firm intention to retain the events of the hypnotic state in his mind upon awakening. Bramwell, in a passage which anticipates a good deal of modern thinking on the subject, goes so far as to question whether post-hypnotic amnesia ever appears without either direct or indirect suggestion. Indirect suggestion is always present because to most people "the idea of hypnosis represents a kind of sleep with subsequent forgetfulness."[71]

When post-hypnotic amnesia fails to occur spontaneously, it can be induced by post-hypnotic suggestion, though Loewenfeld adds[72] that his own experience has been that in most cases this does not work. However, he does not seem to doubt that one can use hypnotic suggestion to produce selective artificial amnesias for ideas and groups of ideas, and for general and particular skills. We can stop subjects remembering a word, a name, a language. We can prevent recollection of greater or lesser parts of life, including recent ones. Such phenomena had long been a standby of hypnotic entertainments, and were increasingly thought of as having possible medical applications.[73]

Hypnotic regression into childhood

Possibly related to hypnotic manipulation of memory is the ostensible regression of a hypnotized subject to an earlier phase of life. Such demonstrations had figured in the shows of popular magnetizers and hypnotists since the mid-century. Good subjects could readily be induced to speak and act like a child, to write in large round childish handwriting, even to take persons present for individuals known to them in childhood days. Opinions differed, Loewenfeld notes,[74] as to whether or not this is simply a variant of hypnotic rôle-playing, and there was further dispute as to what is actually going on when when a hypnotized subject enacts a rôle. Some, like Vincent, traced the changed behaviour back to hallucinations and illusions.[75] Moll attributes it to an inhibition of memory leading to an enhanced creation of images.[76] Lehmann and Sidis think that the subject loses the sense of self and can therefore easily be led to assume any other personality of the hypnotist's choosing.[77]

Some writers, however, took a different kind of view of certain cases of ostensible regression into childhood. In particular, Krafft-Ebing conducted a number of experiments with a gifted somnambule, supposedly free of hysterical symptoms, who was particularly adept at hypnotic age regression.[78] Her dramatic rendition of a return to an earlier phase of her life was particularly striking, and her handwriting and spelling changed appropriately (specimens of her earlier writing were procured for comparison). Krafft-Ebing formed the opinion that there had been an actual reversion to an earlier phase of life still latent within the subject. This rather startling claim attracted a good deal of newspaper publicity with the result that Krafft-Ebing was pounced on by the egregious Benedikt, who described the case as a comedy and hinted that it was a swindle.[79] More sober and more rational were the criticisms of Jolly and of

Köhler, which were based on their own observations.[80] Both reach essentially the same conclusion. There is no question of deliberate simulation by the subject, and though the personation may be fleshed out by actual childhood memories, it is mainly to be ascribed to creative use of imagination and general knowledge. Köhler's subject, a man of twenty-two, put on some excellent performances, but it was apparent that they did not represent a literal return to a previous phase of development. Brought back to the age of two, he recited the Lord's Prayer successfully, a feat which he had not been able to perform at that age.[81] Transformed into a shy girl, and told that he needed to urinate, he squatted *ritu feminarum* on a chamber pot.[82] This certainly did not represent any reversion to a previous phase of his life, yet he seemed as much immersed in one sort of rôle as in the other.

Post-hypnotic suggestion

Loewenfeld devotes a whole chapter to post-hypnotic suggestion,[83] which has widely been regarded as one of the most puzzling and potentially significant of hypnotic phenomena. He alleges that suggestions made to hypnotized individuals generally show no tendency to persist into the waking state unless it is expressly stated that they should do so. "Post-hypnotic" suggestion involves the fulfilment subsequent to hypnosis of a suggestion made during hypnosis, the delayed fulfilment being itself part of the suggestion. Amnesia for the post-hypnotic suggestion is usually also suggested.

Post-hypnotic suggestion may involve a simple carrying over from hypnosis to wakefulness of a suggestion to be immediately fulfilled, or the realization of the suggestion may be delayed – the so-called *suggestion à echéance*. Most hypnotic phenomena can be obtained post-hypnotically, and of course a high percentage of therapeutic suggestions are post-hypnotic. The time of fulfilment of a *suggestion à echéance* can be assigned in various ways, for instance by a direct statement of a certain time, by specifying a certain lapse of days, hours and minutes, or by linking fulfilment to the giving of a certain signal.

In general, says Loewenfeld, post-hypnotic sensory deceptions last only a few minutes, but there are exceptions, as in a case of Bernheim's, where a lady was made to see the image of her husband for twenty-four hours, despite the fact that she knew it was a deception. Lloyd Tuckey made a lady see the tail of her spotted cat black for three days.[84] But they were far outstripped by Londe, who made a subject see a suggested portrait on a visiting card for no less than two years.[85]

A kind of rivalry grew up, particularly among members of the Nancy School, as to who could implant a successful post-hypnotic suggestion with the longest delay before fulfilment. Quite early on, Bernheim succeeded with one of sixty-three days. In August 1883, he told a patient, a former army sergeant referred to him by Liébeault, that on the first Wednesday of October he was to visit Liébeault, and that he would find there the President of the Republic, who would give him a medal and a pension. The sergeant duly arrived, and was observed by Liébeault to engage in a highly satisfactory conversation with a non-existent person who, upon inquiry, he understood to have been the President.[86] On 14th July 1884, Beaunis suggested to a female subject that she would see him at 10.00 a.m. on 1st January 1885, that he would wish her a happy new year, and would then vanish. This suggestion was punctually fulfilled, as the subject told Liébeault and other people (Beaunis was in Paris).[87] These contenders were, however, easily vanquished by Liégeois, who successfully managed a *suggestion à echéance* with a delay of a year, 12th October 1885 to 12th October 1886. The subject was instructed to visit Liébeault in the morning of the latter day and say that his eyes had been so much better during the year that he wished to thank Liébeault and Liégeois. He would request permission to embrace them (both these gentlemen were really present). After that he would see come into the room a dog with a monkey on its back, followed five minutes later by a gypsy and a tame bear. This gypsy would be pleased to find his dog and his monkey again, and would make his bear, a big American one, but very tame, dance for the company. Finally, he would beg Liégeois to provide ten centimes to give to the gypsy. The bear did not appear, and the subject did not ask to embrace Liébeault and Liégeois; otherwise the suggstion was punctually realized.[88] During the fulfilment of this suggestion the subject passed into a renewed hypnosis, for which he was afterwards amnesic. Another example of a post-hypnotic suggestion with a delay of a year is mentioned by Wingfield.[89]

An odd aspect of certain cases of *suggestion à echéance* is the heightened time sense which certain somnambules were alleged to exhibit in calculating (presumably unconsciously) the designated moment for response to the suggestion. This phenomenon (for which there were parallels from the animal magnetic movement[90]) was extensively investigated by Delboeuf, Bramwell, Mitchell, and others,[91] but I know of no satisfactory later replications.

We come now to three interlinked questions of some interest, each in turn raised by Loewenfeld.

1.   What is the state of mind of the subject in the interval between awakening from hypnosis and the fulfilment of the suggestion?
2.   What is the state of mind of the subject at the moment of fulfilment of the suggestion?
3.   How are we to conceive that which persists in the subject following the suggestion, and, though usually not accessible to recall, produces the right effect at the right time or at the right signal?

1. In most cases the mental state in this period is quite normal. Many subjects show a certain restlessness and absent-mindedness, which subsides when the

post-hypnotic suggestion has been carried out.[92] They may have a general sense that something is to be done. Köhler is of the opinion – based, he says, on wide experience – that subjects actually remain in a kind of half-hypnotized or hypnoidal state which passes into a true hypnosis with the passage of the suggested idea from the subconscious into the conscious.[93] But this proposal found little support, and its application to some of the longer-delayed post-hypnotic suggestions would present considerable implausibilities.

2. If the subject recollects the suggestion when the time for fulfilment arrives, he is likely not to carry it out; if he does carry it out, he is likely to say simply, "You wished it, so I did it." If, as is more usual, there is amnesia for the suggestion, the idea that arises in consequence of the latter will often seem spontaneous, especially if it does not contradict custom. If it seems to require explanation, the explanation offered will vary with the subject's imagination and intelligence. These rationalizations can be fairly implausible.[94] On the other hand, practised subjects may immediately grasp that they have been the victims of post-hypnotic suggestions.

Sometimes the person who carries out a post-hypnotic suggestion remembers nothing of it subsequently, and believes he has been awakened out of hypnosis. During the period of enactment of the suggestion, the subject stares, accepts new suggestions, etc. Moll says he can even recollect earlier hypnotic states.[95] F. W. H. Myers cites a case in which one of Liébeault's subjects, induced by post-hypnotic suggestion to see an hallucinatory Myers, could recollect this experience during a subsequent hypnosis, but not during the waking state.[96] Köhler thinks that the performance of post-hypnotic suggestions always depends on reentry into hypnosis or half-hypnosis.[97] Loewenfeld believes, however, that during the fulfilment of post-hypnotic suggestions there are all sorts of intermediate grades between a reinstated hypnosis and ordinary wakefulness.[98] Gurney too is of the opinion that different results are obtained with different subjects,[99] and Forel says that it is not possible to force all cases into the same mould.[100]

3. As for the nature of that which persists in the subject pending the fulfilment of a post-hypnotic suggestion, Loewenfeld thinks[101] that we must consider two points:

a. The persistence of the suggested idea despite its having apparently been forgotten.
b. The emergence of this idea at the right time.

Now a, the retention of a suggestion as a latent memory, presents, according to Loewenfeld, no special difficulties (a view with which it would be possible to disagree). It is b, the question of how the latent idea emerges at the right time, that presents difficulties, for it seems to imply an unconscious or subconscious estimation of time, or else an unconscious or subconscious vigilance for the designated cue. Bernheim tries to avoid these difficulties by proposing that in somnambules who have received a time-linked post-hypnotic suggestion, the suggestion returns to consciousness briefly in periodic spontaneous somnambulic states. Loewenfeld thinks there is no evidence for the regular recurrence to consciousness of the hidden time-idea. He holds that we possess a time sense, more or less developed in different individuals, and operating unconsciously or subconsciously, since our ordinary consciousness is generally occupied with other matters. Bramwell's results[102] show how accurate this subconscious time-sense can be. How it is utilized varies with what has been designated to trigger the response. If the response has been specified so as to follow after a certain number of days, hours and minutes, response does not seem to be based upon a calculation of the point of time thus indicated. Gurney and Bramwell both found that subjects who were rehypnotized in the interval between suggestion and fulfilment did not know the projected time of fulfilment, but only the amount of time that had elapsed and the amount that had still to come.[103] The suggested idea comes to mind at the appointed time because that idea has been associated by suggestion with the idea of that lapse of time. When the specified amount of time has elapsed the idea of it acts like a signal, pulling out the suggested idea.

I cannot say that I find these notions very illuminating.

Physiological effects produced by hypnotic suggestion

A long tradition, with roots in the magnetic movement, and transmitted to the Nancy School by Liébeault,[104] held that the activities of the "vegetative" (or autonomic) nervous might become specially amenable to conscious control when a patient was put into the mesmeric or hypnotic state. Much of the data here has a therapeutic context, and concerns the apparently beneficial effects of hypnotic suggestion upon menstruation, excretion, digestion, lactation, perspiration, etc. Purely experimental investigations of hypnotic influence upon autonomic and related functioning were less frequent perhaps because less commonly successful. There were, for instance, various experimental investigations of the effects of hypnotic suggestion upon heart rate. Some experimenters achieved an effect indirectly, by exciting or calming the subject.[105] More interesting were the experiments in which heart rate was apparently influenced by direct suggestions of acceleration and deceleration. In 1884, Beaunis carried out some experiments with Lisa F., an hysterical somnambule whom we shall meet again, and various other subjects, whose pulse rates were recorded by a sphygmograph on a moving drum. Small but sustained and appropriate changes in heart rate were obtained immediately following hypnotic suggestion of faster and slower heartbeat.[106] Similar experiments were carried out by Bramwell and Bérillon.[107] Bramwell obtained sphygmographic recordings from a subject whose pulse rate rose from 80 to 100, dropped from 100 to 60, and returned to 80 again, all in response to direct hypnotic suggestions.[108] Various other workers reported that they had been able to influence pulse rates hypnotically.[109]

A number of experimenters claimed that they had produced both general and localized variations in bodily temperature. With his celebrated subject Ilma Szandor, Krafft-Ebing was able to produce sustained changes of general body temperature by hypnotic suggestions apparently referring directly to temperature rather than to feeling hot, feeling cold, etc. Thus in one experiment, the subject's temperature, being then 40°C. due to emotional excitement, Krafft-Ebing suggested that that evening her temperature should go down to 36°C. and remain there for the following day, which it did. He was similarly able to raise her temperature from 37°C.(normal) to 38.5°C.[110] Marès and Hellich, of Prague, who gave their subjects suggestions of feeling hot or cold, were able to produce marked changes of temperature, including a sustained drop to as little as 34.5°C.[111] By suggestions of warmth and cold, Sanchez Herrero produced in a non-hysterical subject in a matter of minutes a drop of axillary temperature to 36°C and a rise to 38.5°C.[112]

Marked changes of localized skin temperature in response to hypnotic suggestion were also sometimes reported. In 1889, merely by telling an hypnotized subject "your hand will become cold", Burot caused measured drops in local temperature of up to 10°C.[113] Bramwell, however, tells us that even with "an extremely good somnambule", and very delicate measuring apparatus, he was not able to obtain any increase or decrease of skin temperature by suggestions of heat and cold.[114]

It is to be noted that none of the above experimenters instituted control trials with non-hypnotized subjects.

Allied, perhaps, to localized hypnotic changes in skin temperature, were instances of the alleged hypnotic production of skin markings – reddening, ecchymosis, bleeding, blistering.[115] The alleged phenomena of hypnotic skin marking first came into prominence in the middle 1880s, but scattered examples had been reported almost from the beginning of the mesmeric movement.[116] Blistering in response to suggestion was sometimes reported in an animal magnetic context. An example was recorded in 1819 by M. Le Lieurre de l'Aubépin, with his somnambule Manette T., of Ancenis. One day, in the somnambulic state, Manette prescribed herself a mustard plaster on the stomach, and was told that there was no mustard. "Bah!" she said, "take a piece of linen, and magnetize it *en moutarde*; tomorrow morning, when it is taken off, you will see how red and blistered my skin will be." When the linen was removed, it was found to have irritated the skin and even lifted it in several places.[117] A similar case was observed in 1840 by Dr. Préjalmini of Piedmont. His blistering agent consisted of the paper on which he had written the prescription for a vesicatory.[118]

The bridge between the mesmeric and the hypnotic movements was, in this matter as in so many others, almost certainly Liébeault, who stated in 1866 that with somnambulic subjects he had been able to cause by suggestion localized congestions of the skin, and also haemorrhages of the mucosa at a moment fixed in advance.[119] He gives no details, but goes on to talk of stigmatization; this connection had already been made in a mesmeric context by Kieser and Ennemoser.[120]

It will be convenient to consider the later nineteenth century material on hypnotically produced skin markings under two headings: reddening and bleeding of the skin, and blistering.

Reddening and bleeding of the skin.

In 1886 Beaunis reported some influential experiments on "reddening and cutaneous congestion by hypnotic suggestion"[121] with a nineteen-year old anaemic patient, Mlle. A. E. Beaunis would say to Mlle. A. E. during the hypnotic sleep, "After you are woken, you will have a red mark which will appear on the point that I touch now." Then he would very lightly[122] touch a point on the forearm with his finger. About ten minutes after she was awoken, a red mark, at first faint, but soon very apparent, would begin to appear on the point indicated, and would persist for ten or fifteen minutes, or longer if suggested. She was carefully watched to make sure that she did not rub the spot in question.

The claims of Beaunis and other members of the Nancy School to have produced hypnotic skin markings and blisters were sharply criticized by J. S. Morand.[123] Morand alleged that the precautions taken were insufficient to rule out fraud (self-inflicted injury) by the subjects. A. Bonjean, a Belgian lawyer, set out to answer these criticisms. He gave his subject, Mlle. S., post-hypnotic suggestions that skin markings should appear ten minutes after her awakening from hypnosis, and took steps to ensure that she was carefully watched during the whole of the intervening period. In his concluding experiment, 16th July 1889, he suggested not only that she would develop a red mark on the back of each hand, but that blood would also appear on the skin.

> When the ten minutes had elapsed, we ascertained the existence of two bluish-red marks larger in extent than those of the previous experiments, and with a more marked swelling. Moreover, little beads of blood had formed on the skin. The spectacle was striking...[124]

Suggested and self-suggested bleeding had already achieved some notoriety through its appearance in the well-known multiple personality subject, Louis Vivé.[125] As a further example we may take a case described by Artigalas and Rémond in 1892. The patient, Mme F., was a twenty-seven-year old married woman of hysterical tendencies, with a liability to haemorrhage, including latterly the weeping of blood-stained tears, a phenomenon which could readily be produced by suggestion. Hypnotism did not, however, immediately cure the problem, but on 1st December 1891, Professor Artigalas hypnotized the patient again, and

> ... suggested to her that the ocular haemorrhages would not recur, but that she would bleed from the hollow of the left hand. Several minutes after having been woken, she in fact presented a haemorrhage, or rather a bloody sweat, on the palm of her left hand. This phenomenon took place before our eyes, without M. Artigalas leaving the patient and without it being possible to invoke any sort of fraud. The skin was absolutely healthy on the surface of the spot which bled; the blood seemed to well up in the lines of the hand much as a profuse sweat would have done; on wiping it away, no modification of the skin could be found.[126]

The haemorrhages ceased when the hand was washed in cold water.

From the remainder of Artigalas and Rémond's paper it is clear that in causing blood to appear on the palm of Mme. F.'s hand they were proposing a direct comparison between these phenomena and those presented by various stigmatics who had attracted a good deal of interest during the nineteenth century.[127] Discoveries,or supposed discoveries, in the realm of hypnotism were a principal weapon with which free-thinking members of the medical profession, ecpecially in France, attempted at this time covertly to demolish the claims of certain clergymen concerning alleged miraculous events. The great advantage of this weapon was that it could be directed, not at the phenomena, but at their supposed supernatural causation. The Salpêtrière School, with its psycho-pathological interpretations of even the most impressive examples of demonic possession, supernatural cure, etc., was inevitably caught up in this politico-religious battle. In consequence, the School had many clerical enemies.[128]

So far we have been considering instances of reddening of the skin, bleeding, etc., allegedly produced by direct suggestions. Not infrequently, however, reddening of the skin was secondary to hypnotic suggestions of burns,[129] and this was generally true of hypnotic blistering, to which we come next.

Blistering

Particularly influential here were some experiments reported by Beaunis in 1886.[130] The subject was Lisa F., an hystero-epileptic aged thirty-seven, and the hypnotist M. Focachon, an apothecary of Charmes, whose patient she was. To give one example: Lisa was hypnotized by Focachon at Nancy on the morning of 12th May 1885 in the presence of Liébeault and Beaunis. A pretended vesicatory, consisting of postage stamps held down with plaster and a compress, was placed behind her left shoulder in a spot that she could not reach with her hand. On three occasions she was given brief, relevant suggestions. She spent the night in an hypnotic sleep, being woken only for meals. The next morning the dressing was removed. The stamps had not been disturbed. An area of skin about four by five centimetres was found to be thickened, deadened and of a whitish-yellow colour. This area was surrounded by a zone of intense redness, with swelling. Blisters had developed by the time she returned to Charmes that afternoon.

These experiments, which soon became widely known, seem to have started a minor boom in hypnotic blistering and skin markings.[131] A case which excited considerable interest was that of Ilma Szandor, in whom hypnotic blistering was studied by Jendrássik and by Krafft-Ebing. Their usual procedure was to take some distinctively shaped object, e.g. a cut-out metal letter, tell her under hypnosis that it was hot, and press it upon some inaccessible part of her skin. The area concerned would be covered with sealed bandages, and when these were removed, usually the following morning, burns of some severity might be found.[132] She does not seem to have been watched during the night; but Jendrássik states that on one occasion she was strictly watched (she was engaged in work with her hands) during the five hours which it took a blister to develop.[133]

A curious feature of this case was that if objects were placed on Ilma's left side, and suggested as hot, a corresponding mark or blister would appear not on the side of contact but, symmetrical and reversed, on the corresponding place on the right side (which was hysterically anaesthetic).

Another case of hypnotic blistering was described by the Russian, Dr. J. Rybalkin, of St. Petersburg, in 1890.[134] The subject was a youth of sixteen, suffering from extensive hysterical anaesthesia. He was told, as a post-hypnotic suggestion, that on awakening he would feel cold and go to warm himself at the (cold) stove. While there he would burn his forearm along a line already indicated. The subject obeyed, uttering a cry of pain when he touched the door of the stove. A red mark appeared on his arm within a few minutes, and he began to complain of pain. His arm was bandaged and he lay down under the eyes of the observers. Three hours later the bandage was removed. The "burn" had become inflamed and swollen and showed "papulous erythema", which by next morning had developed into several fully-formed blisters.

We now come to a paper of central importance, published by Schrenck-Notzing of Munich in 1896.[135] Schrenck-Notzing had become involved in the investigation of a case of hypnotic blistering (the subject was a cook, aged twenty, and not regarded as hysterical) and had concluded from certain signs (loosened bandages, and the fact that the phenomena ceased when stricter controls were instituted), that fraud had probably been practised. He then embarked upon a critical study of previous studies of hypnotic blistering. He is particularly severe upon Beaunis and Krafft-Ebing, whose subjects seem to have been left alone for considerable periods with only bandages to hinder them from causing the skin damage themselves. When we remember that these subjects may well have been liable to dermatographia, and, being highly susceptible to hypnosis, have felt prompted to obey their hypnotizers at all costs, we can see that these experiments are worthless. There is as yet, Schrenck-Notzing thinks, no satisfactory evidence for hypnotic vesication. It is to be noted, however, that he did not similarly dismiss all claims concerning suggested skin markings. He describes in some detail how he and Rybalkin induced one of Liébeault's subjects, an hysterical girl of twenty, to develop before their eyes, through waking suggestion, a small red spot, the size of a silver 20 pfennig piece, behind her left ear.[136] Another case in which a suggested skin marking was produced in a subject who was not hypnotized (indeed had never been hypnotized) was reported by Dr. A. N. Khovrin of Tambov.[137]

Schrenck-Notzing was not the first to criticize the supposed evidence for hypnotic blistering and skin markings. In fact, in 1895 Bekhterev of St. Petersburg had published a case of hypnotic blistering in which fraud had been detected and confessed.[138] But after the publication of Schrenck-Notzing's article the need for careful control of the subjects in these experiments was more widely recognized.[139]

Obviously the most impressive of all kinds of evidence for hypnotic blistering possible at this period would be that a skin marking should appear shortly after the suggestion is made, or after the subject is wakened, and, after a relatively brief period, during the whole of which the subject is carefuly watched, develop into an unmistakable blister.

Cases of this last kind had occurred well before the publication of Schrenck-Notzing's article, though he does not seem to have known of them. For example, during his visit to the Salpêtrière in December 1885, Delboeuf asked Charcot to demonstrate the production of an hypnotic burn on a female patient known to be susceptible. Charcot pointed to a place on her arm, without touching it, and told her that it was covered with burning wax. She protested strongly, but Charcot refused to "remove" the wax. Eventually he told her that it had cooled and fallen off. And her skin was now perceptibly red. Delboeuf, being colour-blind, did not at first see this, but he could see that the skin was raised in places. Up to this point at least ten minutes, and perhaps twenty, had elapsed. She was next passed to the assisting doctors, who seated her at a table, held her hands, and from time to time drew her attention to the burn. Delboeuf watched them, while Charcot expounded various points to Taine, who was also present. She now exhibited a veritable burn, pronouncedly red and with raised patches, She was woken by blowing on her eyes, and, feeling the pain of the burn, at once began to protest loudly.[140]

The patient in whom the hypnotic burn was observed by Delboeuf was a highly abnormal individual. The subjects of Doswald and Kreibich of Prague

(1906) and of Smirnoff of Moscow (1912) were, however, healthy persons not, so far as was ascertained, suffering from any psychological problems. Doswald and Kreibich give the following account of the third experiment with their first subject, Dr. U., a clinical assistant. Also present was another medical colleague. The experiment did not at first succeed, but hypnosis was deepened, and the suggestion made to the subject that a spot on his left forearm (lightly touched with a matchstick) would blister as quickly as possible.

> After the subject was woken, the persons present did not take their eyes off the suggested spot. It was not touched by anyone during the whole time ... After three minutes the spot, and its immediate surroundings, turned a delicate rose-red. After the lapse of a further three minutes the skin had raised itself into a thin-walled, flabby blister, about the size of a lentil.[141]

Smirnoff's case was that of a nineteen-year-old peasant girl, employed as a cook, and originally hypnotized for toothache. His procedure was to persuade this hypnotized girl that a small object (a button, a metal letter of the alphabet) laid on her skin was very hot. First reddening and then a blister of appropriate shape would become visible, though curiously enough not always on the precise spot where the object had been laid. "Any simulation whatever is excluded", says Smirnoff of one such experiment, "due to the fact that the whole process ran its course before our eyes."[142]

By the early years of the twentieth century, it was quite widely accepted among students of hypnotism – for instance by the writers of the leading textbooks – that, in certain exceptional subjects, skin markings and blisterings can be produced by hypnotic suggestion. Only rather rarely was there a dissentient voice, such as those of Madden and of Babinski,[143] who argued that such phenomena cannot occur because there is no physiological mechanism which provides for them. Of course, the fact that skin changes may be induced by hypnotic suggestion does not necessarily prove that hypnosis is a special state which favours such phenomena, or that hypnotic induction procedures render a subject more able to produce them than do other forms of psychological manipulation or self-suggestion. No properly thought-out experiments were devoted to this question during the period with which we are concerned. It was just assumed that these phenomena can only be obtained with hypnotized subjects, although there were some data which might have been thought to contradict it. For example, among a number of curious cases cited by Hack Tuke[144] is that of a lady known to him who saw a child's ankle about, as she thought, to be crushed by a heavy iron gate. Her own ankle at once became intensely painful, and shortly afterwards she found a circle round it "as if it had been painted with red currant juice, with a large spot of the same on the outer part." By morning, the whole foot was inflamed. In this case, as in others which might be cited, a powerful and unreflective imaginative impression seems to have been sufficient for the effect without any question of hypnosis. Indeed, a number of writers claimed to know of or to have observed persons who could produce such changes in themselves by a concentrated effort of will.[145]

There was naturally some speculation as to possible physiological mechanisms of these suggested or self-suggested skin changes. Most of the speculation – for example that of Delboeuf, Liébeault, Farez, Podyapolsky and Alrutz – centred around the biological functions and physiological repercussions of pain, especially insofar as these are linked to vasodilation and vasoconstriction. For instance, Alrutz[146] proposes that suggested vesication comes about when the hallucinatorily vivid idea of pain in a certain area brings about a vasodilatory effect in the relevant motor pathways, accompanied by an effusion through the vessel walls. He also proposes, quite logically, that suggested analgesia might have the opposite effect – a strong vasoconstriction, with lessened tendency for the skin to bleed when pricked or cut, and inhibition of blister formation following real burns or cauterization or the application of vesicants. There was in fact some evidence for both these things. A number of workers, for instance Vogt and Trömner,[147] claimed to have found anaemia, with reduced tendency to bleed, in areas of skin rendered hypnotically anaesthetic, whilst Delboeuf actually burned the arm of a non-hysterical female subject with a heated iron rod in order to discover whether analgesia would reduce the resultant inflammation. He made a burn on each arm, the rod, 8 mm in diameter, being applied to both for about one and a half seconds. A comparable burn was produced on each arm. One arm was rendered analgesic and kept so; the other was not. The burn on the analgesic arm occasioned much less inflammation, and began to heal earlier, than that on the non-analgesic arm.[148]

Ostensibly paranormal phenomena

Loewenfeld devotes a substantial chapter to the alleged phenomena of telepathy, clairvoyance, mental suggestion, precognition, etc., as apparently manifested by deeply hypnotized individuals. Such phenomena were, as we have seen, integral to mesmeric practice during extended phases of the animal magnetic movement, and at the same time a major cause of unclarity in mesmeric theorizing. They never became integral to hypnotic practice, but a fair number of late-nineteenth-century hypnotists, knowing, of course, the mesmeric traditions, showed a certain curiosity about them. Among them were Liébeault, Richet, Beaunis, Liégeois, Schrenck-Notzing, Moll, Bramwell, Grasset, Lombroso, Morselli, Sanchez Herrero, Dessoir, James, Gurney and the Myers brothers. Not all, I hasten to add, became believers in the alleged phenomena, but all were willing to experiment or investigate. Reports of positive findings were a good deal thinner on the ground than they had been in the heyday of mesmerism. Whether this was because the new hypnotists had a sounder appreciation of the canons of evidence required in these investigations, or because mesmeric techniques produced effects which hypnotic techniques did not, it is impossible to say. Many hypnotists of this time had little conception of how experiments should be set up and recorded. Of those who had some grasp of the problems involved, most, whether British or Continental, were members of or had links to the Society for Psychical Research.

The relevant literature is comprehensively surveyed in the four volumes of Dingwall's classic *Abnormal Hypnotic Phenomena* (1967–8),[149] and I shall briefly summarize only two aspects of it, namely experiments in the telepathic transmission of images and ideas, and experiments on the induction of hypnosis at a distance.

Among the early experimenters on hypnotism and telepathy were Liébeault, Liégeois and Brullard,[150] but the individual who carried out most work on the matter was Schrenck-Notzing. His experiments involved the transmission of "mental orders" to reproduce drawings or perform other kinds of action, and a rather high success rate was achieved.[151] It is not, however, apparent that unconscious cueing of the subject by the hypnotist was in all instances ruled out. Furthermore, Schrenck-Notzing also carried out various series of experiments on thought-transference using unhypnotized subjects. The targets included drawings, and objects or persons thought of. The success rate was comparable with that of the hypnotic experiments, and the experimental conditions rather stricter, some striking hits being obtained with the experimenter and subject in different rooms.[152]

Extensive experiments on thought-transference under hypnosis were also carried out by leading British members of the Society for Psychical Research, notably Professor and Mrs. Sidgwick and Miss Alice Johnson, the hypnotist in most instances being Mr. G. A. Smith, referred to above. These experiments involved both the transference of numbers (drawn at random from a bag) and the transference of mental pictures. High success rates were achieved, which diminished without wholly disappearing when hypnotist and subject were in separate rooms. Successes were not obtained when the subjects were not hypnotized.[153] However, the allegation has been made that Smith was systematically deceiving the experimenters, who were also his employers, and although the basis of these allegations appears to me to be questionable, a proper discussion of them would be very complex and cannot be entered into here.[154]

Experiments on hypnotization at a distance (*sommeil à distance*), descendants of the older experiments on "magnetization" at a distance, were not numerous; but some stir was created by those conducted at Le Havre in 1885–6 by Pierre Janet and various French and British collaborators, on the famous multiple personality subject Léonie (Mme. B.).[155] In three early series of experiments – October 1885, February-March 1886, and April-May 1886 – twenty-five trials were recorded, of which eighteen were classified as successes in that Léonie fell into a sleep state within a few minutes of the hypnotist (whose distance from her varied between a quarter of a mile and a mile) willing her so to do. Sometimes she would then carry out a command to come to the hypnotist's house, which had also been "willed". No "false positives" were reported. The hypnotist was in all instances either Janet or Dr. Gibert of Le Havre, whose patient Léonie was. For the third series of experiments, in April 1886, Janet and Gibert were joined by Paul Janet, Jules Janet, F. W. H. Myers, A. T. Myers, Ochorowicz and L. Marillier, who conjointly supervized and recorded the experiments. Out of eleven trials in this series, eight were rated as successes; and it must be said that these gentlemen were well aware of the need to make trials at irregular intervals and to beware of the possibility that Mme. B. might pick up cues from their own behaviour.

Further series of experiments by Janet alone in the autumn of 1886, and by Richet in 1887, were less successful, though not without interesting results.[156]

In this section I have briefly outlined some examples of what Loewenfeld referred to as the "extraordinary phenomena of somnambulism". I have confined myself largely to experiments carried out by serious and apparently sensible investigators (many of them known for their work on hypnotism in general) on what would now be called extrasensory perception. Hypnotism was, however, mixed up in all sorts of odder and much stranger endeavours, particularly in late-nineteenth-century France. A kind of continuum of loose interrelationships can be made out between, on the one hand, individuals such as P. Joire and E. Boirac, who were somewhat more prone to believe in "human emanations" and the "exteriorisation of sensibility" than were most of the medical hypnotists of the day, and, on the other, Luys and Gérard Encausse, the latter of whom was an adept of the Rosicrucian order. Between these two extremes were to be found individuals such as H. Durville, C. Lancelin, and E. A. Rochas D'Aiglun, who combined in various permutations and proportions beliefs about human magnetism, the double and its instrumental registration, clairvoyance, reincarnation, Eastern mysticism, (there was, no doubt, some input from theosophy), and the role of hypnotism in promoting all these. I shall not undertake to sort out the varieties of view to be found at different parts of this "continuum". It is interesting to note that the connection between mesmerism and ritual magic which we met in certain French masonic lodges in the 1780s still obtained between hypnotism and magic in the France of the 1880s and 1890s.[157] Indeed, it was not confined to France – R. W. Felkin, who in 1890 published one of the earliest British works on hypnotism (quite a sensible little book) became a leading light of the famous magical order of the Golden Dawn.[158] Lloyd Tuckey, too, was a member of this order,[159] perhaps briefly, and was a mason to boot.[160] The relationship between freemasonry and ritual magic is easy to understand, though how hypnotism fitted in is somewhat more obscure.

It was at this period that we find the beginnings of a practice that has caught on extensively in modern times, and has in some quarters become almost a cult, namely hypnotic regression into previous lives. Given the rise in France of Kardecist Spiritism, with its reincarnationist doctrines, and of public demonstrations of mesmerism or hypnotism, in which a standard trick was to regress someone to childhood, it was inevitable that the two would sooner or later come together. The first recorded case seems to date from 1862,[161] and the topic was mooted at the international *Congrès spirite* held at Paris in 1900. The most famous early work in this field was Rochas D'Aiglun's *Les vies successives* (1911), which describes experiments with a fair number of outstandingly good hypnotic subjects. As well as regressing subjects into past lives, Rochas D'Aiglun progressed them into future ones, though I am not aware that anyone, on leafing though the book, has found his present *curriculum vitae* set down therein.

NOTES

1 Cf. F. W. H. Myers (1903), Vol. 1, p. 191: "We have to regret the lamentable scarcity of purely psychological experiments over the whole hypnotic field. We are habitually forced to base our psychological information on therapeutic practice ..."
2 Some examples may be gleaned from Forel (1906), pp. 360–366 (Bleuler's account); Wetterstrand (1895–6b); Marcinowski (1900); Döllken (1895–6), pp. 70, 91; Straaten (1900); Vogt (1896–7a; 1897).
3 Cf. Loewenfeld (1901), p. 128.
4 Loewenfeld (1901), pp. 129–132, etc.; Forel (1906) pp. 79–80.
5 In Forel (1902), p. 252; cf. Forel (1906), p. 361. Compare also Döllken (1895–6), p. 91: "Above everything, I was always in a condition to produce off my own bat whatever series of ideas I wished."
6 Vogt (1897–8); Marcinowski (1900), pp. 29–31; and cf. Döllken (1895–6), pp. 87–88.
7 Wundt (1893), pp. 100ff; Loewenfeld (1901), p. 136.
8 Loewenfeld (1901), p. 132.
9 Bramwell (1930), pp. 98–100.
10 Bramwell (1930), p. 96.
11 Kingsbury (1891), p. 47.
12 For a succinct account of these see Bramwell (1930), pp. 342–344. Bertrand held a basically similar psychological theory of the nature of rapport. See Bertrand (1823), pp. 241–243.
13 Loewenfeld (1901). pp.159–160.
14 Moll (1892).
15 Moll (1892), pp. 427–428.
16 I have not thought it worth giving author by author references for this paragraph. A brief review of the reaction time literature will be found in Claparède and Baade (1908–9), pp. 316–317.
17 Loewenfeld (1901), pp. 170–171.
18 Loewenfeld (1901), pp. 169–170.
19 Liégeois (1889), chapter 8.
20 Delboeuf (1889), p. 58.
21 Beaunis (1886), pp. 169–171.
22 Lehmann (1890b), pp. 108–109; Parish (1897), p. 248.

23 In his Critical Notice of Binet and Féré (1887), *Proceedings of the Society for Psychical Research*, 1885–7, Vol. 4, p. 548.
24 Ottolenghi and Lombroso (1889).
25 Sidgwick, Sidgwick and Johnson (1894); Gurney, Myers and Podmore (1886); Galton (1883).
26 Delboeuf (1885); Liégeois (1889), pp. 312–313.
27 Forel (1906), p. 126; Bernheim (1889b), p. 40.
28 Bernheim (1889b), pp. 45ff.
29 Loewenfeld (1901), pp. 172–173.
30 E.g. Bernheim (1889b), p. 46.
31 E.g. Vincent (1897), pp. 200–202.
32 Krafft-Ebing (1889b), p. 89.
33 Vincent (1897), pp. 200–202; Binet and Féré (1887b), pp. 306–307.
34 E.g. Binet (1896), p. 301 (he used blank cards where Moll used matches).
35 Moll (1890), p. 185.
36 Cf. e.g. Wingfield (1910), pp. 100–101.
37 Martin (1907), p. 380.
38 See for instance Loewenfeld (1901), pp. 162–164, etc.
39 Döllken (1895–6), p. 75.
40 Bechterew (1894).
41 Schiffer (1895), p. 99; Döllken (1895–6), p. 75.
42 Loewenfeld (1901), p. 164.
43 Bechterew (1905–6), pp. 105–106.
44 Loewenfeld (1901), pp. 180–188.
45 Bernheim (1891), p. 104.
46 But with more tremor. Cf. Binet and Féré (1887b), pp. 121–124.
47 Berger (1880b), p. 121.
48 Benedikt (1894), p. 69.
49 The topic is reviewed by Du Prel (1889), Vol. 2, pp. 34–42.
50 Wolfart (1815), p. 283, describes a female somnambule who, on being magnetized, showed a detailed recollection of her last mesmeric treatment thirteen years previously.
51 Du Prel (1889), 2, pp. 39–40.
52 For examples see Binet and Féré (1887b), pp. 136–138; Moll (1890), pp. 126–127; Richet (1884a), p. 194; Liébeault (1866), pp. 92–93.
53 Loewenfeld (1901), pp. 144–157.
54 Beaunis (1886), pp. 130–131.
55 Cf. below. p. 546.
56 Cited by Moll (1890), pp. 123–124.
57 Dichas (1887), p. 13.
58 Loewenfeld (1901), p. 145.
59 Lombroso (1886), p. 4.
60 Claparède and Baade (1908–9); Chojecki (1912).
61 E.g. Bramwell (1930), pp. 101–102; Bramwell (1896–7b), p. 194; Dufay and Azam (1889–90), pp. 408–409.
62 Loewenfeld (1901), p. 146.
63 Benedikt (1894), p. 27.
64 Felkin (1890), p. 25.
65 Loewenfeld (1910).
66 Loewenfeld (1910), pp. 6–7.

67 Pitres (1884), p. 58.
68 Delboeuf (1889), pp. 66–67.
69 Moll (1890), pp. 124–125.
70 Bernheim (1891), pp. 127–128.
71 Bramwell (1930), p. 105.
72 Loewenfeld (1901), p. 148.
73 For instance Stadelmann, under the influence of Breuer and Freud, treated certain hysterical patients by first recovering the traumatic memory which supposedly underlay their symptoms, and then removing it. Stadelmann (1896), pp. 139–196.
74 Loewenfeld (1901), pp. 154–157.
75 Vincent (1897), p. 177. This seems to be Bernheim's opinion; Bernheim (1891), pp. 119–120.
76 Moll (1890), pp. 134ff.
77 Lehmann (1890b), pp. 145–149.
78 Krafft-Ebing (1889b), pp 97–98.
79 Benedikt (1894), chapter 12.
80 Köhler (1897); Jolly (1893).
81 Köhler (1897), p. 366.
82 Köhler (1897), p. 365.
83 Loewenfeld (1901), chapter 10.
84 Bernheim (1884a), pp. 28–29; Tuckey (1900), pp. 70–71.
85 Moll (1890), p. 140.
86 Bernheim (1884a), p. 29.
87 Beaunis (1886), pp. 233–235.
88 Liégeois (1889), pp. 339–342.
89 Wingfield (1910), p. 99.
90 Noizet (1854), pp. 162–164.
91 Bramwell (1930), pp. 114–139; Delboeuf (1891–2b); T. W. Mitchell (1908–9). For early claims as to enhanced time sense in mesmerized subjects see e.g. Kieser (1826), Vol. 2, pp. 181–182; Brandis (1818), pp. 114–116; Bertrand (1823), pp. 313–316.
92 Pitres (1891), Vol. 2. p. 201. Pitres claims that a subject who is prevented from carrying out a post-hypnotic suggestion may become exceedingly agitated, and gives various examples.
93 Köhler (1897), pp. 371–374.
94 E.g. Forel (1906), pp. 123–124; Moll (1890), pp. 152–154.
95 Moll (1890), p. 149.
96 Myers (1885–7b), pp. 14–15.
97 Köhler (1897), pp. 371–374.
98 Loewenfeld (1901), p. 231.
99 Gurney (1885–7a), p. 282. This article is well worth reading on the point in question.
100 Forel (1906), p. 136.
101 Loewenfeld (1901) p. 238.
102 See note 91.
103 Gurney (1885–7a).
104 See especially Liébeault (1866), pp. 140–142.
105 Cf. Report of the committee ... (1893–4).
106 Beaunis (1886), 44–67.
107 Bramwell (1930), pp. 81–82; Bérillon (1902b).
108 Bramwell (1930), pp. 81–82.
109 Bonjean (1890), pp. 87–90; Tuckey (1900), p. 67; Sanchez Herrero (1905), pp. 407–411.

110 Krafft-Ebing (1889b), pp. 77, 80, 83.

111 Marès and Hellich (1889).

112 Sanchez Herrero (1905), pp. 420–422.

113 Burot (1889–90); for curious modern cases see Benson *et al.* (1982).

114 Bramwell (1930), p. 85.

115 I have given a fuller survey of these early cases in Gauld (1990).

116 I have already mentioned the instances of autosuggested skin markings witnessed by Tardy de Montravel in 1785 and by Billot about 1828. Cf. above pp. 61–62, 172.

117 Le Lieurre de l'Aubépin (1819).

118 Préjalmini (1840–1).

119 Liébeault (1866), p. 140.

120 Kieser (1826), Vol. 1, pp. 260–261; Ennemoser (1842), pp. 251–252.

121 Beaunis (1886), pp. 70–71.

122 Myers criticized Beaunis for not trying whether points touched without suggestion would become equally red, but was later able to experiment with Mlle. A. E. for himself, and tried this control without effect. Myers (1885–7c), pp. 167n-168n.

123 Morand (1890), pp. 308–310.

124 Bonjean (1890), p. 104.

125 Berjon (1886), esp. p. 36.

126 Artigalas and Rémond (1891–2), p.252.

127 On the celebrated Louise Lateau (1850–1883) of Bois d'Haine, see e.g. Lefebvre (1873), and on stigmatization in general, Thurston (1952), pp. 32–129.

128 On this area in general see Goldstein (1982).

129 Tuckey (1900), p. 341, and Liébeault (1894–5b), pp. 40–41, cite such cases. One of Liébeault's subjects showed an actual burn..

130 Beaunis (1886), pp. 73–84.

131 On suggested blistering at the Salpêtrière, see below; on blistering at the Pitié under Dumontpallier see Magnin, *apud* Farez (1907–8), p. 186, who states that it was a commonplace phenomenon there, and Dumontpallier (1885), which reports, however, no blisters, but only changes in skin temperature. Some reports of this period are amazingly casual and brief. E.g. Wetterstrand (1891), p. 31; (1897), pp. 30, 32; Wetterstrand reported in Forel (1898) and (1906), pp. 112–113; Backman (1891–2), p. 204; and Belin (1898) in Zielinski (1968), p. 93.

132 Krafft-Ebing (1889b), pp. 78–79.

133 Jendrássik (1888), pp. 322–323.

134 Rybalkin (1889–90).

135 Schrenck-Notzing (1895–6).

136 Schrenck-Notzing (1895–6), p. 223.

137 Chowrin (1919), p. 12; Zielinski (1968), p. 37.

138 Bechterew (1895), pp. 45–46. The patient applied cantharides plaster to herself, and also burned herself with hot water and a heated metal hook.

139 See e.g. Doswald and Kreibich (1906); Sanchez Herrero (1905), p. 420; Podyapolsky (1909); Podiapolsky (1909–10); Wetterstrand (1903) *apud* Alrutz (1914); Hadfield (1917); Podyapolsky (1904–5); Podiapolsky (1909–10), p. 180; Rybalkin (1889–90); Heller and Schultz (1909). But there were still occasional examples of looseness, e.g. Szöllösy (1907) and Kohnstamm and Pinner (1908).

140 Delboeuf (1886), pp. 20–21. Another case in which the formation of suggested burns was directly observed was that of Louis Vivé, though details are lacking (J. Voisin, 1887, pp. 136–137).

141 Doswald and Kreibich (1906), p. 637.

142 Smirnoff (1912) p. 174.

143 Madden (1903); Babinski, cited by Chertok (1981), p. 26. Similar considerations restrained Pattie (1941), in his well-known review of the subject, from giving a definite assent to the phenomenon.

144 Hack Tuke (1884a), Vol. 2, pp. 35–36.

145 Wright (1900) cited by Kreibich and Sobotka (1909), p. 188; Kohnstamm in Kohnstamm and Pinner (1908), p. 344; cf. Dunbar (1954), pp. 612–613. Mantegazza (1887), Vol. 1, p. 17, asserts that during a period of ill-health in his youth he was able thus to produce urticaria and erythema on his own skin, and Woods (1897), p. 257, states as a fact that a blister has been produced just by sustained concentration on a finger tip.

146 Alrutz (1914), p. 6.

147 Verhandlungen ... (1912), pp. 280–281.

148 Delboeuf (1887a). The experiment was repeated by one of Delboeuf's colleagues using a thermocautery, and a very similar result was reported by Hadfield (1917), p. 678.

149 For contemporary reviews see F. W. H. Myers (1903) and Podmore (1894), pp. 1–17; for the present state of play see Honorton and Krippner (1969) and Schechter (1984).

150 Cf. Gurney, Myers and Podmore (1886), Vol. 2, pp. 656–658.

151 Moser (1967), pp. 181–194. For similar claims see Sanchez Herrero (1905), pp. 500–523; Ochorowicz (1891).

152 Schrenck-Notzing (1891–2).

153 Podmore (1894), pp. 65–80; Sidgwick, Sidgwick and Smith (1889–90); E. M. Sidgwick and Johnson (1892).

154 T. H. Hall (1964); Gauld (1965); Nicol (1966).

155 Podmore (1894), pp. 108–116; Dingwall (1967), pp. 264–273; Pierre Janet (1886b; 1886c); F. W. H. Myers (1885–7c).

156 Madame B. occasionally exhibited apparent powers of precognition, clairvoyance and travelling clairvoyance. Cf. Richet (1923), pp. 104, 125–127; Bickford-Smith (1890–1).

157 I wonder whether Liébeault, Liégeois and Brullard knew that the S. de Guaita, a local "man of letters" who was their co-experimenter in experiments in thought-transference which they conducted in January 1886, was in fact a leading French "Rosicrucian" and magician. See Gurney, Myers and Podmore (1886), Vol. 2, pp. 656–658.

158 Howe (1972), passim.

159 Howe (1972), p. 51.

160 Tuckey (1900), p. 362.

161 Delanne (1924), p. 217.

CHAPTER 22

The turn of the century. III – Uses and applications of hypnotism

Under this very broad heading, I shall give some account of turn of the century views and factual claims in the following six areas:

1. Medical uses of hypnotism and suggestion: general.
2. Success of hypnotic treatment in different categories of ailment.
3. Surgical uses of hypnotism.
4. Educational applications of hypnotism.
5. Hypnotism and crime.
6. Suggestion and hypnotism in folk-psychology.

These are not all the areas which might be included under the heading of this chapter. I have omitted, for instance, as insufficiently interesting, material concerning the use of hypnotism as a method of psychological investigation,[1] and various claims and counterclaims concerning the dangers of hypnotism.[2]

MEDICAL USES OF HYPNOTISM AND SUGGESTION: GENERAL

In his valuable chapter on this subject,[3] Loewenfeld divides the therapeutic uses of hypnotism into three categories:

1. The therapeutic uses of hypnotic sleep.
2. The utilization of the heightened suggestibility peculiar to the hypnotic state.
3. The utilization of the hypermnesia peculiar to the hypnotic state.

The therapeutic use of hypnotic sleep

Since ordinary rest and sleep have a beneficial effect in many ailments, it is reasonable to suppose (thinks Loewenfeld) that hypnotic sleep *per se* may do likewise. Many therapists – including Wetterstrand, Moll, Obersteiner, Binswanger and Vogt[4] – have made use of it, and Loewenfeld has himself found it effective in certain cases. Wetterstrand has been foremost in using hypnotic

sleep therapeutically. By means of it, he has successfully tackled cases of
hysteria, morphinism, alcoholism and cocainism, and has obtained results in
cases where hypnotic suggestion has failed (though of course the fact of having
slept may constitute an implicit suggestion). In some cases, Wetterstrand kept
patients in hypnotic sleep for weeks or even months. It is not essential for the
patient to attain the somnambulic state, but if he can do so it is advantageous in
the matter of feeding etc.[5] Other therapists – Vogt, Corval, Rifat, A. Voisin[6] –
have found long daily sleeps sufficient, and of course hypnotic sleep can readily
be combined with the administration of hypnotic suggestions.

Utilization of the heightened suggestibility peculiar to the hypnotic state

According to Loewenfeld, the therapeutic use of hypnotic suggestion goes back
to Liébeault, who combined sleep suggestions with suggestions for the alleviation
of disease. The disadvantage of this method is that it attacks only symptoms, and
furthermore not all symptoms are suitable for suggestive therapy. Often there is
a complex of symptoms, of which some are suitable for hypnotic treatment and
others are not. When there are numerous symptoms requiring treatment by
suggestion, one has to decide which to deal with and in what order, because the
effectiveness of hypnotic suggestions diminishes in proportion to their number.
In making this decision, two important points to be borne in mind are the degree
to which the symptoms concerned are susceptible to suggestive influence, and
the urgency of the need to remove them. Other things being equal, one should
tackle the most easily influenced symptoms first, because success with them
makes the patient more receptive to subsequent suggestions.

Suggestive therapy has evolved since Liébeault's day. It no longer requires an
intellectual sacrifice on the part of the patient. We do not impose suggestions on
him, but try to enlist his cooperation and understanding. Loewenfeld himself has
adopted this approach since the beginning of his hypnotic practice, and it is also
recommended by Grossmann.[7] We tell the patient not bluntly, "your pains will
disappear", but "this sleep (hypnosis) will thoroughly calm your nerves;
because of this calming your pains will cease." These more sophisticated
suggestions bring us near to other psychotherapies. It is not, however, necessary
to reconcile our suggestions too carefully with the demands of logic. Suggestions,
techniques and supplementary gimmickry must be tailored to fit the level of
understanding and the mental individuality of the patient.

In recent years, Loewenfeld adds (and it is an interesting observation), the
heightened suggestibility of the hypnotic state has been increasingly utilized for
the purposes of ordinary psychotherapy. Because of his psychological condition,
a hypnotized subject is more accessible to any kind of psychotherapeutic
influence than a waking one. Thus we can give a patient explanations, advice,
comfort, reproofs, and so on, while he is in hypnosis with more decided results

than we could achieve were he in the waking state. The separation between hypnotherapy and the various methods of waking psychotherapy has to a certain extent been removed.

The utilization of hypnotic hypermnesia

Hypnotic hypermnesia, says Loewenfeld,[8] can be used for two purposes: 1. the removal of amnesia where this may be beneficial to the patient or fill gaps in the therapist's understanding of the case, and 2. the laying bare of pathogenic influences (*Momente*) which are not accessible during the waking state.

1. Hypnotic suggestion has proved very valuable in cases of pathological amnesia. Vogt has succeeded with all kinds of amnesias – hysterical, infectious, post-epileptic, confusional and affective. Especially interesting are the results of Graeter and Muralt with cases of epileptic amnesia,[9] which was previously thought incurable. Each describes a case in which retrograde and anterograde amnesia associated with epileptic attacks were apparently lifted by hypnotic treatment, though whether Graeter's case is one of true epilepsy might be doubted. Events of actual epileptic attacks are rarely recoverable. Hilger has investigated this question in a number of cases, and has come to largely negative conclusions. He succeeded only in obtaining recall of the hallucinations which had occurred before and during the attacks.[10] In a paper published just after Loewenfeld's book appeared,[11] F. Riklin claims in two cases to have recovered by hypnosis memories both of epileptic attacks and of subsequent twilight states.

2. According to Loewenfeld, Breuer and Freud in Vienna were the first to recommend hypnotism for the discovery of pathogenic influences underlying overt symptoms and no longer accessible to memory.[12] Breuer and Freud[13] interpret hysterical symptoms as manifestations of the undischarged and hence still active affect attaching to traumatic memories which have become inaccessible to consciousness through deliberate blocking (repression) or otherwise. A cure can be obtained if, under hypnosis, the memory of the causative event can be brought clearly to light, along with its accompanying affect, and the latter discharged through speech ("abreaction"), perhaps accompanied by physiological signs of emotion.

Loewenfeld goes on to remark that Breuer and Freud's hopes for their "cathartic method" have not been fulfilled. Often the patient cannot be hypnotized, so the exploration under hypnosis does not have satisfactory results. It was for these reasons that Freud was led to develop his psychoanalytic method. Furthermore, even when the memory of a traumatic event is recovered under hypnosis, and abreaction takes place, a satisfactory result may fail to follow. Krafft-Ebing and Loewenfeld himself were the first to show this.[14] Vogt found that the method of Breuer and Freud worked in several cases, but held that autosuggestion was not excluded.[15] The cures were furthermore not permanent, and in a series of cases in which abreaction took place in typical fashion there was an actual deterioration in the state of the patient.[16] The trials which Seif made with the Breuer–Freud method ran a similar course. Hirschlaff criticizes the claims of Breuer and Freud still more strongly.[17]

Even if discovery of the pathogenic memories which underlie certain hysterical symptoms does not lead to their removal, it may give us valuable hints for the further conduct of the case. Stadelmann has successfully treated several cases of hysteria, in which pathogenic memories lay at the root of the troubles, by hypnotically unearthing the memories concerned and then removing them through suggestion.[18] Vogt has also recommended this procedure. But suggestions of forgetting are not always successful. The emotionally charged pathogenic ideas, which have become deeply ensconced through frequent repetition, may prove too strong for affectless verbal suggestion.

In sum: hypnotherapy has certain advantages over general psychotherapeutic methods. Prolonged sleep (as used by Wetterstrand) is only possible through hypnosis, and the soothing effect provided by brief periods of hypnotic sleep cannot be achieved in any other way. Hypnotic suggestion brings help in many cases which waking suggestion fails to touch, and other forms of psychotherapy – explanation, training, diversion, etc. – can conveniently be combined with it. It opens up prospects, but it also has disadvantages and limitations. Even with persons who pass into a state of deep hypnosis, therapeutic suggestions may remain unfulfilled, and often the results may be only temporary. Or new symptoms may appear instead of the old – this is particularly liable to happen with phobias.

SUCCESS OF HYPNOTIC TREATMENT IN DIFFERENT CATEGORIES OF AILMENT

We come now to the crucial question of what ailments hypnotism can successfully be used to treat. Here the accumulation of relevant material was, even by the end of the century, enormous, but drawing conclusions from this material is fraught with problems. It is, of course, very much a literature of individual case histories and of collections of individual case histories. There are no studies in which the progress of groups of patients treated hypnotically is compared with the progress of groups treated by other methods or not treated at all. Whether hypnotic treatment has in fact benefitted each patient was in the last resort always a matter of the clinical judgment of the physician in charge. It is easy under these circumstances retrospectively to criticize claims made as to the therapeutic efficacy of any method of treatment. But one must remember that the workers concerned were by no means wholly unaware of the desirability of controlled investigations.[19] Such investigations were, and remain, exceedingly difficult to set up, and can generally not be set up within the framework of an individual practice, or clinic, in which the numbers of patients with diseases falling under a given nosological category will usually be limited, and the clientèle will be largely self-selected. The hypnotic literature of this period is probably no worse in these respects than that of any other branch of medicine, and of course medicine *can* progress under these conditions, though it progresses only slowly and with many backtrackings.

A very unsatisfactory feature of the late-nineteenth-century clinical literature on hypnotism is the shortage of detail in many of the individual case reports. In this respect, the hypnotic literature falls on balance below the standard of the mesmeric – medical, and indeed non-medical, mesmerists often reported the case histories of their patients in somewhat tedious detail. From hypnotic case reports it is often impossible to determine what method of hypnotic induction was used, what kind of therapeutic suggestions were made, and whether the patient's symptoms were such as to justify the initial diagnosis.[20] Yet all these pieces of information are essential to one's assessment of the case and of any claims based upon it.

The quite numerous lists or tabulations of such case reports which this literature contains are obviously of rather limited value.[21] There are considerable differences between one therapist and another in successes obtained with ailments of the same category, and we have no means of knowing to what these difference should be attributed. Not a few of the therapists concerned report only their successes. The most comprehensive tabulations are those of van Renterghem and van Eeden.[22] These authors give, in respect of each patient, his or her malady, sex, degree of hypnotizability, and upshot of hypnotic treatment. It is unfortunate that they do not also provide information as to duration of previous treatment and of hypnotic treatment, and as to length of follow-up. Table 2[23] gives the gist of the results at their Amsterdam clinic with 1577 patients treated between May 1887 and June 1897.

Loewenfeld presents no such tabulation of cases, but simply takes each category of ailment in turn and asks whether it is amenable to hypnotherapy. His remarks combine his own clinical experience with the accumulated clinical wisdom of his time, and are on balance more illuminating than Renterghem's tables.

Neuroses

Loewenfeld begins by observing that the chief area of application for hypnotherapy is nervous ailments, more particularly functional nervous ailments (the neuroses).[24] There have been many examples of apparent success in the hypnotic treatment of hysteria, which Grasset refers to as the "triumph" of hypnotism.[25] Here it is hypnotic suggestion rather than the hypnotic sleep *per se* that is effective.[26] Occasionally it produces instant "wonder cures".[27] But it does

Table 2. Results from the Amsterdam Clinic 1887–1897

	A. Ailments of the Nervous System					B. Other Ailments						
	I	II	III	IV	V	V	VI	VII	VIII	IX		
						Functional Distur-bances Linked to:						
	Organic illnesses	Major neuroses, hysterical afflictions	Insanity	Neuropathic complaints	Neuralgia, indefinite pains, employment neuroses	Internal ailments	External ailments	Fever	Chlorisis, menstrual disturbances	Anaesthesia for surgery	Total	Percentage
Number and sex												
Male	57	245	97	186	88	69	25	0	0	2	769	48·8
Female	33	342	82	120	128	51	15	1	22	14	808	51·2
Total	90	587	179	306	216	120	40	1	22	16	1577	
Stage of hypnosis attained												
0. Refractory	6	43	25	23	12	1	2	0	0	1	113	7·16
1. Light sleep	42	214	88	140	100	23	14	1	6	0	628	39·82
2. Deep sleep	38	240	61	109	80	85	22	0	9	5	649	41·15
3. Somnambulism	4	90	5	34	24	11	2	0	7	10	187	11·86
Age												
1–10 years	4	12	0	21	2	0	3	0	0	1	43	2·73
11–20 years	4	88	12	73	24	2	0	5	1		211	13·88
21–40 years	37	356	104	131	106	39	26	0	13	12	824	52·25
41–60 years	29	116	55	61	73	60	7	1	4	2	408	25·87
61–80 years	16	15	8	20	11	19	2	0	0	0	91	5·77
Result of treatment												
a. Without effect	33	88	46	42	26	19	13	1	1	9	269	17·06
b. Slight or transitory improvement	25	83	27	59	23	38	7	0	3	0	265	16·8
c. Decided or lasting improvement	16	188	36	60	26	10	0	2	3[a]		402	25·49
d. Cure	2	176	38	107	89	32	7	0	14	13[b]	478	30·31
e. Unknown[c]	14	52	32	38	17	5	3	0	2	0	163	10·34

[a] Incomplete anaesthesia, [b] Complete anaesthesia, [c] Treatment was broken off after one or two sessions.

not always achieve what might be expected in cases of hysteria, given that hysterical symptoms are themselves commonly the product of suggestion. There are various reasons for the failures. For instance, some hysterics suffer from such lability of mind that they cannot hold fast to the suggested ideas and assimilate them. Again, some hysterics may be so autosuggestible that heterosuggestions can have no more than a transient effect.[28]

Neurasthenic states, says Loewenfeld,[29] are likewise a rewarding field for the use of hypnotic suggestion, though it is not equally effective against all neurasthenic phenomena. In a survey of 228 cases treated by himself and twenty other hypnotherapists, Schrenck-Notzing found a 31.57% recovery rate (the failure rate was the same).[30] Van Renterghem and Van Eeden had a cure rate of 20% in over 100 cases (compared with 39% in 124 cases of hysteria).[31]

Loewenfeld goes on to consider anxiety states occurring independently of neurasthenia.[32] Simple, contentless anxiety states are much more favourable objects for hypnotic suggestion than complex ones with definite external causes, as in phobias. The difference arises because simple anxiety states have various causes, mostly somatic, the influence of which largely depends on the momentary state of the nervous system. This gives the anxiety the character of an inconstant and floating symptom, which is advantageous for its removal. On the other hand, with the anxiety arising from phobias, the proximate causal stimuli are ideas, which may recur again and again in an obsessive manner. If this happens, the production of the anxiety state by the idea may become a sort of psychical reflex or automatism, and function independently of whatever underlying cause originally led to the development of the phobia. It is, so to speak, circularly self-sustaining, and the use of somatic means, especially opium, may be necessary to help break the circle. But the essential thing is the elimination of the anxiety-provoking idea, and this can only be achieved by psychotherapeutic methods, not least hypnotic ones. Hypnotic suggestion, however, does not carry with it the emotional force of the phobic automatism, and may require supplementation by other forms of psychotherapy.

With obsessive ideas, according to Loewenfeld, hypnotic treatment reigns supreme, and Bekhterev likewise states, with examples, that his best results with hypnotherapy have come from patients with obsessive ideas, and all kinds of pathological anxieties.[33] None the less, hypnotism is not an infallible method of treatment for obsessions. Obsessive ideas which are continually intruding, and which have powerful emotional overtones, can be very difficult to eradicate. To deal with strongly developed obsessional ideas we need a strictly systematic stepwise procedure. We must first strive to weaken and to eliminate those obsessive thoughts that set in at certain times or in certain circumstances. From the ground thus won we extend our suggestive operations more and more widely. Unfortunately, obsessive patients are often hard to hypnotize, and many suffer from the obsessive idea that they cannot be hypnotized, which is an insurmountable obstacle. However, Milne Bramwell (1903), whose section on this topic is well worth reading, remarks "...few people have any idea how common the disease [obsessions and phobias] really is...Fortunately hypnotic treatment frequently gives brilliant results in such cases: the majority of those I have seen recovered, and relapse has been rare."[34]

In cases of stuttering due to anxiety or obsessive ideas, hypnotherapy can be very beneficial. Loewenfeld points out that Wetterstrand reports fifteen cures and twenty improvements out of forty-eight cases,[35] and that Corval, Ringier, Hirt and Tatzel have likewise had successes.[36] Bramwell, on the other hand, after surveying his own results, and those of others, maintains that "the results as a whole have not been so satisfactory as those obtained in other functional disorders."[37] But a few years later Wingfield states that he has cured six of the twelve cases he has tackled; the difficulty lies in diagnosing the precise cause of the condition in order to make the appropriate suggestions.[38]

Psychoses

In the therapy of psychoses, hypnotic procedures are not, Loewenfeld asserts, of much avail.[39] This is in part because only about ten per cent of psychotics are hypnotizable. With some, especially paranoiacs, attempts at hypnotism are decidedly harmful. But favourable results have been obtained with cases of mild melancholy and of psychoses with an hysterical basis.

These brief remarks somewhat oversimplify a rather more complex situation. The individual who was most noted at this period for his hypnotic treatment of insane or psychotic[40] or highly excited individuals was Dr. Auguste Voisin of the Salpêtrière (he was, it must be remembered, independent of Charcot). Voisin's first attempts at hypnotic treatment of the insane were in 1880, and thereafter he made use of hypnotic methods not just in cases of mania and melancholia, but in many other mental disorders of a florid kind, and in cases of drug addiction, alcoholism, sexual perversion, criminal tendencies, acquired or native defects of memory or intelligence, convulsions, neuralgia, etc. He wrote numerous articles describing his results.[41] His methods of subduing excited or violent patients were sometimes heroic. He would seat a strait-jacketed patient on a chair, approach his own face to the patient's and stare unrelentingly into his or her eyes, sometimes for hours. Even so, only ten patients in a hundred might eventually respond. When the patient finally passed into the hypnotic sleep, Voisin might give therapeutic suggestions, or might leave him to sleep for 12 or 23 hours out of the 24 during a whole week. With patients who were successfully hypnotized, Voisin enjoyed a fair measure of success, especially in cases of mania and melancholia.[42] Some of Voisin's contemporaries doubted that the *aliénés* whom he had cured were maniacs and melancholiacs rather than excited hysterics. This scepticism was expressed, for example, by Crocq and

Bekhterev in the discussions following Voisin's paper to the Third International Congress of Psychology (Munich, 1896).[43] In his reply, Voisin agreed that among those whom he had seen cured by hypnotic suggestion, hysterics were in a majority; but added: "The observations which I have just presented to you, to the number of 47, relate to *aliénés* of whom more than half remain cured after ten or fifteen years; among them there are 34 lypemaniacs (acute melancholics), *hallucinés*, and sufferers from persecution mania and acute mania."[44]

Some at least of Voisin's cases might nowadays be classified not as hysterics (hysteria has become unfashionable with psychiatrists, or with patients, or possibly both) but as examples of severe reactive depression or hypomania. The same is true of cases reported in his wake by some of his contemporaries. Burot, Jules Voisin (Auguste's brother) and Répoud (of Freiburg), for example, report cases treated by much the same methods as Voisin's, and Répoud asserts categorically that his highly disturbed patients were not hysterics.[45] De Jong essayed hypnotism with thirty-two cases of simple melancholia, in which the principal symptoms were sleeplessness, unmotivated sad anxiety, incapacity for business, etc. He succeeded in hypnotizing seventeen of these patients. Hypnotic suggestion did not have a beneficial effect on their condition, but prolonged hypnotic sleep did.[46] In 1897, J. F. Woods reported that he had treated ten melancholics and four cases of mania by hypnosis. Six of the former recovered and three improved, two of the latter recovered and the other two showed an improvement. The number of hypnotic treatments ranged from one to seven.[47] Van Renterghem and van Eeden were rather less successful. In 1893, they reported on seven cases of hypomania, four of mania, and seventeen of melancholia. Only three of these twenty-eight patients were refractory to the hypnotic influence, but only seven are classified as cured.[48] It is clear that many therapists continued to have great difficulty in hypnotizing their insane patients. Bramwell, for instance, whose discussion of this subject is worthy of study, succeeded in hypnotizing only three out of the eight patients he classed as insane.[49] Farez claimed in 1898, on the basis of his own work, that the solution is to insinuate the curative suggestions while the patient is asleep.[50]

Psychosomatic complaints

Having discussed the uses of hypnotism in psychiatry, Loewenfeld moves on to consider its possible applications in psychosomatic complaints. Many disorders of the latter kind were alleged to yield to hypnotic treatment. Thus Loewenfeld remarks[51] that nervous afflictions of the digestive apparatus, for example dyspepsia, gastralgia, may be treated with hypnotherapy, especially if there are emotional causes. Forel has recommended it for constipation and gives examples. He also states that it is often effective against diarrhoea.[52] Loewenfeld, on the other hand, states that his own and Vogt's experiences have not been favourable.[53] However, Delius emphasizes the value of hypnotism in both obstipation and diarrhoea.[54] Many other writers – for example Wetterstrand, Baierlacher, Hamilton Osgood, Lloyd Tuckey, Stadelmann[55] – report occasional instances of apparent success in the hypnotic treatment of these kinds of internal disturbance. Bernheim even has a number of instances of the beneficial hypnotic treatment of dysentery.[56]

Turning to disorders of the respiratory apparatus, Loewenfeld remarks that among them cases of nervous cough are particularly suitable for hypnotic treatment. He cites a case of Hirt's,[57] a boy of fourteen who had suffered for eight years from coughing attacks which had weakened him so much that he was unable to go to school and which kept the rest of his family from sleep. One hypnotic session cured him completely, and at the time of Hirt's report the cure had been maintained for four and a half years. Among the various forms of asthma, it is, according to Brügelmann, neurasthenic asthma, the kind in which anxiety plays a causal rôle, that may most readily be treated by hypnotism.[58] Forel has reported some success with asthma sufferers, including one who additionally suffered from emphysema (the emphysema was also improved).[59] Wetterstrand too claims good success with cases of asthma and nervous cough.[60]

Hypnosis may also be of value in nervous heart afflictions. Some workers have even found hypnosis of value in cases of organic heart disease. Thus Lloyd Tuckey describes two cases of mitral insufficiency in which the heart's action (though not of course the valvular problem) was improved by hypnotic suggestion, and various supervenient nervous symptoms were removed.[61] Wetterstrand states that he has successfully treated by hypnotism twelve cases of "nervous heart beat" and six of organic heart disease. The latter were likewise cases of mitral insufficiency in which the action of the heart was improved.[62]

Of urinary disturbances, says Loewenfeld, we need only consider nocturnal enuresis. Many authors – for example Cullerre, Liébeault, Bérillon, Ringier, Wetterstrand and Tatzel[63] – have found hypnotic suggestion a very effective way of treating this complaint. Loewenfeld's own experience has been favourable. Unfortunately, most of the authors concerned fail to give us details of the suggestions they used. Liébeault, who was something of a specialist in this matter, seems often to have suggested to the afflicted children that they would be unable to urinate lying down and would have to get up several times during the night.[64]

Cases of the hypnotic treatment of skin complaints were not very numerous. Bramwell instances a number of successfully treated cases of eczema, pruritus and hyperhidrosis, Osgood four cases of eczema and one of dermatitis.[65] There are also a number of instances of the cure of warts by waking suggestion, mostly by the use of charms or other props.[66] Farez and Bérillon both report cases of the successful hypnotic treatment of warts; the latter was, so far as I know, the first to try the now well-known experimental technique of suggesting away warts on one hand only, leaving the warts on the other hand as controls.[67]

Disorders of sexual function

Cases of the hypnotic treatment of disorders of sexual function are scattered, though not in great numbers, across most of the case collections referred to above. As noted in an earlier chapter, the end of the nineteenth century saw, particularly on the continent, a marked upsurge of interest in and literature concerning sexual pathology,[68] and several leading specialists in the area – for instance Moll, Forel, Schrenck-Notzing and Fuchs[69] – were also well-known as hypnotic practitioners. The most influential work especially devoted to the hypnotherapeutic treatment of sexual problems was undoubtedly Schrenck-Notzing's *Die Suggestions-Therapie bei krankhaften Erscheinungen des Geschlecht-sinnes* (1892), which reviews cases published by other writers as well as the author's own. Schrenck was, in his day, most "enlightened". He discusses female impotence (frigidity) as well as male, and says that as long as society "is not in a position to offer to every mature individual satisfaction of the sexual appetite in some generally-recognized form, as in marriage, it is senseless to oppose non-marital sexual intercourse,"[70] a concession which he rather guardedly extends to women as well as to men.[71] He advocates the condom as a contraceptive and prophylactic.[72]

Of disorders of the sexual function in men, Loewenfeld states that impotence, contrary sexual feeling, and the habit of masturbation (which many still believed to bring about mental and physical debility) are most amenable to hypnotic treatment. Thus out of eighteen cases of impotence treated by van Renterghem, Liébeault, Bernheim, Moll and himself, Schrenck-Notzing finds five complete failures, two slightly improved, one essentially improved, and ten cured (four with later report and six without).[73] Out of thirty-two cases of sexual perversion in males (mainly homosexuality, but including some cases of sadism and masochism) taken principally from Krafft-Ebing, Wetterstrand and himself, he finds five total failures, four slightly improved, eleven essentially improved, and twelve cured (ten with later report and two without).[74]

Many authors who report cases of these kinds are disappointingly unspecific as to the actual suggestions which they employed, a shortcoming (as we noted above) of a good many hypnotic case reports, but particularly regrettable in an area where the hypnotist's very tone of voice is important, and counter-suggestibility is so easily called into play. Even Schrenck-Notzing gives us only rather general indications of his lines of attack. With some writers this reticence is no doubt due to the nature of the topic, but hardly with Schrenck, who always calls a spade a spade, or at any rate a *rutrum*. Sanchez Herrero gives an interesting account of his treatment of a case of impotence in a man of thirty-five, who had not had an erection in three years. Under hypnosis, the patient rapidly became somnambulic. After a course of generally reassuring and revivifying suggestions, he was brought back to the days of his youth and induced to make love to the hallucinated person of the mistress who had then most pleased him. Finally a transition was effected by post-hypnotic suggestion to love-making with a more substantial partner.[75]

Loewenfeld states that hypnotic treatment is especially efficacious in nervous disorders of female sexual functioning – he instances vaginismus, hysteralgia, genital erythema and sexual anaesthesia, i.e. frigidity. The case-literature, however, is, as might be expected, very much male-oriented, though Schrenck-Notzing shows a notable sensitivity to the female situation.[76] So far as I know, the only British hypnotherapist of this time who speaks sensibly of the treatment of female sexual problems is Wingfield, a writer of admirable commonsense. He describes the case of Mrs. D., married for eight years, who suffered both from frigidity and failure to conceive. After three weeks of treatment by suggestion, "her natural feelings awoke", she achieved orgasm, and "she now has two children".[77]

Disorders of menstruation and lactation

In contrast to reports of the hypnotic treatment of "female impotence", reports of the hypnotic treatment of menstrual disorders are fairly common, though details of the suggestions given are commonly lacking. Among hypnotherapists who have had success with these problems, Loewenfeld cites Liébeault, Voisin, Bernheim, Brunnberg and himself, and he could have added many others.[78] Corval, for instance, stresses the efficacy of hypnotism in cases of menstrual disturbance,[79] and it will be recalled that the reestablishment of suppressed menstruation was the aim of much mesmeric treatment. The most notable results were perhaps those of Brunnberg and of Delius. Brunnberg (1893) succeeded in curing fourteen out of twenty-six cases of menstrual problems which he treated by hypnotic methods, and Delius in 1897 reported only two total failures in over thirty cases (a number which had increased to sixty by 1905).[80] Occasionally serious cases of haemorrhage were treated by hypnotism with apparent success. Bérillon reports a case of uterine haemorrhage so severe as to threaten life in which he succeeded in stopping the bleeding by hypnotic suggestion, and Wetterstrand has three cases of the hypnotic control of bleeding (one uterine) which he describes as "astonishing".[81]

Disorders of lactation, especially shortage of milk, were also sometimes treated with success by hypnotic methods – in fact, Braid describes what might be called an experimental instance.[82] Grossmann, Freud and Osgood Mason each report cases of some interest, and give some details of the suggestions administered.[83]

Headaches and rheumatism

Quite a few authors reported success in the hypnotic treatment of headaches, muscular rheumatism, rheumatoid arthritis, gout, and various forms of neuralgia.[84] Sometimes the claimed success rates were very high. Out of twenty-three cases of muscular rheumatism and rheumatoid arthritis, Sanchez Herrero claims to have cured twenty-two, and improved the remaining one.[85] The most interesting cases here are probably those of trigeminal neuralgia, a most distressing complaint with characteristics which make it generally easy to diagnose. It is, however, often very difficult to treat, and even modern treatments (carbamazepine, surgery) are liable to have unpleasant side effects. Yet some hypnotherapists succeeded (as certain mesmerists had done in earlier times[86]) in curing quite a high percentage even of refractory cases. Thus in 1893, van Renterghem and van Eeden reported cures in eight out of fourteen cases and in 1897 Woods reported recovery with twelve out of thirteen cases and considerable improvement in the remaining one. Several cases were of from two to five years' standing, and two patients had had teeth extracted in the hope of a cure. It would not surprise me if those hypnotists (like the ones just mentioned) who adopted what might be called quasi-mesmeric methods of hypnotizing achieved a greater success rate with this complaint than did those who followed Bernheim's methods more closely, though one should note here that Delboeuf claims to have cured a case of severe facial neuralgia of fifteen years standing by the remarkable expedient of pulling the (unhypnotized) patient's beard and affirming that he now felt and would feel no more pain[87]

Alcoholism and other addictions

Many authors of this period claim to have found hypnotism of considerable value in the treament of alcoholism and dipsomania, conditions of which it is notoriously difficult to obtain a lasting cure. Among the earliest to publish case reports were Voisin, Ladame, Forel, Wetterstrand and Fontan and Ségard.[88] In Britain, Milne Bramwell and Lloyd Tuckey made something of a speciality of the hypnotic treatment of alcoholism. By 1903, Bramwell had treated by hypnotic suggestion seventy-six cases of alcoholism and dipsomania. Of these he rated twenty-eight as recovered, meaning that they had become and remained total abstainers, thirty-six as improved, and twelve as failures.[89] By the end of 1898, Lloyd Tuckey had treated ninety-three cases, of whom twenty-five were cured in that they had remained abstainers for a minimum of two years, and about half of the remainder had improved, but relapsed before the year was up.[90] Lloyd Tuckey observes that his success rate cannot compare with Voisin's. However, all other figures pale into insignificance beside the claim of Dr. A. A. Tokarsky of Moscow, who states that he has had an 80% cure rate (abstinence for at least a year) among more than 700 alcoholics, most of whom during attacks drank almost a litre of vodka a day.[91] Lloyd Tuckey's method of treatment tended towards the heroic. Weaker, or backsliding, brethren and sisters, especially if they became somnambulic, would be told that they would vomit if they touched alcohol, and then made to drink some.[92]

With morphinism and kindred addictions the results reported were, on the whole, not as striking as with alcoholism. Bramwell states that his results have been less than satisfactory, and he has only two successfully treated cases to report.[93] The best results were those of Dizard, Wetterstrand and Auguste Voisin.[94] Using suggestive methods, Wetterstrand cured thirty-seven out of fifty-one cases of drug addiction (forty-one of pure morphinism), many of long standing (up to twenty years). In some cases, he was able to cure the morphinism by hypnotic treatment of the condition (usually chronic pain) that first led to the adoption of the morphine habit. He did not use his method of protracted sleep for treatment of these addictions, but none the less emphasizes that hypnotic sleep *per se* has a markedly beneficial effect.[95] Cures of individual cases were reported by other workers, for instance Marot and Bérillon.[96]

Organic disorders

We come finally to instances of the allegedly successful hypnotic treatment of organic or physical complaints (insofar as these can be meaningfully separated from psychosomatic ones). The earlier hypnotists were, on the whole, rather more willing to tackle these than were the later ones, but there were always some who were prepared to take them on.[97] Thus Stadelmann describes his hypnotic treatment of a case of cancer of the breast in a woman of fifty-three. He obtained marked regression of the tumours. He ceased treatment, and two months later the disease process resumed. Circumstances prevented him from undertaking hypnotism as often as he wished. The patient died.[98] Fontan and Ségard described a case of multiple sclerosis, confirmed at autopsy, in which the hypnotic treatment greatly alleviated the symptoms (the patient died of tuberculosis).[99] However, this is a disease that not infrequently intermits before advancing again.[100] Occasional reports of cases of improvement in vision and hearing are of very doubtful interpretation.[101]

Rather more interesting were certain cases in which hypnotism was used in attempts to relieve post-ictal symptoms, especially paralyses, and to treat epilepsy. There was a surprising number of examples of apparently successful hypnotic treatment of paralyses following a stroke, and in some cases the paralysis was well-established and the amelioration rapid.[102] Bernheim's explanation of one of his own successes (a particularly severe case) is of some interest:

The functional trouble in diseases of the nervous centres often exceeds the field of the anatomical lesion; this latter is reflected by shock or dynamic irritation upon the functions of neighbouring zones. And it is against this modified dynamic state, independently of a direct material alteration, that psychotherapeutics may be all-powerful.[103]

Cases of the hypnotic treatment of the after-effects of cerebral lesions, traumata and vascular accidents are still sometimes reported.[104]

The value of hypnotic treatment in cases of epilepsy was a matter of dispute. Loewenfeld, Hirt, Crocq, Hirschlaff and De Jong think that little can be hoped from it, and van Renterghem and van Eeden had only very limited success.[105] On the other hand, Hilger, Bérillon, Barwise, Forel, Stadelmann, Bramwell, Voisin, Lloyd Tuckey, and Wetterstrand have examples of apparent improvement or cure even of severe cases, whilst Woods obtained cure or improvement in twelve out of fourteen cases, many of them severe and long standing.[106] Sanchez Herrero claims to have cured or greatly improved all seven of the epileptic patients whom he treated by hypnotic suggestion.[107] It will be recalled that some mesmerists reported remarkable successes in cases of ostensible epilepsy, and it may be of significance that Woods and Wetterstrand both used what might be described as quasi-mesmeric methods of hypnotizing, which seem also to have been strikingly successful in cases of trigeminal neuralgia.

The principal problem in assessing these claims is of course that of the correctness of the diagnosis of epilepsy. This problem was widely appreciated at the time. Forel, for instance, can hardly bring himself to believe that Wetterstrand's cases were epilepsy rather than hysteroepilepsy.[108] It is true that many of the case reports concerned are short on details, but it is difficult not to think that in some instances we are confronted with remarkably sudden cures of or improvements in cases of severe and hitherto intractable epilepsy. For example, in the case of a girl of nineteen who had suffered from both grand mal and petit mal attacks over a period of four years, rising in number latterly to about twenty a week, and had several times been severely burned and scalded in consequence, Bramwell reports that the fits ceased after the first induction of hypnosis. Some sporadic minor relapses were treated by further hypnosis, and the patient was left in a stable and immensely improved condition.[109]

Concluding comments

This survey of the results and alleged benefits of hypnotic treatment has been very brief, and in relation to the materials available even in the period with which we are concerned hopelessly brief. Still, such as it is, it must serve as the basis for a short discussion of the important general question of whether by and large hypnotic treatment produced the benefits that its partisans claimed for it.

Few of those who were actively involved in hypnotism around the turn of the century doubted that in many categories of complaint hypnotic treatment could have beneficial effects. Brodmann, for example, seems to think[110] that the therapeutic utility of hypnotism cannot be questioned, because hypnotic treatment is used day after day by hundreds of clinics and doctors. This argument could have been used in a different context to prove the therapeutic efficacy of blood-letting. At the other extreme, Benedikt, a writer not noted for modesty or moderation, asserts that in many (perhaps 90%) of alleged cures, particularly cures of sexual perversion, morphinism and alcoholism, the patient is merely pretending to be better, and pretending to have been hypnotized, either to please the doctor, or to escape pressure from doctor or family.[111] One wonders whether Benedikt had seriously studied the relevant literature. If someone who has been impotent or homosexual for many years, and has suffered great distress on that account, seeks and pays for treatment from an hypnotic specialist, and afterwards marries and fathers a family, it is idle to suppose that he is merely pretending to be better.[112] And when a medical man, who has been hypnotically treated for with complete success for morphine addiction, publishes an article defending the claims of his therapist, it does not seem likely that he is merely pretending to have been cured.[113] Of course, an hysterical or hypochondriacal patient might (for a while) pretend to be cured following hypnotic treatment, but on a broader view one might reasonably remark that it would be extraordinarily difficult for a patient who had been treated hypnotically (or in any other way) for melancholia, enuresis nocturna, asthma, wry-neck, trigeminal neuralgia, alcoholism, stuttering, etc., etc., to deceive his medical adviser and his family into supposing him cured when he was not.

A more moderate position than the two preceding ones was adopted by Hirschlaff,[114] who admits the therapeutic value of hypnotic treatment (he presents a table of his own results), but warns against the confounding effects of misdiagnosis, failure to follow up cases after the end of treatment, spontaneous recovery, *vis medicatrix naturae*, and the concurrent influence of other physical healing factors. These are very real possibilities, which were no doubt sometimes not sufficiently taken into account by the more optimistic tabulators of "cures". But they cannot be pushed very far. Though this period provides us with no examples of properly controlled studies in which the progress of a group of patients treated hypnotically is compared with that of a matched untreated, or differently treated, group (for that matter our own period does not present very many), it is scarcely plausible to suppose that in any large proportion of the cases the cure or improvement was not genuine and was not directly connected with the hypnotic treatment. Take, for instance, the cases tabulated by Woods (1897), a physician of obvious ability, who lists his successes as well as his failures, and shows himself well aware of the need to follow up cases after the end of treatment. Woods achieved good success rates with complaints that are often refractory, for instance alcoholism (twenty recoveries or improvements out of twenty-six cases), sciatica (six out of ten), mental disorders (twelve out of fifteen), gout (six out of seven), rheumatoid arthritis (nine out of ten), neuralgia (twenty-one out of twenty-one), and epilepsy (twelve out of fourteen). He reports the duration of the disease before treatment and the number of hypnotic sessions, and it is quite apparent that many of his cases were long-standing ones in which conventional remedies had not succeeded, but hypnotism gave relatively rapid relief.[115] The same is widely true of other case collections.

To say that in an appreciable proportion of cases cure or improvement is directly related to hypnotic treatment is, however, not necessarily to imply that it is the supposed state of hypnosis that promotes the successes. The hypnotic situation can have many aspects, the disentangling and assessment of which presents problems that have still to be totally resolved.

THE SURGICAL USES OF HYPNOTISM

The idea that hypnotism, like mesmerism, could be used to procure anaesthesia or analgesia for surgical purposes, originated with Braid, who in 1843 reported several minor operations performed upon patients seemingly rendered cataleptic and analgesic by his eye-fixation method of hypnotizing – he appears to have made no verbal suggestions of the absence of pain.[116] Hypnotic analgesia was, as we have seen, occasionally practised in France during the 1860s, the period of "braidism",[117] and in 1866 Liébeault described the occasional use of his own methods of hypnotizing to keep patients in the hypnotic sleep during tooth extractions and other minor operations.[118] Thus the tradition of painless hypnotic operations, fortified by folk-memories of the mesmerists' remarkable achievements in this sphere, passed into the hypnotism of the 1880s. Many, perhaps most, of the well-known figures in the hypnotic movement were led by medical motives, or by plain curiosity, to attempt now and again to induce hypnotic analgesia for small operations. The technique was usually to combine suggestions of sleep with suggestions of the absence of pain; though analgesia could sometimes be attained even when the subject was not very responsive to sleep-suggestions. There is no point in making a list of the hypnotists who attempted to induce hypnotic analgesia for minor surgery. A brief account of the experiments of J. Milne Bramwell will suffice.

Bramwell began to induce hypnosis for purposes of surgical anaesthesia soon after he commenced hypnotic practice. Usually someone else performed the actual surgery. He was shortly persuaded to give a demonstration in Leeds to a group of some sixty local medical men and dentists. A report appeared in *The Lancet* and the *British Medical Journal* in April 1890.[119] Various minor operations

were performed, the most numerous being tooth extractions, usually of some
difficulty. Thus a servant girl, already under hypnotic treatment, was put asleep
by a note from Bramwell, which added that she was to obey the dentist's
commands:

> Sleep was … so profound that, at the end of a lengthy operation in which
> sixteen stumps were removed, she awoke smiling, and insisted that she had
> felt no pain, and, what was remarkable, there was no pain in her mouth. She
> was found after some time, when unobserved, reading the *Graphic* in the
> waiting-room, as if nothing had happened … [120]

It does not seem likely that the patients thus demonstrated were merely
pretending not to have felt pain, but in case anyone contemplates this possibility
I quote from Bramwell the following self-report of sensations during a tooth
extraction. The reporter was himself a medical man and a hypnotist, and was,
it should be noted, only lightly influenced by the hypnotic procedures, which
consisted of passes, eye-fixation in the manner of Braid, and verbal suggestions.

> At this time probably as much as at any other, the fixing of the mind by the
> auto-suggestion "No pain" had a good effect. Sensation was much dulled,
> as proved by the fact that though the forceps were felt distinctly going into
> the mouth and gripping the tooth, their exit was hardly appreciated, only
> the fact of their absence after they were gone, this being the case on both
> occasions, as the roots (second upper molar left) were removed separately.
> Analgesia was quite complete. After the second root was removed, a
> sensation of fluid, which was known to be blood, was felt in the pharynx,
> and at the same time one was told to sit forward and wash out the mouth. [121]

In at least one case from this period a dentist, to whom the hypnotist's "power"
had been "transferred", was able to "lacerate" the tooth-pulp of an hypnotically
analgesic patient without occasioning any response. [122]

That the inhibition of pain in these cases is a real, not an apparent one, is,
Bramwell thinks, "proved by the following facts: – (1) Not only do the patients
remain passive during operation, but the physiological signs of pain are absent,
i.e. there is neither dilatation of pupil, nor increase of pulse. (2) There is no
corneal reflex. (3) Powerful applications of the faradic brush elicit no signs of
pain. (4) Shock is absent. (5) Pain on waking can be prevented by post-hypnotic
suggestion. (6) The healing process is abnormally rapid." [123]

A number of authors of this period claim to have used hypnosis with success
to ease or remove the pains of childbirth, [124] but of the use of hypnosis as an
anaesthetic or analgesic for major surgery there were only a few instances.
Chemical anaesthesia had displaced it almost completely. As Loewenfeld
observes, hypnotic anaesthesia had many disadvantages. Only a few patients are
good enough hypnotic subjects to undergo an operation under hypnosis.
Hypnotic anaesthesia is rarely so deep or so sustained that one can count on it

for a long operation. It can, however, be used to support chemical anaesthesia, permit smaller doses of anaesthetic, promote more rapid action, and so on. It may also be used to lessen the pain of incurable ailments.[125]

In the following examples of major surgery under hypnosis, hypnoanaesthesia was used in the first because of the patient's medical condition; in the remainder it seems to have been tried because the patient was a good hypnotic subject and the surgeon was curious as to the outcome.

1. *Renterghem* (1892).[126] The patient in this case was a married woman of thirty-two, suffering for many years from a complete rent in the perineum. A radical and painful operation was required, but a heart complaint made the use of chlorofrom dangerous for her. She was exceedingly afraid, and on attempting hypnotism van Renterghem could inspire in her only a feeling of peace accompanied by somnolence. He therefore concentrated upon reassuring her that this was sufficient for the suppression of pain. During the whole of the operation, which lasted three-quarters of an hour, van Renterghem sat with his left hand on her forehead, lightly pressing on her eyes, and his right hand holding both of hers together, while he incessantly assured her that she could feel almost no pain. She remained absolutely quiet, made not the slightest movement, uttered not a single cry, and when the operation was over, assured him that she had in fact felt very little pain.

2. *Mesnet* (1889).[127] The patient was a woman of twenty-five suffering from a vaginal cystocoele and associated damage. She was an excellent hypnotic subject, liable to fall into spontaneous hypnosis, and expressed a desire to be hypnotized for the operation. Her desire was acceded to. The surgeon was Mesnet's colleague Tillaux. During a slow dissection, the hypnotized patient made no movement, talked about indifferent matters, such as what she would have for dinner, and what her mother had said to her; and she wondered when she would have the operation, saying that she wished to be chloroformed. When woken up, back in her bed, she asked when she was going to be operated on. It is noteworthy that during this operation, although analgesic, she retained her sense of touch.

3. *Schmeltz* (1895).[128] The patient, a woman of twenty, suffered from a very large sarcoma of the breast. After extirpation it was found to weigh two kilos. The lady, who had been previously hypnotized only twice, was put into the somnambulic state by look, and by a few downward passes. The hypnotist, Dr. Schmeltz of Nice, was also the surgeon. During the operation, the patient laughed and talked and turned herself into more favourable positions as requested. There was no pupillary dilation or change of the pulse. The operation was carried out slowly and lasted about an hour.

4. *Schmeltz* (1896).[129] In this case the patient, a man of thirty suffering from a sarcoma of the left testicle, was hypnotized by repeated opening and closure of the eyes, accompanied by suggestions of sleep. A wound 20 cm in length was

opened up for the removal of the testicle. During the whole procedure, he made no movement and uttered no cry. "He was, in effect, a corpse."

EDUCATIONAL USES OF HYPNOTISM

There is a considerable late-nineteenth-century literature, touched on briefly by Loewenfeld,[130] on what were rather loosely referred to as the "pedagogic" applications of hypnotism. Much of this literature, however, was concerned not with pedagogy as ordinarily conceived (i.e. the science of teaching), but with the therapy of what might nowadays be referred to as "behaviour disorders" in children, disorders which were often supposed due to "hereditary degeneracy". The central figure was Edgar Bérillon, whom we have already met as a pupil of Dumontpallier at the Pitié, and as the editor of the *Revue de l'hypnotisme* and of the transactions of the two international congresses of hypnotism. Bérillon claims that he first experimented with the pedagogic uses of hypnotism in 1884 and was the first to raise them before a learned body, which he did at Nancy in 1886, at the meeting of the French Association for the Advancement of Sciences.[131] Over the next few decades he regularly gave papers at national and international conferences outlining his procedures and results, and he contributed numerous case reports and related articles to the *Revue de l'hypnotisme* and other periodicals.[132] In 1888,[133] he founded a psychotherapeutic clinic in Paris. In 1892, he joined with various other hypnotists and hypnotherapists to form the *Institut psycho-physiologique* in Paris (rue Saint-André-des-Arts, 49), which added to a neurologic and pedagogic dispensary (the heart of the establishment) a psychological laboratory, a museum, a library and a sanatorium.[134]

Bérillon outlines his principles and procedures in a number of pamphlets and articles, any one of which is much like any other. From a selection of these[135] I have assembled the following account. "Suggestive pedagogics" is applied to children in whom the ordinary procedures of education prove insufficient to repress impulsive tendencies – these tendencies being towards, for instance, kleptomania, onanism, laziness, restlessness, deceitfulness, incontinence, disobedience, chronic temper-tantrums, and nailbiting.[136] He says that the "hypnopedogogic method" applied to such cases constitutes a virtual "moral orthopaedics", and by 1898 claimed to have applied it with a high degree of success to several thousand children of both sexes.[137] Bramwell, whose success rates with such cases was likewise good, talks – less pretentiously if also less charitably – of the hypnotic treatment of "vicious and degenerate children".[138]

Bérillon enumerates five "principles of the hypno-pedagogic method".[139] These are in effect five stages through which the treatment should in each case pass, and are as follows:

1. The first stage is that of assessing the suggestibility of the subject. This Bérillon did by the chair test (his own invention) which we have already mentioned.[140] Ready responsiveness means the child is intelligent and docile, easy to instruct and educate.[141]

2. Next one should induce the state of hypnosis, or at any rate a passive state, preferably before suggestions are undertaken. Bérillon claims that eight out of ten children of six to fifteen can be put into a profound sleep, though the more hereditary neuropathic taints a child has, the harder it is to hypnotize.[142]

3. The subject being hypnotized, one must next impose moral direction on him by imperative suggestions, expressed with authority and with clarity.

4. With this imperative verbal suggestion one should by repeated manoeuvres associate a psycho-mechanical discipline. The aim of this is to create a "centre of psychic arrest" which will render him incapable of performing the forbidden act. For instance, with the chronic onanist, the child being hypnotized, "we raise his arms in the air and suggest to him that his arms are the seat of a veritable *psychic paralysis*. We assure him that when the impulse to yield to onanism manifests, the paralysis which he now suffers from will return immediately, and that he will in consequence find it materially impossible to yield to the habit."[143] In short, the devil cannot find work for immobilized hands to do. Of course, where the "vicious habit" to be overcome is one of idleness or inertia, it is a movement rather than a paralysis that will have to be imposed.

5. The subject should be woken quickly and the same phenomena obtained with his conscious participation.

Many other writers of this time likewise brought the methods of hypnotism to bear upon the reprehensible habits of these wicked children. It is to be doubted that many were self-consciously following Bérillon's lead (in fact Liébeault certainly and Voisin quite probably preceded Bérillon in undertaking hypnotic treatment of such cases) but he was generally recognized as the leader in this field. Other contributors to it included Dumontpallier, Bernheim, Bourdon, Farez, Régis, Ladame, De Jong, Bramwell, Osgood Mason, Bekhterev, Tokarsky, Schrenck-Notzing, Forel and Bauer.[144] By 1897, the topic was sufficiently commonplace for the Paris Faculty of Medicine to accept a doctoral thesis devoted to it.[145]

Many of these writers describe rapid and lasting cures of the disturbed behaviour exhibited by troublesome youngsters of both sexes. The problems involved in assessing their therapeutic claims are no different from the problems that arise in attempting to assess the therapeutic efficacy of other clinical applications of hypnotism.[146]

There were some who proposed a rôle not for hypnotism but for scientifically applied suggestion in the education of normal children, not suffering from any special behaviour disorders. For example, père Felix Thomas, in his *La suggestion: son rôle dans l'éducation* (1895), develops within a very simplistic, associationist framework of thought, a concept of suggestion which he unites with ideas as to "unconscious imitation" to reach some practical proposals as to classroom method) he reserves hypnotism for the treatment of children with *penchants mauvais* and *instincts pervers*).

It is interesting to note that van Eeden, reviewing Binet's *La suggestibilité* (1900), which describes numerous experiments on suggestibility in school-children, adopts a position quite the opposite of Thomas', and remarks: "Any one of common sense will see after perusal of these simple experiments that it is absolutely necessary to change our general principles of education, to do away as much as possible with the influence of personal authority or prestige on the side of the teacher, and to teach our children independence of judgment, and the power of using their own eyes instead of those of the master."[147]

In the end, suggestive pedagogics could annex no territory that was clearly its own, and its principal advocate, Bérillon, was left somewhat isolated and increasingly passé. His institute, however, survived, and he himself remained more or less active, until shortly before the outbreak of war in 1939.

An appendix, as it were, to the educational uses of hypnosis, was the development and dissemination of techniques of autosuggestion for mental and moral self-improvement. The main systems of this kind lie outside the period with which we are concerned. What might be described as preliminary sketches for such systems were, however, presented by Coste de Lagrave and Lévy.[148]

HYPNOTISM AND CRIME

In the second half of the 1880s, and for much of the 1890s, no aspect of hypnotism attracted greater interest, popular, medical, scientific and literary, than that of its possible adaptation to criminal ends. Why the interest, at times amounting to positive excitement, should have become so keen, is by no means obvious. The possibility that animal magnetism might be misused for criminal purposes had now and again been mooted by its enemies since the time of Bailly's secret report, or even earlier, but with little support in the way of proven cases. Nor was the sudden interest in hypnotic crime fed by any sudden access of cases in which the rôle of hypnotism in the commission of a crime had been convincingly established. The matter has been the subject of some social historical speculation.[149] No doubt the rather marked contemporary interest in sexual psychopathology had some bearing upon it. Several of the medical hypnotists who became especially involved in the problems of hypnotism and crime (for instance Schrenck-Notzing and Forel) were also well-known for work in the area of sexual pathology. Others took seriously the dark hints of such as Liégeois who apparently supposed that the use of criminal suggestion, with or without hypnosis, might become so widespread as to threaten law and order.

Liégeois even argued, on the basis of some experiments with soldiers, that hypnotism might be used to subvert the army.[150] Perhaps those who could remember the *communards* and the days of civil disturbance and mob rule found such proposals credible. Whatever the explanation, publications on hypnotism and jurisprudence were for some years so numerous that the *Index Medicus* devoted a separate sub-section to them. Books about, or largely about, the topic, appeared in France, Belgium, Holland, Germany and Italy,[151] and many general textbooks on hypnotism contained a substantial section on it,[152] Loewenfeld's being no exception. The whole area has recently been reviewed in a scholarly monograph by Laurence and Perry (1988), which brings the subject down to the present.

It is sometimes stated that the possible use of hypnotism for criminal purposes constituted yet another ground on which battles were fought between the Schools of Nancy and of the Salpêtrière, the latter adopting a sceptical position and being in this instance victorious or at any rate correct. It is true that differences between certain members of the two schools (notably Liégeois and Gilles de La Tourette) were publicly and dramatically aired at the trial of Gabrielle Bompard, who had aided her lover, and supposed hypnotist, to murder an admirer.[153] But there were disagreements within as well as between the two schools. On the Salpêtrière side, Binet and Féré took the possibilities of hypnotic crime much more seriously than did Charcot and Gilles de La Tourette, whose views are usually regarded as typical of their school.[154] Féré, indeed, claimed to have been the first to draw attention to the medico-legal importance of the subject, which he did in 1883.[155] Charcot and Gilles de La Tourette did not doubt that hypnotized persons might become the victims of crime, particularly rape. Female subjects in the lethargic stage of hypnosis are particularly exposed to sexual molestation, for in it they become, as Charcot put it, "so much lifeless matter offered to the lechery of the magnetizer."[156] Gilles de La Tourette cited several cases in support of this assertion, and also two examples of rape committed on hysterics in the state of lethargy.[157] Somnambules, Charcot thought, would be better able to resist; but "by the very fact of the somnambulism there arises often a quite special state of 'affectivity' between the hypnotizer and the hypnotized"[158] which may undermine the subject's virtue.

Somnambulic subjects, being active, might also be prompted to commit crimes themselves at the instigation of their hypnotizers. The crimes could be committed either during the hypnosis, or afterwards, as the result of a post-hypnotic suggestion. Furthermore "theoretically the 'suggestioner' can assure himself immunity by ordering the subject not to recall, on awaking, the name of the one who gave the suggestion."[159] Charcot and Gilles de La Tourette do not deny that certain somnambules might by these methods be led into crime – many subjects at the Salpêtrière had been induced to commit staged "assassinations," etc. What they deny is that a criminal hypnotist would find using

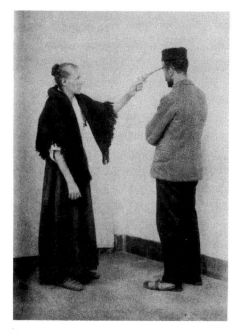

Figure 12. Hypnotic "crimes". (From the Library of the Society for Psychical Research.)

hypnotized proxies either practicable or an effective means of escaping detection. Suitable subjects would be hard to find – one must remember here that for the Salpêtrière School hypnosis was a sign of functional nervous pathology. Even when found they might prove refractory to criminal or immoral suggestions – some somnambules will positively refuse to obey commands which they do not like. Gilles de La Tourette gives an example which, in a slightly distorted form, has been frequently quoted since. Blanche Wittmann, a well-known Salpêtrière subject who, when somnambulic, was perfectly prepared to "slay" all comers at command, was told that it was very hot and that she should take a bath, along with the assembled company. Somewhat reluctantly she began to undress. She got as far as taking off her corset, at which point she showed signs of an imminent hysterical attack, thereby putting an end to the experiment. Less often mentioned is that fact that in comparable circumstances another Salpêtrière subject, Sarah R., had no hesitation in undressing and taking an imaginary bath.[160]

Even if a somnambulic subject could be induced to embark upon the commission of a suggested crime, the result would very probably be the unmasking of the guilty hypnotist. Says Charcot

> The assassin, whose crime has been contrived and planned beforehand by the suggestioner, lies in ambush, with arm raised; he strikes when the victim passes. But if the victim does not pass, what then? Will he put off the crime till the next day? By no means. The victim must be there, else, as I know very

well from repeated experiments, a fit of hysteria will in most cases be the ending of the matter. Or perhaps the subject will have an attack of acute delirium, or of babbling mania, very unfortunate for the magnetizer ... [161]

These arguments, it will be noted, depend heavily upon acceptance of Salpêtrière views about the nature of hypnotic somnambulism.

Members of the Nancy School were at first entirely convinced that hypnotic crime, with hypnotized subjects both as victims and as agents, was a serious possibility. In 1866, Liébeault gave a brief account of his own experiments on hypnotic crime,[162] and expressed himself as follows:

> I have the deep conviction ... that a somnambule who has been given the suggestion to commit bad actions upon waking, will carry them out under the influence of the fixed idea thus imposed. The most wise will become immoral, the most chaste shameless! If a prostitute has been thus forced to abandon her infamous profession why should one not pervert, for the future and by the same means, the most virtuous girl?[163]

Two decades later, Bernheim, Liégeois and Beaunis were expressing similar opinions.[164] The most extreme were those of Liégeois, who published books on the subject in 1884 and 1889, besides a number of articles.[165] Liégeois held that in certain profound hypnotic states subjects may pass into a veritable automatism. In this state, the individual, "deprived of the higher faculties of reason, of judgment, of moral liberty, of coordination of ideas and acts, of conscious and reflective will, is the plaything of a fixed idea, spontaneous or suggested".[166] This is not the ordinary hypnotic state, but a distinct "induced second condition", in which the subject resembles not a sleeping person but a waking one. He can walk, talk, act, come, go, and sustain conversations. Upon waking he is completely ignorant of all that has passed, but if he then executes a post-hypnotic suggestion he will briefly reenter the second state. If a criminal act is firmly suggested to such a subject he will carry it out unhesitatingly and, in general, at once forget about it. Female subjects could be brought by suggestion into a state of insensibility so great that they could be raped without knowing or recollecting it. In a later publication, when he was still licking his wounds after the Bompard trial, Liégeois is careful to point out that he thinks only about four per cent of subjects can be put into this "induced second condition".[167]

Most other members of the Nancy School and most of its immediate followers – for example Brullard, Forel, Moll, Sanchez Herrero, Schrenck-Notzing, Crocq, Bonjean, Dejérine – without going so far as Liégeois, held that a criminal hypnotist might, by a varying mixture of direct suggestion, induced positive and negative hallucination, and suggested amnesia, effect his evil purposes. Similar views were maintained by some prominent individuals who were not exactly followers either of Nancy or of the Salpêtrière – for instance Dumontpallier,

Bérillon, Voisin and Mesnet. The last-named published in 1894 a work somewhat luridly entitled *Outrages à la pudeur*. In it he describes, among other interesting cases, three examples of ladies who were too modest to undergo an internal examination.[168] He was able to go ahead with an examination only after first hypnotizing them. During the examination he was surprised to discover that their genitalia were now anaesthetic. A conclusion widely drawn from his findings was that since hypnosis may spontaneously render ladies insensible to the probings of the speculum, hypnotic rape is a real possibility.

Despite the dramatic pronouncements of so many *savants*, ostensible instances of hypnotic crime remained exceedingly scarce, and cases in which a direct relationship between the hypnotism and the crime could be regarded as satisfactorily established were vanishingly rare. The same small set of ostensible examples is repeated from one textbook to another, and was often supplemented with cases in which the crime was perpetrated by, or upon, individuals in a state of (non-hypnotic) "somnambulism" or "ambulatory automatism". Among the more notable hypnotic examples were the following:

The hypnotized subject as victim

1. *The case of Marguerite C.*, 1858.[169] A girl of eighteen, who had been under treatment by a magnetizer of Marseille, found herself pregnant and accused her magnetizer of having raped her while she was in the magnetic sleep.

2. *The case of Castellan*, 1865.[170] Timothée Castellan, aged twenty-five, was an unsavoury beggar and pretended deaf-mute, of the district of Var, who was given shelter by a farmer and his wife, and rewarded their kindness by carrying off their young daughter (an episode curiously reminiscent of the eighteenth century Scottish ballad of *The Gaberlunzie Man*). Upon her escape from his influence, the daughter returned home. As a result of her revelations, Castellan was accused of having influenced her by certain magnetic procedures, and after expert medical testimony that this was possible, was sentenced to twelve years hard labour. Though Castellan treated her brutally, and exercized a peculiar influence over her, the evidence that he had deliberately practised anything that could be called animal magnetism upon her was of the most tenuous.

3. *The case of the dentist, Lévy*, 1879.[171] Lévy, a dentist of Rouen, was sentenced to ten years' imprisonment for violating a girl patient in the presence of her mother, whose back was turned. He had allegedly placed her under hypnotic influence. He did not deny having engaged in sexual intercourse with her. But his only "hypnotic" manoeuvre seems to have been to get her to hold her lips against her nostrils.

4. *The case of Marie F.*, 1881.[172] A young girl from Zürich, who found herself pregnant, claimed that she had been violated on Xmas Eve by a young man who

had been accustomed to magnetize her. He magnetized her without her permission, and when she half awoke she found herself on his bed and himself on top of her. She lacked the force to repulse him or to cry out. When he noticed she had awoken he sent her more deeply to sleep. She does not seem to have reported the rape until she was six months pregnant, and despite testimony by the Swiss medical hypnotist, Ladame, that such a rape would have been possible, the case against the young man was pronounced not proved.

5. *The Czyński case*, 1894. C. L. Czyński (1858–1932) was an itinerant Polish magnetizer, and the author of various books of an occultist tendency, including one on magnetism and hypnotism published at Cracow in 1889.[173] He seems to have been a person of some plausibility, and in 1894 found himself before a court in Munich accused of having used hypnosis to fascinate and seduce the Baroness Hedwig von Zedlitz auf Luga, a thirty-year-old lady of neurotic tendencies. There was no doubt that Czyński had hypnotized the Baroness and had made passionate declarations of love to her while she was ostensibly in a hypnotic state; but she had not lost consciousness, and he had not suggested that she should be amnesic on waking. Expert witnesses disagreed as to the role which hypnosis had played in her seduction.[174] Grashey, Preyer and Schrenck-Notzing held that Czyński would not have succeeded without it. Hirt, basing himself on the Baroness's actions and utterances, thought that she had been a very willing victim. (Loewenfeld tells us that after seeing both Czyński and the Baroness, he shares Hirt's opinion.[175]) Czyński was acquitted on this charge, but was sentenced to three years' imprisonment on account of a fraudulent marriage ceremony which he went through with her.

The hypnotized subject as criminal

1. *Liébeault's case*, 1887.[176] By post-hypnotic suggestion, Liébeault induced a young lad to steal two statuettes from another patient (in on the plot). Two months later the lad was caught stealing an overcoat. He had in his possession a notebook in which he had recorded a series of crimes, including stealing the statuettes. Liébeault later discovered that the lad had been set upon this course by the hypnotic influence of another doctor, who had been present at the original experiment. He concluded that his own original suggestions had instilled a thieving impulse, and that the same might be done to many others. But, as Loewenfeld points out,[177] the original suggestion might well have failed had the subject been of good moral character.

2. *Voisin's case*, 1888.[178] Auguste Voisin was commissioned to examine the mental state of a girl of twenty, of good background, found guilty of shop-lifting. She proved to be an hysteric liable to fall into an hypnotic sleep if stared at. Interrogating her under hypnosis, Voisin discovered that her crimes had been committed at the hypnotic instigation of an accomplice. She was cured by a

month at the Salpêtrière (i.e. in Voisin's wards). In this case, as in the preceding one, we do not know enough about the girl's ordinary propensities to say what rôle, if any, hypnosis may have played in her career of crime.

3. *The case of Gabrielle Bompard*, 1890. In this case, briefly mentioned above, there was no evidence that the accused girl had been hypnotized into complicity except her own word and the fact that she was known to be a susceptible subject. She seems to have been an amoral and cunning person, and she was on trial for her life. No weight whatever can be attached to her claims that she was under hypnotic influence.

In other contemporary cases often cited – e.g. those of Chambige and of Gabrielle Fenayrou[179] – there was even less evidence that hypnotism or suggestion lay behind the crimes.

It would be going too far, perhaps, to deny that hypnotism may have been a causal factor in one or two of the cases just touched on, but there is almost nothing in any of them to push such a possibility anywhere near to certainty. That this was recognized, at any rate implicitly, at the time, is rather suggested by the great reliance placed by most of the believers in hypnotic crime on staged experiments in which deeply hypnotized subjects were persuaded, either intra- or post-hypnotically to execute all sorts of crimes or immoral acts, the settings for which had been carefully prepared beforehand by the hypnotizers. These hypnotic subjects (often encouraged by appropriate delusory beliefs) knifed, poisoned or shot numerous real or hallucinatory victims with imaginary or surrogate knives, poisons, or revolvers; they heard the explosions and saw the blood and the stiffening corpses; they stole, and secreted their loot; they lied, gave false testimony, signed legal documents in favour of the hypnotist.[180] One or two enterprising experimenters even supplied real revolvers,[181] though they were a little more cautious about real bullets. Experimental hypnotic rape was harder to arrange. But some female subjects were induced to embrace disreputable medical students or even respectable professors with a varying, and no doubt appropriate, degree of passion. Others were placed in delusory situations where they were called upon to undress, and began to do so. Such experiments were a Salpêtrière stock-in-trade, and were diversified and enlarged upon by a good many others, especially by followers of the Nancy School. Bernheim, Bérillon, Beaunis, Crocq, Pitres, Forel and above all Liégeois contributed to this entertaining literature.[182]

The strongest opposition to the views of Liégeois and the central members of the Nancy School came not from the Salpêtrière School, within which there were, as we have seen, differences of opinion on the matter, but from individuals who, in most respects, were in a broad sense themselves adherents of Nancy. The most interesting of these were Delboeuf and Bramwell.

Delboeuf tells us that he was at first convinced by Liégeois' demonstrations, believing that a somnambulic subject is absolutely at the hypnotist's mercy, but by 1889 he had changed his mind, and from then onwards was almost as vociferous in combatting Liégeois' views as the latter was in propounding them.[183] Delboeuf's position is very much that of most modern critics of such experiments. Liégeois' subjects were not idiots. They knew perfectly well that Liégeois could not permit a real crime to take place, and they could tell from the expressions and attitudes of the bystanders that nothing of the kind was in prospect. The calmness with which they executed the "crimes" confirms this view. It is a mistake to suppose that because these subjects were amnesic on awakening (or on emerging from a post-hypnotic episode) they had acted like mere automata. In carrying out their crimes they often behaved intelligently and adaptively. Somnambulic subjects, even the best, are by no means mere playthings of the hypnotizer's will.   There are plenty of examples of highly susceptible subjects none the less resisting commands that are repugnant to them. Delboeuf has an unusually interesting one of his own. He had in his employ a robust and brave country girl, J.,[184] whom he had found to be an excellent subject. J. had already shown herself prepared to use the household revolver against a real nocturnal intruder. Having surreptitiously unloaded this revolver, Delboeuf hypnotized her, and told her that his daughter and another young lady, both of whom were present, were robbers. She was to take the revolver and shoot them. J. took the revolver but refused to shoot despite all exhortations. Putting the revolver down *with precaution*, she backed away. Delboeuf attributes her refusal to shoot to the fact that he allowed her time to reflect. She thought the revolver was still loaded, as usual. Thus the situation was for her not like one of Liégeois' staged crimes.[185]

Loewenfeld is likewise of the opinion that many hypnotic somnambules retain the ability to resist not merely criminal and immoral suggestions but even harmless ones which they happen to find repugnant.[186] He cites numerous experimental instances from Delboeuf, Pitres, De Jong and others, of deeply hypnotized subjects (mainly female) who refused, despite all persuasion, to steal, to hit or alternatively kiss someone, to spill ink over an elegant dress, etc., etc.[187] A nice instance is given by Kingsbury.[188] A young lad who could be made to undergo almost any hallucination or delusion awoke instantly when commanded to kiss his sister.

Liégeois could say, and did,[189] that these subjects were not deeply enough hypnotized, or not good enough subjects, but the only substantial reason for saying so was that they rejected criminal commands, and this is the question at issue. Bramwell takes up this point. He remarks that though he has had frequent opportunities of examining hypnotic subjects, at home and abroad, he had nowhere observed more profound somnambules than among his own patients.[190] None the less, many of them would refuse to carry out suggestions which they disliked,[191] and only one eventually carried out an imaginary crime (which he undoubtedly realized was only imaginary). Both Delboeuf and his opponents have neglected a crucial experiment which he has himself often performed

> A somnambule puts a piece of sugar into her mother's teacup, while her medical man makes various absurd and untruthful assertions as to its composition. Bernheim and Liébeault assert that the subject accepted these absurd statements as true because, being hypnotized, she was unable to distinguish between truth and falsehood, while Delboeuf claims that she had sufficient sense left to know exactly what she was doing. To neither does it seem to have occurred to ask the subject during hypnosis what she thought about the matter herself. If they had done so, she would promptly have solved the difficulty and told them that while they were gravely discussing possibilities, she was quietly laughing in her sleeve at the grotesque absurdity of the whole performance.[192]

Of course Bramwell's subjects were mostly respectable members of the middle class. They were not, like Liébeault's, peasants, or like Bernheim's, poor patients in hospital wards, or like Charcot's, hysterics of dubious antecedents and morality. Hence one cannot simply assume that what held true of them held true of all other hypnotic subjects.

Perhaps none of the parties to the dispute would have totally rejected the possibility of hypnotic crime under certain circumstances and with some subjects. Loewenfeld makes some sensible remarks on the subject.[193] Suggestion is in general no force which overwhelms our intellectual and moral powers. But criminal suggestions may work not only with immoral persons but with those who lack settled moral principles. It is hard to distinguish such persons from the truly moral. Behaving morally can have many non-moral motives. Clearly the amount of moral, or prudential, resistance a suggestion may meet with will vary very much from person to person and from circumstance to circumstance. It would be rash to say that hypnotism might not, for whatever reason, sometimes tip the balance. It would be equally rash to represent hypnotic crime as a serious danger to society. Delboeuf offers the interesting propostion that a hypnotized subject will not carry out an action that he will not do in his dreams.[194] One might remark here that probably many more people dream of seducing or being seduced than of murdering or stealing.

Certain other issues, connected with hypnotism and crime, and much debated at this period, can be touched on only briefly. One is that of the possible bearings of waking suggestion upon crime. Bernheim, as one would expect from his denial that there is anything "special" about the supposed "state" of hypnosis, was particularly interested in this topic, and was prone to accord suggestion an important rôle in some crimes in which there was no question of hypnotism.[195] But it is difficult to state what is involved in this proposal with sufficient clarity to assess it satisfactorily in any given case; there is usually no hint that what might be called the psychological mechanisms of suggestion have been deliberately set in motion, and no way of sorting out the possible workings of suggestion from those of other psychological factors. These issues merge into more general questions about the rôle of suggestion in social and crowd behaviour.

A topic that aroused a fair amount of speculation at this time was the accidental or deliberate use of hypnotism to contaminate eyewitness testimony. This too was a particular interest of Bernheim's, and he was able to illustrate the possibilities by experimental demonstrations with his hospital patients.[196] Some patients could be easily persuaded that they had witnessed events which they had not, and had not witnessed or done things which they had. Bernheim went so far as to cause one of his patients to believe that she had witnessed a particularly nasty rape perpetrated by an elderly man on a little girl, both of whom were known to her. She related these "facts" in detail to a lawyer. Afterwards, Bernheim made her forget it all so completely that she was exceedingly shocked when she was told the "news" of the rape, and had to be calmed by removal of this further false memory.[197] It may be that the ethics committees of modern research institutions have their uses.

Liégeois, Bonjean and Forel were among others who reported examples of hypnotically perverted testimony.[198] Bernheim was even able to arrange demonstrations in which different patients each gave concordant false accounts.[199] What was rather noticeably lacking was any actual court case in which the hypnotic falsification of eyewitness testimony was convincingly shown to have occurred. The examples cited by Bernheim and by Liégeois are mostly of false testimony given by children, who may have been pressurized or terrorized, but had not been hypnotized or even suggestionized in the sense of being deliberately manipulated through suggestions.[200]

It is to be noted that we hear nothing, or almost nothing, at this period of the hypermnesia supposedly produced by hypnosis being utilized to help eye-witnesses of crimes to remember additional details. Nineteenth-century forensic hypnotists were too concerned about the possible hypnotic perversion of testimony to entertain this questionable proposal for a moment. It was, however, occasionally suggested that an accused person might be hypnotized, if he so desired, in the hope of enabling him to remember facts which would lead to his acquittal.[201]

SUGGESTION AND HYPNOTISM IN FOLK-PSYCHOLOGY

Investigations into the history of mesmerism and hypnotism have almost as long a history as mesmerism and hypnotism themselves. We have already noted that both the friends and the opponents of mesmerism were fond of citing supposed early precedents for the use of mesmeric methods of therapy, and of finding supposed instances of mesmeric phenomena in other times or other cultures than our own. Certain later mesmeric writers, for instance Ennemoser and Colquhoun, argued with considerable erudition that our growing knowledge of mesmeric phenomena gives us a fresh understanding of important historical episodes (particularly episodes in religious history) and of religious and magical practices among primitive peoples. Some quotations from Colquhoun will illustrate the sort of position which such writers took:

> The most remarkable instances of the apparently natural occurrence of these extraordinary states – Somnambulistic or ecstatic visions, accompanied, in many cases, with cataleptic insensibility, and the development of the faculty of *clairvoyance* – appear to have occurred among the religious mystics and fanatics of all ages – among the Eastern Brahmins and Bonzes, the Hebrew Prophets, the early Christian Saints and Martyrs, the Mahometan devotees, and the Protestant sectaries in France, Germany, England, Scotland, and America. The same phenomena, under similar circumstances, re-appeared in those remarkable occurrences which took place, towards the middle and end of the last century, at the tomb of the Abbé Paris, at St. Medard ... Similar scenes took place among the early Methodists ... and, to a certain degree, among the Scotch Covenanters; and various attempts have been subsequently made, in different countries and at different times, to renew these extravagances, particularly in Scotland and the United States, by the modern *Revivalists*. Somnambulism may thus arise, in some one or other of its various degrees or modifications, either as an idiopathic affection, or as a symptom in other disorders of the sensitive or intellectual systems. It is not at all surprising that, previous to the great discovery of MESMER, and the subsequent elucidation of the magnetic doctrine, occurrences, such as we have alluded to, should have been generally regarded as miracles ... [202]

Later in the century, many students of hypnotism likewise came to believe that studies of the psychology of hypnotism and suggestion could provide keys to many puzzling historical, and indeed contemporary, episodes, especially ones involving the irrational behaviour of groups and crowds, and the emergence of charismatic leaders, and also to many of the magical, religious and thaumaturgic practices of non-European cultures. The materials surveyed by these hypnotic writers overlapped very considerably with those which their mesmeric predecessors had disinterred. How far the hypnotists were aware of the mesmerists' prior researches I cannot say. Probably they did not, in general, bother to seek them out. If so, it is no more than just that their own scholarly tomes remain equally unread by today's gleaners in the same fields.

One might have expected that the Salpêtrière School, whose members had done so much to reinterpret past instances of possession, ecstasy, and so forth in pathological terms, and had republished or otherwise made available relevant old books and pictures, would likewise have been in the forefront of endeavours to use the latest discoveries in hypnotism and suggestion to enlighten us about other and no doubt analogous historical episodes. This did not, however, happen to any large extent. Hysteria rather than hypnotism remained the Salpêtrière's favourite historical diagnosis, and inevitably so, for on the Salpêtrière view the various hypnotic states were usually produced by specialist manipulations likely to have been hit upon by chance only somewhat rarely. To say this, of course, is not to exclude suggestion and autosuggestion as factors in shaping the hysterical phenomena. However, perhaps oddly, the only work of Salpêtrière tenor known to me that extensively investigates the historical bearings of hypnotism as opposed to hysteria – L. R. Regnier's *Hypnotisme et croyances anciennes* (1891) – denies the rôle of both hypnosis and suggestion in just those historical movements or happenings which other hypnotists were most anxious to interpret in such terms. For instance, Regnier says of cases of possession in which there are apparent in main outline the signs of an "hysterical attack":

> But in these singular aberrations of mind there is nothing of an hypnotic order. Perhaps the exorcists have sometimes induced the sleep by their gestures and thanks to the bright priestly accessories which they make shine in the eyes of the possessed. But ... that would have been involuntarily and without realising what they were doing.[203]

He seems to think that these considerations suffice to exclude all question of hypnotic influence.

It was certain writers broadly in the tradition of Nancy who made the topic of hypnotism and folk-psychology (the psychology of different nations and cultures, which had affiliations in turn to the emerging subject of crowd psychology) particularly their own. It may again have been Liébeault who provided a bridge between the old mesmerism and the new hypnotism. In 1866, he showed himself clearly aware of some of the possibilities,[204] and of course he became very influential after the foundation of the Nancy School. (Braid's brief excursion into these matters, in the shape of comments on Colquhoun's *History of Magic* (1851), seems not to have become widely known.[205]) By the end of the century, the relevant literature was of some substance, and had given rise to various popular offshoots. Loewenfeld devotes a substantial chapter to it under the title of "The significance of suggestion for the mental life of the masses".[206] He draws upon a number of writers, including Sidis, Bekhterev, Scipio Sighele, M. Friedmann, and Otto Stoll, the strongest influence being that of Stoll. Stoll (1849–1922) was professor of geography and ethnology at Zürich, and his *Suggestion und Hypnotismus in der Völkerpsychologie*, a substantial and re-markably erudite work, which first appeared in 1894, came out in a considerably enlarged edition in 1904.

The weakest point of Stoll's book (commented on at the time[207]), and that from which its other weaknesses flow, is his initial definition of suggestion, which runs as follows:

> The term "suggestion" denotes first of all nothing further than an idea, a conception, which is roused in us through various means on the part of the organic and inorganic outer world, and which now creates in us a point of departure for further thought processes, without this causal connection coming clearly before our consciousness.[208]

The problem here is that it is always possible to maintain of any chain of thought that the thinker is not aware of underlying causal links. What exactly is meant by the "causal connection" clause in the definition is left to be established in practice. Stoll's example is of the sight of a lemon arousing the idea of sourness and then salivation. This is "immediate" or "primary" suggestion. Mediate or secondary or autosuggestion is when a suggestion takes effect without a new sense-impression; for example, the image of a lemon (whether or not produced by a verbal suggestion) arouses the ideas of sourness and of salivation.

Now if the arousal of the idea of sourness and of salivation by the sight of a lemon (the subject being in some indefinite sense unaware of the causal relationships involved) is to be taken in practice as a paradigm case of suggestion, we have a dilemma. When a malign small boy disrupts a brass band by ostentatiously sucking a lemon, the players are one and all very well aware of the causal relationship between their seeing the lemon and their salivating excessively. Does their knowledge exclude the effects of sight of the lemon from the category of suggestion? Of course not – a victim of suggestion does not cease to be a victim of suggestion because he realizes as he strugglingly submits that he is one. Then what has not "come clearly to consciousness" when a "suggestion" is taking effect must be, not knowledge of the causal relationship of antecedent and consequent, but some factor (any factor?) that plays a role in the causal chain of thought and activity. But such factors (e.g. neural processes) are no doubt involved in every chain of thought and activity. Stoll appears in theory to embrace the first horn of this dilemma, and to want to exclude from the category of suggestion all cases in which the subject understands the antecedent-consequent relation; in practice he embraces the second horn, and so is able to call any idea which occasions any further idea or chain of ideas and activities a "suggestion". His propensity to do this is constrained only by an understandable preference for the bizarre, for the dramatic over the merely pedestrian.

The great weakness of his book – its catch-all definition of suggestion – leads immediately to its greatest strength. The loose definition gives Stoll an excuse to display a truly awe-inspiring erudition on a great variety of topics of his own choice. Fortifying himself with the observation that many people are susceptible to suggested sensory deceptions even in the ordinary waking state, and that certain suggestions possess an enormous capacity for running like an infection through all the members of a crowd or group, he presents us with an immense array of information concerning what might roughly be described as the collective folly of groups or the socially sanctioned and encouraged follies of individuals. All of these (being incomprehensible to a nineteenth-century rationalist) he readily attributes without further critical analysis to "suggestion". The topics he treats, in no superficial manner, include shamanism, possession by oriental fox-spirits, the powers of yogis and fakirs, possession in the New Testament, analgesia during torture, temple sleep, epidemic suggestive ecstasy, epidemic possession, the crusades and other religious wars, the flagellants, convulsive and dancing epidemics, tarantism, the witch-craze in Western Europe, the tremblers of the Cevennes, revolutionary movements (including the French Revolution), stock market crazes, and the vagaries of fashion. Any modern author who, armed with a refined and more restrictive definition of suggestion, may wish to repeat Stoll's exercize, would be well-advised to begin the venture by consulting his copious materials.

In his chapter on these issues,[209] Loewenfeld charts a fairly circumspect course, taking us neither historically nor geographically so far afield as Stoll. He begins by remarking that every collection of human beings, however assembled or united, and whether organized or not, has a certain degree of suggestibility, which is not just the average of the suggestibility of its members, but greater than that. Suggestion may not only awaken ideas and passions in a collection of people more easily than it would do in the same people separately; it may cause these individuals to behave in ways from which they would normally shrink. To what may we attribute the heightened suggestibility of people *en masse?*

Such suggestibility is in certain directions only, and is of an elective character. Only those suggestions are accepted which are in harmony with group prejudices and preconceptions. People who set off for religious and political meetings, and so forth, leave as it were, part of their personality behind them – for instance the part relating to home, family, business, etc. This narrowed personality has a narrowed mental horizon, and ideas which do not fall within it cease to be effective, leading to enhanced suggestibility. Another relevant circumstance is that participants in gatherings often arrive much preoccupied with leading figures whom they trust and enthusiastically support.

Again, most of the causes which bring together and agitate a gathering are of a kind which occasion strong emotional excitement. A strongly developed emotional excitement may put the crowd into a kind of hypnoid state, with correspondingly heightened suggestibility for ideas in harmony with the predominant mood. Furthermore, this excitement restricts the available range of psychic elements in members of the crowd, and hence diminishes their sense of personal responsibility. They accordingly develop a tendency to subordinate themselves or conform to the crowd, i.e. a tendency towards imitation. Resistance is very hard for individuals in a crowd because they receive hundreds of impressions from other members which powerfully arouse the imitative tendency.

Religious movements provide us with some very instructive examples of the phenomena of mass-suggestion. We should expect this, Loewenfeld thinks, because religious belief consists in holding for true what cannot be rationally proved. The form of the suggestive phenomena concerned varies, of course, with different sorts of religious ideas. If we take the three principal Christian churches we find that the kinds of mass-suggestions to which their followers are susceptible vary according to the differences in their teachings and usages. Catholics are liable to mass-hallucinations of the Blessed Virgin Mary, to mass-illusions relating to sacred pictures and to miraculous healings through the touching of relics and the like. To Protestantism, especially English-American, belong epidemic revivalism and the Salvation Army "with its farcical, noisy piety". The Russian-Greek Church specializes in the creation of sects with tendencies to self-torture and self-murder.

The last century has seen the appearance altogether outside the main churches of religious sects in which the phenomena of mass-suggestion have been exalted to an extraordinary pitch. Loewenfeld cites the Millerites, the followers of Konrat Maljowanny, and the followers of David Lazzarotti. Maljowannism appeared in Russia in the late 1830s, and is decribed in some detail by Bekhterev.[210] Its founder had been a drunkard in the earlier years of his life. He suffered from religious delusions and hallucinations and from a preaching mania which soon brought him numerous followers. At mass-meetings, these followers often exhibited hysterical manifestations reminiscent of those mani-fested in earlier epidemics of demonic possession. They also manifested delusions, hallucinations and preaching tendencies almost exactly mirroring those of their founder. It was as though these phenomena, which Maljowanny himself had experienced as a result of autosuggestions, were in turn suggested by him to his followers whose predispositions and excitement had put them into a receptive, hypnoid state.

The problem with Loewenfeld's treatment of these issues, as with Stoll's, is the vagueness of his concept of suggestion. In an earlier chapter of his book he made it central to his account of the "taking" (as distinct from the giving) of a suggestion that at the relevant moment there is (for whatever reason) a narrowing of the associative activity of the brain, so that ideas which would normally arouse opposing or diverging associations no longer do so, but run their course unimpeded and with corresponding force. They may in consequence lead to bizarre, unwanted or illogical results. Having come to think of suggestion in these terms, he seems, in his chapter on mass-suggestion, to adopt as an implicit rule of thumb the converse notion, namely that whenever we find instances of peculiar mental states or events, irrational behaviour, etc., suggestion or autosuggestion, that is to say the "taking" of a suggestion, must be at work. This is like saying that since shortage of vitamins makes people catch colds, anyone who has caught a cold must be suffering from shortage of vitamins. Obviously poor logic, impaired critical thought, and bizarre behaviour frequently occur in circumstances when it is quite inappropriate to talk of suggestion. Aberrations such as these are found in people who are, for instance, tired, drunk, distracted, excited, psychotic, simple-minded, etc., etc. No doubt they may also occur as a result of suggestion, explicit or implicit, but before we can demarcate the sphere of influence of suggestion in these matters we shall have to clarify our concept of suggestion very considerably.

NOTES

1 Loewenfeld (1901), pp. 465–469. Those principally involved in this were Vogt and his school. See especially Vogt (1896–7a; 1902). Vogt held that in what he called the partial systematic waking state (induced by hypnotism) the neural substrate of one complex of ideas remains awake and as it were super-activated, while all others are dormant, and that self-observation in this state may be greatly enhanced. Ach (1900) believed he had found support for this notion in the enhanced arithmetical performance of one specially good hypnotic subject. See also Farez (1902) and Regnault (1902).

2 Many writers dwelt on the supposed dangers of public or theatrical demonstrations of hypnotism. In one case, that of Ella von Salamon, an Hungarian girl of twenty-three, a hypnotized subject collapsed and died whilst giving a demonstration of clairvoyance. See Schrenck-Notzing (1895a), pp. 9–11; The first recorded death in hypnosis (1894). The relationship between the death and the hypnosis remained quite obscure. Some writers, e.g. Vogt (1897–8), cited cases in which deep hypnosis evolved into so-called hysterical somnambulism. Brügelmann (1896–7), p. 275, and Stadelmann (1895), claimed that such cases involved sexual excitement directed on the hypnotist.

3 Loewenfeld (1901), chapter 15.

4 Wetterstrand (1892–3; 1897b); Renterghem (1896–7), pp. 54–55; Binswanger (1892). In earlier times, the mesmeric sleep *per se* was often thought to have beneficial effects, and this tradition was taken over by Braid and other early hypnotists. Thus Mason (1901), pp. 195–203, describes six cases he treated from 1870 on, without suggestion, and "before the Bernheim period of suggestion had arrived."

5 Hirschlaff (1905), pp. 105–106, points out the advantages of combining prolonged hypnotic sleep with rest and rich diet treatments of the Weir Mitchell kind.

6 A. Voisin (1897b); Rifat (1887–8); Corval (1892–3), p. 199.

7 Grossmann (1892–3). Loewenfeld might have added Corval (1892–3), pp. 193–194.

8 Loewenfeld (1901), p. 355.

9 Graeter (1900); Muralt (1902).

10 Hilger (1900).

11 Riklin (1902–3).

12 Loewenfeld (1901). p. 356. This assertion might be disputed – see for example Janet (1925), I, pp. 589–600, 672–674.

13 Breuer and Freud (1893).

14 Loewenfeld (1901), p. 358.

15 Cf. Ranschburg, cited by Hirschlaff (1899–1900), p. 270.

16 Vogt (1899), p. 74. Benedikt (1894), p. 65, says that Breuer and Freud were lucky to find so many positive cases. He has found them to be much rarer.

17 Seif (1900); Hirschlaff (1905), pp. 190–192. Seif seems later to have gone through a Freudian phase.

18 Stadelmann (1896), especially pp. 139–144, 158–185, 190–199.

19 For instance Forel shows a clear grasp of the need for placebo trials in the evaluation of new drugs and other treatments. Forel (1906), 307–312; Forel (1902), 211–215.
20 For instance Hirschlaff (1899–1900), p. 325, criticizes Stadelmann (1896) on such grounds.
21 Among books or articles which contain, or are, lists or tabulations of cases or case histories we may note the following: Hirt (1890); Felkin (1890); Hirschlaff (1905); Bernheim (1891); Ringier (1891); J.F. Woods (1897); Stadelmann (1896); Tatzel (1894); Janet (1925), Vol. 1, pp. 340–360; Osgood (1890); Tuckey (1900); Bramwell (1903); Wetterstrand (1891; 1897a); Baierlacher (1889); Vogt in Forel (1895), pp. 42–48; and the items listed in the next footnote.
22 Renterghem and Eeden (1889; 1894); Renterghem (1899).
23 Renterghem (1899), p. 9.
24 Loewenfeld (1901), p. 393.
25 Grasset (1903), p. 383.
26 But Wetterstrand (1892–3 and 1897b) contains some striking successes with prolonged sleep in cases of hysteria.
27 E.g. Terrien (1902), reports on an immediate cure, by hypnotic suggestion, of a case of hysterical astasia-abasia (inability to stand or walk) which had lasted for four years.
28 Delius (1898), pp. 37–38, following Ringier, distinguishes between those hysterics in whom heterosuggestibility is stronger than autosuggestibility, and vice versa. Those in whom autosuggestibility is stronger are harder to treat by suggestive methods, especially if, as often happens, they come to believe themselves incurable.
29 Loewenfeld (1901), p. 394.
30 Schrenck-Notzing (1894a).
31 Renterghem and Eeden (1894).
32 Loewenfeld (1901), p. 394.
33 Loewenfeld (1901), p 396; Bechterew (1895), pp. 67–68.
34 Bramwell (1930), p. 246.
35 Wetterstrand, quoted by Bramwell (1930), p. 262.
36 Hirt (1890), p. 1228.
37 Bramwell (1930), p. 262.
38 Wingfield (1910), p. 148.
39 Loewenfeld (1901), p. 397.
40 The term psychosis in its modern sense did not have a wide currency at this time, though Loewenfeld uses it. Insanity, *Geisteskrankheit, Wahnsinn, folie,* were more usual terms.
41 A. Voisin (1884; 1886a; 1886b; 1889; 1897a; 1897c).
42 There is an interesting account by F.W.H. Myers of a visit to Voisin's wards at the Salpêtrière in the *Journal of the Society for Psychical Research*, 1886–7, Vol. 2, pp. 450–451.
43 A. Voisin (1897b).
44 A. Voisin (1897b), p. 382.
45 Burot (1888–9); J. Voisin (1887–8b); Répoud, cited by Bramwell (1930), p. 216.
46 De Jong (1889; 1893–4).
47 Woods (1897), p. 262.
48 Renterghem and Eeden (1893), pp. 56–57.
49 Bramwell (1930), pp. 210–216; but cf. Bramwell (1909), pp. 28–29 (four cases).
50 Farez (1898).
51 Loewenfeld (1901), p.401.
52 Forel (1893–4).
53 Loewenfeld (1901). p. 401.

54 Delius (1896–7), pp. 220–225.

55 Wetterstrand (1897a), pp. 90–94; Baierlacher (1889), pp. 53–56; Tatzel (1894); Osgood (1890); Stadelmann (1896), pp. 15, 118–128; Tuckey (1900), pp. 299–300.

56 Bernheim (1891).

57 Loewenfeld (1901), p. 399; Hirt (1893–4).

58 Brügelmann (1893–4).

59 Forel (1906), pp. 258–259.

60 Wetterstrand (1897a), pp. 80–82.

61 Tuckey (1900), pp. 302–303, 325–326.

62 Wetterstrand (1897a), pp. 82–88.

63 Tatzel (1894), pp. 64–65; Liébeault (1886–7b); Cullerre (1902); Bérillon (1893–4); Ringier (1897).

64 Rolleston (1889), p. 123; cf. above, pp. 321–322.

65 Bramwell (1930), pp. 263–266; Osgood (1894–5).

66 A.T. Myers and F.W.H. Myers (1893–4), p. 196; cf. *Journal of the Society for Psychical Research*, 1897–8, Vol. 8, pp. 7–10, 40–41, 96, 226–227; *Journal of the Society for Psychical Research*, 1899–1900, Vol. 9, pp. 100–104, 121–122; M.H. Mason (1899–1900); Sanchez Herrero (1905), pp. 116–119. Haeberlin (1902–3) reported two cases in which warts were apparently cured by autosuggestion. Bourdon (1898–9), pp. 150–151, reports the disappearance of warts, without special suggestions, in a girl under hypnotic treatment for sciatica.

67 *Apud* Haeberlin (1902–3), pp. 89–90.

68 A useful source of references to this literature at the end of the nineteenth century is the series of surveys contributed by Schrenck-Notzing to volumes 7 to 9 of the *Zeitschrift für Hypnotismus* (1898–1900)

69 Schrenck-Notzing (1892); Fuchs (1899).

70 Schrenck-Notzing (1895b), p. 38.

71 Schrenck-Notzing (1895b), p. 40.

72 Schrenck-Notzing (1895b), p. 41.

73 Schrenck-Notzing (1895b), p. 113.

74 Schrenck-Notzing (1895b). p. 302.

75 Sanchez Herrero (1905), pp. 780–781.

76 See especially Schrenck-Notzing (1895b), p. 89.

77 Wingfield (1910), p. 136.

78 Gascard (1899); A. Voisin (1887); Brunnberg (1896); Bernheim (1889b), pp. 399–404; Bernheim (1891), pp. 465–466; Bramwell (1930), pp. 196–200; Wetterstrand (1897a), pp. 106–107.

79 Corval (1892–3), pp. 258–259.

80 Delius (1896–7), pp. 225–230; Delius (1905).

81 Wetterstrand (1897a), pp. 73–75.

82 Braid (1899), pp. 334–335.

83 R.O.Mason (1901), pp. 117–118; Freud (1892–3); Grossmann (1892–3).

84 The statistical tables of Renterghem and Eeden present a fair number of cases under these headings. Cf. Bernheim (1899b), pp. 370–399; Bernheim (1891), pp. 434–437; Woods (1997); Grossmann (1894–5a); Baierlacher (1889), pp. 37–40; Tatzel (1894), pp. 43–47; Stadelmann (1895); Stadelmann (1896), *passim*.

85 Sanchez Herrero (1905), pp. 607–608.

86 Cf. above, pp. 249–251.

87 Delboeuf (1892–3), pp. 47–48.

88 A. Voisin (1887–8); Ladame (1887–8); Wetterstrand (1897a), pp. 56–61.

89 Bramwell (1930), pp. 221–229.

90 Tuckey (1892; 1902; 1904).

91 Tokarsky (1902).

92 On Tuckey's methods in general see Tuckey (1900), pp. 213–214.

93 Bramwell (1930), p. 229.

94 Wetterstrand (1895–6a); A. Voisin (1886–7b); Dizard (1893).

95 Wetterstrand was defended by a former patient, a medical man, against the charge of being too ready to accept cures without proper investigation, and of taking insufficient account of patients who relapsed. See Landgren (1893–4).

96 Marot (1892–3); Bérillon (1892–3).

97 For alleged cures or ameliorations of tuberulosis see e.g. Wetterstrand (1897a), pp. 74; Liébeault (1866), pp. 472–475; Tuckey (1900), p. 195; Sanchez Herrero (1905), pp. 593–600.

98 Stadelmann (1896), p. 100.

99 Fontan and Ségard (1887).

100 Cf. Bernheim's case, Bernheim (1891), pp. 440–444.

101 See Myers (1903), Vol. 1, p. 474; Fontan (1887); Tatzel (1894), pp. 47–50; Delboeuf (1890a). Cf. Tuckey (1900), pp. 291–292.

102 Tatzel (1894), pp. 50–55; Wetterstrand (1897a), pp. 26–28; Grossmann (1894–5); Delboeuf (1887a), p. 19; Fontan (1889).

103 Bernheim (1889b), p. 229, italics omitted. Bernheim's proposal that the area of functional disturbance exceeds the area of the lesion is confirmed by the results of brain scans.

104 Cf. e.g. Crasilneck and Hall (1970); Crasilneck and Hall (1975), pp. 203–222; Eliseo (1974); Manganiello (1986–7); Radil et al. (1988).

105 Hirt (1890), p. 228; De Jong (1891–2), p. 82; Loewenfeld (1901), p.398; Crocq (1900), p. 575; Renterghem and Eeden (1894); Hirschlaff (1905), p. 131.

106 A. Voisin (1894–5); Tuckey (1900), pp. 178, 290–291; Bérillon (1890–1), pp. 105–6; Stadelmann (1896), pp. 144–157, 186–190, 214–218; Barwise (1888), p. 120; J.F. Woods (1897), pp. 270–271; Hilger (1901); Bramwell (1930), pp. 257–262; Wetterstrand (1897a), pp. 28–30.

107 Sanchez Herrero (1905), pp. 690–695; Braid (1899), pp. 305–306.

108 Forel (1906), p. 222.

109 Bramwell (1930), pp. 257–258.

110 Brodmann (1897), p. 3.

111 Benedikt (1894), pp. 45–46, 66–67.

112 Cf. e.g. Schrenck-Notzing (1895b); Bramwell (1930), pp. 470–471; Wetterstrand (1897a), pp. 50f.

113 Landgren (1893–4).

114 Hirschlaff (1905), pp. 193–195.

115 J.F. Woods (1897).

116 Braid (1899), pp. 310–311.

117 Cf. above, pp. 287–288.

118 Liébeault (1866), pp. 383–387.

119 *The British Medical Journal*, 1890, i, 801–802.

120 *The British Medical Journal*, 1890, i, p. 802.

121 Bramwell (1909), p. 23.

122 Delboeuf (1887a), pp. 35–36.

123 Bramwell (1930), p. 107.

124 Tuckey (1900), p. 284; Auvard and Secheyron (1888); A. Voisin (1895–6); Fanton (1890–1); Delboeuf and Fraipont (1890–1); Dobrovolsky (1890–1); Schrenck-Notzing (1892–3); Dumontpallier (1891–2); Mesnet (1887–8); Kingsbury (1891), pp. 192–194 (cf. *British Medical Journal*, 1891, i, p. 460); Tatzel (1892–3); Grandchamps (1889); Cajal, reported in *The British Medical Journal*, 1889, ii, p. 1053.

125 Loewenfeld (1901), p. 404.

126 Renterghem (1892–3).

127 Mesnet (1894), pp. 140–153.

128 Schmeltz (1894–5).

129 Schmeltz (1895–6).

130 Loewenfeld (1901), p. 405.

131 Bérillon (1898), pp. 1, 8.

132 For partial lists of his articles see Bérillon (1897–8) and Bérillon (1902a).

133 The *Dictionnaire de biographie française* says 1886.

134 On Bérillon and on this establishment see Renterghem (1896–7), pp. 115–120. Photographs of the institution and its activities (supervized by Bérillon in his white apron) will be found in Bérillon and Farez, eds., (1902), pp. 190–199 and 297–300.

135 Bérillon (1887–8; 1889; 1891; 1892; 1896; 1897–8; 1898; 1902a; 1905).

136 Bérillon claims to have invented the term "onychophagia" himself. See Bérillon (1902a), p. 197.

137 Bérillon (1898), p. 9.

138 Bramwell (1930), p. 232.

139 I use mainly the version in Bérillon (1905).

140 See above, p. 427; Bérillon (1898), pp. 12–13.

141 Bérillon differed from Binet on this point.

142 Bérillon (1898), pp. 9–10.

143 Bérillon (1902a), p. 194.

144 A full list of relevant publications would consume too much space. See the articles listed by F.W.H. Myers (1903), Vol. 1, pp. 458–461; also R.O. Mason (1896) and (1901), pp. 448–455; Quackenbos (1900) and (1908), pp. 292–297. Bernheim (1886–7) was one of the earliest contributors.

145 Pigeaud (1897).

146 Hirschlaff (1899) criticizes Bérillon's "hypnotic pedagogics" as though it were in fair part a class-room oriented endeavour and therefore liable to weaken the will of perfectly ordinary school-children. Bérillon, however, makes it perfectly plain that only disturbed children require hypnotic treatment. The same error was made by "Ein Schulmann" (1888) and was repeated by Moll (1890), pp. 331–332.

147 *Proceedings of the Society for Psychical Research*, 1903, Vol. 17, p. 264.

148 Coste de Lagrave (1907–8). And cf. below, pp. 563, on P.E. Lévy.

149 Harris (1985).

150 Liégeois (1892), pp. 262–268.

151 Liégeois (1884; 1889); Lilienthal (1887); Bentivegni (1890); Schrenck-Notzing (1902a and b); Campili (1886); Crocq (1894); Gilles de La Tourette (1887); Mesnet (1894); Ladame (1888); Höfelt (1889); Boekhoudt (1890).

152 Moll (1890); Forel (1906); Felkin (1890); Bonjean (1890); Stucchi (1892); Du Prel (1889a); Crocq (1900); Grasset (1903); Björnström (1889).

153 Harris (1985).

154 Binet and Féré (1887), ch. XIV; Charcot (1890a); Gilles de La Tourette (1887).

155 Binet and Féré (1887), p. 361.

156 Charcot (1890), p. 162.

157 Gilles de La Tourette (1887), pp. 227–231, 322–344.

158 Charcot (1890), p. 162.

159 Charcot (1890), p. 162.

160 Gilles de La Tourette (1887), pp. 139–140.

161 Charcot (1890), p. 164; cf. Gilles de La Tourette (1887), pp. 374–375.

162 Liébeault (1866), p. 525.

163 Liébeault (1866), p. 525.

164 Bernheim (1891), pp. 138–164.

165 See especially Liégeois (1889), pp. 244–278.

166 Liégeois (1889), pp. 713–714.

167 Liégeois (1892), p. 236.

168 Mesnet (1894), pp. 86–98.

169 Charpignon (1860), pp. 52–54; Gilles de La Tourette (1887), pp. 90–92.

170 Gilles de La Tourette (1887), pp. 345–352 (taken from Tardieu , 1878, pp. 92ff.).

171 Gilles de La Tourette (1887), pp. 330ff.

172 Gilles de La Tourette (1887), pp. 339–344.

173 Zielinski (1968), pp. 132–133.

174 Grashey et al. (1895); Hammerschlag (1956), pp. 34–48; Schrenck-Notzing (1895a), pp. 11–14; Grossmann (1894–5b).

175 Loewenfeld (1901), p. 431.

176 Liébeault (1894–5a).

177 Loewenfeld (1901), p. 443.

178 Crocq (1900), pp. 277–278; A. Voisin (1891–2b); Ballet (1891–2)..

179 Bernheim (1891), pp. 147–148 (Fenayrou) and 152–156 (Chambige); Liégeois (1892–3) (Chambige).

180 A leading British hypnotist, Dr. G.C. Kingsbury, was actually accused in court of having exercized, through hypnotism, undue influence upon an elderly lady who had made him her executor and residuary legatee. The case was dismissed. See Tuckey (1900), pp. 368–369.

181 Cf. Forel (1906), p. 326; a photograph of such an experiment is provided by Crocq (1896).

182 See especially Liégeois (1889), pp. 130–171; Forel (1906), pp. 318–349; Höfelt (1889).

183 Delboeuf (1889), pp. 77–128; Delboeuf (1893–4; 1894; 1894–5; 1897).

184 Earlier experiments with J. are described in Delboeuf (1886a).

185 Delboeuf (1893–4), pp. 194–196.

186 Loewenfeld (1901), p. 435.

187 On this topic cf. De Jong (1893–4).

188 Kingsbury (1891), p. 136.

189 Liégeois (1889), p. 359.

190 Bramwell (1896–7c), p. 235.

191 Bramwell (1896–7b), pp. 197–202.

192 Bramwell (1896–7c), p. 236.

193 Loewenfeld (1901), pp. 444–445.

194 Delboeuf (1889), p. 112.

195 See Bernheim (1891), Lesson VIII.

196 Bernheim (1889b), pp. 165–176.

197 Bernheim (1889b), pp. 165–166.

198 Bonjean (1890), pp. 185–188.

199 Bernheim (1889a); Bernheim (1889b), pp. 169–175.

200 Liégeois (1889), pp. 493–501; Bernheim (1897), pp. 80–89.

201 Loewenfeld (1901), pp. 450–451.

202 Colquhoun (1851), Vol. 1, pp. 61–63.

203 Regnier (1891), p. 202.

204 He discusses various examples of what might be called "hysterical epidemics" at some length.

205 Braid (1852).

206 Among important, and also passing, influences on this chapter we may note Sidis (1898); Le Bon (1895); Sighele (1891a and b); Bechterew (1899; 1905); Friedmann (1901); Lehmann (1898); Rossi (1904); and Stoll (1894; 1904).

207 Hirschlaff (1899–1900), p. 258.

208 Stoll (1904), p. 3.

209 Loewenfeld (1901), pp. 470–487.

210 Bechterew (1905), pp. 79–99.

CHAPTER 23

The turn of the century. IV – States
ostensibly related to hypnosis

Loewenfeld's chapter on the theory of hypnosis is preceded by two chapters on states which he regards as related to hypnosis.[1] Since many others of his time likewise thought that hypnosis is related to these states, and since a problem so complex as that of the nature of hypnosis certainly cannot be considered in isolation, I shall follow his example.

The states concerned share with the deeper stages of hypnosis as traditionally conceived some, but not necessarily all, of the following characteristics:

1. A reduced or restricted awareness of the "outer" or "real" world, together, very often, with a growing immersion in an "inner", or fantasy, world.
2. A heightened responsiveness to ideas ("suggestions") implanted from outside or generated internally.
3. Apparent enhancement of certain psychological or physiological functions, together, sometimes, with a restriction of others.
4. Amnesia, or amnesic tendencies, on returning to the ordinary alert state, for the events of the abnormal or unusual state.
5. State-dependent memory for the events of that state and of past states of similar aetiology.

By what criteria one such "state" is to be distinguished from another is not made clear, but for present purposes the question is not crucial, and the "states" which come up for consideration may be assembled for discussion into three groups, as follows:

1. Spontaneously occurring normal or non-pathological states, sleep being the most obvious example.
2. Spontaneously occurring abnormal or pathological states, such as somnambulism, fugue and secondary personality.
3. States (pathological, non-pathological or indeterminate) artificially induced by psychological, physiological or pharmacological means; to which may be added certain states (for instance various kinds of possession state) which present apparent continuities with states of

category 2, but which are, or become, deliberately cultivated, thus qualifying for admission to category 3.

It will also be appropriate to look briefly at "animal hypnosis".

SPONTANEOUSLY OCCURRING STATES: SLEEP

A strong tradition in thinking about hypnosis held that there is a close relationship between hypnosis and sleep. Some workers asserted that it is possible without too much difficulty to transform the one into the other, a fact which seems to indicate that they are nearly akin.[2] This supposition went back at least as far as Bertrand.[3] Bernheim gives an example of the transformation of sleep into hypnosis that was frequently quoted:

> Recently I found in my hospital wards a poor phthisic woman who was asleep; I had never hypnotized her. Lightly touching her hand, I said to her: "Do not wake up. Sleep. You are continuing to sleep. You cannot wake up." After two minutes I lifted both her arms; they remained cataleptic. I left her after having told her that she would wake up at the end of three minutes. Some while after she had woken (which she did at about the time indicated) I returned to talk to her. She remembered nothing .... [4]

Various other workers claimed to have achieved the same effect.[5] Loewenfeld states that he has done so by the use of mesmeric passes.[6] The most extensive discussion of the topic was that by Vogt (1897),[7] to which I shall return when discussing his theory of hypnosis. Interest in the phenomenon, or alleged phenomenon, seems largely in abeyance at the present time.

Sleep shares certain obvious characteristics with hypnosis as traditionally conceived[8] – reduced awareness of the outer world, immersion in an "inner" world, enhancement of some psychological functions and restriction of others, subsequent amnesia for events of the sleep state. It is not generally thought to be a state characterized by responsiveness to suggestion; but in fact occasional instances of apparently successful suggestions during sleep are to be found right through the period with which we are concerned. The idea that the contents of dreams might be influenced by suggestions or by suitable external stimulation was an old one,[9] and has been abundantly confirmed. Rather more interesting were the claims by a number of workers around the turn of the century to have influenced their patients' waking thought and behaviour by suggestions given during sleep, thus paralleling the implanting and working out of post-hypnotic suggestions. The leading exponent of sleep-suggestion was perhaps Paul Farez, a colleague of Bérillon's at the Institut Psycho-Physiologique in Paris, who devoted various articles and a pamphlet (*De la suggestion pendant le sommeil naturel*, 1898) to the topic. Some of his case histories are curious, not to say evocative. He cites, for instance, the cautionary tale of one of his patients, a dramatist, who was surreptitiously hypnotized by an actress, and became her plaything. Meanwhile, his wife (or so he alleged under hypnosis) had taken to whispering suggestions to him during his slumbers. Step by step she would transform his sleep into hypnosis until he was completely at her mercy. Between his wife and his mistress the wretched man, until rescued by Farez, had hardly a part of his mind to call his own.[10] Wetterstrand, on the other hand, was in one instance able to turn sleep-suggestions by a patient's wife to good account. The husband suffered from facial neuralgia, but proved refractory to hypnotism. So Wetterstrand got the wife to suggest to him during his sleep that he would fall into hypnosis when told to do so. The stratagem worked, the patient was hypnotized, and the facial neuralgia was cured.[11]

According to Farez, the principal uses of sleep-suggestion (in his pamphlet he gives full instructions how to administer such suggestions without waking the patient) are with disturbed patients refractory to hypnotism, with persons who take a long time to hypnotize, with cowardly individuals who are afraid of hypnotism, and with children whose parents or teachers object to hypnotism. It seems in fact to have been with children suffering from the sorts of problem that Bérillon might have treated by his "hypno-pedagogical" methods[12] that sleep-suggestion found its readiest applications. The idea was taken up in the United States by Sydney Flower, LL. D., a writer on psychological topics, and editor of the *Journal of Suggestive Therapeutics*, who clearly hoped to initiate a popular movement. In his *Education During Sleep* (1898), Flower asserts that during sleep the subconscious mind is receptive, because fixed on a single idea and freed from opposing autosuggestions, while the conscious mind is in abeyance. He has himself experimented only with children, but thinks that, for example, a wife might cure her husband of drunkenness by sleep-suggestion, provided that he desires to be cured. A parent who wishes to assure himself that his sleeping child is responding to suggestion, should raise the child's arm and tell him it will remain in that position. If it does so, the sleeper's attention is fixed on the parent.[13] A few years later Farez was still speaking with undimmed optimism. Commenting on some case reports presented by Dr. A. S. Rambotis of Corfu, he remarks on the number of cases recently described by other workers, and claims: "It is definitely established that natural sleep constitutes an hypotaxis very favourable to suggestion, just as do the hypnotic sleep and the various narcoses chemically obtained."[14]

Related to sleep is the curious state of consciousness between wakefulness and sleep, characterized by a mainly spontaneous flow of visual and auditory images and hallucinations, which Mavromatis has recently christened "hypnagogia". Mavromatis asserts that such states may be influenced by auto- and hetero-suggestions.[15] There was considerable interest in the later nineteenth century in these "hypnagogic" and "hypnopompic" hallucinations, but not much attempt to relate them to hypnosis. Delboeuf, however, proposes that during this period between sleep and wakefulness the mind has great power over the body. Hypnotism is the art of prolonging this period.[16]

SPONTANEOUSLY OCCURRING ABNORMAL OR PATHOLOGICAL STATES

We come now to spontaneously occurring abnormal or pathological states. Of these, the one that has traditionally been regarded as most similar to deep hypnosis, and most closely related to it, is natural somnambulism, under which I include sleeptalking or somniloquism as well as sleepwalking. How far somnambulism shares with the supposed state of hypnosis the criterial characteristics which I listed above is debatable. Natural somnambulism is commonest among children and adolescents, and the somnambulic performances tend to be of routine and unintelligent activities such as coming downstairs or wandering around the house by well-known routes. Awareness of the outer world is certainly limited, and there is usually subsequent amnesia. Responsiveness to suggestion, however, is quite likely to be absent, and if there is an enhanced inner or fantasy life subjects tend to have forgotten it on waking up. Opportunities to test for state-dependent memory are usually lacking. Nowadays, it is widely assumed that since somnambulic episodes are known to take place in Stages 3 and 4 of NREM sleep, they are not accompanied by dreams, and cannot in any case be related to hypnosis.

Those turn of the century theorists of hypnosis who continued to think that hypnosis was closely related to somnambulism (for example Vogt and Grasset) usually had in mind more complex instances of somnambulism and somniloquism than these run of the mill childhood performances. I have already given examples of some of these more complex cases.[17] The early literature is full of such stories, in some of which we may suspect exaggeration. But if we look at, for example, Hack Tuke's *Sleep-Walking and Hypnotism* (1884) – and Hack Tuke is a most judicious writer – we still find cases that are at any rate not simple. Tuke has examples of somnambulists who (for instance) carried out intelligent and quite complex movements, sometimes continuing the purposes of the day; solved a mathematical problem and wrote down the answer; responded to outside influences; conducted conversations or monologues; sang; ate a meal; and committed murders and other violent assaults. Sometimes (though not frequently) the somnambulists could recall events of the somnambulic episode upon being woken, and sometimes they reported relevant dreams of a kind which makes it seem unlikely that they could have been in Stage 3 or Stage 4 of NREM sleep.[18] With cases such as these, the apparent resemblances to hypnosis become more numerous. We find occasional responsiveness to remarks by other persons; heightened fantasy, as in the vivid accompanying dreams; and heightened faculty, as in the solution of refractory problems from Euclid. In one of the cases cited by Tuke (from Mesnet[19]) there were even signs of state-dependent memory. A woman who made unsuccessful suicide attempts when somnambulic, and had no recollection of them when awake, apparently learned from her failures and changed her techniques accordingly. From other sources, we can find occasional more straightforward instances of apparent state-dependent memory in somnambules, the classic situation being that of the somnambule who hides objects when sleepwalking, forgets about them on waking up, but remembers again when next sleepwalking. Thus Carpenter cites two cases from his own experience,[20] the shorter of which concerns a sleepwalking servant girl, who missed one of her combs and charged a fellow-servant with having appropriated it. One morning, however, she woke up with the missing comb in her hand. In the case of florid somnambulism and somniloquism reported by Burrell (1904), objects hidden during somnambulism were retrieved during later somnambulic episodes.[21] Moll cites from Macario the case of a girl who was outraged during an attack of spontaneous somnambulism, but knew nothing of it when she awoke, and only remembered and told her mother when she next had an attack.[22]

Other considerations made it easy to believe that there are not just marked resemblances between hypnosis and somnambulism, but a fundamental relationship strongly suggesting an identity of underlying mechanism. For example, some writers, such as Vogt,[23] asserted on the basis of their own experiences, that individuals with a history of sleepwalking are likely to make good hypnotic subjects (one or two more recent studies have supported this statement[24]). There was some evidence that hypnotism can be used to tap ordinarily inaccessible memories of somnambulic states. Thus Dr. J. F. C. Dufay, of Blois, reported in the *Revue scientifique* for 1885 the case of a "respectable servant girl" who was accused of having stolen certain missing objects, could give no account of them, and ended up in prison. Dufay, who was the prison doctor, recognized her as someone known to him as a good hypnotic subject, discovered that she had been sleepwalking in her dormitory, hypnotized her himself, and found out what had become of the missing valuables. During somnambulism the girl had for safety put them in a different place – fetching or moving objects is not uncommon among sleepwalkers.[25] A striking case of hypnotic recollection of the contents of somnambulic episodes was presented by Bekhterev to the Neuropathological and Psychiatric Society at Kasan in 1892. The patient made nightly excursions in her natural sleep and sometimes hurt herself, but on awakening she could not remember what had taken place. In hypnosis, however, she could give a detailed account of her nocturnal peregrinations.[26] Vogt mentions a case in which, by the use of hypnosis, he recovered the otherwise inaccessible memory of a "somnambulic dream", i.e. a presumed dream-state accompanied by movements, somniloquism, etc.[27] He also observed a number of instances in which such "somnambulic dreams" recurred under hypnosis; their recurrence was accompanied by a resurgence of a somnambulic monoideism – i.e. it was now possible to communicate with the subjects only by catching the thread of their dreams.[28] There is furthermore some modern evidence that the contents of episodes of somniloquism can be influenced by prior hypnotic suggestion.[29] (Rather surprisingly, hypnotic cures of sleepwalking were, and are, somewhat infrequently reported.[30]) Last, and most impressively, there was a certain amount of evidence, for instance from Vogt and Mesnet, that somnambulic states, or states of "somnambulic dreaming", could readily be turned into a state of hypnosis. This usually involved the hypnotist entering into the thread of a somnambulist's conversation, which tended to be "monoideistic" or at least restricted to a very limited range of themes, and then beginning to insinuate his own suggestions. If the subject started to respond to these suggestions, a state of "hypnosis" was thought to have been initiated, and the scope of the suggestions could gradually be widened.[31] Vogt claimed that, with time, he could invariably transform a "somnambulic dream" into a state of "dreamless hypnosis". I shall cite a couple of his cases later.[32] No wonder that Vogt developed a theory of hypnosis according to which hypnosis is at root just a state of somnambulism, or more usually of somniloquent "somnambulic dreaming", in which a "rapport" (or line of communication) has been established or maintained with the hypnotist. Other writers – for instance Moll, Loewenfeld, Dessoir and Grasset[33] – were likewise of the opinion that natural somnambulism and hypnosis are very nearly related.

In an earlier chapter,[34] I mentioned Azam's proposal that cases of fugue, double consciousness, etc., are essentially developments or elaborations of ordinary somnambulism. An unbroken series of cases can be presented to bridge the gap between ordinary childhood somnambulism and instances of well-developed adult dual personality. If that is so, and if deep hypnosis ("artificial somnambulism") really is closely related to, or at root identical with, natural somnambulism, then cases of fugue, multiple personality, etc., are presumably likewise akin to hypnosis and thus in some sense or in some way hypnotic phenomena. Opinions on this point varied somewhat, but it was widely believed that there must be an important connection of some kind between hypnosis and all the other phenomena mentioned.

This view could be supported by two chief sorts of consideration:

(a) We have seen that there is a close underlying relationship between artificial (or hypnotic) somnambulism and natural somnambulism. Between the latter and cases of dual or multiple personality a series of cases can be set up between any one of which and its neighbours there are no clear-cut grounds for differentiation. Loewenfeld brings forward a good deal of relevant material.[35] He recognizes, for instance, a condition of "hysterical somnambulism", which can have various forms or degrees. In the most elementary, the patient is liable to pass into a kind of hallucinating delirium with no awareness of the outside world. More like hypnosis, and also like Vogt's somnambulic dreams, are cases in which there is a more or less explicit state of hallucination, but the patient remains sensitive to impulses from the outer world (including other individuals) provided that these influences are consonant with his train of thought. It may prove possible to induce him to carry out certain kinds of behaviour by verbal suggestion. There is subsequent amnesia. We noted above that such "delirious" states may supervene upon hypnosis in certain pathological cases. In a case of this kind studied by Charcot and Guinon,[36] the patient, a journalist who exhibited some classical hysterical symptoms, passed during hypnosis into delirious states, whose contents were not the result of direct suggestion, and even began work on a novel, of which he wrote several instalments, with continuity of material, in separate somnambulic episodes, thereby demonstrating undoubted state-dependent memory.

In yet more developed instances of hysterical somnambulism (which Loewenfeld rather vaguely describes as being "above" hypnosis, whereas simple episodes of hysterical delirium are "below" it) the patient's behaviour will only be recognized as abnormal by persons who know his ordinary demeanour well. There are quite a few cases of fugue in the literature of this period (we have already looked at several), and they vary among themselves a good deal in complexity and apparent origins. Some were precipitated by obvious stress and trauma. A case in point is that of a woman of twenty-two, described by Boeteau in 1892.[37] On 24th April 1891, she learned at a Paris hospital that she would have to undergo a serious operation. She left the hospital at 10.00 a.m., and came to herself in another Paris hospital at 6.00 a.m. on 27th April, in a worn-out condition, with no recollection of what she had done in the interval. She proved easily hypnotizable, and at once remembered many of the missing events. She had walked to Versailles seeking the nurse who had looked after her baby, and failing to find her, had walked back again. On reaching Paris, she became increasingly excited and began to be pursued by hallucinatory surgeons endeavouring to perform operations upon her. She was brought to hospital in a state of delirium, which may be compared with the hallucinating deliria of other hysterical patients of this period.

In other instances, the fugue state was without any obvious precipitating cause. In a case described by Dr. W. F. Drewry, of Petersburg, Virginia, in 1896,[38] the patient was a healthy man of fifty, successful in business and happy in his family life. During a business trip he suddenly disappeared, and was eventually presumed dead. Six months later he came suddenly to himself in a distant city in the South where he had been working for some while in a humble capacity. After returning home, and convalescing (he had lost 100 pounds in weight) he was able to resume his old business with his former industry and success.

In the absence of any obvious precipitating cause for such an episode, physicians were apt to look to the patient's heredity. Professor Proust, of the Hôtel-Dieu at Paris did so in the case of Emile X.,[39] a lawyer aged thirty-three. Proust notes that Emile was the son of an eccentric and drunken father and a nervous mother, and that he had a mentally retarded younger brother. He had been earlier treated by Luys, who found him an exceptionally easy hypnotic subject. In fact, he was such an easy subject that he was liable to fall asleep spontaneously, sometimes in embarrassing circumstances, if his gaze became fixated, if he heard a loud sound, or if he was struck by some vivid impression. But more than this: he was prone at times to lose for several days or even weeks all or nearly all memory of who he was and of his past history. When he "came to" he would have no recollection of what he had done in the interim. On several occasions it transpired that he had committed petty crimes and was in trouble with the law. He was also liable to get into debt and to mislay sums of money. Under hypnosis he regained a detailed knowledge of the events of these episodes, and was even enabled to secure the return of a substantial sum of money which he had left in an hotel room more than six months previously. In the interim he had been short of cash.

With this case we come to the border of dual or multiple personality. Emile X. had two pronouncedly different phases of personality. In one he was a responsible if neurotic lawyer; in the other a feckless wanderer given to sponging. It wanted only that the various episodes of his second state should have exhibited clear continuity of state-dependent memory, and that he should have called himself by a new name during them, for him to have become a manifest case of dual personality. Loewenfeld brings us into the realms of such cases with the example of Félida X., which was described earlier.[40] I have adopted a somewhat different route to this case than does Loewenfeld, but the upshot is the same – between on the one hand cases of hypnotic somnambulism and "somnambulic dreams", and on the other hand cases of dual personality, we can find a series of transition cases which make it very tempting to suppose that the latter are but highly elaborated and highly aberrant developments of the former. And indeed we find even in multiple personality cases some of the characteristics which I mentioned at the beginning of this chapter as outlining the traditional concept of hypnosis. There may be state-dependent memory; enhancement of certain psychological functions; if not suggestibility, at any rate a certain lability to suggestion; and amnesia when in one personality phase for the events of other phases – though as with hypnosis this amnesia may be unidirectional, in that just as a hypnotized person may remember both the events of waking life and those of past hypnoses, even though when he is awake he is amnesic for events of the hypnotic state, so personality B in a multiple personality case may remember the events of past A phases as well as those of past B phases, whilst personality A may be able to recall only the events of past A phases.[41]

(b) A second, and related, set of reasons for supposing that there must be some fundamental underlying relationship between hypnosis, somnambulism, hysterical somnambulism, fugue and multiple personality, is the way in which hypnosis runs like a connecting thread through all of them. I have already mentioned the supposed transformation of somnambulism into hypnosis and vice versa, the recovery, under hypnosis, of the memory of somnambulic episodes, and the fact, or alleged fact, that persons predisposed to natural somnambulism make good hypnotic subjects. Analogous claims were advanced with regard to the relationship between hypnosis and the other states under consideration. Individuals liable to states of fugue or of hysterical somnambulism were often good hypnotic subjects, even too good, like Emile X. mentioned above, and the events of the fugue state, of which the patient generally had little or no recollection, would in many cases be recalled under hypnosis. In cases of multiple personality, a transition between one personality or state and another could often be wrought by hypnosis, and in some examples, one personality might represent another of the personalities "in hypnosis", might even have been created, or brought to light, by hypnotizing the patient in one or other of his or her phases.[42] Certainly, all the major symptoms of hysteria, fugue and multiple personality could readily be mimicked by hypnosis.

It was, of course, to dissociationist writers influenced by Janet that the apparent analogies among, and interrelations between, the various kinds of phenomena, or alleged states, that we have been discussing, made their strongest appeal. For one and all involve states of mind the memories of which, though in some sense "existing", are not accessible to the ordinary waking consciousness, but can be retrieved only during recrudescence of the original, unusual or abnormal, state of mind, or of ones nearly related to it. It is as though the missing memories were the "property" of some stream of consciousness other than the workaday one, a stream of consciousness which had come to the fore at the time of the events and activities concerned. We have already examined in some detail the positions adopted by several of these writers. What convinced them, perhaps, more than anything else that they were right in assuming these various states to be closely related, and to be manifestations of a common, underlying stream of consciousness, were cases like Janet's Madame D.[43] This lady, thirty-four years old and hitherto healthy, was informed by a cruel practical joker that her husband was dead. In consequence, she fell into a violent attack of hysteria, from which she emerged with both a retrograde and an anterograde amnesia, the latter of which continued so that for a period of some months she was apparently unable to retain a memory of any new event, however important, but had to keep a notebook continually by her. She became a patient in the Salpêtrière. There it was discovered from her sleeptalking that she in fact retained some memory of the missing events. She was therefore hypnotized, and in the state of hypnosis proved to have full recollection.

Furthermore, by his technique of abstraction and automatic writing, Janet was able to show that the missing memories could be recovered while the subject was in the waking state. It thus seemed that a stream of consciousness which had access to the missing memories could come forward when the subject was in a state of somniloquism, in a state of hypnosis, or in a state of waking abstraction sufficiently pronounced to permit automatic writing. What can these states be but somewhat variant expressions of the same dissociated state of consciousness? It is susceptibility to dissociation – whether induced by trauma, reverie, heterosuggestion, or autosuggestion – that makes the good hypnotic subject, the hysteric, the fugue patient, and the sufferer from multiple personality disorder.

STATES ARTIFICIALLY INDUCED BY PSYCHOLOGICAL, PHYSIOLOGICAL OR PHARMACOLOGICAL MEANS

Our next category of states putatively related to hypnosis is that of states, pathological, non-pathological or indeterminate, induced by psychological, physiological or pharmacological (including toxic) means. To take the last first: I have already noted that towards the close of the century some workers experimented with the use of chloroform, ether and various narcotic agents to "soften up" subjects refractory to hypnotization. There was controversy as to whether their occasional apparent successes were themselves the result of suggestion.[44]

Of the toxins and pharmacological agents which were sometimes said to produce a state resembling hypnosis, that in most widespread use was of course alcohol. It was alleged by some that merely being drunk could enhance suggestibility. Schrenck-Notzing, for instance, experimented on this question with a young man who had never been hypnotized. This young man drank one and a half litres of Munich beer, followed by numerous glasses of Spanish and Rhenish wine. Schrenck-Notzing sent him into an apparently deep sleep by laying a hand on his forehead and closing his eyes with his fingers. He then induced arm catalepsy by appropriate suggestions. The subject was also found to be insensitive to pain and to tickling. It is remarkable that he was not totally paralytic. He awoke after about twenty minutes with amnesia, and carried out a minor post-hypnotic suggestion.[45] No attempt was made to ascertain his normal susceptibility to hypnotism, an omission all the more remarkable because Schrenck-Notzing was more attuned to methodological issues than most other hypnotists of his time.

Rather more striking, perhaps, were the apparent similarities between hypnosis (and also natural somnambulism) and certain instances of alcoholic automatism or "blackout" in which complex chains of actions, often ruled by a single auto- or heterosuggested dominant idea, might be carried out. Early studies of such episodes (which are generally followed by total amnesia) were undertaken by an American expert on alcoholism, T. D. Crothers[46] and were incorporated by James into his Lowell Lectures of 1896.[47] Crimes were not uncommonly committed during blackouts (as occasionally during sleepwalking), and Lloyd Tuckey (an expert on alcoholism) remarks that "there are plenty of cases on record where men have been hanged for murders of which those unhappy criminals had preserved no recollection."[48] Instances of state-dependent memory were (and are) sometimes reported in connection with blackouts,[49] increasing the apparent resemblance to hypnosis, though there does not seem any reason to suppose that persons undergoing such episodes are particularly susceptible to suggestion. Individuals in the throes of alcoholic hallucinosis or delirium tremens, on the other hand, may be markedly suggestible as regards the occurrence and form of their hallucinations and delusions,[50] the traditional instance being that of the nurse who sweeps up the hallucinatory insects which have been troubling the patient and throws them out of the window. The hallucinations and delusions of person delirious with fever may likewise be manipulated by suggestion.[51]

Another state which turn of the century writers sometimes compared with hypnosis was that of hashish or cannabis intoxication; the comparison had, indeed, already been made by Braid.[52] A certain amount of experimental work was carried out later in the century. In a note contributed to the *Revue philosophique* in 1886, Dr. Bonnassies describes experiments on suggested hallucinations under hashish, for some of which he was himself the subject. "I have myself tested the effects of suggestion, with regard to the sense of sight, in a very clear, though transitory, manner ... I was shown a fricassee of potatoes, and told that it was a chicken in white sauce. I saw the chicken in white sauce, cut up, and I put my finger on the drumstick, depressing the flesh."[53] A few years later, Schrenck-Notzing carried out some more extensive experiments.[54] He claimed that suggestions are realized by persons under the influence of cannabis just as they are by persons under hypnosis; that drug states can be transformed into hypnosis by the establishment of rapport and the suppression of contrary ideas; and that hypnosis educed from narcotic states is deeper than that educed from waking states. His experiments, however, are once again not very satisfactory. For example, Frau H., aged twenty-seven, took during the evening of 16th March 1890 a considerable quantity of cannabis extract and two litres of Munich beer. (Schrenck-Notzing assures us that the beer alone could not have been effective because, as a *Münchnerin*, she was well used to it.) She was induced to see hallucinatory colours in the shadow of a curtain, to see and taste various foods wrongly, and to fulfil various other minor suggestions. Attempts to make her perceive the portrait of an old man as that of a young officer, and to make her cataleptic, failed. It is almost impossible to reach any useful conclusions on the basis of experiments such as these.[55] It is true that the lady had proved refractory to hypnotism on several previous occasions, so we might conclude that the limited successes of this occasion should be ascribed to the cannabis. But we have no means of assessing what contributions might have been made by the beer, and what by the subject's knowledge that she had taken cannabis and her expectations based thereon.

Although some experimentation on the major hallucinogens, notably mescalin, had been carried out before the end of the nineteenth century, I do not know of any attempts from that period to establish their effects upon suggestibility and hypnosis. Stoll (1904) mentions the use in religious ceremonies of peyotl and other drugs by the original inhabitants of the West Indies. He speculates that there may be a suggestive element in their effects but adds that the matter requires further proof.[56]

With the rapidly growing knowledge of eastern religious and philosophical systems, it was to be expected that comparisons might be drawn between eastern meditative and related practices and western hypnosis. In fact, both Braid and Esdaile drew such comparisons, and it is somewhat surprising that serious papers on the subject are not more frequent in the later literature. The paper most frequently referred to is one by Schrenck-Notzing on "yoga sleep".[57] It is, however, of very limited interest, being mainly an account of a translation from the Sanscrit of a work on Hatha Yoga. Schrenck-Notzing draws a few obvious comparisons between the phenomena of yoga-sleep (catalepsy, suggest-ibility) and those of hypnosis, and between the methods of induction of the two states (quietness, concentration of thought, fixation, restriction of sensory input). Stoll (1904) has a chapter on the beliefs and practices of yogis and fakirs.

Occasional parallels were drawn during the nineteenth century between the state of hypnosis and the supposedly cognate states that could be produced in individuals or groups by chanting, drumming, dancing and so forth, the assumption being of course that powerful, monotonous stimulation, combined perhaps with physical exhaustion and an atmosphere of mystery and religious exaltation, with all kinds of indirect suggestion, could bring about a kind of "Braidian" hypnosis. The individuals concerned were generally shamans or priests from some non-European culture, and the groups ones engaged in religious rituals. Such practices are widespread historically and geographically. They are occasionally, though briefly, referred to by quite a few late-nineteenth-century writers on hypnotism, particular favourites being the whirling dervishes and the ghost dance of the Sioux.[58] Stoll has a good deal to say on such topics.[59] The whole issue passes into that of possession, since in many instances the aim of the dance-engendered trance is to bring about a state of possession by god, demon, ancestral spirit, etc.

Alleged spirit-possession figured prominently, of course, in the traditions of western spiritualism, and in fact the major studies of the psychology of the trance medium date from the period with which we are principally concerned.

Two related issues attracted attention. The first was that of the status of the entities which ostensibly controlled the entranced medium and spoke to persons present. Were they essentially analogous to the various personalities which manifest in multiple personality cases (the implication being, of course, that if so they could not be "spirits" independent of the medium), or had they to be regarded in some other way? The second issue was that of the relationship between the state of hypnosis and the mediumistic trance, with the associated problem of the rôle played by suggestion in the genesis of the various controlling personalities.

With regard to the first of these questions, no very clear answer was (or is) possible, because of the considerable obscurities which envelop the concepts and the phenomena of both multiple personality and control of a medium by a spirit. Many writers of the time assumed without serious examination that the latter class of phenomena is merely a subcategory of the former. Much, in fact, might depend upon which part of the range of each kind of phenomenon one scrutinized. Some multiple personality cases (for instance Louis Vivé) are most unlike cases of trance mediumship, whilst others (for instance that of Doris Fischer, described by W. F. Prince in 1915–16[60]) resemble certain instances of trance mediumship in a number of respects – in fact one of Doris's personalities claimed to be a spirit independent of her, and at various periods in her career there were manifestations of kinds traditionally associated with mediumship.

The most remarkable, and also the most-studied, trance medium of the late-nineteenth and early-twentieth centuries was Mrs. Leonora E. Piper (1857–1950) of Boston, Massachusetts, who was "discovered" for the learned world by William James,[61] and was studied extensively by members of the American and the British Societies for Psychical Research. Her gifts, whatever may have been their nature, were certainly very remarkable.[62] In her trances, Mrs. Piper both spoke and wrote (sometimes simultaneously) and copious records of her séances remain. These records were extensively analyzed by Mrs. E. M. Sidgwick, a leading member of the SPR, with a view to throwing light on the relationship of the controlling entities to each other and to the normal personality of Mrs. Piper.[63] The analysis is too detailed to be followed here, but its upshot is that there are clear indications that the controls are one and all aspects or phases of Mrs. Piper herself.

Does this mean that the controls are to be looked upon as analogous to the various personalities of multiple personality cases? Mrs. Sidgwick does not say this, but offers a more guarded opinion. She rejects the multiple personality analogy on the grounds of the very large number of personalities that would be involved, and notes that it would be liable to some of the same objections that beset the hypothesis of alien spirits (e.g. the different "personalities" share common associations and peculiarities of language). She proposes instead (and this brings us into a discussion of the second issue mentioned above, that of the relationship between hypnosis and the mediumistic trance) that Mrs. Piper's "hypnotic self" (i.e. presumably Mrs. Piper in a state of autohypnosis) succes- sively personates a number of different characters, but that there is no part of her which has assumed and permanently retains the character of each controlling spirit. The proper analogy is with hypnotic rôle-playing (*objectification des types*):

> Mrs Piper wills to go into trance with a definite idea ... that her own personality is to disappear and its place to be taken by various other spirits whose function is to converse with the sitter, advise him, and put him into communication with the spirit world ... From the deep sleep awakes a consciousness apparently conceiving Mrs. Piper's spirit as absent and itself as another spirit who is there for the purpose of conversing with the sitter. Who this other spirit is to be and also what he is to begin by saying is sometimes foreshadowed in the utterances and gestures of the going-into-trance stage, as though it had been suggested beforehand, – and it is, of course, quite likely that the normal Mrs. Piper does sometimes suggest it to herself consciously or subliminally as she goes off. There is a similar relation between the ideas of the trance proper and of the waking-stage which succeeds the deep sleep that follows the withdrawal of the controls. During this waking-stage the consciousness present again realises itself as Mrs. Piper, who conceives herself as coming back to this world from the spirit world ... [64]

This remains, I think, the most plausible account of the matter. The main snag with it is that William James found Mrs. Piper to be only a moderately good hypnotic subject.[65] The only other in-depth psychological study of a trance medium was T. Flournoy's classic *Des Indes à la planète Mars* (1900), but Flournoy says even less than Mrs. Sidgwick about the psychogenesis of the trance personalities. He talks of the dissociative effects of emotional shocks and psychic traumatisms which his medium, Hélène Smith, had undergone when young. Such shocks may lead to hypnoid states in which secondary personalities may be born and somnambulistic romances elaborated. We must also "take into account the enormous suggestibility and auto-suggestibility of mediums, which render them so sensitive to all the influences of spiritistic reunions, and are so favorable to the play of those brilliant subliminal creations in which, occasionally, the doctrinal ideas of the surrounding environment are reflected together with the latent emotional tendencies of the medium herself."[66] But he adds that "besides this general explanation how many points of detail there are which remain obscure!"

Various other mediums or psychics were subjected to psychological or psychiatric investigation at this period,[67] but their cases are not of comparable interest to those just cited.

Of greater interest for our purposes are the self reports written by two individuals, Mr. C. Hill Tout and "Mr. Le Baron", of their own progress towards mediumship in the setting of Spiritualist "home circles".[68] Mr Hill Tout,

Principal of Buckland College, Vancouver, began to participate in such circles about 1892. Some ladies present at these circles became afflicted with convulsive movements of the fingers and arms. Mr. Hill Tout felt a strong impulse to imitate these movements and occasionally gave way to it. At later sittings he began to feel kindred impulses to assume the personality of a departed spirit, and on a number of occasions yielded, acting the rôles of the characters concerned, and even feeling appropriate emotions. "After about half an hour", he says of one such occasion, "I felt a strange sensation creeping over me. I seemed to be undergoing a change of personality. I seemed to have, as it were, stepped aside, and some other intelligence was now controlling my organism. I was merely a passive spectator interested in what was being done. My second self seemed to be a mother overflowing with feelings of maternal love and solicitude for some one."[69] The nature and direction of these personations seemed very often to be determined by direct or indirect suggestion from persons present, or by other environmental events. At times, Mr. Tout was only partly aware of his actions and surroundings, and he would certainly have been regarded by contemporary writers on hypnotism as in a state of light self-induced hypnosis. He himself makes an interesting remark on his condition. "I know myself – and my susceptibility, even under normal conditions, to suggestion in all sorts of forms, not necessarily verbal, – so well that no alternative remains to me but to believe that what I did was due simply to every-day suggestion in one form and another. Building and peopling *chateaux en Espagne* was a favourite occupation of mine in my early days, and this long-practised faculty is doubtless a potent factor in all my characteristics, and probably also in those of many another full-fledged 'medium.'"[70]

All this is entirely consistent with the speculations of Mrs. Sidgwick and of Flournoy which I have just quoted. Mr. Hill Tout was apparently treading the same path as Mrs. Piper and Mlle. Smith, though he lacked the requisite abilities (whatever these may be) to progress anything like so far along it as did those ladies. It is a great pity that we do not know more of Mrs. Piper's girlhood and early imaginative life.

ANIMAL "HYPNOSIS"

Our final subject for discussion is that of animal "hypnosis". The use of scare quotes here is appropriate since it might well be doubted that the phenomenon shares with hypnosis any of the characteristics listed at the beginning of the chapter. Loewenfeld, however, along with a number of his contemporaries, was keenly interested in the topic and devotes a whole chapter to it.

Loewenfeld cites some early materials from Preyer's book on the subject,[71] going back as far as Athanasius Kircher in the seventeenth century, whose experiments involved placing hens with bound feet on the floor and drawing a chalk mark away from their beaks. The birds would remain immobile even when their feet were freed. In 1872–3, Czermak demonstrated the same phenomenon with other species of bird.[72] He would hold the birds in his hands and gently press their heads and necks on the ground. Immobility ensued. It ensued also if, while holding them, he put small objects (corks, glass balls) before their eyes. He thinks that what put them into an "hypnotic" state was staring at these objects, but he does not explain how the state was achieved without staring.

Preyer himself found that the objects and the chalk line were irrelevant, handling alone being sufficient. Seeing signs of fear in his animals, he attributed the "hypnosis" to sudden fright consequent upon being seized. He termed the state "cataplexy", and argued against Heubel, who had claimed on the basis of his work with frogs that the phenomenon was one of sleep. Preyer replied that sleep comes on gradually, whereas his animals became suddenly motionless, with eyes open. Forel objects to Preyer that the "hypnosis" can be induced in tame animals, which are not afraid of being handled, much more easily than in wild ones, which are. He follows Braid and Liébeault in comparing the state to hibernation. He reports some charming experiments on hibernation, or rather artificially induced aestivation, in dormice. He links the lethargic state of these animals to hypnosis and catalepsy.

Loewenfeld now turns to the numerous experiments conducted by Danilewsky of Kharkov on a large number of species of animal[73] – hens, guinea-pigs, snakes, crabs, lobsters, frogs, and a young crocodile. Danilewsky thinks that the crucial factors are the placing of the animal in an abnormal position, and the persistent mild but firm retaining of it in this position by the experimenter. His animal subjects were for the most part not afraid, and he attributes their immobility to suggestion. The animal understands the procedures of the experimenter as a command,which then subdues it. Danilewsky even succeeded in producing anaesthesia of up to 30 seconds duration, though it is not quite clear how he squares this with his hypothesis.

A number of investigators who worked principally upon the contact cataplexy of frogs came to different conclusions. With these animals, Gley[74] observed cessation of voluntary movement, catalepsy, slowing and even cessation of the breath, slowing of the heart, abolition of reflexes and of responsiveness to stimulation. He regarded their condition as one of hypnosis without suggestibility. Verworn,[75] on the other hand, linked the phenomenon to tonic excitation of reflex centres combined with lack of activity of the cortex, and drew comparisons with the changes in reflex responses characteristic of decerebrate animals.

Loewenfeld now comes to his own ideas and experiments. Not all investigators, he thinks, have examined the same kind of state. The kind of state obtained often depends upon the procedures of the experiment, which affect the degree of fear aroused in the animals. As for the question whether cases in which the animals show fear (Preyer's cataplexy) can be distinguished from hypnosis, we may note that fear may call out an hypnosis-like (hypnoid) state. So cataplexy is not really a state different from hypnosis, though the fact of its being accompanied by the bodily effects of fear distinguishes it from states called out by suggestive means. In both cases, however, we find anaesthesia, catalepsy and suppression of spontaneous movement.

The (supposed) effect of the gaze of snakes on small animals is probably due to terror rather than to fixation, and so are instances of Preyer's cataplexy. But perhaps fixation also has an hypnosigenic effect.

Finally, Loewenfeld comes to his own experiments with hens.[76] Not all birds are affected. Age may play a rôle, and the chalk line is not as useless as Preyer and Czermak made out. Without it, animals tended to run off as soon as released. Once the experiment has succeeded with the use of a chalk line, it will succeed in those same individuals without. Loewenfeld finds this is also true of pigeons, which require a chalk line even after long holding down. But how the line exerts its effect is obscure. That fixation of it plays as such no rôle is shown by the fact that hens hypnotized without a line leave their head and neck in the original position when their bodies are turned, just as do those hypnotized with a chalk mark. Loewenfeld thinks that the effect of the chalk mark occurs too quickly for it to be due to tiring the eyes, which is known to be hypnosigenic in man. The line most probably awakens ideas which increase distress. He thinks that seizing the animals produces a state of anxious excitation which heightens suggestibility. Holding them gives them the idea that they cannot move. So we really are dealing here with a hypnosis-like state brought about by suggestion. Loewenfeld cannot agree with Preyer's cataplexy theory because, apart from accelerated heart-beat, he observed in his animals no signs of fear. Furthermore, treatment designed to frighten them (e.g. hitting the beak) did not bring on the state.

Animal "hypnosis" has been intermittently a topic of interest during the twentieth century,[77] particularly in Eastern Europe. The interested reader should consult *Hypnosis of Men and Animals* by F. A. Völgyesi of Budapest, a pupil of Pavlov. Völgyesi (an advocate of electrical methods of hypnotizing) probably "hypnotized" a larger variety of species of animal, and at greater personal risk, than anyone else before or since. In recent years, the most popular experimental animal has been the rabbit. The most widely held viewpoint has probably been that expressed by Woodruff and Baisden (1985) who say: "Animal hypnosis, also called tonic immobility, is a misnomer for a behavioral response that functions in a large number of species as the terminal response in a series of distance-related predator-defense behaviors consisting of freezing in position when the predator is relatively far away from the prey, then fleeing as the predator approaches, fighting when the predator makes initial contact, and, finally, becoming immobile as a last defense."[78] The period of immobility has sometimes been found to be accompanied by analgesia. Also, some workers have found that the period of immobility/analgesia is truncated by naloxone and extended by morphine, the implication being that the state of immobility is controlled, at least in part, by the release of endogenous opioids.[79] This claim is of interest because similar claims have been made in connection with acupuncture analgesia,[80] and because some writers have speculated that the release of endogenous opioids may play a rôle in the shamanic trance and in the mesmeric state (or magnetic sleep) as distinct from the state of hypnosis induced by the customary verbal suggestions.[81] The further proposal has been advanced that there may be a distinction between animal hypnosis and pressure immobility, but this distinction seems by no means clear.[82]

In this chapter I have discussed various states of human beings or of animals which seemed to hypnotists of the end of last century and the beginning of this to be obviously, or probably, related to the supposed state of hypnosis. The relationship was postulated either because the state concerned shared certain leading characteristics with hypnosis, or because these states seemed in certain respects interchangeable with hypnosis, so that they could, for example, be turned into hypnosis (and vice versa), or events from them (inaccessible to waking consciousness) could be recovered during hypnosis. An examination of these "states related to hypnosis" is important to any discussion of theories of hypnosis, as was perhaps realized at the turn of the century even more clearly than today. For the study of one or other of these states may provide some clue which may enable us to make a start upon unravelling the problems of hypnosis, and any theory of hypnosis which can give some plausible account of these kindred states, and of their relationship to hypnosis, is bound, other things being equal, to have the edge over any theory which can not, or which can give an account of fewer of them. That is, of course, so long as it is agreed that there *is* a state of hypnosis whose relationship to other psychophysiological states can usefully be discussed.

NOTES

1 Cf. Moll (1890), chapter IV.
2 However Hirschlaff (1905), p. 218, thinks this argues for their difference.
3 Bertrand (1823), pp. 499–501.
4 Bernheim (1884b), p. 11.
5 E.g. Bramwell (1930), pp. 47–48; Moll (1890), p. 46; Loewenfeld (1901), p. 91.
6 Loewenfeld (1901), p. 91. Tuckey (1900), p. 43, says Eeden told him of finding a patient asleep in his consulting room. He said to him, "Don't wake, but come with me into my consulting room." The patient did so, received his treatment, and was returned to his seat in the waiting room. He finished his sleep, and expressed surprise that his turn for treatment had not yet come.
7 Vogt (1897–8).
8 For modern views see Evans (1982).

 9 E.g. Schmidt (1799–1800), Part I, p. 38.
10 Farez (1901), p. 674.
11 Wetterstrand (1897a), pp. 22–23.
12 See above pp. 492–494.
13 Cf. Hilger [n.d.], pp. 33–34.
14 Rambotis (1905–6), p. 155.
15 Mavromatis (1987).
16 Delboeuf (1889), pp. 71, 75.
17 See above pp. 260–261.
18 Cf. Arkin (1981), pp. 111–112, who has a case of somnambulism with speech, accompanied by a dream. The somnambulic actions were consistent with the dream content. He has other cases of somniloquism accompanied by relevant dreams. He also notes (p. 284), cases in which there is simple sleep-speech plus report of mentation, and yet no relation between the two. He suggests this may be evidence for multiple, concurrent streams of mentation. Spitta (1882), p. 382, has a case of somnambulism with complete recollection. Du Prel (1889b), Vol. 2, pp. 81–86, has many examples of the somnambulic state being apparently recalled in dreams, suggesting a continuity between the two.
19 Mesnet (1860), pp. 149–164.
20 Carpenter (1881), pp. 596–598.
21 Arkin (1981), p. 324.
22 Moll (1890), p. 240, from Macario (1857). Cf above, p. 260.
23 Vogt (1897–8), pp. 85–86.
24 E.g. Evans (1977), following unpublished work by J. P. Sutcliffe in 1958, finds a relation between frequency of sleepwalking as determined by a questionnaire, and hypnotizability.
25 F. W. H. Myers (1885–7a), pp. 228–229.
26 Bechterew (1906–7), p. 24.
27 Vogt (1897–8), p. 309.
28 Vogt (1897–8), pp. 91–93.
29 Cf. Arkin (1981).
30 Tuckey (1900), p. 99, has one. For a modern study see Reid *et al.* (1981).
31 Carpenter (1881, p. 596) remarks of one case of somniloquism known to him that by "a little adroitness … she might be led to talk upon almost any subject, a transition being *gradually* made from one to another, by means of leading ideas."
32 See p. 541 below.
33 Moll (1890), p. 200; Loewenfeld (1901), pp. 285–291; Dessoir (1896), pp. 13, 19, etc; Grasset (1903), p. 23, etc.
34 Chapter 17 above.
35 Loewenfeld (1901), pp. 291–301. Another useful review is that by Guinon (1892).
36 Binet (1896), pp. 65–73; Guinon (1891).
37 Boeteau (1892); *Journal of the Society for Psychical Research*, 1891–2, Vol. 5, pp. 260–261.
38 Drewry (1896); *Journal of the Society for Psychical Research*, 1899–1900, Vol. 9, 265–267.
39 Proust (1889–90).
40 Pp. 366–367.
41 Cf. above, pp. 366–367. Interesting accounts of the relationships between different secondary personalities are given by T. W. Mitchell (1912–13; 1922).
42 Loewenfeld (1901), p. 299.
43 Janet (1901), pp. 90–91; Janet (1898), pp. 116ff.
44 F. W. H. Myers (1903), Vol. 1, p. 440.
45 Schrenck-Notzing (1891).

46 Crothers (1882; 1886). Cf. Francotte (1896–7; 1897); Lentz (1897).

47 Taylor (1984), pp. 22–24.

48 Tuckey (1900). p. 89.

49 Myers (1885–7a), p. 228; Lishman (1987), p. 510.

50 Lishman (1987), p. 515.

51 Carpenter (1881), pp. 654–655, gives a curious instance.

52 Braid (1850); Braid (1899), pp. 321–322. Also by Richet (see above pp. 299–300).

53 Bonnassies (1886), pp. 673–674.

54 Schrenck-Notzing (1891).

55 Schrenck-Notzing (1891), pp. 59–62.

56 Stoll (1904), pp. 140–141.

57 Schrenck-Notzing (1894–5).

58 Moll (1890), p. 32; Tuckey (1900), p. 14; Kingsbury (1891), p. 111; Hickmet (of Constantinople), *apud* Regnault (1902), p. 95.

59 Especially Stoll (1904), pp. 28–35 (on Siberian shamanism) and 103–105 (on kindred practices among Polynesian and Australian tribes).

60 W. F. Prince (1915–16).

61 James (1889–90).

62 Kenny (1986), chapter 3, seems to think that Mrs. Piper relied for her information upon knowledge of a wide range of likely sitters from a certain milieu; but this seems to me a hopelessly over-simple view. His account of her is useful as it publishes for the first time the names of certain leading sitters hitherto disguised under pseudonyms.

63 E. M. Sidgwick (1915).

64 E. M. Sidgwick (1915), pp. 327–328.

65 James (1889–90), pp. 653–654.

66 Flournoy (1963), p. 443.

67 E.g. Forel (1891).

68 Tout (1895); "Le Baron" (1896–7). "M. Le Baron" was H. G. Waters.

69 Tout (1895), p. 310.

70 Tout (1895), p. 315.

71 Preyer (1878).

72 Czermak (1873).

73 Danilewsky (1890).

74 Gley (1895; 1896).

75 Verworn (1898).

76 Loewenfeld (1901), pp. 306–307.

77 Maser and Gallup (1977).

78 Woodruff and Baisden (1985).

79 E.g. Carli (1978); Fanselow and Sigmundi (1986).

80 On acupuncture and endorphins in general see Pomeranz (1982).

81 See e.g. R. Prince (1982).

82 Carli *et al.* (1984).

CHAPTER 24

The turn of the century. V – Theories of hypnosis

We have looked in some detail at the phenomena of hypnotism as these presented themselves to interested students in the early years of this century, and also at certain phenomena presumed to be closely related to them. It remains for us to examine some of the then current theories of hypnosis and kindred phenomena. The field here is fairly large – larger than it would be today – and we shall have to look for ways of dividing it up. Common subdivisions have been into theories which regard hypnosis as a form of sleep versus those that do not, and into attention-theories, suggestion-theories and dissociation-theories. However, the classification which seems to me most successfully to capture the aims and outlook of the various theorists is one which refers to the kind of explanatory context within which the theorists concerned sought to exhibit hypnotic and related phenomena. Some found a physiological (or physiological-psychological) context more appropriate, some a psychological (or even social psychological), and some a psychopathological. I shall consider in turn a representative theory of each of these classes. No theory belongs absolutely in one or other of these classes, for most theories are to an extent physiological, psychological and psychopathological; assignment of a theory to a class is determined by the relative emphasis which that theory places upon the physiological, psychological or psychopathological context of the phenomena. Of course, some theories (for instance that of Sidis) could with equal justice be assigned to any of these categories.

PHYSIOLOGICALLY ORIENTED THEORIES: VOGT

The class with the largest membership was that of the physiologically oriented theories.[1] Many of these sound fairly odd to modern ears. Krarup, for example, explains in detail how expansion of certain arteries causes, through mechanical pressure on certain nerves, the reflex catalepsy supposedly characteristic of hypnosis.[2] Van de Lanoitte and Pupin[3] explain hypnosis (and sleep) by supposing an amoeba-like withdrawal of the protuberances which connect one brain-cell with another. Schleich[4] invokes the glial cells of the brain, which he thinks can function as an isolation mechanism. Cerebral hyperaemia causes plasma-filling of the neuroglia, which so to speak tighten their grip upon the axons of neurones, diminishing cerebral excitation and inhibiting associative pathways. Sleep, hypnosis and somnambulism all represent a regression to an earlier stage of development, a regression brought about by this means. Preyer proposes that sustained eye fixation leads to a localized accumulation of waste products in the brain, with partial loss of activity in the regions concerned, and heightened activity elsewhere. He was also one of the first to suggest that hypnosis and double consciousness may involve a shift in the preponderance of activity from left to right hemisphere.[5] More interesting than any of these are the brief speculations by J. F. Woods,[6] which attempt to set hypnosis in the context of Hughlings Jackson's hierarchical view of the structure and functioning of the central nervous system.

Many of the physiologically oriented theories suppose that hypnosis is in some way or another linked to local or general changes in the cerebral blood supply. The theorists who attained the highest reputation amongst their contemporaries were Döllken[7] and Vogt. Vogt's theory is not fully expounded in any one place, and remains incomplete.[8] No doubt it was put aside as his interests moved away from hypnotism. What we have of it reveals a powerful and wide-ranging, if somewhat blinkered, mind, and constitutes a not unimpressive attempt at systematization.

In philosophy, Vogt, like his mentor Forel, is a materialistic monist,[9] though for psychological purposes he confines himself to psychophysical parallelism. In psychology, he is an associationist for whom the formation of associations consists in the opening up (*Bahnung*) of connections between different brain centres or functional complexes of brain cells. His most distinctive neuro-physiological doctrine was the concept of *neurokyme* (literally "nerve-wave"), a special kind of energy liberated in nerve-cells, in the first instance through sensory stimulation, and irradiating through established pathways to cortical centres. It acts there as a "functional stimulus", producing increased metabo-lism in cortical elements. If the metabolism of a given centre or location remains above a certain level, that centre will make a contribution to the individual's consciousness; if not, the mental phenomena linked to activation of that region disappear, though the region's metabolic activity may continue to have some influence on the activity of other brain regions. Any increase in local brain metabolism will be accompanied by a local increase in the supply of materials. This is achieved by localized changes in the flow of blood in capillaries, which in turn is brought about by reflex nervous action.

Vogt now deploys these notions to give an account of attention. The direction or focus of attention (he implicitly assumes) is determined by the relative balance of metabolic activity in different brain centres. Now the metabolic activity of the brain has an upper limit, so we must suppose that an increase in attention or a

Figure 13. Oskar Vogt. (From the C. and O. Vogt-Institut für Hirnforschung, Düsseldorf.)

change in the direction of attention is accompanied by an increase in metabolism in one brain region and a decrease in others. This involves the conduction of neurokyme from centres of lesser activity (this is the basis of "inhibition") to centres of greater activity.

Vogt's theory of hypnosis derives from his theory of sleep. Sleep involves a lessening of brain excitability, due to reflexly produced anaemia with consequent diminution in metabolism. There ensue a feebleness and uniformity of neural events, a slowed transmission of neurokyme and a loss of available pathways. If, however, a brain region does become excited, it may be excited to an unusual degree, for two reasons. Conduction in general may be stronger because limited to many fewer localities, and "functional stimulation" (neurokyme) becomes as it were trapped in those centres which it has reached. Contrary or conflicting ideas are thus not activated, and memory images once aroused may acquire an exceptional liveliness.[10] In these terms can be understood certain features of the mental states that are precursors to sleep, and also certain features of sleep itself.

Sleep is promoted or unleashed by reflex centres which remain active when the cortex is becoming exhausted (this exhaustion may be due to toxic by-products of metabolic processes). In accordance with the principle just enunciated, more and more neurokyme will now be conducted to the still-active centres, activating them yet further. One such centre is the reflex centre for the closing of the eyelids; another (more important) is a vaso-motor centre in the medulla,[11] excitation of which causes anaemia of the brain and hence bodily and mental sluggishness, and sleep. To function effectively, this medullary centre must first be freed from the rival activity of all concurrent centres which might rob it of incoming neurokyme. Hence monotonous or diminished stimulation, which lowers the excitability of cortical centres, is conducive to sleep.

Also conducive to sleep is the arousal of the idea of sleep, or rather (since we tend to have few recollections of actually being asleep) of the ideas of the events or scenes which commonly lead to sleep. The brain centres corresponding to these ideas acquire (through ordinary processes of simultaneous and successive association) direct or indirect pathways connecting them with the sleep centre. Hence arousal of ideas associated with sleep may cause neurokyme to be drawn first to the sleep-idea centre, and from there to the sleep-centre itself, heightening its activity.

Vogt distinguishes with Liébeault two kinds of sleep, superficial and deep, which are differentiated principally by the kinds of dreams which characterize them. The dreams of superficial sleep exhibit a diffuse dissociation of consciousness, with totally illogical associations. They have only a loose connection with the core of the dreamer's personality, may be morally dubious, do not express themselves in movement, and may be remembered upon waking. The dreams of deep sleep manifest a circumscribed systematically narrowed waking state. Most elements of consciousness are not excitable, but a smaller more or less logically coherent system shows a more or less normal excitability. The dream-content may correspondingly be more or less logical but, because of the highly circumscribed range of the ideas remaining active, quite inappropriate to the dreamer's circumstances. It retains, however, a close connection with the dreamer's personality, and frequently represents a continuing preoccupation with his daytime concerns. The moral tone is less defective, and there is amnesia on waking. The dream-content finds motor expression, in which one may distinguish three grades: expressive movements; sleeptalking (sometimes in the form of a dialogue with an hallucinated other person); and sleepwalking.[12]

The dreams of deep sleep evince a more or less full waking state in certain cortical centres, together with a deep sleep-inhibition of the rest.

There is one further important difference between the dreams of superficial and of deep sleep. It is difficult to obtain a rapport with, i.e. intelligent responses from, a dreamer in a state of superficial sleep, without waking him up. But with a dreamer who is in a state of deep sleep, a "somnambulic" dreamer, rapport may be easily established provided that one can recognize the thought-content of the dream. One may then, by cautious trial and error, so to speak enter the dream and begin to manipulate its course and the consequent behavioural manifestations.

Vogt cites several cases in which he has himself succeeded in establishing rapport with a somnambulic dreamer. In one case, the dreamer was a woman who began to cry loudly in her sleep. Vogt was summoned by her husband, and succeeded in entering into conversation with the sleeping lady. It appeared that she fancied she was still at a place in the mountains she had lately visited, and that her husband had become buried in the snow. Vogt succeeded in convincing her that he had been rescued. Then Vogt set about persuading her that he was himself, and not the neighbour for whom she took him. By various logical arguments he succeeded in this endeavour. Then (by getting her to grasp pillow and bedclothes) he induced her to admit that she was at home in bed, and that she had been dreaming. Finally, he gave her the suggestion that she would wake up when he counted to three, and would have full recollection of her dream without any accompanying distress. These suggestions were fulfilled.[13]

We have now, by a route seemingly indirect but really direct, reached the heart of Vogt's theory of hypnosis. For this last case represents for him a clear example of the transformation of a somnambulic dream into hypnosis, something which, given time, he has succeeded in achieving in all cases. And a hypnotized person is, in essence, a somnambulic dreamer with whom someone has established rapport. Most of that person's brain centres are asleep, but a connected subset of them (a "system") is still active, and the "hypnotist" has established communication with it and is able to influence it to a greater or lesser extent. The subject is in a state of "systematic partial waking", with a residual or reestablished line of communication to another person.

Of course, it is not usual to induce hypnosis by entering a subject's somnambulic dreams. Though Vogt remarks that a tendency towards somnambulic dreaming is a good prognostic of hypnotizability,[14] it would be impracticable for a hypnotist to wait by his subject's bedside watching for signs of a somnambulic dream. Hypnosis (as partial systematic waking) is usually induced in the reverse way – by lulling the subject into sleep (arousal of centres for sleep-related ideas will in turn arouse the reflex sleep-centre) while keeping certain of his brain centres (those for the idea of the hypnotist who is working upon the centres for sleep-related ideas) in a state of alertness. This will give the hypnotist a foothold, so to speak, in the mental life of the subject, a foothold which he can then enlarge and exploit.

There are, as Vogt admits, differences between natural sleep and hypnosis – for example in the extent and depth of the inhibition of cortical centres, the preceding psychophysiological "constellation", the rapidity of onset of the sleep-inhibition, and the characteristic sensations which accompany this inhibition. But these differences are only secondary, they are quantitative not qualitative, and do not outweigh the numerous similarities and signs of close relationship between the two.[15] Vogt claims, for instance, that he and others have shown that subjects in spontaneous and in suggested sleep exhibit similar series of changes in muscle tone, skin sensitivity and volume of the arm (measured plethysmographically). In both natural and suggested sleep, these first increase and then diminish again. He states also that he has hypnotized a series of patients who suffered from involuntary choreic or athetotic movements of organic origin. These movements disappeared during natural sleep – and also during suggested sleep. Individuals who have somnambulic dreams in deep natural sleep tend to have them also in suggested sleep, a fact of which Vogt gives a number of examples from his own patients, and sometimes the dream-content is the same in both cases.[16] If one accepts Vogt's theory of hypnosis, or at any rate of deep hypnosis (there are various other states which he classifies as hypnotic, and he has also an account of waking suggestibility in terms of the hyperexcitability of certain centres, the others remaining normal), it becomes, from his point of view, not too difficult to give some sort of explanation of most of the ordinary phenomena of hypnosis. In practice, he says little or nothing about many of them, but it is fairly easy to work out what he might have said. The abnormally intense general responsiveness to suggestion, and the occasional apparent heightening of mental capacities, may be attributed to heightened arousal of those brain centres still active (which will receive most of the available neurokyme). Ideas instilled by the hypnotist through exploitation of the centres open to him may have an abnormal liveliness and effectiveness and will be unconstrained by any rivals from other centres, most of which will be in a state of sleep-inhibition. Hence these ideas may in the right circumstances attain an hallucinatory vividness. Negative hallucination, the failure to perceive something that is actually there, may be explained as the suppression of the perception of a real object (e.g. a person sitting on a chair) by an hallucinatory perception of an object in the same field of view (e.g. the seat and back of the chair in which the person is sitting). Here "as the result of suggestion the centre for the physiological events paralleling the positive hallucination, in contrast to that for the physiological events paralleling the perception which has to be suppressed, shows a materially stronger excitation. Hence the neurokyme arriving at the latter centre from the periphery will be conducted away to the former."[17]

This very exiguous outline of Vogt's ideas leaves out many aspects of his thinking and does not do justice to the wide range of physiological information with which he supported his theory. The theory itself is too constrained by the rather crude elementarist and associationist psychology and physiological psychology in which he had been brought up to have much value today. But Vogt was an acute observer and clinician and some of the case histories and clinical distinctions (e.g. that between the dreams of superficial and of deep sleep) with which he supports his theory remain of interest.[18]

PSYCHOLOGICALLY ORIENTED THEORIES: BERNHEIM

I come now to theories which considered hypnosis in a mainly psychological context. Such theories were not especially numerous, partly because academic psychologists interested in hypnotism – for example Lehmann, Wundt and McDougall – tended to develop physiological theories. A theory which might be described as purely psychological is presented by Theodore Lipps.[19] But even Lipps develops a curious notion of "mental energy".

It is hard to know whether theories of hypnosis utilizing Freudian concepts should be regarded as having a psychological or a psychopathological context. The question is not of great moment because the most influential neo-Freudian approaches to hypnosis were developed after the period with which this book is principally concerned. The most important Freudian contribution to the theory of hypnosis which falls within our period is that of Sándor Ferenczi (1873–1933), of Budapest, who in 1909 published a paper developing hints dropped by Freud a few years earlier.[20] According to Ferenczi, suggestion and hypnosis are to be understood in terms of the Oedipus complex. The subject regresses, and directs upon the hypnotist the infantile libidinal attitude which he or she formerly directed upon the parent of the opposite sex. A child obeys the mother willingly because it believes it will be rewarded by her; it obeys the formidable father in order to avoid punishment. This blind obedience transfers to the hypnotist, in a manner dependent (one supposes) either upon that hypnotist's sex, or upon his or her mode of hypnotizing (frightening into submission versus lulling into receptivity). Contemporary hypnotists for the most part looked upon this theory as a myth within a myth.[21]

The most prominent turn-of-the-century theorist who placed the phenomena of hypnosis in a mainly psychological setting was of course Bernheim. Bernheim's ideas about hypnotism had undergone gradual but profound changes,[22] and his eventual position was very different from his initial one. Three key assumptions, however, changed very little.

The first was a very general and excessively simple concept of suggestion. According to this, suggestion is "the act by which an idea is introduced into the brain and accepted".[23] The second was that there is a general law of *ideo-dynamism*, or suggestibility, "the tendency of an idea accepted by the brain, whether coming from without through the senses, or evoked by the organism itself (auto-suggestion), to become active, to realize itself, to translate itself into sensation, movement, emotion, response of the organs, sensory image, illusion or hallucination, etc."[24]

The third assumption was that normally the realization of a suggested idea is "moderated by the higher reasoning faculties of the brain, which constitute the controlling power; reason struggles with the tendency to credence and cerebral automatism."[25] There are, however, certain factors which may undermine the influence of the reasoning faculties and promote the actualization of the suggested idea. In particular, there is *credivité*, which has been translated as "credence", the tendency to unreflecting belief in whatever is said, suggested, or implied. People have different natural levels of *credivité*, so that some are more suggestible than others. Other factors which suspend the proper exercising of reason, and thus heighten *credivité*, are, for instance, religious faith (especially if accompanied by strong emotions), and faith in quack medicines.

Now in the earlier phases of his theorizing, Bernheim assigned considerable importance to two further factors which he supposed to heighten *credivité* and hinder the force of reason; later he changed his mind about both of them, thus setting himself in opposition to many adherents of the movement of which he had once been a leader. These two factors were hypnosis and sleep. Whereas Liébeault had regarded hypnosis as essentially identical with sleep, Bernheim, mainly on account of his successes in producing "hypnotic" phenomena with waking subjects, separated the two, and defined hypnosis as "a peculiar psychological condition which may be artificially produced, and which, if brought into play, increases in various degrees the suggestibility – i.e. the tendency – to be influenced by an idea which is accepted and realised by the brain."[26] Sleep is just one among the phenomena that can be produced by suggestion, and it can be produced more readily in persons who are susceptible to hypnosis, but in such persons it need not be the first phenomenon produced. The other phenomena do not depend on it, and may be obtained before sleep is induced. None the less, "although sleep is not necessary for suggestion, nevertheless, it may be said to facilitate it. Whether it is natural or induced, it suppresses or attenuates intellectual initiative and concentrates cerebral activity on the phenomena of automatism. It frees the imagination of the moderating brake of the faculties of reason."[27]

The later phases of Bernheim's theorizing may best be considered in terms of his grounds for retreating from these two earlier proposals, that sleep enhances the effectiveness of suggestions, and that hypnosis is a "special state" of heightened responsiveness to suggestion. At the twelfth International Congress of Medicine, held at Moscow in August 1897, he created something of a sensation by asserting that "Il n'y a pas d'hypnotisme",[28] "there is no such thing as hypnotism". But for a convenient summary of his final views on sleep and suggestibility and on the "special state" of hypnosis, we may turn to a paper which he presented in September 1911 to the International Society for Medical Psychology and Psychotherapy.[29]

He begins this paper by discussing the older view, held by Braid and by Liébeault, that hypnosis is an artificial sleep, with subsequent amnesia. His own experience has not favoured this view. It is true that some subjects present all the appearances of sleep, and that one can provoke in them all the phenomena commonly called "hypnotic". Left to themselves, they may snore, dream, or wake up spontaneously rubbing their eyes. But others on being hypnotized will feel only a certain drowsiness, or nothing at all, and will afterwards remember everything. Nevertheless, they can be influenced, will manifest catalepsy, anaesthesia, hallucinations, therapeutic effects, and will admit that they could not resist the suggestions. Must one say, despite their denials, that they were asleep? Still other subjects will believe they are asleep if told that they are, and will remember nothing on awakening. However, they will come, go, reply to all questions; they may do so even though they fail to respond to suggestions of catalepsy, anaesthesia, and hallucination. Nothing differentiates them from a waking person, and if told during their pseudo-sleep that they are not asleep, they will reply that they are not asleep and will later remember everything. Such subjects have not slept; they have only the illusion of having slept. It is even possible retroactively to create the illusion of having slept in subjects who have not slept at all.

In short, there is no relationship between whether or not subjects have slept or seemed to sleep, and whether or not they exhibit the customary hypnotic phenomena; nor is the intensity of the hypnotic phenomena related to the depth of sleep. Suggestibility is not exalted in induced sleep; nor yet in natural sleep. If there is a "state" of hypnosis, characterized by enhanced suggestibility, it is not any kind of sleep-state.

Bernheim does not in fact deny that among the various factors or states of mind that enhance suggestibility, *certain phases* of sleep may be included. He thinks that the first phase of sleep, when consciousness still lingers, but rational control is dulled, and the last phase of sleep, before complete awakening, may favour *certain sorts* of suggestions, especially ones of, for example, hallucinations, which fulfil themselves automatically, while the subject remains passive.[30] But this is counterbalanced by the fact that other kinds of suggestion are better realized in the waking state, especially ones involving complex motor and intellectual acts, which require the active collaboration of the faculties of control. Subjects in somnambulism, who can open their eyes, walk about, converse intelligently, etc., may appear to all intents and purposes awake. Bernheim seems to think that such subjects *are* awake. Their mental processes are monopolized by a dominant idea, just as may happen to any one of us in the waking state. If, subsequently, they are amnesic for what they have been doing, it is because they suffered from the illusion of being asleep and still suffer from the illusion of having been asleep. There is nothing here to force us to assign a special status to sleep in the matter of promoting responsiveness to suggestion.

Is there a "special state" of enhanced suggestibility at all? Bernheim emphasizes that all the phenomena called hypnotic can be obtained with subjects who are in a condition of complete wakefulness, and have not been previously subjected to hypnotization. It might be objected that if waking suggestion can influence subjects so as to transform the suggested idea into actuality, waking suggestion must have brought about some special state of the brain, which is precisely the state of hypnosis. "If this definition and this conception were adopted", says Bernheim, "every cerebral effect determined by speech, by persuasion addressing itself to sentiment as much as to reason, by an impression, by an emotion, everything which influences the brain and decides it to act would be hypnotism."[31] Of course, it would all be suggestion (on Bernheim's wide definition). "But it would be an abuse of words to call it hypnotism, a word which involves the idea of sleep and the idea of a special state which would not be our ordinary state."[32]

Perhaps, then, we should narrow the concept of suggestion, and talk of suggestion only when what is evoked is an unusual idea or action which does not accord with the subject's normal state of mind, but implies a sort of dissociation of the mental processes, a disaggregation of the mental synthesis? The trouble with this approach is that such subconscious processes play a part in many of our actions. Consider the following:

> Without saying anything, I record [a waking subject's] pulse on a cardiograph ... While the trace is forming I count the pulse loudly, at first correctly. After a while I count more beats than there really are; and then less. Studying the trace later, I discover that the pulse rate increased during the accelerated counting, and diminished during the slowed counting, and that without the knowledge of the subject ...[33]

This is an "unusual" phenomenon of suggestion executed subconsciously by a waking subject unaware of what was happening. A similar effect can be induced by counting out loud the steps of a walking person. His steps will quicken as the counting quickens. "We see that these so-called unusual phenomena of mental dissociation or inhibition, are in reality normal phenomena which happen every day in ordinary life."[34] So it appears that the term suggestion, like the term hypnotism, should not be taken to imply a special state, requiring special manoeuvres, and giving rise to special phenomena.

Another line of argument that could be deployed to show that there is a "special state" of hypnosis involves the supposed exaltation of the cerebral faculties of subjects in an hypnotic state. To this Bernheim replies that in former years he conducted many experiments on the question, especially on the alleged hypnotic enhancement of memory, always with negative results.[35]

So, there is, according to Bernheim, no "special state" of hypnosis, characterized by enhanced suggestibility, whether that state be conceived as sleep-like or otherwise. What, then, of the numerous apparently remarkable effects of suggestion, effects that can be obtained both with subjects in "artificial" sleep and with waking subjects? How are these to be explained, and, if no special state is required for their production, why are they not obtainable with more subjects more of the time? Bernheim's answers (such as they are) to these questions can be obtained from another of his late publications, and they are at root no different from the answers he had been giving since he first became interested in the subject:

> The divers ideo-dynamisms, the divers suggestibilities are variable according to the idea, to the dynamism, which they must activate, according to the *credivité* which causes their acceptance, according to the way the *credivité* is activated, according to the circumstances which may reinforce them, according to the state of mind which can increase or diminish cerebral control, favour or attenuate the tendency of the idea to be actualized. The same idea expressed in eloquent, vibrant and persuasive speech, can be suggestive, whereas handicapped by colourless and monotonous speech, it will remain sterile and a dead letter ... All emotions, all passions, generous or vile, love, pity, hate, anger, enthusiasm, fanaticism political, religious, socialist and other, blind certain judgments, exalt certain *credivités*, and favour certain modes of suggestibility. The impulsive person is an enthusiast who unreflectingly obeys certain mental impulses and almost automatically transforms his impressions into dynamisms.[36]

The hypnotist emerges here as being like a demagogue, a revivalist, a confidence-trickster, or an actor who can be any one of these. He plays upon the susceptibilities of his hearer so as to minimize all rational scepticism and nagging doubts and permit the desired ideas to rush through unopposed with all the impetus he can give them. The rôle of specifically hypnotic or magnetic manipulations is simply to work upon the minds of those persons who antecedently have faith in them.

Psychotherapy too has become for Bernheim largely a matter of therapeutic suggestion given to subjects in the waking state. Only in certain limited kinds of case is it beneficial to put the subject into a state of sleep. These are cases – night terrors, somnambulism, nocturnal enuresis – in which the disorder itself manifests during sleep. Inducing sleep may also be of value in cases where there is a great deal of agitation or excitement.

Bernheim's revised ideas on hypnotism and psychotherapy won him some supporters among psychotherapists of the rising generation to whom we shall come. They also made him enemies, and those members of the old guard who did not become enemies stuck for the most part to their guns. Forel, for example, responded to Bernheim's paper to the Society for Clinical Psychology and

Psychotherapy with a firm statement of his own position (which is close to Vogt's),[37] and Vogt asserted categorically that suggestibility is heightened in hypnosis, and heightened in proportion to the depth of hypnosis. Contrary findings are to be explained on the assumption that a partial waking state had not been attained, but only a diffuse general sleep, or that the conscious elements which were tested were outside the rapport relation.[38]

Of course, what is central to Bernheim's later views is his claim that waking suggestions in subjects who are not and never have been hypnotized can be just as effective as suggestions given to subjects who are hypnotized or are in some other suppositious special state. He had been interested in the topic of "waking suggestion" as early as 1883,[39] and by the end of the 1880s, the phenomenon was regarded as well-established. In fact, there are sections on it in several of the textbooks of the time,[40] though the examples of waking suggestion cited were mostly of suggestions which took effect upon subjects who had on previous occasions responded to similar suggestions whilst hypnotized. However, before the end of the 1880s, and from then on for the remainder of his life, Bernheim found that by mere verbal affirmation he was able to produce phenomena of kinds usually regarded as hypnotic (anaesthesias, contractures, hallucinations, and so forth) in waking subjects not previously hypnotized and not suffering from neurotic or mental ailments. (Braid had to some extent anticipated him in this by about thirty years.[41]) Here, by way of example, is a case that he described to the First International Congress of Hypnotism in 1889:

> On the first occasion, and without his having taken part in any experiment of this kind, I lifted the arm of such a subject, and said to him: "You are unable to lower it." He could no longer do so. I said: "Your body is without feeling," and I pricked him without his showing any pain. I said to him: "You are compelled to get up and walk." And he walked without being able to resist. I said: "Look, here is a great dog! It barks." He saw it, and drew back in fright. I added, "*Sleep!*" He closed his eyes and *slept, whilst walking.*[42]

Bernheim was by no means alone in reporting such cases. For instance, in 1891, Dejérine gave the following account of two subjects of his own:

> I have studied two subjects, never before hypnotized, who were suggestible in the waking state to a degree as pronounced as that which the most suggestible hypnotic subject can attain. They were young countrymen ... recently arrived at Paris on leaving military service. On the day when I examined them for the first time, I obtained with them, by simple suggestion, in the mental, sensory or motor domain, results not often obtained at the outset even with highly suggestible hypnotic subjects (contractures, paralyses, anaesthesias, sensory hallucinations, doubling of the personality, posthypnotic suggestions after more or less of a delay, which were realized at a determined time, etc.). In other words, I obtained with them on the first day, by waking suggestion, the same results as I obtained the next day, employing suggestion during the [hypnotic] sleep.[43]

It is a great pity that medical men as distinguished as Dejérine and Bernheim did not find it worth giving, in a matter of such importance, many further details concerning their procedures, their subjects, and their relations with those subjects. But even if we could be sure that Bernheim's cases are all that he claims, mere citation of them could only establish that waking suggestion can *sometimes*, and with some subjects, be as effective as hypnotic suggestion; it could not establish what it is essential for him to establish, namely that waking suggestion is *generally* and *on balance* no less effective than hypnotic. This could only be established by systematic comparison of a series of hypnotized with a series of unhypnotized subjects. The proper design and execution of such experiments is by no means easy, but so far as I know there was during the period with which we are concerned not even an attempt at it. The best we can do is mention three studies which seemed to show that a high degree of responsiveness to suggestion is at any rate fairly widespread among waking and hitherto unhypnotized subjects not selected for special suggestibility.

The earliest of these studies is that by Richet (1884). Richet, a pioneer in this as in so many matters, had been interested in the phenomena of suggestion without hypnosis for some while, and had devised a test for assessing it. The test consisted in inviting the subject to close his hand on some commonplace object, such as a coin or a knife, and then affirming that he would be unable to open it. In a paper of 1884,[44] Richet notes that this test had succeeded with two out of seven or eight subjects, and goes on to describe his experiments with one of them, Mlle. B., aged about forty-five. This lady had never been magnetized or hypnotized before. Richet turned her to face him, and by gestures "pushed" her away from him or "pulled" her towards him, and drew her left or right. He told her the sleeve of her black dress was blue. She denied it but said that it had blue reflections. He told her that she would be unable to play on the piano any tune other than *J'ai du bon tabac*. Though a good pianist, all she could manage was this tune in various keys. He told her that she could not count beyond ten, and she stuck on eleven. He told her that she would forget the name (Gaston) of an old friend who was present. He told her further that the name was really Jules. There ensued the following dialogue: "Ah! that; I defy you. – Very well, but then what is his name? – He is called … he is called … I know his name perfectly well, I can't say it, though I would recognize it; anyway, it is not the one which you want to make me say. – Very well, but he is called Jules. – No. – What is his name? – He is called Ju … Ju … no, I don't want to say Jules, it is not his name."[45]

A somewhat wider range of phenomena was obtained with waking suggestion by Wingfield (1888).[46] Wingfield's subjects were healthy young men, mostly Cambridge undergraduates, aged between eighteen and twenty-four. All were in the waking state; most had not been previously put into "the hypnotic sleep"; most of the experiments were experiments in self-suggestion, sometimes assisted by self-administered mesmeric passes. Among the phenomena produced were:

arm rigidity, inability to part joined hands, eye closure with inability to reopen the lids, anaesthesia, crystal visions, automatic writing (but in every case but one the subject had previously been sent to sleep), and (in one subject only) "delusions".

Also of interest in the present connection is Wingfield's normal method of hypnotizing:

> With many – in fact, with most of the Cambridge subjects whom I sent into somnambulism – I induced it in the first instance by this method, making no suggestion of sleep. I suggested as strongly as I could a visual hallucination, generally of a bright star, and then changed it for another, and so on through a series of different hallucinations. In almost every case the subject experienced complete amnesia on awaking, followed by a revival of memory when again hypnotized.[47]

Here we have a complete reversal of what would generally be considered the normal order of events, with hallucinations being provoked before hypnosis, and apparently inducing it. It is a claim that merits further investigation.

Other workers carried out experiments in which the "suggestions" were what might be described as misleading expectations instilled in gullible waking subjects by the pronouncements of authority figures. For instance, by making false statements about the likely effects of magnets, electrical machines, prolonged exposure to light, etc., Ventra[48] was able to induce subjects to feel the predicted sensations, see a black point on a white paper as a cross, etc. On one occasion he gave a somewhat excitable patient (insusceptible to hypnotism) bread pills, telling her that they would have a cholera-like effect. In ten minutes she was suffering from sickness, diarrhoea, cramps, cyanosis, cold of the extremities, and prostration. Similar techniques of suggestion by purveying misleading information were adopted by Binet in his *La suggestibilité* (1900) and by Professor Emile Yung of Geneva from 1882 onwards.[49] In an extensive series of experiments, Yung made use of a simple card trick to convince a student audience (or some part of it) that he could detect by moving his finger over the cards that card which had, in his absence, been impregnated with the magnetic fluid of one of the persons present. After a few successful trials, he invited students to try their hands. Well over half the students turned out to be "suggestible" in that they experienced hallucinations or sensory illusions in connection with the card that they supposed to have been magnetized. These included an "attraction" or "repulsion" in the muscles of the finger, hand or arm, shocks, cramps, feelings of heat or cold, prickings, a wind, various smells (mentioned by Yung himself) and movements of the card. Similar results were obtained with coins in which Yung pretended to recognize the characteristic odour of the persons who owned them.

Experiments like these raise rather acutely the question of the meaning of the terms "suggestion" and "expectation", which we shall have to consider again later. It is also (as Yung recognizes) not easy to assess the possible rôle of compliance in producing the results, for compliance is most to be anticipated where the experimenter is a person of authority vis-à-vis the subject. However, compliance does not seem a likely explanation of the findings of A. Chojecki, when at Geneva.[50] Chojecki applied three tests of suggestibility to sixty unhypnotized university students, thirty male and thirty female. Two of these tests involved pieces of pseudo-apparatus, which subjects were led to believe would respectively produce a feeling of warmth in one finger, or anaesthesia of a finger (as assessed by an aesthesiometer or a needle). The third utilized a suggestibility test of Binet's, the object of which was to affect through the order of presentation a subject's judgment of the length of lines. 31.8% of subjects were responsive to each of the first two tests, and 60% to the third. There was almost no correlation between suggestibility as measured by any one of these tests and suggestibility as measured by either of the others, a fact which argues against compliance by the subjects as an explanation of their responsiveness to suggestion.

To return to Bernheim: I think it would be fair to say that while the materials just outlined fell a long way short of proving that waking suggestion is generally and on balance as effective as hypnotic, they warranted continuance of the debate. Bernheim's rejection of a "special state" of hypnosis, and his insistence that all "hypnotic" phenomena can be obtained with unhypnotized waking subjects, made him widely unpopular in his own time, but they are in accord with the dominant tendencies of today's thought.

PSYCHOPATHOLOGICALLY ORIENTED THEORIES: GRASSET

The theorists who set their theoretical speculations about hypnosis in the context of certain psychopathological phenomena were principally those whom I described in earlier chapters as having to a greater or lesser extent come under the influence of Pierre Janet. Janet himself did not make much impact as a theorist of hypnosis, partly because he remained firmly wedded to the idea that hypnosis is itself a pathological state, and partly because his fullest exposition of his ideas on the subject did not appear until 1919. But others took, as we have seen, their cue, or their materials, from him, and produced accounts of hypnosis in terms of a "dissociation" of consciousness, or in terms of multiple streams of consciousness. Obvious examples are F. W. H. Myers, James, Sidis and Morton Prince. For our immediate purpose – that of discussing a representative theorist whose initial standpoint is psychopathological – a suitable person is Jules Grasset (1848–1918), professor of clinical medicine at Montpellier.

Like Janet, Grasset came at first under the influence of Charcot and his school,[51] and remained for a long time convinced that the Charcot "stages" might in some cases manifest spontaneously. Before long he fell under the sway of Janet (who looked upon him, not altogether justly, as a mere regurgitator of his own ideas[52]). However, Grasset also recognized the force of some of Bernheim's claims, with the result that, although he never abandoned the notion that hypnosis is an abnormal and probably undesirable state, he was less prone than Janet to set it down as definitely pathological. Grasset had some influence upon other contemporary workers, most notably Crocq.[53] His proposals are, indeed, recognizably similar to those of present day "neo-dissociationists", not least because they are illustrated by a comprehensive diagram of the imaginary input–output relations between putative "centres" at different hypothetical "levels" of the nervous system.

Grasset's mature theory of hypnosis is set forth in a number of articles,[54] and in a substantial book, L'hypnotisme et la suggestion (1903), which I shall principally follow here. Grasset is a clear expositor, but his presentation is curiously a priori, in that although his book contains a good deal of clinical material, his theory is in large part developed without direct reference to it, but as it were abstractly and schematically. He begins by distinguishing between an inferior and a superior "psychism". The inferior psychism may also be looked on as a superior automatism, and has its centres not (as do other automatisms) in the midbrain and basal ganglia, but in the grey matter of the cerebral cortex. It is "an automatic function which is not the ordinary reflex arc since it results in acts that are coordinated, intelligent and within limits spontaneous." But it is also "psychic" or mental because, though an "automatic" psychism in that it is not freely willed, it may exhibit memory and intelligence. Consciousness is none the less not essential to it. Grasset does not elaborate on this distinction between the conscious and the psychic (or mental).

The superior automatism (or inferior psychism) "must be carefully distinguished from the superior psychic function, seat of superior intelligence, of full and true personality, of complete and moral consciousness, of freedom and responsibility."[55]

On the basis, presumably, of neuropathological findings, Grasset recognizes six "inferior psychic centres or centres of psychological automatism". He refers to each by a letter of the alphabet. A is the auditory centre in the temporal cortex; V the visual centre in the calcarine fissure; T is the perirolandic tactile centre; K the kinetic (movement) centre, also perirolandic; M the centre of spoken language, at the foot of the left third frontal convolution; and E the centre for writing (at the centre of the left second frontal convolution). In the tradition of the neurological "diagram-makers" of his time, he represents the centres AVTEMK by the six corners of a polygon. Each centre has a pathway connecting it to every other centre. The first three centres (the sensory ones) have pathways coming in from the sense-organs; the last three (the motor ones) have pathways heading for the musculature. The activities carried out by these centres of psychological automatism Grasset refers to as "polygonal" or "functions of the polygon".

Above the polygon, and regulating its functions, Grasset places a further centre, O, the "superior psychic centre of conscious personality", which he tentatively locates in the prefrontal cortex. O is represented as having connections to each of the centres of psychological automatism. One cannot study the functions of the polygon unless it has been dissociated from O, as in distraction or sleep. Distraction is when one thinks of one thing and does another; sleep is the repose (or near-repose) of the centre O, whilst the activity of the polygon persists. In natural somnambulism, there is total suppression of communication between O and the polygon, and the motor centres, K, of the latter, thus emancipated, exhibit bursts of activity, modulated, however, by the still functioning intra-polygonal connections.[56] Hysteria too, with its anaesthesias, restrictions of the field of consciousness, troubles of attention, paralyses and amyosthenias, contractures, circumscribed amnesias, *idées fixes*, etc., is primarily a malady of the emancipated polygon.[57]

Anatomical lesions in different brain regions will also have characteristic effects upon the integrity of the O-polygon complex. For instance, if we take just the aphasias, we can distinguish "polygonal aphasias (lesion seated in the polygonal language centres), supra-polygonal aphasias (lesion seated between the polygon and O), transpolygonal aphasias (lesion seated in the fibres connecting the various polygonal centres) and sub-polygonal aphasias (lesions seated below the polygon)".[58]

We come now to the state of hypnosis. The only constant characteristic of hypnosis is heightened suggestibility. Hypnosis must be regarded as pathological, or at any rate morbid or abnormal. The proof of this is that not everyone can be put into it. The suggestibility of a hypnotized subject must not be confused with responsiveness to persuasion, advice, command, conversation, example, discussion or demonstration. In the latter cases he *accepts* the injunction, he *consents* to obey; in the former he obeys uncritically, without reflection or reasoning.[59]

In terms of the polygon scheme, two factors together constitute the state of suggestibility, that is of hypnosis:

I. The first is a dissociation between O and the polygon. O no longer has control over the polygon, the lines of communication have been interrupted. However, this is true of some somnambules and cataleptics, who are not suggestible. So we must add a second element, viz:

2. A state of pliability of the polygon; that is to say the immediate obedience of the polygon not to the subject's centre O but to that of the hypnotist. The hypnotist so to speak takes possession of the subject's polygon, even of parts of

it not normally under the control of the subject's O. (We may contrast this state of affairs with that which obtains in persuasion, advice, etc. – in these it is the subject's centre O, not his polygon, that is influenced by the O of his interlocutor.)

What remains obscure here is how precisely the standard methods of hypnotizing, magnetizing, etc., operate to dissociate a subject's O from his polygon, or to still his O's activities, and how his polygon comes to be at the disposal of the hypnotizer's O. Grasset seems vaguely and conventionally to suppose that methods of hypnotizing work by putting the idea of sleep into the subject's mind, that is, presumably, into his O, which must then somehow close down the links between itself and the polygon, dissociating itself from the latter. If this is so, there is a kind of suggestibility (the susceptibility of O to sleep-suggestions) that cannot be explained in terms of the dissociation of O from the polygon. Perhaps for this reason, Grasset continues to suspect that monotonous sensory stimulation may have an hypnotic effect *per se*, and that there may be something in Pitres' claims concerning hypnogenic zones. But how these zones might operate is not made clear.

As for waking suggestion: Grasset agrees that such suggestions may be effective with highly susceptible subjects, but maintains that these subjects are "in hypnosis in the waking state".[60] Their polygons have become (partially?) dissociated from their O's and they are in a state of partial hypnosis.

One must not underestimate the capabilities of a polygon freed from the control of its centre O. The polygon has a characteristic unity, an individuality determined by a mind (*psychisme*), a memory (as exhibited in post-hypnotic suggestion), etc. We can classify types of hypnosis, or of hypnotic phenomena, in terms of the way in which, or the extent to which, the polygon operates upon the suggestions which it receives. Three major classes of phenomena can be distinguished. In the first, the polygon simply transmits a suggestion from a sensory to a motor centre; in the second the polygon transforms the suggestion into a complex act of the whole polygon; and in the third the polygon adds new elements up to and including the development of a wholly new polygonal personality.[61] Of course as we have already seen, the polygon can in some circumstances (e.g. sleepwalking) be freed from its O without the O of some other person taking control. Then we have not hypnosis, but various kinds of more or less pathological state. Grasset distinguishes[62] two modes of such troubles of personality: *doubling*, which is the appearance of the polygonal personality alongside the personality of the centre O; and *transformations*, which are at root transformations of the emancipated polygonal personality. The polygon autosuggests these transformations (a process of which Grasset gives no useful account), and the result is mediumistic controls, certain cases of multiple personality, and so forth.

Seen, as it were, from a distance, Grasset's theory seems of wide scope – wider than I can convey in this brief summary – and creates an air of scientific exactitude. But the more closely one studies it, the less satisfactory it appears. It stands at several removes from any detailed consideration of the phenomena it purports to explain; important issues (such as the way in which verbal suggestion operates to separate O from the polygon to produce a state of enhanced suggestibility) are largely passed over; key assumptions are not submitted to critical scrutiny – could there, for instance, possibly be a single centre (A) for audition which, among all its other functions, interprets certain patterns of sound as speech, extracts the "meaning" of the speech, and transmits appropriate "orders" to motor centres (K)? And, as Bernheim remarks, can one seriously suppose that in suggestion "there is only passive automatic cerebration, only inferior mentality ... "[63] Consider, Bernheim goes on, the following instance of hypnotic suggestion, for which the subject was a highly intelligent and educated, but very suggestible, young lady, whom he had several times made see an hallucinatory rose in her hand by waking suggestion. One day he placed a real rose in one of her hands, and an hallucinatory one in the other, and invited her to tell him which was which. She looked, touched, fingered, sniffed the scent, made use of all her critical faculties, but was unable to reach a decision. As Bernheim pointedly observes, during this experimental hallucination the young lady "had all her active intelligence, all her judgment; she could argue all the issues with the resources of her superior mentality [*psychisme supérieur*]; but she could not neutralize the suggested image."[64] In short, the young lady's O was in excellent working order and in no wise divorced from the activities of her polygon, and yet she manifested a high degree of suggestibility. Other writers reported cases of similarly suggestible subjects who seemed fully capable of exerting their critical faculties even while in the grip of apparently successful hypnotic suggestions.[65] Cases of such a kind, if reported with sufficient frequency, would make it very difficult to regard hypnosis as a state of enhanced suggestibility due to the functional separation of the "higher" (O) from the "automatic" (polygonal) faculties of the mind.

CONCLUDING COMMENTS

There are, broadly speaking, three categories of phenomena which a theory of hypnosis needs to tackle. These are the phenomena of hypnosis itself (including allegedly enhanced suggestibility and alleged enhancements of faculty); the alleged therapeutic benefits of hypnosis; and the phenomena of states supposed (by some) to be related to hypnosis. In this chapter, I have considered three representative theories of hypnosis, those of Vogt, the later Bernheim, and Grasset. It is hard to say which one of these covers the most of the designated phenomena. They all suffer from certain lacunae. None gives a satisfactory, or even useful, analysis of the concepts of suggestion and suggestibility, and this leaves their accounts of the corresponding phenomena imperfectly anchored. None of them gives any special account of why hypnotic suggestion may have therapeutic efficacy, though indeed it could be argued that no special account is necessary. Their approaches to the phenomena of waking suggestion are curiously contrasted. Grasset in effect denies that there are such phenomena, for those who exhibit them are *ex hypothesi* hypnotized. Bernheim *per contra* denies the existence of hypnosis as a state of heightened suggestibility. Only Vogt tries to accomodate both hypnotic and waking suggestibility within his scheme. For handling the various states ostensibly related to hypnosis (fugue, multiple personality, drug states, etc.) Grasset's theory has perhaps the greatest flexibility, though he could well, as Binet remarks, have some difficulty with cases of multiple personality in which more than two different personalities are apparently co-conscious.[66] Bernheim has very little to say about the related states; not surprisingly, since he does not look on hypnosis itself as a special state. Vogt does not say a great deal about these states (other than somnambulism) but would probably have little difficulty in accomodating them.

At a certain level, Grasset's theory is the most comprehensive. It has some kind of a niche for most things. But it is a superficial and uncritical piece of work. The theories of Vogt and Bernheim are less comprehensive; but it is easy to see how they could be extended, and each has certain strengths. Vogt's strength lies in the convincing evidence that he cites, from his own careful observations, for a close relationship between hypnosis and what he calls "somnambulic dreams". I do not think that this evidence merits the immediate dismissal it would nowadays usually be accorded. Bernheim's strength lies in keeping his theorizing at all times close to his practical experience so that as the latter enlarged the former altered *pari passu*. In the end, he was led by gradual stages to a position radically different from that which he had originally occupied, and close to positions that are widely advocated today.

NOTES

1 Theories of hypnosis, especially physiological ones, are reviewed at some length by Hirschlaff (1899–1900), pp. 65–97, and (1905), pp. 212–240.
2 Schrenck-Notzing (1894b), p. 12.
3 Pupin (1895–6); Van de Lanoitte (1895–6).
4 Schleich (1897).
5 Preyer (1890), pp. 144–145.
6 Woods (1897), pp. 252–258.
7 Döllken (1895–6).
8 Vogt (1894–6; 1896–7a; 1896–7b; 1897–8). Cf. Vogt's notes in Forel (1895).
9 Vogt (1894–6), pp. 277–314.
10 Vogt (1894–6), p. 313.
11 Vogt (1894–6), p. 320.
12 Vogt (1894–6), pp. 336–339; Vogt (1897–8), pp. 80–83.

13 Vogt (1897–8), pp. 84–85.

14 Vogt (1897–8), pp. 85–86.

15 Vogt (1897–8), pp. 87–91.

16 E.g. Vogt (1897–8), pp. 89–90.

17 Vogt (1896–7a), p. 183.

18 A few years afterwards, William McDougall in England came forward with a theory of hypnosis very like Vogt's – he even adopts the term "neurokyme". But to account for the phenomena of suggestibility and also of rapport he invokes a "submissive propensity" which he believes all human beings to possess. This leaves him with certain problems over autosuggestion. McDougall's various writings on hypnotism and suggestion are conveniently assembled in McDougall (1967). His contributions on the subject to the 11th (1910–11) edition of the *Encyclopaedia Britannica* were of some influence.

19 Lipps (1897a; 1897b; 1898). On "mental energy" see especially (1897b), pp. 398–399, 405–406, 411, 498–499, 502.

20 Ferenczi (1916), pp. 30–79 (originally published in the *Jahrbuch der Psychoanalyse* for 1909); Freud (1905).

21 Cf. the discussion in Verhandlungen ... (1912), pp. 289–293.

22 Cf. Barrucand (1967), pp. 119–128.

23 Bernheim (1891–2), p. 87.

24 Bernheim (1917), p. 52.

25 Bernheim (1892b), p. 1214.

26 Bernheim (1892b), p. 1214.

27 Bernheim (1980), p. 57.

28 Bernheim (1897), p. 9; cf. Bernheim (1897–8).

29 Bernheim (1911).

30 Cf. Bernheim (1917), p. 148.

31 Bernheim (1911), p. 472.

32 Bernheim (1911), p. 472.

33 Bernheim (1911), p. 473.

34 Bernheim (1911), p. 473.

35 Bernheim (1911), p. 476.

36 Bernheim (1917), pp. 81–82.

37 Verhandlungen ... (1912), pp. 276–277.

38 Verhandlungen ... (1912), p. 284.

39 Bernheim (1883a; 1883b).

40 E.g. Bottey (1886), pp. 119–135; Cullerre (1887), pp. 219–225.

41 Braid (1855), pp. 15–16.

42 Bernheim (1889c), p. 85.

43 Dejérine (1890–1), pp. 227–228.

44 Richet (1884c).

45 Richet (1884c), p. 555.

46 Wingfield (1888–9).

47 Wingfield (1910), p. 9.

48 Ventra (1891); Schrenck-Notzing (1895a), pp. 20–21.

49 Yung (1908–9).

50 Chojecki (1911–12); cf. Guidi (1908–9).

51 Cf. Grasset (1888–9).

52 Janet (1925), Vol. 1, p. 230.

53 Crocq (1902).

54 Cf. Grasset (1896; 1902–4); Barrucand (1967), pp. 159–165.
55 Grasset (1903), p. 6.
56 Grasset (1903), p. 23.
57 Grasset (1903), pp. 24–25.
58 Grasset (1903), p. 26.
59 Grasset (1903), pp. 64–65.
60 Grasset (1903), p. 74.
61 Grasset (1903), p. 165.
62 Grasset (1903), p. 265.
63 Bernheim (1917), pp. 70–71.
64 Bernheim (1917), p. 74.
65 Moll (1890), pp. 161–164; Bramwell (1930), pp. 99–100.
66 Binet (1900), p. 12; cf. Grasset (1903), pp. 269–271.

CHAPTER 25

The decline of hypnotism

It is a commonplace of historical sketches of hypnotism that the hypnotic movement, which had advanced and diversified so strikingly in the decade following 1882, had by the turn of the century halted and begun to disintegrate rapidly. Ellenberger speaks of its "swift decline",[1] and Janet of its "rapid decline" after about 1896.[2] Janet makes the decline sound quite dramatic:

> No one repudiated hypnotism, no one denied the power of suggestion; people simply ceased to talk about hypnotism and suggestion. The number of publications devoted to these topics declined enormously...Whereas in previous years the books and articles concerning hypnotism, suggestion, and allied subjects had been numbered by thousands per annum, the number now fell to a few dozens.[3]

But he exaggerates.[4] Loewenfeld (1901) and Milne Bramwell (1903) seem unaware of any decline. The number (211) of ordinary members attending the Second International Congress of Hypnotism, held at Paris in 1900, noticeably exceeded the number (171) of those attending the similar congress held there in 1889.[5]

Table 3 gives the number of books and articles on hypnotism and suggestion listed in the indices[6] to the *Index Medicus* from 1885 to 1899 and from 1903 to 1914.[7] There is no sudden drop here, but only a slow and irregular decline from a peak in the later 1880s. A symptom of the decline (but in part perhaps also a cause) was that the two leading hypnotic periodicals, the *Zeitschrift für Hypnotismus*, and the *Revue de l'hypnotisme*, turned themselves, in 1902 and 1910 respectively, into journals of more general scope, dropping the term "hypnotism".[8] I hestitate to give precise figures concerning the distribution of these items by country of origin – the language in which an item is written is by no means a safe guide – but I am satisfied that the approximate percentages shown in table 4 are sufficiently close for our purposes. Austria is included with Germany. Totals are not comparable across columns, since the periods of time involved differ.

It will be seen that France maintains preeminence, albeit a diminishing one, throughout most of the three decades. One must add, however, that this is a numerical preeminence, not an intellectual one. The latter probably belonged on

Table 3. Publications on hypnotism and suggestion

Vol. no.	Year	Hypnotism	Suggestion	Total
7	1885	29		
8	1886	37	14	51
9	1887	77	20	97
10	1888	92	26	118
11	1889	80	33	113
12	1890	69	30	99
13	1891	68	23	91
14	1892	35	18	53
15	1893	41	16	57
16	1894	35	10	45
17	1895 (Jan)	unindexed		
18	1895–6	61	6	67
19	1896–7	59	29	88
20	1897–8	47	24	71
21 (New Series)	1898–9	45	52	97
1	1903	24	24	48
2	1904	31	22	53
3	1905	24	16	40
4	1906	39	20	59
5	1907	23	16	39
6	1908	21	11	32
7	1909	27	16	43
8	1910	18	17	35
9	1911	19	13	32
10	1912	22	8	30
11	1913	19	9	28
12	1914	13	12	25

balance to Germany after the mid-1890s. The numerical preponderance of France may have been even greater than the figures indicate because of the growth there by the end of the century of a curious semi-popular, semi-occult literature, embracing aspects of hypnotism, animal magnetism, spiritualism and psychical research, which found its way only to a limited extent into the *Index Medicus*.[9]

The most obvious changes in the relative positions of the different countries are the quick descent of Italy from the middle of the order to the bottom, probably linked to the strongly somaticist tendencies of Italian psychiatry, and the marked improvement in the position of the United States. The very rapid rise of the United States to the top of the table in the later 1890s may be partly due to

Table 4. Percentage of total publications by country

1885–9 N = 408		1890–4 N = 345		1895–9 N = 323		1903–8 N = 271		1909–14 N = 193	
France	37	France	30	USA	35	France	33	France	22
Other	19	Germany	16	France	29	Germany	21	USA	20
Germany	18	Other	16	Germany	18	USA	14	Britain	19
Italy	15	USA	14	Other	8	Other	13	Germany	17
Russia	5	Britain	11	Britain	5	Britain	10	Other	14
Britain	3	Italy	8	Italy	3	Russia	6	Italy	5
USA	3	Russia	5	Russia	2	Italy	3	Russia	3

the fact that the *Index Medicus* is an American publication – a good many of the American articles covered are derivative and somewhat trifling; but it was certainly not wholly or even largely due to this. Interest there declined a good deal after about 1900, though there was a brief resurgence in 1909 and 1910, linked to the rise of the Emmanuel movement.

Conversely, the *Index* may well underestimate the amount of Russian work on hypnotism – as we have seen medical interest there was not inconsiderable.

In Britain, after a mild flurry of interest in the early 1890s, the medical profession, traditionally conservative, was slow to take to hypnotism, and interest (though limited) was still stable or rising there when it was falling in most other countries.

REASONS FOR THE DECLINE

The factors which determined the rate and date of the decline of interest in hypnotism in these various countries could only be elucidated by detailed studies of the social and intellectual climate in each individual country. But certain general factors were at work in most of the countries which we are considering. I refer especially to the activities, amounting at times almost to propagandizing, of two groups of persons whom we may refer to for short as the sceptics and the moralizers.

By "sceptics" I do not mean individuals who totally disbelieved in the alleged phenomena of hypnotism and suggestion, but those writers who denied that there is anything "special" about the supposed state of hypnosis. The starting-point for these writers was the claim, forcefully put forward by Bernheim, that most of the so-called phenomena of hypnosis, and many of its therapeutic benefits, can be obtained through suggestion alone, in subjects who are in a state of ordinary wakefulness and have not been put through an hypnotic induction procedure. Similar claims were made by Delboeuf in a paper of 1892, provocatively entitled "Comme quoi il n'y a pas d'hypnotisme."[10] He cites various cases in which he wrought cures himself by mere strong suggestion or affirmation.[11] As we saw in the previous chapter, Bernheim himself took up the phrase "il n'y a pas d'hypnotisme" at a conference held in Moscow in 1897,[12] a move which seems to have caused some confusion among the supporters of hypnotism. A more aggressive and somewhat propagandist stance was taken by Bernheim's pupils, P. Hartenberg (1871–1949) and P. E. Valentin, who had a joint clinic in Paris. Hartenberg, too, took the catch-phrase "il n'y a pas d'hypnotisme" as the title of an article[13] in which he expounded what he thought to be Bernheim's meaning. Bernheim is not denying the curative properties of induced sleep, only that the sleep itself is an extraordinary and unique state. What has led certain observers into the error of supposing that induced sleep is an exceptional phenomenon is that during it the suggestibility of the sleeper may be enhanced. However, this is not always the case. For instance, a lady patient whom he and Valentin are treating at their clinic by the method of prolonged hypnotic sleep passes into a state of the most profound sleep, but is then infinitely less responsive to suggestion than in the waking state.

In other papers Hartenberg and Valentin give examples of cases treated by waking suggestion,[14] for the administration of which they developed some special techniques. These methods included dummy treatments formulated to impress the patient, and carrying out activities designed to counteract the symptoms. For example, Hartenberg, treating an actor of forty-one, who was suffering from headache, depression and insomnia, gave him the usual waking suggestions of improvement (during which he simply reclined on a sofa with his eyes closed, concentrating on the words), but in addition made him (quite voluntarily) act out the part of a happy and confident man.[15] They developed a rationale for this treatment, which they expounded before the Fourth International Congress of Psychology, held at Paris in 1900. Much of their rationale derives from earlier papers by Bernheim.[16]

Modern clinical psychology, says Valentin,[17] has shown that there are as many possible kinds of suggestion as there are varieties of images, and that every sort of manoeuvre capable of evoking a curative image in the patient's brain belongs to the domain of therapeutic psychology. Thus we can make use of affective images, sensory images, movement images, as in treatment by action, and verbal images, as in suggestive therapy by speech. Hartenberg then takes up these ideas.[18] He asserts that every function of the brain – sensibility, affectivity, language – can be used for therapeutic purposes. It was Bernheim who established as a principle that the brain can have a curative action without the subject being hypnotized, and who thought of reinforcing suggested verbal images by the corresponding motor ones and by actual performance of the movements involved. He has himself used similar methods with success. But of all psychotherapeutic procedures the most important is pure verbal suggestion, the production of a curative effect through arousal of the corresponding verbal images. For this he has devised the term "logotherapy".[19]

Another of Bernheim's disciples was P. E. Lévy (b. 1869), whose *L'éducation rationnelle de la volonté* first appeared in 1898. The term "education" in the title is hardly appropriate. The book is in effect a manual, on Bernheim's lines, of the principles of autosuggestion. If, as Bernheim holds, the idea of cure carries with it the real cure, why should not the patient cure himself by affirming the therapeutic idea and attending to it to the exclusion of other ideas? This may best be done by catching or creating a condition of half-sleep or drowsiness. Lévy claims by these means to have cured a considerable diversity of ailments.

These followers of Bernheim did not deny that subjects put through an hypnotic induction procedure may pass into a sleep-like condition, or even into true sleep. They simply denied that they passed into a special state of hypnosis. They had, however, a general inclination to disparage any hypnotic phenomenon (such as sleep) which even hinted at a special state of trance. Thus Hartenberg says:

> As for the nature of induced sleep ... each subject realizes in his own fashion the idea of sleep which is suggested to him. Certain of them, in spite of the appearance of sleep which they present ... do not sleep at all. They realize, often out of complaisance for the doctor, a form of sleep which Dr. Valentin and I are accustomed to call conventional sleep. They are not therefore less suggestible, and one can obtain with them hallucinations, anaesthesias and contractures.[20]

Hartenberg and Valentin were not the first to express the idea that some "hypnotized" subjects may simply be complying with the wishes of the hypnotist. Delboeuf had come to the same conclusion nearly fifteen years before.[21]

Bernheim's change of mind as to the central place of the hypnotic sleep in the theory and practice of hypnotherapy disturbed some of his former supporters, particularly the venerable pioneer Liébeault, who argued in 1898 that "the so-called waking state, in which suggestion is effective, is simply a partial and spontaneous condition of sleep, and ... the therapeutic results obtained in it are not so satisfactory as those gained in the condition of deep somnambulism."[22] Dumontpallier, Liégeois, Bérillon, Bonjour, and others, rallied to his support.[23] Discussing Hartenberg's paper, "Il n'y a pas d'hypnotisme", Dumontpallier maintained, "Suggestion in the waking state has an undeniable therapeutic action; hypnotic suggestion has a still greater therapeutic effect",[24] and he cited cases to prove his point. Furthermore, there is no denying that he *has* a point. The sceptics were much given to citing examples of successful waking suggestion, but none of them set up the systematic experiments which alone could prove that waking suggestion is just as effective as suggestion under hypnosis.

In 1910, G. Preda summarized the situation in France somewhat pointedly:

> At present, there are in France two whole schools which explain the inner mechanism of induced sleep in a slightly different way:
>
> 1. One, represented by Bernheim, Babinski, Dejérine, etc., says that, for a subject, the possibility of being hypnotized proves precisely the existence in him of the state of suggestibility;
>
> 2. The second, represented by Grasset, Bérillon, Janet, Farez, [J.] Voisin, Magnin, etc., maintains the inverse thesis ... that the possibility in a subject of responding to suggestion proves precisely in him a state of hypnosis.[25]

The division between the traditionalists, for whom the supposed hypnotic state was pivotal, and the sceptics, for whom it was not, found its counterpart in other countries, though it is hard to say how far the division abroad reflected that in France. In German-speaking countries, writers on hypnotism, though aware of the state of play in France, tended towards traditionalism, so that we find Vogt and his disciple Hilger,[26] for instance, strongly opposing the French sceptics; but in these countries psychotherapy in general was being gradually infiltrated by psychoanalysis, leading to quite different kinds of conflicts. In Britain and the United States, particularly the latter, a largely indigenous scepticism with regard to the utility of, or necessity for, hypnotic treatment, gained ground for a variety of reasons, most of them, I think, largely unrelated to developments in France. Particularly surprising, and little-known even at the time, was the partial defection of J. Milne Bramwell, who in 1903 had published what was for several decades the best-known of all English-language hypnotic textbooks. In a new book, *Hypnotism and Treatment by Suggestion* (1909) Bramwell tells us that he has now practically abandoned the use of hypnotism. The essence of the whole "hypnotic" situation is, he thinks, an increased suggestibility; "the production of a preliminary imitation sleep is not necessary, and is simply waste of time."[27] In earlier days, he might have to try a hundred times before he could induce "so-called hypnosis". Now he can commence treatment at once and obtain quicker results. His new method is to sit his patient in an arm-chair, and tell him to close his eyes, and try to concentrate his mind on some restful mental picture. He is then given suggestions of two kinds, those intended to facilitate repose, and those that are directly curative. The patient often passes into a drowsy, day-dreamy state, and sometimes into a condition of slight natural sleep.[28] Bramwell's procedure bears a marked resemblance to Sidis's "hypnoidisation", which we have already met. Sidis described the method in a paper contributed to an American Therapeutic Society symposium in 1909. The symposium was published under the editorship of F. H. Gerrish in 1910 as *Psychotherapeutics*, and reveals a rather marked retreat from hypnotism. Only Gerrish himself seems unequivocally in its favour. Morton Prince accords it only a modest place, Sidis has retreated into "hypnoidisation", and E. W. Taylor and G. A. Waterman expound doctrines of "re-education", derived from Dubois of Berne, which sound very like suggestion under another name.

A similar shift is evident in a work at once more popular and on a larger scale, the three volumed *Psychotherapy* (1909) edited by W. B. Parker in the wake of the Emmanuel Movement. In the second volume, Dr. Beatrice M. Hinkle[29] divides the methods of psychotherapy into four categories: hypnotism, which may be of value in selected cases refractory to other methods; suggestion, the most generally applicable method, in which the patient sits comfortably, with his mind fixed on some relaxing and monotonous thought, while the doctor explains what is wrong and how the symptoms will disappear; persuasion, or psychic reeducation, in which the symptoms and their causes are reviewed with the patient, and his good sense and logic appealed to; and psychoanalysis, in which relevant but inaccessible material is dredged up by the method of free association.

That suggestive methods of treatment should have been preferred to hypnotic ones, given that the two are potentially of equal effectiveness, or almost, permits of ready explanation. Simple suggestion can be used with almost any patient, is less liable to disruption by chance factors than hypnotism, and is much quicker and easier to administer.

Hypnotism, attacked by sceptics as having no special effectiveness and numerous practical difficulties, was also attacked during this period by persons who fully believed in its effectiveness, but condemned it on that very account. Some of them even disapproved of all forms of suggestive therapy. I refer to the critics whom I earlier termed the "moralists".

The idea that the practice of hypnotism, or of animal magnetism, is fraught with moral dangers, that it may render the patient liable to seduction, exploitation, domination or demonic invasion, was, as we have seen, already old.[30] However, the "moralists" with whom we are currently concerned objected to hypnotism not as opening the patient to demons or seducers, but as being likely to enfeeble his will and his self-reliance, and as causing an unhealthy dependence on the hypnotizer, amounting in extreme cases to a sort of slavery.

It is hard to trace the ancestry of this kind of criticism of hypnotism, or to see why it had the appeal which in certain quarters it obviously had. In 1892, we find Wundt, always the *Sittenrichter* as well as the *Gelehrte*, describing the dependence of the subject upon the hypnotist as a transient slavery. It is a slavery all the worse in that it removes from the slave not just the right, but the capacity to act freely.[31] Writing in 1904, Hirschlaff notes[32] that many authors, for instance Goldscheider,[33] have taken an ethical standpoint and declared that hypnosis is an unworthy slavery which puts the patient into a slavish dependence on the hypnotizer. He adds that certain hysterical patients can develop a highly undesirable craving for hypnosis. Another writer who thought that hypnotism, where effective, might demoralize the patient and sap his will-power was Moritz Benedikt.[34]

The most influential opponent of hypnotherapy and suggestion therapy was Professor Paul Dubois of Berne (1848–1918), whose *Les psychonévroses et leur traitement moral* appeared in 1904.[35] Dubois had visited Bernheim in 1888, and for a while had applied Bernheim's methods himself. But soon he abandoned them to strike the path which he had left, the path of "rational therapeutics". From *Les psychonévroses*, Dubois emerges as an evangelizing, over-simplifying, moralizing, rationalistic and engagingly naive free-thinker. He is a great advocate of the treatment of psychoneuroses by prolonged rest, isolation, overfeeding, and above all education through "persuasion", which in his terminology means rational demonstration to the patient of the irrationality of his problems and of his responses to them. A therapist who "persuades" his patients believes in what he is saying; one who makes "suggestions" to them does not. "Here we have a professional lie, a justifiable lie, to which I would only have recourse in the event of my bona-fide methods of persuasion not succeeding. I have never found myself in this situation."[36] The aim in therapy of the psychoneuroses is often to cure the patient of his autosuggestibility. How then can it be sensible or right to cultivate his suggestibility and credulity through suggestive methods of treatment?

Les psychonévroses drew some strong responses from hypnotherapists, including Dubois' Swiss compatriots Bonjour and Forel,[37] both of whom are in substantial agreement. Patients are not turned under hypnosis into will-less automata. It is absolutely false, says Forel, "to insinuate that we render people less reasonable and more suggestible by therapeutic hypnotizing: On the contrary, we remove pathological brain dynamisms, and thus render the will and reason freer."[38] Furthermore, Dubois errs in supposing that when, after rational persuasion, he succeeds in curing a patient, the cure is to be attributed to his rational procedures. "On the contrary". says Bonjour, "we are convinced that it is the patient himself who autosuggests the cure under the influence of Dubois' rational suggestions, under the influence of what we call waking suggestion."[39] Reasoning would have no effect on the patient had he not in advance given his complete confidence to the doctor, a confidence produced by his fame or by intense need for a cure. Forel makes similar remarks.[40] Bonjour points out further that in any case many ailments which may be treated by hypnotism or suggestion – for instance *enuresis nocturna* – are inaccessible to reason, so that Dubois' assertions have no relevance to them.

Despite these, and other, criticisms and objections, Dubois' ideas achieved a certain vogue. Dejérine, who had visited him at Berne, was influenced by his teachings, and contributed a preface to *Les psychonévroses*. He contributed a preface also to another work heavily influenced by Dubois, Camus and Pagniez'

Isolement et psychothérapie (1904), and, together with E. Gauckler, himself published a book of somewhat similar tendencies.[41] But it was in the United States that Dubois enjoyed his greatest influence. The English translation of *Les psychonévroses*, published in New York in 1908, had gone through six editions by the following year, whereas the French original had only achieved three. Dubois also contributed a series of articles to Parker's *Psychotherapy*,[42] and his influence was marked on the Gerrish symposium, *Psychotherapeutics*, in which the contributions by Taylor and Waterman bear his stamp,[43] and even the paper by Morton Prince says a good deal about "reeducation".[44]

As reeducation rose into favour as a method of therapy, so hypnotism and suggestion sank more and more into a subsidiary position, or a position of last or limited resort. In the America of Theodore Roosevelt, self-help and individual enterprise were highly valued. The suspicion that hypnotism involved an abnegation by the subject of his self-determination, and a passing of control into the hands of another, was hard to eradicate, and hypnotherapy was therefore in some respects not congenial to the temper of the times. Voluntary "reeducation" was far more acceptable. But high-minded therapy by persuasion and rational reeducation did not enjoy its vogue for long. Soon it was in its turn displaced and almost overwhelmed by a set of doctrines which emphasized the irrational and non-moral foundations of human nature, namely the doctrines of psycho-analysis.

A FINAL SCENE

The state of hypnotism in Europe in the years immediately preceding the First World War may be usefully gauged from the proceedings at the second annual assembly of the International Society for Medical Psychology and Psychotherapy held at Munich on 25th and 26th September 1911.[45] Oskar Vogt was in the chair, a good many distinguished persons were present, and the topic for discussion was "The definition, psychological interpretation and therapeutic value of hypnotism". The discussions were preceded by two position papers by Bernheim and Claparède. I have already used Bernheim's paper to illustrate his later theoretical views.[46] Claparède[47] begins (in opposition to Bernheim) by defending the idea that there is a special state of hypnosis. He thinks that some experiments of his own, to which we shall come, on memory in hypnosis, strongly support this view. He propounds three central questions: What is the subject's state of mind during hypnosis? How is hypnosis produced? And what is its biological significance? He handles the first two of these questions by briefly reviewing current opinions. His approach to the third question is more unusual and more interesting. He begins by pointing out that in a previous work[48] he was able to illuminate many phenomena of sleep by adopting a biological (evolutionary) approach in terms of which sleep is an active function, like a reflex or an instinct of defence. Can we understand hypnosis in the same way? The hypnotized subject may certainly be described as selectively active rather than as merely passive. Now if hypnosis is an active state, what is its function? To progress with this question we have to adopt the comparative method; we have to look at "animal hypnosis". The various interpretations of animal hypnosis all lead to the idea that the immobility is a sort of adaptive reaction, a state of fear, or respect, or resignation, *vis-à-vis* another individual organism. It may be not only fear, but phenomena linked to love, that provoke hypnoidal states in animals, a supposition which, as Claparède notes, brings us close to the views of Ferenczi. He cites his own experiments with a female monkey, which he could put into a state resembling catalepsy by eye-fixation or passes. Perhaps, he says, "these caresses and light touches satisfy her voluptuousness (*volupté*): What is striking, however, is the state of passivity into which the animal, who the moment before was jumping in her cage, is suddenly plunged."[49] We can begin to see human hypnosis as belonging to a group of quite general biological processes, set in action when two beings find themselves face to face, and having the function of regulating the attitude of the one with respect to the other. Another question of biological and evolutionary significance is that of why the state of hypnosis has a curative, or at any rate reparatory, nature, and how it acts on vegetative processes ordinarily outside the influence of the will. Here, Claparède mentions Delboeuf's proposal[50] that hypnosis represents a kind of primitive state of the organism, in which energies that, in the course of evolution, have become increasingly directed upon the outer world, are once again turned inwards. But he holds that such an interior "dynamogenisation" might also be a secondary consequence of the attainment of a state of passivity.

Delboeuf's speculations also provide one possible answer to a further biological question, namely what is the evolutionary origin of the state of hypnosis? Myers voiced a similar view when he regarded hypnosis as a return to a state of "primitive plasticity", whilst Sidis postulates an "hypnoidal state", which is a positive state of repose from which normal sleep is derived.[51] Anastay expresses an analogous opinion.[52]

The discussions (in three languages) were only somewhat loosely related to the contents of the preceding position papers. The first general discussion was on the nature of hypnosis. It had two parts, the first of which was about the reality of hypnosis and objective proofs for this. Bernheim asserted that he who denies the characteristic phenomena of hypnosis denies facts, and Loewenfeld maintained that the weight of evidence is now so great that it is no longer necessary to prove that there is such a thing as hypnosis. The rest of the participants devoted themselves to citing hypnotic phenomena which they thought could not be simulated.

The most interesting contribution was Claparède's. Claparède had performed an experiment (which he had described in his position paper) using two deeply hypnotized subjects. He read these subjects ten words while they were hypnotized, and ten words while they were awake. Then he mixed these words into a list of thirty, which he again read out, asking "Which words do you recognize?" When waking, his subjects recognized the ten words they had heard when awake, and only those ten. Claparède believed it would be impossible for a simulating subject to remember with such accuracy which words had come from which list. Hence the posthypnotic amnesia (and by implication the hypnotic state) was genuine.

Ernst Trömner of Hamburg and L. Frank of Zürich were not convinced that simulation would have been impossible. Claparède replied that, while longer series of experiments were necessary, control experiments showed that people are ordinarily much less precise in distinguishing the words belonging to each list.

Bonjour raised the question of hypnotic anaesthesia. He mentioned a case of a uterine operation in which hypnotic anaesthesia had lasted 45 minutes. This could not be simulated. Bernheim cited hypnotic tooth extractions which he had carried out himself, and Renterghem described a case of extirpation of the womb through the vagina for which he was himself the hypnotist.

Trömner cited the following phenomena as objective proofs of the existence of the hypnotic state: The ability of a hypnotized child who had had an arm rendered cataleptic to hold that arm horizontally outstretched for 20 minutes when five is difficult for an ordinary healthy person; the production under hypnosis of certain kinds of skin phenomena, for example the anaemia of skin rendered hypnotically analgesic, the absence of bleeding from such skin, the appearance on the skin of suggested erythema and oedema; the suppression of certain skin reflexes, for example conjunctival and plantar reflexes; and the suppression of normal pain reactions during hypnotic analgesia. He did not claim that these signs were always present, but when they are present, they cannot be simulated. C. de Montet, of Paris, replied that with a susceptible individual these phenomena could be obtained in the waking state, as with stigmatics.

Frank remarked that in many, though not all, individuals in hypnotic sleep the eyeballs are rolled upwards, and exhibit even, slow, horizontal movements. Vogt disagreed but claimed that alterations of muscle tone provide an objective proof of the reality of hypnosis. It first increases and then decreases. Also he agrees with Trömner about the absence of bleeding from skin rendered hypnotically anaesthetic. He tested it himself in an earlier study with thirty subjects.

The discussion on the nature of hypnosis then moved on to the question of whether hypnosis is ordinary sleep, a new peculiar sleep, or a special state of consciousness.

Bernheim reaffirmed his view that hypnosis and ordinary sleep are fundamentally the same, but no one else fully agreed with him, though de Montet
remarked that those subjects who pass into an ordinary sleep are above all those
hypnotized by means of passes.

Vogt launched into a long exposition of his personal theory. He thought that
the concept of hypnosis should be confined to that of a sleep-like state with
rapport. But what is sleep and what is rapport? Psychologically, sleep is an
inhibition of consciousness. It can, however, take on the special form that only
ideas associated with the hypnotizer remain more or less unrestrained. Then we
have suggested sleep with rapport, the state which he (Vogt) calls hypnosis.
There can be many variations of this state. For Vogt the deepest ideal-hypnosis
is a complete monoideistic waking state of those conscious elements which have
currently some function to perform, and a complete deep sleep of all others. This
he calls a partial systematic waking state.

Trömner maintained that fundamentally, hypnosis involves a one-sided
direction of active attention; the narrowing and concentration of consciousness
on the idea of sleep produces hypnosis. Ernest Jones (1879–1958), best-known
as Freud's biographer, proposed instead that what is crucial is the concentration
of attention on the hypnotizer.

L. Seif, of Munich, observed that in hypnosis one sees produced under the
influence of the hypnotizer phenomena which appear spontaneously in hysteria.
This suggests that a common mechanism underlies both, and it is this
mechanism that requires investigation.

Forel claimed that hypnosis is pathological suggestibility.

Vogt thought that there is a distinction to be made between the suggestive
working of relatively unemotional ideas, and the dissociated working of powerful
feeling-complexes. The same psychophysical event can be produced by either.
But in the former case the event can be modified by all sorts of other ideas, in the
latter case not, and there is not the heightened emotivity characteristic of
hysteria. Not the dissociation state as such, but its origin and (linked to this) the
degree of its modifiablilty distinguishes hypnosis from hysteria.

The next general topic to be taken up was that of modifications of
consciousness and of the nervous system in hypnosis. The first question
addressed was that of whether or not suggestibility is heightened in hypnosis.
Bernheim claimed that hypnosis as such does not heighten suggestibility. Hilger
alleged that Bernheim had misled himself by his own outstanding abilities as a
hypnotist. He could do things that others could not, and so had overestimated
the possibilities of waking suggestion.

Vogt maintained that suggestibility is heightened in hypnosis, the more so the
deeper the hypnosis. Contrary observations can be explained in terms of his
theory. Heightened suggestibility is displayed in hypnosis only in the setting of
a partial waking state, i.e. when some conscious elements are actively

functioning but the remainder are restrained or quiescent. When tests of suggestibility fail it is either because the elements tapped are quiescent, or because they were not comprehended in the rapport relationship.

The next part of the discussion of modifications of consciousness and the nervous system in hypnosis was concerned with hypnotically induced catalepsy and was not very interesting, though it was enlivened by Claparède's account of some experiments which he had conducted with his monkey. When he held this beast, stroked it several times (or made "passes"), he succeeded in evoking a state of catalepsy, in which, if one bent or stretched the legs of the hitherto wild animal, they remained in the position concerned. He thought the phenomenon could hardly be explained through suggestion.

A concluding part of the discussion of modifications of consciousness and the nervous system under hypnosis took up the issue of heightened mental capacities. Trömner described some of his own (rather crude) experiments on the matter.[53] With children of ten to twelve years, he has found a halving under hypnosis of the time taken to do addition problems. He found, too, with uneducated persons knowing nothing of the purpose of the experiment, a marked lowering under hypnosis of the absolute visual threshold, and a smaller, but distinct, lowering of the threshold for auditory stimuli. On the other hand, Chojecki stated that experiments which he had carried out to compare mental processes in the waking state and in hypnosis seemed to prove that mental activity in hypnosis is enfeebled. He found that repetition of series of numbers, and learning of series of nonsense syllables, were poorer in hypnosis than in the waking state.[54]

To explain these disagreements, Trömner distinguished with Vogt between two different kinds of sleep-inhibition, a general diffuse falling asleep, which leads into natural sleep, and a dissociated falling asleep (i.e. one in which the activity of parts of the brain only is inhibited), which occurs spontaneously in hysteria, and is artificially induced in hypnosis. In the latter, mental functions which are still waking can be heightened. Vogt agreed. Those experimenters who have found diminished capacity have been experimenting with a different hypnotic state from the one he has termed a partial systematic waking state. He warned that to reach this state a subject may require extensive training.

The next main topic of discussion was the induction of hypnosis, with particular reference to the significance of fear and liking. This subject gave the two Freudians, Seif and Ernest Jones, scope to air their convictions. Seif adopts from Freud and Ferenczi the view that what is important in bringing about hypnosis is the re-creation between subject and hypnotist of the subject's relationship to his parents. This relationship has two aspects, the maternal and the paternal, love and fear. Forel objected that very young children are difficult subjects, that it is difficult for men to hypnotize their wives, and that strangers, who possess neither fear nor love for the hypnotist, may be good subjects. To this, Seif produced the classic Freudian counter that peoples' unconscious mental attitudes may be the reverse of their conscious ones.

Trömner asked what proof the Freudians had for their assertions. He objected to the claims of the Freudian school that suggestion and love are identical; no proof of this assertion has he found in any writing of the school in question. And there are good reasons for rejecting it. Those who are closest to us are the hardest to hypnotize; on the other hand, the more objective the relationship one stands in to someone, the better one can hypnotize him; fear is an especial hindrance to hypnosis. Vogt agreed with Trömner. Hypnotizing people in 1895, he had a very high success rate. He used group procedures. Afterwards he gave these up, and even gave up showing each patient another patient being hypnotized. His success rate fell markedly. He can only conclude that the later patients were more sceptical. Patients on whom he has spent much trouble, and who have therefore become more friendly towards him, have not become correspondingly easier to hypnotize. Furthermore, he does not see how the Freudians can derive from the father-complex the fact that ordinary sleep can be turned into hypnosis.

Seif, who had all the enthusiasm, and the absence of critical spirit, of a fresh religious convert, replied with a long reassertion of the Freudian position, and the discussion turned more and more into a rather scrappy series of arguments about Freudian claims in general. Seif asserted that although hypnosis can obtain quick cures by setting the unconscious aside, the psychoanalyst wants to know the causes of the symptoms. He seeks to enlighten the patient as to the causes of his illness. Vogt pointedly asked where the statistics were that proved the better therapeutic results of psychoanalysis. He had never had a patient whom a Freudian could help, but he could not. On the other hand, he knows of many who were not helped by Freudians but improved under the care of non-Freudians.

A final period was consigned to the therapeutic value of hypnosis. Unfortunately it was turned by the Freudians largely into a discussion of the supposed dangers of hypnotism, and the possible psycho-sexual dependence of the subject on the hypnotist.

This brief sketch has shown the more traditional hypnotists, Forel and Vogt and their allies, under some pressure from the psychoanalysts in Germany and the advocates of waking suggestion in France.[55] Although well aware of these pressures, and aware of their own diminishing constituency, they respond vigorously and do not seem to regard themselves as seriously beleaguered. And indeed, few of the participants challenged the reality of hypnotic phenomena and the possibility (as distinct from desirability) of hypnotic cures. It is abundantly clear that the hypnotic movement, though no longer the force it had been, had not subsided anything like as drastically as Janet implies. In hindsight, the most significant contributions were those of Claparède. In a small, but interesting, way he had been making serious and systematic attempts, not unique but without many parallels, to compare the performances of hypnotized subjects with those of waking controls in standardized experimental situations. Here, had the participants but known it, was a sign of things to come. Some aspects of those things to come will be touched on in the concluding chapter.

NOTES

1 Ellenberger (1970), p. 171.
2 Janet (1925), Vol. 1, p. 200.
3 Janet (1925), Vol. 1, p. 200.
4 Janet remarks that the current indices compiled by Baldwin and Ebbinghaus show a sharp drop in hypnotic items. He is referring to *The Psychological Index*, published as an adjunct to the *Psychological Review*, and the *Zeitschrift für Psychologie und Physiologie*. The drops seem to have been a matter of editorial policy rather than of a drop in papers published.
5 The figures for the 1900 Congress by country (figures for the 1889 Congress in brackets) were: Germany 9(5); Austria-Hungary 2(2); Belgium 3(6); Brazil 2(2); Canada 4(1); Ecuador 1(1); Spain 2(2); USA 23 (6); Egypt 2(0); Cuba 1(0); Colombia 1(1); Great Britain 6(5); Haiti 1(1); France 101(103); Greece 3(5); Holland 2(3); Iceland 1(0); Ireland 0(1); Italy 6(2); Mexico 2(1); Persia 1(1); Poland 8(0); Roumania 3(1); Russia 15(15); Sweden and Norway 3(2); Switzerland 3(4); Turkey 5(1); Venezuela 1(1).
6 It is quite clear that in the first series not all possibly relevant entries have been indexed under the headings of hypnotism and suggestion. The system of indexing has changed somewhat in the second series, and I have reached my figures by examining and categorizing the actual entries rather than by counting items in the index.
7 I have not included the French volumes which partially bridged the gap in the sequence.
8 They became, respectively, the *Journal für Psychologie und Neurologie*, and the *Revue de psychothérapie et de psychologie appliquée*.
9 Cf. e.g. Boirac (1908); Joire (1908); Rochas d'Aiglun (1892); Filiatre [1906].
10 Delboeuf (1891–2).
11 For instance Delboeuf describes a very troublesome hysterical lady who had remained in her bed for months, not eating or sleeping, and the scourge of those around her. She had herself practised hypnotism, and laughed in the faces of the doctors who tried to hypnotize her. Delboeuf was summoned from the country, and was shortly installed at her bedside. He told her point blank that there was nothing the matter with her, and that her muscles were in excellent working order. The infuriated patient proposed to demonstrate the completeness of her paralysis. "She placed herself on her bottom at the edge of the bed, raised herself and got ready to fall. Her husband and the nurse rushed to her to hold her. For my part, I pushed them aside brusquely, and plunging my gaze into her, said, with imperious voice and gesture: 'No comedy, madame, walk.' And she walked, mastered. That very evening she ate a steak and the next day again shared the conjugal bed." Delboeuf (1891–2), p. 131.
12 Bernheim (1897a), p. 9.
13 Hartenberg (1897–8c).
14 For examples of cases treated by waking suggestion by Hartenberg and Valentin and others see Crocq (1900), pp. 582–603.
15 Hartenberg (1897–8a).
16 E.g. Bernheim (1897–8b).
17 Valentin (1901).

18 Hartenberg (1901), p. 666.

19 For Hartenberg's more developed views see Hartenberg (1897–8b; 1908; 1910; 1912).

20 Hartenberg (1897–8c), p. 214.

21 See above, pp. 314–315.

22 Quoted in Hilger (n.d.) p. 9.

23 Bérillon was particularly bitter about Bernheim's apostasy. In a history of hypnotism which he presented to the Second International Congress of Hypnotism in 1900, he mentions Bernheim only once. Barrucand (1967), p. 181, quotes a diatribe by Bérillon against Bernheim, but gives an erroneous reference. I have not been able to trace the original.

24 Hartenberg (1897–8c), p. 221.

25 Preda (1910–11); Barrucand (1967), p. 181.

26 Hilger (n.d.) is indignant at Bernheim, Hartenberg and Valentin, but reserves his strongest denunciations for Dejérine and Dubois and the advocates of "persuasion".

27 Bramwell (1909), p. 164.

28 Bramwell (1909), p. 168.

29 Hinkle (1909).

30 Cf. pp. 118, 168, 420 above.

31 Wundt (1893), p. 157.

32 Hirschlaff (1905), pp. 305–306.

33 Presumably A. Goldscheider (1858–1935), the eminent neurologist.

34 Benedikt (1894), pp. 36, 54, etc.

35 I have used the English translation (Dubois, 1909b). On Dubois and his influence, see Janet (1925), Vol. 1, pp. 98–120.

36 Dubois (1909b), p. xi.

37 Bonjour (1905–7); Forel (1906), pp. 234–238.

38 Forel (1906), p. 237.

39 Bonjour (1905–7), p. 359.

40 Forel (1906), p. 235.

41 Dejérine and Gauckler (1911).

42 Dubois (1909a).

43 See above p. 564.

44 On the origin of reeducation as a method of therapy, see Janet (1925), Vol. 2, pp. 710–732.

45 Verhandlungen ... (1912).

46 Bernheim (1911). See above, pp. 545–547.

47 Claparède (1911).

48 Claparède (1905).

49 Claparède (1911), pp. 510–511.

50 Delboeuf (1887a).

51 Sidis (1909).

52 Anastay (1908–9).

53 See Trömner (1913).

54 See Chojecki (1912).

55 Similar lines of division were to be found in other psychological and psychiatric congresses of the time. Cf. Ellenberger (1970), chapter 10.

Epilogue: Barber and beyond

It is, at a certain level, easy enough to chart the fluctuating fortunes of the mesmeric and hypnotic movements. Those fluctuations can be traced through indicators (books, public demonstrations, newspaper reports) that would have been obvious at the time. Harder to pursue, because complex and multifaceted and indefinitely ramifying, are the interactions of the mesmeric and hypnotic movements with other intellectual and social movements and tendencies of their times. I have touched on some of these interactions – for instance those between mesmerism and nature-philosophy and between mesmerism and the radical movements associated with phrenology – but detailed consideration of them has lain outside the purposes of this book, and would, indeed, require a separate, larger and differently oriented volume. In its heyday, from the mid-1880s to the turn of the century, hypnotism was a social and intellectual force of some significance. The possible bearings of hypnotism and suggestion upon crime, upon eyewitness testimony, upon crowd behaviour and public order were, as we have seen, of concern to lawyers, criminologists, sociologists, and hypnotists themselves. In France, hypnotic phenomena were among the munitions deployed in the frequent sniping and occasional overt skirmishes that went on between the Church and liberal anticlericals disposed to put a psycho-pathological interpretation, fortified by Salpêtrière case histories, on certain exotic religious phenomena.[1] On contemporary psychiatry and psychology, the influence of the hypnotic movement, and of the abnormal phenomena with which it was associated, were profound. Psychotherapy as a self-conscious endeavour emerged from hypnotic practices in the later 1880s; indeed at first psychotherapy and hypnotherapy were synonyms. Psychoanalysis and other non-hypnotic psychotherapies developed in the following decade in part as a reaction against hypnotherapy. In psychology, the period between 1890 (when James's *Principles* was published) and 1914 (by which time McDougall's earlier works had appeared) was one of unprecedented movement and sense of intellectual adventure, and hypnotism and the odder kinds of hysterical phenomena forced a radical reappraisal of previously accepted views as to the organization of human personality upon all who were not prisoners of a narrow structuralism.

All these developments are susceptible of documentation, though disinterring that documentation might in some instances be a considerable task. But there is an important aspect of the history of the mesmeric and hypnotic movements on which it is almost impossible to get an adequate grip, not from shortage of materials, but because the materials are scattered, fragmentary, and dubiously representative of that of which they are fragments. I refer to the ideas about mesmerism and hypnotism entertained by members of the general population, including those of very limited educational attainments. It was recognized even by early mesmerists that a subject's ideas about animal magnetism might have an important influence on the phenomena which he would be likely to exhibit. That subjects' beliefs about hypnotism might function as covert autosuggestions was well understood by turn-of-the-century hypnotists and has been often pointed out since. So clearly the question is one of some significance for the understanding of hypnotic phenomena and of the supposed state of hypnosis. But unfortunately we can only guess at popular ideas about hypnotism – mainly from examining the sources that may be supposed to have reflected or inspired them.

There were, of course, a good many literary sources of very varying degrees of "literariness". There were periods – most notably in early-nineteenth-century Germany, with the rise of Romanticism, and in *fin-de-siècle* France and Vienna – when some writers of fiction and drama systematically exploited current psychopathological notions, including ones linked to mesmerism and hypnosis, to lay bare the hidden roots of human conduct.[2] But this was a somewhat intellectual literature. More widely influential were the excursions into mesmerism and hypnotism of authors who were tellers of tales rather than self-conscious delvers into the recesses of the psyche. Dumas and Balzac each introduced mesmerism into a novel (*Mémoires d'un médecin* and *Ursule Mirouet*). Poe's "The facts in the case of M. Valdemar" (1845), which narrates the ill consequences of mesmerizing a dying man, and the even worse consequences of waking him up,[3] made a considerable impact, and was even taken for a genuine case report.[4] His "Mesmeric revelation" is an essay, in the guise of fiction, on the significance of mesmeric phenomena for our view of man's place in the cosmos. Frédéric Soulié, in *Le magnétiseur* (1834), and the anonymous author of *The Power of Mesmerism, A Highly Erotic Narrative of Voluptuous Facts* (1880),[5] present us with the possibility of turning mesmerism to evil uses.

Among novels in which hypnotism plays a central part, the most famous was George Du Maurier's *Trilby* (1894), which enjoyed huge success both as a book and as a play. It conveys, however, some very peculiar ideas about hypnotism, as does John Buchan's well-known adventure novel *The Three Hostages* (1924). In works of fiction, the emphasis is generally upon the coercive powers of the hypnotist, and to an extent on the occult powers of the entranced subject. Probably most people took these avowed fictions with a pinch of salt. What is of interest, however, is the fact that many story writers could introduce mesmerism or hypnotism, often quite casually, with no need of explanation.[6]

In part, of course, ideas about mesmerism and hypnotism came from popular and semi-popular books and articles, which were numerous. They varied very much in critical spirit, but it was no doubt the more startling claims which stuck in readers' minds.

Knowledge of mesmerism and hypnotism, however, was to be found among persons unlikely to have done much reading. Mesmerists and hypnotists alike looked upon subjects (usually peasants) who had no antecedent knowledge of the subject as quite a find.[7] How the relatively unlettered acquired their ideas about mesmerism and hypnotism must remain largely conjectural.[8] In early mesmeric days, word of the goings-on at magnetic clinics probably spread by word of mouth. Then came the period of the itinerant magnetic demonstrators, some of whom attracted large audiences and a good deal of publicity. It goes without saying that the claims about mesmerism made by these demonstrators, though not always wild, did not err on the side of caution.

Caution was largely abandoned by the stage hypnotists who began to appear in increasing numbers towards the end of the century. They ranged from Madam Card, before whom Cambridge undergraduates fell like flies in the late 1880s,[9] to "Doctor" Bodie, who combined electrical wizardry with hypnotism, and claimed if challenged that the letters "M. D." which he had assumed stood for "Merry Devil".[10] Stage hypnotism was closely associated with conjuring,[11] and it early became *de rigueur* for conjurers to "hypnotize" their assistants in second-sight or levitation acts.

We may reasonably conclude, I think, that some knowledge of mesmerism and of hypnotism, though vague and sensationalized, probably became quite quickly diffused across wide sections of the populace in Europe and America, including sections that were low in literacy. Among the ideas about hypnotism current towards the end of the nineteenth century would probably have been: that a hypnotized subject is in a peculiar trance state resembling somnambulism; that afterwards he will remember nothing of what has transpired; that he may exhibit clairvoyance; that he is a mere automaton under the hypnotist's complete control; that there is a peculiar, almost mystical, relationship between hypnotist and hypnotized subject; that a hypnotist's eyes, like those of the Ancient Mariner, may have a strange and spell-binding power (depicted in cartoons even today by emergent lines of quasi-magnetic force); that hypnotic induction is largely a matter of the powerful will of the hypnotizer overwhelming the feebler will of a delicate or neuropathic and therefore vulnerable subject.

Now some of these ideas – particularly the ones about automatism and the post-hypnotic amnesia of the hypnotized subject, and about the affinity of hypnosis with somnambulism and neuropathy – were in fact held by some medical and academic hypnotists, especially in the period 1875–1885. But by

1900 many moderate and able hypnotists, for example Bramwell and Loewenfeld, had, as we have seen, developed considerable reservations about them, and Bernheim had ceased to believe in the "state" of hypnosis at all. Hypnotherapists found it desirable to give each patient at the outset a reassuring chat about hypnotism to dispel erroneous and anxiety-provoking notions that might have interfered with treatment. Yet when today's hypnotists set out to characterize late-nineteenth-century, or "traditonal", ideas about hypnotism, the views that they set up tend to be those of the general populace rather than the much more cautious views expressed by leading hypnotists of the period.

HYPNOTISM SINCE THE FIRST WORLD WAR

Among the ups and downs of the mesmeric and hypnotic movements there were one or two, for example the last two decades of the nineteenth century, which one might refer to as "golden ages". "Ormolu" would probably be nearer the mark, but for present purposes I shall stick to the more extravagant term. The period from the end of the First World War to the end of the 1950s was not a golden age. During this period, hypnotism remained in some respects curiously static. It had fallen somewhat out of fashion as a method of treatment, and progressed only at a moderate, though accelerating, pace as a subject of experimental study. Despite some inputs from behaviourism and psychoanalysis, the most commonly adopted conceptual frameworks reflected late-nineteenth-century ways of thought.

In passing this period largely by, I do not wish to imply that it was devoid of interest and importance. On the experimental side, there were advances of lasting significance, particularly in methodological matters. A growing sophistication in experimental design and the application of statistics was greatly accelerated by the publication in 1933 of C. L. Hull's *Hypnosis and Suggestibility*, while the later part of the period saw the evolution of the scales of hypnotic susceptibility which, in more developed form, are so much used by hypnotic experimentalists and indeed clinicians today.

Clinical hypnotism, after a brief upsurge during and immediately after the First World War, when its use for the treatment of shell-shock and kindred disorders received some notice,[12] was, like other psychotherapeutic methods, heavily overshadowed for several decades by psychoanalysis. A revival of interest in hypnotherapy began with World War Two. It centred around post-traumatic stress, but it diversified quite adventurously, so that we find hypnotism used as an adjunct to analysis by promoting dreams, age regression, etc., or by eliciting fantasy material and the symbolic acting-out of conflicts.

Materials for a history of hypnotism during this period have been ably presented by a number of writers,[13] and, though I do not wish to belittle such important contributions as those of (say) Hull, Erickson, Sutcliffe, Pattie, R. W. White, Weitzenhoffer, and Eysenck and Furneaux, I shall not deal with them here. Instead, I shall plunge straight into the period from 1960 to the present. For this is a period which has some claims to be considered a "golden age". Certainly it has been an age of striking productivity, of methodological innovations and theoretical turmoil, and a very brief sketch of its leading features will serve as a convenient basis from which to consider certain general and historical issues.

If ours is a golden age, it has to be said that the gold has been somewhat unevenly distributed. The most striking developments and controversies, though not without clinical repercussions, have been on the experimental side. On the clinical side, though there has been a great diversification in the variety, ingenuity and theoretical background of the techniques used,[14] much of the literature continues to take the form of studies of individual cases, or of a few cases, without untreated, or non-hypnotically treated, controls, and has changed little in this respect since the end of the nineteenth century. To the experimentally orientated, such slackness seems deplorable. A clinician could perhaps reply that setting up proper control groups is very difficult. Most hypnotherapists do not have access to large subject pools, and their clientèle, which is largely self-selected, has come specially for a certain sort of treatment, and is not properly comparable to any likely control group. And indeed those writers who reviewed the clinical literature in search of adequately controlled studies could at first find little evidence that hypnotic treatment as such brings any benefits at all to patients, once methodological errors have been allowed for. These errors included failure to assign patients randomly to groups, failure to control against placebo effects, failure to assess the relationship of hypnotic susceptibility to treatment outcome, failure to give details of the induction procedures and suggestions used, and failure to set up procedures for independent assessment of therapeutic benefit. Doubts were cast on the utility even of such favourites as hypnotic treatments of smoking, of skin disorders, headaches and obesity, and of alcoholism.[15] It was not that the patients did not improve under treatment; it was rather that the rôle, if any, of the specifically hypnotic aspects of the treatment remained obscure. In this respect, the situation had not been greatly clarified since the nineteenth century.

However, in a classic paper of 1982, reviewing a fair number of studies in six different areas of hypnotherapy, Wadden and Anderton came to more complex conclusions. With regard to the hypnotic treatment of obesity, smoking and alcoholism, they found that although benefits followed the treatment, these were probably related to non-hypnotic factors, especially the motivational level of the patients. On the other hand, hypnotic treatment did seem to be effective in cases of clinical pain, warts and asthma. Thus their review confirmed an earlier speculation by Perry and collaborators in 1979 that hypnotism is less effective in disorders involving self-initiated behaviour than in largely psycho-somatic disorders.

That hypnotism may be especially effective in the treatment of psychosomatic disorders receives a certain support from recent successes in the treatment of external and internal haemorrhages, burns, stress symptoms, allergies, and other problems linked to autonomic and immune system functioning.[16] There is now evidence that the nervous system and the immune system are closely interconnected both by direct innervation of organs specially involved in immune responses (thymus, bone marrow, lymph glands),[17] and by a great variety of neurohumoral and neurochemical pathways.[18] Thus lymphocytes have recently been shown to possess receptor sites for various hormones and neurotransmitters, including catecholamines and other stress-related sub-stances, and for a variety of neuropeptides, including the endogenous opioids. Some of the substances have immuno-suppressive and some immuno-en-hancing effects. It has even proved possible to condition certain immune system responses.[19] All this brings us a very small step, but perhaps a very important step, towards understanding how it may be that forms of psychotherapy, including hypnotherapy, can bring about effects which must, seemingly, be mediated by the immune system.[20] Further clarification of these important issues may in addition throw some retrospective light on the therapeutic *modus operandi* of certain mesmeric procedures.

Traditionally, hypnotism has been regarded as a form of treatment especially applicable in cases of hysterical and neurotic complaints, a view commonly propounded being that disorders such as hysteria, in which suggestion and self-suggestion play important causal rôles, ought to be particularly amenable to hypnotic treatment. Furthermore, hysterics should be unusually suggestible. More recently, it has been claimed both that hypnotism is a viable method of treatment in cases of phobia and that phobics as a class are more than usually suggestible.[21] However, hysteria as a diagnostic category has gone out of fashion, and disorders that in the nineteenth century would have been labelled hysterical are divided out by DSM-IIIR into or among a number of different categories: post-traumatic stress disorder (PTSD), somatoform disorder (Bri-quet's syndrome), conversion disorder, and dissociative disorder. All in one way or another involve responses to psychological stress, for example conversion of the stress into physical symptoms, or excluding it from the main stream of psychological functioning. There is evidence that in PTSD, and also in an extreme form of dissociative disorder, multiple personality disorder (MPD), the sufferers are unusually suggestible;[22] and for MPD hypnotism remains, as it has traditionally been, the commonest form of treatment.

Though we may not be in a golden age of hypnotherapy, we are undoubtedly in not just a golden age but *the* golden age of multiple personality disorder. A modest increase in the number of MPD cases during the 1960s has become a positive flood since 1980. However, this flood has been largely a North American phenomenon,[23] and even in America it is much more apparent to some psychiatrists than to others, a fact which has occasioned some controversy.[24] Whatever the cause of the very uneven distribution of cases, there is no doubt that sufferers from MPD have often undergone childhood abuse or trauma,[25] in response to which they have retreated into fantasies involving imaginary companions or change of identity. Bliss and others have argued that what underlies MPD is "the patient's unrecognized abuse of self-hypnosis" whilst endeavouring to cope with intolerable stress.[26]

I said above that the present "golden" age of hypnotism is golden mainly in respect of its experimental side. To that side we must now turn. The materials are very copious, and all I can do is outline some of the important issues. For our purposes, it will be best to do so by relating them to certain current theoretical positions. To this end, I shall consider firstly the views and impact on the field of T. X. Barber; then the "neodissociationist" position of E. R. Hilgard and his followers; and finally the social psychological approach of N. P. Spanos and his collaborators. It is of interest to note that each of these three approaches is associated with characteristic views as to the nature of multiple personality and other dissociative disorders.

THE BARBER REVOLUTION

The theoretical and methodological debates which we have to discuss began rather suddenly in the early 1960s, and were principally, though not exclusively, initiated by T. X. Barber of the Medfield Foundation. Barber adopted a strongly positivistic or operationalistic view of psychology.[27] All important variables are to be specified by concepts that are close to observable events, and the aim of a science is to establish empirical relationships between dependent and independent variables that have been thus specified. He also talks of "mediating variables", which appear to be hypothetical dependent variables that can in turn act as independent variables related to the observable dependent variables under investigation.

Now from this point of view the concept of hypnosis, or trance, as interpreted by Barber, is a very unsatisfactory one. Hypnosis is a state indexed by the hypnotized subject's high level of responsiveness to suggestion; yet a subject's being in that state is supposed to give rise to that responsiveness. Hypnosis itself is not indexed independently of the phenomena of which it is the alleged antecedent – no stable physiological concomitants of hypnosis have been observed. As a mediating variable, it contributes nothing and can be dispensed with, particularly if we can show (as Barber believed he could) that the phenomena can be fully accounted for in terms of their functional relationships to adequately specified independent and mediating variables, especially attitudes, motivations and expectancies.

In pursuit of this last aim, Barber undertook an intensive research programme, becoming probably the most prolific author in the history of hypnotism prior to the advent of N. P. Spanos. With respect to certain hypnotic phenomena (amnesia, enhanced muscular performance, certain physiological effects, arm levitation), he tried to show that they may equally well be obtained with unhypnotized subjects in whom positive attitudes, motivations, expectancies, etc., concerning the hypnotic task have been inculcated. With respect to certain other phenomena (the more startling or bizarre ones, such as hallucinations, age regression, enhanced recall, hypnotic blistering), Barber tried to show either that they do not occur as reported, or that they have been greatly exaggerated and may be obtained with unhypnotized subjects as above.

Central to Barber's pursuit of the first of these aims were questions of experimental design. He was particularly critical of designs which involve the comparison of the same group of subjects under hypnotic and non-hypnotic conditions. Subjects may guess what the experimenter expects or hopes for; they may even "hold back" under the non-hypnotic conditions. Barber's own preferred design was to assign independent groups of subjects, drawn at random from the same population (and usually not selected for hypnotic susceptibility), to different treatment conditions, "hypnotic", "task-motivated", and "base level". Tape-recorded instructions and test-suggestions minimized the possibilities of influence from the hypnotist's tone of voice and from hypnotist-subject interactions. For his test suggestions, Barber often made use of the Barber Suggestibility Scale which contains eight graded suggestions – arm-lowering, arm levitation, hand lock, thirst hallucination, verbal inhibition, bodily immobility, "post hypnotic-like" response and selective amnesia.[28] It is scored both objectively in terms of the number of suggestions observably responded to, and subjectively in terms of whether or not the subject "felt" the suggested effect or simply complied. Barber's findings[29] may be summarized as follows:

1. Under base-level conditions most subjects responded to some of the test suggestions.

2. A small proportion of subjects (about one sixth to one fourth) showed a high level of base-rate response to test suggestions, both on the objective and the subjective scoring systems.

3. Standardized hypnotic induction procedures raised objective and subjective scores on the BSS by about 2.5 points (out of 8) above the base level.

4. Task-motivated instructions facilitated responsiveness to test suggestions by about the same amount.

The conclusions that might be drawn are obvious. For decade after decade, hypnotists and mesmerists were deceiving themselves as to how they produced their supposed results. They thought they produced them by inducing in their subjects a special state of trance. In fact, however, they may have produced them by (more or less) unwitting manipulation of their subjects' motivations, attitudes, expectancies, and so forth. It certainly is more parsimonious to suppose this, than to postulate a mysterious state of "hypnosis".

Barber has had a stronger influence on both conceptual and methodological aspects of contemporary hypnotism that any other worker, and that influence has on the whole been salutary. But is is impossible to contemplate the views which he developed during the 1960s (he has himself moved away from them, though they still figure among the basis assumptions of some of his former associates) without developing considerable reservations.

In the first place, it is a great over-simplification to allege that the older hypnotists defined hypnosis circularly as a state characterized by enhanced responsiveness to suggestion. Many of them thought that the state of mind of a hypnotized person has subjective characteristics – relaxation, dreamy mentation, a sense of drifting with the tide of events, a diminished awareness of outer events, a lowering of associative activity combined with an increased vividness or forcefulness of certain systems of ideas – which *explain* his increased suggestibility. Indeed, suggestibility might sometimes diminish with increased "depth" of hypnosis. The relationship between the two was thus open to empirical study and might vary from one person to another.[30] We may perhaps suspect that the accounts given by these workers of the subjective state of hypnotized persons were contaminated by theoretical presuppositions, but that is beside the point. Certain modern theorists – for example Shor, Orne and Tart[31] – have likewise claimed that hypnosis is a distinct psychophysiological state linked only contingently to enhanced responsiveness to suggestion.

In the second place, Barber's frequent (not invariable) use of groups of unselected subjects means that most of his conclusions are reached on the basis of experiments in which only a small number of really good hypnotic subjects has been tested. The influence of these subjects will of course be heavily diluted, thus stacking the odds against the theory that there is something special about the state of hypnosis.

The most widely-voiced criticisms of Barber[32] relate, however, to the use of task-motivational control groups. Barber seemed to suppose that if, on some conventional "hypnotic" task or test, the performance of a task-motivated group does not differ significantly from that of an "hypnotic" group we are justified, on some principle of parsimony, in assuming that the hypnotic group was actually task-motivated. But obviously the same end-result might be produced in different ways in the two groups. It has been forcefully suggested,[33] with supporting experimental data, that Barber's task-motivational instructions bring such social pressure to bear upon subjects to comply with the demands of the situation that few would not bend a little. It is not at all obvious that subjects who are hypnotized by conventional induction procedures are put under comparable pressure.

Barber pursued the second part of his campaign, that of more or less eliminating certain rather startling alleged findings (for instance ones involving hypnotic blistering and skin markings, and hypnotically engendered hallucinations) partly by sharp criticism of previous experiments (his criticisms of hypnotic blistering[34] are very like those advanced by Schrenck-Notzing in 1895) and partly by experiments of his own.[35] However, more recently Barber has very considerably modified his stance. He now accepts the occurrence of hypnotic blistering and skin marking, and accepts that some exceptional persons may have the ability to hallucinate "as real as real"; though he would still, I imagine, deny that these phenomena require a special state of "hypnosis" for their production.

BARBER: THE POST-REVOLUTIONARY PHASE

In the early 1970s Barber began to move from what Fellows[36] has called his "essentially destructive period" into a "mainly constructive phase". He allied himself to an extent with other contemporary trends in hypnotism, especially a tendency to emphasize the importance in the production of hypnotic phenomena of a subject's capacity for "imaginative involvement".[37] He developed a "cognitive-behavioural" theory of hypnosis,[38] which postulated that subjects respond to test-suggestions "to the extent that they think along with and imagine the themes that are suggested."[39] He devised a new "think-with" set of instructions to accompany test-suggestions (it has affinities with other "active-alert" methods of inducing a high level of responsiveness to suggestion[40]), and (with Sheryl C. Wilson) a new Creative Imagination Scale (CIS).[41] He showed a growing interest in the therapeutic applications of these ideas. And he became rather more receptive to reports of certain of the odder phenomena of hypnosis with regard to which he had previously had considerable reservations.[42]

Out of this combination of therapeutic work with interest in imaginative involvement has arisen what may prove to be the most interesting of all the lines of inquiry in which Barber has been involved. Wilson and Barber encountered two patients each of whom scored highly on the CIS and in addition proved to be an outstandingly good hypnotic subject. In the course of therapy, it became apparent that these two ladies shared a number of unusual characteristics – they had "a profound involvement in fantasy, spent much of their life in fantasizing, and typically experienced their fantasies 'as real as real' (at hallucinatory intensities)."[43] This led Wilson and Barber to conjecture that other persons who score highly on the CIS, and are in addition excellent hypnotic subjects, may share these characteristics. They located twenty-seven subjects (all female) who rated highly both on the CIS and as hypnotic subjects, and administered to them (as to twenty-five control subjects) a Memory, Imagination, and Creativity Interview Schedule of 100 questions. All but one of the twenty-seven subjects, and none of the control group, turned out to be "fantasy-prone personalities" or "gifted fantasizers". As children they had lived in a make-believe world much of the time, often in reaction to an oppressive or even abusive home life. Many thought that their dolls and toy animals were alive. Most believed in fairies, leprechauns, elves, guardian angels, etc., and some had actually seen such beings "as real as real". More than half had had imaginary companions, whom they reported having clearly seen, heard and felt.

Most continued their fantasizing into adulthood, and regarded it as an integral part of their lives. Almost all had vivid sexual fantasies "that they experience 'as real as real' with all the sights, sounds, smells, emotions, feelings and visual sensations."[44] Their memories might also attain an hallucinatory vividness. Fantasies were at times accompanied by appropriate physical symptoms. Many of the fantasizers regarded themselves as psychic, as having mediumistic gifts, etc., and had had out-of-the-body experiences, religious visions, etc.

Out of this study two obvious lines of reflection emerged. The first was that there could very well be a relationship between being a gifted fantasizer and becoming a victim of multiple personality disorder. As Rhue and Lynn point out, "there are many parallels between the attributes and developmental characteristics of fantasizers and multiples."[45] Multiple personality patients as a class are highly hypnotizable,[46] and tend to have had dreadful childhoods, from which they escaped into realms of fantasy. Many cases of multiple personality seem in fact to have begun in childhood.[47] For fantasizers to respond to unbearable stress by totally immersing themselves in imagined or suggested other identities seems a sensible and possibly life-saving strategy, and there is no doubt that many MPD patients belong in the class of gifted fantasizers. Certain cases of fugue may represent a less developed version of the same tendencies.[48]

The second (advanced by Wilson and Barber) was that these subjects constantly underwent experiences resembling the classic hypnotic phenomena. Perhaps good hypnotic subjects simply *are* fantasy-prone individuals who are capable of performing these feats anyway, without benefit of hypnotic induction. Hypnotic induction encourages them to do and talk about things concerning which they would normally have remained silent.[49]

However, Wilson and Barber go on to note that not all highly hypnotizable subjects are gifted fantasizers. Furthermore, it seems unlikely that if hypnotic subjects "good" enough to hallucinate, rôle-play, etc., had been doing so all their lives, this fact would not have emerged long ago. Although the findings of Wilson and Barber as to the existence and characteristics of a class of gifted fantasizers have received considerable, though not complete, confirmation from the later studies of Myers and Austrin and of Lynn and Rhue,[50] the link between fantasy proneness and hypnotic susceptibility has not emerged as strongly as was originally thought. In general, we may say that while there is a substantial overlap between fantasy proneness and high hypnotic susceptibility, there are none the less not a few individuals who are highly hypnotizable but not fantasy prone, and vice versa, so that a complete account of hypnotic susceptibility is not to be derived from the study of fantasy proneness.

THE NEODISSOCIATIONISTS

The longest-running debate in current experimental hypnotism has been that between the "neodissociationists", headed by E. R. Hilgard, of Stanford University, and the advocates of a social psychological approach to hypnotism, of whom the most active has been N. P. Spanos of Carleton University. Each side has advanced numerous experimental claims and counter-claims in support of its position.

Hilgard's *Hypnotic Susceptibility* (1965) and *Hypnosis in the Relief of Pain* (1975), written with Josephine Hilgard, are generally ranked among the classics of hypnotic literature. From about 1973, he has expounded what he calls a "neodissociationist" view of the nature of hypnosis.[51] A number of other workers – for example J. F. Kihlstrom, F. J. Evans and K. S. Bowers – have adopted a similar stance, and together with Hilgard may loosely be referred to as a neodissociationist "school".

Hilgard's approach is a catholic one. He thinks that a straightforward "special state" view of hypnosis cannot be sustained because often a hypnotized person, despite exhibiting many of the standard hypnotic phenomena, does not undergo any profound change in subjective experience.[52] He proposes, before developing a theory, to identify the "domain" of hypnosis in general terms, and to "place" that domain in relation to different areas of psychology. The domain of hypnosis is constituted by the kinds of phenomena (subjective and objective) and the kinds of clinical interventions which we have already discussed so extensively. It has affinities with many parts of psychology – with social, cognitive and physiological psychology, but above all with the psychology of personality and individual differences (Hilgard thinks hypnotic susceptibility is a stable trait), and with certain "dissociative" phenomena of the kinds we looked at when considering Pierre Janet and his successors. These included hysterical anaesthesias, contractures and paralyses in which the afflicted limb, as if guarded by an intelligence, comes to no harm; automatic writing; fugue states and multiple personality disorder; dreams, sleepwalking and sleeptalking. For all of these, hypnosis can provide analogies; and it adds a few further categories of apparently dissociative phenomena of its own – for instance, post-hypnotic suggestion, post-hypnotic amnesia and negative hallucination. According to Kihlstrom, all these phenomena

> ...involve deliberate, intelligent, behavioral and cognitive activities of such complexity and extent as to require representation in phenomenal awareness, if not the full commitment of the person's attentional resources... Yet these very activities appear to occur involuntarily; or the person has little awareness of having engaged in them; or, if the person is aware of them, they are not integrated into the other activities of which he or she is also aware.[53]

It is to handle this "family" of phenomena, but most particularly the hypnotic ones, that Hilgard has developed his "neodissociationist" theory. He begins from the assumption that consciousness "can be considered in two major modes. The first of these is the mode of awareness, appreciation, esthetic sensitivity. The second mode is that of active agent, planner and controller."[54] Though he recognizes that hypnosis may bring about changes in the quality of awareness, most of his speculations relate to the active mode of consciousness. A person's life, he notes, "is made up of an almost infinite number of activities, from trivial responses to stimuli (brushing a fly off the face), to those consuming more time, still with definable beginnings and endings (writing a letter, playing a game of golf, listening to a symphony), and on to activities that are enduring over many years and complexly interacting with each other (raising a family, doing the housework, saving for retirement)."[55] Each of these activities Hilgard refers to as a "subsystem". At any given moment, some will be "latent"; others will be "activated". But even those few that are activated may pull different ways. Clearly, to avert a chaos of conflicting activities there must be some factor or factors that determine at any time which subsystems are activated, and which will predominate. Hilgard proposes an "executive ego" or "central control structure" which monitors and controls the activities of an hierarchy of subsystems (subordinate cognitive control structures).

The central control structure must obviously have both a planning and executive function and a monitoring function that is alert to the progress of the plans being implemented. The subordinate control structures likewise have their own inputs and outputs (monitoring and control functions). The subordinate control structures are complexly linked to each other, and normally have input to and receive input from the central control structure. Hilgard illustrates all this with a diagram that has many points of similarity with Grasset's "polygon" diagram which we discussed in an earlier chapter.

Once subordinate control structures are activated, their activity may achieve a high degree of autonomy, as when a skilled act is carried on with little conscious attention. When subordinate systems lose connection with each other, or with the central control system, we have "dissociation" of functions. Dissociation may be relative or absolute, and of one subsystem or several linked subsystems. In cases of extensive dissociation, the result may be a developed dissociated consciousness, of the kind that may emerge in automatic writing or multiple personality. Hypnotic induction procedures facilitate dissociative experiences. The emphasis on muscular relaxation has a generally disorienting effect. The metaphor of sleep, so commonly urged upon hypnotic subjects, has a numbing effect upon memory, which in turn impedes the ability to formulate critical judgments. Reality testing is reduced. Under such circumstances, "response to the stimulation provided by the hypnotist takes precedence over planned or self-initiated action, and the voice of the hypnotist becomes unusually persuasive."[56]

Hypnotic induction procedures may thus produce a readiness for dissociative experiences. What remains relatively obscure, however, is how specific suggestions bring about particular dissociations in the hypnotized subject. The first step in hypnosis is that part of the central executive functions is handed over to the hypnotist (this is comparable to the centre O in Grasset's scheme coming under the control of the O of the hypnotist). Then, within the "hypnotic contract", the subject, who to a greater or lesser extent loses his normal power of initiative, will do what the hypnotist suggests. (In self-hypnosis, the central executive function "divides itself into two parts, representing the rôle of hypnotist and hypnotized."[57]) The subject "remains essentially the hypnotist's assistant in producing the phenomena."[58] In superficial hypnosis, mild dissociations can be brought about with little alteration of the general state of consciousness. But a more massive dissociation "may result from the summing up of many specific subsystems for which control has been relinquished."[59]

The monitoring aspect of the central control structure also undergoes changes. The executive "issues an order" to the monitor to reduce the amount of cortical scanning. Thus critical discrimination is no longer exercised concerning the outcome of the hypnotist's suggestions. This may express itself in a certain detachment of the observing part from what is happening. "The monitoring system knows what is going on, but some information is concealed from it. The executive system, in collaboration with the hypnotist, succeeds in giving rise to the activated experiences. How this has been done may be concealed from the monitor or lost by the fractionation of memory."[60]

How such things are done is also, I fear, to a fair extent concealed from the student of Hilgard's neodissociationist model of hypnosis. What is meant by a "division" of the executive functions between subject and hypnotist? Or between part playing the rôle of subject and part playing the rôle of hypnotist? What is an "executive ego" such that it can divide itself into two? How can it "issue orders" to another part of itself, the "monitor"? How can it conceal information from that other part of itself? The underlying model here seems to be not that of an "executive ego" but of an anthropomorphically conceived chief executive's office, with various functionaries who are capable of cooperating with each other or of being set against each other or of being suborned from without. The details of how they fulfil their various functions are left to the reader's imagination.

Equally obscure is Hilgard's hierarchical model of superordinate and subordinate and coordinate cognitive control structures, represented in his flow diagram by boxes and interconnecting arrows, with the central control structure at their head. Dissociations, such as those allegedly brought about by hypnosis, involve severance of the links between boxes. How many superordinate and subordinate control structures are there? Is there one for each function that can be logically or empirically distinguished in terms of some set of agreed criteria? Suppose a subject's arm is hypnotically paralyzed, or made to rise in the air. Presumably the cognitive control structure for the arm has been dissociated from the main part of the central control structure, and is controlled by the hypnotist through that part of the central control structure that has been handed over to him. What if just a forefinger is involved? Has a forefinger control structure, subordinate to arm and hand control structures, alone been dissociated? Suppose now that the dissociated forefinger is made to tap out automatically the answers to addition sums the components of which are indicated by pricks administered to the opposite hand, which has been rendered hypnotically anaesthetic. Then we could say that the forefinger centre, an arithmetic centre, and the monitoring aspect of the hand centre, though dissociated from the central control structure, are all still linked together and have conspired, at the hypnotist's suggestion, to produce the observed phenomenon. The point is not whether or not such phenomena actually occur (similar things have occasionally been reported), but that one can go on playing the game of "explaining" hypnotic phenomena in this way almost indefinitely. Any conceivable hypnotic phenomenon can be given some sort of interpretation in terms of Hilgard's schema of hierarchically organized but dissociable control centres. But there is no point at all in engaging in this game unless we have some independent idea how many control centres there are, what their interrelations are, and what would be the lines of cleavage (each having distinctive behavioural or experiential consequences) along which they would dissociate and divide. Grasset, in fact, made some suggestions on this matter, and until Hilgard does the same his speculations will remain – untestable speculations.

Consider, for example, Hilgard's now famous investigations of the so-called "hidden observer" phenomenon.[61] In its modern form, the phenomenon involves the establishment, with highly hypnotizable subjects, of analgesia of an arm or hand which is then subjected to sustained painful stimulation (immersion in iced water or restriction of circulation whilst muscular activity is engaged in). The hypnotized subject rates the level of pain he is experiencing (non-existent or much reduced as compared to what he reported when not hypnotized). A hypothetical "hidden part" of him is then addressed, and is invited to rate by means of automatic writing or speaking or signalling the degree of pain it is undergoing. In an appreciable percentage of highly hypnotizable subjects, the "hidden observer" acknowledges feeling sensory pain (as distinct from "suffering"[62]), though usually to a somewhat reduced degree.

Hilgard's interpretation of the "hidden observer" phenomenon is as follows. "The inaccessible information about felt pain in hypnotic analgesia ... must be stored in a manner similar to that in which amnesic memories are stored. If we understood amnesia better, part of our problem would be solved. One additional question would still have to be answered. In the usual posthypnotic amnesia, the original registration of perception is over; only the temporary inability to recall creates the problem. In covert pain an additional first step is involved, in which the perception of the pain is diverted to an amnesic condition *before it has entered consciousness.*"[63] The diversionary process may perhaps involve the subject engaging in fantasy, imagining his arm as numb, or "whatever has succeeded in the past". The pain information is now behind an "amnesia-like" barrier created by the central control structure and is thus inaccessible to the consciousness of the hypnotized person. However, the pain still exists covertly behind the barrier. Hilgard seems to think it is felt by part of the monitoring aspect of the central executive control structure divided off at the command of the control aspect (he says "separating the central control from the central monitor may seem arbitrary, but phenomenally that is what takes place"[64]). But the pain cannot find overt expression because a second barrier has been erected between the covert hypnotic recognition of the pain and all channels of communication. This second barrier is breached when the hidden observer (detached portion of the monitor) is induced to communicate by automatic speech or automatic writing. There is then a fully-fledged dissociation between the parts of the overall cognitive hierarchy of the hypnotized subject which report "no pain" to and through the ordinary monitor, and the "hidden observer" which reports "pain" through outlets which have been put at the disposal of the detached portion of the monitor. The amnesic barrier may later be lifted by post-hypnotic suggestion, and the subject may thereafter have some recollection of the experience of the "hidden observer", but usually as appertaining to a "separate part" of the mind.[65]

Of these speculations one might make the following remarks:

1. Hilgard's account of the "diversion" of pain behind the amnesic barrier before it is felt is highly implausible for all but minor discomforts.

2. He offers no account of the nature of the amnesic barrier behind which the covert pain and the hidden observer are supposed to be concealed, except to indicate that like hypnotic amnesia it can be removed by a signal or a command. Now the nature of hypnotic amnesia is controversial, and its very existence is

questioned by some. Under these circumstances it is unrealistic to suppose that any useful tests of the theory's applicability to anaesthesia can be devised.

3. The "amnesic" barrier in Hilgard's theorizing appears to be not just a barrier to recall of pain, but a barrier which prevents a continuously occurring pain from penetrating the consciousness of an hypnotized subject. Now the shutting out from ordinary consciousness of certain ongoing mental events is not what is ordinarily understood by amnesia; by "amnesia" is normally meant an inability to activate memories that are latent and currently unexpressed. It is by no means obvious that the latter can profitably be taken as any kind of model for the former.

4. No useful account is given of the nature and origins of the part-personality or quasi-personality (a personality adequately endowed with linguistic and conceptual capacities) that communicates information about what has gone on "behind" the amnesic barrier.

I must not create the impression that Hilgard, who is one of the most moderate, as well as one of the most distinguished, of current writers on hypnotism is not well aware that his theory is incomplete and highly speculative. He would regard his model, I think, not as a developed theory, but as a mapping out of the direction which a theory of hypnosis and of dissociative phenomena should take. Some other neodissociationists have advanced more specific theories to accommodate circumscribed regions of the phenomena,[66] but I cannot say that these theories seem to me to represent any genuine advance in precision.

Of course all these theories share the dubious *a priori* assumptions that pervade modern cognitive psychology. It may be useful to glance briefly at a basic or Ur-form of dissociation theory not far removed from those current at the beginning of the century. An initial problem is what is meant by "dissociation", a concept that awaits adequate explication.[67] Thus some writers talk as though even the phenomena of subliminal perception are evidence for dissociated mental functioning; but clearly such phenomena are not evidence for an organized "secondary consciousness" of the kind that might be postulated to explain, say, automatic writing. Still others talk as though tasks which are performed "automatically", without the need for much conscious attention, involve the "dissociation" of a subsystem from the main stream of psychoneural activity, which may or may not be a helpful way of talking. For still others, someone is in a "dissociated" state if he has become "dissociated" from his surroundings, i.e. lost in his own thoughts, or if he has become "dissociated" from his own past and is not sure who or where he is. And these are perfectly understandable and atheoretical descriptive uses of the terms, provided that we beware of letting them entangle us in empty speculations about the severance of hypothetical subordinate control systems from each other. It is clear, however, that the only use of the term that can be turned to effect in the explanation of major hypnotic phenomena involves the co-existence in connection with the same organism of two separate streams of consciousness (or, to adopt Braude's rather more appropriate usage, "apperceptive centres"), which are, at least potentially, co-active and which pursue their courses not necessarily without mutual interference, but with limited mutual cognizance and a large measure of independence. Some such scheme of things has been an assumption of most dissociationist explanations of such key hypnotic phenomena as post-hypnotic amnesia, post-hypnotic suggestions, the seemingly automatic and involuntary nature of some responses to hypnotic suggestions, state-dependent memory in hypnosis, the hidden observer, positive and negative hypnotic hallucinations, and the hypersuggestibility of the hypnotized individual. I shall suggest that the implications of these dissociationist explanations have only to be spelled out to be seen to be unacceptable, and that the kinds of phenomena usually cited as evidence for dissociated and potentially co-active apperceptive centres do not as a rule constitute such evidence. I shall further suggest alternative ways of looking at some of these phenomena.

Consider first the phenomena of hypnotic (and hysterical) anaesthesia and negative hallucination. According to classical dissociation theory, an awareness of the events extruded from the main consciousness is diverted to and maintained in the secondary consciousness, which can then influence the motor system so as to keep the subject out of any danger consequent upon the sensory lacunae in his primary consciousness. The secondary consciousness must therefore possess a vigilant awareness of events in the external world, together with a sophisticated ability to control the motor system appropriately. What is the nature of the process of extrusion? It would be very difficult (perhaps self-contradictory) to argue that there is a conscious process (conscious, that is, in terms of the primary consciousness) of expulsion. The secondary consciousness must therefore somehow come forward and engulf the relevant material. What does this mean? How can it be done? How is the missing part of the visual field filled in? How does the secondary consciousness know how to do it? These are questions which it is difficult to make sense of, and impossible to answer.

Very similar problems arise with regard to post-hypnotic suggestions. A suggestion is implanted into a hypnotized subject's secondary consciousness (or apperceptive centre). On being awoken the subject can no longer recall the suggestion, but it persists in the now submerged secondary consciousness and in due course is implemented. The secondary consciousness must therefore have carried out the calculations or other intellectual work required for its fulfilment at the appointed time, or have monitored the environment for the designated cue. Let us say that the suggestion (as has so often been the case) involves seeing an hallucinatory figure. Since this figure is not deliberately constructed by the primary consciousness, but appears in it suddenly to the subject's surprise, it must have been constructed by the secondary consciousness, and injected into the primary consciousness at the appropriate moment. The secondary consciousness must possess the skills needed to construct the hallucination, and such knowledge of the relationship between itself and the primary consciousness as will enable it to transfer material from itself to the latter as required.

One could adduce further examples, but the message is in all instances the same. To fill the theoretical rôle which dissociationists have assigned to it, a secondary stream of consciousness or second apperceptive centre would have to possess gifts, insights and information greatly in excess of those of the vast majority of primary consciousnesses. Perhaps it may be so – this was F. W. H. Myers's opinion. But most dissociationists have looked upon secondary streams of consciousness as being in general rather impoverished, as containing mostly the detritus or ejecta of the primary stream. This makes it easy to explain the supposed hypersuggestibility of hypnotized subjects. The rather feeble stream of consciousness that is tapped in deep hypnosis lacks the critical resources to combat suggestions, which consequently "take" in a quasi-automatic manner. But if secondary streams of consciousness are not thus impoverished, but are in many respects richer and more skilful than primary ones, this explanation breaks down, and no account, or at any rate only a highly fanciful account, can be given of what is widely supposed to be the central phenomenon of hypnosis.

My further proposal, that the phenomena most often cited as evidence for dissociated streams of consciousness do not constitute such evidence, would, if accepted, undercut all attempts at dissociationist explanations of hypnotic phenomena. But I can do no more than touch on the heads of the arguments that might be deployed. It is obvious, to begin with, that cases where different phases of personality simply alternate, as in classic early examples of MPD, for example Félida X., do not supply such evidence, though they were often misleadingly referred to as cases of "double consciousness".[68] Of such cases, as of cases of fugue, one can perfectly well propose that there is just a sudden sharp change in what remains at root one and the same apperceptive centre. The aetiology of the changes may be quite different in different categories of case, with epilepsy a possible factor in some instances.[69] But often, especially as we move into the domain of multiple personality, we may regard the changes as essentially constructive and creative responses to stress, as an immersing of oneself in a fantasy in which pain and anxiety are forgotten or become more tolerable. I am not saying that we have any proper understanding of the psychodynamics or neurodynamics involved; only that we are not required to postulate supernumerary apperceptive centres.

There are, however, some features of some multiple personality cases which do seem to constitute *prima facie* evidence for disconnected streams of consciousness. It occasionally happens that an alter when "in control" may claim to remember a "submerged" existence during the periods when another personality held sway, and to remember that during such periods it was aware of the deeds and experiences of the other as those of a separate individual. Rather more rarely a submerged personality will manifest surreptitiously, for example by automatic writing, while the normal waking consciousness, or some other apperceptive centre, holds the stage. Both these categories of phenomena seem to indicate the separate and simultaneous functioning of more than one apperceptive centre in connection with the same organism.

Now as regards the claimed memories of one alter concerning a submerged existence during the predominance of another, these claims are surely most readily interpreted as a shift in imaginative perspective from first-person to third-person (or vice versa) towards events that are still, at root, the memories of one individual. Such shifts (which are not difficult to adopt towards particular periods of one's life[70]) are put to use in maintaining and heightening the fantasy-drama of alternating personalities. What the underlying psychodynamics or neurodynamics may be remains obscure; but hardly as obscure as the neurodynamics of multiple apperceptive centres linked to the same nervous system. To make just one point about the latter: alter personalities sometimes give very peculiar accounts of the way in which (while dormant) they learn the thoughts etc., of the controlling personality, and even influence them. They talk rather vaguely as though they can see and hear the visual and verbal images passing before the controlling personality's mind.[71] But it makes no sense to talk of directly "perceiving" images in a mind other than one's own (or indeed in one's own mind), and even if one could perceive them, they would not suffice to tell one what the other is thinking, for it is the meaning or intentionality with which the experient invests an image that yields the direction of his thought. This story too has to be simply a plausible elaboration of a fantasy drama.

As for those peculiar occasions on which a submerged personality appears to manifest surreptitiously, for example by automatic writing, during the dominance of an entirely different personality, often the "normal" one: The problem here is no different from the problems raised by more commonplace examples of automatic writing in which the automatist's hand, acting outside the deliberate control of the waking consciousness, apparently develops, and maintains over different sessions, a characteristic personality of its own, with a view-point and even fund of information different from that of the normal personality. When automatic writing has developed to this degree of complexity it has often provoked the following line of argument: (a) Such writing has as good a claim to be regarded as the manifestation of a conscious intelligence as any other form of human conceptual activity; (b) it is not a deliberate manifestation of the automatist's own conscious intelligence; (c) therefore it must be a manifestation of some other conscious intelligence (apperceptive centre) linked to his organism.

Now there have undoubtedly been some very remarkable cases of intelligent automatic writing being executed whilst the automatist is apparently fully aware of his or her surroundings, yet completely unaware of what his or her had is writing or has just written.[72] And I can see no reason for arguing that the concept of a secondary stream of consciousness, or second "apperceptive centre", is incoherent or totally without instantiation.[73] So I shall simply urge caution with regard to any proposal that takes even well-developed cases of automatic writing as evidence for the occurrence of secondary streams of consciousness such as might figure in dissociationist explanations of hypnotic phenomena. The fact is that we have very little detailed information about the state of mind of waking individuals engaged in automatic writing. Even Anita Mühl's most interesting monograph on automatic writing does not tell us much.[74] A question therefore arises as to how far an automatist – and many gifted automatists probably belong in the category of gifted fantasizers – could be deceiving himself as to the extent of his influence on the ongoing or recently completed automatic writing. There is a curious tendency to suppose that whatever goes on "in" someone's mind (and may influence behaviour) must be as patent to his observation and description as events in an "inner theatre" or private what-the-butler-saw machine. But knowledge of events "in" one's mind is not like that. States of mind are not like sequences of scenes packed with observable characters and events. They are, so to speak, successive phases of oneself, each fleeting and pregnant with tendencies (some conflicting with others) which it would often be very hard to pin down and put into words. How easy it can be, under these circumstances, to deceive oneself into believing that one is thinking or doing one thing, when really one is thinking or doing quite another.

Take such a basic form of automatic writing as operating a ouija board, an example which is of particular relevance because of the readiness with which one can deceive oneself as to whether or not one had some inkling of what was about to be spelled out, and as to whether or not one gave the pointer a slight but sufficient nudge. Self-deception here would involve one's being aware "in the back of one's mind", or "in one's heart", as Fingarette puts it,[75] that one had surreptitiously helped the pointer along by just a very slight exploration of a movement in a certain direction, but shying away from verbalizing this incriminating fact even to oneself.

My point is not that all examples of complex automatic writing can be satisfactorily accorded this treatment, but that much work of a conceptual, phenomenological and practical kind remains to be done before we can contemplate accepting such cases as evidence for the occurrence in some individuals of secondary streams of consciousness or second apperceptive centres.

THE SOCIAL PSYCHOLOGICAL APPROACH

The social psychological approach to the problems of hypnotism, most prominently represented by N. P. Spanos and his associates, cannot boast so long and distinguished an ancestry as the approach of the neodissociationists. Spanos himself traces his position back[76] to a celebrated article by R. W. White, published in 1941,[77] though there were occasional earlier anticipations. Nearer influences have been T. R. Sarbin, of the University of California, Santa Cruz, and his collaborator, W. C. Coe. Sarbin had been publishing views on hypnotism not unlike White's since about 1950, and these culminated in Sarbin and Coe's *Hypnosis* (1972), which presents a "rôle theory" analysis of hypnotic phenomena. "Rôle" in the Sarbin and Coe theory is a metaphor drawn from the stage. The fact that the concept is metaphorical does not, they hold, detract from its value, since all scientific theories involve metaphors. The "hypnotized" subject may be regarded as striving to "enact" the rôle of a hypnotized person, a rôle defined for him both culturally and through hints thrown out by the hypnotist. The subject's success in playing the rôle will be determined by such factors as his "self-rôle congruence", his rôle-playing skills, his rôle expectations, and the feedback from his audience. In certain cases, he may (as some actors do) become totally immersed in his rôle, show appropriate physiological changes ("organismic involvement"), vividly imagine a fictitious setting and surroundings (hallucinations are "believed-in imaginings") and lose awareness of himself. Although a hypnotized subject may act like, indeed feel like, an automaton if doing so is part of this rôle expectations, he never *is* an automaton. However much absorbed in his rôle, he is always an actor.

Spanos's own theoretical statements tend to be quite short.[78] It is as though he is so keen to be at the practical work of testing hypotheses, and fustigating the neodissociationists, that he has not the patience for over-much theorizing. The essence of his position is this:

> [A social psychological formulation implies] that hypnotic behavior is, first and foremost, social behavior. It can be understood only by taking into consideration (a) subjects' motivations to respond in terms of the rôle requirements specified by the hypnotic context and (b) the understandings they develop concerning those requirements ... Social psychological accounts of hypnotic responding are not alternatives to cognitive accounts; they *are* cognitive accounts. More specifically, social psychological accounts construe hypnotic subjects as continuously modifying their cognitive activities and behavior in terms of the changing social context that constitutes the ongoing hypnotic situation.[79]

A "hypnotized" subject uses whatever "cognitive strategies" he is capable of or can devise to maintain the socially prescribed rôle requirements of a "good hypnotic subject". One must not, however, suppose that such subjects are "faking" or "shamming". Good subjects often state that their responses to suggestions were involuntary. They manage to convey not just to the audience but to themselves as well the impression that they could not help doing or experiencing what they did or experienced. A satisfactory account of how subjects can thus deceive themselves is a key problem for Spanos. He seems to think that what may mediate the experience of non-volition is the "strategic enactment" of engaging in appropriate "goal-directed fantasies".[80]

I shall now proceed to outline the application of Spanos's theoretical views to three centrally important categories of phenomena, viz. hypnotic analgesia, hypnotic amnesia, and multiple personality and related matters. Since Spanos is probably the most prolific writer in the whole history of hypnotism, and since each of these three topics has in recent years generated a large controversial literature, it will be understood that I can offer only indications of tendency, not adequate reviews.[81]

I begin with Spanos's views on hypnotic analgesia.[82] Spanos does not deny that hypnotic suggestions may be effective in reducing pain. But whereas the neodissociationists regard hypnotic analgesia as an involuntary response to suggestion, resulting in the dissociation of the felt pain from consciousness, social psychologists

> ... view the subject as an active agent who can deliberately initiate cognitive activity in order to cope more effectively with noxious events ... A large number of studies on this tradition have demonstrated that treatments designed to encourage the use of strategies such as imagery, self-distraction, and coping verbalization (e.g. "this isn't so bad") are effective in raising pain thresholds and pain tolerance and in reducing ratings of pain intensity.[83]

If appropriate strategies are used, there is no reason why pain reduction should not be as successful in the waking state as during "hypnosis". Spanos cites numerous studies, mostly by himself and his collaborators, to show that this is so, at least so long as the experiments compare the performance of matched hypnotic and waking groups. If a within-group design is used, so that waking and hypnotic analgesia is compared in the same subjects, a difference in favour of hypnosis is likely to emerge. This has been Hilgard's preferred design. Spanos attributes these results to the tendency of highly suggestible persons to strive to present themselves as "good" subjects.[84]

It seems to be generally agreed that there are significant positive correlations between pretested level of hypnotic susceptibility and degree of suggested analgesia. Such findings are compatible with the view that both measured suggestibility and analgesia test performance reflect a stable capacity for hypnotic dissociation. Spanos, however, proposes that low suggestibles may connect the analgesia test situation with their previous hypnotic susceptibility testing and be correspondingly pessimistic about their prospects. As a result, they are unlikely to initiate and sustain the attention diversion or cognitive reinterpretation required to reduce pain.[85] However, given suitable instruction and encouragement, low suggestibles ought to be able to bring their analgesic responses up to the level of those of high suggestibles. Spanos cites several studies which show that this is indeed so. For example, Spanos, Ollerhead and Gwynn (1985–6) compared the analgesic effects of four treatments, to each of which a separate group of twenty-one subjects was assigned. The four treatments were: hypnotic analgesia; "Do whatever you can to reduce pain"; stress inoculation (i.e. instruction in strategies for minimizing pain); and no instructions. The first three types of treatment produced significant, and very similar, degrees of pain reduction. Only in the hypnotic condition did pretested hypnotic susceptibility correlate significantly with degree of pain reduction. Participants in all three experimental treatments showed a significant baseline to test increase in the percentage of time during which coping strategies were engaged in, and differences between treatments in this respect were not significant, though the hypnotic group spent less time in coping strategies than did the others. Spanos and collaborators conclude that the pain reduction seen in hypnotic as well as non-hypnotic subjects may be attributed to the use of coping strategies, and that the correlation between hypnotic susceptibility and pain reduction in the hypnotic condition is probably due to the subjects bringing to the analgesic situation the strategies which they had successfully adopted during the previous susceptibility testing.

Miller and Bowers (1986) obtained different results. They assigned ten pretested high and ten pretested low susceptibles to each of three treatment conditions: stress inoculation, stress inoculation defined as hypnosis, and hypnotic analgesia. Hypnotic susceptibility was related to pain reduction only for the hypnotic analgesia group. Strategy use was related to pain reduction only for the two stress-inoculation groups. The authors conclude: "All in all it appears that the pain reductions occurring in hypnotic and nonhypnotic treatments are brought about by different means. The use of cognitive strategies and one's hypnotic responsiveness seem to operate quite distinctly and independently from one another."[86]

In reply, Spanos argues as follows. Miller and Bowers' subjects were only counted as using a cognitive strategy "if they explicitly indicated that they *deliberately* carried out cognitive activity to reduce their pain ... statements like 'My arm felt like it turned into a piece of wood' were *not* classified as instances of strategy use because such statements do not explicitly indicate deliberateness on the subject's part."[87] From a socio-cognitive viewpoint, however, both groups of subjects may be regarded as having used coping strategies. "Hypnotized" subjects, who had been subjected to procedures and propaganda that emphasized passivity and automaticity, were led by this context to interpret their coping cognitions as events which occurred without their active intervention, whereas the stress-inoculation procedures encouraged subjects to regard their coping cognitions and pain reduction as self-generated and deliberate activities.

This view is supported by further experimental data which show that whether or not highly hypnotizable subjects rate their pain reductions as effortless or as the result of "coping" depends very much upon whether they are given "passive" or "active" instructions, and not at all upon whether or not they were put through an hypnotic induction procedure, what degree of hypnotic "depth" they attained, and how much use they made of "coping" imagery. The conclusion drawn is that we do not have evidence that suggested analgesia in hypnotic and non-hypnotic subjects results from different mechanisms.[88]

All the work on analgesia which we have been discussing used standardized tests of susceptibility, standardized hypnotic inductions, and standardized pain situations (usually immersion of the arm in iced water for 60 seconds). These procedures are relatively brief, and most of the subjects had not been extensively hypnotized before participation. It is quite unclear how far work of this kind could bear upon the central problem of hypnoanalgesia – the use of hypnotism in the relief of severe clinical and surgical pain. There continues to be a trickle of surgical operations, some quite severe, performed with unsupplemented hypnotic analgesia.[89] Spanos, it appears, thinks that these phenomena too can be accounted for in his terms; for my part I remain quite uncertain as to how such cases should be assessed.

We come next to Spanos's views about hypnotic amnesia. We are talking here only about suggested hypnotic amnesia.[90] Suggested amnesia has most often been investigated in the form of the amnesia suggestion incorporated in many scales of hypnotic susceptibility (the amnesia relates to previous items on the scale), but sometimes hypnotized subjects have been required to learn standardized verbal learning materials and have been rendered post-hyp-notically amnesic for them. There is no doubt that under these conditions "good" hypnotic subjects show a considerable degree of amnesia, more, in general, than that shown by "task-motivated" subjects put through comparable procedures. Such findings have of course given heart to neodissociationists. However, Spanos has an interpretation of his own.[91] Subjects in hypnotic amnesia experiments achieve amnesia by refraining from attending consistently to target-related retrieval cues. Large individual differences arise because instructions in these experiments have often been ambiguous – it has not been made clear to subjects whether they are actively to do something in order to forget or to wait passively for the fog of amnesia to descend upon them. The challenge to recall which precedes the cancellation of the amnesia suggestion is also likely to be ambiguous – subjects do not know whether they are to refocus attention on the recall of target material, or to recall target material that happens to come to mind while attention is maintained elsewhere. If it is made clear to subjects what they are expected to do, and how they should set about doing it, even low susceptible subjects and unhypnotized subjects can be led to show a high degree of amnesia, and they may even feel that their amnesia is involuntary.

A new kind of experimental paradigm was introduced by Kihlstrom and collaborators.[92] They gave hypnotic subjects (and of course control subjects also) two trials on which to try to recall target items. Between the first trial and the second or "breaching" one, subjects are given various kinds of instructions (to be honest, to try very hard) which pressurize them to breach. Breaching is measured in terms of the number of extra items recalled on the second trial. In a considerable number of studies based on this paradigm, highly hypnotizable subjects weathered all these pressures and continued to show a residual amnesia as compared to controls. In some experiments, a "lie detector" has been used to check the honesty of subjects' claims to be doing their best; in others even the playback of a video of the proceedings failed to lift amnesia completely.[93] Neodissociationists have tended to look upon these failures, or partial failures, to breach as one of their strongest cards. But Spanos has ways of explaining these matters. The high hypnotizables fail to breach because of their concern to present themselves as high hypnotizables. "Instead of refocusing their attention on the recall task, the subjects [sc. some subjects] reported experiences such as 'I concentrated on driving very fast down an open highway repeating "I cannot remember" and just concentrating on moving ahead quickly ...' or 'I went over and over in my mind saying "forget" over and over ...'"[94] If we manage to persuade highly susceptible subjects, by a crafty manoeuvre, that breaching amnesia is a sign of being a "good" subject, even high hypnotizables should breach. To do this, Spanos, Radtke and Bertrand (1985) made use of a variant of Hilgard's "hidden observer" technique. Highly hypnotizable subjects learned under hypnosis a list containing both abstract and concrete words. They were told that abstract words are stored in the right (or left) hemisphere and concrete ones in the left (or right). They were also informed that they possess a hidden part that remains aware of all that occurs in the left hemisphere, and one that remains aware of all that occurs in the right. After an amnesia suggestion, all subjects showed a high level of amnesia. But then the experimenter proceeded to contact the "hidden part" of each hemisphere, and the subjects at once recalled the appropriate words.

The problem with this experiment, as Silva and Kirsch (1987) point out, is that for highly hypnotizable subjects the preliminary instructions and obvious misinformation could have had the force of hypnotic suggestions. The subjects would in that case have breached amnesia because it had been suggested that they should, and not because by so doing they thought they were acting as "good" subjects. However, Silva and Kirsch's own experiment, designed to overcome this problem by administering the misinformation (that deepening the hypnotic state would increase/lessen the amnesia) before the hypnotic induction, seems to me not really to overcome it. An expectation implanted before hypnosis could easily become attached to related ideas conveyed during hypnosis. A stronger experiment is that by Radtke and collaborators (1987). Hypnotized subjects learned a forty-eight-item word list, and on the breaching trial were given twelve category names (which corresponded to the arrangement of items on the list) and were required to produced four words under each. The subjects who had been amnesic now all breached. The experimenters argue that this was because the task demands prevented the subjects from engaging in the cognitive strategies they would otherwise have used to avoid breaching. But against this it might be pointed out that the retrieval cues were very powerful and that the unselected subjects (113) contained only a sprinkling of high hypnotizables.

I come finally to Spanos's views on multiple personality and related matters.[95] These are readily predictable from his general position with regard to hypnosis. He notes the remarkable recent upsurge in reported cases of MPD,[96] and points out that this upsurge is largely North American, and is much more apparent to some psychiatrists than to others. Certain psychiatrists have been uncovering multiple personalities with quite startling frequency in patients who would previously have been diagnosed as schizophrenic, psychopathic, depressive, etc. The scale of their discoveries has inspired scepticism in various of their professional colleagues.[97] Spanos thinks that this multiplication of multiples may best be understood in social psychological terms, and rejects the view of Bliss and others that the cause of MPD is "the patient's unrecognized abuse of self-hypnosis".[98] MPD patients have simply learned to enact the social rôle of a multiple personality patient "in the same sense that good subjects learn to enact the hypnotic role".[99] To this end, they use whatever strategies they can and whatever information they can obtain concerning the requirements of the rôle. It is probable that psychotherapists "often play a particularly important part in the generation and maintenance of the role enactment. Although it may be done unwittingly, therapists sometimes encourage patients to adopt the multiple personality role, provide them with information about how to enact that role convincingly, and, perhaps most important, provide 'official' validation for the different identities that their patients enact."[100] Spanos has some amusing examples of the ways in which certain over-eager psychiatrists may have elicited second personalities by interventions based on the assumption that such personalities must be there, and have made them more definite by asking their names, histories and so forth. (But suggestions as to the iatrogenesis of MPD have been strongly contested by psychiatrists who point to cases in which symptoms clearly pre-dated resort of psychiatric help,[101] to cases of childhood MPD,[102] and to the similarity of cases studied by psychiatrists in widely separated localities.[103]) Another readily available source of information about the ways in which the rôle of multiple personality patient should be enacted is the numerous popular books on the subject that have lately been published.

The gains which may accrue to the patient from enacting the multiple personality rôle(s) are fairly obvious – sympathy and attention from high-status persons, the satisfaction of knowing oneself to be someone rather unusual and special, a touch of drama in one's life and in the recounting of one's life, an excuse for shedding the blame for some of one's less worthy actions. At times, Spanos seems to teeter on the brink of proposing that MPD patients commonly enact their symptoms with conscious and deliberate guile for the sake of these gains. But he also asserts that he does not hold that such patients are necessarily faking. He says that the multiple identities created by these individuals "are both implicitly encouraged and then consistently validated by high-status experts. Thus, it should come as no surprise that those who carry out such enactments sometimes adopt the interpretation of their behavior that is so strongly and consistently reinforced by significant others."[104] This is not exactly a full or clear account of how patients who are in effect play-acting manage to convince themselves that they are not. It is like proposing that a malingerer who has succeeded in gulling his doctor is very likely to be persuaded by the sympathetic attention of that doctor that he really must be suffering from an ailment after all. However, Spanos accepts that sufferers from MPD are likely to be gifted fantasizers (hence their high scores on suggestibility tests) who have spent much time in rehearsing "make believe" rôles; this may account for their total absorption in the new rôle of MPD patient.

THE SOCIAL PSYCHOLOGICAL APPROACH: CONCLUDING COMMENTS

Comments on the social psychological approach to hypnosis fall naturally under two headings, namely comments on the rôle-playing metaphor or analogy which forms a background to so much of that approach, and comments on the proposal that the characteristic phenomena of hypnosis are brought about by the deployment of "cognitive strategies".

The rôle-playing analogy has been widely influential in general social psychology; but we are concerned only with its applications to hypnotism and related matters. Of course, when we talk ordinarily, without any special psychological theory in mind, of somebody playing the rôle of husband, father, mentor, financial adviser, tour guide, etc., etc., it is in a general way perfectly clear what we mean. We are talking of contextually situated social behaviour which, though indefinitely variable and flexible, is constrained by, yet at the same time partially generated by, certain subsets of stricter or looser social or institutional rules or conventions. The force of the theatrical analogy is simply that just as an actor can take on many different rôles, each of us behaves in different more or less conventionalized ways in different social settings. The analogy cannot be pressed very far, for in playing a rôle in a play one may be enacting the part of someone who plays many "rôles" in the rôle theory sense (Hamlet is son, lover, fellow-student, avenger of blood, etc.); but such as it is one can see its point.

Now if this is the force of the rôle-playing analogy, the analogy cannot be extended to the supposed rôle of "good hypnotic subject". There is no such rôle. And it is clear why this is so. According to the notions of hypnosis current in our society, being a good hypnotic subject is not something one can do or achieve. It is something one is, and hypnosis with its consequent phenomena is something that happens to one. *Mutatis mutandis*, rather similar considerations apply to MPD.

However, the rôle-playing analogy can in some circumstances have a somewhat different force. To describe someone as "playing the rôle of father" can be a slightly peculiar form of speech. It could imply something a little unusual, different from just *being* a good father. It might be used, for instance, of someone who had to spend most of his time in other capacities, and is now for a while throwing himself into fatherhood; or of someone who is honestly, and from the best of motives, trying to act as father to a step-child, an orphan, a pupil, etc.; or it might be used of somebody who is trying to give the world the impression that he is what he is not – a good father. Through all these instances runs a common thread – the activities concerned are somewhat forced and artificial, and at worst they are a downright sham, a piece of play-acting.

The rôle-playing analogy as developed by Spanos and his collaborators has this latter kind of force. One *pretends* to be a good hypnotic subject as one understands the concept, and then indeed one is play-acting. But many such subjects would deny any conscious deceit, and in most (not all) cases there is no good reason for not taking their words. They accept "honesty" demands, show an appreciation of the duties of subjects in scientific experiments, etc. So the rôle-playing metaphor can only be maintained by supposing that persons who play the rôle of "good" hypnotic subject somehow manage to deceive not just other people but also themselves as to what they are really doing. This is not at all a novel or revolutionary notion. Someone who is briefly playing the rôle of good father, but is more often neglectful, might very well be reluctant to admit even to himself that he is pretending to be what he is not. He might even become so absorbed in his rôle that he forgets that he is merely pretending. Similarly, Spanos proposes that "good" hyponotic subjects may become so absorbed in the "goal-directed fantasies" which support their rôle-playing that they deceive themselves as to the genuineness of their hypnotic state and reactions.

Introducing the notion of self-deception into a psychological theory opens, however, pitfalls for the theorist, of which the most important is that it now becomes very easy for him to deceive himself. For unless he manages to lay down workable criteria by adhering to which we can decide whether or not self-deception has taken place (and this has not yet been accomplished and would certainly be very difficult to accomplish) he will find it impossible to set any limitations on the applicability of his theory. He will be able to apply it to whatever phenomena he wishes, unrestrained by the denials of the hypnotized subjects that they were rôle-playing. Extending the theory to embrace some new, and at first sight refractory, observation, becomes no more than a sort of game that one can always play if one has sufficient ingenuity. To take just one example:[105] Kirsch and collaborators (1989) compared the responsiveness to tape-recorded hypnotic suggestions of high-hypnotizable subjects and low-hypnotizable "simulators" (a) when each subject was on his own but surreptitiously observed, and (b) when an experimenter, unaware of group assignment, was present. Simulators who believed themselves to be alone were much less responsive to suggestion than high-hypnotizables who believed themselves to be alone. The presence or absence of an experimenter made no difference to the scores of high hypnotizables, but made a considerable difference to those of simulators, which were much lower in the absence of a hypnotist. Here we find hypnotically suggested responses made by subjects who are not deliberately rôle-playing and who do not have an audience before whom rôle-playing could appropriately be engaged in. Does not this press the rôle-playing analogy very hard? Not at all, for the "good" hypnotic subjects were *ex hypothesi* persons who became deeply, indeed self-deceptively immersed in their rôles;[106] they did not fully realize that the curtain had come down, and they may besides have had an unformulated intuition that their status as good subjects would at any rate not suffer by their playing out their rôles to the bitter end.

I do not say that the rôle-playing analogy *never* has application, only that one should resist the temptation to stretch it very far when stretching it relies upon the so far imperfectly elucidated concept of self-deception. Far from pursuing the (disguised) desire to *present themselves* as good hypnotic subjects, the ostensible motive of many cooperative subjects is simply *to be hypnotized*, whether from curiosity, from helpfulness, from the scientific spirit, or from the desire to be cured. It may, indeed, be the case that when such subjects suppose themselves to be hypnotized, they are really deluding themselves by the self-deceptive use of certain cognitive strategies. The point, however, is that the "cognitive strategy" aspect of the social psychological approach (and this is by far its most interesting and productive aspect) need not be tied to social motives, rôle-playing, etc. Cognitive strategies could be pressed into the service of many different kinds of motives.

We come now to the proposal that the characteristic phenomena of hypnosis are brought about by the deployment of cognitive strategies. Those who support this proposal are usually contrasting the active, and at root voluntary, deployment of such strategies, with the more traditional idea that the hypnotized subject responds passively and involuntarily to the hypnotist's suggestions.

We might distinguish three relevant possible kinds or levels of responsiveness to suggestions or autosuggestions. The suggestions concerned could be waking suggestions or hypnotic suggestions. Waking suggestibility is a topic much underrepresented in the current literature.[107] I do not know whether a socio-cognitive theorist would think that waking responsiveness to the body sway test of suggestibility is mediated by cognitive strategies aimed at self-presentation as a "suggestible" subject. If so, he might find suggested or self-suggested impotence, a very common and distressing condition, an interesting case to consider. Waking subjects tend to report that in response to standard suggestions they simply felt as though they were being pulled backwards, felt the held object getting warmer and warmer, and so forth; but perhaps closer questioning would reveal hints of the employment of "cognitive strategies". In view of the uncertainty here, the kinds of responsiveness to suggestion that I shall discuss will primarily involve responsiveness to hypnotic suggestions.

1. A subject is led to think, or leads himself to think, that he may or will undergo or carry out a certain experience, movement or action, and his so thinking brings about the event thought of without his further attempting to promote it.

2. A subject, aware of the reputed efficacy of suggestion and hypnotic suggestion, but finding himself uninfluenced by them, decides to find out for himself as nearly as he can what it would be like to be hypnotized, be responsive to suggestion, etc. Accordingly, whilst being put through an hypnotic induction procedure, or even while on his own, he relaxes, turns his thoughts "inwards" as though drowsy, listens to or gives himself an arm levitation suggestion, and imagines as vividly as he can that his arm is very light, is floating upwards willy-nilly, etc. At the same time, he gently raises his arm, so gently that there is almost no sensation of muscular effort. It is almost as though his arm really is floating spontaneously upwards. Then he imagines that one of his hands is dead, numb, does not belong to him, etc., sets it mentally aside, and concentrates very hard on pleasant and distracting imagery. By this means he manages to avoid paying too much attention to moderately painful stimuli applied to his "analgesic" hand. And so on and so on. At the end of it all he tells himself a fantasy story in terms of which he has been "hypnotized" and hence cannot remember what has just happened to him. By thinking determinedly of other things he manages to suppress the recollection of what happened to him during his "hypnotic trance".[108]

3. A subject pursues essentially the same cognitive strategies as in strategy 2, but somehow manages to disguise from himself the fact that this is what he is doing. Perhaps he becomes so absorbed in his fantasies that he loses sight of the fact that they are fantasies, especially if he is operating in a context (e.g. an old-

fashioned hypnotic induction) that encourages him to believe that they are not. Perhaps an element of motivated self-deception enters into it, and he finds ways (further cognitive strategies?) of avoiding spelling out to himself the true nature of what he is doing.

The first thing to be noted about these three kinds or levels of responsiveness to suggestions is that they do not exclude each other. The same individual might respond in each of the ways on different occasions, and there might even be elements of each way of responding on the same occasion. My own view is that both strategy 1 (traditional passive responsiveness to suggestion) and strategy 3 (absorption in, or self-deceptive use of, cognitive strategies) are quite commonly found; and of course strategy 2 (the deliberate use of cognitive strategies) is coming increasingly into prominence with the development of skills training "packages" for hypnotically related skills.[109] There is no need whatever to try to force all instances of responsiveness to suggestion into one mould.

The second thing to be noted is that it is extraordinarily difficult to be clear what the differences between strategies 1, 2 and 3 might in practice amount to. Yet unless we can become clear on this issue we shall remain unable to reach any firm conclusions as to the extent to which cognitive strategies underlie hypnotic phenomena. The difficulty has three kinds of source:

1. Partly it is due to the fact that we run once again upon the slippery notion of self-deception, and as yet we lack, at least in this context, any adequately thought-out and generally agreed criteria for distinguishing self-deceptive but at root cognitive strategical responsiveness to suggestion from direct responsiveness unmediated by cognitive strategies.

2. Partly it arises because experiments intended to distinguish between strategies 1 and 3 present many difficulties of design and interpretation. One cannot exactly instruct subjects to deceive themselves as to whether or not they are utilizing cognitive strategies, though there may be ways of encouraging them to do so. On the other hand, there is always a danger that any mental activity that a hypnotized subject is engaged in at the relevant time will be taken as the unacknowledged pursuit of a cognitive strategy; for since *ex hypothesi* the subject has lost sight of the fact that he is utilizing strategies, or has somehow disguised it from himself, no one can decisively refute this proposal.[110] This opens up the possibility that we might claim that any subject whose mind was not a blank during the experiment was operating a cognitive strategy.

Of course one can give subjects instructions that enjoin, promote or encourage the use of cognitive strategies to obtain the desired effects. This may at any rate illustrate what it is possible to achieve; but such instructions may themselves function as direct suggestions,[111] as may the subject's knowledge that he is utilizing a strategy that is supposed to facilitate a certain effect. For example, it is by no means apparent why certain "coping strategies" are successful in reducing pain. It seems to be generally supposed that such strategies work by diverting attention away from pain. But of course instructions (or decisions) to use coping strategies are also strong implicit suggestions of the form "do X and the pain will diminish". Could the suggestive aspect of the instructions rather than the strategies be fundamental? Would the instructions be as effective if X were an activity that directed attention towards the pain instead of away from it, e.g. "yell as loudly as you can the moment it begins to hurt."[112] Answers to these questions will not be easily found.

3. Partly, the difficulty may be due to the fact that subjects for apparently relevant experiments are likely to find themselves out of their depths when confronted with the questions they are required to answer and the conceptual distinctions they are required to make (a situation often not helped by the experimenters' own uncertainty on these issues). Should one admit to being hallucinated only if one was deceived into believing something was really there? (One well-known experiment on hypnotic hallucinations did not allow subjects to say they were hallucinated but knew there was nothing there. Such hallucinations were counted as vivid images, which certainly improved the case against the view that subjects may hallucinate as real as real.[113]) If one has a transparent hallucination should one call it an "image" (a perennial source of confusion)? Should one talk of analgesia and of the diminution of pain if one has simply managed to think of something else for a while? Doesn't true analgesia involve so to speak looking for the pain and not being able to find it, or concentrating on the region of skin which the needle is impaling and feeling nothing? At what level should questions as to whether one complied with suggestions voluntarily or involuntarily be taken? Experimenters have thought the question of the voluntariness or involuntariness of responses to suggestion of great importance because the traditional view of an hypnotized subject is supposed to have been that he is a passively receptive automaton. But there has been continual confusion of the distinction between the voluntary and the involuntary with the distinction between action and non-action.[114] Experimenters have not provided (and very likely could not provide) their subjects with adequate clarification, and controversy seems set to rumble on indefinitely.[115] What is needed, it seems to me, is intensive and detailed investigation, by conceptually sensitive experimenters, of a small number of highly scrupulous and highly sophisticated subjects – the work of the Würzburg School comes to mind.

My own guess is that the adherents of the socio-cognitive school will prove to have gone about things in partly the wrong way. They have developed a social psychological approach to the various different categories of hypnotic phenomena. Thus they have propounded a social psychological approach to hypnotic amnesia, to hypnotic analgesia, and so forth, and have found that the behaviour and self-reports of an appreciable proportion of subjects are at any rate compatible with the socio-cognitive position. And they have concluded that in each of these areas their account must in fact be correct for all subjects. It seems to me, however, more probable than not that there is more than one road to high hypnotic susceptibility. We can now say with fair certainty, for example, that whilst many "gifted fantasizers" (individuals who must certainly be rated skilled cognitive strategists) are good hypnotic subjects, not all are; and not all good hypnotic subjects are gifted fantasizers.[116] It seems possible, therefore, that there may be a subclass of "good hypnotic subjects" who achieve their characteristic performances in *all* hypnotic areas principally by the sorts of strategies which the socio-cognitive school propose; and other subclasses of "good hypnotic subjects" for whose performances in *some or all* areas we may need other explanations.

A WAY OF LOOKING AT HYPNOTIC PHENOMENA

It is among those hypnotists whose background and training has been in experimental psychology that scepticism about the supposed "special state" of hypnosis is most widespread. Hypnotists whose orientation is medical and clinical are much more prone to believe in a special state – no doubt hypnotherapy proceeds more effectively from such a basis. Some of the arguments commonly advanced by the anti-special-state theorists are a good deal less than compelling. These theorists frequently repeat the suggestion that it is circular to define hypnosis as a state of heightened suggestibility whilst explaining the heightened suggestibility in terms of the state. But this was not how such late-nineteenth-century theorists as Vogt proceeded. For Vogt, hypnosis was a partial systematic waking state; it was a state in which most cognitive and motor functions had subsided into an inert or sleep-like condition, leaving a circumscribed system or systems (probably directed upon the hypnotist) active or even hyper-active (the general idea goes back at least to Wienholt). Such a state could be characterized introspectively without reference to the heightened suggestibility, or the post-hypnotic amnesia and state-dependent memory, supposedly consequent upon it. It might be achieved by meditational techniques as well as by hypnotic induction procedures. Not a few modern writers have given a somewhat similar account of the hypnotic consciousness.[117]

Now of course if anyone wishes to apply the label "hypnosis" to states of this kind, I do not see why he should not do so. His decision would be based not on the presence or absence in such states of some special quality or characteristic which would make the label appropriate – we have no idea what such a characteristic would be – but simply on whether or not it is for theoretical and practical purposes useful to distinguish these states of mind from more commonly occurring states.

Some opponents of a "special state" position argue that states of hypnosis are "really" no different from waking states because the EEGs of hypnotized individuals have been found not to differ from those of waking persons. But this argument is misconceived for two reasons. Firstly, the data are by no means clear-cut – some workers claim to have found that hypnotized persons may show a pre-sleep EEG or a shift in balance towards the right hemisphere, or that persons susceptible to hypnosis show an unusual amount of frontal theta.[118] Secondly, the EEG is a fairly weak indicator of changes in mental state – it is perfectly possible for someone under the influence of a major hallucinogen to register a normal EEG.[119]

Granted, however, that classifying certain mental states as "hypnosis" is a perfectly legitimate move, the question still remains, is it in the end going to prove a useful move? And I am inclined to answer, probably not. The principal reason for saying this is essentially that offered by Bernheim and his fellow sceptics nearly a century ago. "Hypnotic" phenomena are readily obtainable from persons who are not in a state of sleep-like passivity and have not been put through any form of traditional hypnotic induction procedure. Dupotet obtained them by exposing waking subjects to mysterious "magical" drawings; the electrobiologists by having waking subjects hold and look at pseudoscientific charms; Schrenck-Notzing by making subjects drunk or giving them cannabis; Delboeuf, Dejérine, Bernheim, Hartenberg and Valentin by forceful affirmative suggestions. More recently, we have accumulated evidence (admittedly of somewhat variable quality, mainly due to the lack in some instances of a placebo control group) that a measurable increase in responsiveness to suggestion can be brought about by cannabis intoxication, by nitrous oxide inhalation, by the administration of hallucinogens, by excessive exercise, by meditational prac-tices, by guided fantasies leading to a sort of euphoria.[120] Very likely it could also be induced by an indefinitely large range of other slightly disorienting experiences, let us say vertigo, religious ecstasy, sexual encounters or ritualized drumming and dancing.[121] And some gifted individuals can generate many of the phenomena concerned without any such external assistance. It seems most unlikely that the various psychophysiological states that may lead to enhanced suggestibility have any common denominator that could usefully be accommo-dated by some such rubric as "partial systematic waking state". It begins to look as though the "state" of hypnosis, even if one can mark one out, can have at best a very limited part to play in accounting for the phenomena of enhanced suggestibility, many of which, I must none the less emphasize – and this includes the "hypnotic" ones – appear to me to be genuine.

A secondary reason for doubting that such a move would be useful is that one would obviously like the postulated properties of the supposed state to be such as would help us understand the enhanced suggestibility of the hypnotized subject. And generally speaking this has not been the case. Take, for instance, Vogt's "partial systematic waking state" and the state of "partial systematic dissociation" with which he supplemented it.[122] These are essentially "monoideistic" states in the tradition of Braid (the general stance has quite a few contemporary representatives). But monoideism yields no satisfactory account of the phenomena of hypnotism and suggestion. To take just one point: The theory of monoideism seems inextricably tied to the ideomotor theory of action. Hypnotized subjects carry out motor suggestions because of the force and vividness attained by the supposed image or idea of movement. But the ideomotor theory was shown to be unacceptable many years ago,[123] and recently it has been demonstrated that hypnotic subjects will respond to motor suggestions even when these are contrary to their imaginings.[124] I am far from sure that any of the concepts of "hypnosis" or "hypnotic state" developed since Vogt have any greater explanatory validity than his. A truly viable concept of hypnosis remains to be achieved.

Should we then dismiss the concept of hypnosis as superfluous? The answer to this is both yes and no. The yes part of the answer has just been given. I shall return to the no part shortly. Meantime, we need to look more closely at the topics of suggestion and suggestibility. For clearly the problems of enhanced suggestibility are part of the general problem of suggestibility; and the problems of hypnosis are part of (though they may also extend beyond[125]) the problems of enhancement of suggestibility.

Suggestibility, as a topic in its own right, has not received the attention that its importance and pervasiveness merit. Fortunately, two recent books on the subject, *Human Suggestibility* (1991) edited by J.F. Schumaker, and *Suggestion and Suggestibility* (1989) edited by V. A. Gheorghiu and collaborators, have done quite a lot to improve this situation.[126] Our present concerns are with the enhancement of suggestibility, but before pursuing them we must ask what is meant by suggestibility, which in turn requires an investigation, however cursory, of the concept of suggestion.

In chapter 20, I briefly discussed some turn-of-the-century definitions of suggestion, and we may begin by distinguishing, as there, between the making of a suggestion, and the "taking" of a suggestion (as when Janet's subject, quoted above, would say of hypnotic suggestions made to her that they had or had not "taken"). Clearly, it is the "taking" of a suggestion that is important for our purposes. Under what circumstances would it be appropriate to say that a suggestion had "taken"? I think we could say this when

1. The subject formulates, receives, or otherwise has implanted in his mind the idea of carrying out an action, or piece of behaviour (or of not carrying it out), or the idea of undergoing an experience or feeling (or of not undergoing it when he would normally be expected to); or he receives the idea of feeling a powerful urge to carry out the action concerned.

2. The possibility further occurs to him, or is put to him, that he may be unable to refrain from carrying out the action or behaviour (or feeling the urge to carry

it out), or having the experience (or that he will be unable to carry out the action or have the experience). This constitutes the "suggestion" or suggested idea that he receives or as it were puts to himself.

3. He does not consciously decide to implement or attempt to implement the suggested idea, or deliberately engage in activities ("cognitive strategies") designed to achieve its implementation.

4. The subject carries out (or is unable to carry out) the action, or undergoes (or excludes) the experience either fully or in part, whether or not as the result of feeling a powerful urge to carry it out; or at least comes to feel a strong if frustrated urge to carry it out.

5. *Ceteris paribus*, the action would not have been carried out (or would not have been not carried out) or the experience undergone (or excluded) or the urge felt, had the conditions outlined in 1 and 2 not obtained.

6. The action was carried out (etc.) because of the meaning or content of the idea and not because of its mere occurrence. For example, a golfer who conceives the idea that he is going to mis-hit (autosuggestion), and then mis-hits, must mis-hit because he thought he would and not because the mere occurrence of the idea was a distraction to him.

Note:

(i) I have not stipulated that the action- or experience-idea is implemented because the subject *believes* or expects he will be unable to refrain from implementing it.[127] I have only stipulated that it is implemented because the idea (whether believed in or not) that he will or may be unable to refrain from implementing it has entered the subject's mind.

(ii) I have not said that the suggested idea must necessarily be in the subject's mind at or immediately before the moment when the suggestion is implemented; all that is required is that the idea was in his mind at some time prior to its implementation.

(iii) I have not said that when the subject carries out or undergoes the suggested action or experience, that action or experience must at the time of occurrence seem to him to be something he passively undergoes rather than brings about himself.

I hope that this brief preliminary definition is tolerably theory-neutral. For instance, 3 above excludes cases in which there is deliberate compliance with a suggestion or the deliberate engagement in strategies thought likely to promote its implementation; it would not, however, rule out cases (prominent in Spanos's theory) in which a person who had received a suggestion carried out strategies likely to promote the fulfilment of that suggestion whilst at the same time successfully disguising from himself that that was what he was doing.

One can at any rate see how certain straightforward examples of the "taking" of a suggestion would fit in with the proposals I have just made. Consider a recent experience of one of my colleagues. It fell to him to raise with a student a topic that was bound to cause considerable embarrassment on both sides. As he sat with the student, the thought occurred to him that he might not come out with what he had to say (1), and indeed that he might simply be unable to come out with it (2). He dismissed this idea as quite irrational, and certainly made no conscious effort to implement it (3) – he was indeed quite determined not to be influenced. None the less, he found himself for several minutes quite unable to come out with what he had meant to say (4). He had to approach the topic by a circuitous route, and when he finally came to the point, his words burst out in an odd way after some seconds of struggling. He is quite sure that had he not dwelt upon this thought he would not have found himself thus impotent to act (5).

One can also readily see why certain categories of event which might be, but should not be, classified as the "taking" of a suggestion do not qualify in terms of this definition. One is obedience to a command. When I set out to obey a command or comply with a request, I conceive my relation to the destined action or event in a certain way (involving notions of duty, obligation, prudence, etc.), and it is a different way to the way in which I conceive it when a suggestion "takes" (then I think of myself as possibly unable to intervene effectively in what is to follow). Even if I obeyed a command thinking myself bound by moral or prudential reasons, or did it simply from habit, and would in fact have been unable to refrain from obeying, this would not make my obeying into the "taking" of a suggestion. For the idea that I could not help obeying it did not enter my mind, or if it did it was still not a cause of my obeying.

Another is responsiveness to suggestion in cases of what has been called "secondary suggestibility".[128] Secondary suggestibility involves susceptibility (in the tradition of Binet's *La suggestibilité*) to manoeuvres (presentation of certain sequences of stimuli, pressure from other persons present, confusion by the assertions of authority figures) designed to warp judgment about features of external stimulus objects (the length of lines, the temperature or weight of objects held in the hand). Responsiveness to such "suggestions" does not count as those suggestions "taking" as "taking" has just been defined, for the "ideas" in the subject's mind are not, as such, ideas about what he will do or experience, but ideas as to the state or properties of certain external physical objects. It is, I think, desirable that cases of secondary suggestion should not be counted as the "taking" of suggestions, both on the empirical grounds that the two are poorly correlated,[129] and on the theoretical grounds that gulling a subject into making an erroneous judgment is psychologically very different from directly influencing his actions and experiences.

In addition, however, to excluding phenomena that ought to be excluded, the proposals under consideration would exclude from the concept of the "taking" of a suggestion certain phenomena which one might think should be included. A doctor who, so to speak, insinuated himself into the hallucinatory world of a patient suffering from delirium tremens, and was then able to shape and influence the hallucinations through conversation with the patient and pretended participation in his fantasies,[130] would not, generally speaking, be utilizing suggestions that then "took" in the relevant sense, any more than would the nurse who swept up the hallucinatory fauna infesting the patient's bed. The doctor's and the nurse's success in influencing the patient's mind is not due to their instilling the idea that he will be unable to refrain from seeing or not seeing certain things. It is due to a more direct activation of transiently exalted imaginative gifts in someone whose critical faculties have been dulled. The same was probably true of, for instance, those occasions, already noted, on which Janet (like others before and since) was able to "enter" the "hysterical" states of deluded and hallucinating patients and influence those states by assuming and playing out a rôle within them, and also true of episodes in which someone (for instance Oskar Vogt) has succeeded in communicating with a somnambule or sleepwalker and altering his actions or experiences.

If we are to talk of episodes such as these as involving the "taking" of a suggestion, we must distinguish two senses in which or levels on which a suggestion may be said to "take". We might call the kind of "takings" we first attempted to characterize "reflexive takings", because conditions 1, 2, 4 and 5 together imply that the subject's idea that the suggestion may "take" is part of the cause of its "taking". And we might call the kind of "taking" manifested by a delirious patient whose actions and thoughts are manipulated by his attendants "non-reflexive takings", because the manipulations concerned do not involve working upon his ideas of what he will or will not find himself able to do or experience, etc. Of the criteria by means of which I attempted to mark out "reflexive takings", "non-reflexive takings" exhibit only conditions 1, 4 and 5, with 3 not necessarily applicable. Obedience to a command understood as a command, or compliance with a request understood as a request, would be excluded from non-reflexive as well as from reflexive "takings" because obedience and compliance would require a new and different condition 2, mentioning the subject's ideas of obligation, prudence, status, politeness, etc.

Reflexive and non-reflexive "takings" are in practice likely to be very difficult to distinguish from one another because while the form of many suggestions is such as to indicate only a non-reflexive "taking" ("your arm is becoming heavier and heavier"), subjects in our society, because of their knowledge of hypnotism and related matters, are likely to put a reflexive gloss on them ("am I going to be unable to hold my arm up?"). Furthermore, it is possible that a subject in whom suggestions had "taken" reflexively, might on subsequent occasions find them begin to take non-reflexively, as if responding to suggestion had become a kind of habit that no longer required reflection.

Given definitions of the reflexive and non-reflexive "takings" of a suggestion, it becomes, one might suppose, a relatively simple matter to reach at least preliminary definitions of "suggestibility" and "enhanced" suggestibility. An individual's "suggestibility" would be the degree to which suggestions given him are likely to "take"; and "enhanced" suggestibility would be an enhancement of this degree above some established base level. Unfortunately, nothing in this field is ever wholly simple. "Enhanced suggestibility" has in practice two rather different aspects, which are confounded in most scales of hypnotic susceptibility, and which it is almost impossible to disentangle, namely an enhanced readiness to respond to suggestions and an enhanced capacity to respond to suggestions the fulfilment of which might be thought beyond the subject's ordinary faculties. The latter might include suggestions of heightened muscular strength, skin marking and blistering, certain illusions, hallucinating "as real as real", and changes in autonomic or immune system functioning. It would certainly be convenient for theoretical purposes if (as some have tried to do) we could dismiss all alleged happenings of these kinds as the products of experimental errors, compliance, etc. But as I have already indicated I do not think that suggested skin markings and autonomic system changes are so easily dismissed.

With regard to suggested hallucinations (perhaps the most dramatic of all hypnotic phenomena) the issue has been muddied by conceptual confusions forced upon the experimental subjects, and by the use of experimental situations so barren and simplistic that whatever the subject's imaginative gifts they must surely have withered and died at the very outset.[131] There is no doubt that some very careful and highly respected investigators – for example E. R. Hilgard and M. Orne[132] – have convinced themselves, although aware of the problems involved, that a few highly susceptible subjects may, in response to hypnotic suggestions, hallucinate "as real as real". A recent variant of Orne's work by Marks, Baird and McKellar (1989) is to the point. These experimenters compared the responses of a group of highly susceptible hypnotized subjects with those of a group of highly susceptible unhypnotized subjects instructed to simulate hypnosis. Eight out of eighteen hypnotized subjects reported having seen the suggested visual hallucination of a target person, and twelve out of eighteen simulators did likewise – one and only one of whom stated on post-experimental interview that she had actually had an hallucination. Six out of the nine hallucinators reported their hallucinations in terms that indicated transparency or lack of solidity; three reported that the figure was initially solid but became transparent after a while. Hallucinated subjects expressed surprise at seeing the real co-experimenter. The authors claim that the doubled person hallucination "is definitely more than an ordinary mental image and that it is indeed an hallucination."[133] The conclusion that might be drawn (and this is supported by other studies[134]) is that a smallish number of highly suggestible subjects can be induced to hallucinate "as real as real", and that a somewhat larger number will hallucinate incompletely, for example by seeing a transparent object, or part of an object, or something out of the corner of the eye. Even the incomplete hallucinations are not simply describable as vivid images, and the aspect of "enhanced suggestibility" that involves apparently enhanced capacity to hallucinate still remains. Further developments may be hoped for from investigations somewhat in the Würzburg tradition of the hypnotic experiences of highly sophisticated and scrupulous subjects,[135] and from studies of the regional brain metabolism of subjects undergoing hypnotic and other hallucinations.[136]

Enhanced suggestibility, then, definitely includes an enhanced readiness to respond, and it may well also involve an enhanced capacity to perform various feats or to undergo various kinds of experience. At least, if we are to rule out the latter, it has to be on the grounds not that the phenomena do not occur, but that they occur as readily when the subject has not been subjected to suggestion-enhancing techniques as when he has. And this is a matter on which, with respect to all the major categories of phenomena concerned, we have insufficient evidence to pronounce.

I remarked earlier that the problems of hypnosis are part of (though they may also extend beyond) the problems of enhanced suggestibility. We can now rephrase this as follows. A suggestion may or may not "take" (reflexively or non-reflexively) as defined above. If it does take, we may ask why it takes, meaning how, speaking psychologically, or psychophysiologically, are we to understand the peculiar and repeated transitions between someone's receiving or conceiving the idea that he may or will do, undergo or experience so and so, and his doing, undergoing or experiencing it. And if in some individual under some circumstances suggestions begin to "take" with exceptional readiness, or force, we may ask what are the factors which have promoted this "enhanced suggestibility" and why?

With regard to the question of what psychologically or psychophysiologically, happens when suggestions (hypnotic or other) "take" (whether reflexively or non-reflexively), the short answer is that we have virtually no idea. Most of the proposals that have been made, for instance Vogt's speculations about the flow of neurokyme, or Hilgard's about the handing over of central executive functions to the hypnotist, or Spanos's about the self-deceptive use of cognitive strategies in the service of social rôle-playing, hover uneasily on the margins of testability. A question which can be answered somewhat more firmly is that of what background factors may promote an enhanced responsiveness to suggestion in general, though even here there is a good deal of uncertainty. We may group the varied factors that have been supposed conducive to the reflexive or the non-reflexive "taking" of suggestions (we do not have sufficient data to separate these classes of "taking") into three categories. It should be noted that owing to the relative scarcity of studies of waking suggestibility, much of the literature (which it is impossible here to review in detail) concerns work done in a hypnotic context.

Factors to do with the enduring personal characteristics of the subject

These would include the subject's age (persons in older age-groups have generally been found to be less suggestible than those in younger ones), his capacity for absorption and imaginative involvement, his fanatasy proneness and openness to experiences, his educational level, and whether his education was arts- or science-based.

Factors to do with the current psychophysiological state of the subject

These might include drug states, somnambulic states, meditational states, pre-sleep states, hypnagogic states, states of daydream and fantasizing, "flashbacks" and states in which prior episodes of traumatic stress are relived and reenacted, states produced by drumming and dancing, states of relaxation, drunkenness, nitrous oxide intoxication, fatigue, hypoxia, hypercarbia, "hyperempiria", euphoria, etc.,[137] and of course the state of "hypnosis" if it proves useful and viable to mark one out. These states are usually supposed (with varying degrees of supporting evidence) to enhance suggestibility directly. But there is another way in which the subject's psychophysiological state might favour the "taking" of suggestions, namely if certain of its experienced features are especially consonant with the suggestions concerned. For instance, in their extremely interesting study of an hysterical epidemic, *The June Bug*, Kerckhoff and Back propose that the contagion spread as it did because some of the victims were already suffering from stress symptoms which they could readily interpret as consequences of the bite of a rumoured but non-existent poisonous insect.

Factors to do with the subject's physical and social environment

1. *His immediate surroundings.* The possibly relevant factors here are so numerous that it would be pointless to try to enumerate them. The confidence and persuasiveness with which suggestions are made have traditionally been of great importance, and so have the real or presumed status or authority of the person making them. The setting in which a suggestion is given is of obvious relevance. It would, for instance, be difficult to persuade even a nervous and susceptible person while he is watching a football match that he is likely to see a ghost. But matters may be very different the same night if his road home passes by a graveyard. The reactions of other persons present may be of importance. A crucial factor in the spreading of hysterical epidemics is the witnessing by each potential new victim of the antics of present sufferers, and the first move in halting such epidemics is to separate all patients and endangered persons from each other.

2. *The general ambience in which he lives.* I refer here especially to the social ambience and institutionalized frameworks of social concepts against which as background so many suggestions "take" and are enacted. One might propose – though one is here broaching issues which it would be difficult to investigate experimentally – that suggestions which harmonize with a conceptual system which is already deeply embedded in a subject's conceptual functioning (have "penetrated" it, to adapt Pylyshyn's term,[138] right up to the hilt), are, *ceteris paribus*, more likely to take than others which have not. Particularly interesting are suggestions which "take" because of their consonance with conceptual frameworks which have become institutionalized and pervasive within a given culture. Many members of the culture may then prove to be suggestible with respect to suggestions that cohere with, or derive their meaning from, these institutionalized or semi-institutionalized frameworks of thought.

To take just one example, from a wide range of possible examples, consider the healing practises of the !Kung bushmen, of which there have been several recent studies.[139] Those who aspire to heal dance around a fire to the sound of drumming. The purpose of the dance is to activate a healing energy, *n/um*, a distant cousin of the mesmeric fluid, which resides in the pit of the stomach and becomes heated during the dance. When it boils, the vapours rise to the brain, and the dancer passes into a trance state, *!kia*, in which he achieves power of clairvoyant diagnosis, spirit seeing, or travel out of the body. Above all, he acquires the power to heal by transmitting *n/um* to the patient through laying on of hands (*n/um* is held to exude with the sweat) and by pulling the sickness out and transferring it to his own body, whence it is expelled by shaking and shrieking. Novice healers, for whom the boiling and the *!kia* may be too powerful, are restrained, guided and encouraged by the more experienced. The first-hand accounts of those who have become healers in this way leave no doubt that the boiling of *n/um* and the transition to *!kia* occasion real and vivid sensations of heat, pain, dissociation, etc., as well as certain sorts of hallucinations. These sensations, and the various kinds of correlated behaviours, might be looked upon as the outcome of (mainly reflexive) auto- and heterosuggestions, communicated by the setting, tribal traditions, the more experienced participants in the dance, etc., suggestions which take effect as they do because the dancer's cognitive functioning has been rendered cognitively penetrable to them by the prior penetration of a system of ideas which grants them privileged entry. There may also be some assistance from the use of psychoactive drugs.[140] It is of interest to note further that Katz, who has worked extensively with the !Kung, has found that TAT-type tests show that !Kung healers "have easier access to a richer fantasy life. The healer's distinctive access to their own fantasy life can be seen as conducive to experiencing *!kia* and as being affected by it."[141] And we must bear in mind that it may not just be cognitive functions that are "cognitively penetrable", a point most dramatically shown up, perhaps, by the phenomenon of "Voodoo death",[142] which occurs particularly in persons brought up in cultures where that concept plays a significant rôle.

We are now in a position to return again to the problems of hypnotism and hypnosis and their place among the problems of the enhancement of suggestibility. And by now it will be quite obvious what I am going to say. It may or may not be appropriate to place among my second category of factors conducive to the "taking" of suggestions (factors to do with the current psychophysiological state of the agent) a state that may usefully be called "hypnosis". Personally, I doubt that such a state may profitably be distinguished; certainly not one intrinsically characterized by unsuggested posthypnotic amnesia[143] and state-dependent memory; but I am prepared to be convinced. "Hypnosis" may none the less have powerful suggestibility-enhancing effects, not as a state which subjects enter, but as a concept, or rather socially transmitted conceptual system, which works within their minds. The hypnotic state, its characteristics, and the manner of inducing it, have become almost institutionalized in our society. I briefly sketched the history of this process at the beginning of this chapter and shall not repeat myself. Just as a !Kung healer, steeped from childhood in the concepts of his culture, and imaginatively gifted to boot, is soil in which certain suggestions will readily "take", so that he actually feels the n/um boil within him, and sees the world swirl as he passes into !kia, so certain persons in our own culture, knowing as we all do something about hypnosis, and fearing or suspecting a good deal more, are primed for the (mainly reflexive) "taking" of suggestions presented to them in an hypnotic context, and will experience, feel or act correspondingly. And whereas the conceptual framework of the !Kung healers facilitates only the "taking" of suggestions relating to the boiling of n/um, the experience of !kia, etc., the system of hypnosis-related concepts current in our society will, if it sinks deeply enough into a person's mind, facilitate the "taking" of a very large range of kinds of suggestion, for the concept of hypnosis is the concept of a state that facilitates the "taking" of suggestions in general.[144]

If it is indeed the case that what is important in promoting "hypnotic" phenomena – the phenomena of enhanced suggestibility and enlarged faculty ostensibly brought on through procedures of "hypnotic induction" – is not so much subjects passing into an "hypnotic state", as suggestions working upon and through their antecedent concepts of "hypnosis", we are left with two closely interrelated problems which are bound to occur with special force to anyone who has delved into the history of mesmerism and hypnotism. They are the problem of accounting for the remarkable power which these mesmeric and hypnotic concepts have exercized over the minds of the public at large in general and of mesmeric and hypnotic practitioners and subjects in particular, and the problem of accounting for the remarkable resilience and durability of the mesmeric and hypnotic movements in different countries and localities, and in differing social and economic conditions, despite decades of official and establishment neglect and discouragement.

I think that the answer to both problems clearly has to be this. N/um, the boiling energy of the !Kung bushmen, does not exist. It is imaginary, or at best metaphorical. It works, produces felt effects and genuine benefits, not because it is really there, but because those educated into bushman culture believe that it is or might be. It fits into an institutionalized set of concepts which they all share, and which influences the "penetrable" aspects of their cognitive functioning. "Hypnosis" too does not exist, at least in the strong traditional sense involving (in addition to hypersuggestibility) spontaneous post-hypnotic amnesia and state-dependent memory. But the concept of hypnosis in which the majority of hypnotic subjects in our society participate may likewise have powerful effects on the minds and behaviour of some among those who possess it. Of course, the concept of hypnosis which emerged from the mesmeric movement, and has evolved over the last century and a half, has many facets, and is very variously represented in the minds of different persons, educated and uneducated, lay and professional. But in most of its manifestations it might be described as "polysynthetic", synthesized from divers elements. Into the *tout ensemble*, there have become blended elements derived from ideas concerning (for instance) fascination, fixity of gaze, sleep, somnambulism, somniloquy, personal magnetism, healing touch, suggestibility, fugue, trance and allied states of consciousness, multiple personality, the unconscious, epilepsy and psychomotor epilepsy. How these ideas became incorporated is in some instances clear (Puységur brought in the notion of somnambulism) and in others quite obscure (fascination and the evil eye are long established superstitions, and surely they paved the way for and helped promote the spread of mesmerism and hypnotism.) The immediate point, however, is that though the concept of hypnosis, like the concept of n/um, may be an artefact, corresponding to no reality that it has not itself engendered, the elements of the concept are not all factitious, are not all derived from folk-supersitions, socially inculcated practices, etc. Some are genuine in the sense that the phenomena in question occur independently of whether or not the persons to whom they occur antecedently know anything about them. One of the best known of these phenomena, and the one most strongly linked with hypnosis in the popular imagination, is somnambulism in the loose and extended sense accepted by mesmerists and early hypnotists. Others are suggestibility and state-dependent memory. The fact that into the concept of hypnosis there enter elements (further concepts) that are thus "genuine", and are widely known or believed to be genuine, has helped to give hypnotism as a concept and a set of practices its durability and also its powerful and suggestibility enhancing influence on the minds of those about to be hypnotized.[145]

Of all the "elements" in, or supporting, the traditional concept of hypnosis, the most significant is probably the concept of "natural" somnambulism. As we have seen,[146] animal magnetists often cited cases of "natural" somnambulism as providing parallels for and vindications of their proposed state of mesmeric somnambulism. The cases concerned were rare, but instances continued to occur throughout the nineteenth century.[147] Very deep hypnosis was often referred to as the "somnambulic" stage. Modern textbooks tend to play such cases down, stating that somnambulism is mainly a disorder of adolescence, involves simple and somewhat stereotyped behaviour, arises from NREM sleep, and is not accompanied by corresponding consciousness or dream. But in fact cases of complex adult somnambulism continue to be reported, especially, but by no means exclusively, when they result in injury to the patient or to some other person.[148] For example, a case described by Rice and Fisher seems to lie in a no-man's-land between somniloquism, somnambulism and fugue. The patient was a man of forty-two whose problems stemmed from his inability to tolerate the death of his father. He suffered from daytime ambulatory fugue states and from nocturnal somniloquism on topics (the death of his father) related to those of the somniloquism of his daylight fugues. He was studied for four nights in a sleep laboratory and exhibited a number of somniloquent episodes. The authors conclude on the basis of this case that "fugue states may originate out of at least three different, but related, states of consciousness, all of them characterized by waking alpha rhythm. These three states are: (a) waking; (b) the predormescent period immediately preceding the onset of stage I sleep; and (c) the post-REMP waking state artificially produced by sounding a buzzer."[149] They note that in many fugue patients the fugue states are preceded by drowsiness, sleep, fainting spells, or even periods of stupor, and speculate that the fugues may represent periods of dissociation characterized by alpha rhythm.

Schenck and collaborators[150] have recently published polysomnographic data concerning eight adult patients whose presenting diagnosis was somnambulism, and who had suffered injuries in the course of sleep-related dissociative episodes. These episodes (which in some instances occurred during the day as well as at night) were characterized by a waking or drowsy EEG, but subsequent amnesia for the actions carried out (one patient recalled relevant dreams). In several cases, the behaviour concerned was quite complex. One patient regularly underwent an impressive "transformation" into a large and athletic jungle cat with an appropriate behavioural repertoire (which did not however, extend to catching live prey). Another patient underwent two episodes in which she got up, drove to an airport (appropriately dressed), purchased a ticket, and boarded a long-distance flight, in the middle of which she "woke up". Many of the patients had had traumatic childhoods of a kind often found in cases of dissociative disorders.

Schenck and collaborators regard these cases as examples of sleep-related dissociative disorder, and discuss at some length the differential diagnosis of this complaint from sleepwalking, night terrors and REM sleep behavioural disorder (which involves the acting out of dream experiences, sometimes vigorously, and sometimes with somniloquy). In their view, as in that of other recent writers, the term sleepwalking or somnambulism should be confined to cases in which the aberrant behaviour emerges from NREM sleep, though they agree that sleepwalking and sleep-related dissociative disorder share some common distinctive features. But questions of differential diagnosis are not important for present purposes. All these cases are cases of what, following an old but appropriate usage, one might loosely call "sleep-waking", and with them we are clearly approaching the sorts of case that the mesmerists and early hypnotists cited as providing "natural" parallels to "artificial" somnambulism or hypnosis.

It seems that cases of complex somnambulism (including REM sleep behaviour disorder and merging into fugue and ambulatory automatism) continue to occur, and they they may possess features (subsequent amnesia, occasional signs of state-dependent memory) that could give heart to writers disposed to believe that hypnosis is a special suggestibility-enhancing state with many of the same characteristics. What, then, happens if a person liable to somnambulism, or even undergoing a somnambulic episode, is hypnotized? There is some evidence that sufferers from sleepwalking are more than usually hypnotically susceptible, and we have noted Vogt's claim, backed by experiment, that by discovering and following a subject's dream-theme, he could turn "somnambulic dreaming" (surely to be equated with REM sleep behavioural disorder) into "hypnosis" (this probably meant simply that the subject became non-reflexively responsive to suggestions). In his most interesting book on sleeptalking, Arkin has this to say about one of his more complex cases:

> ... the findings demonstrate the possibility that under conditions where brain function is neither altered by drugs nor by other physical interventions, it is possible for a subject to elaborate complex, hallucinatory, dramatic mentation; respond to it; and "live in it" in the presence of sustained alpha rhythm in the EEG. This is succeeded by almost total wakeful amnesia immediately afterward. Furthermore this amnesia is capable of removal many hours later when the subject is deeply hypnotized, at which time the unavailable material is retrievable.[151]

It is as though we have not just similarity, but a deep underlying relationship, between the somniloquent or fugue state and the "state", whatever it may be, brought about by hypnotic induction. Obviously, we are getting close to cases of dual consciousness and multiple personality which, both in the popular imagination and in psychiatric practice, are so closely linked with hypnotism. This topic is too large to go into here, and I will restrict myself to a related matter. The First World War (like all subsequent major ones) produced a great many examples of what was then known as "shell-shock", but would now be categorized as forms of post-traumatic stress disorder (PTSD).[152] These were cases in which the symptoms – nightmares, flashbacks, trembling, paralyses, amnesia, fugues, precipitation of terrors by environmental events reminiscent of the traumata – though caused by prolonged exposure to battle, were functional or psychological in nature, and not due to neural lesions. A case of McDougall's provides a very simple example.[153] A soldier who was in hospital after being dazed by a shell exhibited a series of curious somnambulic episodes in which he would walk over to the bed of the only sergeant in the ward and stand there until led back to his bed. It later transpired that he was reliving the scene of his accident. A shell had exploded, killing and wounding several comrades. He rushed off to the sergeant to report, and as he did so a second shell exploded, dazing him. McDougall treated the case by hypnosis, through which the soldier recovered memory of the episode. During and after the Second World War, recovery of memory in such cases was often assisted by subanaesthetic doses of barbiturates.[154] The similarities to hypnosis could be quite striking, though in fact the administration of these small doses of barbiturates produces an EEG very different from that of waking or hypnotized persons.[155] In this area, we find no less than three "states" which approximate to "hypnosis" as traditionally conceived: the subject's initial fugue state, the state of hypnosis produced by the hypnotist if hypnotic treatment is undertaken, and the state produced by the barbiturate. Such cases and methods of treatment, it may be added, became very well known, and have figured in several films.

If one moves back in time to Charcot's Salpêtrière and considers (say) some of the "hysterical" states exhibited by certain of Janet's classic cases, one can see in several of them (as van der Kolk and van der Hart have argued[156]) similarities to the fugues and flashbacks of patients now classified as suffering from PTSD (such patients, it may be noted, are as a class highly suggestible[157]). For Janet, as we have seen, hypnosis simply is the artificial recreation of such states in persons liable to them, often, apparently, by the hypnotist so to speak insinuating himself into the patient's self-generated drama and thereafter manipulating it by suggestions which would presumably have "taken" non-reflexively. I do not see any reason for doubting that some of these hypnotic/hysterical cases were genuine in the sense of not being modelled upon cases of which the patient had antecedent knowledge,[158] and I think, correspondingly, that Janet was almost certainly right in maintaining, against Bernheim, that perfectly normal and healthy individuals cannot be "hypnotized" *in this sense*. Janet's accounts of his cases, which became very well known, none the less, and despite his own reservations, continued to feed the concept of hypnosis in general.

As further examples of states with features which may have helped engender, or have lent some support to, the concept of hypnosis, I could go on to mention states induced by drugs other than barbiturates, alcoholic states, various forms of delirium, the states induced by shamanic practices and by certain meditative techniques, and so forth. I could even propose once again that mesmeric passes (perhaps also Braid's very stressful eye-fixation technique) may in some subjects have produced effects (possibly due to the activation of endogenous opioids) which did not simply occur because the subjects expected them to occur. It might even be that a genuine state of opioid-caused mesmeric "sleep-waking" helped to engender the concept of a fictional "hypnosis". But by now the general drift should be clear. We do not know what, psychophysiologically, goes on when a suggestion "takes", and quite possibly no one psychophysiological account will cover all cases. It may be, for instance, that there are both "active-agent" and "passive-subject" ways in which suggestions reflexively "take".[159] We do know of various background factors which can enhance suggestibility, that is, increase the likelihood of a suggestion "taking". Among them seem to be certain enduring personal characteristics or personality traits.[160] Among them *may* be (though I rather doubt it) a state of relaxation and "loss of generalized reality orientation" which it would be appropriate to call "hypnosis" (at any rate I would not wish to deny that the mental state of a deeply hypnotized person, turned to the "inner" rather than the outer world, may develop threads of state-dependent memory linking it to some kinds of dissociated state). Among them undoubtedly is possession of, or participation in, the more or less institutionalized concept of hypnosis current in our society. This concept draws part of its power and its durability from the fact that although no reality precisely corresponds to it, it has significant elements or aspects which reflect or parallel certain genuine phenomena, and certain fairly well-known abnormal states which some individuals are liable in some circumstances to enter. And we must not neglect the possibility that if these individuals are subjected to hypnotic or mesmeric induction procedures they may pass into or towards the abnormal states concerned, thus opening up another route by which the traditional concepts of hypnosis largely current in our society may be strengthened, first among certain hypnotists, and then derivatively among the public at large. Indeed it is hard to see how some features of the traditional concept of hypnosis could have arisen unless something like this had occasionally occurred. Such individuals will be found only very rarely in the student populations upon which modern hypnotic experimenters largely draw; but it is arguable that they were more frequently encountered in nineteenth-century clinical practice, from which so many common ideas about hypnosis derive, and that they may occasionally be encountered today.[161]

The mysterious domain of hypnosis emerges, then, as a kind of fairy palace, less than real, but more than illusion. It has, one might say, sufficient substance in its foundations to have deceived mortals rather well. Especially has it ensnared savants of past generations, who in turn have misled the media and the public at large. But from our vantage point at the end of the twentieth century we can begin to see that there is no one path by which it may be reached, no one material of which it is built, no one hidden chamber containing all its secrets, no one key which will open all its doors, and no simple formula by which it may be dispelled. Those who set out to investigate it should beware of the bafflements to come.

NOTES

1 Cf. Goldstein (1982).
2 See Ellenberger (1970), pp. 158–170, 278–284.
3 The magnetization of dying persons was not in fact attended with the frightful results which Poe envisaged. See Siemers (1835), p. 249.
4 Ellenberger (1970), p. 162.
5 I know this work only from the reference in Dingwall (1968a), p. 129.
6 E.g. Lytton's *The Haunted and the Haunters* (1859) and *A Strange Story* (1862), Mrs Amelia Edwards' *The Phantom Coach* (1864), Conan Doyle's *The Great Keinplatz Experiment* (1885), Bram Stoker's *Dracula* (1897) and Richard Marsh's *The Beetle* (1897).
7 See e.g. Dejérine (1890–1).
8 No doubt preexisting superstitions about fascination and the evil eye prepared the ground for the ready reception of mesmerism and hypnotism.
9 Wingfield (1910), pp. 165–166.
10 It was Bodie's medical claims which seem to have inspired the "Bodie riots" among Glasgow medical students in 1909.
11 On the methods of stage hypnotists see Leonidas (1901); Burrows (1912).
12 See e.g. Rivers (1920); Brown (1918; 1919; 1920); McDougall (1920–1).
13 Weitzenhoffer (1963); Wolberg (1948); Schneck (1959); Kline (1958); Scott (1984).
14 See e.g. Kroger (1977); Udolf (1988); Heap (ed.) (1988); Heap and Gibson (1990).
15 Smoking: Perry and Muller (1975), Perry, Gelfand and Marcovitch (1979), Holroyd (1980); skin disorders, headaches and obesity: De Piano and Salzberg (1979); obesity: Wadden and Flaxman (1981); alcoholism: Wadden and Penrod (1981–2).
16 Particularly well-known are the studies of the hypnotic treatment of burns by Ewin (1983–4; 1986–7). But van der Does and van Dyck (1989) in a review of the literature on hypnotic treatment of burns urge caution and the need for control studies.
17 See review by Blalock (1984).
18 Jemmott and Locke (1984); Ader *et al.* (1990).
19 See reviews by Ader (1981); Bovbjerg *et al.* (1982).
20 H. Hall (1982–3; 1989); Goldberg (1988); O'Leary (1990).
21 Frankel (1974); Frankel and Orne (1976); McGuinness (1983–4); but contrast Owens *et al.* (1989).
22 Bliss (1984a; 1986); Spiegel, Hunt and Dondershine (1988).
23 On this, and on this topic in general, see Kluft (1987); Ross, Norton and Wozney (1989); Boor and Coons (1983). Ross, Norton and Wozney state (p. 413) that whereas about 200 cases of MPD had been diagnosed up to 1980, 6000 have been so diagnosed since. Yet Japan is apparently immune to the disorder! Cf. Takahashi (1990).
24 See Bliss (1988); Dell (1988).
25 Ross, Norton and Wozney (1989) in a sample of 236 cases found that 79.2% of patients had been sexually abused as children, and 74.9% physically abused. In a sample of 100 cases analyzed by F. W. Putnam *et al.* (1986) the figures were even higher.

26 Bliss (1984b), p. 138.
27 The most convenient sources are T. X. Barber (1969; 1970; 1973).
28 It is thus appreciably shorter than the better-known Stanford Scale of Hypnotic Susceptibility (SSHS) of Hilgard and Weitzenhoffer, and the Harvard Group Scale of Hypnotic Susceptibility (HGSHS) of Shor and Emily Orne.
29 T. X. Barber and Calverly (1962–1965).
30 Cf. above pp. 440–442.
31 Tart (1970; 1979); Shor (1979).
32 Cf. Sheehan and Dolby (1974); Sheehan and Perry (1976).
33 Sheehan and Dolby (1974).
34 T. X. Barber (1973), p. 950.
35 See for example his negative findings on hypnotic hallucinations (Barber and Calverly, 1964), and compare the reply by Bowers (1968) and the further experiment of Spanos and Barber (1968).
36 Fellows (1986), p. 52.
37 Josephine Hilgard (1970).
38 T. X. Barber, Spanos and Chaves (1974).
39 T. X. Barber and Wilson (1977).
40 Banyai and Hilgard (1976), cf. Callen (1983–4); Gibbons (1974; 1975; 1979); Vingoe (1973); Malott (1984).
41 Wilson and Barber (1977–8); T. X. Barber and Wilson (1978–9).
42 E.g. the suggested production of skin markings and blistering (of which he obtained some evidence himself) which he was now prepared to attribute to the influence of imagination on capillary blood flow (T. X. Barber, 1984).
43 Wilson and Barber (1983), p. 342.
44 Wilson and Barber (1983), p. 351.
45 Rhue and Lynn (1991), p. 213; cf. Lynn, Rhue and Green (1988).
46 Bliss (1984a; 1986, pp. 166–169); cf. Spiegel, Hunt and Dondershine (1988).
47 Vincent and Pickering (1983); Albini and Pease (1989).
48 Benson et al. (1986); Schenk and Bear (1981); but contrast Devinsky et al. (1989). A connection between the two was proposed as early as Rieger (1884).
49 Wilson and Barber (1983), pp. 376–377.
50 S. A. Myers and Austrin (1985); cf. the review of their own and others' work by Lynn and Rhue (1988).
51 E. R. Hilgard (1973b; 1976; 1977; 1978). He calls his theory "neodissociationist" because earlier "dissociationists" have been supposed by some to have held that "dissociated" functions may be carried out together without interference, which experiment shows not to be the case. However, some data described by Bowers (1991) might be taken to suggest that in hypnotized subjects "dissociated" functions can be carried on without interfering with the main stream of consciousness.
52 E. R. Hilgard (1979), p. 46.
53 Kihlstrom (1984), p. 188.
54 E. R. Hilgard (1978), p. 29.
55 E. R. Hilgard (1977), p. 222.
56 E. R. Hilgard (1977), p. 227.
57 E. R. Hilgard (1977), pp. 228–229.
58 E. R. Hilgard (1977), p. 230.
59 E. R. Hilgard (1977), p. 230.
60 E. R. Hilgard (1977), p. 232.

61   The literature on this topic is very large. See especially E. R. Hilgard and J. Hilgard (1983),
     chapter 9.
62   E. R. Hilgard and J. Hilgard (1983), pp. 166–176.
63   E. R. Hilgard and J. Hilgard (1983), p. 178.
64   E. R. Hilgard (1976), p. 163.
65   See the subjects' reports quoted in E. R. Hilgard and J. Hilgard (1983), p. 173.
66   For instance, Kihlstrom and Evans have both recently offered accounts of the nature of post-
     hypnotic amnesia (Kihlstrom 1984; Evans 1988). Bowers (1991) presents data which he
     regards as supporting the neodissociationist position in general. Hypnotized highly
     susceptible subjects were found able to perform without mutual interference two concurrent
     tasks each of which required high-level cognitive effort. In unhypnotized and low-
     susceptible subjects the two tasks interfered with each other. Bowers regards this finding as
     a "straightforward prediction" from Hilgard's neodissociationist theory. The findings are
     certainly *compatible with* the neodissociationist position, and may force supporters of the
     social psychological position into further *ad hoc* supplementary hypotheses. But I do not see
     that they would have constituted any particular difficulty for (say) Vogt or Bernheim.
67   For an analysis of the concept of dissociation see Braude (forthcoming).
68   Perhaps following Wigan (1844) on the dual functions of the cerebral hemispheres. Cf.
     Bérillon (1884).
69   Cf. footnote 48 above and Loewenstein and Putnam (1988).
70   Consider, for example, Samuel Butler's *The Way of All Flesh*, which describes from a third
     person perspective, in the form of a novel, his own childhood sufferings.
71   Cf. W. F. Prince (1915–16), p. 839 etc., and M. Prince (1908), chapter xxiii.
72   I have witnessed this phenomenon several times myself.
73   But even in what once seemed the clearest cases of such concurrent streams of
     consciousness – those of certain "split brain" cases – the proper interpretation is becoming
     less clear. Cf. Gazzaniga (1987); J. J. Myers and Sperry (1985).
74   Mühl (1961).
75   Fingarette (1969), p. 136.
76   Spanos (1982b), pp. 231–232.
77   R. W. White (1941).
78   See the theoretical passages in two summary articles, Spanos (1982a and b).
79   Spanos (1982b), p. 258.
80   Cf. Spanos and Gorassini (1984).
81   Views generally similar to those of Spanos are expressed by G. Wagstaff of Liverpool
     University (e.g. Wagstaff 1981; 1986). Wagstaff is more prone than Spanos to allege that
     "hypnotized" subjects may simply be exhibiting "compliance" in response to the demands
     of a social situation. But he also holds that they may engage in a kind of motivated believing
     that their experiences are concordant with being "hypnotized", a believing facilitated by
     the ambiguous nature of the sensations commonly undergone during hypnotic induction
     procedures.
82   For summaries see Spanos (1982b), pp. 246–252; Spanos (1986c), pp. 457–462; Spanos
     (1989).
83   Spanos (1986c), p. 457.
84   Cf. Stam and Spanos (1980).
85   Spanos (1986c), p. 460.
86   Miller and Bowers (1986), pp. 11–12.
87   Spanos (1989), p. 218.
88   Spanos and Katsanis (1989).

89 See Hilgard and Hilgard (1983), pp. 120–163; Morse *et al.* (1984); Morse and Wilcko (1978–9).

90 It has recently been reported by Simon and Salzberg (1985). Cf. Kunzendorf and Benoit (1985–6). Like most persons who have practised hypnotism to any extent, I have encountered occasional examples of apparent spontaneous post-hypnotic amnesia. In two cases, the amnesia could not be reversed by waking suggestion. It was hard to convince these subjects that they had been hypnotized, and they invented plausible reasons for changes which had "suddenly" taken place in their surroundings. Both belonged to that interesting class of subjects who "fall asleep" quietly when someone else is being worked on and is therefore playing the "rôle" of hypnotic subject.

91 See Spanos (1982b), pp. 252–257; Spanos (1986c), pp. 450–457. A seminal article is that by Coe (1978).

92 E.g. Kihlstrom *et al.* (1980).

93 McConkey, Sheehan and Cross (1980).

94 Spanos (1982b), p. 254.

95 See especially Spanos, Weekes and Bertrand (1985); Spanos (1986b). For articles on related topics see Spanos (1978); Spanos and Gottlieb (1976; 1979); Spanos and Hewitt (1979); Spanos (1983); Stam and Spanos (1982).

96 Cf. footnote 23 above.

97 See Bliss (1988); Dell (1988).

98 Bliss (1984b), p. 138.

99 Spanos (1986b), p. 39.

100 Spanos, Weekes and Bertrand (1985), p. 364.

101 Kluft (1987), pp. 367–368.

102 Weiss, Sutton and Utecht (1985); Fagan and McMahon (1984); Vincent and Pickering (1988); Albini and Pease (1989).

103 Cf. especially Ross, Norton and Fraser.

104 Spanos, Weekes and Bertrand (1985), p. 365.

105 As one example of how one might play the explaining-away game, consider a young man, X, observed by me on both the relevant occasions, who had proved an excellent subject at a demonstration given to a university society. He returned a year later to a demonstration by the same hypnotist (not myself), expressing the firm intention to be only a spectator this time, so as to observe the phenomena for himself. While he stood at the back of the hall, watching another subject being worked on, a brief tape-recorded message from the hypnotist was played, "Mr X, go to sleep now." Within a second his expression of surprise turned into one of sleep, and I was just in time to catch him as he slumped to the floor. He appeared to have had insufficient time to realize what had happened to him, let alone to change his mind about playing a rôle which he had consciously set out not to play. But no doubt he was very quick-thinking and instantly grasped (and then disguised from himself) the fact that the reputation which he had built up the previous year as a "good" hypnotic subject would be shattered if he did not again respond as required.

106 Cf. Marks *et al.* (1989).

107 But see Gheorghiu *et al.* (eds.)(1989); Schumaker (ed.)(1991); the most useful review of the earlier literature is Evans (1966).

108 On hypnotic "skill" training see Spanos (1986a); Bertrand (1989), A central article is that by Gorassini and Spanos (1986).

109 For references see footnote 108.

110 Cf. Spanos's reply to Miller and Bowers, above pp. 598–599.

111 Cf. Weitzenhoffer (1974).

112 A surgical patient known to me practised and taught a form of coping with post-operative pain that involved "going into" the pain, "taking hold of it", and "coming out the other side".

113 Spanos *et al.* (1982–3).

114 Lynn, Rhue and Weekes (1989), in their valuable review of the literature on this issue, are very clear (p. 99) about the conceptual difficulties which confront subjects in these experiments.

115 Cf. the experiments recently reported by Levitt, Baker and Finch (1989–90), and the accompanying comments by Coe, Lynn, Perry, Spiegel and Weitzenhoffer, with rejoinder by the authors.

116 Cf. Lynn and Rhue (1988).

117 E.g. Cardeña and Spiegel (1991), pp. 99–101.

118 There is a large literature on these topics. See e.g. Barker and Bergwin (1949); Chertok and Kramarz (1959); Gorton (1949); Engstrom (1976); Saletu *et al.* (1975); Sabourin (1982); Gabel (1988); Bick (1989); De Pascalis et al (1987); De Pascalis, Silveri and Palumbo (1988); Sabourin *et al.* (1990).

119 T. X. Barber (1970), pp. 21–22; Kiloh *et al.* (1981), p. 226.

120 Cf. e.g. Fenwick *et al.* (1977); Mavromatis (1987), p. 187; Kiloh *et al.* (1981), p. 199; Kelly *et al.* (1978); T. X. Barber (1970), p. 36; Sjoberg and Hollister (1965); J. Barber *et al.* (1979); H. Benson *et al.* (1982) and note 40 above.

121 Cf. R. Prince (1968; 1982); Jilek (1982); Winkelman (1986); Ward and Kemp (1991).

122 By "partial systematic dissociation" Vogt meant a state in which one idea or idea-complex is hyperactivated, all else being normal.

123 Woodworth (1906).

124 Wagstaff (1991), p. 134.

125 Factor analytic studies suggest that not all dimensions of hypnotizability correlate with the dimensions of waking suggestibility. Cf. Evans (1989).

126 Cf. note 107.

127 There is a large literature on the rôle of subject expectancy in promoting hypnotic response. See especially Council *et al.* (1983).

128 Eysenck and Furneaux (1945).

129 Cf. Eysenck (1991), pp. 86–87.

130 Lishman (1987), p. 515.

131 E.g. Spanos *et al.* (1982–3); Rhue and Lynn (1987).

132 E. R. Hilgard (1965), pp. 133, 135; Orne (1959; 1979).

133 The analysis of the language of hallucinations by Sarbin and Juhasz (1978) seems to me to leave the issues more confused than before.

134 Sheehan and McConkey(1982), pp. 24–25 and 168.

135 McConkey, Glisky and Kihlstrom (1989) is a modern example of something like what I have in mind.

136 Spiegel and his collaborators (Spiegel and Barabasz, 1988–9; Spiegel *et al.*, 1985; Spiegel *et al.*, 1989) report with highly susceptible subjects that an hypnotic hallucination may interfere with evoked potentials in the same way as would perception of a real object. However, Walter *et al.* (1990), investigating regional cerebral blood flow during auditory hallucinations, found differences between hypnotized subjects and psychotic ones; for instance, the hypnotized subjects showed relatively increased blood flow in thalamic regions, especially on the right.

137 Cf. notes 40 and 120 above.

138 On "cognitive penetrability" see Pylyshyn (1984), pp. 130–145, 217–229.

139 Lee (1968); Katz (1982; 1989).
140 Winkelmann and De Rios (1989).
141 Katz (1982), p. 365.
142 Cf. Cohen (1985); H. Hall (1986).
143 But cf. note 90 above.
144 All experiments purporting to test the effects of putting a subject into hypnosis thus require a placebo control group in which some non-hypnotic procedure (e.g. in hypnotic analgesia administration of a dummy pill) is used to work powerfully on subjects' waking suggestibility. Cf. Evans (1989), pp. 151–152.
145 The nearest parallel I can find for the situation with regard to the "polysynthetic" concept of hypnosis is that which obtains over the concept of *piblokto* or arctic hysteria, of which an interesting account is given by Locke and Kelly (1985). Aspects of *piblokto* are "genuine" and according to Locke and Kelly originate from calcium deficiency. Yet undoubtedly the concept (which has been taken up by shamans for their own ends) shapes the behaviour of those who hold it.
146 Cf. above pp. 259–261.
147 Cf. for example Tuke (1884); Spitta (1882).
148 Cf. Crisp *et al.* (1990); Schenck *et al.* (1989a and b); Oswald and Evans (1985); Hartmann (1983); Mahowald *et al.* (1990).
149 Rice and Fisher (1976), p. 85.
150 Schenck *et al.* (1989a and b); Mahowald and Schenck (1991).
151 Arkin (1981), p. 185.
152 On PTSD see Horowitz (1986). Cf. Loewenstein and Putnam (1988).
153 McDougall (1926), p. 259.
154 For a brief history of the technique see Haward (1972). A classic treatment is in Sargent and Slater (1963). There are some signs of a revival, e.g. Perry and Jacobs (1982); Waller *et al.* (1985).
155 Kooi (1971), p. 227; Kiloh *et al.* (1981), p. 224.
156 Van der Kolk and van der Hart (1989).
157 Spiegel, Hunt and Dondershine (1988).
158 There are quite a few cases of spontaneous complex somnambulism or "double consciousness" in the pre-mesmeric literature. The patients (mostly young) would pass into aberrant states, often characterized by delusory perceptions and unawareness of painful stimuli. These phases were generally ushered in and terminated by a brief catalepsy or period of insensibility, and there was always subsequent amnesia and sometimes state-dependent memory. I have treated this topic in greater detail in Gauld (1992). For some eighteenth-century cases see Du Prel (1889b), Vol. 2, pp. 67–98. Cf. Boissier de Sauvages (1768), Vol. 2, p. 207; Haen (1760), Vol. 4, pp. 199–210; Pomme (1778), pp. 39–49; Lorry (1770), Vol. 1, pp. 101–105; Sauvages de la Croix (1742); Held (1722); J.C. Müller (1725). A similar case is described in Erasmus Darwin's poem *Zoonomia* (1794–6). For even earlier cases see Thurston 1952, pp. 104–117; Montalvo (1719), chapter 64; Bourneville (ed.) (1886), pp. 25–27. Crabtree's suggestion (1985a), pp. 138–139, that there were no MPD cases prior to the mesmeric movement is at best a great oversimplification.
159 Cf. the different "roads to hypnosis" apparently adopted by the two hypnotic "virtuosos" described by McConkey, Glisky and Kihlstrom (1989).
160 It is worth bearing in mind here the point made by Nadon *et al.* (1987) following an experimental investigation of the personality correlates of hypnotic suggestibility: " ... the experiments in this study support the notion that hypnotizability reflects various abilities

and preferences and that univariate analyses alone cannot capture the complexity that underlies hypnotic talent." Cf. Nadon, Laurence and Perry (1991), Pekala (1991).

161   Various late-nineteenth-century writers described cases in which hypnosis evolved into "hysterical somnambulism'. Vogt wrote a celebrated article on such cases (Vogt, 1897–8), and Loewenfeld (1901, pp. 207–208) cites two from his own experience. The first is of an hysterical lady of thirty, who had formerly had convulsive attacks, and now suffered from headaches and obsessional ideas. This lady passed into hypnotic sleep readily enough, but proved unresponsive to conventional test suggestions. Instead she began to conduct a conversation with Loewenfeld, casting him in the rôle of her husband. In a second case, a hypnotized lady began to hold a conversation with the hallucination of her husband, and proved very difficult to wake up. These patients (Loewenfeld asserts) were not just undergoing somnambulic dreams, but were in states of mind satisfying Vogt's criteria for hysterical somnambulism. There was motor expression of the contents of consciousness, a chain of ideas which appeared spontaneously and was not easily influenced by suggestion, and subsequent amnesia. Brügelmann (1896–7, p. 275) cites a case described to him by Vogt, in which a deeply hypnotized lady took the hypnotist (Vogt) for her husband, and engaged in sustained copulatory movements.

References: Animal magnetism

Abrégé de l'histoire des magnétiseurs de Lyon, par un nouveau converti (1784). N. p., n. pub.

Ackerknecht, E. H. (1967). *Medicine at the Paris Hospital (1794–1848)*. Baltimore, the Johns Hopkins Press.

Acland, –. (1852–3). The galvanic disc delusion dispelled. *The Zoist*, 10, 48–67.

Adams, –. (1849–50). Remarkable mesmeric phenomena. *The Zoist*, 7, 79–80.

Adelon, N. P., *et al* (eds.) (1812–22). *Dictionnaire des sciences médicales* ... 60 vols., Paris, C. L. F. Pancoucke.

Adelon, N. P., *et al* (eds.) (1821–28). *Dictionnaire de médecin.* 21 vols., Paris, Béchet le jeune.

Ader, R. (1981). A historical account of conditioned immunobiologic responses. In R. Ader (ed.), *Psychoneuroimmunology.* New York, Academic Press. Pp. 321–352.

Akil, H., *et al* (1984). Endogenous opioids: Biology and function. *Annual Review of Neuroscience*, 7, 223–256.

Allison, J. (1844). *Mesmerism: Its Pretensions as a Science Physiologically Considered.* London, Whitaker.

Amadou, R. (1953). Esquisse d'une histoire philosophique du fluide. *Revue métapsychique*, No. 21, 5–33.

Amadou, R. (1969). *Trésor Martiniste.* Paris, Villain et Belhomme.

Amadou, R., and Joly, A. (1962). *De l'agent inconnu au Philosophe inconnu.* Paris, Éditions Denoël.

Angoff, A. (1968). Hypnotism in the United States of America. In E. J. Dingwall (ed.), *Abnormal Hypnotic Phenomena: A Survey of Nineteenth-Century Cases*, vol. 4. London, J. and A. Churchill. Pp. 1–78.

Animal magnetism (1830a). *The Foreign Review and Continental Miscellany*, 5, 96–124.

Animal magnetism (1830b). *Fraser's Magazine*, 1, 673–684.

Annales de la Société harmonique des amis réunis de Strasbourg ou cures que des membres de cette société ont opérées par le magnétisme animal (1789). Strasbourg, n. pub.

L'antimagnétisme Martiniste ou Barbériniste ... (1784). Lyon, n. pub.

Arkin, A. M. (1981). *Sleep-Talking: Psychology and Psychophysiology.* Hillsdale, N. J., L. Erlbaum.

Arndt, W. (1816). *Beyträge zu den durch animalischen Magnetismus zeither bewirkten Erscheinungen. Aus eigner Erfahrung.* Breslau, C. Cnobloch.

Artelt, W. (1951). Der Mesmerismus im deutschen Geistesleben. *Gesnerus*, 8, 4–14.

Artelt, W. (1965). *Der Mesmerismus in Berlin.* Wiesbaden, F. Steiner.

Ashburner, J. (1847–8). On the silent influence of the will. *The Zoist*, 5, 260–273.

Ashburner, J. (1848–9). Facts in clairvoyance. *The Zoist*, 6, 96–110.

Ashburner, J. (1867). *Notes and Studies in the Philosophy of Animal Magnetism and Spiritualism* ... London, H. Baillière.

Atkinson, H. G., and Martineau, H. (1851). *Letters on the Laws of Man's Nature and Development*. London, J. Chapman.

Audry, J. (1924). Le mesmérisme et le somnambulisme à Lyon avant la révolution. *Mémoires de l'Académie des sciences, belles lettres et arts de Lyon*, sér. 3, 18, 57–100.

Autographe de Mesmer (1849). *Journal du magnétisme*, 8, 653–655.

Baader, F. (1818). *Ueber die Ekstase oder das Verzuckseyn der magnetischen Schlafredner ...* Nürnberg, Monath und Kussler.

Baas, J. H. (1971). *Outlines of the History of Medicine and the Medical Profession*. 2 vols. Huntington, N. Y., R. E. Krieger.

Bähr, J. K. (1853). *Der animalische Magnetismus und die experimentirende Naturwissenschaft*. Dresden, W. Türk.

[Bähr, J. K., and Kohlschütter, R.] (1843). *Mittheilungen aus dem magnetischen Schlafleben der Somnambüle Auguste K. in Dresden*. Dresden, Arnold.

Bährens, J. C. F. (1816). *Der animalische Magnetismus und die durch ihm bewirkten Kuren*. Elberfeld, Mannes.

Bakker, G., Wolthers, H., and Hendriksz, P. (1814–18). *Bijdragen tot den tegenwoordigen staat van het animale magnetismus in ons vaderland*. 2 vols. Groningen, Schierbeek en van Boekeren.

Bakker, G., Wolthers, H., and Hendriksz, P. (1818). *Beobachtungen über die Heilkraft des thierischen Magnetismus*. Halle, Renger.

Baldock, T. (1845–6). Cure of severe affection of the stomach with mesmerism. *The Zoist*, 3, 337–338.

Baldwin, G. [1810?]. *La Prima Musa Clio. Or, the Divine Traveller; Exhibiting a Series of Writings obtained in the Extasy of Magnetic Sleep ...* London, privately printed.

Baldwin, G. (1818–19). Extrait d'un ouvrage sur le magnétisme animal. *Bibliothèque du magnétisme animal*, 3, 212–230; 5, 279–285; 6, 70–75, 184–188; 7, 52–55, 146–164.

Ballou, A. (1852). *An Exposition of the Views respecting the Principal Facts, Causes and Peculiarities involved in Spirit Manifestations ...* Boston, Bela Marsh.

Banks, Sir J. (1958). *The Banks Letters ...* Ed. W. R. Dawson, London, The Trustees of the British Museum.

Baraduc, H. (1896). *L'âme humaine: Ses mouvements, ses lumières et l'iconographie de l'invisible fluidique*. Paris, G. Carré.

Baraduc, H. (1897). *Photographie des états hypervibratoires de la vitalité humaine*. Paris, A. Quelquejeu.

Baraduc, H. (1904). *Les vibrations de la vitalité humaine*. Paris, J. B. Baillière.

Barberin, – . (1818). Auszug aus dem magnetistischen Tagebuch des Ritters von Barberin. *Blätter für höhere Wahrheit*, 1, 208–242.

Baréty, A. (1882). *Contribution à l'étude des propriétés du système nerveux. Des propriétés physiques d'une force particulière du corps humain ... connue vulgairement sous le nom de magnétisme animal*. Paris, J. Lechevalier.

Baréty, A. (1897). *Le magnétisme animal étudié sous le nom de force neurique rayonnante et circulante ...* Paris, O. Doin.

[Barré, P. Y., and Radet, J. B.] (1785). *Les docteurs modernes, comédie-parade, en un acte et en vaudevilles, suivie du baquet de santé ...* Paris, Brunet.

Barrett, W. F. (1884). Note on the existence of a "magnetic sense". *Proceedings of the Society for Psychical Research*, 2, 56–60.

Barrett, W. F., *et al.* (1882–3). First report of the "Reichenbach" committee. *Proceedings of the Society for Psychical Research*, 1, 230–237.

Barrett, W. F. (1884). Note on the existence of a "magnetic sense." Proceedings of the Society for Psychical Research, 2, 56–60.

Barrow, L. (1986). *Independent Spirits: Spiritualism and English Plebeians 1850–1910.* London, Routledge and Kegan Paul.

Bartels, E. (1812). *Grundzüge einer Physiologie und Physik des animalischen Magnetismus.* Frankfurt-am-Main, Varrentrapp und Sohn.

Barth, G. (1853). *What is Mesmerism?* ... London, H. Baillière.

Basbaum, A. I., and Fields, H. L. (1984). Endogenous pain control systems: Brainstem spinal pathways and endorphin circuitry. *Annual Review of Neurosciences*, 7, 309–328.

Bauer, E. (1989). Exkursionen in "Nachtgebiete der Natur" – Justinus Kerner und die historische Spukforschung. *Zeitschrift für Parapsychologie und Grenzgebiete der Psychologie*, 31, 3–19.

Becht, H. G. (1876). *De mesmerische tooverlantaarn. Eene bladzijde wit mijne praktijk...* Entschede, Van der Loeff.

Beer, M. de, Fourie, D. P., and Niehaus, E. C. (1985–6). Hypnotic analgesia: Endorphins or situations? *British Journal of Experimental and Clinical Hypnosis*, 3, 139–145.

Belden, L. W. (1834). *Somnambulism. The Extraordinary Case of Jane C. Rider, the Springfield Somnambulist.* London, Simpkin and Marshall.

Bendsen, B. (1821–2). Tagebuch einer lebensmagnetischen Behandlung der Wittwe Petersen zu Arröeskjöping. *Archiv für den thierischen Magnetismus*, 9(1), 61–168; 9(2), 42–203; 10(1), 83–150; 10(2), 1–55.

Bendsen, B. (1822–3). Nachtrag zu der Krankheitsgeschichte der Wittwe Petersen... *Archiv für den thierischen Magnetismus*, 11(1), 66–164; 11(2), 34–138; 11(3), 36–143; 12(1), 59–80.

Bennett, J. H. (1851). *The Mesmeric Mania of 1851.* Edinburgh, Sutherland and Knox.

Benoit, J. T., and Biat, C. (1850). *Communication universelle et instantée de la pensée à quelque distance que ce soit, à l'aide d'un appareil portatif appelé Boussole pasilalinique sympathique.* Paris, Bureau de l'Institut polytechnique.

Benor, D. J. (1984). Fields and energies related to healing: A review of Soviet and Western studies. *Psi Research*, 3, 21–35.

Benz, E. (1968). *Les sources mystiques de la philosophie romantique allemande.* Paris, J. Vrin.

Benz, E. (1977). *Franz Anton Mesmer und die philosophischen Grundlagen des "animalischen Magnetismus".* Mainz, Akademie der Wissenschaften und der Literatur.

Bergasse, N (1784). *Considérations sur le magnétisme animal, ou sur la théorie du monde et des êtres organisés, d'après les principes de M. Mesmer*... La Haye, n. pub.

[Bergasse, N.] (1785a). *Dialogue entre un docteur de toute les universités & académies du monde connu... & un homme de bon sens, ancien malade du docteur.* In *Recueil des pièces les plus intéressantes pour & contre le magnétisme animal*, Vol. 1. Lyon, n. pub. Pp. 175–198.

[Bergasse, N.] (1785b). *Lettre d'un médecin de la Faculté de Paris à un médecin du College de Londres.* In *Recueil des pièces les plus intéressantes pour & contre le magnétisme animal*, Vol 1. Lyon, n. pub. Pp. 199–267.

Bergasse, N. (1786). Observations de M. Bergasse sur un écrit du docteur Mesmer, ayant pour titre: Lettre de l'inventeur du magnétisme animal, à l'auteur des Réflexions preliminaires. In *Recueil de piéces pour servir à l'histoire du magnétisme.* N.p., n. pub. Pp. 28–155.

Berger-Fix, A. (ed.) (1986). *Justinus Kerner. Nur wenn man von Geistern spricht: Briefe und Klecksographien.* Stuttgart-Wien, Edition Erdmann in K. Thienemanns Verlag.

Bérillon, E. (1884). *Hypnotisme experimental: La dualité cérébrale et l'indépendance fonctionelle des deux hémisphères cerebraux.* Paris: A. Delahaye and E. Lecrosnier.

Bersot, E. (1853). *Mesmer et le magnétisme animal.* Paris, L. Hachette.

Bertrand, A. J. F. (1820). Cours public sur le magnétisme animal. *Archives du magnétisme animal,* 2, 57–93.

Bertrand, A. J. F. (1823). *Traité du somnambulisme et des différentes modifications qu'il présente.* Paris, J. G. Dentu.

Bertrand, A. J. F. (1826). *Du magnétisme animal en France, et des jugements qu'en ont portés les sociétés savantes* ... Paris, J. B. Bailliére.

Bertrand, A. J. F. (1829). Opération d'un cancer faite sur une extatique insensible. *Le Globe,* 7, 255–256.

Bicker, G. (1787a). Herrn Doctor Bickers zu Bremen Brief an den Herrn Hofrath Baldinger über Lavaters Magnetismus. *Hannoverisches Magazin,* Pt. 3, 34–48. Also in *Magnetistisches Magazin für Niederteutschland,* 1787, 1, 31–39.

Bicker, G. (1787b). Zweiter Brief, an den Herrn Hofrath Baldinger, über den thierischen Magnetismus. *Hannoverisches Magazin,* Pt. 19, 290–304. Also in *Magnetistisches Magazin für Niederteutschland,* 1787, 3, 247–264.

Bicker, G. (1788). Erklärung des Hrn. D. Bickers über seine beyden an Hrn. Hofrath Baldinger geschriebene Briefe, den animalischen Magnetismus betreffend. *Archiv für Magnetismus und Somnambulismus,* Pt. 6, 85–95.

Billot G. P. (1839). *Recherches psychologiques sur la cause des phénomènes extraordinaires observés chez les modernes voyans, improprement dits somnambules magnétiques, ou correspondance sur le magnétisme vital, entre un solitaire et M. Deleuze* ... 2 vols. Paris, Albanel et Martin.

Birot, – . (1814). Suite des recherches sur la faculté de se magnétiser soi-même. *Annales du magnétisme animal,* 2, 261–268.

Bjelfvenstam, E. (1967). Hypnotism in Scandinavia. In E. J. Dingwall (ed.), *Abnormal Hypnotic Phenomena; A Survey of Nineteenth-Century Cases,* vol. 2. London, J. and A. Churchill.

Blalock, J. E. (1984). The immune system as a sensory organ. *The Journal of Immunology,* 132, 1067–1070.

Blankenburg M. (1985). F. A. Mesmer – Aufklärer und Citoyen. In H. Schott (ed.), *Franz Anton Mesmer und die Geschichte des Mesmerismus.* Wiesbaden, F. Steiner. Pp. 68–87.

Bonnaymé, E. (1908). *La force psychique.* Paris, Librairie du magnétisme.

Bonnefoy, J. B. (1784). *Analyse raisonnée des rapports des commissaires chargés par le roi de l'examen du magnétisme animal.* Lyon, Prault.

Bonnefoy, J. B. (1785). *Examen du comte rendu par M. Thouret, sous le titre de Correspondance de la Société royale de médecine, relativement au magnétisme animal.* N. p., n. pub.

[Bouvier, M. A. J.] (1784). *Lettres sur le magnétisme animal* ... Brussels, n. pub.

Bovbjerg, D., Cohen, N. and Ader, R. (1982). The central nervous system and learning: A strategy for immune regulation. *Immunology Today,* 3, 287–291.

[Brack, – ., attr.] (1784). *Histoire du magnétisme en France* ... Vienna, Royez.

Braid, J. See References: Hypnotism.

Bramwell, J. Milne (1930). *Hypnotism; Its History, Practice and Theory.* 3rd edn., London, Rider.

Brandis, J.D. (1818). *Ueber psychische Heilmittel und Magnetismus.* Copenhagen, Gyldenal.

Brazier, M.A.B. (1984). *A History of Neurophysiology in the 17th and 18th Centuries.* New York, Raven Press.

Brendel, F. (1840). *Kritik der commissarischen Berichte und Protokolle über die ärztliche Beobachtung der Somnambule Christiane Höhne in Dresden.* Freiberg, J.G.. Engelhardt.

Brice, J.M.P.A. (1823). Journal de la maladie de Mme G.**, traitée par les procédés du magnétisme animal, sans l'intervention du somnambulisme. *Archives du magnétisme animal,* 7, 151–175, 179–186, 215–245.

Brice de Beauregard, − . (1844–5). Némesis magnétique. *Revue magnétique,* 1. 180–192.

Brosse, P.T. (1818). Sur quelques effets du magnétisme observés à Berlin. *Bibliothèque du magnétisme animal,* 5, 64–75.

Brosse, P.T. (1819). Description du baquet magnétique composé de M. Wolfart, à Berlin. *Bibliothèque du magnétisme animal,* 6, 225–232.

Brown, R.F. (1977). *The Later Philosophy of Schelling.* London, Associated University Presses.

Brown, S. (1970). *The Heyday of Spiritualism.* New York, Hawthorn Books.

Brownson, O.A. (1854). *The Spirit-Rapper: An Autobiography.* Boston, Little, Brown and Co.

Buchanan, A. (1851). *On Darlingism, Misnamed Electro-Biology ...* London, J.J. Griffin.

Buchner, E. (1922). *Ärzte und Kurpfuscher.* Munich, A. Langen.

Buckland, T. (1850). *The Hand-Book of Mesmerism ...* London, H. Baillière.

Buranelli, V. (1976). *The Wizard from Vienna.* London, P. Owen.

Burdin, C., *jeune,* and Dubois, F. (1841). *Histoire académique du magnétisme animal ...* Paris, J.B. Baillière.

Burney, F. (1975). *The Letters and Journals of Fanny Burney (Madame D'Arblay),* Vol. 6., ed. J. Hemlow *et al.* Oxford, Clarendon Press.

Burq, V. (1852–3). Nervous affections, metallo-therapy, or metal cure; New properties of metals illustrated through mesmerism. *The Zoist,* 10, 121–140, 230–278.

Burq, V. (1853). *Métallothérapie ...* Paris, Germer Baillière.

Bursy, C. (1818). Ein Beitrag zur Geschichte des Selbstmagnetisirens. *Archiv für den thierischen Magnetismus,* 3(1), 163–165.

Bush, G. (1847). *Mesmer and Swedenborg ...* New York, J. Allen.

C***, G. (1784). *Observations sur le Rapport des commissaires chargés par le roi de l'examen du magnétisme animal.* Vienna, n. pub.

Cahagnet, L.A. (1848–54). *Magnétisme. Arcanes de la vie future devoilés ...* 3 vols. Paris, Germer Baillière.

Cahagnet, L.A. [1850?]. *The Celestial Telegraph; Or, Secrets of the Life to Come Revealed through Magnetism ...* 2 vols. London, G. Peirce.

Cailliet, E. (1980). *The Themes of Magic in Nineteenth Century French Fiction.* Philadelphia, Porcupine Press.

Caldwell, C. (1842). *Facts in Mesmerism and Thoughts on its Causes and Uses.* Louisville, Ky., Prentice and Weissinger.

Capern, T. (1851). *The Mighty Curative Powers of Mesmerism ...* London, H. Baillière.

Capron, E. W. (1855). *Modern Spiritualism: Its Facts and Fanaticisms, its Consistencies and Contradictions.* Boston, B. Marsh.

Carpenter, W. B. (1881). *Principles of Mental Physiology* ... 6th edn., London, C. Kegan Paul.

Carra, J. (1785). *Examen physique du magnétisme animal* ... London, n. pub.

Carré de Montgeron, L. B. (1745-7). *La verité des miracles opérés par l'intervention de M. de Pâris et autres appelans* ... 3 vols. Cologne, Librairies de la Compagnie.

Carrington, H. [1939]. *Laboratory Investigations into Psychic Phenomena.* London, Rider.

Carus, C. G. (1857). *Ueber Lebensmagnetismus und über die magischen Wirkungen überhaupt.* Leipzig, F. A. Brockhaus.

Cederschjöld, P. G. (1815). Erfahrungen über den animalischen Magnetismus in Schweden. *Journal der practischen Arzneykunde und Wundarzneykunst,* 41(3), 86-122; 41(4), 119-136.

Chardel, C. (1828). Note relative au testament du docteur Georget. *L'Hermès,* 3, 105-108.

Chardel, C. (1844). *Essai de psychologie physiologique.* 3rd. edn. Paris, Germer Baillière.

Charpignon, J. (1848). *Physiologie, médecine et métaphysique du magnétisme.* 2nd edn. Paris, Germer Baillière.

Charpignon, J. (1860). *Rapports du magnétisme avec la jurisprudence et la médecine legale.* Paris, Germer Baillière.

Chastenet de Puységur, A. H. de (1783). *Lettre de M. le C** de C** P** à M. le P** E** de S**.* N. p., n. pub.

[Chastenet de Puységur, A. M. J. de] (1784a). *Détail des cures opérées à Buzancy, près Soissons, par le magnétisme animal.* Soissons, n. pub.

Chastenet de Puységur, A. M. J. de (1784b). *Mémoires pour servir à l'histoire et à l'établissement du magnétisme animal.* N. p., n. pub.

Chastenet de Puységur, A. M. J. de (1785). *Détail des cures opérées à Buzancy, près Soissons, par le magnétisme animal.* In *Recueil des pièces les plus interessantes pour & contre le magnétisme animal,* Vol. 1. Lyon, n. pub. Pp. 317-360.

Chastenet de Puységur, A. M. J. de (1786). *Mémoires pour servir à l'histoire et a l'établissement du magnétisme animal.* London, n. pub.

Chastenet de Puységur, A. M. J. de (1807). *Du magnétisme animal, considéré dans ses rapports avec diverses branches de la physique générale.* Paris, Cellot.

Chastenet de Puységur, A. M. J. de (1809a). *Mémoires pour servir à l'histoire et à l'établissement du magnétisme animal.* 2nd edn. Paris, Cellot.

Chastenet de Puységur, A. M. J. de (1809b). *Suite des mémoires pour servir à l'histoire et à l'établissement du magnétisme animal.* 2nd edn. Paris, Cellot.

Chastenet de Puységur, A. M. J. de (1811). *Recherches, expériences et observations physiologiques sur l'homme dans l'état de somnambulisme naturel, et dans le somnambulisme provoqué par l'acte magnétique.* Paris, J. G. Dentu.

Chastenet de Puységur, A. M. J. de (1812a). *Continuation du traitement magnétique du jeune Hébert, (mois de septembre).* Paris, J. G. Dentu.

Chastenet de Puységur, A. M. J. de (1812b). *Les fous, les insensés, les maniaques et les frénétiques ne seraient-ils que des somnambules désordonnés?* Paris, J. G. Dentu.

Chastenet de Puységur, A. M. J. de (1813). *Appel aux savans observateurs du dix-neuvième siècle, de la décision portée par leurs prédécesseurs contre le magnétisme animal, et fin du traitement du jeune Hébert.* Paris, J. G. Dentu.

Chastenet de Puységur, A. M. J. de (1817). Lettre de M. le Mis de Puységur à M. Lamy Senart, à Saint-Quentin. *Bibliothèque du magnétisme animal*, 2, 165–168.

Chastenet de Puységur, A. M. J. de (1818). Lettre de M. le Mis de Puységur ... *Bibliothèque du magnétisme animal*, 4, 135–142.

Chastenet de Puységur, J. M. P. de (1784). *Rapport des cures opérées à Bayonne par le magnétisme animal* ... Bayonne, Prault.

Chaves, J. F., and Barber, T. X. (1975–6). Hypnotic procedures and surgery: A critical analysis with applications to acupuncture analgesia. *The American Journal of Clinical Hypnosis*, 18, 217–236.

Chenevix, R. (1829). On mesmerism, improperly denominated animal magnetism. *London Medical and Physical Journal*, 61, 219–230, 491–501; 62, 114–125, 210–221, 315–329.

[Cheron, – .] (1817). Mémoire sur le fluide vital, ou magnétisme animal. *Bibliothèque du magnétisme animal*, 1, 148–156, 247–262; 2, 25–72.

"Christian, A" (1852). *The Veil Uplifted and Mesmerism Traced to its Source*. London, B. L. Green.

Clement-Martin, E. (1926). Essais expérimentaux sur les effets moteurs du supposé fluide magnétique humain. *Revue métapsychique*, No. 2, 125–134.

[Cloquet, J.] (1829). Ablation d'un cancer du sein pendant un sommeil magnétique. *Archives générales de médecine*, 20, 131–134.

Cogevina, A., and Orioli, F. (1842). *Fatti relativi a mesmerismo e cure mesmeriche*. Corfu, Tipografia del Governo.

Coll, – . (1817). Cure d'une maladie chronique compliquée. *Bibliothéque du magnétisme animal*, 1, 101–147.

Collyer, R. H. (1843). *Psychography, or, the Embodiment of Thought* ... Philadelphia, Zieber.

Colquhoun, J. C., tr. (1833). *Report of the Experiments on Animal Magnetism, Made by a Committee of the Medical Section of the French Royal Academy of Science* ... Edinburgh, R. Cadell.

Colquhoun, J. C. (1836). *Isis Revelata: An Enquiry into the Origin, Progress, and Present State of Animal Magnetism*. 2nd edn. 2 vols. Edinburgh, Maclachlan and Stewart.

Colquhoun, J. C. (1843). *The Fallacy of Phreno-Magnetism Detected and Exposed*. Edinburgh, W. Wilson.

Colquhoun, J. C. (1851). *An History of Magic, Witchcraft, and Animal Magnetism*. 2 vols. London, Longman, Brown, Green, and Longmans.

Comellas, R. (1846). *Reseña sobre el magnetismo animal*. Madrid.

Commissarische Berichte und Protokolle über die auf allerhöchsten Befehl stattgefundene ärztliche Beobachtung der in Dresden anwesenden sogenannten Somnambule Johanne Christiane Höhnin aus Draschwitz (1840). Dresden, C. Bromme.

Conn, J. H. (1982). Nature of magnetic treatment. *Journal of the American Society of Psychosomatic Dentistry and Medicine*, 29, 44–53.

Cooter, R. (1984). *The Cultural Meaning of Popular Science: Phrenology and the Organization of Consent in Nineteenth-Century Britain*. Cambridge, Cambridge University Press.

Cooter, R. (1985). The history of mesmerism in Britain: Poverty and promise. In H. Schott (ed.), *Franz Anton Mesmer und die Geschichte des Mesmerismus*. Wiesbaden, F. Steiner. Pp. 152–162.

Corbaux, F. (1816). [Lettre de M. Corbaux à M. Deleuze.] *Annales du magnétisme animal*, 7, 59–62.

Corfe, G. (1848). *Mesmerism Tried by the Touchstone of Truth ...* London, Hatchard.

Cotter, T. C. (1829). An account of the total failure of Mr. Chenevix's mesmeric operations in Dublin. *The London Medical and Surgical Journal*, 3, 484–485.

Cour royale de Paris. Procès des dames Burckart et Couturier ... (1828). *L'Hermès*, 3, 108–129.

Court de Gébelin, A. (1784). *Lettre de l'auteur du* Monde primitif, *à messieurs ses souscripteurs, sur le magnétisme animal*. Paris, n. pub.

Court de Gébelin, A. (1785). *Lettre de l'auteur du* Monde primitif, *à messieurs ses souscripteurs. Sur le magnétisme animal*. In *Recueil des pièces les plus intéressantes pour & contre le magnétisme animal*, Vol. 1. Lyon, n. pub. Pp. 65–167.

Courtier, J. (1905). Sur quelques effets de passes dites magnétiques. In S. de Sanctis (ed.), *Atti del V congresso internazionale di psicologia tenuto in Roma del 26 al 30 aprile 1905*. Rome, Forzani. Pp. 536–540.

Crabtree, A. (1985). Mesmerism, divided consciousness, and multiple personality. In H. Schott (ed.), *Franz Anton Mesmer und die Geschichte des Mesmerismus*. Wiesbaden, F. Steiner. Pp. 133–143.

Crabtree, A. (1988). *Animal Magnetism, Early Hypnotism, and Psychical Research, 1766–1925. An Annotated Bibliography*. White Plains, N.Y., Kraus International Publications.

Cremmens, D. and Tarte, J. (1841). *Le propagateur du magnétisme animal*. Brussels.

Cunningham, A., and Jardine, N. (eds.) (1990). *Romanticism and the Sciences*. Cambridge, Cambridge University Press. .

Cure opérée par M. Mesmer sur le fils de M. Kornmann, enfant âgé de deux ans (1785). In *Recueil des pièces les plus intéressantes pour & contre le magnétisme animal*, vol 1. Lyon, n. pub. Pp. 356–360.

Cuvier, G. (1805). *Leçons d'anatomie comparée*. 5 vols. Paris, Crochard.

Dalloz, A. L. J. (1823). *Entretiens sur le magnétisme animal et le sommeil magnétique ...* Paris, Deschamps.

Daly, R. W. (1964). Andrew Jackson Davis: Prophet of American Spiritualism. *The Journal of American History*, 44, 43–56.

[Dampierre, A. E. de] (1784). *Réflexions impartiales sur le magnétisme animal ...* Geneva, B. Chirol.

Darnton, R. (1968). *Mesmerism and the End of the Enlightenment in France*. Cambridge, Mass., Harvard University Press.

Davey, W. (1854). *The Illustrated Practical Mesmerist Curative and Scientific*. Edinburgh, the author.

Davis, A. J. (1847). *The Principles of Nature, her Divine Revelations, and a Voice to Mankind ...* 35th edn. New York, S. S. Lyon and W. Fishbough.

Davis, A. J. (1857). *The Magic Staff ...* Boston, Colby and Rich.

Davis, A. J. (1917). *The Harmonial Philosophy: A Compendium and Digest of the Works of Andrew Jackson Davis*. London, Rider.

Dean, D. (1983). Infrared measurements of healer treated water. In W. G. Roll, J. Beloff and R. White (eds.), *Research in Parapsychology 1982*. Metuchen, N. J., The Scarecrow Press. Pp. 100–101.

Dean, D., and Brame, E. (1972). The effect of healers on biologically significant molecules. *New Horizons*, 1(5), 187–251.

De Giustino, D. (1975). *Conquest of Mind: Phrenology and Victorian Social Thought.* London, Croom Helm.

Delandine, A. F. (1785). *De la philosophie corpusculaire, ou des connaissances et des procédés magnétiques chez les divers peuples.* Paris, Cuchet.

Delatour, –. (1826). Des avantages de l'insensibilité des somnambules dans les traitements et les opérations. *L'Hermès*, 1, 143–146.

Deleuze, J. P. F. (1813). *Histoire critique du magnétisme animal.* 2 vols. Paris, Mame.

Deleuze, J. P. F. (1814). Réflexions sur l'analogie des phénomènes du magnétisme avec les autres phénomènes de la nature, et conjectures sur le principe de l'action magnétique. *Annales du magnétisme animal*, 1, 225–240.

Deleuze, J. P. F. (1815). Sur quelques principes et quelques faits dont la publication peut nuire à la doctrine du magnétisme. *Annales du magnétisme animal*, 3, 31–48.

Deleuze, J. P. F. (1817). *Réponse aux objections contre le magnétisme.* Paris, J. G. Dentu.

Deleuze, J. P. F. (1818a). *Lettre de J.-P.-F. Deleuze à l'auteur d'un ouvrage intitulé: Superstitions et prestiges des philosophes du dix-huitième siècle, ou les démonolâtres du siècle des lumières...* Paris, J. G. Dentu.

Deleuze, J. P. F. (1818b). Réponse de M. Deleuze. *Bibliothèque du magnétisme animal*, 5, 13–63.

Deleuze, J. P. F. (1819). *Défense du magnétisme animal contre les attaques dont il est l'objet dans le Dictionnaire des sciences médicales.* Paris, Belin-Leprieur.

Deleuze, J. P. F. (1825). *Instruction pratique sur le magnétisme animal.* Paris, J. G. Dentu.

Deleuze, J. P. F. (1826a). Du magnétisme animal en France,.... par Alexandre Bertrand. *L'Hermès*, 1, 31–36, 157–172.

Deleuze, J. P. F. (1826b). *Lettre à messieurs les membres de l'Académie de médecine...* Paris, Béchet jeune.

Deleuze, J. P. F. (1828). Lettre de M. Deleuze à M.*** de la Marne... *L'Hermès*, 3, 233–252.

Deleuze, J. P. F. (1836). *Mémoire sur la faculté de prévision: suivi des notes et pièces justificatives recueillies par M. Mialle.* Paris, Crochard.

Deleuze, J. P. F. (1846). *Practical Instruction in Animal Magnetism.* Tr. T. C. Hartshorn. Revised edn. New York, D. Appleton.

[De R******] (1787). *Lettre à madame la comtesse de L*** contenant une observation magnétique faite par une somnambule sur un enfant de six mois.* N. p., n. pub.

D'Eslon, C. (1781). *Observations sur le magnétisme animal.* London, M. Maklot.

D'Eslon, C. (1782). *Lettre de M. D'Eslon, docteur-régent de la Faculté de médecine de Paris ... à M. Philip, doyen en charge de la même Faculté.* La Haye, n. pub.

D'Eslon, C. (1784). *Observations sur les deux rapports de MM. les commissaires nommés par sa majesté pour l'examen du magnétisme animal...* Philadelphia, Clousier.

Despine, A. (1840). *De l'emploi du magnétisme animal et des eaux minérales dans le traitement des maladies nerveuses, suivi d'une observation très curieuse de guérison de névropathie.* Paris, Germer Baillière.

Devillers, C. (1784). *Le colosse aux pieds d'argille.* [Rouen], n. pub.

Dickerson, K. D. D. (1843). *The Philosophy of Mesmerism, or Animal Magnetism...* Concord, N. H., Morrill, Silsby and Co.

Didier, Adolphe (1860). *Animal Magnetism and Somnambulism*. 3rd edn. London, the author.

Didier, Alexis (1856). *Le sommeil magnétique expliqué par le somnambule Alexis en état de lucidité*. Paris, J. G. Dentu.

Dingwall, E. J. (1967). Hypnotism in France. In E. J. Dingwall (ed.), *Abnormal Hypnotic Phenomena: A Survey of Nineteenth-Century cases*, vol. 1. London, J. and A. Churchill. Pp. 1–328.

Dingwall, E. J. (ed.) (1967–8). *Abnormal Hypnotic Phenomena: A Survey of Nineteenth-Century Cases*. 4 vols. London, J. and A. Churchill.

Dingwall, E. J. (1968). Hypnotism in Great Britain. In E. J. Dingwall (ed.), *Abnormal Hypnotic Phenomena: A Survey of Nineteenth-Century Cases*. Vol. 4. London, J. and A. Churchill. Pp. 79–164.

Dods, J. B. (1849). *Six Lectures on the Philosophy of Mesmerism* ... London, Simpkin, Marshall and Co.

Dods, J. B. (1854). *Spirit Manifestations Examined and Explained* ... New York, De Witt and Davenport.

Dods, J. B. (1876). *The Philosophy of Mesmerism and Electrical Psychology*. London, J. Burns.

Dods, J. B., and Grimes, J. S. (1851). *Electrical-Psychology: Or the Electrical Philosophy of Mental Impressions* ... London, J. J. Griffin.

Doerner, K. (1981). *Madmen and the Bourgeoisie: A Social History of Insanity and Psychiatry*. Oxford, Blackwell.

[Doppet, F. A., attr.] (1784a). *La mesmériade, ou le triomphe du magnétisme, poème en trois chants, dédié à la lune*. Geneva, Coutourier.

Doppet, F. A. (1784b). *Traité théorique et pratique du magnétisme animal*. Turin, J. M. Briolo.

[Doppet, F. A., attr.] (1785). *Oraison funèbre du célèbre Mesmer, auteur du Magnétisme animal, & président de la Loge de l'harmonie*. Grenoble, n. pub.

Drake, D. (1844). *Analytical Report of a Series of Experiments in Mesmeric Somniloquism* ... Louisville, F. W. Prescott.

Dresser, H. W. (1921). *The Quimby Manuscripts*. New York, T. Crowell.

[Drouault, – .] (1816). Certificats des cures magnétiques opérées par M. Drouault, à Châtellerault, Poitiers, etc. *Annales du magnétisme animal*, 8, 5–25.

Du Commun, J. (1829). *Three Lectures on Animal Magnetism* ... New York, J. Desnoues.

Dugas, L. A. (1845). Extirpation of the mamma of a female in the mesmeric sleep, without any evidence of sensibility during the operation. *The Southern Medical and Surgical Journal*, NS, 1, 122–125.

Dupau, J. A. (1826). *Lettres physiologiques et morales sur le magnétisme animal* ... Paris, Gabon.

Dupotet de Sennevoy, J. D. (1821). *Exposé des expériences sur le magnétisme animal faites à l'Hôtel Dieu de Paris pendant les mois d'octobre, novembre et décembre 1820*. Paris, Béchet jeune.

Dupotet de Sennevoy, J. D. (1834). *Cours de magnétisme animal*. Paris, the author.

Dupotet de Sennevoy, J. D. (1838). *An Introduction to the Study of Animal Magnetism*. London, Saunders and Otley.

Dupotet de Sennevoy, J. D. (1840). *Le magnétisme opposé à la médecine. Mémoire pour servir à l'histoire du magnétisme en France et en Angleterre*. Paris, A. Réné

Dupotet de Sennevoy, J. D. (1846). *Manuel de l'étudiant magnétiseur* ... Paris, Germer Baillière.

Dupotet de Sennevoy, J. D. (1852). *La magie devoilée, ou principes de science occulte*. Paris, Pommeret.

Dupotet de Sennevoy, J. D. (1927). *Magnetism and Magic*. London, Allen and Unwin.

Dupotet de Sennevoy, J. D. (1930). *Traité complet de magnétisme animal*. 8th edn. Paris, F. Alcan.

Du Prel, C. (1889). *The Philosophy of Mysticism*. 2 vols. London, G. Redway.

Durand, L. (1845). Procès-verbal d'une amputation de la jambe ... *Journal du magnétisme*, 1, 492–496.

Durant, C. F. (1837). *Exposition, or a New Theory of Animal Magnetism* ... New York, Wiley and Putnam

Dureau, A. (1869). *Notes bibliographiques pour servir à l'histoire du magnétisme animal* ... Paris, J. Lechevalier.

Dürr, F. (1822). Das siderische unmagnetisirte Baquet als Heilmittel gegen den Veitstanz. *Archiv für den thierischen Magnetismus*, 10(3), 1–72.

[Duval d'Espréménil, A. A., attr.] (1784). *Réflexions préliminaires à l'occasion de la pièce intitulée* Les docteurs modernes ... N. p., n. pub.

[Duval d'Esréménil, A. A., attr.] (1786). *Sommes versées entre les mains de Monsieur Mesmer, pour acquérir le droit de publier sa découverte*. In *Recueil de pièces pour servir à l'histoire du magnétisme*. N.p., n. pub. Pp. 3–8.

E. W. C. N. (1844–5). The medical journals and medical men. *The Zoist*, 2, 273–288.

Ebhardt, G. F. (1818). *Theologische und philosophische Raisonnements in Bezug auf den animalischen Magnetismus* ... Leipzig, C. F. Kollmann.

Eckhartshausen, K. von (1788–91). *Aufschlüsse der Magie* ... 4 vols. Munich, Lentner.

Edge, H. L., Morris, R. L., Rush, J. H. and Palmer, J. (1986). *Foundations of Parapsychology: Exploring the Boundaries of Human Capacity*. London, Routledge and Kegan Paul.

Edwards, D. (1789). *A Treatise on Animal Magnetism* ... London, J. Wagstaff.

Ehman, O. C. (1819). Thèse sur le magnétisme animal soutenue en Suède, par un jeune médecin, traduite du Latin par M. le comte du Crouseilhe ... *Bibliothèque du magnétisme animal*, 8, 189–213.

Ehrenwald, J. (1957). The telepathic hypothesis and doctrinal compliance in psychotherapy. *American Journal of Psychotherapy*, 11, 359–379.

Eich, J. E. (1977). State-dependent retrieval of information in human episodic memory. In I. M. Birnbaum and E. S. Parker (eds.), *Alcohol and Human Memory*. Hillsdale, N. J., L. Erlbaum.

Eich, J. E. (1980). The cue-dependent nature of state-dependent retrieval. *Memory and Cognition*, 8, 157–173.

Ellenberger, H. F. (1970). *The Discovery of the Unconscious: The History and Evolution of Dynamic Psychiatry*. New York, Basic Books.

Elliotson, J. (1840). *Human Physiology*. 5th edn. 2 vols. London, Longman, Orne, Brown, Green and Longmans.

Elliotson, J. (1843). *Numerous Cases of Surgical Operations without Pain in the Mesmeric State* ... London, H. Baillière.

Elliotson, J. (1843–4a). Address to the Phrenological Association. *The Zoist*, 1, 227–246.

Elliotson, J. (1843–4b). Cures of epileptic and other fits with mesmerism. *The Zoist*, 1, 407–460.

Elliotson, J. (1843–4c). Cures of palsy by mesmerism. *The Zoist*, 1, 300–349.

Elliotson, J. (1843–4d). Dr. Elliotson's cases of cures by mesmerism. *The Zoist*, 1, 161–208.

Elliotson, J. (1844–5a). Case of epilepsy cured with mesmerism. *The Zoist*, 2, 194–238.

Elliotson, J. (1844–5b). A collection of more instances of surgical operations rendered painless by means of mesmerism. *The Zoist*, 2, 90–123.

Elliotson, J. (1844–5c). Postcript by Dr. Elliotson to the account of the late amputation at Leicester. *The Zoist*, 2, 410–429.

Elliotson, J. (1844–5d). Reports of various trials of the clairvoyance of Alexis Didier, last summer, in London. *The Zoist*, 2, 477–529.

Elliotson, J. (1845–6a). Case of a contracted foot with severe pain, cured with mesmerism. *The Zoist*, 3, 339–379, 446–485.

Elliotson, J. (1845–6b). Cure of hysterical epilepsy, somnambulism, etc., with mesmerism. *The Zoist*, 3, 39–79.

Elliotson, J. (1845–6c). More of Alexis Didier. *The Zoist*, 3, 389–398.

Elliotson, J. (1845–6d). More painless amputations and other surgical operations in the mesmeric state. *The Zoist*, 3, 490–508.

Elliotson, J. (1845–6e). Surgical operations without pain in the mesmeric state. *The Zoist*, 3, 380–388.

Elliotson, J. (1846). *The Principles and Practice of Medicine*. 2nd edn., ed. N. Rogers and A. Lee. London, J. Butler.

Elliotson, J. (1846–7a). Accounts of more painless surgical operations. *The Zoist*, 4, 1–59.

Elliotson, J. (1846–7b). Instances of double states of consciousness independent of mesmerism. *The Zoist*, 4, 157–187.

Elliotson, J. (1846–7c). More painless surgical operations in the mesmeric state. *The Zoist*, 4, 193–218.

Elliotson, J. (1846–7d). More painless surgical operations. Report of a committee at Calcutta in favour of the truth and utility of mesmerism. *The Zoist*, 4, 563–583.

Elliotson, J. (1846–7e). Review of An Abstract of Researches on Magnetic and certain Allied Subjects ... By Baron von Reichenbach ... *The Zoist*, 4, 104–124, 277–284, 346–361.

Elliotson, J. (1847–8a). Remarkable cure of intense nervous affections ... *The Zoist*, 5, 234–253.

[Elliotson, J.] (1847–8b). Report of the committee appointed by Government to observe and report upon surgical operations by Dr. J. Esdaile ... *The Zoist*, 5, 50–69.

Elliotson, J. (1847–8c). Three more painless surgical operations performed at Cherbourg in France. *The Zoist*, 5, 197–201.

Elliotson, J. (1848). *Cure of a True Cancer of the Female Breast with Mesmerism*. London, Walton and Mitchell. [From *The Zoist* 1848–9, 6, 213–237.]

Elliotson, J. (1848–9a). Account of a "Record of Cases treated in the Mesmeric Hospital, from June to December, 1847 ... " *The Zoist*, 6, 121–158.

Elliotson, J. (1848–9b). Account of "A Review of my Reviewers. By James Esdaile, M. D. Calcutta, January 26, 1848." *The Zoist*, 6, 158–173.

[Elliotson, J.] (1848–9c). Appendix to Dr. Elliotson's case of the cure of cancer. *The Zoist*, 6, 312–315.

Elliotson, J. (1848–9d). Report by Dr. Elliotson upon "A Record of Cases treated in the Mesmeric Hospital, from November, 1846, to May, 1847 ... " *The Zoist*, 6, 1–42.

Elliotson, J. (1849–50a). Death of Miss Barber. *The Zoist*, 7, 323–326.

Elliotson, J. (1849–50b). Mesmerism in the East ... *The Zoist*, 7, 121–137.

Elliotson, J. (1849–50c). "Second half-yearly Report of the Calcutta Mesmeric Hospital ... " *The Zoist*, 7, 353–363.

Elliotson, J. (1850–1). Mesmeric phenomena in brutes, as effected by the Duke of Marlborough and the Rev. Mr. Bartlett. *The Zoist*, 8, 295–299.

Elliotson, J. (1851–2a). Cure of convulsive and rigid fits ... together with some peculiar mesmeric phenomena. *The Zoist*, 9, 157–195.

Elliotson, J. (1851–2b). Submesmerism and imagination. *The Zoist*, 9, 106–112, 424–436.

Elliotson, J. (1851–2c). Successful result of two cases of lock-jaw or tetanus in horses, treated with mesmerism ... *The Zoist*, 9, 49–51.

Elliotson, J. (1852–3a). An account of the mesmeric hospital in Bengal since Dr. Esdaile's departure from India. *The Zoist*, 10, 278–293.

Elliotson, J. (1852–3b). Dr. Esdaile and mesmerism in Perth. *The Zoist*, 10, 419–427.

[Elliotson, J.] (1853–4). Introduction to Volume XI of *The Zoist*. *The Zoist*, 11, 1–17.

Elliotson, J. (1854–5a). An account of the perfectly painless and successful removal of a female breast in mesmeric sleep-waking ... *The Zoist*, 12, 113–141.

Elliotson, J. (1854–5b). An instance of sleep and cure by imagination only. *The Zoist*, 12, 396–403.

Elliotson, J. (1854–5c). Some rapid cures by Dr. Esdaile. *The Zoist*, 12, 413–415.

[Elliotson, J.] (1855–6). Conclusion of *The Zoist*. *The Zoist*, 13, 441–444.

Elliotson, J. (1982). *John Elliotson on Mesmerism*. Ed. F. Kaplan. New York, Da Capo Press.

Ellis, A. J. (1854–6). Reichenbach and his researches ... *The Zoist*, 12, 317–334; 13, 1–27, 111–131, 218–242, 321–350.

Emerson, R. W. (1971). *The Journals and Miscellaneous Notebooks of Ralph Waldo Emerson, Vol. 9, 1843–1847*. Ed. R. H. Orth and A. R. Ferguson. Cambridge, Mass., The Belknap Press of Harvard University Press.

Engledue, W. C. (1842). *Cerebral Physiology and Materialism*. London, H. Baillière.

Engledue, W. C. (1844–5). Cases of mesmeric clairvoyance and sympathy of feeling. *The Zoist*, 2, 269–273.

Engledue, W. C. (1851–2). Review of "Letters to a Candid Inquirer, on Animal Magnetism." By William Gregory, M. D., ... *The Zoist*, 9, 215–223.

Engelhardt, D. von (1985). Mesmer in der Naturforschung und Medezin der Romantik. In H. Schott (ed.) *Franz Anton Mesmer und die Geschichte des Mesmerismus*. Wiesbaden, F. Steiner. Pp. 88–107.

Ennemoser, J. (1819). *Der Magnetismus nach der allseitigen Beziehung seines Wesens, seiner Erscheinungen, seiner Anwendung und Enträthselung*. Leipzig, F. Brockhaus.

Ennemoser, J. (1842). *Der Magnetismus im Verhältnisse zur Natur und Religion*. Stuttgart, J. G. Cotta.

Ennemoser, J. (1844). *Geschichte des thierischen Magnetismus. Erster Theil: Geschichte der Magie*. 2nd. edn. Leipzig, Brockhaus.

Ennemoser, J. (1854). *The History of Magic*. 2 vols. London, H. G. Bohn.

Erdmann, J. E. (1924). *A History of Philosophy. Vol II. Modern Philosophy*. London, Allen and Unwin.

Erman, W. (1925). *Der tierische Magnetismus in Preussen vor und nach den Freiheitskriegen*. Munich, R. Oldenbourg.

Eschenmayer, C. A. von (1816). *Versuch die scheinbare Magie des thierischen Magnetismus aus physiologischen und psychischen Gesetzen zu erklären*. Vienna, Haas.

Eschenmayer, C. A. von (1817). Allegemeine Reflexionen über den thierischen Magnetismus und den organischen Aether. *Archiv für den thierischen Magnetismus*, 1(1), 11–34.

Eschenmayer, C. A. von (1830). *Mysterien des innern Lebens erläutert aus der Geschichte der Seherin von Prevorst ...* Tübingen, Zu-Guttenberg.

Eschenmayer, C. A. von (1837). *Conflict zwischen Himmel und Hölle ...* Tübingen, Zu-Guttenberg.

Esdaile, J. (1846). *Mesmerism in India, and its Practical Applications in Surgery and Medicine*. London, Longman, Brown, Green and Longmans.

Esdaile, J. (1847). *A Record of Cases Treated in the Mesmeric Hospital, from November 1846, to May 1847, with the Reports of the Official Visitors*. Calcutta, W. Ridsdale.

Esdaile, J. (1847–8). Dr. Esdaile's first monthly report of the Calcutta Mesmeric Hospital ... *The Zoist*, 5, 178–187.

Esdaile, J. (1848). *A Record of Cases Treated in the Mesmeric Hospital, from June to December 1847, with Reports of the Official Visitors*. Calcutta, W. Ridsdale.

Esdaile, J. (1850a). *Mesmerism in India. Second Half-Yearly Report of the Calcutta Mesmeric Hospital, from 1st March to 1st September 1849 ...* London, H. Baillière.

Esdaile, J. (1850b). On the operation for the removal of scrotal tumours. *The London Medical Gazette*, new ser., 11, 449–454.

Esdaile, J. (1852a). *The Introduction of Mesmerism as an Anaesthetic and Curative Agent, into the Hospitals of India*. Perth, Dewar.

Esdaile, J (1852b). *Natural and Mesmeric Clairvoyance with the Practical Application of Mesmerism in Surgery and Medicine*. London, H. Baillière.

Esdaile, J. (1854–5). Mesmeric cases. *The Zoist*, 12, 74–80.

Esdaile, J. (1856). *The Introduction of Mesmerism (with the Sanction of the Government) into the Public Hospitals of India*. 2nd edn. London, W. Kent.

Esposito, J. L. (1977). *Schelling's Idealism and Philosophy of Nature*. London, Associated University Presses.

L'évangile du jour. Pour servir d'éclaircissement aux Doutes d'un provincial *proposés à MM. les médecins commissaires ...* (1785). N.p., n. pub.

Ewin, D. M. (1983–4). Emergency room hypnosis for the burned patient. *The American Journal of Clinical Hypnosis*, 26, 5–8.

Exposé de différentes cures opérées depuis le 25. d'août 1785, époque de la formation de la société, fondée à Strasbourg, sous la dénomination de Société harmonique des amis réunis, jusqu'au 13 du mois de juin 1786, par des membres de cette société (1786). [Strasbourg], Lorenz et Schouler.

Extrait du journal d'une cure magnétique. Traduit de l'allemand (1787). Rastadt, J. W. Dorner.

F., R. (1840). *Die in Dresden sich aufhaltende Somnambule, Johanne Christiane Höhne, ihr Zustand und ihre Kräfte*. Neustadt-Dresden, C. Heinrich.

Fahnestock, W. B. (1844). Clairvoyance. *The Magnet*, 3, 15–17.

Fauchier-Magnin, A. (1958). *The Small German Courts in the Eighteenth Century*. London, Methuen.

Favre, L. (1905). Hypnotisme et magnétisme animal. In *Atti del V congresso internazionale di psicologia tenuto in Roma del 26 al 30 aprile 1905*. Rome, Forzani. Pp. 552–553.

Fenwick, P., and Hopkins, R. (1986). An examination of the effect of healing on water. *Journal of the Society for Psychical Research*, 53, 387–390.

Figuier, L. (1881). *Histoire du merveilleux dans les temps modernes*. Vol. 3. *Le magnétisme animal*. 3rd edn. Paris, L. Hachette.

Fillassier, A. (1832). *Quelques faits et considérations pour servir à l'histoire du magnétisme animal*. Paris, Didot le jeune.

Fischer, F. (1805). Einige Beobachtungen über thierischen Magnetismus und Somnambulismus. *Journal für die Physiologie*, 6, 264–281.

Fischer, F. (1839). *Der Somnambulismus*. 3 vols. Basel, Schweighauser.

Fischer, W. (1984). Die Psychiatrie des Mediziners J. C. Reil – eine kritische Betrachtung. *Psychiatrie, Neurologie und medizinische Psychologie*, 36, 229–235.

Foissac, P. (1827). Pleuro-pneumonie chronique avec épanchement; palpitation du coeur, guéris par le magnétisme et le somnambulisme. *L'Hermès*, 2, 196–200.

Foissac, P. (1828). Des progrès du magnétisme dans les pays étrangers. *L'Hermès*, 3, 165–168.

Foissac, P. (1833). *Rapport et discussions de l'Académie royale de médecine sur le magnétisme animal...* Paris, J. B. Baillière.

Forbes, J. (1845). *Illustrations of Modern Mesmerism from Personal Investigation*. London, J. Churchill.

[Fournel, J. F.] (1785a). *Essai sur les probabilités du somnambulisme magnétique, pour servir à l'histoire du magnétisme animal*. Amsterdam, n. pub.

[Fournel, J. F.] (1785b). *Remontrances des malades aux médecins de la Faculté de Paris*. Amsterdam, n. pub.

[Fournier-Michel, – .] (1781). *Lettre à Monsieur Mesmer, et autres pièces concernant la maladie de la Dlle. Berlancourt de Beauvais*. Beauvais, P. Desjardins.

Frapart, N. N. (1839). *Lettre sur le magnétisme et le somnambulisme, à l'occasion de Mademoiselle Pigeaire...* Paris, J. G. Dentu.

Frid, M., and Singer, G. (1979). Hypnotic analgesia in conditions of stress is partially reversed by naloxone. *Psychopharmacology*, 63, 211–215.

Friedlander, C. M. (1852). *Case of Paralysis and Mania Cured by means of Animal Magnetism, in a Letter to Professor Gregory*. Edinburgh, the author.

Friedlander, M. (1817). Note sur l'état actuel du magnétisme animal en Allemagne. *Bibliothèque du magnétisme animal*, 1, 169–180.

Freschi, G. (1849). Du magnétisme animal en Italie. *Journal du magnétisme*, 8, 120–126.

Fuller, R. (1982). *Mesmerism and the American Cure of Souls*. Philadelphia, University of Philadelphia Press.

Fuller, R. (1985). The American mesmerists. In H. Schott (ed.), *Franz Anton Mesmer und die Geschichte des Mesmerismus*. Wiesbaden, F. Steiner. Pp. 163–173.

Funck, W. (1894). *Der Magnetismus und Somnambulismus in der Badischen Markgrafschaft*. Freiburg im Breisgau, J. C. B. Mohr.

Galart de Montjoye, – [i.e. C. F. L. Ventre de Toulabre]. (1784). *Lettre sur le magnétisme animal, où l'on examine la conformité des opinions des peuples anciens & modernes, des sçavans, et notamment de M. Bailly, avec celles de M. Mesmer*. Philadelphia, P. J. Duplain.

Gandon, F. A. (1849). *La seconde vue dévoilée*. Paris, chez l'auteur.

Garrison, F. H. (1929). *An Introduction to the History of Medicine*. 4th edn. Philadelphia, W. B. Saunders.

Gauld, A. O. (1988). Reflections on mesmeric analgesia. *British Journal of Experimental and Clinical Hypnosis*, 5, 17–24.

Gauthier, A. (1842). *Histoire du somnambulisme chez tous les peuples* ... 2 vols. Paris, F. Malteste.

Gauthier, A. (1844). *Le magnétisme catholique* ... Paris, Bureau de la Revue magnétique.

Gauthier, A. (1845). *Traité pratique du magnétisme et du somnambulisme* ... Paris, Germer Baillière.

Gehrts, H. (1966). *Das Mädchen von Orlach*. Stuttgart, E. Klett.

Gentil, J. A. (1848). *Initiation aux mystères secrets de la théorie et de la pratique du magnétisme* ... Paris, Robert.

Georget, E. J. (1821). *De la physiologie du système nerveux et spécialement du cerveau*. 2 vols. Paris, J. B. Baillière.

Gerber, U. (1844). *Das Nachtgebiet der Natur im Verhältniss zur Wissenschaft, zur Aufklärung und zum Christentum*. Augsburg, J. A. Schlosser.

Gerboin, A. C. (1808). *Recherches expérimentales sur un nouveau mode de l'action électrique*. Strasbourg, F. G. Levrault.

The German somnambulist and Miss M'Avoy (1817–18). *Blackwoods Edinburgh Magazine*, 2, 437–443.

Ghert, P. G. van (1814). *Dagboek eener magnetische behandelung*. Amsterdam, J. Van der Hey.

Ghert, P. G. van (1815). *Mnemosyne of aanteekeningen van merkwaardige verschijnsels van het animalisch magnetismus*. Amsterdam, J. Van der Hey.

Ghert, P. G. van (1817). Tagebuch einer magnetischen Behandlung. *Archiv für den thierischen Magnetismus*, 2(1), 3–188; 2(2), 3–51.

Ghert, P. G. van (1818). Sammlung merkwürdige Erscheinungen des thierischen Magnetismus. *Archiv für den thierischen Magnetismus*, 3(3), 1–97.

Gibson, H. B. (1982). *Pain and its Conquest*. London, P. Owen.

Gilibert, J. E. (1784). *Apperçu sur le magnétisme animal, ou résultat des observations faites à Lyon sur ce nouvel agent*. Geneva, n. pub.

Gillispie, C. C. (1980). *Science and Polity in France at the End of the Old Regime*. Princeton, N. J., Princeton University Press.

[Girard, – .] (1784). *Mesmer blessé, ou réponse à la lettre du R. P. Hervier, sur le magnétisme animal*. London, Coutourier.

[Girardin, S.] (1784). *Observations adressés à Mrs. les commissaires* ... London, Royez.

Girardin, S. (1785). Lettre d'un anglois à un françois sur la découverte du magnétisme animal. In *Recueil des pièces les plus intéressantes pour & contre le magnétisme animal*, Vol 1. Lyon, n. pub. Pp. 269–289.

Giraud, – . (1785). Compte rendu à M. Mesmer de l'état des malades admis au traitement gratuit par lui établi, ancien hôtel de Coigny, rue du Coqhéron. In J. L. N. Tissart de Rouvre, *Nouvelles cures opérées par le magnétisme animal* (see below). Pp. 413–415.

Glowatzski, G. (1983). Der Mesmerismus – und seine Spuren in der heutigen Medizin. *Zeitschrift für Allgemeinmedizin*, 59, 1360–1366.

Gmelin, E. (1787a). Beytrag zur Realität des thierischen Magnetismus. *Archiv für Magnetismus und Somnambulismus*, Pt. 3, 49–84.

Gmelin, E. (1787b). *Ueber thierischen Magnetismus in einem Brief an Herrn geheimen Rath Hoffmann in Mainz*. Tübingen, J. F. Heerbrandt.

Gmelin, E. (1789). *Neue Untersuchungen über den thierischen Magnetismus*. Tübingen, Cotta.

Gmelin, E. (1791). *Materialen für die Anthropologie*, Vol. 1. Tübingen, Cotta.

Gmelin, E. (1793). *Materialen für die Anthropologie*, Vol. 2. Heilbronn, J. D. Class.

Goldstein, A., and Hilgard, E. R. (1975). Lack of influence of the morphine antagonist naloxone on hypnotic analgesia. *Proceedings of the National Academy of Sciences*, 72, 2041–2043.

Görwitz, H. (1851). *Idiosomnambulismus oder näturlich-magnetischer Schlaf Richard's ...* Leipzig, C. E. Kollmann.

Gould, G. M., and Pyle, W. L. (1962). *Anomalies and Curiosities of Medicine ...* New York, The Julian Press.

Grad, B. (1965). Some biological effects of the "laying on of hands": A review of experiments with animals and plants. *Journal of the American Society for Psychical Research*, 59, 95–127.

Grad, B. (1976). The biological effect of the "laying on of hands": A review of experiments with animals and plants. In G. R. Schmeidler (ed.), *Parapsychology: Its Relations to Physics, Biology, Psychology and Psychiatry*. Metuchen, N. J., The Scarecrow Press. Pp. 76–89.

Grad, B. (1989). The healer phenomenon: Implications for parapsychology. In L. A. Henkel and R. E. Berger (eds.), *Research in Parapsychology 1988*. Metuchen, N. J., The Scarecrow Press. Pp. 107–120.

Grad, B., Cadoret, R., and Paul, G. (1961). The influence of an unorthodox method of treatment on wound healing in mice. *International Journal of Parapsychology*, 3(2), 5–24.

Grad, B., and Dean, D. (1984). Independent confirmation of infrared healer effects. In R. A. White and R. S. Broughton (eds.), *Research in Parapsychology 1983*. Metuchen, N. J., The Scarecrow Press. Pp. 81–83.

Grässe, J. G. T. (1843). *Bibliotheca magica et pneumatica*. Leipzig, W. Engelmann.

Gravitz, M. A. (1987). Two centuries of hypnosis specialty journals. *The International Journal of Clinical and Experimental Hypnosis*, 35, 265–276.

Gravitz, M. A. (1987–8). Early uses of hypnosis as surgical anaesthesia. *American Journal of Clinical Hypnosis*, 30, 201–208.

Gravitz, M. A., and Gerton, I. (1984–5). Origin of the term hypnotism prior to Braid. *The American Journal of Clinical Hypnosis*, 27, 107–110.

Greenhow, T. M. (1845). *Medical Report of the Case of Miss H − M −*. London, S. Highley.

Gregory, S. (1843). *Mesmerism, or Animal Magnetism and its Uses ...* Boston, Redding.

Gregory, W. (1846). *Abstract of "Researches on Magnetism and on Certain Allied Subjects," including a Supposed New Imponderable*. London, Taylor and Walton.

Gregory, W. (1851). *Letters to a Candid Inquirer on Animal Magnetism*. London, Taylor, Walter and Moberly.

Gregory, W. (1851–2). Case of vision at a distance. *The Zoist*, 9, 422–424.

[Grézie, C. B. de la] (1784). *Magnétisme animal devoilé ...* Geneva, n. pub.

Grimes, J. S. (1850). *Etherology, and the Phreno-Philosophy of Mesmerism and Magic Eloquence ...* Boston, Mass., J. Munroe.

Grüsser, O. J. (1987) *Justinus Kerner 1786–1862 ...* Berlin, Springer-Verlag.

Guéritaut, – . (1814). Histoire d'une maladie nerveuse fort singulière, observée à Mar, et communiquée à la Société des sciences physiques d'Orléans. *Annales du magnétisme animal*, 2, 123–144.

Guidi, F. (1851). *Magnetismo animale e somnambulismo magnetico.* Turin, Favale.

Guidi, F. (1854). *Trattato teorico-pratico di magnetismo animale considerato sotto il punto di vista fisiologico e psicologico.* Milan, Turati.

Guidi, F. (1860). *Il magnetismo animale considerato secondo le leggi della natura e principalmente diretto alla cura delle malattie.* Milan, F. Sanvito.

[Guigoud-Pigale, P.] (1784). *Le baquet magnétique, comédie en vers et en deux actes.* London, n. pub.

Haddock, J. W. (1851). *Somnolism & Psycheism; or the Science of the Soul and the Phenomena of Nervation as Revealed by Vital Magnetism or Mesmerism...* 2nd edn. London, J. S. Hodson.

Haggard, H. W. (1932). *The Lame, the Halt and the Blind.* London, Heinemann.

Hall, S. T. (1845). *Mesmeric Experiences.* London, H. Baillière.

Hall, S. T. (1873). *Biographical Sketches of Remarkable People...* London, Simpkin, Marshall, and Co.

Halldin, J. G. (1816). Lettre de M. Halldin ... à M. Mouillesaux. *Annales du magnétisme animal*, 5, 7–26.

Hamel, J. (1818). Extrait d'un lettre de M. le docteur Joseph Hamel, médecin de St. Petersbourg ... à M. le baron d'Henin de Cuvillers. *Bibliothèque du magnétisme animal*, 4, 252–254.

Hanák, M. (1833). *Geschichte eines natürlichen, durch sich selbst entwickelte Somnambulismus...* Kaschau, Wigand.

Hardy, H. H. (1852). *Analytic Researches in Spirit Magnetism.* London, G. Peirce.

Hare, R. (1856). *Experimental Investigation of the Spirit Manifestations...* 4th edn. New York, Partridge and Brittan.

Harrington, A. (1988). See References: Hypnotism.

Harte, R. (1902–3). *Hypnotism and the Doctors.* 2 vols. London, L. N. Fowler.

Hartmann, E. (1983). Two case reports: Night terrors with sleepwalking – a potentially lethal disorder. *The Journal of Nervous and Mental Disease*, 181, 503–505.

Hartmann, E. von (1868). *Philosophie des Unbewussten.* Berlin, Duncker.

Hartmann, E. von (1891). *Die Geisterhypothese des Spiritismus und seine Phantome.* Leipzig, W. Friedrich.

Hartmann, E. von (1931). *Philosophy of the Unconscious.* London, Kegan Paul, Trench, Trübner and Co.

Hastings, H. (1854). *Medicina Mentis, or Spiritualism, Commonly Called Animal Magnetism, or Mesmerism, Considered Entirely as a Curative Agent.* London, Aylott.

Heineken, J. (1800). *Ideen und Beobachtungen den thierischen Magnetismus und dessen Anwendung betreffend.* Bremen, F. Wilmans.

Heineken, P. (1818). Geschichte einer merkwürdigen Entzundungskrankheit. *Archiv für den thierischen Magnetismus*, 2(3), 3–71.

Hénin de Cuvillers, E. F. d' (1820). *Le magnétisme eclairé, ou introduction aux Archives du magnétisme animal.* Paris, Barrois.

Hénin de Cuvillers, E. F. d' (1821). *Magnétisme animal retrouvé dans l'antiquité...* Paris, Barrois.

Hénin de Cuvillers, E. F. d' (1822). *Exposition critique du système et de la doctrine mystique des magnétistes*. Paris, Barrois.

[Hénin de Cuvillers, E. F. d'] (1823a). Deuxième annonce faisant suite au catalogue des ouvrages imprimés et des articles et analyses insérés dans les journaux, à commencer de l'an 1800, concernant le phantasiéxoussisme, improprement appelé magnétisme animal. *Archives du magnétisme animal*, 8, 173–192.

[Hénin de Cuvillers, E. F. d'] (1823b). Première annonce d'ouvrages imprimés et d'articles insérés dans les journaux … concernant le magnétisme animal. *Archives du magnétisme animal*, 7, 187–192.

[Hénin de Cuvillers, E. F. d'] (1823c). Troisième annonce faisant suite au catalogue des ouvrages imprimés et des articles et analyses insérés dans les journaux, à commencer de l'an 1800, concernant le phantasiéxoussisme, improprement appelé magnétisme animal. *Archives du magnétisme animal*, 8, 233–273.

Hensler, P. I. (1833). *Ueber die verschiedenen Arten des thierischen Magnetismus, und ihre verschiedenen Wirkungen auf den Menschen im kranken Zustände*. Würzburg, Stahel.

Hensler, P. I. (1837). *Der Menschen-Magnetismus in seinen Wirkungen auf Gesundheit und Leben*. Würzburg, Stahel.

Hervier, – . (1784). *Lettre sur la découverte du magnétisme animal, à M. Court de Gebelin* … Pekin, Coutourier.

Hilgard, E. R., and Hilgard, J. R. (1983). *Hypnosis in the Relief of Pain*. 2nd edn. Los Altos, Calif., W. Kaufmann.

Histoire véritable du magnétisme animal, ou nouvel preuves de la realité de cet agent, tirées de l'ancien ouvrage d'un vieux docteur (1785). La Haye, n. pub.

Ho, B., Chute, D., and Richards, D. (eds.) (1977). *Drug Discrimination and State-Dependent Learning*. New York, Academic Press.

Hodgson, R. (1884–5). On vision with sealed and bandaged eyes. *Journal of the Society for Psychical Research*, 1, 84–86.

Hoek, A. (1852). *Electrobiologie en levens-magnetismus*. s'Gravenhage, van Cleef.

Hoek, A. (1854). *De helderziendheid, een verschijnsel, dat nu'en dan in het levensmagnetismus wordt aagetroffen*. s'Gravenhage, van Cleef.

Holland, G. C. (1848). *The Philosophy of Animated Nature*. London, J. Churchill.

Honigfeld, G. (1964). Non-specific factors in treatment. I. Review of placebo reactions and placebo reactors. *Diseases of the Nervous System*, 25, 145–156.

Huapaya, L. U. M. (1979). Seven cases of somnambulism induced by drugs. *The American Journal of Psychiatry*, 136, 985–986.

Hufeland, C. W. (1794). *Gemeinnützige Aufsätze zur Beförderung der Gesundheit, des Wohlseyns und vernünftiger medicinischer Aufklärung*, vol 1. Leipzig, G. J. Göschen.

Hufeland, C. W. (1809). Ueber den Magnetismus, nebst der Geschichte einer merkwürdigen vollkommen Tageblindheit … welche … durch den Magnetismus völlig geheilt würde. *Journal der practischen Arzneykunde und Wundarzneykunst*, 29(8), 1–68.

Hufeland, C. W. (1816). *Auszug und Anzeig der Schrift des Herrn Leibmedikus Stieglitz über den thierischen Magnetismus, nebst Zusätzen*. Berlin, Reimer.

Hufeland, F. (1805). Ausserordentliche Erhöhung der Sensibilität, ein Beitrag zu den Erfahrungen über Somnambulismus und thierischen Magnetismus. *Archiv für die Physiologie*, 6, 225–264.

Hufeland, F. (1822). *Ueber Sympathie*. 2nd edn. Weimar, Verlag des Landes-Industrie-Comptoirs.

Humboldt, F. A. von (1797). *Versuche über die gereizte Muskel- und Nervenfaser*, vol 1. Posen, Decker.

Husson, H. M. [1831]. *Rapport sur les expériences magnétiques faites par la commission de l'Académie royale de médecine, lu dans les séances des 21 et 28 juin* ... [Paris], n. pub.

Jackson, J. W. (1851). *Lectures on Mesmerism, Delivered at the Rotunda, Dublin*. Dublin, James McGlashan.

Janin de Combe-Blanche, J. (1784). *Réponse au discours de M. O-Rian sur le magnétisme animal*. Geneva, n. pub.

Jastrow, J., and Nuttall, G. F. H. (1885–9). On the existence of a magnetic sense. *Proceedings of the American Society for Psychical Research*, 1, 116–126.

Joly, A. (1938). *Un mystique lyonnais et les secrets de la franc-maçonnerie 1730–1824*. Mâcon, Protat frères.

Jones, J. (1861). *The Natural and Supernatural* ... London, H. Baillière.

Josipovici, J. (1982). *Franz Anton Mesmer, magnétiseur, médecin et franc-maçon*. Monaco, Éditions du Rocher.

Jung-Stilling, J. H. (1808). *Theorie der Geister-Kunde* ... Nürnberg, Raw.

Jung-Stilling, J. H. (1834). *Theory of Pneumatology* ... London, Longman, Rees, Orme, Brown, Green and Longman.

[Jussieu, A. L. de] (1784). *Rapport de l'un des commissaires chargés par le Roi, de l'examen du magnétisme animal*. Paris, veuve Hérissant.

Kales, A., *et al.* (1966). Somnambulism: Psychophysiological correlates. I. All-night EEG studies. *Archives of General Psychiatry*, 14, 586–604.

Kales, A., *et al.* (1980). Somnambulism: Clinical characteristics and personality patterns. *Archives of General Psychiatry*, 37, 1406–1410.

Kales, J. D. *et al.* (1979). Sleepwalking and night terrors related to febrile illness. *The American Journal of Psychiatry*, 136, 1214–1215.

Kaplan, F. (1974). "The mesmeric mania": The early Victorians and animal magnetism. *Journal for the History of Ideas*, 35, 691–702.

Katz, S., Jz. (1863). *Het levensmagnetismus. Was het is en hoe het werkt*. Amsterdam, M. Lobo.

Kellner, R. (1975). Psychotherapy in psychosomatic disorders: A survey of controlled studies. *Archives of General Psychiatry*, 32, 1021–1028.

Kenny, M. (1986). *The Passion of Ansel Bourne: Multiple Personality in American Culture*. Washington, D. C., Smithsonian Institution Press.

Kerner, J. (1819). *Das Bilderbuch aus meiner Knabenzeit* ... Braunschweig, Vieweg.

Kerner, J. (1824). *Geschichte zweyer Somnambülen* ... Karlsruhe, G. Braun

Kerner, J. (1829). *Die Seherin von Prevorst: Eröffnungen über das innere Leben des Menschen und über das Hereinragen einer Geisterwelt in die unsere*. 2 vols. Stuttgart, Cotta.

Kerner, J. (1833). *Ueber das Besessenseyn oder das Daseyn und den Einfluss des bösen Geisterreiche in der alten Zeit. Mit Berücksichtigung dämonischen Besitzungen der neuen Zeit*. Heilbronn, C. Drechsler.

Kerner, J. (1834). *Geschichten Besessener neuerer Zeit. Beobachtungen aus dem Gebiete kakodämonisch-magnetischer Erscheinungen* ... Karlsruhe, Braun.

Kerner, J. (1847). *Revelations of the Invisible World, by a Somnambulist; being the Life of the Seeress of Prevorst* ... London, C. Moore.

Kerner, J. (1856). *Franz Anton Mesmer aus Schwaben, Entdecker des thierischen Magnetismus* ... Frankfurt am Main, J. Rütten

Kerner, J. (1904). *Sämtliche Werke.* 2 vols. Berlin, A. Weichert.

Kiernan, J. G. (1895). Hypnotism in American psychiatry fifty years ago. *The American Journal of Insanity,* 51, 336–345.

Kieser, D. G. (1817a). Ausbreitung des thierischen Magnetismus ausserhalb Deutschland. *Archiv für den thierischen Magnetismus,* 1(2), 155–158.

Kieser. D. G. (1817b). Rhapsodien aus dem Gebiete des thierischen Magnetismus. *Archiv für den thierischen Magnetismus,* 2(3), 63–147.

Kieser, D. G. (1818a). [Review of] Hochst merkwürdige Geschichte der magnetisch-hellsehenden Auguste Müller in Karlsruhe; von Dr. Meier ... *Archiv für den thierischen Magnetismus,* 3(3), 110–125.

Kieser, D. G. (1818b). Das magnetische Behältniss (Baquet) und der durch dasselbe erzeugte Somnambulismus. *Archiv für den thierischen Magnetismus,* 3(2), 1–180.

Kieser. D. G. (1820). Daemonophania, bei einem wachenden Somnambul beobachtet. *Archiv für den thierischen Magnetismus,* 6(1), 56–147.

Kieser, D. G. (1826). *System des Tellurismus oder thierischen Magnetismus: ein Handbuch für Naturforscher und Aerzte.* 2 vols. Leipzig, F. L. Herbig.

Kieserwetter, K. (1909). *Geschichte des neueren Occultismus.* 2nd edn. Leipzig, M. Altmann.

King, L. S. (1971). *The Medical World of the Eighteenth Century.* Huntington, N. Y., R. E. Krieger.

Kiste, A. (1845). *Mesmerism: Or Facts against Fallacy.* London, H. Baillière.

Klein, C. L. von (1819). Geschichte einer durch Magnetismus in 27 Tagen bewirkten Heilung eines 15 monatlichen Nervenleidens. *Archiv für den thierischen Magnetismus,* 5(1), 1–72.

Kluge, C. A. F. (1811). *Versuch einer Darstellung des animalischen Magnetismus als Heilmittel.* Berlin, C. Salfeld.

Kluge, C. A. F. (1812). *Proeve eener voorstelling van het dierlijk magnetismus als geneesmiddel.* Amsterdam, Maaskamp.

Kolb, B., and Whishaw, I. Q. (1990). *Fundamentals of Human Neuropsychology.* 3rd edn. New York, W. H. Freeman.

Köttgen, A. (1819). Maria Rübel, die Hellseherin in Langenberg. *Archiv für den thierischen Magnetismus,* 4(3), 1–279.

Krieger, D. (1979). Therapeutic touch: Searching for evidence of physiological change. *American Journal of Nursing,* 79, 660–662.

Kroger, W. S. (1977). *Clinical and Experimental Hypnosis in Medicine, Dentistry, and Psychology.* 2nd edn. Philadelphia, J. B. Lippincott.

L. U. G. E. (1845–6). Miss Martineau and her traducers. *The Zoist,* 3, 86–96.

L., l'abbé J. B. [i.e. J. B. Loubert] (1844). *Le magnétisme et le somnambulisme devant les corps savants, la cour de Rome et les théologiens.* Paris, Germer Baillière.

L., l'abbé J. B. [i.e. J. B. Loubert] (1846). *Défense théologique du magnétisme humain ...* Paris, Poussielgue-Rusand.

Lafontaine, C. (1847). *L'art de magnétiser ...* Paris, Germer Baillière.

Lafontaine, C. (1866). *Mémoires d'un magnétiseur ...* 2 vols. Paris, Germer Baillière.

Lagerweij, E., Nelis, P. C., Wiegant, V. M., and Ree, J. M. van (1984). The twitch in horses: A variant of acupuncture. *Science,* 225, 1172–1173.

Lamy Senart, – . (1817). Traitement de M. Baron, fils. *Bibliothèque du magnétisme animal,* 2, 1–24.

Landauer, A.A. (1980). Karl Christian Wolfart – Germany's first professor of psychology. Paper presented at the 22nd International Congress of Psychology, Leipzig.

Landauer, A.A. (1981). A note on the role of Karl Christian Wolfart (1778–1832) in the study of animal magnetism. *Journal of the History of the Behavioral Sciences*, 17, 206–208.

Lang, W. (1843). *Mesmerism, its History, Phenomena, and Practice: With Reports of Cases Developed in Scotland*. Edinburgh, Fraser.

Lange, V. (1982). *The Classical Age of German Literature*. London, Edward Arnold.

Lassaigne, A. (1851). *Mémoires d'un magnétiseur ...* Paris, Dentu.

Lausanne, M. de [i.e. A.A.V. Sarrazin de Montferrier] (1815). Traitement magnétique de Madame de Villeneuve. *Annales du magnétisme animal*, 4, 97–123, 198–218.

Lausanne, M. de [i.e. A.A.V. Sarrazin de Montferrier] (1818). *Élémens du magnétisme animal ...* Paris, J.G. Dentu.

Lausanne, M. de [i.e. A.A.V. Sarrazin de Montferrier] (1819). *Des principes et des procédés du magnétisme animal ...* 2 vols. Paris, J.G. Dentu.

Lavater, J.C. (1787). Briefwechsel zwischen Lavater und Marcard. *Magnetistisches Magazin für Niederteutschland*, 1(1), 1–19.

Lavater, J.C. (1821). J.C. Lavaters bisher ungedruckte Briefe und Aufsätze über den thierischen Magnetismus. *Archiv für den thierischen Magnetismus*, 8(3), 1–59; 9(1), 1–60.

Lee, E. (1843) *Animal Magnetism and Homoeopathy ...* 3rd edn. London, J. Churchill.

Lee, E. (1866). *Animal Magnetism and Magnetic Lucid Somnambulism ...* London, Longmans, Green and Co.

Léger, T. (1846). *Animal Magnetism; or Psycodunamy*. New York, D. Appleton.

Léger, T. (1852). *The Magnetoscope ...* London, Baillière.

Lehmann, F. (1819). Vermittelst des thierischen Magnetismus unternommene Kur eines complicirten Nervenübels. *Archiv für den thierischen Magnetismus*, 4(1), 1–57.

Le Lieurre de l'Aubépin, – . (1819). Lettre de M. Le Lieurre de l'Aubépin, à M. Deleuze. *Bibliothéque du magnétisme animal*, 8, 93–128.

Lentin, J.F.L. (1800). Etwas vom thierischen Magnetismus. *Journal der practischen Arzneykunde und Wundarzneykunst*, 11(2), 130–142.

Léonard, – . (1834). *Magnétisme, son histoire, sa théorie, son application au traitement des maladies ...* Paris, Duvignau.

Lettre d'un Bordelais au Père Hervier ... (1784). Amsterdam, n. pub.

Lettre de Mme ***, à M. Deleuze (1817). *Bibliothèque du magnétisme animal*, 2, 228–241.

Lettre d'un médecin de la Faculté de Paris, à M. Court de Jebelin ... (1784). Bordeaux, Bergeret.

Lettre sur la mort de M. Court de Gebelin (1785). In *Recueil des pièces les plus intéressantes pour & contre le magnétisme animal*, Vol I. Lyon, n. pub. Pp. 169–172.

Lettre sur la seule explication satisfaisante des phénomènes du magnétisme animal et du somnambulisme ... adressée à la Société des amis réunis de Strasbourg, par la Société exégétique et philanthropique de Stockholm ... (1788). Stockholm, n. pub.

Levin, J.D., Gorden, N.C., and Fields, H.L. (1978). The mechanism of placebo analgesia. *The Lancet*, 1978, ii, 654–657.

Levitan, A.R., and Johnson, J.M. (1985–6). The role of touch in healing and hypnotherapy. *American Journal of Clinical Hypnosis*, 28, 218–233.

Lévy, J. (1976). Magnétisme animal et médecine à Strasbourg à la fin du XVIIIᵉ siècle.

In G. Livet and G. Schaff (eds.), *Médecine et assistance en Alsace XVIe-XXe siècle*. Strasbourg, Istra.

Lewis, E. E. (1848). *A Report of the Mysterious Noises Heard in the House of Mr. John D. Fox in Hydesville, Arcadia, Wayne County*. Canandaigua, E. E. Lewis.

Lichtenstaedt, J. R. (1816). *Untersuchungen über den thierischen Magnetismus*. St. Petersburg, Imperial Academy of Sciences.

Lichtenstaedt, J. R. (1819a). Commentaires et réflexions sur le magnétisme animal. *Bibliothèque du magnétisme animal*, 7, 93–115.

Lichtenstaedt, J. R. (1819b). *Erfahrungen im Gebiete des Lebensmagnetismus*. Berlin, Sander.

Lishman, W. A. (1987). *Organic Psychiatry: The Psychological Consequences of Cerebral Disorder*. Oxford, Blackwell Scientific Publications.

Loew[enhjelm], –. (1819). Traitement magnétique de mademoiselle de S***, à Saint-Pétersbourg. *Bibliothèque du magnétisme animal*, 8, 129–140.

Loisson de Guinaumont, C. M. L. (1846). *Somnologie magnétique...* Paris, Germer Baillière.

Lòpez Piñero, J. N. (1983). *Historical Origins of the Concept of Neurosis*. Cambridge, Cambridge University Press.

Loubert, J. B. See L., J. B.

Lowes, J. L. (1927). *The Road to Xanadu*. London, Constable.

Luchins, D. J., *et al.* (1978). Filicide during psychotropic induced somnambulism: A case report. *The American Journal of Psychiatry*, 135, 1404–1405.

[Luetzelbourg, –.] (1786). *Extrait des journaux d'un magnétiseur, attaché à la Société des amis réunis de Strasbourg...* 2nd edn. Strasbourg, Lorenz et Schouler.

[Luetzelbourg, –.] (1788). *Dieu, l'homme et la nature: Tableau philosophique d'une somnambule*. London, n. pub.

Mabru, G. (1858). *Les magnétiseurs jugés par eux mêmes*. Paris, Mallet-Bachelier.

McGregor, D. (1964). *The Dream World of Dion McGregor*. New York, Geis.

M'Neile, H. (1842). *Satanic Agency and Mesmerism. A Sermon*. Liverpool [an offprint, pp. 141–152].

MacQuitty, B. (1971). *Victory over Pain: Morton's Discovery of Anaesthesia*. New York, Taplinger.

La magnétisme en Belgique (1840). *Journal du magnétisme animal*, 2, 149–152.

Magnétisme. Insensibilité produite au moyen du sommeil magnétique. Nouvelle opération chirurgicale faite à Cherbourg (1846). Cherbourg, n. pub.

Het magnetismus beschouwd bij het licht der naturkunde in zijnen aard (1858). s'Gravenhage, J. A. De la Vieter.

Mahan, A. (1855). *Modern Mysteries, Explained and Exposed...* Boston, Mass., J. P. Jewett.

[Mahon, P. A. O.] (1784). *Examen sérieux et impartial du magnétisme animal*. London, Royez.

Mainauduc, J. B. de (1798). *The Lectures of J. B. de Mainauduc. Part the First*. London, for the executrix.

Maitland, S. R. (1849). *Illustrations and Enquiries relating to Mesmerism. Part I*. London, W. Stephenson.

Marne, – de La (1828). *Étude raisonnée du magnétisme animal, et preuves de l'intervention des puissances infernales dans les phénomènes du somnambulisme magnétique*. Paris, Bureau de l'éclair.

Martin, J. (1790). *Animal Magnetism Examined, in a Letter to a Country Gentleman.* London, J. Stockdale.

Martineau, H. (1845). *Letters on Mesmerism.* 2nd edn. London, E. Moxon.

Martineau, H. (1851–2). Mesmeric cure of a cow. *The Zoist*, 8, 300–303.

Mattison, H. (1853). *Spirit-Rapping Unveiled!*... New York, Mason Brothers.

Mayer, D. J., Price, D. D., and Raffii, A. (1977). Antagonism of acupuncture analgesia in man by the narcoantagonist naloxone. *Brain Research*, 121, 368–372.

Mayo, H. (1851). *On the Truths Contained in Popular Superstitions with an Account of Mesmerism.* 3rd edn. Edinburgh, Blackwood.

Meier, – . (1818). *Höchst merkwürdige Geschichte der magnetisch hellsehenden Auguste Müller in Karlsruhe* ... Stuttgart, J. B. Metzler.

Meissner, E. (1819). *Bemerkungen aus den Taschenbuche eines Arztes während einer Reise von Odessa durch einen Theil von Deutschland, Holland, England und Schottland.* Halle, Renger.

Mello, N. K., and Overton, D. A. (1972). Behavioral studies of alcoholism. In B. Kissin and H. Begleiter (eds.), *The Biology of Alcoholism.* Vol 2. *Physiology and Behavior.* New York, Plenum Press.

[Meltier, – .] (1787). *Lettre adressée à monsieur le marquis de Puiségur sur une observation faite à la lune, précédée d'un système nouveau sur le mécanisme de la vue.* Amsterdam, n. pub.

Menager, – . (1926). Essais expérimentaux sur ls effets actiniques du supposé fluide magnétique humain. *Revue Métapsychique*, 1926, No. 2, 135–138.

Mesmer, F. A. (1766). *Dissertatio physico-medica de planetarum influxu.* Vienna, Ghelen.

Mesmer, F. A. (1775). *Schreiben über die Magnetkur von Herrn A. Mesmer, Doktor der Arzneygelährtheit, an einen auswärtigen Arzt.* Vienna, Kurzböck.

Mesmer, F. A. (1779). *Mémoire sur la découverte du magnétisme animal.* Geneva, P. F. Didot le jeune

Mesmer, F. A. (1781). *Précis historique des faits relatifs au magnétisme animal jusques en avril 1781.* London, n. pub.

Mesmer, F. A. (1785a). *Aphorismes de M. Mesmer* ... 3rd edn. Paris, n. pub.

Mesmer, F. A. (1785b). *Mémoire sur la découverte du magnétisme animal.* In *Recueil des pièces les plus intéressantes pour & contre le magnétisme animal*, Vol. 1. Lyon, n. pub. Pp. 1–63.

[Mesmer, F. A.] (1786). *Lettre de l'auteur de la découverte du magnétisme animal, à l'auteur des réflexions préliminaires; pour servir de réponse à un imprimé ayant pour titre: Sommes versées entre les mains de M. Mesmer pour acquérir le droit de publier sa découverte.* In *Recueil de pièces pour servir à l'histoire du magnétisme.* N. p., n. pub. Pp. 9–28.

Mesmer, F. A. (1799). *Mémoire de F. A. Mesmer, docteur en médecine, sur ses découvertes.* Paris, Fuchs.

Mesmer, F. A. (1800). *Lettre de F. A. Mesmer, docteur en médecine, sur l'origine de la petite vérole* ... Paris, Imprimerie des sciences et arts.

Mesmer, F. A. (1812a). Allgemeine Erläuterungen über den Magnetismus und den Somnambulismus. *ΑΣΚΛΗΠΙΕΙΟΝ*, 3, 247–302; 4, 3–25.

Mesmer, F. A. (1812b). *Allgemeine Erläuterungen über den Magnetismus und den Somnambulismus.* Halle, Buchhandlungen des Hallischen Waisenhauses.

Mesmer, F. A. (1812c). Über den Ursprung und die wahre Natur der Pocken ... *ΑΣΚΛΗΠΙΕΙΟΝ*, 3, 203–217.

Mesmer, F. A. (1814). *Mesmerismus. Oder System der Wechselwirkungen* ... Berlin, Nikolai.

Mesmer, F. A. (1971). *Le magnétisme animal: oeuvres publiées par Robert Amadou*. Paris, Payot.

Mesmer, F. A. (1980). *Mesmerism: A Translation of the Original Scientific and Medical Writings of F. A. Mesmer*. By G. Bloch. Los Altos, Calif., W. Kaufmann.

Mesmerism Considered (1852). Glasgow, W. Mackenzie.

Mesmerism and Media, with Full Instructions how to Develop the Alleged Spiritual Rappings in every Family (1855). London, H. Baillière.

Mesmerism Solved ... (1853). London, Jones.

Meylink, B. (1837). *Iets over het dierlijk magnetismus* ... Deventer, A. ter Gunne.

[Mialle, S.] (1826). *Exposé par ordre alphabétique des cures opérées en France par le magnétisme animal, depuis Mesmer jusqu'à nos jours (1774–1826)* ... 2 vols. Paris, J. G. Dentu.

Mialle, S. See also Scobardi, – .

Miles, F. (1854). *Mesmerism and the Diseases to which it is most applicable* ... Dublin, the author.

Miller, J. (1983). A Gower Street scandal. *Journal of the Royal College of Physicians of London*, 17, 181–191.

Milt, B. (1953). Franz Anton Mesmer und seine Beziehungen zur Schweiz: Magie und Heilkunde zu Lavaters Zeit. *Mitteilungen der antiquarischen Gesellschaft in Zürich*, 38(1), 1–139.

Mirville, J. E. de (1854). *Des esprits et de leurs manifestations fluidiques*, Vol. 1. 3rd edn. Paris, H. Vrayet de Surcy.

Mitchell, W., Falconer, M. H., and Hill, D. (1954). Epilepsy with fetishism relieved by temporal lobectomy. *The Lancet*, 1954, ii, 626–630.

Monspey, – de, and Barberin, – de (1784). *Expérience magnétique*. Lyon, n. pub.

Montadan, R. (1927). *Les radiations humaines*. Paris, Alcan.

Montègre, A. F. J. de (1812). *Du magnétisme animal et de ses partisans* ... Paris, Colas.

Moore, R. L. (1977). *In Search of White Crows*. New York, Oxford University Press.

Moret, V., *et al.* (1991). Mechanisms of analgesia induced by hypnosis and acupuncture: Is there a difference? *Pain*, 45, 135–140.

Morin, A. S. (1860). *Du magnétisme et des sciences occultes*. Paris, Germer Baillière.

Moser, L. (1967). Hypnotism in Germany. In E. J. Dingwall (ed.), *Abnormal Hypnotic Phenomena: A Survey of Nineteenth-Century Cases*, Vol 2. London, J. and A. Churchill. Pp. 101–199.

[Mouillesaux, – de] (1787). *Appel au public sur le magnétisme animal, ou projet d'un journal pour le seul avantage du public, et dont il serait le coopérateur*. [Strasbourg], n. pub.

Mouillesaux, – de (1789). Rapport sur une somnambule magnétique ... *Annales de la Société harmonique des amis réunis de Strasbourg*, [3], 1–87.

Moulinié, C. (1785). *Lettre sur le magnétisme animal, adressée à Monsieur Perdriau* ... In *Recueil des piéces les plus intéressantes pour & contre le magnétisme animal*, Vol 1. Lyon, n. pub. Pp. 291–316.

Muck, F. (1818). Sur l'état du magnétisme en Allemagne ... *Bibliothèque du magnétisme animal*, 5, 76–87.

Müller-Funk, W. (1985). E. T. A. Hoffmanns Erzählung *Der Magnetiseur*, ein poetisches Lehrstück zwischen Dämonisierung und neuzeitlicher Wissenschaftkritik. In H.

Schott (ed.), *Franz Anton Mesmer und die Geschichte des Mesmerismus.* Wiesbaden, F. Steiner. Pp. 200–214.

Muratori, L. B. (1740). *Della forza della fantasia umane.* Venice, G. Pasquali.

Myers, F. W. H. (1903). *Human Personality and its Survival of Bodily Death.* 2 vols. London, Longmans, Green and Co.

Nani, G. D. (1850). *Trattato teorico-practico sul magnetismo animale.* Turin, F. E. Franco.

Nasse, F. (1809). Untersuchungen über das Verhältniss des thierischen Magnetismus zur Elektricität. *Journal für die Physiologie,* 9, 237–312.

Neilson, W. (1855). *Mesmerism in its Relation to Health and Disease ...* Edinburgh, Shepherd and Elliot.

Neuberth, J. (1841). *Original-Beiträge zur Geschichte des Somnambulismus.* Leipzig, O. Wigand.

Neue Beobachtungen über den Thier-Magnetismus aus dem Tagebuch eines Reisenden (1788). *Journal des Luxus und der Moden,* 3, 153–184.

Newnham, W. (1845). *Human Magnetism; its Claims to Dispassionate Inquiry.* London, J. Churchill.

Nick, F. A. (1817). Darstellung einer merkwürdigen Geschichte durch den thierischen Magnetismus veranlasst. *Archiv für den thierischen Magnetismus,* 1(2), 3–162.

Nick, F. A. (1818). *Darstellung der sehr merkwürdigen durch den thierischen Magnetismus veranlassten Geschichte der C. Krämerin in Stuttgart.* Leipzig, Brockhaus.

Nicolai, J. D. (1787). Fragment einer Predigt am Schluss des Jahres 1787 über Apost. Gesch. 15 v. 12 gehalten. *Magnetistisches Magazin für Niederteutschland,* 1, 39–49.

Nicolas, – . (1784). *Histoire de l'établissement du magnétisme animal, fait à Grenoble le 1er octobre 1784.* Geneva, n. pub.

Noizet, F. J. (1854). *Mémoire sur le somnambulisme et le magnétisme animal adressé en 1820 à l'Académie royale de Berlin.* Paris, Plon frères.

Numan, A. (1815). *Verhandelingen over het dierlijk magnetismus als den grondslag ter verklaring der physische levensbetrekkingen of sympathie tussen de dierlijke lichamen.* Groningen, R. J. Schierbeek.

Oberkirch, H. L. d' (1970). *Mémoires de la Baronne d'Oberkirch.* Paris, Mercure de France.

Observations on animal magnetism (1817–18). *Blackwoods Edinburgh Magazine,* 1, 563–567.

Olbers, H. W. M. (1787a). Hrn. D. Olbers Erklärung über die in Bremen durch Magnetismus vorgenommene Kuren. *Archiv für Magnetismus und Somnambulismus,* Pt. 5, 69–82.

Olbers, H. W. M. (1787b). Erklärung über die in Bremen, durch den sogenannten Magnetismus, vorgenommenen Curen. *Magnetistisches Magazin für Niederteutschland,* 6, 495–514.

Olbers, H. W. M. (1788). Dr. Olbers abermalige Erklärung über die in Bremen durch den sogenannten Magnetismus vorgenommene Kuren. *Magnetistisches Magazin für Niederteutschland,* 7, 631–657.

Olivier, J. (1849). *Traité du magnétisme, suivi des paroles d'un somnambule et d'un recueil de traitements magnétiques.* Toulouse, L. Jougla.

Olson, G. A., Olson. R. D., and Kastin, A. J. (1990). Endogenous opioids: 1989. *Peptides,* 11, 1277–1304.

On the present state of animal magnetism in Germany (1817–18). *Blackwoods Edinburgh Magazine,* 2, 36–38.

Operations without pain in the mesmeric state (1844–5). *The Zoist*, 2, 390–393.

Oppenheim, J. (1985). *The Other World: Spiritualism and Psychical Research in England, 1850–1914*. Cambridge, Cambridge University Press.

Oppert, C. (1817). Observations relatives à la lettre de M. Friedlander sur l'état actuel du magnétisme en Allemagne. *Bibliothèque du magnétisme animal*, 1, 181–196.

Ordinaire, – . (1841). *Le magnétisme et le somnambulisme du docteur Laurent; une somnambule mâconnaise*. Mâcon, De Jussieu.

Orelut, – . (1785). *Détail des cures opérées à Lyon par le magnétisme animal, selon les principes de M. Mesmer*. In *Recueil des pièces les plus intéressantes pour & contre le magnétisme animal*, Vol. 1. Lyon, n. pub. Pp. 361–384.

O'Ryan, – . (1784). *Discours sur le magnétisme animal, lu dans une assemblée du Collège des médecins de Lyon...* Dublin, n. pub.

Oswald, I., and Evans, J. (1785). On serious violence during sleep-walking. *British Journal of Psychiatry*, 147, 688–691.

Overton, D. A. (1972). State-dependent learning produced by alcohol and its relevance to alcoholism. In B. Kissin and H. Begleiter (eds.), *The Biology of Alcoholism*. Vol. 2. *Physiology and Behavior*. New York, Plenum Press.

Palfreman, J. (1977). Mesmerism and the English medical profession: A study of conflict. *Ethics in Science and Medicine*, 4, 51–66.

Pam, V. (1816). Au rédacteur des Annales du magnétisme animal. *Annales du magnétisme animal*, 6, 193–201.

Panin, N. P. (1818). Lettre de S. Ex. M. le comte Panin, adressée aux membres de la Société du magnétisme, séante à Paris... *Bibliothèque du magnètisme animal*, 3, 126–146.

Parker, J. L. (1979). The use of hypnosis to reinstate a child's impaired writing performance resulting from neurological damage. *Australian Journal of Clinical and Experimental Hypnosis*, 7, 225–230.

Parrot, G. F. von (1816). *Coup d'oeil sur le magnétisme animal*. St. Petersburg, Bonnet.

Parsinnen, T. M. (1977–8). Mesmeric performers. *Victorian Studies*, 21, 87–104.

Parsinnen, T. M. (1979). Professional deviants and the history of medicine: Medical mesmerists in Victorian Britain. In R. Wallis (ed.), *On the Margins of Science: The Social Construction of Rejected Knowledge. Sociological Review Monograph 27*. Keele, Staffs., University of Keele.

Parsons, – . (1849–50). Conclusion of Mr. Parsons' case of cataleptic insanity treated mesmerically. *The Zoist*, 7, 9–25.

Pascal, E. (1936). La question du "fluide magnétique." *Revue métapsychique*, No. 5, 375–417.

Pasley, T. H. (1848). *The Philosophy which Shows the Physiology of Mesmerism and Explains the Phenomena of Clairvoyance*. London, Longman, Brown, Green and Longmans.

Passavant, J. C. (1821). *Untersuchungen über den Lebensmagnetismus und das Hellsehen*. Frankfort am Main, H. L. Brönner.

Passavant, J. C. (1837). *Untersuchungen über den Lebensmagnetismus und das Hellsehen*. 2nd edn. Frankfort am Main, H. L. Brönner.

Pattie, F. A. (1956). Mesmer's medical dissertation and its debt to Mead's *De imperio solis ac lunae*. *Journal of the History of Medicine and Allied Sciences*, 11, 275–287.

Pattie, F. A. (1980–1). A Mesmer-Paradis myth dispelled. *American Journal of Clinical Hypnosis*, 23, 29–31.

[Paulet, J. J.] (1780). *Les miracles de Mesmer*. Paris, n. pub.

[Paulet, J. J.] (1784a). *L'antimagnétisme ou origine, progrès, décadence, renouvellement et réfutation du magnétisme animal*. London, n. pub.

[Paulet, J. J.] (1784b). *Mesmer justifié*. Constance, n. pub.

Paulet, J. J. (1785). *Réponse à l'auteur des* Doutes d'un provincial ... London, n. pub.

[Pearson, J., attr.] (1790). *A Plain and Rational Account of the Nature and Effects of Animal Magnetism*. London, W. and J. Stratford.

[Périer, – .] (1814–15). Traitement de fistules avec complications. *Annales du magnétisme animal*, 2, 203–229; 3, 3–18, 55–68.

Perry, C., and Laurence, J. R. (1983). Hypnosis, surgery and mind-body interaction. *Canadian Journal of Behavioral Science*, 15, 351–372.

Petetin, J. H. D. (1787a). *Mémoire sur la découverte des phénomènes de l'affection hysterique essentielle* ... N. p., n. pub.

Petetin, J. H. D. (1787b). *Mémoire sur la découverte des phénomènes qui présentent la catalepsie et le somnambulisme* ... N. p., n. pub.

Petetin, J. H. D. (1808). *Électricité animale* ... Paris, Brunot-Labbé.

Pezold, J. N. (1797). Versuche mit den thierischen Magnetismus. *Archiv für die Physiologie*, 2, 1–18.

Pfaff, C. H. (1817). *Ueber und gegen den thierischen Magnetismus und die jetzt vorherrschende Tendenz auf dem Gebiete desselben*. Hamburg, Perthes und Besser.

Philips, A. J. P. See References: Hypnotism.

"Physician, A" (1838). *A Short Sketch of Animal Magnetism* ... London, J. Hatchard.

Pierer, J. F. (1823). Mesmerismus, Mesmerthum, Naturmagnetismus. In J. F. Pierer and D. L. Choulant (eds.), *Medizinisches Realwörterbuch*, vol. 5. Altenburg, Literatur-Comptoir. Pp. 178–203.

Pigault-Lebrun, C. A. G. (1829). Traitement magnétique de Madame Plantin, redigé d'après les notes de M. le docteur Chapelain ... *L'Hermès*, 4, 173–204.

Pigeaire, J. (1839). *Puissance de l'électricité animale* ... Paris, J. G. Dentu

Podmore, F. (1894). *Apparitions and Thought-Transference: An Examination of the Evidence for Telepathy*. London, Walter Scott.

Podmore, F. (1897). *Studies in Psychical Research*. London, Kegan Paul, Trench, Trübner and Co.

Podmore, F. (1902). *Modern Spiritualism: A History and a Criticism*. 2 vols. London, Methuen.

Podmore, F. (1909). *Mesmerism and Christian Science*. London, Methuen.

Poe, E. A. (1948). *The Letters of Edgar Allen Poe*. Ed. J. W. Ostrom. Cambridge, Mass., Harvard University Press.

Poeti, M. (1848). *Saggio sull'azione curativa del magnetismo animale nelle malattie nervose*. Turin, Bocca.

Poyen St. Sauveur, C. (1837a). *A Letter to Col. Wm. L. Stone of New York* ... Boston, Weekes, Jordan and Co.

Poyen St. Sauveur, C. (1837b). *Progress of Animal Magnetism in New England* ... Boston, Weekes, Jordan and Co.

A Practical Display of the Philosophical System Called Animal Magnetism ... (1790). London, n. pub.

Pratt, M. (1789). *A List of a Few Cures Performed by Mr. and Mrs. de Loutherbourg of Hammersmith Terrace, without Medicine.* London, J. P. Cooke.

Pressavin, J. B. (1784). *Lettres sur le magnétisme ...* N.p., n. pub.

Prideaux, T. S. (1845–6). On the application of phrenology in the choice of Parliamentary candidates. *The Zoist,* 3, 399–416.

Priestley, P. (1985). *Victorian Prison Lives.* London, Methuen.

Procès de Mme Fructus (1826). *L'Hermès,* 1, 99–120.

Procès en police correctionelle ... (1828). *L'Hermès,* 3, 56–70.

Prophétie du douzième siècle [1785]. N.p., n. pub.

Pulos, L. (1979–80). Mesmerism revisited: The effectiveness of Esdaile's techniques in the production of deep hypnosis and total body hypnoanesthesia. *American Journal of Clinical Hypnosis,* 22, 206–211.

Pyne, T. (1844). *Vital Magnetism: A Remedy.* London, S. Highley.

Quen, J. M. (1975). Case studies in nineteenth century scientific rejection: Mesmerism, Perkinsism and acupuncture. *Journal of the History of the Behavioral Sciences,* 11, 149–156.

Quen, J. M. (1976). Mesmerism, medicine, and professional prejudice. *New York State Journal of Medicine,* 76, 2218–2222.

Quinn, J. F. (1987). Therapeutic touch: Report of research in progress. In D. H. Weiner and R. D. Nelson (eds.). *Research in Parapsychology 1986.* Metuchen, N. J., The Scarecrow Press. Pp. 187–202.

R******, M. de (1787). *Lettre à madame la comtesse de L*** contenant une observation magnétique fait par une somnambule sur un enfant de six mois.* [Besançon], n. pub.

Rahn, J. H. (1790). *Physische Abhandlungen von den Ursachen der Sympathie von dem Magnetismus und Schlafwandeln.* Leipzig, F. G. Jacobaer.

Rahn, J. H., and Scherb, C. (1787). Briefwechsel zwischen Hrn Doktor Scherb in Bischofzell, und Dr. und Canonicus Rahn in Zürich über die Heilkräfte des thierischen Magnetismus. *Archiv gemeinnütziger physischer und medizinischer Kenntnisse,* 1, 595–665, 665–688; 2, 217–221, 221–408, 409–420.

Ramm, A. (1967). *Germany 1789–1919: A Political History.* London, Methuen.

Ramsey, M. (1988). *Professional and Popular Medicine in France, 1770–1830. The Social World of Medical Practice.* Cambridge, Cambridge University Press.

Rapport de l'Académie des sciences sur une annonce de guérison des sourds-muets par le magnétisme animal (1840). *Journal du magnétisme animal,* 2, 500–513.

Rapport des commissaires chargeés par le roi de l'examen du magnétisme animal (1784). Paris, Imprimerie Royale.

Rapport des commissaires de la Société royale de médecine, nommés par le Roi, pour faire l'examen du magnétisme animal (1784). Paris, Moutard.

Rapport fait à l'Académie royale de médecine ... par M. Jules Cloquet ... (1829). *L'Hermès,* 4, 132–134.

Razy, – . (1815). Traitement d'une phlegmasie chronique. *Annales du magnétisme animal,* 3, 193–226.

Recueil d'observations et de faits relatifs au magnétisme animal, présenté à l'auteur de cette découverte, & publiée par la Société de Guienne (1785). Philadelphia, n. pub.

Recueil de pièces pour servir à l'histoire du magnétisme (1786). N. p., n. pub.

Recueil des pièces les plus intéressantes pour & contre le magnétisme animal, Vol. 1. (1785). Lyon, n. pub.

Redern, – de (1818). Analyse de l'ouvrage intitulé Versuch einer Darstellung des animalischen Magnetismus... *Bibliothèque du magnétisme animal*, 5, 97–143.

Reich, [la baronne de] (1786). Somnambule magnétique: Cure de convulsions cataleptiques, suivies de faiblesses et d'un engorgement général et squirrheux des viscères. In *Exposé de différentes cures*... Strasbourg, L. Schouler. Pp. 162–250.

Reichel, W. (1829). *Ueber das Entwicklungsgesetz des magnetischen Lebens im Menschen. Nebst der Geschichte zweier merkwürdiger Somnambulen*. Leipzig, Lehnhold.

Reichenbach, K. L. von (1849). *Physikalisch-physiologische Untersuchungen über die Dynamide des Magnetismus*... 2nd edn. 2 vols. Braunschweig, F. Vieweg

Reichenbach, K. L. von (1850). *Physico-Physiological Researches on the Dynamides or Imponderables*... Tr. W. Gregory. London, Taylor, Walton and Moberly.

Reichenbach, K. L. von (1850–1). *Physico-Physiological Researches on the Dynamics of Magnetism*... Tr. J. Ashburner. 2 vols. London, H. Baillière.

Reichenbach, K. L. von (1853–4). Popular letters on the odic force. *The Zoist*, 11, 101–128, 274–294, 329–349

Reil, J. C. (1795), Von der Lebenskraft. *Archiv für die Physiologie*, 1(1), 8–162.

Reil, J. C. (1807). Ueber die Eigenschaften des Ganglien-Systems und sein Verhältniss zum Cerebral-System. *Journal für die Physiologie*, 7(2), 189–256

Reisen in den Mond, in mehrere Sterne und in die Sonne... (1834). Stuttgart, Sonnewald.

Remarques sur le conduite du sieur Mesmer et de son commis le P. Hervier... (1784). N.p., n. pub.

Renard, J. C. (1815). Zwey höchst merkwürdige Beobachtungen über den Somnambulismus oder das Traumleben ohne magnetische Einwirkung. *Journal der practischen Arzneykunde und Wundarzneykunst*, 40(2), 3–101.

Report of the Committee Appointed by Government to Observe and Report upon Surgical Operations by Dr. J. Esdaile upon Patients under the Influence of Alleged Mesmeric Agency (1846). Calcutta, W. Ridsdale.

Ribault, – . (1819). Rapport fait à la Société du magnétisme... d'un traitement par le magnétisme... *Bibliothèque du magnétisme animal*, 7, 35–51.

Ricard, J. J. A. (1840). Cas de sourd-mutité guéri au moyen du magnétisme. *Journal du magnétisme animal*, 2, 251–256.

Ricard, J. J. A. (1841). *Traité théorique et pratique du magnétisme animal*... Paris, Germer Baillière.

Ricard, J. J. A. (1844). *Physiologie et hygiène du magnétiseur*... Paris, Germer Baillière.

Ricard, J. J. A. (1845). *Le magnétisme traduit en cour d'assises: Acquittement*. Paris, Giraudet et Jouaust

Rice, E., and Fisher, C. (1976). Fugue states in sleep and wakefulness: A psychophysiological study. *The Journal of Nervous and Mental Disease*, 163, 79–87.

Robiano, L. M. G. de (1845). *Mesmer, Galvani et les théologiens*... Brussels, Wouters.

Rogers, E. C. (1853). *The Philosophy of Mysterious Agents, Human and Mundane*... Boston, Mass., J. P. Jewett.

Römer, C. (1821). *Ausführliche historische Darstellung einer höchst merkwürdigen Somnambüle*... Stuttgart, J. B. Metzler.

Rosen, G. (1936). John Elliotson, physician and hypnotist. *Bulletin of the Institute of the History of Medicine*, 4, 600–603.

Rosenthal, R. (1966). *Experimenter Effects in Behavioral Research*. New York, Appleton-Century-Crofts.

Rostan, L. (1825). Magnétisme. In N. P. Adelon *et al.* (eds.), *Dictionnaire de médecine*, vol. 13. Paris, Béchet jeune. Pp. 421–469.

Rouillier, A. (1817a). Cure d'une ophtalmie. *Bibliothèque du magnétisme animal*, 1, 17–24.

Rouillier, A. (1817b). *Exposition physiologique des phénomènes du magnétisme animal et du somnambulism* ... Paris, J. G. Dentu.

Rouillier, A. (1818). Exposition physiologique des phénomènes du magnétisme animal. *Bibliothèque du magnétisme animal*, 3, 189–205; 4, 53–73, 188–199.

Rouxel, – . [i.e. A. Leroux] (1892). *Rapports du magnétisme et du spiritisme*. Paris, Librairie des sciences psychologiques.

Rumpelt, F. (1840). *Die Höhne und der animalische Magnetismus* ... Dresden, Walther.

Russell, M., and Goldfarb, C. R. (1978). *Spiritualism and Nineteenth-Century Letters*. Cranbury, N. J., Associated University Presses.

Rutter, J. O. N. (1851). *Magnetoid Currents, their Forces and Directions; with a Description of the Magnetoscope*. London, J. W. Parker.

Rutter, J. O. N. (1854). *Human Electricity: The Means of its Development Illustrated by Experiments*. London, J. W. Parker.

S. du M. (1815). Cure d'une dislocation de l'avant-bras. *Annales du magnétisme animal*, 4, 241–247.

S. du M. (1816a). Du démon de Socrate, etc. *Annales du magnétisme animal*, 5, 109–119.

S. du M. (1816b). Des séances publiques de magnétisme, qui on lieu chez M. l'abbé Faria. *Annales du magnétisme animal*, 5, 186–191.

S. I. L. E. (1843–4). Mesmerism. *The Zoist*, 1, 58–94.

Sabbatini, M. (1846). *Sul magnetismo animale: memoria storico-critica*. Modena, A. Rossi.

Sammlung der neuesten gedruckten und geschriebenen Nachrichten von Magnet-Curen vorzüglich der mesmerischen (1778). Leipzig, C. G. Hilschern.

Sandby, G. (1843). *Mesmerism the Gift of God*. London, W. E. Painter.

Sandby, G. (1844). *Mesmerism and its Opponents: With a Narrative of Cases*. London, Longman, Brown, Green and Longmans.

Sandby, G. (1848). *Mesmerism and its Opponents*. 2nd edn. London, Longman, Brown, Green and Longmans.

Sarbin, T. R., and Slagle, R. W. (1979). Hypnosis and psychophysiological outcomes. In E. Fromm and R. E. Shor (eds.), *Hypnosis: Developments in Research and New Perspectives*. 2nd edn. New York, Aldine Publishing Co. Pp. 273–303.

Sarrazin de Montferrier, A. A. V. See Lausanne, M. de.

Saunders, S. D. (1852). *The Mesmeric Guide for Family Use* ... London, H. Baillière.

Saunders, S. D. (1852–3). Cases by Mr. Saunders of Bristol. *The Zoist*, 10, 167–170

Saunders, S. D. (1855). *Errors Dispelled; or, Mesmerism without Sleep and Mesmerism without Medicine*. London, H. Baillière.

Schelling, F. W. J. von (1978). *System of Transcendental Idealism* (1800). Charlottesville, University Press of Virginia.

Schindler, H. B. (1857). *Das magische Geistesleben: Ein Beitrag zur Psychologie*. Breslau, W. G. Korn.

Schmidt, A. (1846). *Bericht von der Heilung der Frau Marnitz durch Somnambulismus* ... Berlin, Stuhr.

Schott, H. (1982). Die Mitteilung des Lebensfeuers: zum therapeutischen Konzept von Franz Anton Mesmer (1734–1815). *Medizinhistorisches Journal*, 17, 195–214.

Schott, H. (ed.) (1985). *Franz Anton Mesmer und die Geschichte des Mesmerismus.* Wiesbaden, F. Steiner.

Schott, H. (1985). Bibliographie: Der Mesmerismus im Schriftum des 20. Jahrhunderts (1900–1984). In H. Schott (ed.), *Franz Anton Mesmer und die Geschichte des Mesmerismus.* Wiesbaden, F. Steiner.

Schubert, G. H. von (1838). *Die Geschichte der Seele.* 2nd edn. Stuttgart, Cotta.

Schubert, G. H. von (1840a). *Ansichten von der Nachtseite der Naturwissenschaft.* 4th edn. Dresden, Arnold.

Schubert, G. H. von (1840b). *Die Symbolik des Traumes ...* Leipzig, Brockhaus.

Schuerer von Waldheim, F. (1930). *Anton Mesmer, ein Naturforscher ersten Ranges ...* Vienna, Selbstverlag.

Schwartz, S. A. *et al.* (1987). Infrared spectra alteration in water proximate to the palms of therapeutic practitioners. In D. H. Weiner and R. D. Nelson (eds.), *Research in Parapsychology* 1986. Metuchen, N. J., The Scarecrow Press. Pp. 24–29.

Scobardi, – . [i.e. S. Mialle] (1839). *Rapport confidentiel sur le magnétisme animal ...* Paris, J. G. Dentu.

Scoresby, W. (1849). *Zoistic Magnetism.* London, Longman, Brown, Green and Longmans.

Segretier, – . (1819). Relation des cures magnétiques opérées à Nantes par M. Segretier. *Bibliothèque du magnétisme animal*, 7, 189–217.

Semler, J. S. (1776). *Samlungen von Briefen und Aufsätzen über die Gassnerischen und Schröfferischen Geisterbeschwörungen ...* 2 vols. Halle,

[Servan, J. M. A.] (1784a). *Doutes d'un provincial proposés à messieurs les médecins-commissaires chargés par le roi de l'examen du magnétisme animal.* Lyon, Prault.

[Servan, J. M. A.] (1784b). *Questions du jeune docteur Rhubarbini de Purgandis ...* Padua, n. pub.

Shapiro, A. K. (1977). Placebos in psychiatric therapy. *Current Psychiatric Therapies*, 17, 157–163.

Siemers, J. F. (1835). *Erfahrungen über den Lebensmagnetismus und Somnambulismus ...* Hamburg, A. Campe.

Sinnett, A. P. (1892). *The Rationale of Mesmerism.* London, Kegan Paul, Trench, Trübner and Co.

Sloman, – ., and Mayhew, – . (1851–2) An instance of introvision, with the verification after death. *The Zoist*, 9, 289–294.

Sloman, – ., and Mayhew, – ., (1852–3). An instance of introvision with verification after death. *The Zoist*, 10, 307–309.

Snewing, W. (1852–3). An instance of the effect of maternal impressions upon the offspring before its birth. *The Zoist*, 10, 379–385.

Société magnétique du Cap-Français, formée par M. le comte Chastenet de Puységur, en 1784 (1828). *L'Hermès*, 3, 361–370.

Solfvin, J. (1984). Mental healing. In S. Krippner (ed.), *Advances in Parapsychological Research*, vol. 4. Jefferson, N. C., McFarland. Pp. 31–63.

Spanos, N. P. (1986). Hypnotic behavior: A social psychological interpretation of amnesia, analgesia and "trance logic". *The Behavioral and Brain Sciences*, 9, 449–502.

Spanos, N. P., and Gilbert, J. (1979). Demonic possession, mesmerism, and hysteria: A social psychological perspective on their historical interrelations. *Journal of Abnormal Psychology*, 88, 527–546.

Spiegel, D., and Albert, L.H. (1983). Naloxone fails to reverse hypnotic alleviation of chronic pain. *Psychopharmacology*, 81, 140–143.

Spiritus, – . (1819). Beobachtungen über die Heilkraft des animalischen Magnetismus. *Archiv für den thierischen Magnetismus*, 5(3), 78–87.

Stam, H.J., and Spanos, N. (1982). The Asclepian dream healings and hypnosis: A critique. *The International Journal of Clinical and Experimental Hypnosis*, 30, 9–22.

Stieglitz, J. (1814). *Ueber den thierischen Magnetismus*. Hannover, Hahn.

Stone, G.W. (1850). *Electrobiology; or, the Electrical Science of Life*. Liverpool, Willmer and Smith.

Stone, M.H. (1974). Mesmer and his followers: The beginnings of sympathetic treatment of childhood emotional disorders. *History of Childhood Quarterly*, 1, 659–679.

Stone, W.L. (1837). *Letter to Doctor A. Brigham on Animal Magnetism* ... New York, G. Dearborn.

Storer, – . (1846–7). Three cures of epilepsy. *The Zoist*, 4, 447–450.

Storer, – . (1847–8). Cases of tic douloureux and other nervous affections cured with mesmerism. *The Zoist*, 5, 16–20.

Straumann, H. (1928). *Justinus Kerner und der Okkultismus in der deutschen Romantik*. Horgen-Zürich, Münster-Presse.

Strombeck, F.C. de (1814). *Histoire de la guérison d'une jeune personne par le magnétisme animal* ... Paris, Librairie Grecque, Latine, Allemande.

Suite des cures faites par différentes magnétiseurs, membres de la Société harmonique des amis réunis de Strasbourg (1787). Strasbourg, Lourenz et Schouler.

Sulla causa dei fenomeni mesmerici (1856). 2 vols. Bergamo, Mazzolini.

Sunderland, L. (1847). *Pathetism: Man Considered in respect to his Form, Life, Sensation, Soul, Mind, Spirit* ... Boston, Mass., White and Potter.

Supplément aux deux rapports de Mrs. les commissaires ... (1784). Amsterdam, Gueffier.

Sur les faits qui semblent prouver une communication des somnambules avec les êtres spirituels ... (1818). *Bibliothèque du magnétisme animal*, 5, 1–12.

Système raisonné du magnétisme universel d'après les principes de M. Mesmer ... (1786). Paris, Gastelier.

Szapáry, F. von (1838). *Ein Wort über animalischen Magnetismus, Seelenkörper und Lebensessenz* ... Leipzig, F. Brockhaus.

Tandel, E. (1841–2). Nouvel examen d'un phénomène psychologique du somnambulisme. *Mémoires publiées par l'Académie royale des sciences et belles-lettres de Bruxelles*, 15(2), 1–39.

Tanton, – . (1818). Traitemens magnétiques, par M. Tanton, officier de la gendarmerie royale, présentément en résidence à Béziers ... *Bibliothéque du magnétisme animal*, 4, 1–10.

[Tardy de Montravel, A.A.] (1785). *Essai sur la théorie du somnambulisme magnétique*. London, n. pub.

[Tardy de Montravel, A.A.] (1786a). *Journal du traitement magnétique de la Demoiselle N.* London, n. pub.

[Tardy de Montravel, A.A.] (1786b). *Suite du traitement magnétique de la Demoiselle N.* London, n. pub.

[Tardy de Montravel, A.A.] (1787a). *Essai sur la théorie du somnambulisme magnétique*. New edn. London, n. pub.

[Tardy de Montravel, A. A.] (1787b), *Lettres pour servir de suite à l'essai sur la théorie du somnambulisme magnétique*. London, n. pub.

[Tardy de Montravel, A. A.] (1787c). *Journal du traitement magnétique de Mme B.* Strasbourg, Librairie académique.

Tedinngarov, J. A. [i.e. Grandvoinet] (1843). *Esquisse d'une théorie des phénomènes magnétiques*. Paris, Dentu.

Tepperberg, K. (1935). *"Die Lebenskraft" bei Johann Christian Reil*. Halle, C. Nieft.

Teste. A. (1843). *A Practical Manual of Animal Magnetism* ... London, H. Baillière.

Teste, A. (1845). *Le magnétisme animal expliqué* ... Paris, J. B. Baillière.

Thomas d'Onglée, F. L. (1785). *Rapport au public, de quelques abus auxquels le magnétisme animal a donné lieu*. Paris, veuve Hérissant.

Thompson, H. S. (1847–8). On the silent power of the will of one person over another. *The Zoist*, 5, 253–260.

Thouret, M. A. (1784). *Recherches et doutes sur le magnétisme animal*. Paris, Prault.

Thouret, M. A. (1785). *Extrait de la correspondance de la Société royale de médecine, relativement au magnétisme animal*. Paris, Imprimerie royale.

Tinterow, M. M. (ed.) (1970). *Foundations of Hypnosis: From Mesmer to Freud*. Springfield, Ill., C. C. Thomas.

Tischner, R. (1928). *Franz Anton Mesmer: Leben, Werk und Wirkungen*. Munich, Verlag der Münchner Drucke.

Tischner, R., and Bittel, K. (1941). *Mesmer und sein Problem: Magnetismus – Suggestion – Hypnose*. Stuttgart, Hippocrates-Verlag Marquardt & Cie.

[Tissart de Rouvre, J. L. N.] (1785). *Nouvelles cures opérées par le magnétisme animal*. In *Recueil des pièces les plus intéressantes pour & contre le magnétisme animal*, vol 1. Lyon, n. pub. Pp. 385–468.

Tomlinson, W. K., and Perret, J. J. (1974). Mesmerism in New Orleans, 1845–1851. *The American Journal of Psychiatry*, 131, 1402–1404.

Tommasi, M. (1851). *Il magnetismo animale considerato sotto un nuovo punto di vista*. Turin, Pomba.

Topham, W., and Ward, W. Squire (1842). *Account of a Case of Successful Amputation of the Thigh during the Mesmeric State, without the Knowledge of the Patient*. London, H. Baillière.

Tournier, C. (1911). *Le mesmérisme à Toulouse* ... Toulouse, Saint-Cyprien.

Townshend, C. H. (1844). *Facts in Mesmerism* ... 2nd edn. London, H. Baillière.

Townshend, C. H. (1851–2). Recent clairvoyance of Alexis Didier. *The Zoist*, 9, 402–414.

Townshend, C. H. (1854). *Mesmerism Proved True, and the Quarterly Review Reviewed*. London, T. Bosworth.

Traitmens magnétiques de mesdemoiselles Anastasie et Rose, opérées en 1817, à Saint-Quentin, par M.*** (1818). *Bibliothéque du magnétisme animal*, 3, 1–38, 93–125.

Tritschler, J. C. S. (1817). Sonderbare, mit glücklichem Erfolg animal-magnetisch behandelte, Entwicklungskrankheit eines dreizehnjahrigen Knaben. *Archiv für den thierischen Magnetismus*, 1(1), 51–137.

Tschiffeli, [Mme de] (1789). Journal de Mad.^e de Tschiffeli, de la cure magnétique de Marianne V.**** ... pour surabondance d'écoulement menstruel ... *Annales de la Société harmonique des amis réunis de Strasbourg* ..., [3], 132–144.

Tweedie, A. C. (1857). *Mesmerism and its Realities* ... Edinburgh, Paton and Ritchie.

Tymms, R. (1955). *German Romantic Literature*. London, Methuen.

Ueber das Magnetisiren in Strasbourg (1787). *Berlinische Monatsschrift*, 10, 458–477.

Ueber den Magnetismus am Oberrhein und in Niederdeutschland (1787). *Magnetistisches Magazin für Niederteutschland*, 2, 188–191.

Uhlmann, J. (1853). *Blicke in das Jenseits, geoffenbart durch die Hellseherin Magdalene Wenger*. Bern, Haller.

Valentin, J. C. (1820). Beschreibung einer magnetischen Cur, als Beitrag zur Geschichte des Magnetismus. *Archiv für den thierischen Magnetismus*, 7(3), 49–120.

Valleton de Boissière, – . (1785). *Lettre de M. Valleton de Boissière, médecin à Bergerac, à M. Thouret* ... Philadelphia, n. pub.

Van der Hart, O., and Van der Velden, K. (1986–7). The hypnotherapy of Dr. Andries Hoek: Uncovering hypnotherapy before Janet, Breuer, and Freud. *American Journal of Clinical Hypnosis*, 29, 264–271.

Verati, L. [i.e. G. Pellegrino] (1845–6). *Sulla storia, teoria e practica del magnetismo animale*. 4 vols. Florence, V. Bellagambi.

Verordnung über die Ausübung des Magnetismus in den königl. Preussischen Staaten (1812). *Journal der practischen Arzneykunde und Wundarzneykunst*, 35(7), 125–126.

Vess, D. M. *Medical Revolution in France 1789–1796*. Gainesville, University Press of Florida.

Viatte, A. (1928). *Les sources occultes du romantisme* ... 2 vols. Paris, H. Champion.

Villers, C. de (1978). *Le magnétiseur amoureux*. Introduction et notes par François Azouvi. Paris, J. Vrin.

Vinchon, J. (1936). *Mesmer et son secret*. Paris, A. Legrand.

Virey, J. J. (1818). Magnétisme animal. In N. P. Adelon al. (eds.), *Dictionnaire des sciences médicales* ... , vol. 29. Paris, C. L. F. Pancoucke. Pp. 463–558.

Vogel, J. L. A. (1818). *Die Wunder des Magnetismus*. Erfurt, Hennings.

Voss, J. H. (1819). *Der thierische Magnetismus als Wirkung der höchsten Naturkraft*. Cöln, H. Rommerskirschen.

Wadden, T. A., and Anderton, C. H. (1982). The clinical use of hypnosis. *Psychological Bulletin*, 91, 215–243.

Wagstaff, G. F. (1981). *Hypnosis, Compliance and Belief*. Brighton, The Harvester Press.

Walmsley, D. M. (1967). *Anton Mesmer*. London, R. Hale.

Watts, A. M. H. (1883). *The Pioneers of the Spiritual Reformation* ... London, The Psychological Press.

Weber, J. (1816). *Der thierische Magnetismus* ... Landshut, Weber.

Weber, J. (1817). *Über Naturerklärung überhaupt und über die Erklärung der thierisch-magnetisch Erscheinungen* ... Landshut, Weber.

Wecter, D. (1952). *Sam Clemens of Hannibal*. Boston, Houghton Miflin.

Weisse, J. F. (1819). *Erfahrungen über arzneiverständige Somnambulen* ... Berlin, Flittner.

Werner, H. (1839). *Die Schutzgeister oder merkwürdige Blicke zweier Seherinnen in die Geisterwelt* ... Stuttgart, J. G. Cotta.

Werner, H. (1841). *Die Symbolik der Sprache mit besondere Berücksichtigung des Somnambulismus*. Stuttgart, J. G. Cotta.

Werner, H. (1847). *Guardian Spirits, a Case of Vision into the Spiritual World*. New York, J. Allen.

Wesermann, H. M. (1822). *Der Magnetismus und die allgemeine Weltsprache*. Creveld, J. H. Funcke.

West, B.H. (1836). Experiments in animal magnetism. *Boston Medical and Surgical Journal*, 14, 349–351.

Wester, W.C. (1975–6). The Phreno-Magnetic Society of Cincinnati – 1842. *American Journal of Clinical Hypnosis*, 18, 277–281.

Wetzler, J.E. (1833). Meine wunderbare Heilung von beispeilloser Hautschwäche und Geneigtheit zu Erkältungen durch eine Somnambule... Augsburg, Kollmann.

White, W.A. (1916). Critical historical review of Reil's Rhapsodieen. *The Journal of Nervous and Mental Disease*, 43, 1–22.

Wideck, J. (1848). *Der Clairvoyant oder Geschichte eines prophetischen somnambülen Knaben in Oelfe bei Striegau*. Schweidnitz, Weigmann.

Wiener, M. (1838). *Selma, die jüdische Seherin...* Berlin, L. Fernbach.

Wienholt, A. (1787). *Beytrag zu den Erfahrungen über den thierischen Magnetismus.* Hamburg, B.G. Hoffmann.

Wienholt, A. (1788a). Beytrag zu den Erfahrungen über animalischen Magnetismus. *Archiv für Magnetismus und Somnambulismus*, Pt. 6, 53–80

Wienholt, A. (1788b). Krankheits Geschichte eines durch Magnetismus von einer schweren Nervenkrankheit geheilten Frauenzimmer. *Archiv für Magnetismus und Somnambulismus*, Pt. 8, 3–35.

Wienholt, A. (1802–6). *Heilkraft des thierischen Magnetismus nach eigenen Beobachtungen.* 3 vols. Lemgo, Meyer.

Wienholt, A. (1845). *Seven Lectures on Somnambulism.* Tr., with preface, notes, introduction and appendix, by J.C. Colquhoun. Edinburgh, A. and C. Black

Williams, J.H.H. (1952). *Doctors Differ: Five Studies in Contrast.* Springfield, Ill., C.C. Thomas.

Wilson, J. (1839). *Trials of Animal Magnetism.* London, Sherwood, Gilbert and Piper.

Winter, G. (1801). *Animal Magnetism...* Bristol, Routh.

Wirth, D.P. (1990). Unorthodox healing: The effect of noncontact therapeutic touch on the healing rate of full thickness dermal wounds. In L.A. Henkel and J. Palmer (eds.), (1990). *Research in Parapsychology 1989.* Metuchen, N.J., The Scarecrow Press.

Wirth, J.U. (1836). *Theorie des Somnambulismus oder des thierischen Magnetismus...* Leipzig, J. Scheible.

Wohleb, J.L. (1939). Franz Anton Mesmer: Biographischer Sachstandsbericht. *Zeitschrift für die Geschichte des Oberrheins*, 53(1), 33–130.

Wolfart, K.C. (1815). *Erläuterungen zum Mesmerismus.* Berlin, Nikolai.

Wolfart, K.C. (1816). *Der Magnetismus gegen die Stieglitz-Hufelandische Schrift über den thierischen Magnetismus in seinem wahren Werth behauptet.* Berlin, Nikolai.

Wolfart, K.C. (1823). Darlegung über meine magnetisch-ärztliche Wirksamkeit und Behandlungsart. Nebst untermischten einzelnen Krankheitsfäller. *Jahrbücher für den Lebens-Magnetismus oder neues Askläpieion*, 5(1), 1–51.

Wonders and Mysteries of Animal Magnetism Displayed... (1791). London, J. Sudbury.

Wood, A.(1851). *What is Mesmerism? An Attempt to Explain its Phenomena on the Admitted Principles of Physiological and Psychical Science.* Edinburgh, Sutherland and Knox.

Younger, D. (1887). *Instructions in Mesmerism, &c. The Magnetic and Botanic Family Physician...* London, E.W. Allen.

Zerffi, G.G. (1871). *Spiritualism and Animal Magnetism...* London, K. Hardwicke.

Zielinski, L. (1968). Hypnotism in Russia. In E.J. Dingwall (ed.), *Abnormal Hypnotic*

Phenomena: A Survey of Nineteenth-Century Cases, vol 3. London, J. and A. Churchill. Pp 1–105.

Zorab, G. (1967). Hypnotism in Belgium and the Netherlands. In E. J. Dingwall (ed.), *Abnormal Hypnotic Phenomena: A Survey of Nineteenth-Century Cases*, vol. 2. London, J. and A. Churchill. Pp. 1–100.

Zugenbühler, J. A. (1809). Nachricht von Mesmers jetzigem Leben und Aufenthalte. *Journal der practischen Arzneykunde und Wundarzneykunst*, 28(4), 122–125.

Zwelling, S. (1982). Spiritualist perspectives on antebellum experiences. *The Journal of Psychohistory*, 10, 3–25.

References: Hypnotism

Ach, N. (1900). Ueber geistige Leistungsfähigkeit im Zustande des eingeengten Bewusstseins. *Zeitschrift für Hypnotismus*, 9, 1–4.

Ach, N. (1905). *Ueber die Willenstätigheit und das Denken*. Göttingen, Vandenhoeck und Ruprecht.

Ader, R., Felten, D., and Cohen, N. (1990). Interactions between the brain and the immune system. *Annual Review of Pharmacology and Toxicology*, 30, 561–602.

Albini, T. K., and Pease T. E. (1989). Normal and pathological dissociations of early childhood. *Dissociation*, 2(3), 144–150.

Alrutz, S. (1914). Die suggestive Vesikation. *Journal für Psychologie und Neurologie*, 21, 1–10.

Anastay, E. (1908–9). L'origine biologique du sommeil et de l'hypnose. *Archives de Psychologie*, 8, 63–76.

Arkin, A. M. (1981). *Sleep-Talking: Psychology and Psychophysiology*. Hillsdale, N. J., L. Erlbaum

Artigalas, C., and Rémond, A. (1891–2). Note sur un cas d'hémorrhagies auriculaires, oculaires et pulmonaires, provoquées par suggestion. *Revue de l'hypnotisme*, 6, 250–254.

Ash, E. (1906). *Hypnotism and Suggestion. A Practical Handbook*. London, J. Jacobs.

Auvard, – ., and Secheyron, – . (1888). *L'hypnotisme et la suggestion en obstétrique*. Paris, A. Delahaye and E. Lecrosnier.

Azam, E. (1860). Note sur le sommeil nerveux ou hypnotisme. *Archives générales de médecine*, 5 sér., 15, 5–24.

Azam, E. (1887). *Hypnotisme, double conscience et altérations de la personnalité*. Paris, J. B. Baillière et fils.

Azam, E. (1892). Double consciousness. In D. Hack Tuke (ed.), *A Dictionary of Psychological Medicine*, vol 1. London, J. and A. Churchill. Pp. 401–406.

Babinski, J. (1886). Recherches servant à établir que certaines manifestations hystériques peuvent être transférées d'un sujet à un autre sans l'influence de l'aimant. *Revue philosophique*, 22, 697–700.

Babinski, J. (1889). Grand et petit hypnotisme. *Archives de Neurologie*, 17, 92–108, 253–269.

Babinski, J. (1890). La suggestion dans l'hypnotisme. In *Congrès international de psychologie physiologique: première session. Paris 1890*. Paris, Bureau des revues.

Babinski, J. (1891). *Hypnotisme et hystérie: du rôle de l'hypnotisme en thérapeutique*. Paris, G. Masson.

Backman, A. (1891–2). Experiments in clairvoyance. *Proceedings of the Society for Psychical Research*, 7, 199–220.

Baierlacher, E. (1889). *Die Suggestions-Therapie und ihre Technik*. Stuttgart, F. Enke.

Baldwin, J. M. (1892). Hypnotism with Dr. Bernheim at Nancy. *Nation*, 55, 101–103.

Baldwin, J. M. (n.d.). *The Mind.* London, Hodder and Stoughton.

Ballet, G. (1891–2). Observations présentées à propos de la communication faite à la Société par M. A. Voisin. *Revue de l'hypnotisme*, 6, 326–329.

Banyai, E. I., and Hilgard, E. R. (1976). A comparison of active-alert hypnotic induction with traditional relaxation induction. *Journal of Abnormal Psychology*, 85, 218–224.

Barber, J. *et al.* (1979). The relation between nitrous oxide conscious sedation and the hypnotic state. *Journal of the American Dental Association*, 99, 624–626.

Barber, T. X. (1969). *Hypnosis: A Scientific Approach.* New York, Van Nostrand Reinhold.

Barber, T. X. (1970). *LSD, Marihuana, Yoga, and Hypnosis.* Chicago, Aldine Publishing Co.

Barber, T. X. (1973). Experimental hypnosis. In B. B. Wolman (ed.), *Handbook of General Psychology.* Englewood Cliffs, N. J., Prentice-Hall. Pp. 942–963.

Barber, T. X. (1984). Changing "unchangeable" bodily processes by (hypnotic) suggestion: A new look at hypnosis, cognitions, imagining, and the mind-body problem. *Advances*, 1(2), 6–40.

Barber, T. X., and Calverley, D. S. (1964). An experimental study of "hypnotic" (auditory and visual) hallucinations. *Journal of Abnormal and Social Psychology*, 68, 13–20.

Barber, T. X., Spanos, N. P., and Chaves, J. F. (1974). *Hypnosis, Imagination, and Human Potentialities.* New York, Pergamon Press.

Barber, T. X., and Wilson, S. C. (1977). Hypnosis, suggestions, and altered states of consciousness: Experimental evaluation of the new cognitive behavioral theory and the traditional trance state theory of "hypnosis". *Annals of the New York Academy of Sciences*, 296, 34–47.

Barber, T. X., and Wilson, S. C. (1978–9). The Barber Suggestibility Scale and the Creative Imagination Scale: Experimental and clinical applications. *American Journal of Clinical Hypnosis*, 21, 84–108.

Baréty, A. (1887). *Le magnétisme animal étudié sous le nom de force neurique rayonnante et circulante* ... Paris, O. Doin.

Barker, W., and Burgwin, S. (1949). Brain wave patterns during hypnosis, hypnotic sleep and normal sleep. *Archives of Neurology and Psychiatry*, 62, 412–420.

Barkworth, T. (1889–90). Duplex personality. An essay on the analogy between hypnotic phenomena and certain experiences of the normal consciousness. *Proceedings of the Society for Psychical Research*, 6, 84–97.

Barrett, W. F. *et al.* (1882–3). Second report of the committee on mesmerism. *Proceedings of the Society for Psychical Research*, 1, 251–252.

Barrucand, D. (1967). *Histoire de l'hypnose en France.* Paris, Presses Universitaires de France.

Barrucand, D. (1978). Bernheim. Importance historique et actualité de l'école de Nancy. In 103e *Congrès national des sociétés savantes, Nancy, 1978.* Sciences, fasc. V, 75–86.

Baruk, H. (1978). L'école de Nancy et la découverte de la psychothérapie. In 103e *Congrès national des sociétés savantes, Nancy, 1978.* Sciences, fasc. V, 71–74.

Barwise, S. (1888). *Mesmerism (Hypnotism): Its Possibilities, its Uses and Abuses.* Birmingham, Cornish Brothers.

Barzun, J. (1983). *A Stroll with William James.* New York, Harper and Row.

Bates, B. C. *et al.* (1988). Modifying hypnotic suggestibility with the Carleton Skills Training Program. *Journal of Personality and Social Psychology*, 5, 120–127.

Bauer, E. (1991). Periods of historical development of parapsychology in Germany – an overview. *The Parapsychological Association 34th Annual Convention, August 8–11 1991, Heidelberg. Proceedings of Presented Papers.* Pp. 18–34.

Bäumler, C. (1881). *Der sogenannte animalische Magnetismus oder Hypnotismus ...* Leipzig, F. C. W. Vogel.

Beahrs, J. O. (1983–4). Co-consciousnss: A common denominator in hypnosis, multiple personality, and normalcy. *American Journal of Clinical Hypnosis,* 26, 100–113.

Beard, G. M. (1881). *Nature and Phenomena of Trance ("Hypnotism," or "Somnambulism").* New York, G. P. Putnam's sons.

Beaunis, H. (1886). *Le somnambulisme provoqué: études physiologiques et psychologiques.* Paris, J. B. Baillière.

Bechterew, W. von [i.e. Bekhterev, V. M.] (1894). Ueber die Wechselbeziehung zwischen der gewöhnlichen und sensoriellen Anästhesie ... *Neurologisches Zentralblatt,* 13, 252, 297.

Bechterew, W. von (1895). Die Hypnose und ihre Bedeutung als Heilmittel. *Therapeutische Wochenschrift,* 11, 21–27, 42–46, 63–69.

Bechterew, W. von (1899). *Suggestion und ihre sociale Bedeutung.* Leipzig, A. Georgi.

Bechterew, W. von (1904). Was ist suggestion? *Journal für Psychologie und Neurologie,* 3, 100–111.

Bechterew, W. von (1905). *Die Bedeutung der Suggestion im sozialen Leben.* Wiesbaden, J. F. Bergmann.

Bechterew, W. von (1905–6). Des signes objectifs de la suggestion pendant le sommeil hypnotique. *Archives de psychologie,* 5, 103–107.

Bechterew, W. von (1906–7). What is hypnosis? *The Journal of Abnormal Psychology,* 1, 18–25.

Belfiore, G. (1888) *L'ipnotismo e gli stati affini.* Naples, L. Pierro.

Belfiore, G. (1922). *Magnetismo e ipnotismo.* 6th edn. Milan, W. Hoepli.

Bell, C. (1889). Hypnotism. *The Medico-Legal Journal,* 7, 363–371.

Benedikt, M. (1880). Ueber Katalepsie und Mesmerismus. *Wiener Klinik: Vorträge aus der gesammten Heilkunde,* 6, 73–92.

Benedikt, M. (1894). *Hypnotismus und Suggestion. Eine klinisch-psychologische Studie.* Leipzig, M. Breitenstein.

Benn, A. W. (1962). *The History of English Rationalism in the Nineteenth Century.* 2 vols. New York, Russell and Russell.

Bennett, J. H. (1851). *The Mesmeric Mania of 1851, with a Physiological Explanation of the Phenomena Produced.* Edinburgh, Sutherland and Knox.

Benson, D. F., Miller, B. L., and Signer, S. F. (1986). Dual personality associated with epilepsy. *Archives of Neurology,* 43, 471–474.

Benson, H., et al. (1982) Body temperature changes during the practice of g Tum-mo yoga. *Nature,* 295, 234–236.

Bentivegni, A. von (1890). *Die Hypnose und ihre civilrechtliche Bedeutung.* Leipzig, E. Gunther.

Berger, O. (1880a). Experimentelle Katalepsie (Hypnotismus). Neue Beiträge. *Deutsche medicinischer Wochenschrift,* 6, 116–118.

Berger, O. (1880b). Hypnotische Zustände und ihre Genese. *Breslauer aerztliche Zeitschrift,* 2, 109–111, 121–123, 133–138.

Bergmann, – . (1894–5). Ist die Hypnose eine physiologischer Zustand? *Zeitschrift für Hypnotismus*, 3, 169–176.

Bérillon, E. (1884). *Hypnotisme expérimental: la dualité cérébrale et l'indépendance fonctionnelle des deux hémisphères cérébraux*. Paris, A. Delahaye et E. Lecrosnier.

Bérillon, E. (1887–8). De la suggestion et de ses applications à la pédagogie. *Revue de l'hypnotisme*, 2, 169–180.

Bérillon, E. (1889). Les applications de la suggestion à la pédiatrie et à l'éducation mentale des enfants vicieux ou dégénérés. In E. Bérillon (ed.), *Premier congrès internationale de l'hypnotisme expérimentale et thérapeutique*... Paris, O. Doin. Pp. 157–181.

Bérillon, E. (ed.) (1889). *Premier congrès international de l'hypnotisme expérimentale et thérapeutique. Tenu à l'Hôtel-Dieu de Paris, du 8 au 12 août 1889*... Paris, O. Doin.

Bérillon, E. (1890–1). Les indications formelles de la suggestion hypnotique en psychiatrie et en neuropathologie. *Revue de l'hypnotisme*, 5, 97–111.

Bérillon, E. (1891). *Hypnotisme et suggestion: théorie et applications pratiques*... Paris, Société d'éditions scientifiques.

Bérillon, E. (1892). Les applications de la suggestion hypnotique à l'éducation. In *International Congress of Experimental Psychology. Second Session. London 1892*. London, Williams and Norgate. Pp. 166–168.

Bérillon, E. (1892–3). Le traitement psychothérapeutique de la morphinomanie. *Revue de l'hypnotisme*, 7, 129–137.

Bérillon, E. (1893–4). Le traitement psychique de l'incontinence nocturne d'urine. *Revue de l'hypnotisme*, 8, 359–365.

Bérillon, E. (1897). Les principes de la pédagogie suggestive et préventive. In *Dritter internationaler Congress für Psychologie in München von 4. bis 7. August 1896*. Munich, J. F. Lehmann. Pp. 474–475.

Bérillon, E. (1897–8). Les principes de la pédagogie suggestive. *Revue de l'hypnotisme*, 12, 161–167.

Bérillon, E. (1898). *L'hypnotisme et l'orthopédie mentale*. Paris, Rueff.

Bérillon, E. (1898–9). Mal de mer et vertiges de la locomotion. *Revue de l'hypnotisme*, 13, 244–247.

Bérillon, E. (1902a). Les applications de l'hypnotisme à la pédagogie et à l'orthopédie mentale. In E. Bérillon and P. Farez (eds.), *Deuxième congrès international de l'hypnotisme expérimental et thérapeutique*... Paris, Revue d l'hypnotisme. Pp. 190–199.

Bérillon, E. (1902b). L'hypnotisme et la méthode graphique. In E. Bérillon and P. Farez (eds.), *Deuxième congrès international de l'hypnotisme expérimental et thérapeutique*... Paris, Revue de l'hypnotisme. Pp. 262–268.

Bérillon, E. (1905). La méthode hypno-pédagogique. (La suggestion hypnotique envisagée comme procédé de rééducation.) In *Atti del V congresso internazionale di psicologia tenuto in Roma del 26 al 30 aprile 1905*... Rome, Forzani. Pp. 659–661.

Bérillon, E., and Farez, P. (eds.) (1902). *Deuxième congrès international de l'hypnotisme expérimental et thérapeutique tenu à Paris du 12 au 18 août 1902 [1900]. Comptes rendus*... Paris, Revue de l'hypnotisme.

Berjon, A. (1886). *La grande hystérie chez l'homme*... Paris, J. B. Baillière et fils.

Bernheim, H. (1883a). De la suggestion à l'état de veille. *Journal de thérapeutique*, 10, 641–646.

Bernheim, H. (1883b). De la suggestion dans l'état hypnotique et dans l'état de veille. *Revue médicale de l'est*, 15, 513, 545, 577, 610, 641.

Bernheim, H. (1884a). *De la suggestion dans l'état hypnotique et dans l'état de veille*. Paris, O. Doin.

Bernheim, H. (1884b). *De la suggestion dans l'état hypnotique. Réponse à M. Paul Janet*. Paris, O. Doin.

Bernheim, H. (1885). L'hypnotisme chez les hystériques. *Revue philosophique*, 19, 311–316.

Bernheim, H. (1886). *De la suggestion, et de ses applications à la thérapeutique*. Paris, O. Doin.

Bernheim, H. (1886–7). De la suggestion envisagée au point de vue pédagogique [continued as l'hypnotisme et la pédagogie]. *Revue de l'hypnotisme*, 1, 129–139, 332–338, 354–370.

Bernheim, H. (1887). De la suggestion et de ses applications thérapeutiques. *Revue philosophique*, 23, 93–98.

Bernheim, H. (1887–8). De l'action médicamenteuse à distance. *Revue de l'hypnotisme*, 2, 161–165.

Bernheim, H. (1888). *Die Suggestion und ihre Heilwirkung*. Leipzig, Deuticke.

Bernheim (1889a). Les hallucinations rétroactives suggerées dans le sommeil naturel ou artificiel. In E. Bérillon (ed.), *Premier congrès international de l'hypnotisme expérimental et thérapeutique* ... Paris, O. Doin. Pp. 291–294.

Bernheim, H. (1889b). *Suggestive Therapeutics: A Treatise on the Nature and Uses of Hypnotism*. New York, G. P. Putnam's sons.

Bernheim, H. (1889c). Valeur relative des divers procédés destinés à provoquer l'hypnose et à augmenter la suggestibilité au point de vue thérapeutique. In E. Bérillon (ed.), *Premier congrès international de l'hypnotisme expérimental et thérapeutique* ... Paris, O. Doin. Pp. 79–111.

Bernheim, H. (1891). *Hypnotisme, suggestion, psychothérapie: études nouvelles*. Paris, O. Doin.

Bernheim H. (1891–2). Définition et conception des mots suggestion et hypnotisme. *Revue de l'hypnotisme*, 6, 86–99, 109–114.

Bernheim, H. (1892a). *Neue Studien ueber Hypnotismus, Suggestion und Psychotherapie*. Leipzig, F. Deuticke.

Bernheim, H. (1892b). Suggestion and hypnotism. In D. Hack Tuke (ed.), *A Dictionary of Psychological Medicine*, vol. 2. London, J. and A. Churchill. Pp. 1213–1217.

Bernheim, H. (1892–3). Hypnotismus und Suggestion. *Zeitschrift für Hypnotismus*, 1, 115–122.

Bernheim, H. (1895). Conférences cliniques sur la suggestion thérapeutique. *Revue medicale de l'est*, 27, 417–433.

Bernheim, H. (1897). *L'hypnotisme et la suggestion*. Paris, O. Doin.

Bernheim, H. (1897–8a). À propos de l'étude sur James Braid par le Dr. Milne Bramwell et de son rapport lu au Congrès de Bruxelles. *Revue de l'hypnotisme*, 11, 137–145.

Bernheim, H. (1897–8b). Suggestion et hypnotisme. *Revue de psychologie clinique et thérapeutique*, 1, 40–44.

Bernheim, H. (1911). Définition et valeur thérapeutique de l'hypnotisme. *Journal für Psychologie und Neurologie*, 18, 468–477.

Bernheim, H. (1913). *L'hystérie*. Paris, O. Doin.

Bernheim, H. (1917). *Automatisme et suggestion.* Paris, F. Alcan.

Bernheim, H. (1980). *New Studies in Hypnotism.* New York, International Universities Press.

Bertrand, A. J. F. See References: Animal magnetism.

Bertrand, L. D. (1989). The assessment and modification of hypnotic susceptibility. In N. P. Spanos and J. F. Chaves (eds.), *Hypnosis: The Cognitive-Behavioral Perspective.* Buffalo, N. Y., Prometheus Books. Pp. 18–31.

Bick, C. H. (1989). EEG mapping including patients with normal and altered states of hypnotic consciousness under the parameter of posthypnosis. *International Journal of Neuroscience, 47,* 15–30.

Bickford-Smith, R. A. H. (1889–90). Experiments with Madame B. in September, 1889. *Journal of the Society for Psychical Research, 4,* 186–188.

Billot, G. P. See References: Animal magnetism.

Binet, A. (1884a). Hallucinations. *Revue philosophique, 17,* 377–412, 473–503.

Binet, A. (1884b). Visual hallucinations in hypnotism. *Mind, 9,* 413–415.

Binet, A. (1889). Recherches sur les altérations de la conscience chez les hystériques. *Revue philosophique, 27,* 135–170.

Binet, A. (1892). *Les altérations de la personnalité.* Paris, Germer Baillière.

Binet, A. (1896). *Alterations of Personality.* London, Chapman and Hall.

Binet, A. (1900). *La suggestibilité.* Paris, Schleicher frères.

Binet, A., and Féré, C. (1885a). L'hypnotisme chez les hystériques. *Revue philosophique, 19,* 1–25.

Binet, A., and Féré, C. (1885b). La polarisation psychique. *Revue philosophique, 19,* 369–402.

Binet, A., and Féré, C. (1885c). La théorie physiologique des hallucinations. *Revue scientifique, 35,* 49–53. .

Binet, A., and Féré, C. (1887a). *Animal Magnetism.* London, Kegan Paul, Trench and Co.

Binet, A., and Féré, C. (1887b). *Le magnétisme animal.* Paris, Alcan.

Binswanger, R. (1892a). Ueber die Erfolge der Suggestiv-Therapie. *Verhandlungen des Congress für innere Medizin, Wiesbaden 1892, 11,* 299–312.

Binswanger, R. (1892b). Ueber die therapeutische Verwerthung der Hypnose in Irrenanstalten. *Therapeutische Monatshefte, 6,* 105, 163.

Bjelfvenstam, E. (1967). Hypnotism in Scandinavia. In E. J. Dingwall (ed.), *Abnormal Hypnotic Phenomena: A Survey of Nineteenth-Century Cases,* Vol. 2. London, J. and A. Churchill. Pp. 201–246.

Björnström, F. (1887). *Hypnotismen: dess utveckling och nuvarande ståndpunkt.* Stockholm, Geber.

Björnström, F. [1889]. *Hypnotism: Its History and Present Development.* New York, The Humboldt Publishing Co.

Bliss, E. L. (1984a). Hysteria and hypnosis. *The Journal of Nervous and Mental Disease, 172,* 203–206.

Bliss, E. L. (1984b). Spontaneous self-hypnosis in multiple personality disorder. *Psychiatric Clinics of North America, 7,* 135–148.

Bliss, E. L. (1986). *Multiple Personality, Allied Disorders, and Hypnosis.* New York, Oxford University Press.

Bliss, E. L. (1988). Commentary: Professional scepticism about multiple personality. *The Journal of Nervous and Mental Disease, 176.* 533–534.

Boekhoudt, W. (1890). *De beteekenis van hypnotisme en suggestie in ons strafrecht en strafprocess.* Leeuwarden, Coöperative Handelsdrukkerij.

Boeteau, J. M. (1892). Automatisme somnambulique avec dédoublement de la personnalité. *Annales médico-psychologiques,* 7 sér, 15, 63–79.

Boirac, E. (1908). *La psychologie inconnue ...* Paris, F. Alcan.

Boissier de Sauvages, F. (1768). *Nosologia methodica sistens morborum classes ...,* vol 2. Amsterdam, Fratrum de Tournes.

Bonamaison, L. (1889–90). Un cas remarquable d'hypnose spontané, grande hystérie, et grand hypnotisme. *Revue de l'hypnotisme,* 4, 234–243.

Bonjean, A. (1890). *L'hypnotisme: ses rapports avec le droit et la thérapeutique, la suggestion mentale.* Paris, Germer Baillière.

Bonjour, – . (1905–7). Critique du livre de M. le professeur Dubois (de Berne), sur "les Psychonévroses". *Revue de l'hypnotisme,* 20, 357–365; 21, 7–17, 50–60.

Bonnassies, – . (1886). La suggestion dans le hachisch. *Revue philosophique,* 21, 673–674.

Bonnet, G. (1905). *Traité pratique de l'hypnotisme ...* Paris, J. Rousset.

Boor, M. and Coons, P. M. (1983). A comprehensive bibliography of literature pertaining to multiple personality. *Psychological Reports,* 53, 295–310.

Boring, E. G. (1950). *A History of Experimental Psychology.* 2nd edn. New York, Appleton-Century-Crofts.

Bottey, F. (1886). *Le "magnétisme animal". Étude critique et expérimentale sur l'hypnotisme ...* 2nd edn. Paris, Nourrit.

Bourdon, – . (1889). Applications diverses de l'hypnotisme à la thérapeutique. In E. Bérillon (ed.), *Premier congrès international de l'hypnotisme expérimental et thérapeutique ...* Paris, O. Doin. Pp. 206–217.

Bourdon, – . (1898–9). La psychothérapie envisagée comme complément de la thérapeutique générale. *Revue de l'hypnotisme,* 13, 146–153.

Bourneville, D. M. (ed.) (1886). *La possession de Jeanne Fery ...* Paris, A. Delahaye et E. Lecrosnier.

Bourneville, D. M., and Regnard, P. (1877–80). *Iconographie photographique de la Salpêtrière.* 3 vols. Paris, Progrès medicale.

Bourru, H., and Burot, P. (1885). Un cas de la multiplicité des états de conscience chez un hystéro-epileptique. *Revue philosophique,* 20, 411–416.

Bourru, H., and Burot, P. (1886). Sur les variations de la personnalité. *Revue philosophique,* 21, 73–74.

Bourru, H., and Burot, P. (1887). *La suggestion mentale et l'action à distance des substances toxiques et médicamenteuses.* Paris, J. B. Baillière.

Bourru, H., and Burot, P. (1888) *Variations de la personnalité.* Paris, J. B. Baillière.

Bowers, K. S. (1967). The effect of demands for honesty on reports of visual and auditory hallucinations. *The International Journal of Clinical and Experimental Hypnosis,* 15, 31–36.

Bowers, K. S., and Kelly, P. (1979). Stress, disease, psychotherapy and hypnosis. *Journal of Abnormal Psychology,* 88, 490–505.

Bowers, K. S., and LeBaron, S. (1986). Hypnosis and hypnotizability: Implications for clinical interventions. *Hospital and Community Psychiatry,* 37, 457–467.

Bowers, K. S., and Meichenbaum, D. (eds.) (1984). *The Unconscious Reconsidered.* New York, Wiley.

Braid, J. (1842). *Satanic Agency and Mesmerism Reviewed*. Manchester, Simms and Dinham.

Braid, J. (1843). *Neurypnology or the Rationale of Nervous Sleep* ... London, J. Churchill.

Braid, J. (1846). *The Power of the Mind over the Body* ... London, J. Churchill.

Braid, J. (1850). *Observations on Trance: Or, Human Hybernation*. London, J. Churchill.

Braid, J. (1851). *Electro-Biological Phenomena Considered Physiologically and Psychologically*. Edinburgh, Sutherland and Knox.

Braid, J. (1852). *Magic, Witchcraft, Animal Magnetism, Hypnotism, and Electro-Biology* ... 3rd edn. London, J. Churchill.

Braid, J. (1853). *Hypnotic Therapeutics, Illustrated by Cases. With an Appendix on Table-Moving and Spirit-Rapping*. N. p., n. pub. [Reprinted from *The Monthly Journal of Medical Science*, July 1853.]

Braid, J. (1855). *The Physiology of Fascination, and the Critics Criticized*. Manchester, Grant.

Braid, J. (1899). *Braid on Hypnotism. Neurypnology or the Rationale of Nervous Sleep* ... ed. A. E. Waite. London, G. Redway.

Bramwell, J. Milne (1896-7a). James Braid; his work and writings. *Proceedings of the Society for Psychical Research*, 12, 127–166.

Bramwell, J. Milne (1896-7b). Personally observed hypnotic phenomena. *Proceedings of the Society for Psychical Research*, 12, 176–203.

Bramwell, J. Milne (1896-7c). What is hypnotism? *Proceedings of the Society for Psychical Research*, 12, 204–258.

Bramwell, J. Milne (1897). On the so-called automatism of the hypnotized subject. In *Dritter internationaler Congress für Psychologie in München von 4. bis 7. August 1896*. Munich, J. F. Lehmann. Pp. 358–361.

Bramwell, J. Milne (1903). *Hypnotism: Its History, Practice and Theory*. London, G. Richards.

Bramwell, J. Milne (1909). *Hypnotism and Treatment by Suggestion*. London, Cassell.

Bramwell, J. Milne (1930). *Hypnotism: Its History, Practice, and Theory*. 3rd edn. London, Rider.

Brandis, J. D. See References: Animal magnetism.

Braude, S. (1988). Mediumship and multiple personality. *Journal of the Society for Psychical Research*, 55, 177–195.

Braude, S. (forthcoming). *First Person Plural: Multiple Personality and the Philosophy of Mind*. London, Routledge.

Brémaud, P. (1884). *Des différentes phases de l'hypnotisme et en particulier de la fascination* ... Paris, L. Cerf.

Breuer, J., and Freud, S. (1893). Ueber den psychischen Mechanismus hysterischen Phänomene (vorläufige Mittheilung). *Neurologisches Centralblatt*, 12, 4–10, 43–47.

Breuer, J., and Freud, S. (1974). *Studies on Hysteria*. Harmondsworth, Penguin.

Briquet, P. (1859). *Traité de l'hystérie clinique et thérapeutique*. Paris, J. B. Baillière.

Broca, P. (1859). Note sur une nouvelle méthode anesthésique. *Comptes rendus hebdomadaires des séances de l'Académie des sciences*, 49, 902–911.

Broca, P. (1869). Sur l'anesthésie chirurgicale provoquée par l'hypnotisme. *Bulletin de la Société de chirurgerie de Paris*, 2 sér, 10, 247–270

Brock, H. (1880). Ueber stoffliche Veränderungen bei der Hypnose. *Deutsche medicinische Wochenschrift*, 6, 598–599.

Brodmann, K. (1897–8). Zur Methodik der hypnotischen Behandlung. *Zeitschrift für Hypnotismus*, 6, 1–10, 194–214; 7, 1–35, 228–284.

Brown, W. (1918). Treatment of cases of shellshock in an advanced neurological centre. *The Lancet*, 96, ii, 197–200.

Brown, W. (1919). War neuroses. *Proceedings of the Royal Society of Medicine*, 12, 53–61.

Brown, W. (1920). Psychopathology and dissociation. *British Medical Journal*, 1920, i, 139–142; 1920, ii, 847–851.

Brügelmann, W. (1889). *Über den Hypnotismus und seine Verwertung in der Praxis*. Berlin, L. Heuser.

Brügelmann, W. (1893–4). Psychotherapie und Asthma. *Zeitschrift für Hypnotismus*, 2, 84–93.

Brügelmann, W. (1896–7). Suggestive Erfahrungen und Beobachtungen. *Zeitschrift für Hypnotismus*, 5, 256–276.

Brullard, J. (1886). *Considérations générales sur l'état hypnotique*. Nancy, privately printed.

Brunnberg, T. (1896). *Menstruationsstörungen und ihrer Behandlung mittels hypnotischer Suggestion*. Berlin, H. Brieger.

Bucknill, J. C., and Tuke, D. Hack (1858). *A Manual of Psychological Medicine ...* London, J. Churchill.

Bunnemann, – . (1913). Ueber die Erklärbarkeit suggestiver Erscheinungen. *Monatsschrift für Psychiatrie und Neurologie*, 24, 349–369.

Burot, P. (1888–9). Manie hystérique avec impulsions et hallucinations guérie par suggestion. *Revue de l'hypnotisme*, 3, 336–339.

Burot, P. (1889–90). Asphyxie locale des extrêmités chez un hystérique. *Revue de l'hypnotisme*, 4, 197–201.

Burq, V. (1882). *Des origines de la métallothérapie ...* Paris, A. Delahaye et E. Lecrosnier.

Burrows, J. F. [1912]. *Secrets of Stage Hypnotism: Stage Electricity: and Bloodless Surgery*. London, The Magician Ltd.

Callen, K. E. (1983–4). Auto-hypnosis in long distance runners. *American Journal of Clinical Hypnosis*, 26, 30–36.

Campili, G. (1886). *Il grande ipnotismo e la suggestione ipnotica nei rapporti col diretto penale e civile*. Turin, Bocca.

Camus, J., and Pagniez, P. (1904). *Isolement et psychothérapie: traitement de l'hystérie et de la neurasthénie, pratique de la rééducation morale et physique*. Paris, F. Alcan.

Camuset, L. (1882). Un cas de dédoublement de la personnalité. Période amnésique d'une année chez un jeune homme. *Annales médico-psychologiques*, 6 sér, 7, 75–86.

Cardeña, E., and Spiegel, D. (1991). Suggestion, absorption, and dissociation: An integrative model of hypnosis. In J. F. Schumaker (ed.) *Human Suggestibility: Advances in Theory, Research, and Application*. New York, Routledge.

Carli, G. (1978). Animal hypnosis and pain. In F. H. Frankel and H. S. Zamansky (eds.), *Hypnosis at its Bicentennial*. New York, Plenum Press. Pp. 68–79.

Carli, G. *et al.* (1884). Physiological characteristics of pressure immobility. Effects of morphine, naloxone and pain. *Behavioral Brain Research*, 12, 55–63.

Carlson, E. T. (1981). The history of multiple personality in the United States: I. The beginnings. *The American Journal of Psychiatry*, 138, 666–668.

[Carpenter, W. B.] (1853). Electro-biology and mesmerism. *The Quarterly Review*, 93, 501–557.

Carpenter, W.B. (1877). *Mesmerism, Spiritualism, &c. Historically & Scientifically Considered*. London, Longmans, Green and Co.

Carpenter, W.B. (1881). *Principles of Mental Physiology ...* 6th edn. London, C. Kegan Paul.

Carré de Montgeron, L.B. (1745–7). *La verité des miracles opérés par l'intercession de M. de Pâris et autres appelans ...* 3 vols. Cologne, Librairies de la compagnie.

Carroy-Thirard, J. (1980). Hypnose et expérimentation. *Bulletin de psychologie*, 34, 42–50.

Chambard, E. (1881). *Du somnambulisme en générale ...* Paris, O. Doin.

Charcot, J.M. (1877). *Lectures on the Diseases of the Nervous System Delivered at La Salpêtrière*. London, The New Sydenham Society.

Charcot, J.M. (1886–90). *Oeuvres complètes de J.-M. Charcot*. 9 vols. Paris, A. Delahaye et E. Lecrosnier.

Charcot, J.M. (1886). *Oeuvres complètes de J.-M Charcot. Leçons sur les maladies du système nerveux*. I. Paris, A. Delahaye et E. Lecrosnier.

Charcot, J.M. (1887). *Oeuvres complètes de J.-M Charcot. Leçons sur les maladies du système nerveux*. III. Paris, A. Delahaye et E. Lecrosnier.

Charcot, J.M. (1889). *Clinical Lectures on Diseases of the Nervous System Delivered at the Infirmary of La Salpêtrière*. III. London, The New Sydenham Society.

Charcot, J.M. (1890a). Hypnotism and crime. *The Forum*, 9, 159–168.

Charcot, J.M. (1890b). *Oeuvres complètes de J.-M Charcot. IX. Hémorragie et ramollissement du cerveau. Métallothérapie et hypnotisme. Électrothérapie*. Paris, A. Delahaye et E. Lecrosnier.

Charcot, J.M. (1893). The faith-cure. *The New Review*, 8, 18–31.

Charcot, J.M. (1897). *La foi qui guérit*. Paris, Progrès medicale.

Charcot, J.M., and Gilles de La Tourette, G. (1892). Hypnotism in the hysterical. In D. Hack Tuke (ed.), *A Dictionary of Psychological Medicine*, vol. 1. London, J. and A. Churchill. Pp. 607–610.

Charcot, J.M., Luys, J., and Dumontpallier, A. (1878). Second rapport fait à la Société de biologie sur la métalloscopie et la métallothérapie du docteur Burq. *Gazette medicale de Paris*, 4 sér, 7, 419–423, 436–440, 450–452.

Charcot, J.M., Luys, J., and Dumontpallier, A. (1879). Rapport fait à la Société de biologie sur la métallothérapie du docteur Burq ... *Comptes rendus de la Société de biologie 1877*, 6 sér, 4, 1–24.

Charcot, J.M., and Marie, P. (1892). Hysteria mainly hystero-epilepsy. In D. Hack Tuke (ed.), *A Dictionary of Psychological Medicine*, vol. 1. London, J. and A. Churchill. Pp. 627–641.

Charcot, J.M., and Richer, P. (1885). *Les démoniaques dans l'art*. Paris, A. Delahaye et E. Lecrosnier.

Charpignon, J. (1860). *Rapports du magnétisme avec la jurisprudence et la médecine légale*. Paris, Germer Baillière.

Chateaubriand, F.R. (1947). *Mémoires d'outre-tombe*, vol. 2. Paris, Flammarion.

Chaves, J.F. (1989). Hypnotic control of clinical pain. In N.P. Spanos and J.F. Chaves (eds.), *Hypnosis: The Cognitive-Behavioral Perspective*. Buffalo, N.Y., Prometheus Books. Pp. 242–272.

Chertok, L. (1966). Centenaire de la publication "Du sommeil et des états analogues" de

A. A. Liébeault: de la suggestion à la métapsychologie. *Évolution psychiatrique*, 31, 869–901.

Chertok, L. (1981). *Sense and Nonsense in Psychotherapy: The Challenge of Hypnosis*. Oxford, Pergamon Press.

Chertok, L. (1984). On the centenary of Charcot: Hysteria, suggestibility and hypnosis. *British Journal of Medical Psychology*, 57, 111–120.

Chertok, L., and Kramarz, P. (1959). Hypnosis, sleep and electroencephalography. *The Journal of Nervous and Mental Disease*, 128, 227–238.

Chertok, L., Michaux, D., and Droin, M. C. (1977). Dynamics of hypnotic analgesia: Some new data. *The Journal of Nervous and Mental Disease*, 164, 88–96.

Chertok, L., and Saussure, R. de (1979). *The Therapeutic Revolution from Mesmer to Freud*. New York, Brunner/Mazel.

Chojecki, A. (1911–12). Contribution à l'étude de la suggestibilité. *Archives de psychologie*, 11, 182–186.

Chojecki, A. (1912). Comparaison de quelques processus psychiques dans l'hypnose et dans la veille. *Archives de psychologie*, 12, 61–67.

Chowrin, A. N. [i.e. Khovrin, A. N.] (1919). *Experimentelle Untersuchungen auf dem Gebiete des räumlichen Hellsehens (der Kryptoskopie und inadaequaten Sinneserregung)*. Munich, E. Reinhardt.

Claparède, E. (1905). *Esquisse d'une théorie biologique du sommeil* ... Geneva, H. Kündig.

Claparède, E. (1911). Interprétation psychologique de l'hypnose. *Journal für Psychologie und Neurologie*, 18, 501–512.

Claparède, E., and Baade, W. (1908–9). Recherches expérimentales sur quelques processus psychiques simples dans un cas d'hypnose. *Archives de psychologie*, 8, 298–394.

Coconnier, M. T. (1897). *L'hypnotisme franc*. Paris, V. Lecoffre.

Coe, W. C. (1978). The credibility of posthypnotic amnesia: A contextualist's view. *The International Journal of Clinical and Experimental Hypnosis*, 26, 218–245.

Coe, W. C., and Sluis, A. S. E. (1989). Increasing contextual pressures to breach posthypnotic amnesia. *The Journal of Personality and Social Psychology*, 57, 885–894.

Coe, W. C., and Yashinski, E. (1985). Volitional experiences associated with breaching posthypnotic amnesia. *The Journal of Personality and Social Psychology*, 48, 716–722.

Cohen, S. I. (1985). Psychosomatic death: Voodoo death in a modern perspective. *Integrative Psychiatry*, 3, 46–51.

Colquhoun, J. C. (1836). *Isis Revelata: An Inquiry into the Origin, Progress and Present State of Animal Magnetism*. 2nd edn. 2 vols. Edinburgh, Maclachlan and Stewart.

Colquhoun, J. C. (1851). *An History of Magic, Witchcraft, and Animal Magnetism*. 2 vols. London, Longman, Brown, Green and Longmans.

Confer, W. N., and Ables, B. S. (1983). *Multiple Personality: Etiology, Diagnosis, and Treatment*. New York, Human Sciences Press.

Congdon, M. H., Hain, J., and Stevenson, I. (1961). A case of multiple personality illustrating the transition from role-playing. *The Journal of Nervous and Mental Disease*, 132, 497–504.

Congrès internationale de psychologie physiologique: première session. Paris 1890 [1889] ... Paris, Bureau des revues.

Coons, P. M. (1988). Psychophysiologic aspects of multiple personality disorder: A review. *Dissociation*, 1(1), 47–53.

Corval, – von (1892–3). Suggestiv-Therapie. *Zeitschrift für Hypnotismus*, 1, 143–147, 164–171, 193–201, 238–244, 255–264.

Cory, C. B. (1888). *Hypnotism or Mesmerism*. Boston, Mass., A. Mudge.

Coste de Lagrave, L. (1888). *Hypnotisme: états intermédiaires entre le sommeil et la veille*. Paris, J. B. Baillière.

Coste de Lagrave, L. (1889). Quelques expériences d'auto-hypnotisme et d'auto-suggestion. In E. Bérillon (ed.), *Premier congrès international de l'hypnotisme expérimentale et thérapeutique* ... Paris, O. Doin. Pp. 299–311.

Coste de Lagrave, L. (1907–8). Éducation de la volonté et de l'intelligence par l'autosuggestion. *Revue de l'hypnotisme*, 22, 214–220, 238–242.

Council, J. R., *et al.* (1988). "Trance" versus "skill" hypnotic inductions: The effects of credibility, expectancy and experimenter modeling. *Journal of Consulting and Clinical Psychology*, 51, 432–440.

Courtier, J. (1905). Sur quelques effets de passes dites magnétiques. In *Atti del V congresso internazionale di psicologia tenuto in Roma de 26 al 30 aprile 1905* ... Rome, Forzani. Pp. 536–540.

Crabtree, A. (1985a). Mesmerism, divided consciousness, and multiple personality. In H. Schott (ed.), *Franz Anton Mesmer und die Geschichte des Mesmerismus*. Wiesbaden, F. Steiner. Pp. 133–143.

Crabtree, A. (1985b). *Multiple Man: Explorations in Possession and Multiple Personality*. London, Holt, Rinehart and Winston.

Crabtree, A. (1988). *Animal Magnetism, Early Hypnotism, and Psychical Research, 1766–1925. An Annotated Bibliography*. White Plains, N. Y., Kraus International.

Crasilneck, H. B., and Hall, J. A. (1970). The use of hypnosis in the rehabilitation of complicated vascular and post-traumatic neurological patients. *The International Journal of Clinical and Experimental Hypnosis*, 18, 145–159.

Crasilneck, H. B., and Hall, J. A. (1975). *Clinical Hypnosis: Principles and Applications*. New York, Grune and Stratton.

Crawford, H. J., *et al.* (1986). Eidetic-like imagery in hypnosis: Rare but there. *American Journal of Psychology*, 99, 527–546.

Crichton, A. (1798). *An Inquiry into the Nature and Origin of Mental Derangement* ... , vol 1. London, T. Cadell and W. Davies.

Crisp, A. H., *et al.* (1990). Sleepwalking, night terrors, and consciousness. *The British Medical Journal*, 300, 360–362.

Crocq, J., fils (1892–3). Sur quelques phénomènes de l'hypnose. II. – Des zones hypnogènes. *Revue de l'hypnotisme*, 7, 353–362.

Crocq, J., fils (1894). *L'hypnotisme et le crime*. Brussels, H. Lamertin.

Crocq, J., fils (1896). *L'hypnotisme scientifique*. Paris, Société d'éditions scientifiques.

Crocq, J., fils (1900). *L'hypnotisme scientifique*. 2nd edn. Paris, Société d'éditions scientifiques.

Crocq, J., fils (1902). Les rapports de l'hystérie avec l'hypnotisme. In E. Bérillon and P. Farez (eds.), *Deuxième congrès international de l'hypnotisme expérimental et thérapeutique* ... Paris, Revue de l'hypnotisme. Pp. 152–168.

Crothers, T. D. (1882). *The Trance State in Inebriety: Its Medico-Legal Relations*. Hartford, Conn., Case, Lockwood and Brainard.

Crothers, T. D. (1886). Trance states in inebriety. *The Journal of Nervous and Mental Disease*, new ser. 11, 565–571.

Cruise, F. R. (1891). *Hypnotism.* Dublin, J. Falconer.

Cuddon, E. (1955). *Hypnosis: Its Meaning and Practice.* London, Bell.

Cullerre, A. (1887). *Magnétisme et hypnotisme* ... Paris, J. B. Baillière et fils.

Cullerre, A. (1902). Note sur le traitement de l'incontinence d'urine par la suggestion. In E. Bérillon and P. Farez (eds.), *Deuxième congrès international de l'hypnotisme expérimental et thérapeutique* ... Paris, Revue de l'hypnotisme. Pp. 200–202.

Cybulski, N. N. (1887). *O hypnotyzmie ze stanowiska fizyologicz.* Cracow, Jagellonian University.

Cybulski, N. N. (1894). *Spirytyzm i hipnotyzm.* Cracow, Czasu.

Czermak, J. N. (1873). Beobachtungen und Versuche über "hypnotische" Zustände bei Thieren. *Archiv für die gesammte Physiologie des Menschen und der Thiere,* 7, 107–121.

Dalgado, D. G. (1906). *Mémoire sur la vie de l'abbé de Faria* ... Paris, H. Jouve.

Dana, C. L. (1894). The study of a case of amnesia or "double consciousness". *The Psychological Review,* 1, 570–580.

Danilewsky, B. (1890). Recherches physiologiques sur l'hypnotisme des animaux. In *Congrès international de psychologie physiologique: première session. Paris 1890* ... Paris, Bureau des revues. Pp. 79–92.

Decker, H. S. (1986). The lure of nonmaterialism in materialist Europe: Investigations of dissociative phenomena, 1880–1918. In J. M. Quen (ed.), *Split Minds/Split Brains: Historical and Current Perspectives.* New York, New York University Press. Pp. 31–62.

Dejérine, J. (1890–91). Hypnotisme et suggestion. *Revue de l'hypnotisme,* 5, 225–231.

Dejérine, J., and Gauckler, E. (1911). *Les manifestations fonctionelles des psychonévroses, leur traitement par la psychothérapie.* Paris, Masson.

De Jong, A. (1889). Valeur thérapeutique de la suggestion dans quelques psychoses. In E. Bérillon (ed.), *Premier congrès international de l'hypnotisme expérimental et thérapeutique* ... Paris, O. Doin. Pp. 196–201.

De Jong, A. (1891–2). Quelques observations sur le valeur médical de la psychothérapie. *Revue de l'hypnotisme,* 6, 78–86.

De Jong, A. (1892–3). Die Suggestibilität bei Melancolie. *Zeitschrift für Hypnotismus,* 1, 178–181.

De Jong, A. (1893–4). Der Hypnotismus und der Widerstand gegen die Suggestion. *Zeitschrift für Hypnotismus,* 2, 269–275.

Delanne, G. (1924). *Documents pour servir à l'étude de la reincarnation.* Paris, Éditions de la B. P. S.

Delatour, – . (1826). Des avantages de l'insensibilité des somnambules dans les traitements et les opérations. *L'Hermès,* 1, 143–146.

Delboeuf, J. (1885). Une hallucination à l'état normal et conscient. *Revue philosophique,* 20, 513–514

Delboeuf, J. (1886a). La mémoire chez les hypnotisés. *Revue philosophique,* 21, 441–472.

Delboeuf, J. (1886b). *Une visite à la Salpêtrière.* Brussels, C. Muquardt.

Delboeuf, J. (1887a). *De l'origine des effets curatifs de l'hypnotisme.* Paris, F. Alcan.

Delboeuf, J. (1887b). De la prétendue veille somnambulique. *Revue philosophique,* 23, 113–142, 262–285.

Delboeuf, J. (1889). *Le magnétisme animal: à propos d'une visite à l'école de Nancy.* Paris, F. Alcan.

Delboeuf, J. (1890a). *L'hypnotisme appliqué aux altérations de l'organe visuel.* Paris, F. Alcan.

Delboeuf, J. (1890b). *Magnétiseurs et médecins*. Paris, Germer Baillière.

Delboeuf, J. (1891–2). Comme quoi il n'y a pas d'hypnotisme. *Revue de l'hypnotisme*, 6, 129–135.

Delboeuf, J. (1892). De l'appréciation du temps par les somnambules. *Proceedings of the Society for Psychical Research*, 8, 414–422.

Delboeuf, J. (1892–3). Einige psychologische Betrachtungen über den Hypnotismus gelegentlich eines durch Suggestion geheilten Falles von Mordmanie. *Zeitschrift für Hypnotismus*, 1, 43–48, 84–90.

Delboeuf, J. (1893–4). Die verbrecherischen Suggestionen. *Zeitschrift für Hypnotismus*, 2, 177–198, 221–240, 247–268.

Delboeuf, J. (1894). L'hypnose et les suggestions criminelles. *Bulletin de l'Académie royale de Belgique*, 3 sér, 28, 521–553.

Delboeuf, J. (1894–5). L'hypnose et les suggestions criminelles. *Revue de l'hypnotisme*, 9, 225–240, 260–266.

Delboeuf, J. (1897). Les suggestions criminelles. In *Dritter internationaler Congress für Psychologie in München von 4. bis 7. August 1896*. Munich, J. F. Lehmann. Pp. 335–337.

Delboeuf, J., and Fraipont, F. (1890–91). Accouchement dans l'hypnotisme. *Revue de l'hypnotisme*, 5, 289–298.

Deleuze, J. P. F. (1813). *Histoire critique du magnétisme animal*. 2 vols. Paris, Mame.

Delius, H. (1896–7). Erfolge der hypnotischen Suggestiv-Behandlung in der Praxis. I. *Zeitschrift für Hypnotismus*, 5, 219–238.

Delius, H. (1898). Erfolge der hypnotischen Suggestiv-Behandlung in der Praxis. II. *Zeitschrift für Hypnotismus*, 7, 36–53.

Delius, H. (1905). Der Einfluss zerebraler Momente auf die Menstruation und die Behandlung von Menstruationsstörungen durch hypnotische Suggestion. *Wiener klinische Rundschau*, 19, 181, 202.

Dell, P. F. (1988). Professional scepticism about multiple personality. *The Journal of Nervous and Mental Disease*, 176, 528–531.

Delmonte, M. M. (1981). Suggestibility and meditation. *Psychological Reports*, 48, 727–237.

Delmonte, M. M. (1984a). Electrocortical activity and related phenomena associated with meditation practice: A literature review. *International Journal of Neuroscience*, 24, 217–231.

Delmonte, M. M. (1984b). Meditation: similarities with hypnoidal states and hypnosis. *International Journal of Psychosomatics*, 31(3), 24–38.

Demarquay, – ., and Giraud-Teulon, – . (1860). *Recherches sur l'hypnotisme ou sommeil nerveux ...* Paris, J. B. Baillière et fils.

DePascalis, V., Silveri, A., and Palumbo, G. (1988). EEG asymmetry during current mental activity and its relationship with hypnotizability. *The International Journal of Clinical and Experimental Hypnosis*, 36, 38–52.

DePascalis, V., *et al.* (1987). Hemisphere activity of 40Hz EEG during recall of emotional events: Differences between high and low hypnotizables. *International Journal of Psychophysiology*, 5, 167–180.

De Piano, F. A., and Salzberg, H. C. (1979). Clinical applications of hypnosis to three psychosomatic disorders. *Psychological Bulletin*, 86, 1223–1235.

Descourtis, G. (1882). *Du fractionnement des opérations cérébrales et en particulier de leur dédoublement dans les psychopathies.* Paris, privately printed.

Desmartis, T. (1860). *De l'hypnotisme.* Bordeaux, Dupuy.

Despine, P. (1880). *Étude scientifique sur le somnambulisme* ... Paris, F. Savy.

Dessoir, M. (1888). *Bibliographie des modernen Hypnotismus.* Berlin, C. Duncker.

Dessoir, M. (1889). *Das Doppel-Ich.* Berlin, K. Siegismund.

Dessoir, M. (1890a). Le double moi. In *Congrès international de psychologie physiologique: première session. Paris 1890* ... Paris, Bureau des revues. Pp. 146–151.

Dessoir, M. (1890b). *Erster Nachtrag zur Bibliographie des modernen Hypnotismus.* Berlin, C. Duncker.

Dessoir, M. (1896). *Das Doppel-Ich.* 2nd edn. London, Williams and Norgate.

Dessoir, M. (1946). *Buch der Erinnerung.* Stuttgart, F. Enke.

Devinsky, D., *et al.* (1989). Dissociative states and epilepsy. *Neurology*, 39, 835–840.

Diagnostic and Statistical Manual of Mental Disorders (1987). 3rd edn revised [DSMIIIR]. Washington, D. C., American Psychiatric Association.

Dichas, A. (1887). *Étude de la mémoire dans ses rapports avec le sommeil hypnotique (spontané ou provoqué).* Paris, O. Doin.

Diehl, B. J., *et al.* (1989). Mean hemispheric blood perfusion during autogenic training and hypnosis. *Psychiatry Research*, 29, 317–318.

Dingwall, E. J. (1967). Hypnotism in France. In E. J. Dingwall (ed.), *Abnormal Hypnotic Phenomena: A Survey of Nineteenth-Century Cases*, Vol. 1. London, J. and A. Churchill. Pp 1–328.

Dingwall, E. J. (ed.) (1967–8). *Abnormal Hypnotic Phenomena: A Survey of Nineteenth-Century Cases.* 4 vols. London, J. and A. Churchill.

Dingwall, E. J. (1968a). Hypnotism in Great Britain. In E. J. Dingwall (ed.), *Abnormal Hypnotic Phenomena: A Survey of Nineteenth-Century Cases*, Vol. 4. London, J. and A. Churchill. Pp. 79–164.

Dingwall, E. J. (1968b). Hypnotism in Spain, Portugal and Latin America. In E. J. Dingwall, (ed.), *Abnormal Hypnotic Phenomena: A Survey of Nineteenth-Century Cases*, Vol. 3. London, J. and A. Churchill. Pp. 191–203.

Dizard. F. (1893). *Étude sur le morphinisme et son traitement par le suppression totale et définitive des narcotiques et des boissons alcooliques.* Geneva, W. Kündig et fils.

Dobrovolsky, M. (1890–91). Huit observations d'accouchement sans douleur sous l'influence de l'hypnotisme. *Revue de l'hypnotisme*, 5, 274–277.

Döllken, A. (1895–6). Beiträge zur Physiologie der Hypnose. *Zeitschrift für Hypnotismus*, 4, 65–101.

Domangue, B., *et al.* (1985). Biochemical correlates of hypnoanalgesia in arthritis pain patients. *Journal of Clinical Psychiatry*, 46, 235–238.

Doswald, C., and Kreibich, C. (1906). Zur Frage der posthypnotischen Hautphänomene. *Monatshefte für praktischen Dermatologie*, 43, 634–640.

Dreher, E. (1889). *Der Hypnotismus, seine Stellung zum Aberglauben und zur Wissenschaft.* Berlin, L. Heuser.

Drewry, W. F. (1896). Duplex personality: Report of a case. *Medico-Legal Journal*, 14, 39–47.

Dubois, P. (1909a). The method of persuasion. In W. B. Parker (ed.), *Psychotherapy* ... New York, Center Publishing Co., Vol. 2(3), pp. 5–13; 2(4), pp. 22–32; 3(1), pp. 33–45; 3(2), pp. 31–43.

Dubois, P. (1909b). *The Psychic Treatment of Nervous Disorders.* 6th edn. New York, Funk and Wagnalls.

Dufay, J. F. C. (1876). La notion de la personnalité. *Revue scientifique*, 11, 69–70.

Dufay, J. F. C., and Azam, E. (1889–90). Observations on clairvoyance, etc. *Proceedings of the Society for Psychical Research*, 6, 407–427.

Dujardin-Beaumetz, – . (1888). Sur l'action des médicaments à distance. *Bulletin général de thérapeutique medicale et chirurgicale*, 114, 241–261.

Dumontpallier, V. A. A. (1885). De l'action vaso-motrice de la suggestion chez les hystériques hypnotisables. *Gazette des hôpitaux*, 58, 619.

Dumontpallier, V. A. A. (1891–2). De l'analgésie hypnotique dans le travail de l'accouchement. *Revue de l'hypnotisme*, 6, 257–261.

Dunand, – . (1860). *Magnétisme. – Somnambulisme. – Hypnotism .... .* Paris, Ledoyen.

Dunbar, H. F. (1954). *Emotions and Bodily Changes: A Survey of Literature on Psychosomatic Interrelationships 1910–1953.* 4th edn. New York, Columbia University Press.

Du Prel, C. (1889a). *Das hypnotische Verbrechen und seine Entdeckung.* Munich, Verlag der Academischen Monatshefte.

Du Prel, C. (1889b). *The Philosophy of Mysticism.* 2 vols. London, G. Redway.

Durand de Gros, J. P. (1893). *Le merveilleux scientifique.* Paris, F. Alcan.

Durand de Gros, J. P. (1901). Pluralité animale et animique chez l'homme. In *IVe congrès internationale de psychologie tenu à Paris du 20 au 26 août 1900 ...* Paris, F. Alcan. Pp. 648–652.

Durand de Gros, J. P. See also Philips, A. J. P.

Durville, H. (1895–6). *Traité expérimentale de magnétisme ...* 2 vols. Paris., Librairie du magnétisme.

Edmonston, W. E. (1986). *The Induction of Hypnosis.* New York, Wiley.

Edmonston, W. E., and Moscovitch, H. C. (1990). Hypnosis and lateralized brain function. *The International Journal of Clinical and Experimental Hypnosis*, 38, 70–84.

Eeden, F. van (1892). The theory of psycho-therapeutics. In *International Congress of Experimental Psychology. Second Session. London, 1892.* London, Williams and Norgate. Pp. 150–154.

Eeden, F. van (1892–3). Die Grundzüge der Suggestionstherapie. *Zeitschrift für Hypnotismus*, 1, 53–65, 91–101.

Eliseo, T. S. (1974). Three examples of hypnosis in the treatment of organic brain syndrome with psychosis. *The International Journal of Clinical and Experimental Hypnosis*, 22, 9–19.

Ellenberger, H. F. (1970). *The Discovery of the Unconscious: The History and Evolution of Dynamic Psychiatry.* New York, Basic Books.

Empson, J. (1986). *Human Brainwaves: The Psychological Significance of the Electroencephalogram.* Basingstoke, Macmillan.

Engstrom, D. R. (1976). Hypnotic susceptibility, EEG-alpha, and self-regulation. In G. E. Schwartz and D. Shapiro (eds.), *Consciousness and Self-Regulation: Advances in Research.* I. London, Wiley.

Ennemoser, J. (1842). *Der Magnetismus in Verhältnisse zur Natur und Religion.* Stuttgart, J. G. Cotta.

Erasmus, D. (1703). *De laude medicinae.* In *Opera omnia*, vol. 1. Leyden, P. Vander.

Erickson, M. H. (1980a). *The Collected Papers of Milton H. Erickson on Hypnosis.* 4 vols. New York, Irvington.

Erickson, M. H. (1980b). A special inquiry with Aldous Huxley into the nature and character of various states of consciousness. In E. L. Rossi (ed.), *The Nature of Hypnosis and Suggestion: The Collected Papers of Milton H. Erickson on Hypnosis*. Vol. I. New York, Irvington. Pp. 83–107.

Evans, F. J. (1967). Suggestibility in the normal waking state. *Psychological Bulletin*, 67, 114–129.

Evans, F. J. (1977). Hypnosis and sleep: The control of altered states of awareness. *Annals of the New York Academy of Sciences*, 296, 162–174.

Evans, F. J. (1979). Hypnosis and sleep. In E. Fromm and R. E. Shor (eds.), *Hypnosis: Developments in Research and New Perspectives*. New York, Aldine. Pp. 139–183.

Evans, F. J. (1982). Hypnosis and sleep. *Research Communications in Psychology, Psychiatry and Behavior*, 7, 241–256.

Evans, F. J. (1988). Posthypnotic amnesia: Dissociation of content and context. In H. M. Pettinati (ed.), *Hypnosis and Memory*. New York, The Guilford Press. Pp. 157–192.

Evans, F. J. (1989). The independence of suggestibility, placebo response, and hypnotizability. In V. A. Gheorghiu *et al.* (eds.), *Suggestion and Suggestibility: Theory and Research*. Berlin Heidelberg, Springer-Verlag. Pp. 145–154.

Ewin, D. M. (1983–4). Emergency room hypnosis for the burned patient. *American Journal of Clinical Hypnosis*, 26, 5–8.

Ewin, D. M. (1986–7). Emergency room hypnosis for the burned patient. *American Journal of Clinical Hypnosis*, 29, 7–12.

Eysenck, H. J., and Furneaux, W. D. (1945). Primary and secondary suggestibility: An experimental and statistical study. *Journal of Experimental Psychology*, 35, 485–503.

Fagan, J., and McMahon, P. P. (1984). Incipient multiple personality in children: Four cases. *The Journal of Nervous and Mental Disease*, 172, 26–36.

Fahnestock, W. B. (1846). Artificial somnambulism in Pennsylvania. *The Boston Medical and Surgical Journal*, 35, 184–203.

Fahnestock, W. B. (1871). *Statuvolism; or, Artificial Somnambulism, hitherto Called Mesmerism ...* 2nd edn Chicago, Religio-Philosophical Publishing House.

Fajardo, F. (1889). *Hypnotismo*. Rio de Janeiro, Laemmert.

Fanselow, M. S., and Sigmundi, R. A. (1986). Species-specific danger signals, endogenous opioid analgesia, and defensive behavior. *Journal of Experimental Psychology: Animal Behavior Processes*, 12, 301–309.

Fanton, – . (1890–1). Un accouchement sans douleur sous l'influence de l'hypnotisation. *Revue de l'hypnotisme*, 5, 150–155.

Farez, P. (1897–8). Rapport sur la candidature de Dr. Vogt (de Berlin). *Revue de l'hypnotisme*, 12, 248–250.

Farez, P. (1898). *De la suggestion pendant le sommeil naturel*. Paris, A. Maloine.

Farez, P. (1901). L'hypnotisme et l'évocation du subconscient. In *IVᵉ congrès international de psychologie tenu à Paris du 20 au 26 août 1900 ...* Paris, F. Alcan. Pp. 670–674.

Farez, P. (1902a). L'hypnotisme comme moyen d'investigation psychologique. In E. Bérillon and P. Farez (eds.) *Deuxième congrès international de l'hypnotisme expérimental et thérapeutique ...* Paris, Revue de l'hypnotisme. Pp. 72–77.

Farez, P. (1902b). Sommeil naturel et suggestion. In E. Bérillon and P. Farez (eds.),

Deuxième congrès international de l'hypnotisme expérimental et thérapeutique... Paris, Revue de l'hypnotisme. Pp. 203–207.

Farez, P. (1908–9). Les troubles trophiques dans l'hystérie, brûlures suggerées. *Revue de L'hypnotisme*, 22, 179–187.

Faria, J.C. de (1876). Diario da viagem que fez o brigadiero José Custodiò de Sá e Faria de cidade de S. Paulo à praca de Nossa Senhoro dos prazeres de Rio Igatomy, 1774–5. *Rivista trimensal do Instituto historico, geographico e ethnographico do Brasil*, 39(1), 218–291.

Faria, J.C. de (1906). *De la cause du sommeil lucide ou étude de la nature de l'homme.* Paris, H. Jouve.

Favre, L. (1905). Hypnotisme et magnétisme animal. In *Atti del V congresso internazionale di psicologia tenuto in Roma del 26 al 30 aprile 1905*... Rome, Forzani. Pp. 552–553.

Felkin, R.W. (1890). *Hypnotism or Psycho-Therapeutics.* Edinburgh, Y.J. Pentland.

Fellows, B.J. (1986). The concept of trance. In P.L.N. Naish (ed.), *What is Hypnosis? Current Theories and Research.* Milton Keynes, Open University Press. Pp. 37–58.

Fenwick, P.B., *et al.* (1977). Metabolic and EEG changes during transcendental meditation: An exploration. *Biological Psychiatry*, 5, 101–118.

Féré, C. (1883). Les hypnotiques hystériques considérés comme sujets d'expérience en médecine mentale. *Archives de Neurologie*, 6, 122–135.

Ferenczi, S. (1916). The role of transference in hypnosis and suggestion. In S. Ferenczi, *Contributions to Psycho-Analysis.* Boston, R.G. Badger.

Figuier, L. (1881). *Histoire du merveilleux dans les temps modernes.* Vol. 3. *Le magnétisme animal.* 3rd edn. Paris, L. Hachette.

Filiâtre, J. [1906]. *Hypnotisme et magnétisme*... Bourbon-L'Archambault, Genest.

Fingarette, H. (1969). *Self-Deception.* London, Routledge and Kegan Paul.

The first recorded death in hypnosis (1894). *The Boston Medical and Surgical Journal*, 131, 474–475.

Flournoy, T. (1900). *Des Indes à la planète Mars: étude sur un cas de somnambulisme avec glossolalie.* Paris, F. Alcan.

Flournoy, T. (1963). *From India to the Planet Mars*... New York, University Books.

Flower, S. (1898). *Education during Sleep.* Chicago, C.H. Kerr.

Fontan, J. (1887). *La suggestion hypnotique appliquée aux maladies des yeux.* Paris, F. Alcan.

Fontan, J. (1889). Les effets de la suggestion hypnotique dans les affections *cum materiâ* du système nerveux. In E. Bérillon (ed.), *Premier congrès international de l'hypnotisme expérimental et thérapeutique*... Paris, O. Doin. Pp. 112–121.

Fontan, J., and Ségard, C. (1887). *Éléments de médecine suggestive*... Paris, O. Doin

Fontan, J., and Ségard, C. (1888–9). Effets de la suggestion dans un cas de sclérose en plaques. *Revue de l'hypnotisme*, 3, 230–240.

Forel, A. (1887). Einige therapeutische Versuche mit dem Hypnotismus (Braidismus) bei Geisteskranken. *Correspondenz-Blatt für Schweizer Aerzte*, 17, 481–488.

Forel, A. (1888–9). Un cas d'autohypnotisation. *Revue de l'hypnotisme*, 3, 277–278.

Forel, A. (1889a). *Der Hypnotismus: seine Bedeutung und seine Handhabung.* Stuttgart, F. Enke.

Forel., A. (1889b). *Zu den Gefahren und dem Nutzen des Hypnotismus.* [Offprint from the *Münchener medicinische Wochenschrift*, 1889, No. 38.]

Forel, A. (1891). Ein Gutachten über einen Fall von spontanem Somnambulismus mit angeblicher Wahrsagerei und Hellseherei. *Schriften der Gesellschaft für psychologische Forschung*, 1(1), 77–80.

Forel, A. (1893–4). Die Heilung der Stuhlverstopfung durch Suggestion. *Zeitschrift für Hypnotismus*, 2, 55–64.

Forel, A. (1895). *Der Hypnotismus, seine psycho-physiologische, medicinische, strafrechtliche Bedeutung, und seine Handhabung*. 3rd edn., with notes by O. Vogt. Stuttgart, F. Enke.

Forel, A. (1895–6). Der Hypnotismus in der Hochschule. *Zeitschrift für Hypnotismus*, 4, 1–8.

Forel, A. (1897). Der Unterschied zwischen der Suggestibilität und der Hypnose. Was ist Hysterie? In *Dritter internationaler Congress für Psychologie in München von 4. bis 7. August 1896* ... Munich, J.F. Lehmann. Pp. 367–374.

Forel, A. (1898). Ueber suggestive Hauterscheinungen. *Zeitschrift für Hypnotismus*, 7, 137–139.

Forel, A. (1902). *Der Hypnotismus und die Suggestive Psychotherapie*. 4th edn. Stuttgart, F. Enke.

Forel, A. (1906). *Hypnotism or Suggestion and Psychotherapy* ... London, Rebman.

Forel, A. (1935). *Rückblick auf mein Leben*. Zürich, Europa-Verlag.

Fortineau, E. (1985). Bernheim face à Charcot et Freud: l'école de Nancy. *L'information psychiatrique*, 61, 415–420.

Foveau de Courmelles, F. V. (1891). *Hypnotism*. London, G. Routledge.

Fraenkel, E. (1889). *Magnetisme og Hypnotisme*. Copenhagen, Gjällerup.

Franco, G. G. (1886). *L'ipnotismo tornato di moda: storia e disquisizione scientifico*. Prato, Giachetti.

Franco, G. G. (1888). *L'hypnotisme revenu à la mode* ... Le Mans, Leguicheux.

Francotte, X. (1896–7). Du somnambulisme alcoölique considéré surtout au point de vue médico-légal. *Revue de l'hypnotisme*, 11, 129–138.

Francotte, X. (1897). [The above also in *Journal de neurologie et d'hypnologie*, 2, 24–31.]

Frankel, F. H. (1974). Trance capacity and the genesis of phobic behavior. *Archives of General Psychiatry*, 31, 261–263.

Frankel, F. H. (1976). *Hypnosis: Trance as a Coping Mechanism*. New York, Plenum Medical Book Co.

Frankel, F. H., and Orne, M. T. (1976). Hypnotizability and phobic behavior. *Archives of General Psychiatry*, 33, 1259–1261.

Frankel, F. H., and Zamansky, H. S. (eds.) (1978). *Hypnosis at its Bicentennial*. New York, Plenum Press.

Franzolini, F. (1882). *Del somnambulismo* ... Udine, G. B. Doretti e soni.

Freud, S. (1892–3). Ein Fall von hypnotischer Heilung nebst Bemerkungen über die Entstehung hysterischer Symptome durch den "Gegenwillen." *Zeitschrift für Hypnotismus*, 1, 102–107, 123–129.

Freud, S. (1905). *Drei Abhandlungen zur Sexualtheorie*. Leipzig, F. Deuticke.

Freud, S. (1924). Charcot. In S. Freud, *Collected Papers*, vol. I. London, International Psycho-Analytical Press. Pp. 9–23.

Freud, S. (1954). *The Interpretation of Dreams*. London, Allen and Unwin.

Frid, M., and Singer, G. (1979). Hypnotic analgesia in conditions of stress is partially reversed by naloxone. *Psychopharmacology*, 63, 211–215.

Friedmann, M. (1901). *Über Wahnideen im Völkerleben*. Wiesbaden, J. F. Bergmann.

Fromm, E., and Shor, R. E. (eds.) (1979). *Hypnosis: Developments in Research and New Perspectives*. 2nd edn., New York, Aldine.

Funck, H. (1894). *Magnetismus und Somnambulismus in der Badischen Markgrafschaft*. Freiburg im Breisgau, J. C. B. Mohr.

Gabel, S. (1988). The right hemisphere in imagery, hypnosis, rapid-eye-movement sleep and dreaming. Empirical studies and tentative conclusions. *The Journal of Nervous and Mental Disease*, 176, 323–331.

Galton F. (1883). *Inquiries into Human Faculty and its Development*. London, Macmillan.

Gardner, H. (1977). *The Shattered Mind: The Person after Brain Damage*. London, Routledge and Kegan Paul.

Garrison, F. H. (1929). *An Introduction to the History of Medicine*. Philadelphia, W. B. Saunders.

Gascard, A. (1889). Influence de la suggestion sur certains troubles de la menstruation. In E. Bérillon (ed.). *Premier congrès international de l'hypnotisme expérimental et thérapeutique*... Paris, O. Doin. Pp. 129–137.

Gasc-Desfossés, E. (1897). *Magnétisme vital*... Paris, Société d'éditions scientifiques.

Gauld, A. (1965–6). Mr. Hall and the SPR. *Journal of the Society for Psychical Research*, 43, 53–62.

Gauld, A. (1968). *The Founders of Psychical Research*. London, Routledge and Kegan Paul.

Gauld, A. (1990). The early history of hypnotic skin marking and blistering. *British Journal of Experimental and Clinical Hypnosis*, 7, 139–152.

Gauld, A. (1992). Hypnosis, somnambulism and double consciousness. *Contemporary Hypnosis*.

Gazzaniga, M. S. (1987). Perceptual and attentional processes following callosal section in humans. *Neuropsychologia*, 25, 119–133.

Gheorghiu, V. A., and Kruse, P. (1991). The psychology of suggestion: An integrative perspective. In J. F. Schumaker (ed.), *Human Suggestibility: Advances in Theory, Research, and Application*. New York, Routledge. Pp. 59–75.

Gheorghiu, V. A., Netter, P., Eysenck, H. J., and Rosenthal, R. (eds.)(1989). *Suggestion and Suggestibility: Theory and Research*. Berlin Heidelberg, Springer-Verlag.

Gibbons, D. E. (1974). Hyperempiria: A new "altered state of consciousness" induced by suggestion. *Perceptual and Motor Skills*, 39, 47–53.

Gibbons, D. E. (1979). *Applied Hypnosis and Hyperempiria*. New York, Plenum Press.

Gibson, H. B., and Heap, M. (1990). *Hypnosis in Therapy*. London: L. Erlbaum.

Gilbert, J. A. (1902). A case of multiple personality. *The Medical Record*, 37, 207, 254.

Gilles de La Tourette, G. (1887). *L'hypnotisme et les états analogues au point de vue médico-légal*... Paris, E. Plon, Nourrit et cie.

Gilles de La Tourette, G. (1891–5). *Traité clinique et thérapeutique de l'hystérie d'après l'enseignment de la Salpêtrière*. 3 vols. Paris, Plon et Nourrit.

Glaskin, G. M. (1974). *Windows of the Mind: The Christos Experiment*. London, Wildwood House.

Gley, E. (1895). De quelques conditions favorisant l'hypnotisme chez les grenouilles. *Comptes rendus de la Société de biologie*, 10 sér, 2, 518–521.

Gley, E. (1896). Étude sur quelques conditions favorisants l'hypnose chez les animaux. *L'année psychologique*, 2, 70–78.

Goldberg, B. (1988). Hypnosis and the immune response. *International Journal of Psychosomatics,* 32, 34–36.

Goldstein, J. (1982). The hysteria diagnosis and the politics of anticlericalism in late nineteenth-century France. *The Journal of Modern History,* 54, 209–239.

Goodwin, J. (1987). Mary Reynolds: A post-traumatic reinterpretation of a case of multiple personality disorder. *The Hillside Journal of Clinical Psychiatry,* 9, 89–99.

Gorassini, D. R., and Spanos, N. P. (1986). A social-cognitive skills approach to the successful modification of hypnotic susceptibility. *Journal of Personality and Social Psychology,* 50, 1004–1012.

Gorton, B. E. (1949). The physiology of hypnosis. I. *Psychiatric Quarterly,* 23, 317–343.

Graeter, C. (1899). Ein Fall von epileptischer Amnesie, durch hypnotische Hypermnesie beseitigt. *Zeitschrift für Hypnotismus,* 8, 129–163.

Grandchamps, – . de (1889). Accouchement en état de fascination... *Gazette des hôpitaux civils et militaires,* 62, 857.

Grashey, – ., Hirt, L ., Schrenck-Notzing, A. von, and Preyer, W. (1895). *Der Process Czynski...* Stuttgart, F. Enke.

Grasset, J. (1888–9). Leçons sur le grand et le petit hypnotisme. *Revue de l'hypnotisme,* 3, 321–335, 356–364.

Grasset, J. (1896). De l'automatisme psychologique (psychisme inférieur; polygone corticale) à l'état physiologique et pathologique. *Nouveau Montpellier médicale,* supp. v, 47–162.

Grasset, J. (1902–4). L'hypnotisme et la suggestion. *Revue de l'hypnotisme,* 17, 257–268, 295–303, 329–334, 358–362; 18, 3–10.

Grasset, J. (1903). *L'hypnotisme et la suggestion.* Paris, O. Doin.

Gravitz, M. A. (1985–6). An 1846 report of tumour remission associated with hypnosis. *American Journal of Clinical Hypnosis,* 28, 16–19.

Grossmann, J. (1892–3a). Die Suggestion, speciell die hypnotische Suggestion, ihr Wesen und ihr Heilwerth. *Zeitschrift für Hypnotismus,* 1, 355–377, 398–433.

Grossmann, J. (1892–3b). Suggestion und Milchsecretion, vorläufige Mittheilung. *Zeitschrift für Hypnotismus,* 1, 71–72

Grossmann, J. (ed.) (1894). *Die Bedeutung der hypnotischen Suggestion als Heilmittel...* Berlin, Bong.

Grossmann, J. (1894–5a). Die Erfolge der Suggestionstherapie (Hypnose) bei organischen Lähmungen und Paralysen. *Zeitschrift für Hypnotismus,* 3, 54–64, 76–80.

Grossmann, J. (1894–5b). Zum Fall Czynski. *Zeitschrift für Hypnotismus,* 3, 185–192.

Grossmann, J. (1894–5c). Zur suggestiven Behandlung der Gelenkskrankheiten mit besonderer Berücksichtigung des chronischen Gelenksrheumatismus und der Gicht. *Zeitschrift für Hypnotismus,* 3, 245–259.

Guarnieri, P. (1988). Between soma and psyche: Morselli and psychiatry in late-nineteenth-century Italy. In W. F. Bynum, R. Porter and M. Shepherd (eds.), *The Anatomy of Madness: Essays in the History of Psychiatry. Vol. III. The Asylum and its Psychiatry.* London, Tavistock Publications. Pp. 102–124.

Guérineau, A. (1860). Amputation de cuisse pratiquée sans douleur sous l'influence des manoeuvres hypnotiques. *Gazette médicale de Paris,* 15, 21–23.

Guidi, G. (1908–9). Recherches expérimentales sur la suggestibilité. *Archives de Psychologie,* 8, 49–54.

Guillain, G. (1955). *J.-M. Charcot: sa vie, son oeuvre.* Paris, Masson.

Guillain, G. (1959). *J.-M. Charcot 1825–1893: His Life, his Work*. London, Putnam.

Guinon, G. (1891). Documents pour servir à l'histoire des somnambulismes: du somnambulisme hystérique (phase passionnelle de l'attaque, attaque délirante, attaque de somnambulisme). *Progrès médicale*, 2 sér, 13, 401–404, 425–429, 459–466, 513–517; 14, 41–48, 137–141.

Guinon, G. (1892). Du dédoublement de la personnalité d'origine hystérique (vigil-ambulisme hystérique). *Progrès médicale*, 2 sér, 15, 193–196, 236–238, 361–364, 401–402; 16, 1–4, 17–21, 73–76, 131–134.

Gurney, E. (1884a). An account of some experiments in mesmerism. *Proceedings of the Society for Psychical Research*, 2, 201–206.

Gurney, E. (1884b). The problems of hypnotism. *Proceedings of the Society for Psychical Research*, 2, 265–292.

Gurney, E. (1884c). The stages of hypnotism. *Proceedings of the Society for Psychical Research*, 2, 61–72.

Gurney, E. (1885). Local anaesthesia induced in the normal state by mesmeric passes. *Proceedings of the Society for Psychical Research*, 3, 453–459.

Gurney, E. (1885–7a). Peculiarities of certain post-hypnotic states. *Proceedings of the Society for Psychical Research*, 4, 268–323.

Gurney, E. (1885–7b). The stages of hypnotic memory. *Proceedings of the Society for Psychical Research*, 4, 515–531.

Gurney, E. (1887). Further problems of hypnotism. *Mind*, 12, 212–232, 397–422.

Gurney, E. (1888–9a). Hypnotism and telepathy. *Proceedings of the Society for Psychical Research*, 5, 216–259.

Gurney, E. (1888–9b). Recent experiments in hypnotism. *Proceedings of the Society for Psychical Research*, 5, 3–17.

Gurney, E., Myers, F. W. H., and Podmore, F. (1886). *Phantasms of the Living*. 2 vols. London, Trübner.

Gurney, E., *et al.* (1882–3). First report of the committee on mesmerism. *Proceedings of the Society for Psychical Research*, 1, 217–229.

Hadfield, J. A. (1917). The influence of hypnotic suggestion on inflammatory conditions. *The Lancet*, 95, ii, 678–679.

Haeberlin, – . (1902–3). Action vasomotrice de la suggestion dans le traitement des verrues. *Revue de l'hypnotisme*, 17, 84–90.

Haen, A. de (1760). *Ratio medendi in noscomio practico …*, vol. 4. Vienna, Trattner.

Hale, N. G. (1971). *Freud and the Americans: The Beginnings of Psychoanalysis in the United States, 1876–1917*. New York, Oxford University Press.

Hall, G. S. (1881). Recent researches on hypnotism. *Mind*, 6, 98–104.

Hall, G. S. (1883). Reaction time and attention in the hypnotic state. *Mind*, 8, 170–182.

Hall, H. R. (1982–3). Hypnosis and the immune system: A review with implications for cancer. *American Journal of Clinical Hypnosis*, 25, 92–103.

Hall, H. (1986). Suggestion and illness. *International Journal of Psychosomatics*, 33(2), 24–27.

Hall, H. R. (1989). Research in the area of voluntary immunomodulation: Complexities, consistencies and future research considerations. *International Journal of Neuroscience*, 47, 81–89.

Hall, T. H. (1964). *The Strange Case of Edmund Gurney*. London, Duckworth.

Hammerschlag, H. E. (1956). *Hypnotism and Crime*. London, Rider.

Hammond, W. A. (1887). The medico-legal relations of hypnotism or syggignoscism. *The New York Medical Journal*, 30th July 1887, 115–120.

Harrington, A. (1988). Hysteria, hypnosis, and the lure of the invisible: The rise of neo-mesmerism in fin-de-siècle French psychiatry. In W. F. Bynum, R. Porter and M. Shepherd (eds.), *The Anatomy of Madness: Essays in the History of Psychiatry*. Vol. III. *The Asylum and its Psychiatry*. London, Tavistock Publications. Pp. 226–246.

Harris, R. (1985). Murder under hypnosis in the case of Gabrielle Bompard: Psychiatry in the courtroom in Belle époque Paris. In W. F. Bynum, R. Porter and M. Shepherd (eds.), *The Anatomy of Madness: Essays in the History of Psychiatry*. Vol. II. *Institutions and Society*. London, Tavistock Publications. Pp. 197–241.

Hart, E. (1893). *Hypnotism, Mesmerism and the New Witchcraft*. London, Smith, Elder and Co.

Harte, R. (1902–3). *Hypnotism and the Doctors*. 2 vols. London, L. N. Fowler.

Hartenberg, P. (1897–8a). Un cas de neurasthénie psychique guéri par la dynamogénie suggestive. *Revue de L'hypnotisme*, 12, 42–46.

Hartenberg, P. (1897–8b). Essai d'une psychologie de la suggestion. *Revue de psychologie clinique et thérapeutique*, 1, 264–278.

Hartenberg, P. (1897–8c). Il n'y a pas d'hypnotisme. *Revue de l'hypnotisme*, 12, 211–221.

Hartenberg, P. (1908). *Psychologie des neurasthéniques*. Paris, Alcan.

Hartenberg, P. (1910). *L'hystérie et les hystériques*. Paris, Alcan.

Hartenberg, P. (1912). *Traitement des neurasthéniques*. Paris, Alcan.

Hartmann, E. (1983). Two case reports: night terrors with sleep-walking – a potentially lethal disorder. *The Journal of Nervous and Mental Disease*, 171, 503–505.

Haward, L. R. C. (1972). Barbiturate-induced hypnosis: A golden jubilee review. *Bulletin of the British Psychological Society*, 25, 23–24.

Heap, M. (ed.) (1988). *Hypnosis: Current Clinical, Experimental and Forensic Practises*. London, Croom Helm.

Hébert, H. J. (1861). *Recherches sur l'hypnotisme et ses causes ...* Poissy, Arbieu.

Hécaen, H., and Lanteri-Laura, G. (1977). *Évolution des connaissances et des doctrines sur les localisations cérébrals*. Paris, Desclee de Brouwer.

Hecker, E. (1893). *Hypnose und Suggestion im Dienste der Heilkunde. Ein Vortrag*. Wiesbaden J. F. Bergmann.

Heidenhain, R. (1880). *Der sogenannte thierische Magnetismus. Physiologische Beobachtungen*. Leipzig, Breitkopf und Härtel.

Heidenhain, R. (1888). *Hypnotism or Animal Magnetism. Physiological Observations*. London, Kegan Paul, Trench and Co.

Held, G. (1722). Rarer casus einer Jungfer, welche allerhand actiones im Schlafe vorgenommen, die eine andre fast wachend nicht wird verrichten können. *Sammlung von Natur- und Medicin-, wie auch hierzu gehörigen Kunst- und Literatur- Geschichten ... von einigen Academ. Naturae Curios. in Breslau*. Pp. 192–194.

Heller, F., and Schultz, J. H. (1909). Ueber einen Fall hypnotisch erzeugte Blasenbildung. *Muenchener medizinischer Wochenschrift*, 56, 2112.

Hermas, M. (1938). On recovery of memory in psychogenic amnesia and fugue states. *Psychiatric Quarterly*, 12, 738–742.

Herter, C. A. (1888a). Hypnotism: What it is, and what it is not. *Popular Science Monthly*, 33, 755–771.

Herter, C. A. (1888b). The therapeutic uses of hypnotism. *Medical News*, 53, 456–458.

Hicks, W. H. (1910). The present status of hypnotism. *Journal of the Medical Society of New Jersey*, 6, 483–493.

Hilgard, E. R. (1962). Impulsive versus realistic thinking: An examination of the distinction between primary and secondary processes in thought. *Psychological Bulletin*, 59, 477–488.

Hilgard, E. R. (1965). *Hypnotic Susceptibility*. New York, Harcourt, Brace and World.

Hilgard, E. R. (1971). Hypnotic phenomena: The struggle for scientific acceptance. *American Scientist*, 59, 567–577.

Hilgard, E. R. (1973a). The domain of hypnosis with some comments on alternative paradigms. *American Psychologist*, 28, 972–982.

Hilgard, E. R. (1973b). A neodissociation interpretation of pain reduction in hypnosis. *Psychological Review*, 80, 396–411.

Hilgard, E. R. (1976). Neodissociation theory of multiple cognitive control systems. In G. E. Schwartz and D. Shapiro (eds.), *Consciousness and Self-Regulation: Advances in Research*. Vol. 1. New York, Plenum Press. Pp. 137–171.

Hilgard, E. R. (1977). *Divided Consciousness: Multiple Controls in Human Thought and Action*. New York, Wiley.

Hilgard, E. R. (1978). States of consciousness in hypnosis: Divisions or levels? In F. H. Frankel and H. S. Zamansky (eds.), *Hypnosis at its Bicentennial*. New York, Plenum Press. Pp. 15–36.

Hilgard, E. R. (1979). Divided consciousness in hypnosis: The implications of the hidden observer. In E. Fromm and R. E. Shor (eds.), *Hypnosis: Developments in Research and New Perspectives*. New York, Aldine. Pp. 45–79.

Hilgard, E. R. (1987). Research and advances in hypnosis: Issues and methods. *The International Journal of Clinical and Experimental Hypnosis*, 35, 248–264.

Hilgard, E. R., and Hilgard, J. R. (1983). *Hypnosis in the Relief of Pain*. Rev. edn. Lost Altos, Calif., W. Kaufmann.

Hilgard, J. R. (1970). *Personality and Hypnosis: A Study of Imaginative Involvement*. Chicago, University of Chicago Press.

Hilger, W. (1899). Zur Kasuistik der hypnotischen Behandlung der Epilepsie. *Zeitschrift für Hypnotismus*, 8, 17–61,

Hilger, W. (1902). Beitrag zur Frage der Hypnotisirbarkeit. *Zeitschrift für Hypnotismus*, 10, 190–201.

Hilger, W. [1912]. *Hypnosis and Suggestion ...* London, Rebman.

Hillairet, J. (1958). *Les 200 cimetières du vieux Paris*. Paris, Les Éditions de minuit.

Hinkle, B. M. (1909). The methods of psychotherapy. In W. B. Parker (ed.) *Psychotherapy ...* 3 vols. New York, Center Publishing Co. Vol. 2, No. 1, pp. 5–14.

Hirsch, M. (1895). Zur Begriffsbestimmung der Hypnose. *Deutsche Medizinal-Zeitung*, 16, 1029.

Hirschlaff, L. (1899). Die angebliche Bedeutung des Hypnotismus für die Pädagogik. *Zeitschrift für pädagogische Psychologie*, 1, 127–132.

Hirschlaff, L. (1899–1900). Kritische Bemerkungen über den gegenwärtigen Stand der Lehre vom Hypnotismus. *Zeitschrift für Hypnotismus*, 8, 257–274, 321–341; 9, 65–97, 202–228.

Hirschlaff, L. (1905). *Hypnotismus und Suggestivtherapie. Ein kurzen Lehrbuch für Ärzte und Studierende nach der I. Auflage des Dr. Max Hirsch ...* Leipzig, J. A. Barth.

Hirt, L. (1890). Ueber die Bedeutung der Suggestionstherapie (vulgo "Hypnotismus") für die ärztliche Praxis. *Wiener medizinische Wochenschrift*, 40, 1137–1142, 1182–1186, 1225–1228, 1273–1276.

Hirt, L. (1893–4). Ueber die Bedeutung der Verbalsuggestion für die Neurotherapie. *Zeitschrift für Hypnotismus*, 2, 287–294.

Hodgson, R. (1891–2). A case of double consciousness. *Proceedings of the Society for Psychical Research*, 7, 221–257.

Hoefelt, I. A. (1889). *Het hypnotisme in verband met het strafrecht*. Leiden, privately printed.

Holland, H. (1852). *Chapters on Mental Physiology*. London, Longman, Brown, Green and Longmans.

Hollander, B. (1910). *Hypnotism and Suggestion in Daily Life, Education and Medical Practice*. New York, G. P. Putnam's sons.

Holroyd, J. (1980). Hypnosis treatment for smoking: An evaluative review. *The International Journal of Clinical and Experimental Hypnosis*, 28, 341–357.

Honorton, C., and Krippner, S. (1969). Hypnosis and ESP performance: A review of the experimental literature. *Journal of the American Society for Psychical Research*, 63, 214–252.

Horowitz, M. J. (1986). *Stress Response Syndromes*. 2nd edn. Northvale, N. J., Jason Aronson.

Horstius, J. (1593). De natura, differentiis et causis eorum qui dormientes ambulant. In J. Horstius, *De aureo dente maxillari puerisilesii* ... Leipzig, M. Lantzenberger.

Houge, D. R., and Hunter, R. E. (1988). The use of hypnosis in orthopaedic surgery. *Contemporary Orthopaedics*, 16, 65–68.

Hövelmann, G. H. (1987). Max Dessoir and the origin of the word "parapsychology." *Journal of the Society for Psychical Research*, 54, 61–63.

Howarth, C. I. (1989). Psychotherapy: Who benefits? *The Psychologist*, 12, 150–152.

Howe, E. (1972). *The Magicians of the Golden Dawn*. London, Routledge and Kegan Paul.

Hückel, A. (1888). *Die Rolle der Suggestion bei gewissen Erscheinungen der Hysterie und des Hypnotismus*. Jena, G. Fischer.

Hull, C. L. (1933). *Hypnosis and Suggestibility: An Experimental Approach*. New York, Appleton-Century.

Humphrey, G. (1963). *Thinking: An Introduction to its Experimental Psychology*. New York, Wiley.

Huss, M. (1888). *Om hypnotismen. De vådor den innebär och kan innebära*. Stockholm, C. E. Fritze.

Ireland, W. W. (1893). *The Blot upon the Brain: Studies in History and Psychology*. 2nd edn. Edinburgh, Bell and Bradfute.

Jaguaribe, D. (1902). L'hypnotisme au Brésil. In E. Bérillon and P. Farez (eds.), *Deuxième congrès international de l'hypnotisme expérimental et thérapeutique* ... Paris, Revue de l'hypnotisme. Pp. 258–259.

James, W. (1885–9). Reaction-time in the hypnotic trance. *Proceedings of the American Society for Psychical Research*, 1, 246–248.

James, W. (1888–90). A record of observations of certain phenomena of trance. Part III. *Proceedings of the Society for Psychical Research*, 6, 651–659.

James, W. (1890). *The Principles of Psychology*. 2 vols. London, Macmillan.

James, W. (1892). *Text-Book of Psychology*. London, Macmillan.

James, W. (1903–4). [Review of] *Human Personality and its Survival of Bodily Death*. By Frederic W. H. Myers... *Proceedings of the Society for Psychical Research*, 18, 22–33.

James, W. (1925). *The Varieties of Religious Experience: A Study in Human Nature*. London, Longmans, Green and Co.

James, W. (1961). *William James on Psychical Research*. Ed. G. Murphy and R. O. Ballou. London, Chatto and Windus.

James, W. (1986). *Essays in Psychical Research*. Cambridge, Mass., Harvard University Press.

Janet, Paul (1884). De la suggestion dans l'état d'hypnotisme. *Revue politique et littéraire*, 3 sér, 8, 100–104, 129–132, 179–185, 198–203.

Janet, Pierre, (1886a). Les actes inconscientes et la dédoublement de la personnalité pendant le somnambulisme provoqué. *Revue philosophique*, 22, 577–592.

Janet, Pierre (1886b). Deuxième note sur le sommeil provoqué à distance et la suggestion mentale pendant l'état somnambule. *Revue philosophique*, 22, 212–223.

Janet, Pierre (1886c). Note sur quelques phénomènes de somnambulisme. *Revue philosophique*, 21, 190–198.

Janet, Pierre (1886d). Les phases intermédiaires de l'hypnotisme. *Revue scientifique*, 23, 577–587.

Janet, Pierre (1887). L'anesthésie systematisée et la dissociation des phénomènes psychologiques. *Revue philosophique*, 23, 449–472.

Janet, Pierre (1888). Les actes inconscients et la mémoire pendant le somnambulisme provoqué. *Revue philosophique*, 25, 238–279.

Janet, Pierre (1889). *L'automatisme psychologique: essai de psychologie expérimentale sur les formes inférieures de l'activité humaine*. Paris, F. Alcan.

Janet, Pierre (1892). Étude sur quelques cas d'amnésie antérograd dans la maladie de la désagrégation psychologique. In *International Congress of Experimental Psychology. Second Session. London 1892*. London, Williams and Norgate. Pp. 26–30.

Janet, Pierre (1893–4). *État mental des hystériques I.* ... 2 vols. Paris, Rueff.

Janet, Pierre (1898). *Névroses et idées fixes.* .... Paris, F. Alcan.

Janet, Pierre (1901). *The Mental State of Hystericals* ... New York, G. P. Putnam's sons.

Janet, Pierre (1903). *Les obsessions et la psychasthénie. I.* ... Paris, Alcan.

Janet, Pierre (1905). Les oscillations du niveau mental. In *Atti del V congresso internazionale di psicologia tenuto in Roma del 26 al 30 aprile 1905* ... Rome, Forzani. Pp. 110–126.

Janet, Pierre (1907). The Major Symptoms of Hysteria ... New York, Macmillan

Janet, Pierre (1909). Les problèmes du subconscient. In *VIme congrès international de psychologie, tenu à Genève du 2 à 7 août 1909* ... Geneva, Kündig, 57–70

Janet, Pierre (1911). Les problémes de la suggestion. *Journal für Psychologie und Neurologie*, 17, 323–343.

Janet, Pierre (1919). *Les médications psychologiques*. 3 vols. Paris, F. Alcan.

Janet, Pierre (1924). *Principles of Psychotherapy*. New York, Macmillan.

Janet, Pierre (1925). *Psychological Healing: A Historical and Clinical Study*. 2 vols. New York, Macmillan. [Translation of Janet (1919).]

Janet, Pierre (1936). *Les névroses*. Paris, E. Flammarion.

Janet, Pierre, and Raymond, F. (1898). *Névroses et idées fixes. II* ... . Paris, F. Alcan.

Janet, Pierre, and Raymond, F. (1903). *Les obsessions et la psychasthénie. II* .... Paris, F. Alcan.

Jeanniard du Dot, A. (1913). *Où en est l'hypnotisme?* Paris, Bloud.

Jemmott, J. B., and Locke, S. E.. (1984). Psychosocial factors, immunologic medication, and human susceptibility to infectious diseases. How much do we know? *Psychological Bulletin*, 95, 78–108.

Jendrássik, E. (1886). De l'hypnotisme. *Archives de neurologie*, 11, 362–380; 12, 43–53.

Jendrássik, E. (1888). Einiges über Suggestion. *Neurologisches Centralblatt*, 7, 281–283, 321–330.

Jilek, W. (1982) Altered states of consciousness in North American indian ceremonials. *Ethos*, 10, 326–343.

Johnson, R. F. Q. (1989). Hypnosis, suggestion and dermatological changes: A consideration of the production and diminution of dermatological entities. In N. P. Spanos and J. F. Chaves (eds.), *Hypnosis: The Cognitive-Behavioral Perspective*. Buffalo, N. Y., Prometheus Books. Pp. 297–312.

Joire, P. (1908). *Traité de l'hypnotisme expérimental et thérapeutique* ... Paris, Vigot frères.

Jolly, F. (1893). Ueber Hypnotismus und Geistesstörung. *Archiv für Psychiatrie und Nervenkrankheiten*, 25, 599–616.

Kaan, H. (1885). *Über Beziehungen zwischen Hypnotismus und cerebraler Blutfüllung*. Wiesbaden, J. F. Bergmann.

Katz, R. (1982). Accepting "boiling energy": The experience of !Kia healing among the !Kung. *Ethos*, 10, 344–368.

Katz, R. (1989). Healing and transformation: Perspectives on development, education, and community. In C. A. Ward (ed.), *Altered States of Consciousness and Mental Health: A Cross-Cultural Perspective*. Newbury Park, Calif., Sage Publications. Pp. 207–227.

Kelly, S. F., Fisher, S., and Kelly, R. J. (1978). Effects of cannabis intoxication on primary suggestibility. *Psychopharmacology*, 56, 217–219.

Kenny, M. G. (1986). *The Passion of Ansel Bourne: Multiple Personality in American Culture*. Washington, D. C., Smithsonian Institution Press.

Kerckhoff, A. C., and Back, K. W. (1968). *The June Bug: A Study of Hysterical Contagion*. New York, Appleton-Century-Crofts.

Kieser, D. G. (1826). *System des Tellurismus oder thierischen Magnetismus: ein Handbuch für Naturforscher und Aerzte*. New edn. 2 vols. Leipzig, F. L. Herbig.

Kieserwetter, K. (1909). *Geschichte des neueren Occultismus* ... 2nd edn. Leipzig, M. Altmann.

Kihlstrom, J. F. (1984). Conscious, subconscious, unconscious: A cognitive perspective. In K. S. Bowers and D. Meichenbaum (eds.), *The Unconscious Reconsidered*. New York, Wiley.

Kihlstrom, J. F., and Evans, F. J. (1979). Memory retrieval processes during posthypnotic amnesia. In J. F. Kihlstrom and F. J. Evans (eds.), *Functional Disorders of Memory*. Hillsdale, N. J., L. Erlbaum. Pp. 179–218.

Kihlstrom, J. F., Evans, F. J., Orne, M. T., and Orne, E. C. (1980). Attempting to breach posthypnotic amnesia. *Journal of Abnormal Psychology*, 89, 603–626.

Kihlstrom, J. F., and McConkey, K. M. (1990). William James and hypnosis: A centennial reflection. *Psychological Science*, 1, 174–178.

Kiloh, L. G., McComes, A. J., Osselton, J. W., and Upton, A. R. M. (1981). *Clinical Electroencephalography*. 4th edn. London, Butterworths.

Kingsbury, G. C. (1891). *The Practice of Hypnotic Suggestion* ... Bristol, J. Wright.

Kirsch, I., *et al.* (1989). The surreptitious observation design: An experimental paradigm for distinguishing artifact from essence in hypnosis. *Journal of Abnormal Psychology*, 98, 132–136.

Kissel, P., and Barrucand, D. (1964). Le sommeil hypnotique, d'après L' "école de Nancy"... *Encéphale*, 53, 571–588.

Kline, M. V. (1958). *Freud and Hypnosis*... New York, The Julian Press.

Kluft, R. P. (1987). An update on multiple personality disorder. *Hospital and Community Psychiatry*, 38, 363–373.

Kluge, C. A. F. (1811). *Versuch einer Darstellung des animalischen Magnetismus als Heilmittel*. Berlin, C. Salfeld.

Köhler, F. (1897). Experimentelle Studien auf dem Gebiete des hypnotischen Somnambulismus. *Zeitschrift für Hypnotismus*, 6, 357–374.

Kohnstamm, O., and Pinner, R. (1908). Blasenbildung durch hypnotische Suggestion und Gesichtpunkte zu ihrer Erklärung. *Verhandlungen der deutschen dermatologischen Gesellschaft*, 10, 342–345.

Kooi, K. A. (1971). *Foundations of Electroencephalography*. New York, Harper and Row.

Krafft-Ebing, R. von (1889a). *Eine experimentelle Studie auf dem Gebiete des Hypnotismus nebst Bemerkungen über Suggestion und Suggestionstherapie*. 2nd. edn. Stuttgart, F. Enke.

Krafft-Ebing, R. von (1889b). *An Experimental Study in the Domain of Hypnotism*. New York, G. P. Putnam's sons.

Krafft-Ebing, R. von (1891). Zur Verwerthung der Suggestionstherapie (Hypnose) bei Psychosen und Neurosen. *Wiener klinische Wochenschrift*, 4, 795–799.

Krafft-Ebing, R. von (1893). *Hypnotische Experimente*. Stuttgart, F. Enke.

Kravis, N. M. (1988). James Braid's psychophysiology: A turning point in the history of dynamic psychiatry. *The American Journal of Psychiatry*, 145, 1191–1206.

Kreibich, C., and Sobotka, P. (1909). Experimenteller Beitrag zur psychischen Urticaria. *Archiv für Dermatologie und Syphilologie*, 97, 187–192.

Kroger, W. S. (1977). *Clinical and Experimental Hypnosis*. 2nd. edn. Philadelphia, Lippincott.

Kunzendorf, R. G., and Benoit, M. (1985–6). Spontaneous post-hypnotic amnesia and spontaneous recovery in repressors. *Imagination, Cognition and Personality*, 5, 303–310.

Ladame, P. L. (1881). *La névrose hypnotique ou le magnétisme dévoilé*. Paris, Sandoz et Fischbacher.

Ladame, P. L. (1887–8) Le traitement des buveurs et des dipsomanes par l'hypnotisme. *Revue de l'hypnotisme*, 2, 129–135, 165–168.

Ladame, P. L. (1888). *L'hypnotisme et la médecine légale*. Lyon, A. Storck,

Lallart, C. (1864). *Essai sur l'hypnotisme*... Soissons, Lallart.

Landgren, S. (1893–4). Offener Brief. *Zeitschrift für Hypnotismus*, 2, 23–26.

Langley, J. N. (1884). *The Physiological Aspect of Mesmerism*. London, [a lecture to the Royal Institution].

Lasègue, C. (1865). Des catalepsies partielles et passagères. *Archives générales de médecine*, NS 2, 385–402.

Laurence, J. R., and Perry, C. (1988). *Hypnosis, Will, and Memory: A Psycho-Legal History*. New York, The Guilford Press.

Laurent, L. (1892). *Des états secondes: variations pathologiques du champ de la conscience*. Bordeaux, Cadoret.

Le Bon, G. (1895). *Psychologie des foules*. Paris, F. Alcan.

Lee, R. B. (1968). The sociology of !Kung Bushman trance performances. In R. Prince (ed.), *Trance and Possession States*. Montreal, R. M. Bucke Memorial Society. Pp. 35–54.

Lefebvre, F. (1873). *Louise Lateau of Bois d'Haine*. London, Burns and Oates.

Lehmann, A. (1890a). *Hypnosen og de dermed beslaegtede normale Tilstande*. Copenhagen, Philipsen.

Lehmann, A. (1890b). *Die Hypnose und die damit verwandte normalen Zustände*. Leipzig, O. R. Reisland.

Lehmann, A. (1898). *Aberglaube und Zauberei von den ältesten Zeiten an bis in die Gegenwart*. Stuttgart, F. Enke.

"Le Baron, A." (1896–7). A case of psychic automatism, including "speaking with tongues". *Proceedings of the Society for Psychical Research*, 12, 277–297.

Le Lieurre de l'Aubépin, –. (1819). Lettre de M. le Lieurre de l'Aubépin, à M. Deleuze. *Bibliothèque du magnétisme animal*, 8, 93–128.

Lentz, –. (1897). L'automatisme alcoolique. *Journal de neurologie et d'hypnologie*, 3, 42–46.

"Leonidas, Professor" (1901). *Stage Hypnotism. A Text Book of Occult Entertainments*. Chicago, Bureau of Stage Hypnotism.

Leppo, L. (1968). Hypnotism in Italy. In E. J. Dingwall (ed.), *Abnormal Hypnotic Phenomena: A Survey of Nineteenth-Century Cases*, Vol. 3. London, J. and A. Churchill. Pp. 137–189.

Levitt, E. E., Baker, E. L., and Fish, R. C. (1989–90). Some conditions of compliance and resistance among hypnotic subjects. *American Journal of Clinical Hypnosis*, 32, 225–236.

Lévy, P. E. (1898). *L'éducation rationnelle de la volonté, son emploi thérapeutique*. Paris, F. Alcan.

Liébeault, A. A. (1866). *Du sommeil et des états analogues considérés au point de vue de l'action du moral sur le physique*. Paris, V. Masson et fils.

Liébeault, A. A. (1883). *Étude sur le zoomagnétisme*. Paris, G. Masson.

Liébeault, A. A. (1886–7a). Confession d'un médecin hypnotiseur. *Revue de l'hypnotisme*, 1, 105–110, 143–148.

Liébeault, A. A. (1886–7b). Traitement par suggestion hypnotique de l'incontinence d'urine chez les adultes et les enfants au dessus de trois ans. *Revue de l'hypnotisme*, 1, 71–77.

Liébeault, A. A. (1889). *Le sommeil provoqué et les états analogues*. Paris, O. Doin.

Liébeault, A. A. (1891). *Thérapeutique suggestive: son mécanisme. Propriétés diverses du sommeil provoqué et des états analogues*. Paris, O. Doin.

Liébeault, A. A. (1891–2). De l'influence psychique exercée par la mère sur le foetus. *Revue de l'hypnotisme*, 6, 52–54.

Liébeault, A. A. (1894–5a). Criminelle hypnotische Suggestionen. Gründe und Tatsachen welche für dieselben sprechen. *Zeitschrift für Hypnotismus*, 3, 193–206, 225–229.

Liébeault, A. A. (1894–5b). Das Wachen, ein activer Seelenzustand. – Der Schlaf ein passiver Seelenzustand. – Physiologische passive Zustände, beziehentlich pathologische, welche dem Schlaf analog sind. – Suggestion. *Zeitschrift für Hypnotismus*, 3, 22–28.

Liégeois, J. (1884). *De la suggestion hypnotique dans ses rapports avec le droit civil et le droit criminel*. Paris, A. Picard.

Liégeois, J. (1889). *De la suggestion et du somnambulisme dans leurs rapports avec la jurisprudence et la médecine légale.* Paris, O. Doin.

Liégeois, J. (1892). Hypnotisme et criminalité. *Revue philosophique,* 33, 233–272.

Liégeois, J. (1892–3). Der Fall Chambige vor den Schwurgerichtshof in Constantine (Algier) 1888. Eine Studie zur criminellen Psychologie. *Zeitschrift für Hypnotismus,* 1, 212–216, 234–238.

Lilienthal, K. von (1887). *Der Hypnotismus und das Strafrecht.* Berlin, Guttentag.

Lipps, T. (1897a). Zur Psychologie der Suggestion. *Zeitschrift für Hypnotismus,* 6, 94–199, 154–159.

Lipps, T. (1897b). Suggestion und Hypnose: eine psychologische Untersuchung. *Sitzungsberichte der philosophisch-philologischen und der historischen Class der k.b. Akademie der Wissenschaften zu München,* 2, 391–522.

Lipps. T. (1898). *Suggestion und Hypnose* ... Munich, F. Straub.

Lishman, W. A. (1987). *Organic Psychiatry: The Psychological Consequences of Cerebral Disorder.* 2nd edn. Oxford, Blackwell Scientific Publications.

Locke, R. G., and Kelly, E. F. (1985). A preliminary model for the cross-cultural analysis of altered states of consciousness. *Ethos,* 13, 3–55.

Loewenfeld, L. (1893–4). *Pathologie und Therapie der Neurasthenie und Hysterie.* Wiesbaden, J. F. Bergmann.

Loewenfeld, L. (1899). *Sexualleben und Nervenleben.* Wiesbaden, J. F. Bergmann.

Loewenfeld, L. (1901). *Der Hypnotismus: Handbuch der Lehre von der Hypnose und der Suggestion, mit besonderer Berücksichtigung ihre Bedeutung für Medicin und Rechtspflege.* Wiesbaden, J. F. Bergmann.

Loewenfeld, L. (1910). Ueber die hypermnestischen Leistungen in der Hypnose in bezug auf Kindheitserrinerungen. *Zeitschrift für Psychotherapie und medizinische Psychologie,* 2, 1–29.

Loewenstein, R. J., and Putnam, F. W. (1988). A comparison study of dissociative symptoms in patients with complex partial seizures, MPD and posttraumatic stress disorder. *Dissociation,* 1(4), 17–23.

Lombroso, C. (1886). *Studi sull'ipnotismo con ricerche oftalmoscopiche del prof. Reymond.* Rome, Fratelli Bocca.

Lombroso, C. (1909). *Richerchi sui fenomeni ipnotici e spiritici.* Turin, Unione tipografico-editrice Torinese.

Lorry, A. C. (1770). *Von der Melancholie und den melancholischen Krankheiten,* vol. 1. Frankfurt, Andrea.

Ludwig, A. M. (1969). Altered states of consciousness. In C. Tart (ed.), *Altered States of Consciousness.* New York, Wiley. Pp. 9–22.

Luria, A. (1968). *The Mind of the Mnemonist.* New York, Basic Books.

Luys, J. (1887). *Les émotions chez les sujets en état d'hypnotisme* ... Paris, J. B. Baillière et fils.

Luys, J. (1888). De la transmission à distance des émotions d'un sujet hypnotisé à un autre. *Comptes rendus de la Société de biologie,* 8 sér, 5, 545–547.

Luys, J. (1890a). *Les émotions dans l'état d'hypnotisme et l'action à distance des substances médicamenteuses ou toxiques.* Paris, J. B. Baillière et fils.

Luys, J. (1890b). The latest discoveries in hypnotism. *Fortnightly Review,* 47, 896–921; 48, 168–183.

Luys, J. (1890c). *Leçons cliniques sur les principaux phénomènes de l'hypnotisme dans leurs rapports avec la pathologie mentale.* Paris, G. Carré.

Luys, J., and Encausse, G. (1891). *Du transport à distance à l'aide d'une couronne de fer aimanté d'états névro-pathiques variés, d'un sujet à l'état de veille sur un sujet à l'état hypnotique.* Clermont, Daix frères.

Lynn, S. J., and Rhue, J. W. (1988). Fantasy proneness: Hypnosis, developmental antecedents, and psychopathology. *American Psychologist*, 43, 35–44.

Lynn, S. J., and Rhue, J. W. (eds.) (1991). *Theories of Hypnosis.* New York, Guilford Publications.

Lynn, S. J., Rhue, J. W., and Green, J. P. (1988). Multiple personality and fantasy proneness: Is there an association or dissociation? *British Journal of Experimental and Clinical Hypnosis*, 5, 138–142.

Lynn, S. J., Rhue, J. W., and Weekes, J. R. (1989). Hypnosis and experienced nonvolition: A social-cognitive integrative model. In N. P. Spanos and J. F. Chaves (eds.), *Hypnosis: The Cognitive-Behavioral Perspective.* Buffalo, N. Y., Prometheus Books. Pp. 78–109.

Macario, M. M. A. (1857). *Du sommeil des rêves et du somnambulisme dans l'état de santé et de maladie.* Lyon, Perisse.

McConkey, K. M., Glisky, M. L., and Kihlstrom, J. F. (1989). Individual differences among hypnotic virtuosos: A case comparison. *Australian Journal of Clinical and Experimental Hypnosis*, 17, 131–140.

McConkey, K. W., Sheehan, P. W., and Cross, D. G. (1980). Posthypnotic amnesia: Seeing is not remembering. *British Journal of Social and Clinical Psychology*, 19, 99–107.

McDougall, W. (1911). Hypnotism. *Encyclopaedia Britannica*, vol. 14. New York, The Encyclopaedia Britannica Co. Pp. 200–207.

McDougall, W. (1920–1). Four cases of "regression" in soldiers. *The Journal of Abnormal Psychology*, 15, 136–156.

McDougall, W. (1926). *An Outline of Abnormal Psychology.* London, Methuen.

McDougall, W. (1967). *William McDougall: Explorer of the Mind. Studies in Psychical Research.* Ed. R. van Over and L. Oteri. New York, Helix Press.

McGuinness, T. P. (1983–4). Hypnosis in the treatment of phobias: A review of the literature. *American Journal of Clinical Hypnosis*, 26, 261–269.

McKellar, P. (1979). *Mindsplit: The Psychology of Multiple Personality and the Dissociated Self.* London, J. M. Dent and Sons.

Macleod-Morgan, C. (1983). Hypnosis is a right-hemisphere task: A model based on arguments from the laboratory and from literature. *Svensk tidskrift för hypnos*, 10, 84–90.

Macnish, R. (1830). *The Philosophy of Sleep.* Glasgow, W. R. M'Phun.

Madden, J. (1903). Wounds produced by suggestion. *American Medicine*, 5, 288–290.

Magini, G. (1887). *Le meraviglie dell'ipnotismo* ... Turin, E. Loescher

Magnin, P. (1883). Rémarques générales sur l'hypnotisme ... *Mémoires de la Société de biologie*, 7 sér, 5, 43–49.

Magnin, P. (1884). *Étude clinique et expérimentale sur l'hypnotisme* ... Paris, A. Delahaye et E. Lecrosnier.

Magnin, P. (1889). Des effets opposés déterminés par un même agent physique sur l'hystéro-épileptique hypnotisable. In E. Bérillon (ed.), *Premier congrès international de l'hypnotisme expérimental et thérapeutique* ... Paris, O. Doin. Pp. 279–288.

Mahowald, M. W., and Schenck, C. H. (1991). Status dissociatus – a perspective on states of being. *Sleep*, 14, 69–79.

Mahowald, M. W., *et al.* (1990). Sleep violence: Forensic science implications: Polygraphic and video documentation. *Journal of Forensic Sciences*, 35, 413–432.

Malott, J. M. (1984). Active-alert hypnosis: Replication and extension of previous research. *Journal of Abnormal Psychology*, 93, 246–249.

Manganiello, A. S. (1986–7). Hypnotherapy in the rehabilitation of a stroke victim. *American Journal of Clinical Hypnosis*, 29, 64–68.

Mantegazza, P. (1887). *Le estasi umane*. 2 vols. Milan, P. Mantegazza editore.

Marcinowski, – . (1900). Selbstbeobachtungen in der Hypnose. *Zeitschrift für Hypnotismus*, 9, 5–46, 177–192.

Marès, J., and Hellich, B. (1889). L'abaissement de la température chez l'homme après perte de la sensibilité pour le froid et le chaud, suggérée dans l'état hypnotique. *Comptes rendus de la Société de biologie*, sér 9, 1, 410–415.

Marin, P. [1890]. *L'hypnotisme théorique et pratique* ... Paris, E. Kolb.

Marks, D. F., Baird, J. M., and McKellar, P. (1989). Replication of trance logic using a modified experimental design; highly hypnotizable subjects in both real and simulator groups. *The International Journal of Clinical and Experimental Hypnosis*, 37, 232–248.

Marot, F. (1892–3). Morphinomanie et suggestion: guérison datant de trois ans et demi. *Revue de l'hypnotisme*, 7, 233–236.

Martin, L. J. (1907). Zur Begründung und Anwendung der Suggestionsmethode in der Normalpsychologie. Vorläufige Mitteilung. *Archiv für die gesamte Psychologie*, 10, 321–402.

Maser, J. D., and Gallup, G. J. (1977). Tonic immobility and related phenomena: A partially annotated, tricentennial bibliography, 1636–1972. *Psychological Record*, 1, 177–217.

Mason, M. K. (1899–1900). The cure of warts by suggestion. *Journal of the Society for Psychical Research*, 9, 225–227.

Mason, R. Osgood (1893). Duplex personality. *The Journal for Nervous and Mental Disease*, 18, 594–598.

Mason, R. Osgood (1896). Educational uses of hypnotism. *The North American Review*, 158, 448–455.

Mason, R. Osgood (1897). *Telepathy and the Subliminal Self*. New York, H. Holt.

Mason, R. Osgood (1901). *Hypnotism and Suggestion in Therapeutics, Education and Reform*. London, Kegan Paul, Trench, Trübner and Co.

Mavromatis, A. (1987). *Hypnagogia: The Unique State of Consciousness between Wakefulness and Sleep*. London, Routledge and Kegan Paul.

Meares, R., *et al.* (1985). Whose hysteria: Briquet's, Janet's or Freud's? *Australian and New Zealand Journal of Psychiatry*, 19, 256–263.

A medical society for the study of hypnotism (1907–8). *Journal of the Society for Psychical Research*, 13, 14–15.

Mesnet, E. (1860). Études sur le somnambulisme, envisagé au point de vue pathologique. *Archives générales de médecine*, 5 sér, 15(1), 147–173,

Mesnet, E. (1874). *De l'automatisme de la mémoire et du souvenir dans le somnambulisme pathologique. Considérations médico-légales*. Paris, F. Malteste. [From *L'union médical*, July 21st and 28th 1874.]

Mesnet, E. (1887–8). Un accouchement dans le somnambulisme provoqué. *Revue de l'hypnotisme*, 2, 33–42.

Mesnet, E. (1889–90). Autographisme et stigmates. *Revue de l'hypnotisme*, 4, 321–335.

Mesnet, E. (1894). *Outrages à la pudeur. Violences sur les organes sexuels de la femme dans le somnambulisme provoqué et la fascination. Étude médico-légale.* Paris, Rueff et cie.

Miller, M. E., and Bowers, K. S. (1986). Hypnotic analgesia and stress inoculation in the reduction of pain. *Journal of Abnormal Psychology*, 95, 6–14.

[Mills, C. K.] (1882). Review on hypnotism. *American Journal of Medical Sciences*, 83, 143–163.

Mitchell, J. K. (1859). *Five Essays*. Philadelphia, J. B. Lippincott and Co.

Mitchell, S. L. (1816). A double consciousness or a duality of person in the same individual. *Medical Repository*, 3, 185–186.

Mitchell, S. Weir (1889). *Mary Reynolds: A Case of Double Consciousness*. Philadelphia, W. J. Dorman.

Mitchell, T. W (1908). *Hysteria and Hypnosis*. Croydon, n. pub. [Offprint from *The General Practitioner*, 9th and 16th May 1908.]

Mitchell, T. W. (1908–9). The appreciation of time by somnambules. *Proceedings of the Society for Psychical Research*, 21, 2–59.

Mitchell, T. W. (1912–13). Some types of multiple personality. *Proceedings of the Society for Psychical Research*, 26, 257–285.

Mitchell, T. W [1922]. *Medical Psychology and Psychical Research*. New York, E. P. Dutton and Co.

Moll, A. (1889). *Der Hypnotismus*. Berlin, Fischer.

Moll, A. (1890). *Hypnotism*. 2nd edn. London, Walter Scott.

Moll, A. (1892). Der Rapport in der Hypnose: Untersuchungen über den thierischen Magnetismus. *Schriften der Gesellschaft für psychologische Forschung*, 1(3,4), 273–514.

Moll, A. (1936). *Ein Leben als Arzt der Seele; Erinnerungen*. Dresden, C. Reissner.

Mondeil, G. (1927). *Le fluide humain devant la physique révélatrice et la métapsychique objective* ... Paris, Berger-Levrault.

Montalvo, T. de (1719). *Vita prodigiosa de la extatica virgen y venerable madre Sor Beatriz Maria de Jesus* ... Granada, F. Dominguez.

Morand, J.-S. (1889). *Le magnétisme animal (hypnotisme et suggestion): Étude historique et critique*. Paris, Garnier frères.

Morse, D. R., Schoor, R. S., and Cohen, B. B. (1984). Surgical and non-surgical dental treatments with meditation-hypnosis as the sole anesthetic: Case report. *International Journal of Psychosomatics*, 31, 27–33.

Morse, D. R., and Wilcko, J. M. (1978–9). Nonsurgical endodontic therapy for a vital tooth with meditation-hypnosis as the sole anesthetic: A case report. *American Journal of Clinical Hypnosis*, 21, 258–262.

Morselli, E. (1886). *Il magnetismo animale: la fascinazione e gli stati affini*. Turin, Roux e Favale.

Morselli, E. (1894–5). Les indications cliniques de l'hypnotisme. *Revue de l'hypnotisme*, 9, 180–182.

Morselli, E., and Tanzi, E. (1890). Contributo sperimentale alla fisiopsicologia dell'ipnotismo – polso e respiro negli stati suggestivi dell'ipnose. *Rivista di filosofia scientifica*, 9, 705–729.

Moser, L. (1967). Hypnotism in Germany. In E. J. Dingwall (ed.), *Abnormal Hypnotic*

Phenomena: A Survey of Nineteenth-Century Cases, Vol. 2. London, J. and A. Churchill. Pp. 101–199.

Mühl, A. M. (1963). *Automatic Writing: An Approach to the Unconscious.* 2nd. edn. New York, Helix Press.

Müller, J. C. (1725). Somnambulo diurnus, oder von einer Manns-Person, so bey hellen Tage darjenige gethan, was ein Mond-süchtiger bey Nacht. *Sammlung von Natur- und Medecin-, wie auch hierzu gehörigen Kunst- und Literatur-Geschichten ... von einigen Academ. Naturae Curios. in Breslau.* Pp. 653–658.

Munro, H. S. (1911). *Handbook of Suggestive Therapeutics, Applied Hypnotism, Psychic Science ...* St. Louis, C. V. Mosby.

Münsterberg, H. *et al.* (n.d.). *Subconscious Phenomena.* London, Rebman.

Muralt, L. von (1902). Zur Frage der epileptischen Amnesie. *Zeitschrift für Hypnotismus,* 10, 75–90.

Myers, A. T. (1885–6). On the action of drugs at a distance. *Journal of the Society for Psychical Research,* 2, 58–62.

Myers, A. T. (1886). *The Life-History of a Case of Double or Multiple Personality.* [Offprint from *The Journal of Mental Science,* January 1886.]

Myers, A. T. (1889–90). International Congress of Experimental Psychology. *Proceedings of the Society for Psychical Research,* 6, 171–182.

Myers, A. T. (1890). Hypnotism at home and abroad. *The Practitioner,* March, 197–206.

Myers, A. T. (1892). Hypnotism, history of. In D. Hack Tuke (ed.), *A Dictionary of Psychological Medicine,* vol. 1. London, J. and A. Churchill. Pp. 603–606.

Myers, A. T., and Myers, F. W. H. (1893–4). Mind-cure, faith-cure, and the miracles of Lourdes. *Proceedings of the Society for Psychical Research,* 9, 160–209.

Myers, F. W. H. (1884). On a telepathic explanation of some so-called spiritualistic phenomena [= Automatic writing. – I.]. *Proceedings of the Society for Psychical Research,* 2, 217–237.

Myers, F. W. H. (1885). Automatic writing. – II. *Proceedings of the Society for Psychical Research,* 3, 1–63.

Myers, F. W. H. (1885–7a). Automatic writing. – III. *Proceedings of the Society for Psychical Research,* 4, 209–261.

Myers, F. W. H. (1885–7b). Human personality in the light of hypnotic suggestion. *Proceedings of the Society for Psychical Research,* 4, 1–24.

Myers, F. W. H. (1885–7c). On telepathic hypnotism and its relation to other forms of hypnotic suggestion. *Proceedings of the Society for Psychical Research.* 4, 127–188.

Myers, F. W. H. (1888–9a). Automatic writing. – IV. – The daemon of Socrates. *Proceedings of the Society for Psychical Research,* 5, 522–547.

Myers, F. W. H. (1888–9b). French experiments on strata of personality. *Proceedings of the Society for Psychical Research,* 5, 374–397.

Myers, F. W. H. (1888–9c). The work of Edmund Gurney in experimental psychology. *Proceedings of the Society for Psychical Research,* 5, 359–373.

Myers, F. W. H. (1889–90a). Binet on the consciousness of hysterical subjects. *Proceedings of the Society for Psychical Research,* 6, 200–206.

Myers, F. W. H. (1889–90b). [Review of Dessoir's] "Das Doppel-Ich." *Proceedings of the Society for Psychical Research,* 6, 207–215.

Myers, F. W. H. (1891–2a). [Review of James's] "The Principles of Psychology." *Proceedings of the Society for Psychical Research,* 7, 111–133.

Myers, F. W. H. (1891–2b). The subliminal consciousness [chs. I and II]. *Proceedings of the Society for Psychical Research*, 7, 298–355.

Myers, F. W. H. (1892). The subliminal consciousness [chs. III to V]. *Proceedings of the Society for Psychical Research*, 8, 333–404, 436–535.

Myers, F. W. H. (1893). Professor Wundt on hypnotism and suggestion. *Mind*, new ser. 2, 95–101.

Myers, F. W. H. (1893–4). The subliminal consciousness [chs. VI and VII]. *Proceedings of the Society for Psychical Research*, 9, 3–128.

Myers, F. W. H. (1895). The subliminal self [chs. VIII and IX]. *Proceedings of the Society for Psychical Research*, 11, 334–593.

Myers, F. W. H. (1898–9). The psychology of hypnotism. *Proceedings of the Society for Psychical Research*, 14, 100–108.

Myers, F. W. H. (1903). *Human Personality and its Survival of Bodily Death*. 2 vols. London, Longmans, Green and Co.

Myers, J. J., and Sperry, R. W. (1985). Interhemispheric communication after section of the forebrain commissures. *Cortex*, 21, 248–260.

Myers, S. A., and Austrin, H. R. (1985). Distal eidetic technology: Further characteristics of the fantasy-prone personality. *Journal of Mental Imagery*, 9, 57–66.

Nadon, R., Laurence, J. R., and Perry C. (1987). Multiple predictors of hypnotic susceptibility. *Journal of Personality and Social Psychology*, 53, 948–960.

Nadon, R., Laurence, J. R., and Perry, C. (1991). The two disciplines of scientific hypnosis: A synergistic model. In S. J. Lynn and J. W. Rhue (eds.), *Theories of Hypnosis*. Guilford Publications, New York. Pp. 485–519.

Naples, M., and Hackett, T. P. (1978). The amytal interview: History and current use. *Psychosomatics*, 19, 98–105.

Nash, M. R., Lynn, S. J., and Givens, D. L. (1984). Adult hypnotic susceptibility, childhood punishment, and child abuse: A brief communication. *The International Journal of Clinical and Experimental Hypnosis*, 32, 6–11.

Newbold, W. R. (1895). Experimental induction of automatic processes. *The Psychological Review*, 2, 348–362.

Newbold, W. R. (1895–6a). Suggestibility, automatism and kindred phenomena. *Popular Science Monthly*, 48, 193–198.

Newbold, W. R. (1895–6b). Normal and heightened suggestibility. *Popular Science Monthly*, 48, 641–651.

Newbold, W. R. (1895–6c). Hypnotic states: trance and ecstasy. *Popular Science Monthly*, 48, 804–815.

Newbold, W. R. (1896a). Post-hypnotic and criminal suggestion. *Popular Science Monthly*, 49, 230–241.

Newbold, W. R. (1896b). Suggestion in therapeutics. *Popular Science Monthly*, 49, 342–353.

Newbold, W. R. (1896c). Spirit writing and speaking with tongues. *Popular Science Monthly*, 49, 508–522.

Newbold, W. R. (1896d). Illusions and hallucinations. *Popular Science Monthly*, 49, 630–641.

Newbold, W. R. (1896e). The self and its derangements. *Popular Science Monthly*, 49, 810–821.

Newbold, W. R. (1896–7a). Double personality. *Popular Science Monthly*, 50, 67–79.

Newbold, W. R. (1896–7b). Possession and mediumship. *Popular Science Monthly*, 50, 220–231.

Nicol, F. (1966). The silences of Mr. Trevor Hall. *International Journal of Parapsychology*, 8, 3–59.

Nicoll, A. (1890). *Hypnotism*. London, H. Renshaw.

Noizet, F. J. (1854). *Mémoire sur le somnambulisme et le magnétisme animal adressé en 1820 à l'Académie royale de Berlin*. Paris, Plon frères.

Nonne, M. (1888) Zur therapeutischen Verwerthung der Hypnose. *Neurologisches Zentralblatt*, 7, 185, 226.

Obersteiner, H. (1887). *Der Hypnotismus ...* Vienna, M. Breitenstein.

Ochorowicz, J. (1884–5). A new hypnoscope. *Journal of the Society for Psychical Research*, 1, 277–282.

Ochorowicz, J. (1890). Tous les phénomènes de l'hypnotisme, peuvent-ils être attribués à la suggestion. In *Congrès international de psychologie physiologique: première session. Paris 1890 ...* Paris, Bureau des revues. Pp. 74–76.

Ochorowicz, J. (1891). *Mental Suggestion*. New York, Humboldt.

O'Leary, A. (1990). Stress, emotion and human immune function. *Psychological Bulletin*, 108, 363–382.

Orne, M. T. (1959). The nature of hypnosis: artifact and essence. *Journal of Abnormal and Social Psychology*, 58, 277–299.

Orne, M. T. (1979). On the simulating subject as a quasi-control group in hypnosis research: What, why, and how. In E. Fromm and R. E. Shor (eds.), *Hypnosis: Developments in Research and New Perspectives*. 2nd edn. New York, Aldine. Pp. 519–565

Orne, M. T., Dinges, D. F., and Orne, E. C. (1984). On the differential diagnosis of multiple personality in the forensic context. *The International Journal of Clinical and Experimental Hypnosis*, 32, 118–169.

Osborne, A. E. (1894–5). People who drop out of sight. *The Medico-Legal Journal*, 12, 22–41.

Osgood, H. (1890). The therapeutic value of suggestion during the hypnotic state, with an historical sketch of hypnotism and a report of thirty-four cases. *The Boston Medical and Surgical Journal*, 122, 411–417, 441–447.

Osgood, H. (1891). The outcome of personal experience in the application of hypnotism and hypnotic suggestion. *The Boston Medical and Surgical Journal*, 124, 277–280, 303–307.

Osgood, H. (1894–5). Quatre cas d'eczéma et un de dermatite traités par suggestion. *Revue de l'hypnotisme*, 9, 300

Oswald, I., and Evans, J. (1985). On serious violence during sleep-walking. *British Journal of Psychiatry*, 147, 688–691.

Ottolenghi, S., and Lombroso, C. (1889). *Nuovi studi sull'ipnotismo e sulla credulità*. Turin, stamperia dell'unione tipografico-editrice. [Offprint from the *Giornale della Accádemia di medicina di Torino*, 1889.]

Owen, A. R. G. (1971). *Hysteria, Hypnosis and Healing: The Work of J.-M. Charcot*. New York, Garrett Publications.

Owens, M. E., *et al.* (1989). Phobias and hypnotizability: A reexamination. *The International Journal of Clinical and Experimental Hypnosis*, 37, 207–216

Parish, E. (1897). *Hallucinations and Illusions: A Study of the Fallacies of Perception.* London, Walter Scott.

Parker, J.L. (1979). The use of hypnosis to reinstate a child's impaired writing performance resulting from neurological damage. *Australian Journal of Clinical and Experimental Hypnosis,* 7, 225–230.

Parker, W.B. (ed.) (1909). *Psychotherapy: A Course of Reading in Sound Psychology, Sound Medicine and Sound Religion.* 3 vols. New York, Center Publishing Co.

Pattie, F.A. (1935). A report of attempts to produce uniocular blindness by hypnotic suggestion. *British Journal of Medical Psychology,* 15, 230–241.

Pattie, F.A. (1941). The production of blisters by hypnotic suggestions: A review. *Journal of Abnormal and Social Psychology,* 36, 62–72.

Paul, G.L. (1963). The production of blisters by hypnotic suggestion: Another look. *Psychosomatic Medicine,* 25, 233–244.

Pekala, R.J. (1991). Hypnotic types: Evidence from a cluster analysis of phenomenal experience. *Contemporary Hypnosis,* 8, 95–104.

Perronet, C. (1886). *Force psychique et suggestion mentale ...* Paris, J. Chevalier.

Perry, C. (1976). The abbé Faria: A neglected figure in the history of hypnosis. In F.H. Frankel and H.S. Zamansky (eds.), *Hypnosis at its Bicentennial: Selected Papers.* New York, Plenum Press. Pp. 37–45.

Perry, C., Gelfand, R., and Marcovitch, P. (1979). The relevance of hypnotic susceptibility in the clinical context. *Journal of Abnormal Psychology,* 88, 592–603.

Perry, J.C., and Jacobs, D. (1982). Overview: Clinical applications of the amytal interview in psychiatric emergency settings. *The American Journal of Psychiatry,* 139, 552–559.

Petetin, J.H.D. (1808). *Électricité animale ...* Paris, Brunot-Labbé.

Philips, A.J.P. [i.e. J.P. Durand de Gros] (1860). *Cours théorique et pratique de braidisme ou hypnotisme nerveux ...* Paris, J.B. Baillière.

Pigeaud, P.E. (1897). *La suggestion en pédagogie: dangers et avantages.* Paris, H. Jouve.

Pitres, A. (1884). *Des suggestions hypnotiques ...* Bordeaux, Feret et fils.

Pitres, A. (1885). *Des zones hystérogènes et hypnogènes, des attaques de sommeil.* Bourdeaux, Gounouilhov. [Offprint from the *Journal de médecine de Bordeaux,* November 1884-January 1885.]

Pitres, A. (1891). *Leçons cliniques sur l'hystérie et l'hypnotisme faites à l'hôpital Saint-André de Bordeaux.* 2 vols. Paris, O. Doin.

Plummer, W.S. [1887]. Mary Reynolds, a case of double consciousness. In E.W. Stevens, *The Watseka Wonder.* Chicago, Office of the Philosophical Journal.

Podiapolsky, P.P. (1904–5). Brûlure suggérée chez une femme ayant présenté du mutisme hystérique. *Revue de l'hypnotisme,* 19, 50–53.

Podyapolsky, P.P. (1909). [Vasomotor disorders caused by hypnotic suggestion.] *Zhurnal nevrologii i psikhiatrii,* 9, 101–109.

Podiapolsky, P.P (1909–10). Des troubles vaso-moteurs par suggestion hypnotique. *Revue de l'hypnotisme,* 24, 178–188

Podmore, F. (1894). *Apparitions and Thought-Transference: An Examination of the Evidence for Telepathy.* London, Walter Scott.

Podmore, F. (1897). *Studies in Psychical Research.* London, Kegan Paul, Trench, Trübner and Co.

Podmore, F. (1902). *Modern Spiritualism: A History and a Criticism.* 2 vols. London, Methuen.

Podmore, F. (1909). *Mesmerism and Christian Science.* London, Methuen.

Pomeranz, B. (1982). Acupuncture and the endorphins. *Ethos,* 10, 385–393.

Pomme, P. (1778). *A Treatise on Hysterical and Hypochondriacal Disease* ... London, P. Elmsly.

Preda, G. (1910–11). L'hypnotisme d'hier et l'hypnotisme d'aujourd'hui. *Revue de psychothérapie et de psychologie appliquée,* 25, 210–213.

Préjalmini, L. (1840–1). Faits. *Journal du magnétisme,* 2, 151$_2$-152$_2$.

Preyer, W. (1878). *Die Kataplexie und der thierische Hypnotismus.* Jena, G. Fischer.

Preyer, W. (1881). *Die Entdeckung des Hypnotismus. Nebst einer ungedruckten Original-Abhandlung von Braid in deutscher Uebersetzung.* Berlin, Gebrüder Paetel.

Preyer, W. (1890). *Der Hypnotismus. Nebst Anmerkungen und einer nachgelassenen Abhandlung von Braid aus dem Jahre* 1845. Vienna, Urban und Schwarzenberg.

Prince, M. (1890). Some of the revelations of hypnotism: Post-hypnotic suggestion, automatic writing and double personality. *The Boston Medical and Surgical Journal,* 122, 463–467, 475–476, 493–495.

Prince, M. (1898–9). A contribution to the study of hysteria and hypnosis; being some experiments on two cases of hysteria and a physiologico-anatomical theory of the nature of these neuroses. *Proceedings of the Society for Psychical Research,* 14, 79–97.

Prince, M. (1900–1). The development and genealogy of the Misses Beauchamp: A preliminary report of a case of multiple personality. *Proceedings of the Society for Psychical Research,* 15, 466–483.

Prince, M. (1908). *The Dissociation of a Personality: A Biographical Study in Abnormal Psychology.* 2nd edn. New York, Longmans, Green and Co.

Prince, M. (1908–10). The unconscious. *The Journal of Abnormal Psychology,* 3, 261–297, 335–353, 391–426; 4, 36–62.

Prince, M. (1909). The subconscious. In *VIᵉ congrès international de psychologie, tenu à Genève du 2 à 7 août 1909* ... Geneva, Kündig. Pp. 71–97.

Prince, M. (1909–10). The psychological principles and field of psychotherapy. *The Journal of Abnormal Psychology,* 4, 72–98.

Prince, M. (1914). *The Unconscious: The Fundamentals of Human Personality.* New York, The Macmillan Company.

Prince, M. (1919–20). The psychogenesis of multiple personality. *The Journal of Abnormal Psychology,* 14, 225–280.

Prince, M. (1975). *Psychotherapy and Multiple Personality: Selected Essays.* Ed. N. G. Hale. Cambridge, Mass., Harvard University Press.

Prince, M., *et al.* (1910). *Psychotherapeutics: A Symposium.* Boston, Mass., R. Badger.

Prince, R. (1968). Can the EEG be used in the study of possession states. In R. Prince (ed.), *Trance and Possession States.* Montreal, R. M. Bucke Memorial Society. Pp. 121–137.

Prince, R. (1982). Shamans and endorphins: Hypotheses for a synthesis. *Ethos,* 10, 409–423.

Prince, W. F. (1915–16). The Doris case of multiple personality. *Proceedings of the American Society for Psychical Research,* 9, 1–700; 10, 701–1419.

Proust, – . (1889–90). Automatisme ambulatoire chez un hystérique. *Revue de l'hypnotisme,* 4, 267–269.

Pupin, C. (1895–6). La théorie histologique du sommeil. *Revue de l'hypnotisme*, 10, 289–299.

Putnam, F. W. (1986). The scientific investigation of multiple personality disorder. In J. M. Quen (ed.), *Split Minds/Split Brains: Historical and Current Perspectives*. New York, New York University Press. Pp. 109–125.

Putnam, F. W., Zahn, T. P., and Post, R. M. (1990). Differential autonomic nervous system activity in multiple personality disorder. *Psychiatry Research*, 31, 251–260.

Putnam, F. W., *et al.* (1986). The clinical phenomenology of multiple personality disorder: Review of 100 recent cases. *The Journal of Clinical Psychiatry*, 47, 285–293.

Pylyshyn, Z. W. (1984). *Competition and Cognition: Towards a Foundation for Cognitive Science*. Cambridge, Mass., MIT Press.

Quackenbos, J. D. (1900). *Hypnotism in Mental and Moral Culture*. New York, Harper and Brothers.

Quackenbos, J. D. (1908). *Hypnotic Therapeutics in Theory and Practice...* New York, Harper and Brothers.

Radil, T., *et al.* (1988) Attempts to influence movement disorders in hemiparetics. *Scandinavian Journal of Rehabilitation, Medical Supplement*, 17, 157–161.

Radtke, H. L., Thompson, V. A., and Egger, L. A. (1987). Use of retrieval cues in breaching hypnotic amnesia. *Journal of Abnormal Psychology*, 96, 335–340.

Rambotis, A. S. (1905–6). Suggestion pendant le sommeil naturel. *Revue de l'hypnotisme*, 20, 153–155.

Ranschburg, P. (1895–6). Beiträge zur Frage der hypnotisch-suggestiven Therapie. *Zeitschrift für Hypnotismus*, 4, 269–302.

Regnault, F. (1897). *Hypnotisme, religion*. Paris, Schleicher frères.

Regnault, F. (1902). Valeur de l'hypnotisme comme moyen d'investigation psychologique. In E. Bérillon and P. Farez (eds.), *Deuxième congrès international de l'hypnotisme expérimental et thérapeutique...* Paris, Revue de l'hypnotisme. Pp. 80–92.

Regnier, L. R. (1891). *Hypnotisme et croyances anciennes*. Paris, Lecrosnier et Babé.

Reid, W. H., Ahmed, I., and Levie, C. A. (1981). Treatment of sleepwalking: A controlled study. *American Journal of Psychotherapy*, 35, 27–37.

Remic, M. (1974). L'esplorazione ipnotica dopo i traumi cerebrali. *Rivista internazionale di psicologia e ipnosi*, 15, 409–417.

Renterghem, A. W. van (1892–3). Kompleter seit mehreren Jahren bestehender Dammriss... Radicale und fast schmerzlose Operation unter dem Einfluss der Suggestion ohne Hypnose. *Zeitschrift für Hypnotismus*, 1, 139–142.

Renterghem, A. W. van (1895–6). Liébeault et son école. Causeries. *Zeitschrift für Hypnotismus*, 4, 333–375.

Renterghem, A. W. van (1896–7). Liébeault et son école. Causeries. *Zeitschrift für Hypnotismus*, 5, 46–55, 95–127,

Renterghem, A. W. van (1897). Liébeault et son école. Causeries. *Zeitschrift für Hypnotismus*, 6, 11–44.

Renterghem, A. W. van (1897–8). Résultats obtenus à la Clinique de psychothérapie d'Amsterdam (de 1893 à 1897). *Revue de psychologie clinique et thérapeutique*, 1, 111–118, 139–146.

Renterghem, A. W. van (1898a). Liébeault en zijne school. Amsterdam, F. van Rossem.

Renterghem, A.W. van (1898b). Liébeault et son école. Causeries. *Zeitschrift für Hypnotismus*, 7, 54–96.

Renterghem, A.W. van (1898c). *Liébeault und seine Schule*. Amsterdam, F. van Rossem.

Renterghem, A.W. van (1899). Dritter Bericht über die in der psychotherapeutischen Clinik in Amsterdam erhaltenen Resultate während den Jahren 1893–1897. *Zeitschrift für Hypnotismus*, 8, 1–9.

Renterghem, A.W. van (1902). L'évolution de la psychothérapie en Hollande. In E. Bérillon and P. Farez (eds.), *Deuxième congrès international de l'hypnotisme expérimentale et thérapeutique* ... Paris, Revue de l'hypnotisme. Pp. 54–62.

Renterghem, A.W. van, and Eeden, F. van (1889). Clinique de psycho-thérapie suggestive fondée à Amsterdam le 15 août 1887. Compte rendu des résultats obtenus pendant la première période bisannuelle 1887–1889. In E. Bérillon (ed.), *Premier congrès international de l'hypnotisme expérimental et thérapeutique* ... Paris, O. Doin. Pp. 57–78.

Renterghem, A.W. van, and Eeden F. van (1894). *Psycho-thérapie: communications, statistiques, observations cliniques nouvelles. Compte rendu des résultats obtenus dans la Clinique de psycho-thérapie suggestive d'Amsterdam, pendant la deuxième période 1889–1893*. Paris, Société d'éditions scientifiques.

Report of the committee for the systematic investigation of hypnotic phenomena (1893–4). *Journal of the Society for Psychical Research*, 6, 247–249.

Rezende de Castro Monteire, A. (1976). "Hypnotisme." Analyse de la thèse présentée à la Faculté de médecine de Rio de Janeiro par Francesco de Paula Fajardo junior ... Rio de Janeiro, 1888. Paper presented at the Congrès international de l'histoire de la médecine XXVe, Quebec, 1976.

Rhue, J.W., and Lynn, S.J. (1987). Fantasy proneness: The ability to hallucinate as real as real. *British Journal of Experimental and Clinical Hypnosis*, 4, 173–180.

Rhue, J.W., and Lynn, S.J. (1991). Fantasy proneness, hypnotizability, and multiple personality disorder. In J.F. Schumaker (ed.), *Human Suggestibility: Advances in Theory, Research, and Application*. New York, Routledge. Pp. 200–218.

Ricard, J.J.A. (1844). *Physiologie et hygiène du magnétiseur* ... Paris, Germer Baillière.

Rice, E., and Fisher, C. (1976). Fugue states in sleep and wakefulness: A psychophysiological study. *The Journal of Nervous and Mental Disease*, 163(2), 79–87.

Richer, P. (1885). *Études cliniques sur la grande hystérie ou hystéro-épilepsie* ... 2nd edn. Paris, A. Delahaye et E. Lecrosnier.

Richet, C. (1875). Du somnambulisme provoqué. *Journal d'anatomie et physiologie*, 11, 348–378.

Richet, C. (1880a). Les démoniaques d'aujourd'hui. *Revue des deux mondes*, 37, 340–372.

Richet, C. (1880b). Les démoniaques d'autrefois. *Revue des deux mondes*, 37, 552–583, 828–863.

Richet, C. (1880c). Du somnambulisme provoqué. *Revue philosophique*, 20, 337–374, 462–484.

Richet, C. (1881a). De l'excitabilité réflexe des muscles dans la première période du somnambulisme. *Archives de physiologie*, 2 sér, 8, 155–157.

Richet, C. (1881b). The simulation of somnambulism. *The Lancet*, 59, i, 8–9, 51–52.

Richet, C. (1884a). *L'homme et l'intelligence: fragments de physiologie et de psychologie*. Paris, F. Alcan.

Richet, C. (1884b). Hypnotisme et contracture. *Comptes rendus de la Société de biologie*, 7 sér, 5, 662–664.

Richet, C. (1884c). De la suggestion sans hypnotisme. *Comptes rendus de la Société de biologie*, 8 sér, 1, 553–556.

Richet, C. (1923). *Thirty Years of Psychical Research*. London, Collins.

Rieger, C. (1884). *Der Hypnotismus* ... Jena, G. Fischer.

Rifat, –. (1887–8). Étude sur l'hypnotisme et la suggestion. *Revue de l'hypnotisme*, 2, 297–302.

Riklin, F. (1902–3). Hebung epileptischer Amnesien durch Hypnose. *Journal für Psychologie und Neurologie*, 1, 200–225.

Ringier, G. (1891). *Erfolge des therapeutischen Hypnotismus in der Landpraxis*. Munich, J. F. Lehmann.

Ringier, G. (1897). Zur Redaction der Suggestion bei Enuresis nocturna. *Zeitschrift für Hypnotismus*, 6, 150–153.

Rivers, W. H. R. (1920). *Instinct and the Unconscious: A Contribution to a Biological Theory of the Psycho-Neuroses*. Cambridge, Cambridge University Press.

Rochas d'Aiglun, E. A. de (1892). *Les états profonds de l'hypnose*. Paris, Chamuel.

Rochas d'Aiglun, E. A. de (1924). *Les vies successives* ... Paris, Chacornac frères.

Rogo, D. Scott (1988). *The Infinite Boundary: Spirit Possession, Madness, and Multiple Personality*. Wellingborough, Northants, The Aquarian Press.

Rolleston, H. D. (1889). Treatment by hypnotic suggestion. *St. Bartholomew's Hospital Reports*, 25, 115–126.

Romanes, G. J. (1878). Hypnotism. *Nature*, 18, 492–494.

Romanes, G. J. (1880). Hypnotism. *The Nineteenth Century*, 8, 474–480.

Rosenzweig, S. (1987). Sally Beauchamp's career ... *Genetic, Social, and General Psychology Monographs*, 113, 5–60.

Rosenzweig, S. (1988). The identity and ideodynamics of the multiple personality "Sally Beauchamp". *American Psychologist*, 43, 45–48.

Ross, C. A., and Norton, G. R. (1989–90). Effects of hypnosis on the features of multiple personality disorder. *American Journal of Clinical Hypnosis*, 32, 99–106.

Ross, C. A., Norton, G. R., and Fraser, G. A. (1989). Evidence against the iatrogenesis of multiple personality disorder. *Dissociation*, 2(2), 61–65.

Ross, C. A., Norton, G. R., and Wozney, K. (1989). Multiple personality disorder: An analysis of 236 cases.

Rossi, L. (1984). Enrico Morselli e le scienza dell'uomo nell'età del positivismo. *Rivista sperimentale di freniatria e medicina legale delle alienazioni mentale*, 108, supp. to fasc. 6, 2192–2464.

Rossi, P. (1904). *Les suggestions et la foule* ... Paris, A. Michelon.

Roth, M. (1887). *The Physiological Effects of Artificial Sleep with some Notes on the Treatment by Suggestion*. London, Baillière, Tindall and Cox.

Rumpf, T. (1880). Ueber Reflexe. *Deutsche medicinischer Wochenschrift*, 6, 392–395.

Rutter, J. O. N. (1851). *Magnetoid Currents, their Forces and Directions; with a Description of the Magnetoscope*. London, J. W. Parker and son.

Rybalkin, J. (1889–90). Brûlure du second degré provoquée par suggestion. *Revue de l'hypnotisme*, 4, 361–362.

S. du M. (1816). Des séances publiques de magnétisme, qui ont lieu chez M. l'abbé Faria. *Annales du magnétisme animal*, 5, 186–191.

Sabourin, M. E. (1982). Hypnosis and brain function: EEG correlates of state-trait differences. *Research Communications in Psychology, Psychiatry and Behavior*, 7, 149–168.

Sabourin, M. E., Cutcomb, D. D., Crawford, H. J. and Pribram, K. (1990). EEG correlates of hypnotic susceptibility and hypnotic trance: spectral analysis and coherence. *International Journal of Psychophysiology*, 10, 125–142.

Saletu, B., *et al.* (1975). Hypno-analgesia and acupuncture analgesia: A neuro-physiological reality. *Neuropsychobiology*, 1, 218–242.

Sallis, J. G. (1888). *Ueber hypnotische Suggestionen, deren Wesen, deren klinische und strafrechtliche Bedeutung.* Berlin, L. Heuser.

Sanchez Herrero, A. (1889a). *El hipnotismo y la sugestiòn: estudios de fisio-psicologia y de psico-terapia.* Valladolid, Pastor

Sanchez Herrero, A. (1889b). L'hypnotisation forcé et contre la volonté arrêtée du sujet. In E. Bérillon (ed.), *Premier congrès international de l'hypnotisme expérimental et thérapeutique ...* Paris, O. Doin. Pp. 312–318.

Sanchez Herrero, A. (1905). *El hipnotismo y la sugestiòn: estudios de fisio-psicologia y de psico-terapia.* 3rd edn. Madrid, A. Marzo.

Sarbin, T. R. (1981). On self-deception. *Annals of the New York Academy of Sciences*, 364, 220–235.

Sarbin, T. R., and Coe, W. C. (1972). *Hypnosis: A Social Psychological Analysis of Influence Communication.* New York Holt, Rinehart and Winston.

Sarbin, T. R., and Juhasz, J. B. (1978). The social psychology of hallucinations. *Journal of Mental Imagery*, 2, 117–144.

Sargant, W., and Slater, E. (1963). *An Introduction to Physical Methods of Treatment in Psychiatry*, 4th edn. Edinburgh, E. C. S. Livingstone.

Sarlo, F. de (1893). La teorie moderne sulla psicologia della suggestione. *Rivisita di filosofia*, 8, ii, 172–205.

Sauvages de la Croix, F. (1742). Observation concernant une fille cataleptique & somnambule en même temps. *Mémoires de l'Académie royale des sciences à Paris*, 409–415.

Savage, G. H. (1909). *The Harveian Oration: On Experimental Psychology and Hypnotism.* London, H. Froude.

Schaffer, K. (1895). *Suggestion und Reflex: eine kritisch-experimentelle Studie über die Reflexphaenomene des Hypnotismus.* Jena, G. Fischer.

Schechter, E. I. (1984). Hypnotic induction vs. control conditions: Illustrating an approach to the evaluation of replicability in parapsychological data. *Journal of the American Society for Psychical Research*, 78, 1–27.

Schenck, C. H., *et al.* (1989). Dissociative disorders presenting as somnambulism: Polysomnographic, video, and clinical documentation (8 cases). *Dissociation*, 2(4), 194–204.

Schenck, C. H., *et al.* (1989b). A polysomnographic and clinical report on sleep-related injury in 100 adult patients. *American Journal of Psychiatry*, 146, 1166–1173.

Schenk, L., and Bear, D. (1981). Multiple personality and related dissociative phenomena in patients with temporal lobe epilepsy. *The American Journal of Psychiatry*, 138, 1311–1316.

Schleich, K. L. (1897). *Schmerzlose Operationen. Oertliche Betäubung mit indifferenten Flüssigkeiten ...* Berlin, Springer.

Schmeltz, – . (1890). Amputation du sein faite pendant l'anesthésie hypnotique. *Gazette médicale de Strasbourg*, 49, 79–81.

Schmeltz, – . (1894–5). Opérations faites pendant le sommeil hypnotique. *Revue de l'hypnotisme*, 9, 47–52.

Schmeltz, – . (1895–6). Sarcome du testicle gauche opéré pendant le sommeil hypnotique. *Revue de l'hypnotisme*, 10, 120–121.

Schmidkunz, H. (1892). *Psychologie der Suggestion* ... Stuttgart, F. Enke.

Schmidt, J. G. (1799–1800). *Archytas, oder das Wichtigste für den Menschen* ... 2 parts. Berlin, Maurer.

Schneck, J.M (1963). *Hypnosis in Modern Medicine*, 3rd edn. Springfield, Ill., C. C. Thomas.

Schneider, G. H. (1880). *Die psychologische Ursache der hypnotischen Erscheinungen.* Leipzig, A. Abel.

Schopenhauer, A. (1909). *The World as Will and Idea*, vol 2. London, Kegan Paul, Trench, Trübner and Co.

Schrenck-Notzing, A. von (1888). *Ein Beitrag zur therapeutischen Verwerthung des Hypnotismus.* Leipzig, F. C. W. Vogel.

Schrenck-Notzing, A. von (1891). Die Bedeutung narcotischer Mittel für den Hypnotismus. *Schriften der Gesellschaft für psychologische Forschung*, 1(1), 1–73.

Schrenck-Notzing, A. von. (1891–2). Experimental studies in thought-transference. *Proceedings of the Society for Psychical Research*, 7, 3–22.

Schrenck-Notzing, A. von (1892). *Die Suggestions-Therapie bei krankhaften Erscheinungen des Geschlechtsinnes mit besonderer Berücksichtigungen der conträren Sexualempfindung.* Stuttgart, F. Enke.

Schrenck-Notzing, A. von (1892–3). Eine Geburt in der Hypnose. *Zeitschrift für Hypnotismus*, 1, 49–52.

Schrenck-Notzing, A. von [1893a] Suggestion, Suggestivtherapie. Offprint from A. Eulenberg (ed.), *Real-Encyclopädie der Gesammten Heilkunde. Encyclopädische Jahrbücher*, vol 3. Vienna, Urban and Schwarzenberg.

Schrenck-Notzing, A. von (1893b). *Ueber Suggestion und suggestive Zustände.* Munich, J. F. Lehmann.

Schrenck-Notzing, A. von.(1894a). *Ein Beitrag zur psychischen und suggestiven Behandlung der Neurasthenie.* Berlin, Brieger.

Schrenck-Notzing, A. von [1894b]. Suggestion, Suggestivtherapie. Offprint from A. Eulenberg (ed.), *Real-Encyclopädie der Gesammten Heilkunde. Encyclopädische Jahrbücher*, vol. 4. Vienna, Urban and Schwarzenberg.

Schrenck-Notzing, A. von (1894–5). Ueber den Yoga-Schlaf. *Zeitschrift für Hypnotismus*, 3, 69–75.

Schrenck-Notzing, A. von [1895a]. Suggestion, Suggestivtherapie. Offprint from A. Eulenberg (ed.), *Real-Encyclopädie der Gesammten Heilkunde. Encyclopädische Jahrbücher*, vol. 5. Vienna, Urban and Schwarzenberg.

Schrenck-Notzing, A. von. (1895b). *Therapeutic Suggestion in Psychopathia Sexualis (Pathological Manifestations of the Sexual Sense), with especial reference to Contrary Sexual Instinct.* New York, F. A. Davis.

Schrenck-Notzing, A. von (1895–6). Ein experimenteller und kritischer Beitrag zur Frage der suggestiven Hervorrufung circumscripter vasomotorischer Veränderungen auf der äusserer Haut. *Zeitschrift für Hypnotismus*, 4, 209–228.

Schrenck-Notzing, A. von (1896). *Ueber Spaltung der Persönlichkeit (sogenanntes Doppel-Ich)*. Vienna, A. Hölder.

Schrenck-Notzing, A. von (1898). Zur Frage der suggestiven Hauterscheinungen. *Zeitschrift für Hypnotismus*, 7, 247–249.

Schrenck-Notzing, A. von (1898–1900). Literaturzusammenstellung über die Psychologie und Psychopathologie der vita sexualis. *Zeitschrift für Hypnotismus*, 7, 121–131; 8, 40–53, 275–291; 9, 98–112.

Schrenck-Notzing, A. von (1902a). *Kriminal-psychologische und psychopathologische Studien. Gesammelte Aufsätze aus den Gebieten der Psychopathia sexualis, der gerichtlichen Psychiatrie und der Suggestionslehre*. Leipzig, J. A. Barth.

Schrenck-Notzing, A. von (1902b). La suggestion et l'hypnose dans leurs rapports avec le jurisprudence. In E. Bérillon and P. Farez (eds.), *Deuxième congrès international de l'hypnotisme expérimental et thérapeutique*... Paris, Revue de l'hypnotisme. Pp. 121–141.

"Ein Schulmann," (1888). *Der Hypnotismus in der Pädagogik*. Berlin, Heuser.

Schumaker, J. F. (ed.)(1991). *Human Suggestibility: Advances in Theory, Research, and Application*. New York, Routledge.

Schuyler, B. A., and Coe, W. C. (1989). More on volitional experiences and breaching posthypnotic amnesia. *The International Journal of Clinical and Experimental Hypnosis*, 37, 320–331.

Scott, J. A. (1984). History of medical hypnoanalysis. *Medical Hypnoanalysis*, 5, 124–151.

Seashore, C. E. (1895). Measurement of illusions and hallucinations in normal life. *Studies from the Yale Psychological Laboratory*, 2, 1–67.

Seif, L. (1900). Casuistische Beiträge zur Psychotherapie. *Zeitschrift für Hypnotismus*, 9, 275–282.

Semal, F. (1888). *De l'utilité et des dangers de l'hypnotisme*. Brussels, F. Hayez. [Offprint from the *Bulletin d l'Académie royale de médecine de Belgique*, 1888.]

Senso, M. (1881). *La verité sur le magnétisme animal*. Lausanne, L. Vincent.

Seppilli, G. (1891). Report on the therapeusis of mental diseases by means of suggestion. *The American Journal of Insanity*, 47, 542–556.

Shea, J. D. (1991). Suggestion, placebo, and expectation: Immune effects and other bodily changes. In J. F. Schumaker (ed.), *Human Suggestibility: Advances in Theory, Research, and Application*. New York, Routledge. Pp. 253–276.

Sheehan, P. W., and Dolby, R. M. (1974). Artifact and Barber's model of hypnosis: A logical-empirical analysis. *Journal of Experimental Social Psychology*, 10, 171–187.

Sheehan, P. W., and McConkey, K. M. (1982). *Hypnosis and Experience: The Exploration of Phenomena and Process*. Hillsdale, N. J., L. Erlbaum.

Sheehan, P. W., McConkey, K. M., and Cross, D. (1978). Experiential analysis of hypnosis: Some new observations on hypnotic phenomena. *Journal of Abnormal Psychology*, 87, 570–573.

Sheehan, P. W., and Perry, C. W. (1976). *Methodologies of Hypnosis: A Critical Appraisal of Contemporary Paradigms of Hypnosis*. Hillsdale, N. J., L. Erlbaum.

Shor, R. E. (1959). Hypnosis and the concept of the generalized reality-orientation. *The American Journal of Psychotherapy*, 13, 582–602.

Shor, R. E. (1962). Three dimensions of hypnotic depth. *The International Journal of Clinical and Experimental Hypnosis*, 10, 23–28.

Shor, R. E. (1979). A phenomenological method for the measurement of variables important to an understanding of the nature of hypnosis. In E. Fromm and R. E. Shor (eds.), *Hypnosis: Developments in Research and New Perspectives*. New York, Aldine. Pp. 105–135.

Shor, R. E., and Orne, M. T. (eds.) (1965). *The Nature of Hypnosis: Selected Basic Readings*. New York, Holt, Rinehart and Winston.

Sidgwick, E. M. (1891–2). On the evidence for clairvoyance. Part I. *Proceedings of the Society for Psychical Research*, 7, 30–99.

Sidgwick, E. M. (1915). A contribution to the study of the psychology of Mrs. Piper's trance phenomena. *Proceedings of the Society for Psychical Research*, 27, 1–657.

Sidgwick, E. M., and Johnson, A. (1892). Experiments in thought-transference. *Proceedings of the Society for Psychical Research*, 8, 536–596.

Sidgwick, H., Sidgwick, E. M., and Johnson, A. (1894). Report on the census of hallucinations. *Proceedings of the Society for Psychical Research*, 10, 25–422.

Sidgwick, H., Sidgwick, E. M., and Smith, G. A. (1889–90). Experiments in thought-transference. *Proceedings of the Society for Psychical Research*, 6, 128–170.

Sidis, B. (1896). A study of mental epidemics. *The Century Magazine*, 52, 849–853.

Sidis, B. (1898). *The Psychology of Suggestion …* New York, D. Appleton and Co.

Sidis, B. (1902). *Psychopathological Researches: Studies in Mental Dissociation*. New York, G. E. Stechert.

Sidis, B. (1909). *An Experimental Study of Sleep*. Boston, R. G. Badger.

Sidis, B. (1910). The psychotherapeutic value of the hypnoidal state. In M. Prince *et al.*, *Psychotherapeutics: A Symposium*. London, T. Fisher Unwin. Pp. 121–144.

Sidis, B. (1912–13). The theory of the subconscious. *Proceedings of the Society for Psychical Research*, 26, 319–343.

Sidis, B. (1914). *The Foundations of Normal and Abnormal Psychology*. Boston, R. G. Badger.

Sidis, B., and Goodhart, S. P. (1904). *Multiple Personality: An Experimental Investigation into the Nature of Human Individuality*. New York, D. Appleton and Co.

Siemers, J. F. See References: Animal magnetism.

Sighele, S. (1891a). *La folla deliquente. Studio di psicologia colletiva*. Turin, Fratelli Bocca.

Sighele, S. (1891b). *La foule criminelle, essai de psychologie collective*. Paris, F. Alcan.

Silva, C. E., and Kirsch, I. (1987). Breaching hypnotic amnesia by manipulating expectancy. *Journal of Abnormal Psychology*, 96, 325–329.

Simon, M. J., and Salzberg, H. C. (1985). The effect of manipulated expectancies on posthypnotic amnesia. *The International Journal of Clinical and Experimental Hypnosis*, 33, 40–51.

Singer, O. (1959). *A Short History of Scientific Ideas to 1900*. Oxford, Clarendon Press.

Sjoberg, B. M., and Hollister, L. E. (1965). The effects of psychotomimetic drugs on primary suggestibility. *Psychopharmacologia*, 8, 251–262.

Smirnoff, D. (1910–11). Modifications vaso-motrices produites par la suggestion hypnotique. *Revue de l'hypnotisme*, 24, 185–188.

Smirnoff, D. (1912). Zur Frage der durch hypnotische Suggestion hervorgerufenen vasomotorischen Störungen. *Zeitschrift für Psychotherapie und medizinische Psychologie*, 4, 171–175.

Smith, M. V., Glass, G. W., and Miller, T. I. (1980). *The Benefits of Psychotherapy*. Baltimore, Johns Hopkins Press.

Smith, R. P., and Myers, A. T. (1890). *On the Treatment of Insanity by Hypnotism*. Lewes, South Counties Press. [Offprint from *The Journal of Mental Science*, April 1890.]

Spanos, N. P. (1978). Witchcraft in histories of psychiatry: A critical analysis and an alternative conceptualization. *Psychological Bulletin*, 85, 417–439.

Spanos, N. P. (1982a). Hypnotic behavior: A cognitive, social psychological perspective. *Research Communications in Psychology, Psychiatry and Behavior*, 7, 199–213.

Spanos. N. P. (1982b). A social psychological approach to hypnotic behavior. In G. Weary and H. L. Mirels (eds.), *Integrations of Clinical and Social Psychology*. Oxford, Oxford University Press. Pp. 231- 271.

Spanos, N. P. (1983). Ergotism and the Salem witch panic: A critical analysis and an alternative conceptualization. *Journal of the History of the Behavioral Sciences*, 19, 358–368.

Spanos, N. P. (1986a). Hypnosis and the modification of hypnotic susceptibility: A social psychological perspective. In P. L. N. Naish (ed.), *What is Hypnosis? Current Theories and Research*. Milton Keynes, Open University Press. Pp. 85–120.

Spanos, N. P. (1986b). Hypnosis, nonvolitional responding, and multiple personality: A social psychological perspective. In B. A. Maher and W. A. Maher (eds.), *Progress in Experimental Personality Research*, vol. 14. New York, Academic Press, 1–62.

Spanos, N. P. (1986c). Hypnotic Behavior: A social-psychological interpretation of amnesia, analgesia, and "trance logic." *The Behavioral and Brain Sciences*, 9, 449–502.

Spanos, N. P. (1989). Experimental research on hypnotic analgesia. In N. P. Spanos and J. F. Chaves (eds.), *Hypnosis: The Cognitive-Behavioral Perspective*. Buffalo, N. Y., Prometheus Books. Pp. 206–240.

Spanos, N. P., and Barber, T. X. (1968). "Hypnotic" experiences as inferred from auditory and visual hallucinations. *Journal of Experimental Research in Personality*, 3, 136–150.

Spanos, N. P., and Barber, T. X. (1974). Towards a convergence in hypnosis research. *American Psychologist*, 29, 500–511.

Spanos, N. P., and Chaves, J. F. (eds.) (1989). *Hypnosis: The Cognitive-Behavioral Perspective*. Buffalo, N. Y., Prometheus Books.

Spanos, N. P., and Gorassini, D. R. (1984). Structure of hypnotic test suggestions and attributions of responding involuntarily. *Journal of Personality and Social Psychology*, 46, 688–696.

Spanos, N. P., and Gottlieb, J. (1976). Ergotism and the Salem witch trials, *Science*, 194, 1390–1394.

Spanos, N. P., and Gottlieb, J. (1979). Demonic posssssion, mesmerism, and hysteria: A social psychological perspective on their historical interrelations. *Journal of Abnormal Psychology*, 88, 527–546.

Spanos, N. P., Ham, M. W., and Barber, T. X. (1973). Suggested ("hypnotic") visual hallucinations: Experimental and phenomenological data. *Journal of Abnormal Psychology*, 81, 96–106.

Spanos, N. P., and Hewitt, E. C. (1979). Glossalalia: a test of the "trance" and psychopathology hypotheses. *Journal of Abnormal Psychology*, 88, 427–434.

Spanos, N. P., and Katsanis, J. (1989). Effects of instrumental set on attributions of nonvolition during hypnotic and nonhypnotic analgesia. *Journal of Personality and Social Psychology*, 86, 182–188.

Spanos, N. P., Ollerhead, V. G., and Gwynn, M. I. (1985–6). The effects of three instructional treatments on pain magnitude and pain tolerance: Implications for theories of hypnotic analgesia. *Imagination, Cognition and Personality*, 5, 321–337.

Spanos, N. P., Radtke, H. L., and Bertrand, L. D. (1985). Hypnotic amnesia as a strategic enactment: Breaching amnesia in highly susceptible subjects. *Journal of Personality and Social Psychology*, 47, 1155–1169.

Spanos, N. P., Robertson, L. A., Menary, E. P., and Brett, P. J. (1986). Component analysis of cognitive skill training for the enhancement of hypnotic susceptibility. *Journal of Abnormal Psychology*, 95, 350–357.

Spanos, N. P., Weekes, J. R., and Bertrand, L. D. (1985). Multiple personality: A social psychological perspective. *Journal of Abnormal Psychology*, 94, 362–376.

Spanos, N. P., Williams, V., and Gwynn, M. I. (1990). Effects of hypnotic, placebo, and salicylic acid treatments on wart regression. *Psychosomatic Medicine*, 52(1), 109–114.

Spanos, N. P., *et al.* (1982–3). When seeing is not believing: The effects of contextual variables on the reports of hypnotic hallucinators. *Imagination, Cognition and Personality*, 2, 195–209.

Spiegel, D., and Albert, L. H. (1983). Naloxone fails to reverse hypnotic alleviation of chronic pain. *Psychopharmacology*, 81, 140–143.

Spiegel, D., and Barabasz, A. F. (1988–9). Effects of hypnotic instructions on P^{300} event-related-potential amplitudes: Research and clinical implications. *American Journal of Clinical Hypnosis*, 31, 11–17.

Spiegel, D., Bierre, P., and Rootenberg, J. (1989). Hypnotic alterations of somatosensory perception. *The American Journal of Psychiatry*, 146, 749–754.

Spiegel, D., Cutcomb, S., Ren, C., and Pribram, K. (1985). Hypnotic hallucination alters evoked potential. *Journal of Abnormal Psychology*, 94, 249–255.

Spiegel, D., Hunt, T., and Dondershine, H. E. (1988). Dissociation and hypnotizability in posttraumatic stress disorder. *The American Journal of Psychiatry*, 145, 301–305.

Spiritus, – . (1819). Beobachtungen über die Heilkraft des animalischen Magnetismus. *Archiv für den thierischen Magnetismus*, 5(3), 78–87.

Spitta, H. (1882). *Die Schlaf-und Traumzustände der menschlichen Seele mit besonderer Berücksichtigung ihres Verhältnisses zu den psychischen Alienation.* 2nd edn. Tübingen, F. Fues.

Stadelmann, H. (1895). *Der acute Gelenkrheumatismus und dessen psychische Behandlung.* Würzburg, Stahel.

Stadelmann, H. (1896). *Der Psychotherapeut ...* Würzburg, Stahel.

Stam, H. J., and Spanos, N. P. (1980). Experimental designs, expectancy effects and hypnotic analgesia. *Journal of Abnormal Psychology*, 89, 751–762.

Stam, H. J., and Spanos, N. P. (1982). The Asclepian dream healings and hypnosis: A critique. *The International Journal of Clinical and Experimental Hypnosis*, 30, 9–22.

Stead, W. T. [1891]. *Real Ghost Stories.* London, The Review of Reviews.

Steggles, S., Stam, H. J., Fehr, R., and Aucoin, P. (1986–7). Hypnosis and cancer: An annotated bibliography 1960–1985. *American Journal of Clinical Hypnosis*, 29, 281–290.

Steinen, E. von den (1881). *Über den natürlichen Somnambulismus.* Heidelberg, C. Winter.

Stevenson, I. (1983–4). Cryptomnesia and parapsychology. *Journal of the Society for Psychical Research*, 52, 1–30.

Stoll, O. (1904). *Suggestion und Hypnotismus in der Völkerpsychologie.* 2nd edn. Leipzig, Veit.

Straaten, T. van (1900). Zur Kritik der hypnotischen Technik. *Zeitschrift für Hypnotismus*, 9, 129–176, 193–201.

Stucchi, G. (1892). *I problemi dell'ipnotismo*. Treviso, Turazza.

Sturgis, W. Russell (1894). The use of suggestion during hypnosis of the first degree, as a means of modifying or completely eliminating a fixed idea. *The Medical Record*, 45, 193–197.

Sunderland, La Roy (1843). *Pathetism; with Practical Instructions Demonstrating the Falsity of the hitherto Prevalent Assumptions in regard to what has been called "Mesmerism" and "Neurology"* ... New York, P. P. Good.

Sunderland, La Roy (1847). *Pathetism: Man Considered in respect to his Form, Life, Sensation, Soul, Mind, Spirit* ... Boston, White and Potter.

Sunderland, La Roy (1853). *Book of Psychology. Pathetism, Historical, Philosophical, Practical* ... New York, Stearns and Co.

Sunderland, La Roy (1868). *The Trance and Correlative Phenomena*. Chicago, J. Walker.

Sutcliffe, J. P. (1960). "Credulous" and "skeptical" views of hypnotic phenomena: A review of certain evidence and methodology. *The International Journal of Clinical and Experimental Hypnosis*, 8, 73–101.

Szöllösy, L. von (1907). Ein Fall multipler neurotischer Hautgangrän in ihrer Beziehung zur Hypnose. *Muenchener medizinische Wochenschrift*, 54, 1034–1035.

Tagliarini, A. (1985). Aspects of the history of psychiatry in Italy in the second half of the nineteenth century. In W. F. Bynum, R. Porter and M. Shepherd (eds.), *The Anatomy of Madness: Essays in the History of Psychiatry. Vol. II. Institutions and Society*. London, Tavistock Publications. Pp. 175–196.

Takahashi, Y. (1990). Is multiple personality disorder really rare in Japan? *Dissociation*, 3(2), 57–68.

Tamburini, A. (1891). On the nature of the somatic phenomena in hypnotism. *The Alienist and Neurologist*, 12, 297–321.

Tamburini, E., and Seppilli, G. (1881). *Contribuzioni allo studio sperimentale dell'ipnotismo. I. Communicazione preventiva*. Milan, Fratelli Rechiedei. [Offprint from the *Gazzetta medica Italiana-Lombardia*.]

Tamburini, A., and Seppilli, G. (1882). *Contribuzioni allo studio sperimentale dell'ipnotismo. 2ᵃ communicazione. Ricerche sui fenomeni di moto, di senso, del respiro et del circolo* ... Reggio nell'Emilia, Calderini. [Offprint from *Rivista sperimentale de freniatria*, 1882, 8.]

Tanzi, E. (1887). Studi sull'ipnotismo: la così detta "polarizzazione cerebrale" e le leggi associative. *Rivista di filosofia scientifica*, 6, 548–558.

Tarchanoff [Tarkhanov], J. de (1891). *Hypnotisme, suggestion et lecture des pensées*. Paris, G. Masson.

Tardieu, A. A. (1878). *Étude médico-légale sur les attentats aux moeurs*. 7th edn. Paris, J. B. Baillière.

Tardy de Montravel, A. A. (1787). *Lettres pour servir de suite à l'essai sur la théorie du somnambulisme magnétique*. London, n. pub.

Tart, C. T. (1970). Self-report scales of hypnotic depth. *The International Journal of Clinical and Experimental Hypnosis*, 18, 105–125.

Tart, C. T. (1979). Measuring the depth of an altered state of consciousness with particular reference to self-report scales of hypnotic depth. In E. Fromm and R. E. Shor (eds.), *Hypnosis: Developments in Research and New Perspectives*. New York, Aldine. Pp. 567–601.

Tatzel, R.W. (1892–3). Eine Geburt in der Hypnose. *Zeitschrift für Hypnotismus*, 1, 245–247.

Tatzel, R.W. (1894). *Die Psychotherapie (Hypnose)* ... Berlin, L. Heuser.

Tatzel, R.W. (1899). *Die hypnotische Suggestion und ihre Heilwirkungen*. Leipzig, J. A. Barth.

Taylor, E. (1984). *William James on Exceptional Mental States: The 1896 Lowell Lectures*. Amherst, The University of Massachusetts Press.

Terrien, – . (1902). Un cas complexe d'astasie-abasie. In E. Bérillon and P. Farez (eds.), *Deuxième congrès international de l'hypnotisme expérimental et thérapeutique* ... Paris, Revue de l'hypnotisme. Pp. 237–241.

Theodoridès, J. (1978). Hippolyte Bernheim et la pathologie infectieuse. *103ᵉ congrès national des sociétés savantes, Nancy, 1978*, sciences, fasc. v. Pp. 87–94.

Tholey, P. (1985). Haben Traumgestalten ein eigenes Bewusstsein? Eine experimentell-phänomenologische Klartraumstudie. *Gestalt Theory*, 7, 29–46.

Thomas, F. (1895). *La suggestion, son rôle dans l'éducation*. Paris, F. Alcan.

Thornton, E. M. (1976). *Hypnotism, Hysteria and Epilepsy: An Historical Synthesis*. London, Heinemann.

Thurston, H. (1952). *The Physical Phenomena of Mysticism*. London, Burns and Oates.

Tinterow, M. (1970). *Foundations of Hypnosis from Mesmer to Freud*. Springfield, Ill., C. C. Thomas.

Tissié, P. A. (1890). *Les rêves, physiologie et pathologie*. Paris, Germer Baillière.

Titchener, E. B. (1897). *An Outline of Psychology*. New York, Macmillan.

Tokarsky, A. A. (1902). Les indications de l'hypnotisme et de la suggestion dans le traitement de l'alcoolisme. In E. Bérillon and P. Farez (eds.), *Deuxième congrès international de l'hypnotisme expérimental et thérapeutique* ... Paris, Revue de l'hypnotisme. Pp. 173–177.

Tout, C. Hill (1895). Some psychical phenomena bearing upon the question of spirit control. *Proceedings of the Society for Psychical Research*, 11, 309–316.

Trömner, E. (1908). *Hypnotismus und Suggestion*. Leipzig, Teubner.

Trömner, E. (1913). Steigerung der Leistungsfähigkeit in hypnotische Zustand. *Journal für Psychologie und Neurologie*, 20, 181–184.

Tuckey, C. Lloyd (1890). *Treatment by Suggestion*. London, Baillière, Tindall and Cox.

Tuckey, C. Lloyd (1892). *The Value of Hypnotism in Chronic Alcoholism*. London, J. and A. Churchill.

Tuckey, C. Lloyd (1897). The value of hypnotism in chronic alcoholism. In *Dritter internationaler Congress für Psychologie in München von 4. bis 7. August 1896* ... Munich, J. F. Lehmann. Pp. 384–385.

Tuckey, C. Lloyd (1900). *Treatment by Hypnotism and Suggestion in Psychotherapeutics*. 4th edn. London, Baillière, Tindall and Cox.

Tuckey, C. Lloyd (1902). Les indications de l'hypnotisme et de la suggestion dans le traitement de l'alcoolisme. In E. Bérillon and P. Farez (eds.), *Deuxième congrès international de l'hypnotisme expérimental et thérapeutique* ... Paris, Revue de l'hypnotisme. Pp. 169–172.

Tuckey, C. Lloyd (1904). The position of hypnotic treatment in the cure of chronic alcoholism. *The British Journal of Inebriety*, 1, 268–273.

Tuckey, C. Lloyd (1907). *The Utility of the Study of Suggestion to the Student and Practitioner*. [Offprint from *The General Practitioner*, 6th April 1907.]

Tuckey, C. Lloyd (1909). How suggestion works. In W. B. Parker (ed.), *Psychotherapy* ... New York, Center Publishing Co. Vol 2(2), Pp. 5–19.

Tuke, D. Hack (1881). Hypnosis redivivus. *The Journal of Mental Science*, NS 26, 531–551.

Tuke, D. Hack (1884a). *Illustrations of the Influence of the Mind upon the Body in Health and Disease.* 2nd edn. London, J. and A. Churchill.

Tuke D. Hack. (1884b). *Sleep-Walking and Hypnotism.* London, J. and A. Churchill.

Tuke, D. Hack (ed.) (1892). *A Dictionary of Psychological Medicine.* 2 vols. London, J. and A. Churchill.

Udolf, R. (1981). *Handbook of Hypnosis for Professionals.* New York, Van Nostrand Reinhold.

Valentin, P. (1901). Psychothérapie et logothérapie. In *IVe congrès international de psychologie tenu à Paris du 20 au 26 août 1900* ... Paris, F. Alcan. Pp. 664–666.

Van der Does, A. J. W., and Dyck, R. van (1989). Does hypnosis contribute to the care of burn patients? A review of the evidence. *General Hospital Psychiatry*, 11, 119–124.

Van der Does, A. J. W., Dyck, R. van, and Spijker, R. E. (1988). Hypnosis and pain in patients with severe burns: A pilot study. *Burns*, 14, 399–404.

Van der Hart, O., and Horst, R. (1989). The dissociation theory of Pierre Janet. *Journal of Traumatic Stress*, 2, 397–411.

Van der Kolk, B. A., and Van der Hart, O. (1989). Pierre Janet and the breakdown of adaptation in psychological trauma. *The American Journal of Psychiatry*, 146, 1530–1540.

Van der Lanoitte, – , (1895–6). La suggestion et le fonctionnement du système nerveux. *Revue de l'hypnotisme*, 10, 263–276.

Vaschide, N. (1905). Du dédoublement de la conscience chez certains névropathes et neurasthéniques. In *Atti del V congresso internazionale di psicologia tenuto in Roma del 26 al 30 aprile 1905* ... Rome, Forzani. Pp. 496–497.

Veith, I. (1965). *Hysteria: The History of a Disease.* Chicago, University of Chicago Press.

Velander, – . (1889). Un cas de mutisme mélancolique guéri par suggestion. In E. Bérillon (ed.), *Premier congrès international de l'hypnotisme expérimental et thérapeutique* ... Paris, O. Doin. Pp. 323–325.

Ventra, D. (1891). La suggestione non ipnotica nelle persone sane e nella psicoterapia: studio sperimentale e clinico. *Il manicornio moderna*, 7, 75–122.

Verhandlungen der internationalen Gesellschaft für medizinische Psychologie und Psychotherapie. Zweite Jahresversammlung in München am 25. und 26. September 1911 (1912). *Journal für Psychologie und Neurologie*, 19, 273–299.

Verkroost, C. M. (1980). A. W. van Renterghem und Frederik van Eeden: zwei Holländische Ärzte als Bahnbrecher auf dem Gebiete der psychischen Therapie. *Janus*, 67, 191–199.

Veronesi, A. (1887). *L'ipnotismo e magnetismo davanti alla scienza.* Rome, Armanni.

Verworn, M. (1898). *Beiträge zur Physiologie des Centralnervensystems. Erster Theil. Die sogenannte Hypnose der Thiere.* Jena, G. Fischer.

Vincent, M., and Pickering, M. R. (1988). Multiple personality disorder in childhood. *Canadian Journal of Psychiatry*, 33, 524–529.

Vincent, R. H. (1893). *The Elements of Hypnotism* ... London, Kegan Paul, Trench, Trübner and Co.

Vincent, R. H. (1897). *The Elements of Hypnotism* ... 2nd edn. London, Kegan Paul, Trench, Trübner and Co.

Vingoe, F. J. (1973). Comparison of the Harvard Group Scale of Hypnotic Susceptibility Form A, and the Group Alert Trance Scale in a university population. *The International Journal of Clinical and Experimental Hypnosis*, 21, 169–179.

Vizioli, F. (1885). Del morbo ipnotico (ipnotismo spontaneo, autonomo) e delle suggestioni. *Giornale di neuropatologia*, 3, 289–342.

Vogt, O. (1894–6). Zur Kenntniss des Wesens und der psychologischen Bedeutung des Hypnotismus. *Zeitschrift für Hypnotismus*, 3, 277–340; 4, 32–45, 122–167, 229–244.

Vogt, O. (1896–7a) Die directe psychologische Experimentalmethode in hypnotischen Bewusstseinzuständen. *Zeitschrift für Hypnotismus*, 5, 7–30, 180–218.

Vogt, O. (1896–7b). Die Zielvorstellung der Suggestion. *Zeitschrift für Hypnotismus*, 5, 332–342.

Vogt, O. (1897). Die directe psychologische Experimentalmethode in hypnotischen Bewusstseinzuständen. In *Dritter internationaler Congress für Psychologie in München von 4. bis 7. August 1896*. Munich, J. F. Lehmann. Pp. 250–260.

Vogt, O. (1897–8). Spontane Somnambulie in der Hypnose. *Zeitschrift für Hypnotismus*, 6, 79–93; 7, 285–314.

Vogt, O. (1898). Ueber die Natur der suggerirten Anästhesie. *Zeitschrift für Hypnotismus*, 7, 336–341.

Vogt, O. (1899). Zur Methodik der ätiologischen Erforschung der Hysterie. *Zeitschrift für Hypnotismus*, 8, 65–83.

Vogt, O. (1902). Valeur d'hypnotisme comme moyen d'investigation psychologique. In E. Bérillon and P. Farez (eds.), *Deuxième congrès international de l'hypnotisme expérimental et thérapeutique* ... Paris, Revue de l'hypnotisme. Pp. 63–70.

Voisin, A. (1884). *Étude sur l'hypnotisme et sur les suggestions chez les aliénés*. [Offprint from *Annales médico-psychologiques*, 12th September 1884.]

Voisin, A. (1886a). *De l'hypnotisme et de la suggestion hypnotique dans leur application au traitement des maladies nerveuses et mentales*. Paris, Revue de l'hypnotisme.

Voisin, A (1886b). *De la thérapeutique suggestive chez les aliénés*. Paris, O. Doin. [Offprint from the *Bulletin général de thérapeutique*, 15th April 1886.]

Voisin, A. (1886–7a). De l'hypnotisme et de la suggestion hypnotique dans leur application au traitement des maladies nerveuses et mentales. *Revue de l'hypnotisme*, 1, 41–49.

Voisin, A. (1886–7b). Morphinomanie guérie par suggestion hypnotique. *Revue de l'hypnotisme*, 1, 161–163.

Voisin, A. (1887). *Du traitement de l'aménorrhée par la suggestion hypnotique et de l'influence de la suggestion sur les vaso-moteurs*. Paris, J. Lechevalier. [Offprint from the *Annales médico-psychologiques*, 5, 1887.]

Voisin, A. (1887–8). De la dipsomanie et des habitudes alcooliques et leur traitement par la suggestion hypnotique. *Revue de l'hypnotisme*, 2, 48–52, 65–70.

Voisin, A. (1889). Les indications de l'hypnose et de la suggestion dans le traitement des maladies mentales et des états connexes. In E. Bérillon (ed.), *Premier congrès international de l'hypnotisme expérimentale et thérapeutique* ... Paris, O. Doin. Pp. 134–136.

Voisin, A. (1891–2a). De l'aide donnée par le chloroforme à la production du sommeil hypnotique chez les aliénés et les obsédés. *Revue de l'hypnotisme*, 6, 114–116.

Voisin, A. (1891–2b). Delit de vol commis sur l'influence de la suggestion hypnotique. *Revue de l'hypnotisme*, 6, 219–220.

Voisin, A. (1894–5). Épilepsie Jacksonienne traité par la suggestion. *Revue de l'hypnotisme*, 9, 304–306.

Voisin, A. (1895–6). Un accouchement dans l'état d'hypnotisme. *Revue de l'hypnotisme*, 10, 360–361.

Voisin, A. (1897a). *Emploi de la suggestion hypnotique dans certaines formes d'aliénation mentale*. Paris, J. B. Baillière.

Voisin, A. (1897b). *Traitement de certaines formes d'aliénation mentale par la suggestion hypnotique*. In *Dritter internationaler Congress für Psychologie in München von 4. bis 7. August 1896 ...* Munich, J. F. Lehmann. Pp. 380–382.

Voisin, J. (1885). Note sur un cas de grande hystérie chez l'homme avec dédoublement de la personnalité ... *Archives de neurologie*, 10, 212–225.

Voisin, J. (1887). Séance du 25 octobre 1886. *Annales médico-psychologiques*, 7 sér, 5, 134–160.

Voisin, J. (1887–8a). Action des médicaments à distance. *Revue de l'hypnotisme*, 2, 209–211.

Voisin, J. (1887–8b). Guérison par la suggestion hypnotique d'idées délirantes et de mélancolie avec conscience. *Revue de l'hypnotisme*, 2, 242–244.

Volgyesi, F. A. (1966). *Hypnosis of Man and Animals*, 2nd edn. London, Baillière, Tindall and Cassell.

Wadden, T. A., and Anderton, C. H. (1982) The clinical use of hypnosis. *Psychological Bulletin*, 91, 215–243.

Wadden, T. A., and Flaxman, J. (1981). Hypnosis and weight loss: A preliminary study. *The International Journal of Clinical and Experimental Hypnosis*, 19, 162–173.

Wadden, T. A., and Penrod, J. H. (1981–2). Hypnosis in the treatment of alcoholism: A review and appraisal. *American Journal of Clinical Hypnosis*, 24, 41–47.

Wagstaff, G. F. (1981). *Hypnosis, Compliance and Belief*. Brighton, The Harvester Press.

Wagstaff, G. F. (1986). Hypnosis as compliance and belief: A socio-cognitive view. In P. L. N. Naish (ed.), *What is Hypnosis? Current Theories and Research*. Milton Keynes, Open University Press. Pp. 59–84.

Wagstaff, G. F. (1991). Suggestibility: A social psychological approach. In J. F. Schumaker (ed.), *Human Suggestibility: Advances in Theory, Research and Application*. New York Routledge. Pp. 132–145.

Walser, H. H. (1960). Ambroise-Auguste Liébeault (1823–1904) der Begründer der "École hypnologique de Nancy". *Gesnerus*, 17, 145–162.

Walter, H., *et al.* (1990). A contribution to classification of hallucinations. *Psychopathology*, 23, 97–105.

Ward, C., and Kemp, S. (1991). Religious experiences, altered states of consciousness, and suggestibility: Cross-cultural and historical perspectives. In J. F. Schumaker (ed.), *Human Suggestibility: Advances in Theory, Research, and Application*. New York, Routledge. Pp. 159–182.

Weinhold, A. F. (1880). *Hypnotische Versuche ...*, 3rd edn. Chemnitz, M. Bübz.

Weiss, M., Sutton, P. J., and Utecht, A. J. (1985). Multiple personality in a 10-year old girl. *Journal of the American Academy of Child Psychiatry*, 24, 495–501.

Weitzenhoffer, A. M. (1963). *Hypnotism: An Objective Study in Suggestibility.* New York, Wiley.

Weitzenhoffer, A. M. (1974). When is an "instruction" an "instruction"? *The International Journal of Clinical and Experimental Hypnosis,* 22, 258–269.

Weller, E. B., Weller, R. A., and Fristed, M. A. (1985). Use of sodium amytal interviews in prepubertal children: Indications, procedure and clinical utility. *Journal of the American Academy of Child Psychiatry,* 24, 747–749.

Werner, H. (1839). *Die Schutzgeister oder merkwürdige Blicke zweier Seherinnen in die Geisterwelt ...* Stuttgart, J. G. Cotta.

Wetterstrand, O. G. (1888a). Om den hypnotiska suggetionens använding i den praktiska medicinen. *Hygiea,* 50, 28–50, 130–165, 171–203.

Wetterstrand, O. G. [1888b]. *Om hypnotismens användande i den praktiska medicinen.* Stockholm, J. Seligmann.

Wetterstrand, O. G. (1891). *Der Hypnotismus und seine Anwendung in der praktischen Medicin.* Vienna, Urban and Schwarzenberg.

Wetterstrand, O. G. (1892–3). Ueber den künstlich verlängerten Schlaf, besonders bei Behandlung der Hysterie, Epilepsie und Hystero-Epilepsie. *Zeitschrift für Hypnotismus,* 1, 17–23.

Wetterstrand, O. G. (1895–6a). Die Heilung der chronischen Morphinismus, Opium-genusses, Cocaïnismus und Chloralismus mit Suggestion und Hypnose. *Zeitschrift für Hypnotismus,* 4, 8–26.

Wetterstrand, O. G. (1895–6b). Selbstbeobachtungen während des hypnotischen Zustandes. (Angaben zweier Patienten.) *Zeitschrift für Hypnotismus,* 4, 112–121.

Wetterstrand, O. G. (1897a). *Hypnotism and its Application to Practical Medicine.* New York, G. P. Putnam's sons.

Wetterstrand, O. G. (1897b). Ueber den künstlich verlängerten Schlaf, besonders bei der Behandlung von Hysteria. In *Dritter internationaler Congress für Psychologie in München von 4. bis 7. August 1896 ...* Munich, J. F. Lehmann. Pp. 361–363.

Wettley, A. (1953). *August Forel, ein Arztleben im Zweispalt seiner Zeit.* Salzburg, O. Müller.

White, R. W. (1941). A preface to the theory of hypnotism. *The Journal of Abnormal and Social Psychology,* 36, 477–505.

White, R. W., and Shevach, B. J. (1942). Hypnosis and the concept of dissociation. *The Journal of Abnormal and Social Psychology,* 37, 309–328.

Wiebe, A. (1884). Einige Fälle von therapeutischer Anwendung des Hypnotismus. *Berliner klinische Wochenschrift,* 21, 33–36.

Wiener, M. (1838). *Selma, die jüdische Seherin ...* Berlin, L. Fernbach.

Wienholt, A. (1806). *Heilkraft des thierischen Magnetismus,* vol. 3 Lemgo, Meyer.

Wigan, A. L. (1844). *A New View of Insanity. The Duality of the Mind Proved by the Structure, Function and Diseases of the Brain, and by the Phenomena of Mental Derangement ...* London, Longman, Brown, Green and Longmans.

Wilson, S. C., and Barber, T. X. (1977–8). The Creative Imagination Scale as a measure of hypnotic responsiveness: Applications to experimental and clinical hypnosis. *American Journal of Clinical Hypnosis,* 20, 235–249.

Wilson, S. C., and Barber, T. X. (1983). The fantasy-prone personality: Implications for understanding imagery, hypnosis, and parapsychological phenomena. In A. A. Sheikh (ed.), *Imagery: Current Theory, Research and Application*. New York, Wiley. Pp. 340–387.

[Wingfield, H. E.] (1888–9). The connection of hypnotism with the subjective phenomena of spiritualism. *Proceedings of the Society for Psychical Research*, 5, 279–287.

Wingfield, H. E. (1908). *Some Observations on Hypnotism*. Croydon. [Reprinted from *The General Practitioner*, 11th, 18th and 25th January 1907.]

Wingfield, H. E. (1910). *An Introduction to the Study of Hypnotism Experimental and Therapeutic*. London, Baillière, Tindall and Cox.

Winkelman, M. (1986). Trance states: A theoretical model and cross-cultural analysis. *Ethos*, 14, 174–203.

Winkelman, M., and DeRios, M. D. (1989). Psychoactive properties of !Kung Bushmen medicine plants. *Journal of Psychoactive Drugs*, 21, 51–59.

Winkelmann, O. (1965). Albert Moll (1862–1939) als Wegbereiter der Schule von Nancy in Deutschland. *Praxis der Psychotherapie*, 10(1), 1–7.

Wolberg, L. R. (1948). *Medical Hypnosis*. 2 vols. New York, Grune and Stratton.

Wolfart, K. C. (1815). *Erläuterungen zum Mesmerismus*. Berlin, Nikolai.

Woodruff, M. L., and Baisden, R. H. (1985). Lesions of the dorsal spinal cord decrease the duration of contact defense immobililty (animal hypnosis) in the rabbit. *Behavioral Neuroscience*, 99, 778–783.

Woods, H. F. (1908). Hypnotism in relation to medicine. *The Middlesex Hospital Journal*, 12, 1–15.

Woods, J. F. (1897). The treatment by suggestion, with and without hypnosis. *Journal of Mental Science*, 43, 248–276.

Woodworth, R. S. (1906). The cause of a voluntary movement. In *Studies in Philosophy and Psychology, by Former Students of C.E Garman*. Boston, Houghton Miflin.

Wundt, W. (1892). *Hypnotismus und Suggestion*. Leipzig, W. Engelmann.

Wundt, W. (1893). *Hypnotisme et suggestion: Étude critique*. Paris, F. Alcan.

Wundt, W. (1912). *An Introduction to Psychology*. London, Allen and Unwin.

Yeo, G. F. (1884). *The Nervous Mechanism of Hypnotism*. London, King's College Science Society.

Young, W. C. (1988). Observations on fantasy in the formation of multiple personality disorder. *Dissociation*, 1(3), 13–20.

Yung, E. (1883). *Le sommeil normal et le sommeil pathologique. Magnétisme animal, hypnose, névrose hystérique*. Paris, O. Doin.

Yung, E. (1908–9). Contribution à l'étude de la suggestibilité à l'état de veille. *Archives de psychologie*, 8, 263–285.

Zielinski, L. (1968a). Hypnotism in Poland. In E. J. Dingwall (ed.), *Abnormal Hypnotic Phenomena: A Survey of Nineteenth-Century Cases*, Vol. 3. London, J. and A. Churchill. Pp. 107–135.

Zielinski, L. (1968b). Hypnotism in Russia. In E. J. Dingwall (ed.), *Abnormal Hypnotic Phenomena: A Survey of Nineteenth-Century Cases*, Vol. 3. London, J. and A. Churchill. Pp. 1–105.

Zilboorg, G. (1941). *A History of Medical Psychology*. New York, W. W. Norton.

Zorab, G. (1967). Hypnotism in Belgium and the Netherlands. In E. J. Dingwall (ed.), *Abnormal Hypnotic Phenomena: A Survey of Nineteenth-Century Cases*, Vol. 2. London, J. and A. Churchill. Pp. 1–100.

Index of names

Ségard, C., 338, 429, 485, 486
Seif, L., 476, 570, 571, 572
Semal, F., 339
Senator, H., 304
Seppilli, G., 298, 331, 339, 354, 442
Servan, J. M. A., 29, 32, 34, 35
Shor, R. E., 533
Sidgwick, Eleanor M., 389, 465, 529–30, 531
Sidgwick, H., 389, 390, 465
Sidis, B., 375, 389, 403, 405–11, 453, 505, 537, 551, 564, 568
Siemers, J. F., 146
Sighele, S., 505
Silva, C. E., 600
Sixdeniers, A. V., 171
Smirnoff, D., 463
Smith, G. A., 391, 465
Smith, Hélène, 530
Soulié, F., 576
Spanos, N. P., 581, 582, 586, 596–608, 611, 615
Spiritus, 147, 220
Spurzheim, J., 185
Stadelmann, H., 422, 431, 476, 482, 486, 487
Stahl, G. E., 82, 125, 144
Stanhope, Earl, 213
Steinen, E. von den, 304
Stembo, L. 345
Stieglitz, J., 155, 156–7
Stirling, 348
Stoerck, A., 2, 3
Stoffregen, 91
Stoll, O., 424, 505–7, 508, 528
Stone, G. W., 185, 231
Stone, W. L., 182–3
Storer, 250–1
Straaten, T. van, 422
Strombeck, F. C. de, 147, 150
Stucchi, G., 340
Sunderland, La Roy, 184, 185, 189, 193, 288
Sutcliffe, J. P., 579
Swieten, G. van, 2
Szandor, Ilma, 342, 446, 458, 462
Szapàry, F. von, 152

Tamburini, A., 298, 331, 339, 354, 442
Tandel, E., 167
Tanzi, E., 339
Tardy de Montravel, A. A., 56–7, 59, 60–1, 64, 68–9, 93, 100, 102, 116, 147
Tarkhanov, I., 345
Tart, C. T., 583
Tatzel, 480, 482
Taylor, E., 400
Taylor, E. W., 565, 567
Teste, A., 138, 249, 257
Thomas, F., 493

Thouret, M. A., 8–9, 34
Tillaux, 491
Tillaye, 134
Tissart de Rouvre, J. L. N., Marquis, 55
Tissié, P. A., 367
Tokarsky, A. A., 485–6, 493
Topham, W., 220
Tout, C. Hill, 530–1
Townshend, C. H., 203, 208, 239
Treviranus, G. R., 80
Tritschler, J. C. S., 147, 148
Trömner, E., 443, 464, 569, 570, 571, 572
Tubbs, 213
Tuckey, C. Lloyd, 321, 336, 349, 420, 454, 466, 482, 485, 486, 487, 527
Tuke, D. Hack, 298, 305, 348–9, 364, 368, 463, 520–1

Valentin, J. C., 147–8, 149
Valentin, P., 562, 563, 609
Van der Hart, O., 622
Van der Kolk, B. A., 622
Van der Lanoitte, 537
Velianski, D., 92
Velander, 347
Velpeau, A. A. L. M., 287
Ventra, D., 550
Verati, L. [G. Pellegrino], 168, 169
Veronesi, A., 340
Verworn, M., 532
Viélet, P. H., 45–6, 47, 60
Vigouroux, R., 310
Villagrand, P., 136, 137
Villers, C. de, 70–1
Vincent, R. H., 453
Vinchon, J., 5
Virey, J. J., 127–8, 134
Vivé, L., 365–6, 459, 529
Vizioli, F., 331, 354
Vogel, J. L. A., 146
Vogt, O., 344, 421–2, 423, 424, 427, 433, 439, 440, 441, 443, 464, 473, 474, 475–6, 481, 518, 520, 521, 522, 523, 537–43, 548, 555, 556, 564, 567, 569 570–1, 571, 572, 608, 609, 613, 615, 621
Voisin, A., 338, 346, 347, 350, 421, 434, 473, 480–1, 484, 485, 486, 487, 493, 498, 499
Voisin, J., 335, 365, 481, 564
Volgyesi, F. A., 533

Wadden, T. A., 579
Wagner, N. P., 345
Wagstaff, G. F., 626 n. 81
Wakley, T., 202, 210–11, 212, 264
Ward, W. Squire, 220
Warren, 231

Index of subjects

sleep, 518–19, 544 n. 6, 518–19, 540–1
waking, 61, 340–1, 548–9, 554, 610, 615
see also electrobiologists; fascination
surgery, mesmeric and hypnotic, *see* analgesia
Swedenborgianism, 16, 67, 145, 172, 180,
    190–1, 209
sympathetic system, *see* ganglion system

telepathy, *see* community of sensation; ESP
temple sleep, 245
testimony, hypnotism and, 502–3
theories, *see* hypnosis, mesmerism
time sense, enhanced, 455
traction (of subjects by mesmerizer), 153, 162 n.
    64, 201
trance, 289, 351
    see also hypnosis; mesmeric state
transcendentalism, 179–80
transfer, 333–6
treatment, mesmeric and hypnotic
    ailments
        abscess, retropharyngeal, 34
        addictions (drugs, alcohol), 485–6, 579
        aphasia, 57
        asthma, 13, 249, 482, 579
        burns, 34, 249, 580
        chorea, *see* treatment, St. Vitus dance
        constipation, 249, 286
        consumption, *see* treatment, tuberculosis
        dropsy, 54, 249
        ear problems, 54, 84, 267 n. 15
        enuresis, 200, 321–2, 482, 566
        epilepsy, 12, 54, 84, 201, 249, 487
        erysipelas, 25, 337
        eye problems, 12–13, 54, 84, 249, 283
        fevers, 43, 54, 249
        fistula, rectal, 121
        gout, 33
        haemorrhage, 484, 569
        headaches, 54, 249, 485
        heart problems, 482
        hysteria, 17, 84, 201–2, 249, 286, 374–5,
            478–9
        indefinable, 55, 56–7
        insanity, 113–14, 129, 200, 249, 480–1
        lactation, disordered, 484–5
        menstrual disorders, 13, 54, 249, 286, 484
        migraine, *see* treatment, headaches

neuralgia, trigeminal *see* treatment, tic
    douloureux
neurasthenia, 479
neuroses, 477–80
obesity, 579
obsessive neurosis, 479–80
paralysis, 15–16, 43, 57, 84, 249, 283, 486–7
pneumonia, 84
psychoses, *see* treatment, insanity
rheumatism, 32–3, 34, 54, 249, 485
St. Vitus' dance, 201, 249
scrofula, *see* treatment, tuberculosis
sexual problems, 483–4
skin complaints, 33, 482, 579
smoking, 579
stress, 580
stuttering, 480
tic douloureux, 249, 250–1, 283, 485, 519,
    579
tuberculosis (consumption, scrofula), 30, 121,
    249, 250
tumours, 13, 15, 33, 84, 205, 249, 250,
    289, 486
warts, 482–3, 511 n. 33, 579
wounds, 84, 249
crises in, 13, 30, 42, 58–60
methods of, 12–14, 53, 99–100, 321–2, 474–6
success of, 14–16, 32–4, 53–7, 83–4, 247–9,
    251, 476–8, 487–9
rationale for, 12, 84–5, 105–6, 253–7, 322,
    330, 580
trees, magnetization of, 11, 28, 41, 275

unconscious, the, 159, 402 n. 53
Univercoelum, The, 186
water, magnetized, 3, 5, 12, 13, 28, 54, 56, 61,
    100, 117, 121, 130, 132, 170, 249, 255, 275
will, willing, rôle of, in magnetization, 47, 48–9,
    66, 70–1, 113, 133, 275–6
Würzburg School, 422–3, 447, 451, 607, 615

Zeitschrift für Hypnotismus, 344–5, 421, 559
*Zeitschrift für Psychologie, Kriminal-Anthropologie
    und Hypnotismus*, 421
Zoist, The, 206–8, 210, 212–13, 219–40
zones, hypnogenic, 328, 357 n. 48
zones, hysterogenic, 309, 328, 357 n. 48